CHEMISTRY RESEARCH AND APPLICATIONS

ORGANIC AND MEDICINAL CHEMISTRY

VOLUME 1

CHEMISTRY RESEARCH AND APPLICATIONS

Additional books and e-books in this series can be found
on Nova's website under the Series tab.

CHEMISTRY RESEARCH AND APPLICATIONS

ORGANIC AND MEDICINAL CHEMISTRY

VOLUME 1

BIMAL KRISHNA BANIK
EDITOR

nova science publishers
New York

Library of Congress Cataloging-in-Publication Data

ISBN: 978-1-53614-453-6

Published by Nova Science Publishers, Inc. † New York

CONTENTS

PREFACE

Organic and medicinal chemistry are the crucial branches of chemistry as they are more widely studied than any other subjects in chemistry. In principle, organic and medicinal chemistry have been defined as a hybrid subject of many other disciplines that include biology, pharmacology, medicine and pharmacy. The synthesis of molecules, mechanism of a process, studies of new reagents, natural products, and biological evaluation of compounds against different components of cells are the keys in this subject. As a result of unlimited possibilities, several scopes exist for students and researchers to study and explore organic and medicinal chemistry together. In this book (Volume 1), diverse research topics are included to benefit the readers. This book has nine chapters from active authors.

Bhalla and his group have written three chapters in this volume of the book. In the first chapter, Bhalla et al. have reported recent trends in the nitrone-olefin cycloaddition reaction for the preparation of biologically active molecules. In the second chapter, Bhalla and Narula have reported the synthesis of medicinally important compounds via several organic transformations using DDQ. In the third chapter, Saini and Bhalla have described a benzoxazole scaffold for effective drug design in therapeutic drug design. Sahoo et al. have reported the therapeutic potential of a pyrimidine-related new drug in the fourth chapter. Sahoo et al. have demonstrated new medicinal and pharmacological importance of thiazolidinones in Chapter Five. In Chapter Six, Maiti et al. have investigated synthesis and medicinal chemistry of isoxazolines and their analogues. Chapters Seven through Nine are written by Banik and his group. For example, in Chapter Seven, they have explored the synthesis of biologically active pyrroles through a variety of methods. In Chapter Eight, an exploration of novel polyaromatic compounds is described by this group. In Chapter Nine, Banik et al. have reported the preparation of diverse glycosides using the Ferrier rearrangement.

Organic and medicinal chemistry has unlimited goals. The application of organic and medicinal chemistry is significantly high since they are related to many notable discoveries. On this basis, the book will be useful for chemists, biologists, clinicians, pharmacists, biotechnologists, industrialists and engineers who are working in the field of interdisciplinary science as well as specific chemical and medicinal science.

In: Organic and Medicinal Chemistry, Volume 1
Editor: Bimal Krishna Banik
ISBN: 978-1-53614-453-6

Chapter 1

RECENT TRENDS IN NITRONE-OLEFIN [3+2] CYCLOADDITION REACTION: SYNTHETIC AND BIOLOGICAL POTENTIALS

Jitender Bhalla, Shamsher S. Bari and Aman Bhalla*
Department of Chemistry and Centre of Advanced Studies in Chemistry,
Panjab University, Chandigarh, India

ABSTRACT

Cycloaddition reactions are one of the most important and direct method for construction of heterocyclic rings. Nitrone-olefin [3+2] cycloaddition reaction is a well known example of 1,3-dipolar cycloaddition reaction which are used for the synthesis of many biologically important heterocycles such as isoxazole, isoxazoline and isoxazolidine which can also be transformed into valuable natural products and other heterocyclic compounds. The modification of this reaction (also known as Kinugasa reaction) is used to prepare β-lactam derivatives. These β-lactam derivatives in turn have immense synthetic and biological applications in various fields such as organic synthesis, peptidomimetics and pharmaceutical industry. The nitrone-olefin [3+2] cycloaddition reaction involves formation of new C–C and C–X bond in the products. The advantage of these cycloaddition reactions lies in the fact that the substrates are easily accessible with high diversity, require mild conditions and affords high efficiency, atom economy and selectivity. Therefore, in light of high end utility of nitrone-olefin [3+2] cycloaddition reactions in organic synthesis and medicinal application of various products, review of latest progresses and development on this reaction is of utmost importance. The present chapter aims to provide a comprehensive survey of nitrone-olefin [3+2] cycloaddition reactions in past ten years.

Keywords: 1,3-Dipolar cycloaddition, nitrone-olefin [3+2], diastereoselectivity, enantioselectivity, β-lactams, kinugasa reaction

* Corresponding Author Email: amanbhalla@pu.ac.in.

ABBREVIATIONS

Å	Angstrom
Ac	Acetyl
Ar	Aryl
BARAC	Biaryl-aza-cyclooctynone
Boc	*tert*-Butyloxycarbonyl
Bu	Butyl
i-Bu	*iso*-Butyl
t-Bu	*tert*-Butyl
Bn/Bz	Benzyl
[bmim]	1-Butyl-3-methylimidazolium
°C	Degree centigrade
CAI	Cholesterol absorption inhibitor
CFTR	Cystic fibrosis transmembrane conductance regulator
$CHCl_3$	Chloroform
COX	Cyclooxygenase
CTAB	Cetyltrimethylammonium bromide
Cy	Cyclohexyl
D	Deuterium
DCM	Dichloromethane
DIPEA	*N,N*-Diisopropylethylamine
DMAD	Dimethyl acetylenedicarboxylate
DMPM	3,4-Dimethoxybenzyl
DMF	*N,N*-Dimethylformamide
de	Diastereomeric excess
ee	Enantiomeric excess
Et	Ethyl
EDGs	Electron donating groups
ERGs	Electron releasing groups
EWGs	Electron withdrawing groups
FMOs	Frontier molecular orbitals
g	Gram
h	Hour
HOMO	Highest occupied molecular orbital
HCMV	Human cytomegalovirus
HLE	Human leukocyte elastase
INAC	Intramolecular nitrone-alkene cycloaddition
LUMO	Lowest unoccupied molecular orbital
MeOH	Methyl alcohol
MWI	Microwave irradiation
µ	Micro
min.	Minute
m-CPBA	*meta*-Chloroperoxybenzoic acid
mL	Millilitre

mmol	Millimole
Me	Methyl
Ms	Mesylate
Nu	Nucleophile
Ph	Phenyl
Pr	Propyl
i-Pr	*iso*-Propyl
PMB	*p*-Methoxybenzyl
PMP	*p*-Methoxyphenyl
ppm	Parts per million
RCY	Radiochemical yield
Sat.	Saturated
SACW-3CR	Staudinger/aza-wittig/ugi three-component reaction
SPANC	Strain-promoted alkyne-nitrone cycloaddition
TEA	Triethylamine
Tf	Trifluoromethanesulfonyl
TFA	Trifluoroacetic acid
TMG	1,1,3,3-Tetramethylguanidine
Tr	Trityl
Ts	Tosyl
THF	Tetrahydrofuran

1. INTRODUCTION

Heterocyclic compounds have attracted significant attention of research community ever since the search for chemical compounds of biological importance had started [1]. Heterocyclic rings containing *N*, *O* and *S* of different sizes are well acknowledged in the medicinal field [2]. These also constitute the structure of many biologically active compounds such as alkaloids, steroids and nucleic acids. Therefore, synthesis of heterocyclic compounds has remained an area of continuous exploration and research.

Construction of heterocyclic rings can be achieved via mainly two different strategies; Cycloaddition and Cyclization reactions. Out of these two, cycloaddition reactions are more popular in comparison to cyclization reaction due to their high efficiency, high atom economy, easily accessible substrates and carried out under mild reaction conditions.

In general, cycloaddition reactions involve formation of two new σ bonds. Depending upon the number of atoms contributing towards new ring, these reactions can be classified into various categories i.e., [2+2] cycloaddition reaction, [4+2] cycloaddition reaction and other cycloaddition reactions. A general representation of these cycloaddition reactions is depicted below (Figure 1).

Figure 1. General representation of cycloaddition reactions.

1.1. 1,3-Dipolar Cycloaddition

The 1,3-dipolar cycloaddition or [3+2] cycloaddition reactions are pioneered by Rolf Huisgen, hence also known as Huisgen cycloaddition [3]. These cycloaddition involves reaction between a 1,3-dipolar compounds and a dipolarophile to afford a five membered heterocyclic product such as pyrazoles, pyrazolones, isoxazoles, isoxazolones, 1,2,4-oxadiazolines etc. 1,3-Dipolar compound can be assumed as made up of three atoms (**a–b–c**) and possess delocalized electrons along with charge separation spread over three atom. A general representation of 1,3-dipolar compounds and 1,3-dipolar cycloaddition reactions is presented in Figure 2.

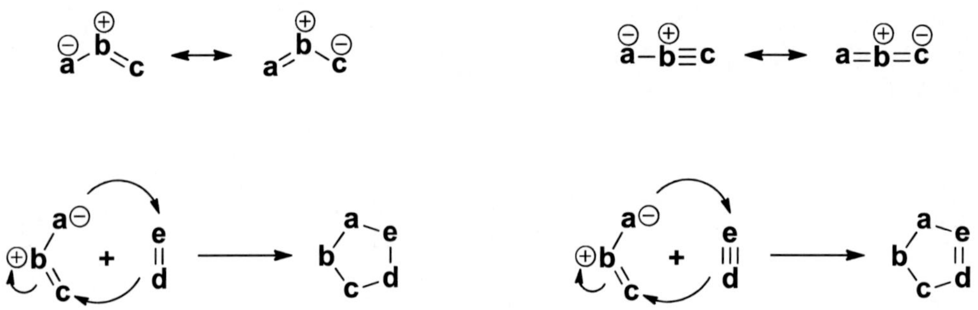

Figure 2. 1,3-Dipolar cycloaddition reaction.

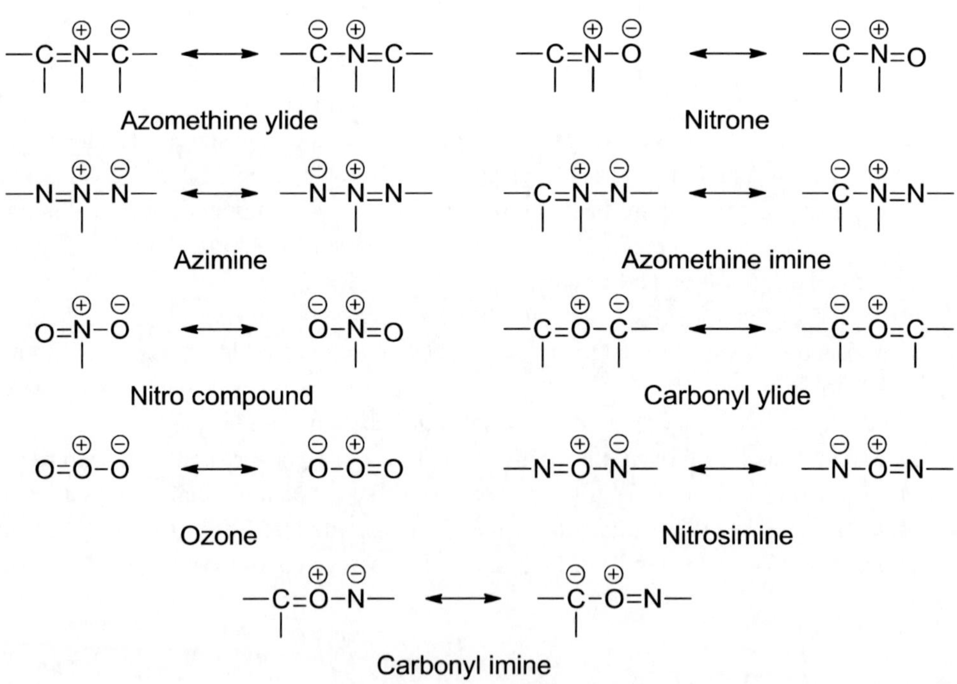

Figure 3. Allyl type 1,3-dipoles.

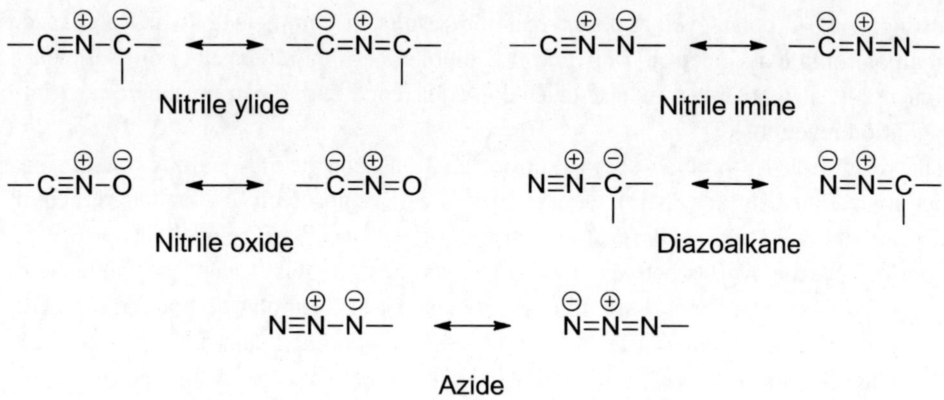

Figure 4. Propargyl-allenyl type 1,3-dipoles.

A wide variety of 1,3-dipolar compounds are present in literature. 1,3-Dipolar compounds are mainly composed of *C*, *N* and *O* in different combinations. The 1,3-dipolar compounds can be broadly divided into two main categories. First category is allyl type 1,3-dipoles (Figure 3) which includes compounds viz. azomethine ylides and imines, carbonyl ylides and imines, nitrones, ozone etc. The second category is propargyl-allenyl type 1,3-dipoles (Figure 4) which constitutes nitrile ylides and imines, diazoalkanes, azides etc. All the 1,3-dipolar compounds exhibit both nucleophilic as well as electrophilic character which plays important role in their reactivity towards dipolarophiles. This means that 1,3-dipolar compounds having high nucleophilic character (e.g., diazomethane, nitrile ylide) reacts faster with electron deficient dipolarophiles while reverse is true for electrophilic 1,3-dipolar compounds (e.g., ozone).

On the other hand, any compound containing double or triple bond can act as dipolarophiles. Most common examples of dipolarophiles include alkenes, alkynes and molecules with functional groups having heteroatom linked multiple bond such as carbonyl, nitrile, imine etc.

1.2. Mechanism of 1,3-Dipolar Cycloaddition

The 1,3-dipolar cycloaddition reaction involves participation of 4π electrons of 1,3-dipolar compound and 2π electrons of dipolarophile to afford the product via pericyclic electron shift. This shift is concerted and suprafacial in nature. Therefore, it is designated as $[\pi^4{}_s + \pi^2{}_s]$ cycloaddition reaction similar to Diels-Alders cycloaddition. These cycloadditions takes place via interactions between highest occupied molecular orbital (HOMO) and lowest unoccupied molecular orbital (LUMO) of both the substrates (1,3-dipole and dipolarophile). In general, presence of electron donating groups on dipolarophile favors the interaction between LUMO of 1,3-dipole and HOMO of dipolarophile and inverse is true for dipolarophile with electron withdrawing groups.

Firestone proposed that it is a two step process and proceed via formation of a diradical intermediate [4]. However, stereospecificity of the reaction indicate the involvement of concerted mechanism [5]. A concerted mechanism provides best explanation of various experimental observations: stereoselectivity in case of isomeric dipolarophiles (*cis-trans*), orientation phenomenon, effects of substitutions and solvent on rate constant etc [6].

Interestingly, 1,3-dipolar cycloaddition reaction results in complete loss of stereospecificity when dipolarophile is electron rich and 1,3-dipolar compound is electron deficient. This phenomenon is also observed in case of Diels-Alder reaction and can be generalized for other cycloaddition reactions [7].

The results of 1,3-dipolar cycloaddition reactions can be rationalized on the basis of frontier molecular orbital (FMO) theory. In general, regioselectivity is determined by the interaction between atoms of largest orbital coefficient on HOMO and LUMO [8,9]. These reactions involve the overlap between HOMO of one substrate with LUMO of another substrate in bonding fashion. The FMO diagram of diazomethane (1,3-dipolar compound) and ethylene (dipolarophile) and interaction of HOMO and LUMO in bonding fashion are depicted in Figure 5 and Figure 6 respectively. The 1,3-dipolar cycloaddition reactions are thermally allowed similar to Diels-Alder cycloaddition.

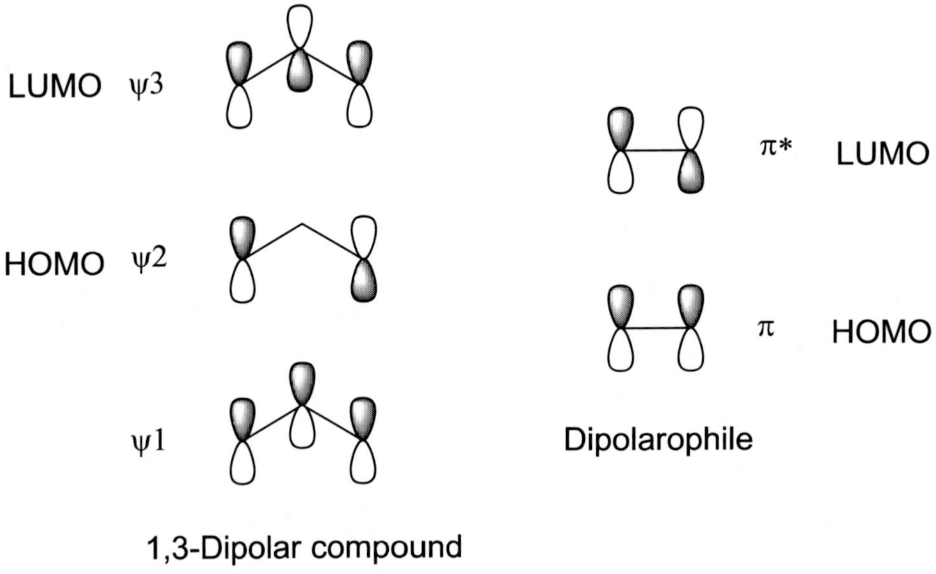

Figure 5. FMO diagram of 1,3-dipolar compound and dipolarophile.

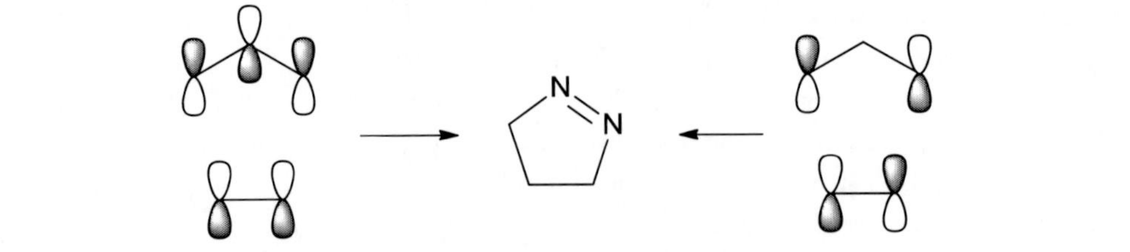

Figure 6. HOMO-LUMO interaction in bonding fashion.

2. NITRONE-OLEFIN CYCLOADDITION REACTIONS

One of the most common and important class of 1,3-dipolar cycloaddition reactions is nitrone-olefin cycloaddition reaction. In this reaction, nitrone acts as 1,3-dipolar compound and olefin acts as dipolarophile. The reaction involves three atoms of nitrone (1,3-dipolar compound) and two atoms of alkene (dipolarophile) add to each other to form a five membered heterocyclic adduct. Therefore, it is also known [3+2] cycloaddition reaction.

Nitrones, an important class of allyl type 1,3-dipolar compound which is frequently used for cycloaddition reactions. Chemically nitrones are N-oxide of differently substituted imine derivatives with general formula $R^1R^2C{=}N^+R^3O^-$ (Figure 7). 1,3-Dipolar cycloaddition of nitrones results in the formation of five membered heterocyclic rings viz. isoxazole, isoxazolines, isoxazolidines and β-lactams. These heterocyclic rings are biologically active in nature and act as potential synthons for compounds of medicinal interest [10, 11].

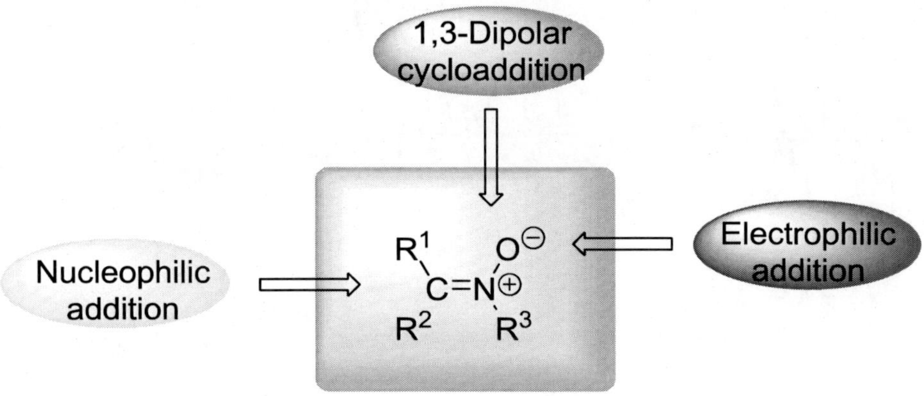

Figure 7. Reactivity of nitrones.

2.1. Intermolecular [3+2] Nitrone-Olefin Cycloaddition Reactions

Shindo and co-workers [12] have reported asymmetric synthesis of 5-isoxazolidinone derivatives **3-4** via 1,3-dipolar cycloaddition reaction between nitrones **1** and lithium ynolates **2** in very good yields (Scheme 1). The chiral nitrone **1** was derived from another chiral intermediate i.e., Garner's aldehyde. The reaction resulted in the formation of different isomeric isoxazolidinones viz. (3*S*,4*R*), (3*S*,4*S*) and (3*R*,4*S*). The (3*S*,4*R*)-**4** was obtained as major product with excellent diastereoselectivities (up to > 99:1) and results revealed that ynolates exhibited better selectivities in comparison to enolates. Furthermore, the alkylation of 5-isoxazolidinone enolates and their transformation into enantiomerically pure β-lactams, γ-lactams and β-amino acids was successfully achieved.

Garcia-Tellado et al. [13] have reported first ever example of regioselective 1,3-dipolar cycloaddition in water (Scheme 2). The reaction was performed between nitrones **5** and conjugated alkynoates **6** in the presence of an organic catalyst. The more reactive nitrones exhibited very good results. However, the less reactive nitrones required the aid of LiCl. Moreover, PPh_3 produce better results with aromatic nitrones while quinuclidine was found to be good in case of aliphatic nitrones. No formation of product was achieved when the reaction was carried out in organic solvents. The reaction was believed to proceed via *in-situ* formation of β-phosphonium or ammonium allenolates.

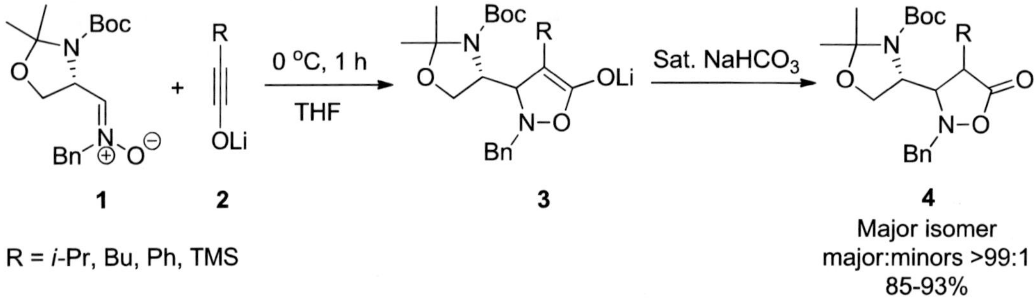

Scheme 1. Asymmetric synthesis of 5-isoxazolidinones.

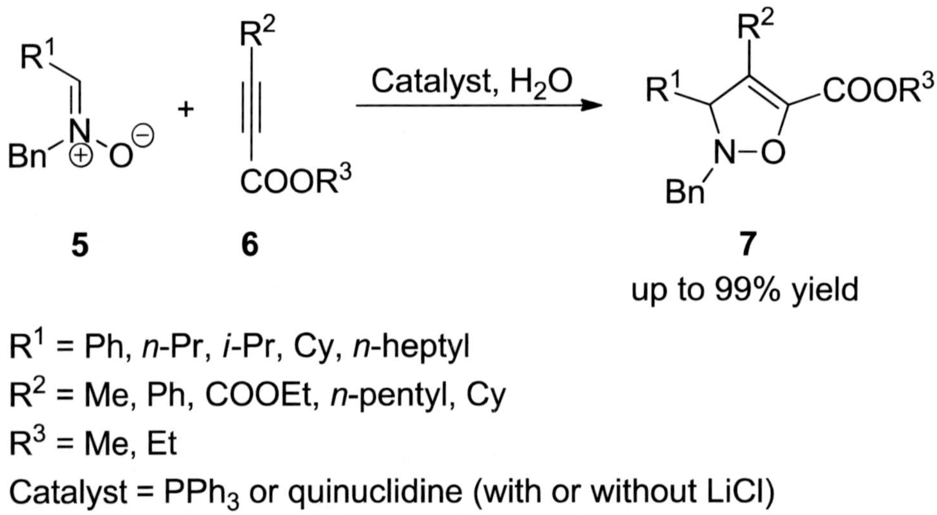

Scheme 2. Organocatalyzed [3+2] cycloaddition reaction in water.

A facile three-step strategy involving 1,3-dipolar nitrone-alkyne cycloaddition followed by oxido-alkylidenation (Scheme 3) has been reported by Houk et al. [14]. Initially, differently substituted nitrones **8** and alkynoates **9** undergo 1,3-dipolar cycloaddition under mild conditions to afford isoxazoline derivatives **10** in good to excellent yields. These isoxazolines **10** were further subjected to oxidation with *m*-CPBA which resulted in the stereoselective extrusion of nitrosomethane to yield isomeric mixture of unsaturated ketones **11-12**. The stereoselectivity and torquoselectivity arises due to steric hindrance between α-substituent of isoxazoline nitrogen (R^1) and oxidizing agent which was further supported by quantum mechanical model. It was found that *in-out* selectivity is reversed when R^1 possess large *ortho*-substituents.

Coutouli-Argyropoulou et al. [15] have carried out 1,3-dipolar cycloadditions of various hydrophobic nitrones **13** with differently substituted acrylates **14** in organic solutions as well as aqueous suspensions (Scheme 4). The reaction was more selective for isoxazolidine derivatives having 1,3-substituents in *trans*- configuration. Moreover, the reaction was much faster in case of water suspension. The best results in terms of rate accelerations were obtained when reaction was performed between D-ribose derived nitrone and methyl acrylate in water suspension (144 h to 6 h). The authors have also concluded that hydrophobic interactions could be the reason for rate accelerations and stereoselectivities.

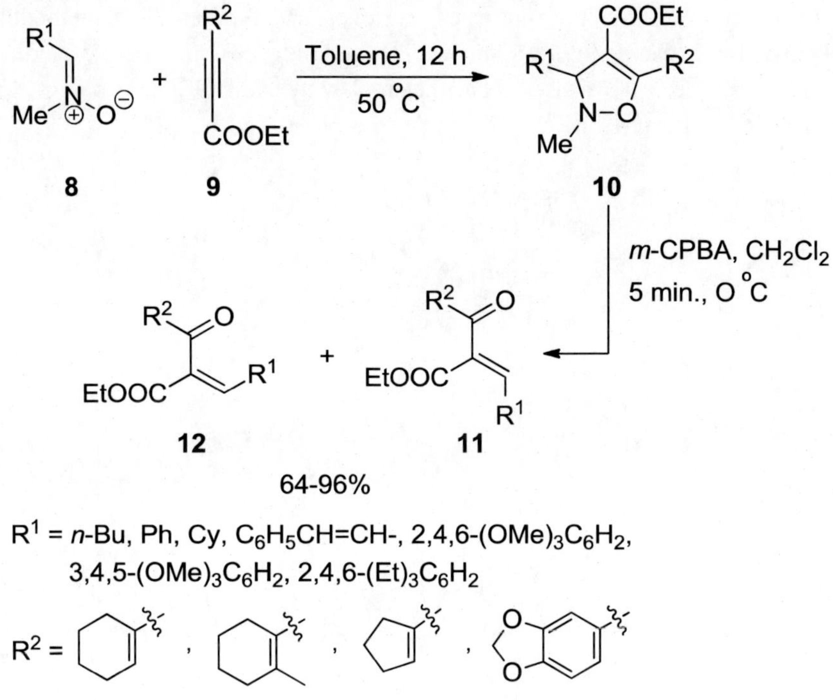

Scheme 3. Stereoselective synthesis of isoxazolines and unsaturated ketones.

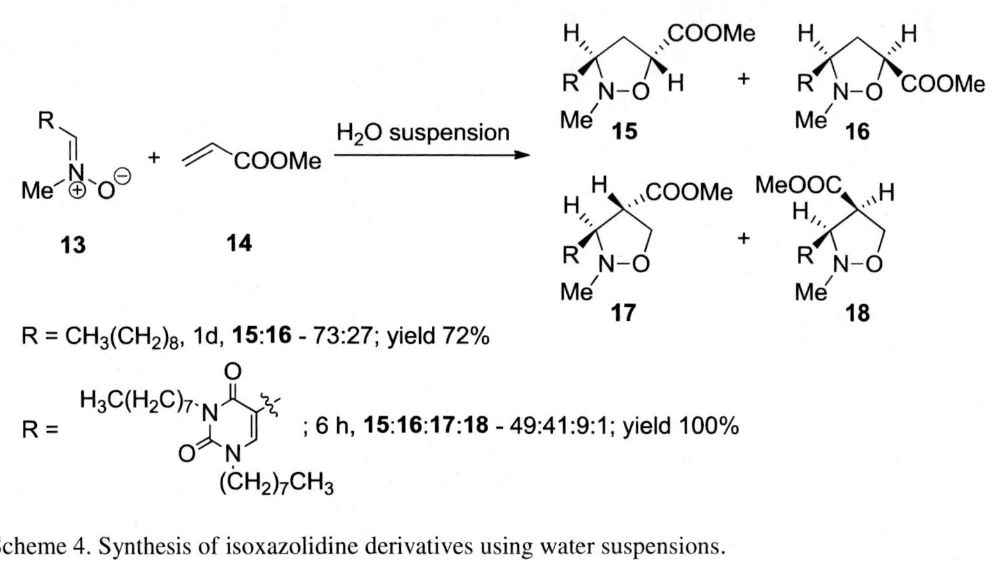

Scheme 4. Synthesis of isoxazolidine derivatives using water suspensions.

A mild and convenient approach for the synthesis of aryl vinyl or divinyl ketones **22-23** (Nazarov cyclisation substrates) involving 1,3-dipolar cycloaddition followed by oxidation (Scheme 5) has been reported by Canterbury et al. [16]. The reaction was carried out between different nitrones **19** and electron deficient alkynes **20** which resulted in the formation of olefin products with moderate to high selectivities. The reaction was applicable with wide range of substrates viz. alkynyl esters, phosphonates, sulfones and amides. The methodology was mainly selective for *in*-isomer with some exceptions and does not require acidic or basic conditions. Furthermore, it was also found that replacement of OEt with larger *t*-Bu group reduces the selectivity for *in*-isomer.

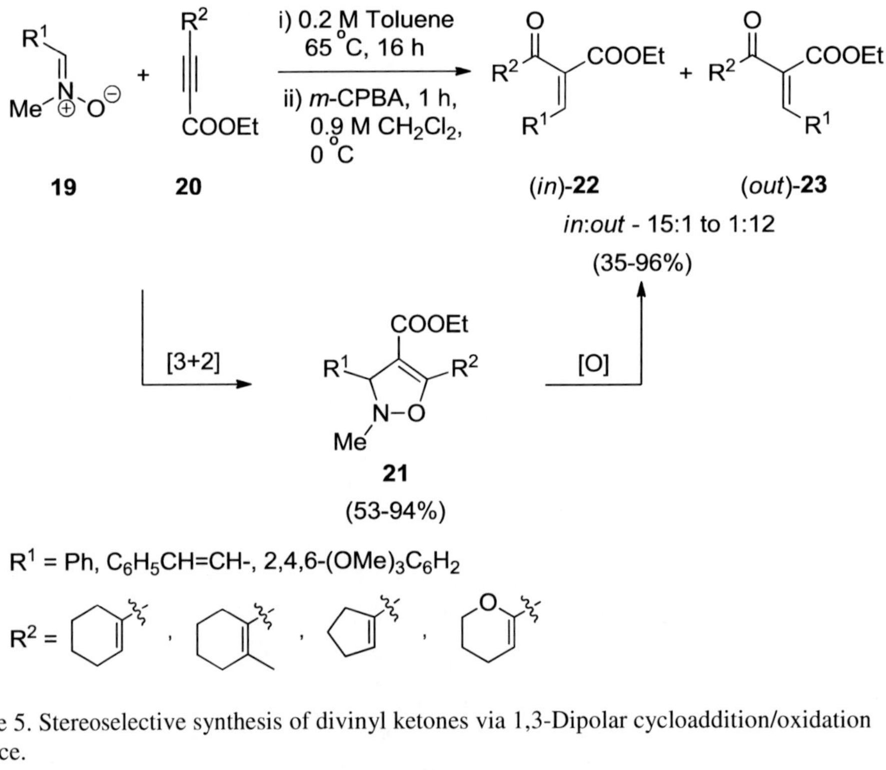

Scheme 5. Stereoselective synthesis of divinyl ketones via 1,3-Dipolar cycloaddition/oxidation sequence.

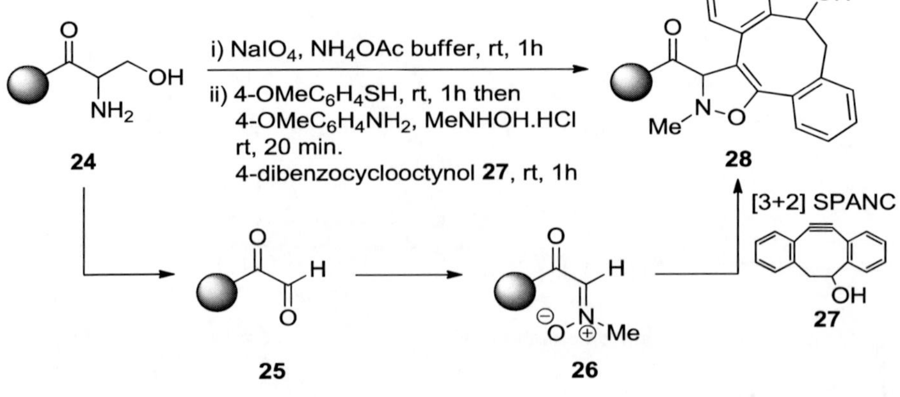

Scheme 6. One pot protein modification by SPANC.

van Delft and co-workers [17] have described novel and fast one-pot strategy involving strain-promoted alkyne-nitrone cycloaddition (SPANC) for site-specific protein modification. The 1,3-dipolar cycloaddition was carried out between various nitrones **26** and cyclooctynes **27** having ester or amide functionalities (Scheme 6). The SPANC reaction was found to be extremely fast in compared to azide-alkyne reactions.

Lopes et al. [18] have synthesized 4-isoxazoline **32** and pyrrole derivatives **33** via regioselective 1,3-dipolar cycloaddition reaction in moderate yields (Scheme 7). The nitrones **30** were derived from differently substituted allenoates **29** and made to react with dimethyl acetylenedicarboxylate (DMAD) **31** and other analogues. It was found out that thermal rearrangement of isoxazoline derivatives **32** yields pyrrole derivatives **33**. Therefore, reaction conditions (temperature) can be controlled to afford the desired product. Moreover, quantum mechanical calculations performed on mechanistic of thermal rearrangement have supported the intermediacy of 2-acylaziridine.

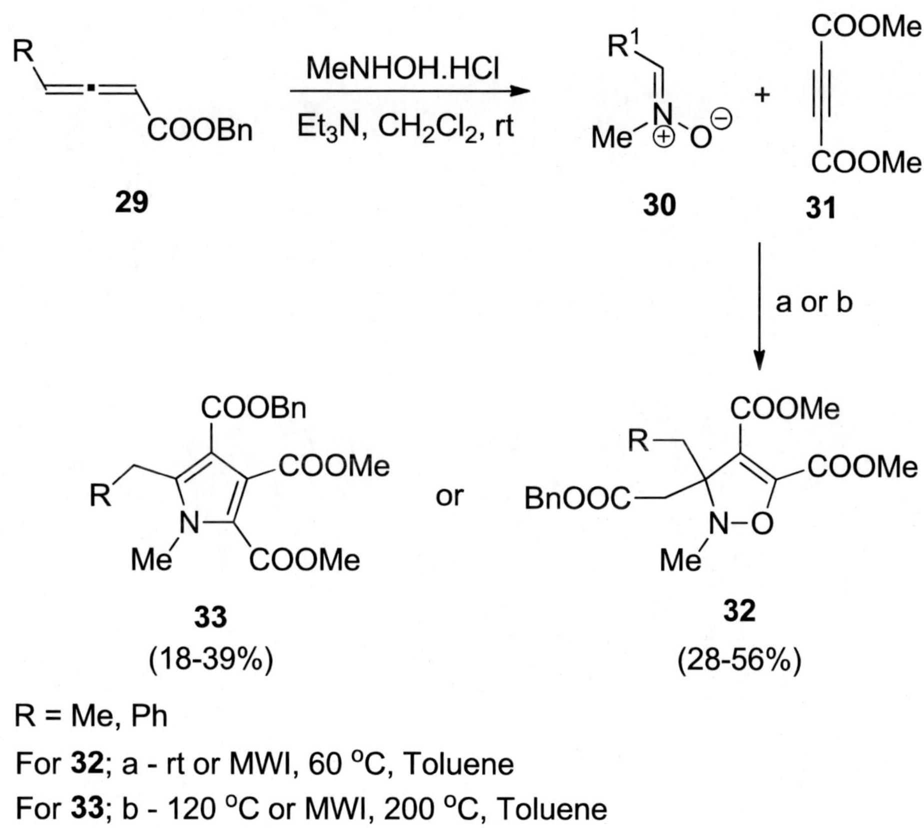

Scheme 7. Synthesis of isoxazoline and pyrrole derivatives.

A facile and efficient synthesis of 4-isoxazolines **36** via $TiCl_4$ catalyzed [3+2] nitrone-alkyne cycloaddition (Scheme 8) has been reported by Valizadeh and Dinparast [19]. The reaction was carried out under solvent-free conditions and resulted in the formation of products in excellent yields. Further, low reaction time and low cost of the catalyst enhances the practical utility of the methodology.

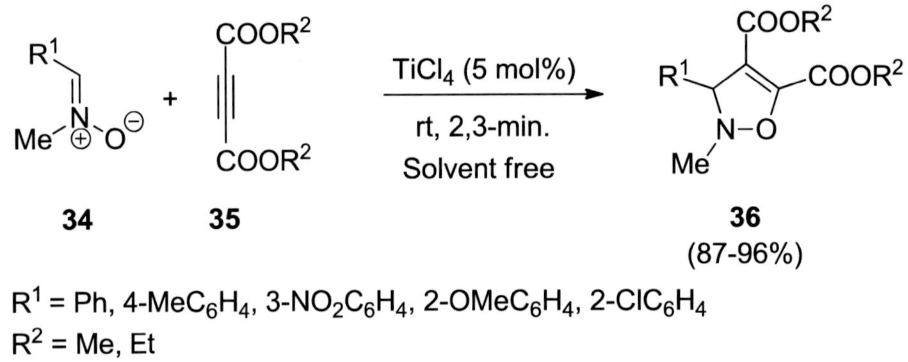

Scheme 8. Solvent free synthesis of isoxazolines via $TiCl_4$ catalyzed 1,3-dipolar cycloaddition.

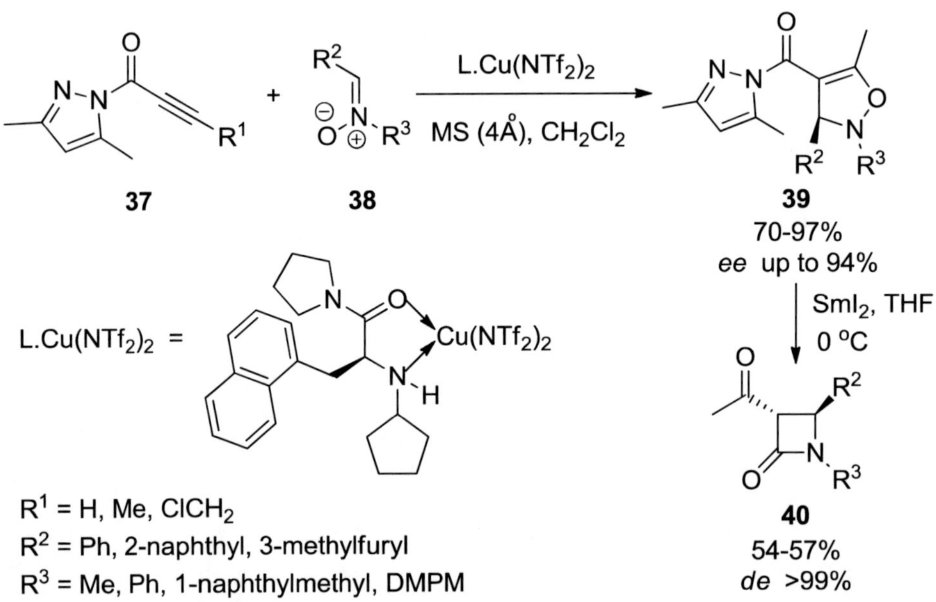

Scheme 9. Enantioselective synthesis of pyrazole linked isoxazoline derivatives.

A novel series of pyrazole linked isoxazoline derivatives has been synthesized by Sakakura et al. [20]. The 1,3-dipolar cycloaddition reaction was carried out between propiolyl pyrazole **37** and various nitrones **38** in the presence of chiral catalyst which furnished isoxazoline derivatives **39** in very good yields (Scheme 9). The isoxazolines were further transformed into β-lactams **40** via samarium halide mediated N-O reductive cleavage.

Worgull and co-workers [21] have reported synthesis of novel bicyclic isoxazole derivatives **44** via intermolecular 1,3-dipolar cycloaddition reaction between various α,β-unsaturated aldehyde **41** and malononitrile derivatives **42** (Scheme 10). This protocol afforded the products in good yields and exhibited high diastereo and enantioselectivity. Further, the octahydrobenzo[*c*]isoxazole derivatives were successfully transformed into various 1,3-amino alcohols.

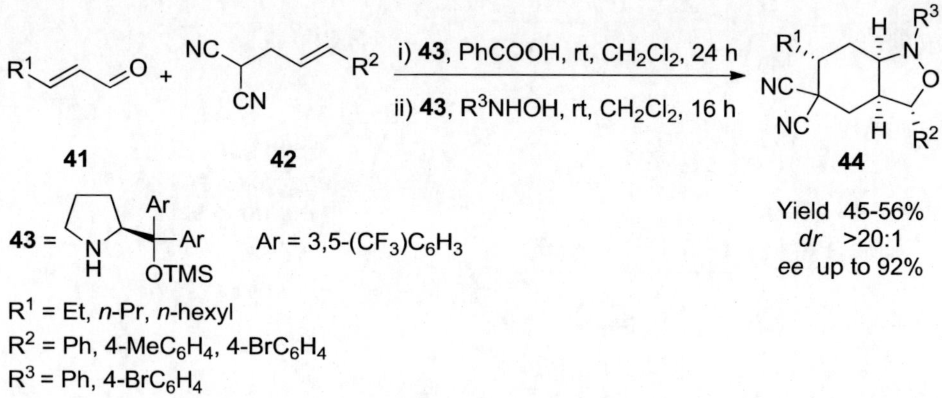

Scheme 10. Diastereo- and enantioselective synthesis of bicyclic isoxazolidine derivatives.

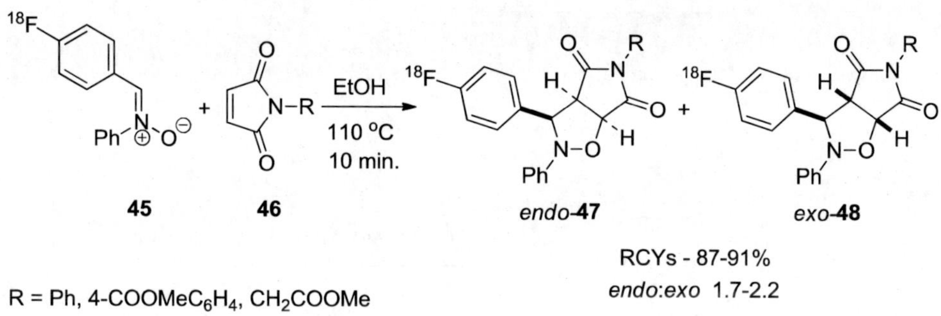

Scheme 11. Synthesis of [^{18}F]-labeled isoxazolidine derivatives.

Zlatopolskiy and co-workers [22] have carried out radiofluorination via 1,3-dipolar cycloaddition reaction using C-(4-[^{18}F]fluorophenyl)-N-phenyl nitrone **45** as novel synthon (Scheme 11). The nitrone **45** was prepared with high radiochemical purity (> 95%) and very good yields (74%). On the other hand, 1,3-dipolar reaction between nitrone **45** and diversely substituted maleimide **46** resulted in the formation of isomeric mixture of isoxazolidines in high RCYs. The reaction was found to be more selective for *endo*-products.

A mild and novel methodology for the synthesis of *N*-alkyl/benzyl isoxazolidine derivatives **51-52** via highly selective 1,3-dipolar cycloaddition reaction (Scheme 12) has been reported by Badoiu and Kundig [23]. The reaction was carried out between *N*-alkyl/benzyl nitrones **49** and various dipolarophiles **50** to afford the products in moderate to very good yields. The nature of the substituents on the nitrone moiety play important role in regulating the diastereoselectivity. The electron withdrawing groups (EWGs) on α-aryl group of nitrone favors *endo* product whereas electron donating groups (EDGs) yields *exo*- as major product. In addition to this, the *N*-alkyl/benzyl nitrones afforded isoxazolidines with *endo*- as major products. Furthermore, the reaction was also tried in the presence of a chiral Ru based catalyst which furnished the products in good yields with high region, diastereo and enantioselectivity.

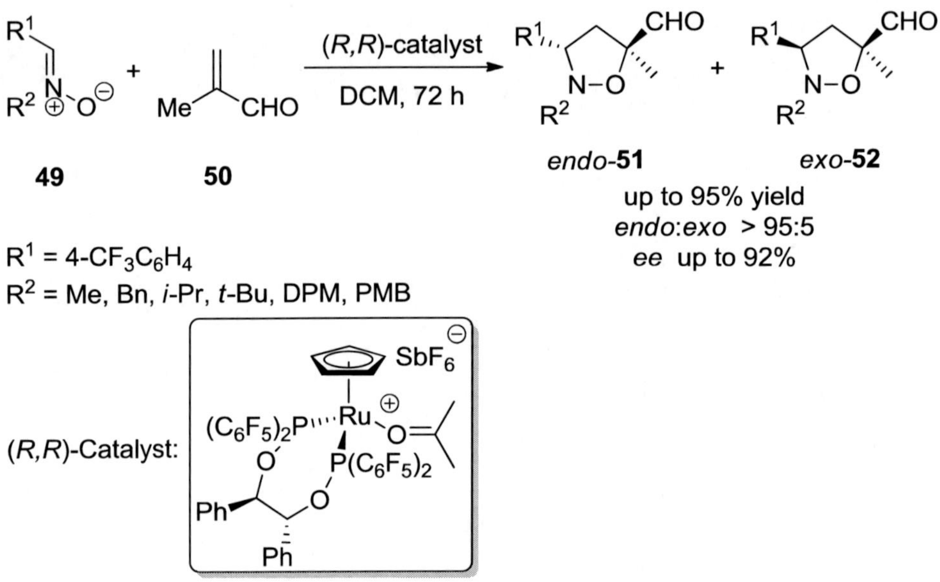

Scheme 12. Regio-, diastereo- and enantioselective synthesis of isoxazolidines.

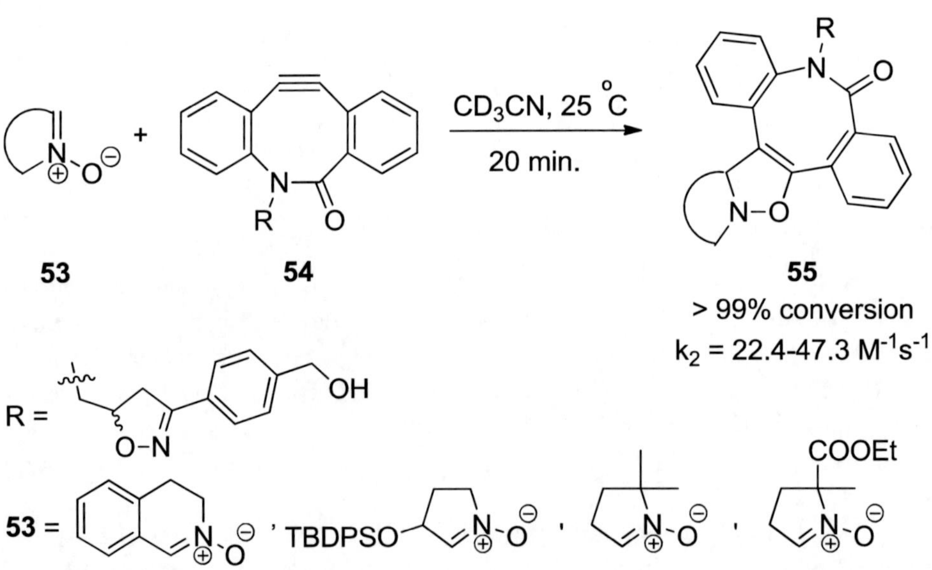

Scheme 13. Kinetic study of strain-promoted alkene-nitrone (SPANC) cycloaddition reaction.

Pezacki and coworkers [24] have carried out studies on strain-promoted 1,3-dipolar cycloaddition of various cyclic nitrones **53** and biaryl-aza-cyclooctynone (BARAC) **54** (Scheme 13). The reaction was found to be extremely fast with rate constant of 47.3 $M^{-1}s^{-1}$. The results concluded that five membered cyclic nitrones exhibited fast rate kinetics in comparison to six membered cyclic nitrones. Further, reaction was sensitive towards group present on nitrogen of nitrone but no significant effect was observed for groups present on α-carbon of nitrone. From linear free energy relationship studies, the rho value was found to be 0.25±0.04 which indicated the low sensitivity to substituents and thus high applicability.

Chakraborty and Luitel [25] have successfully carried out 1,3-dipolar cycloaddition using 1-butyl-3-methyl-imidazolium based ionic liquids. The reaction between *N*-benzyl-fluoro nitrones **56** and various activated alkenes/electron deficient alkynes in the presence of ionic liquid resulted in the formation of isoxazolines **57** and isoxazolidines **58-59** in very good yields (Scheme 14). The authors have also studied the recyclability of ionic liquid catalyst which revealed no loss of activity up to 5th cycle.

Tran and co-workers [26] have reported the synthesis of novel spiro-isoxazolidines **62** via 1,3-dipolar cycloaddition reaction followed by base promoted ring opening-cyclization to afford bicyclic β-lactams **63** in very good yields (Scheme 15). Initially, the cycloaddition reaction was carried out between various *C*-carbamoyl nitrones **60** and dialkyl substituted 2-benzylidenecyclopropane-1,1-dicarboxylates **61**. The reaction afforded spiro[cyclopropane-1,4-isoxazolidine] cycloadduct **62** in moderate to excellent yields with high selectivity.

Mandoli et al. [27] have reported a novel chiral organocatalyst immobilized on a monolithic reactor which can be utilized for carrying out various cycloaddition reactions with high diastereo and enantioselectivities. The flow reactor was prepared by radical copolymerization of divinyl benzene followed by modification with chiral imidazolidinone in the presence of porogens (dodecanol and toluene) inside a steel column. In case of nitrone-olefin 1,3-dipolar cycloaddition reaction, the products **66-67** were obtained in very good yields with high enantioselectivities (Scheme 16).

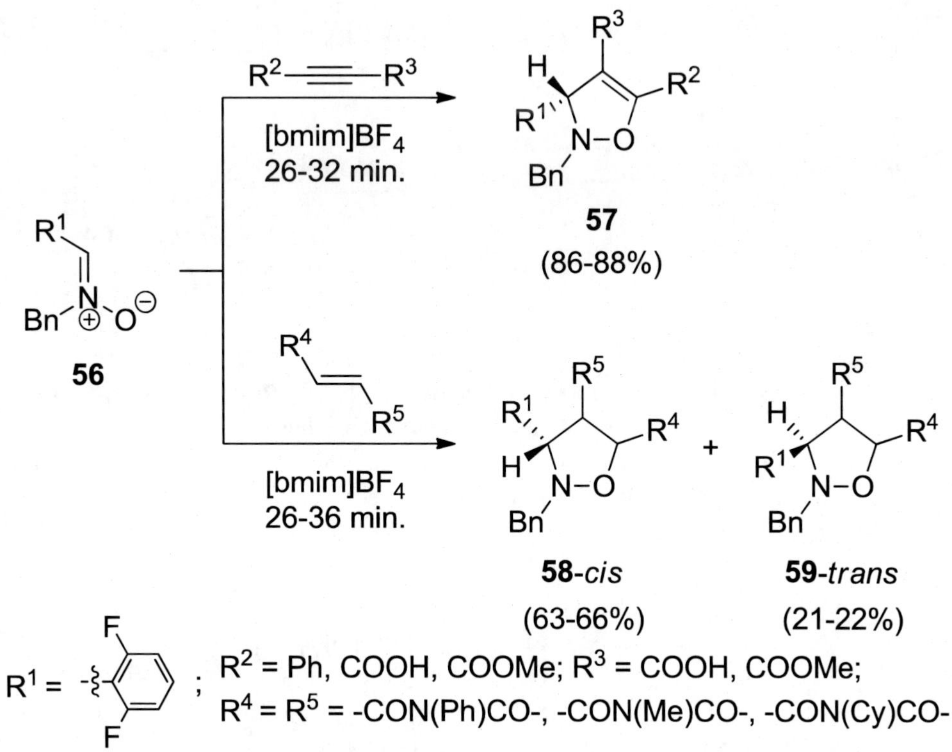

Scheme 14. Preparation of fluoroisoxazolines and fluoroisoxazolidines using ionic liquids.

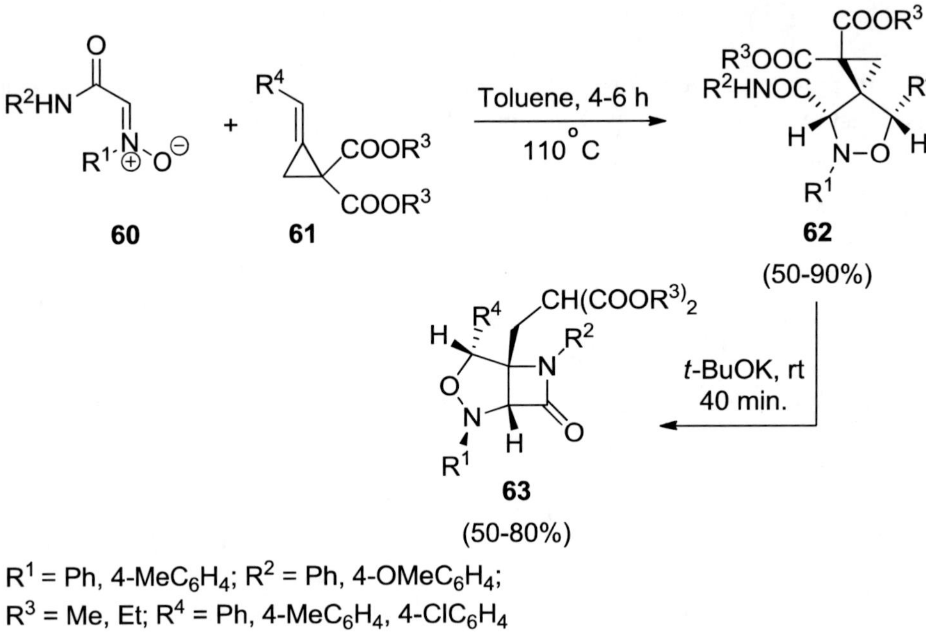

Scheme 15. Efficient synthesis of isoxazole fused bicyclic β-lactams.

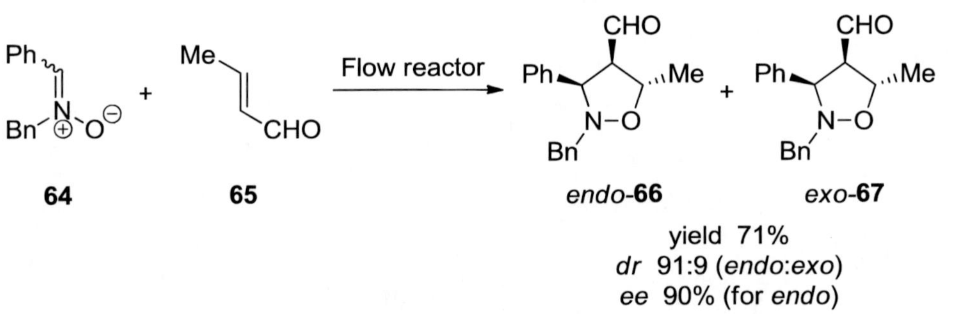

Scheme 16. Enantioselective catalytic synthesis of isoxazolidines under continuous-flow conditions.

A stereoselective synthesis of various cycloadducts using (-)-menthone derived nitrone **68** as a substrate has been reported by Aouadi et al. [28] (Scheme 17). Initially, 1,3-dipolar cycloaddition reaction was carried out between menthone derived nitrone **68** and allyl bromide under microwave irradiation. The reaction resulted in the formation of product **69** in excellent yield. The bicyclic adduct **69** undergo ring opening and click reactions to afford novel heterocyclic adducts **70-72** in moderate to good yields.

Aouadi and co-workers [29] have reported a combined cycloaddition-cyclization strategy for the synthesis of unprecedented 4-hydroxyl proline/pyrrolidine derivatives **77** (Scheme 18). Initially, 1,3-dipolar cycloaddition reaction between (-)-menthone-derived nitrone **73** and (*Z*)-1,4-dichlorobutene **74** furnished dichlorinated isoxazolidine adducts **75** in good yields with high selectivity. This cycloadduct further undergo reductive cleavage of N-O bond of isoxazolidine ring followed by reduction to afford 4-hydroxy prolines **77** in moderate to good yield.

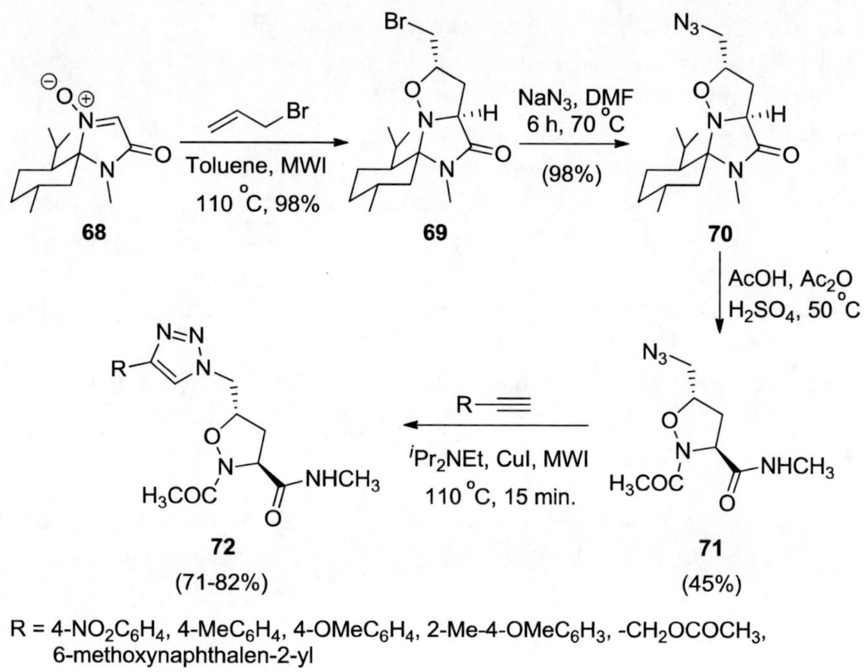

Scheme 17. Stereoselective synthesis of triazolyl substituted isoxazolidine derivatives.

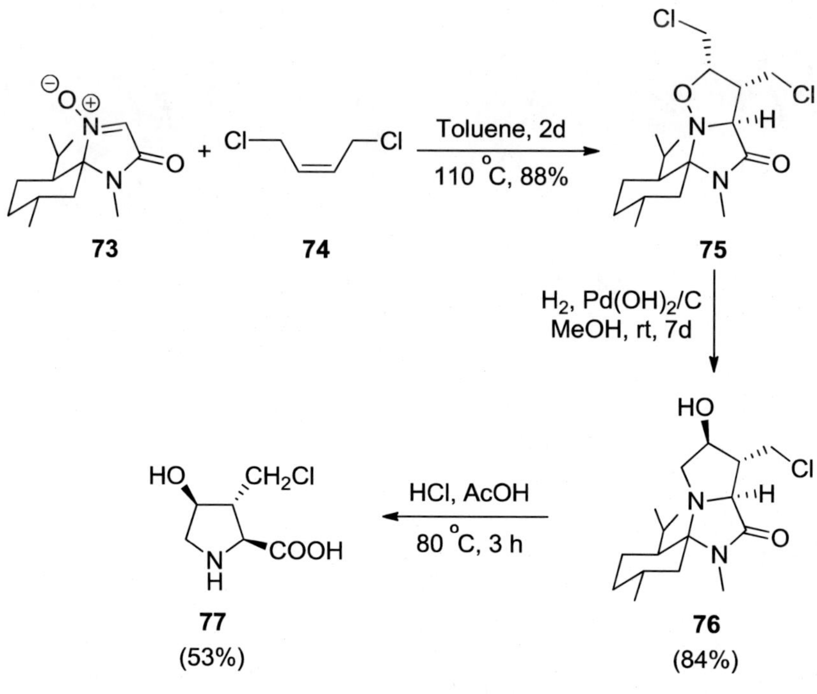

Scheme 18. Stereocontrolled synthesis of 3-substituted 4-hydroxyproline derivatives.

A facile synthesis of novel 1,5-dideoxy-1,5-iminoribitol iminosugar *C*-glycosides **81** using domino strategy involving [3+2] nitrone-olefin cycloaddition has been reported by Siriwardena et al. [30] (Scheme 19). The reaction started with *in-situ* generation of cyclic sugar nitrone **79** followed by its condensation with an alkene dipolarophile to yield corresponding isoxazolidines **80** with high stereoselectivity in one-pot strategy. The isoxazolidines **80** were further transformed into iminosugars **81** in very good yields.

Brahmi et al. [31] have carried out synthesis of isoxazolidine cycloadduct **83** and their transformation into enantiopure 4-hydroxy proline derivatives **87** using a fast and stereoselective approach (Scheme 20). The isoxazolidine derivatives **83** were prepared by [3+2] cycloaddition reaction between menthone derived nitrone **82** and (*Z*)-1,4-dichlorobutene. Further, **82** undergo regioselective azidation followed by azide-alkyne cycloaddition to afford 1,2,3-triazolyl substituted isoxazolidines **85**. Finally, cleavage of chiral auxiliary and reductive cleavage of N–O bond afforded proline derivatives **87** in good yields.

Wennekes and co-workers [32] have successfully synthesized differently substituted bicyclic isoxazolidines **90** starting from enantiopure furanone **89** (Scheme 21). Initially, various nitrones **88** undergo 1,3-dipolar cycloaddition reaction with chiral furanone **89** to furnish bicyclic isoxazolidine derivatives **90** followed by their subsequent modification in good yields. Finally, a series of pipecolic acid derivatives were prepared via one-pot SACW-3CR (Staudinger/aza-wittig/ugi three-component reaction) of bicyclic isoxazolidine cycloadduct **92**.

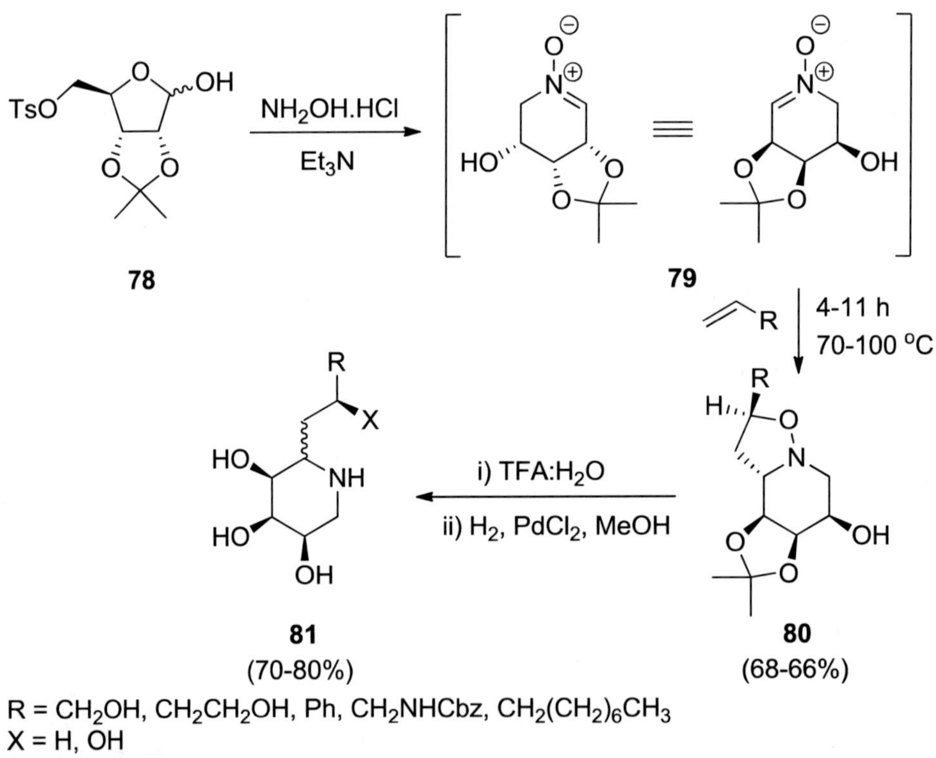

Scheme 19. Mild and efficient synthesis of novel iminosugars.

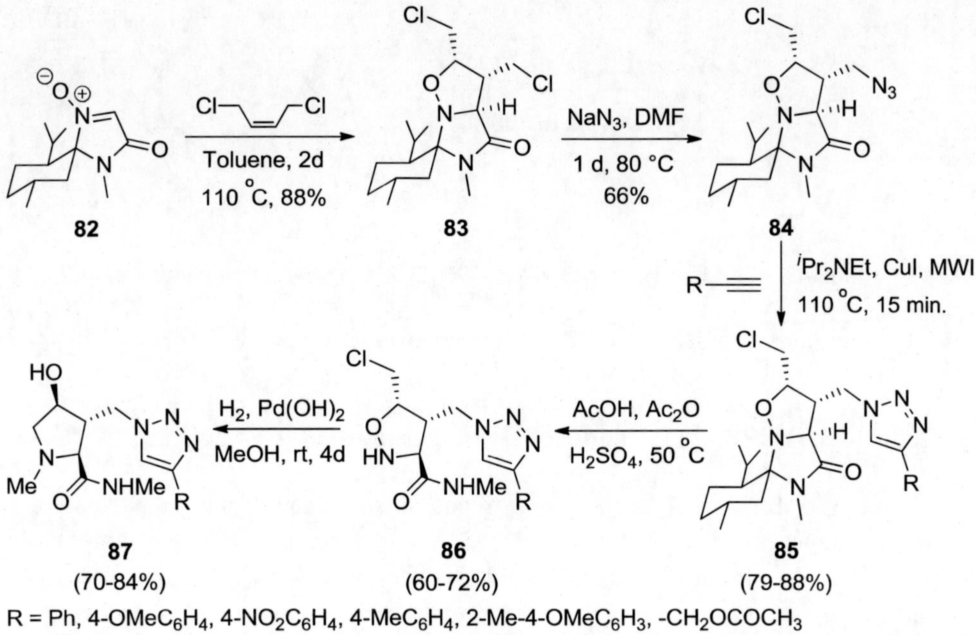

Scheme 20. Enantioselective synthesis of 4-hydroxyproline derivatives.

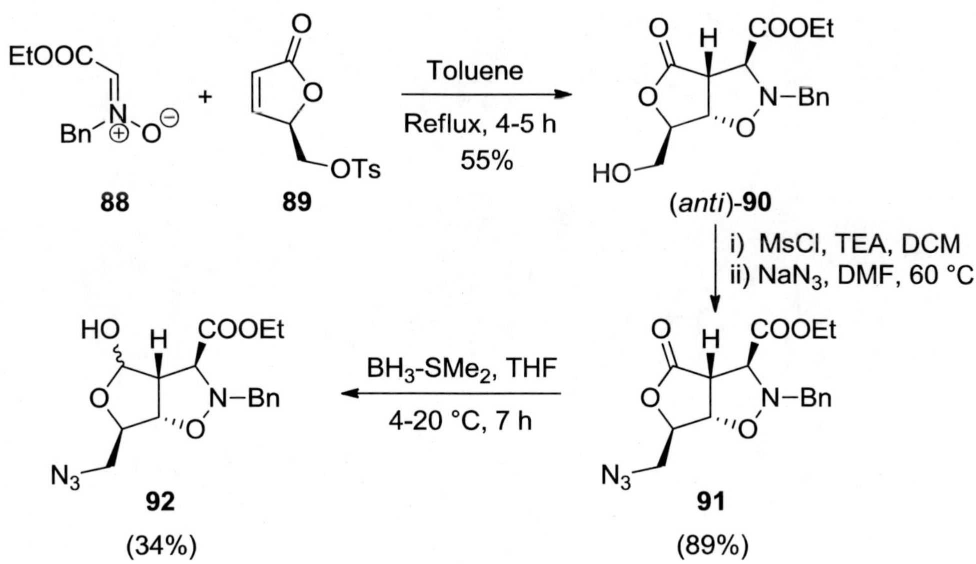

Scheme 21. Synthesis of bicyclic isoxazolidine-derived azidohemiacetal.

Abda et al. [33] has carried out intermolecular [3+2] nitrone-alkyne cycloaddition reaction under microwave irradiation (Scheme 22). The reaction furnished novel series of enantiopure aziridine derivatives **95-96** in moderate to good yields. The use of hindered chiral nitrone **93** was responsible for the formation of aziridine derivatives via thermal pathway.

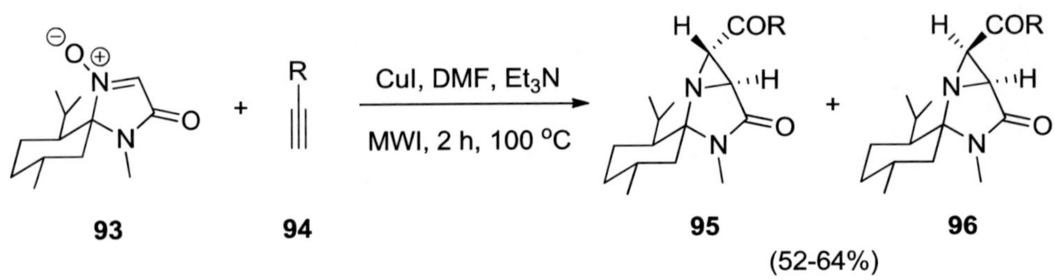

R = Ph, 4-MeC_6H_4, 4-$OMeC_6H_4$, 4-$NO_2C_6H_4$, -CH_2OCOCH_3, 6-methoxynaphthalen-2-yl

Scheme 22. Facile synthesis of enantiopure aziridine derivatives via nitrone-alkyne cycloaddition.

2.2. Intramolecular [3+2] Nitrone-Olefin Cycloaddition Reaction

Chatterjee and Bhattacharya [34] have reported surfactant catalyzed stereoselective synthesis of oxepanes **98** and pyrans **99** via intramolecular 1,3-dipolar nitrone-olefin cycloaddition reaction in aqueous medium in excellent yields (Scheme 23). The reaction resulted in exclusive formation of only one type of product. The nitrones derived from 3-*O*-allyl glucofuranose, corresponding allose derivatives and 3-*O*-crotyl glucofuranose yielded oxepanes whereas nitrones derived from 3-*O*-prenyl carbohydrates afforded pyrans respectively.

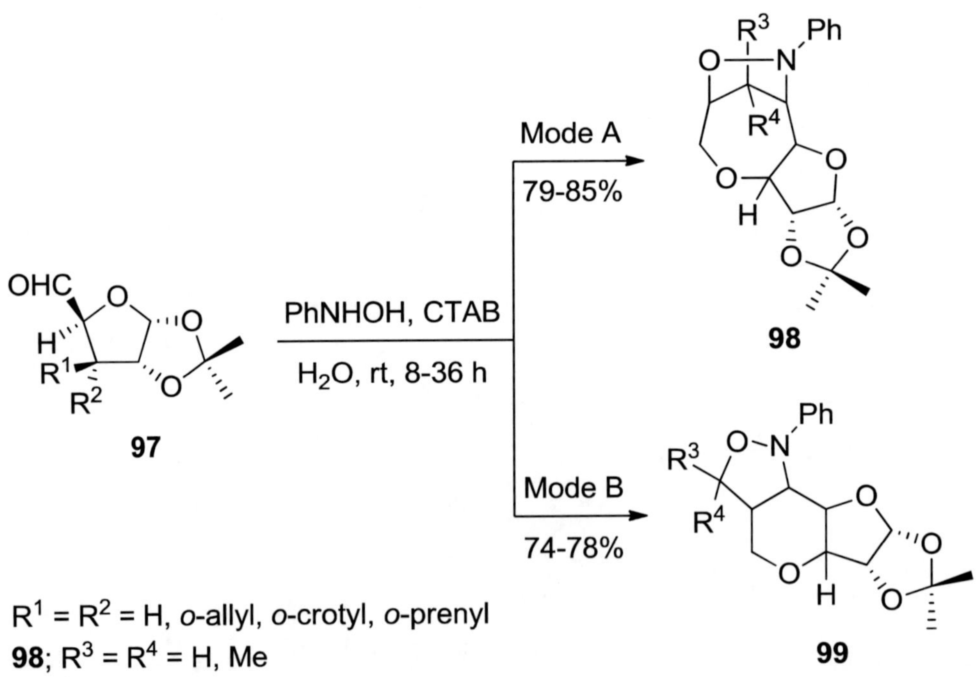

$R^1 = R^2$ = H, *o*-allyl, *o*-crotyl, *o*-prenyl
98; $R^3 = R^4$ = H, Me
99; $R^3 = R^4$ = Me

Scheme 23. Stereoselective synthesis of oxepanes and pyrans via 1,3-dipolar cycloaddition.

Shing et al. [35] have carried out extensive studies on regio and stereoselectivity in intramolecular [3+2] nitrone-olefin cycloaddition reaction of hept-6-enoses **100**. Various hept-6-enoses containing different blocking groups were treated with *N*-alkyl hydroxylamine to afford nitrone which subsequently undergo intramolecular 1,3-dipolar cycloaddition reaction and results in the formation of *cis*- and *trans*- fused isoxazolidines **102-103** (i.e., cyclohexanols) and bridged isoxazolidines **104-105** (i.e., cycloheptanols) (Scheme 24). The cyclohexanols and cycloheptanols were formed via *exo*- or *endo*- mode respectively. The *exo*- cyclisation was more favored when 2,3-*O*-isopropylidene ring is present in the substrate. However, the *endo*-mode of cyclisation was directed by *trans*-diacetal ring and is favored when 2-propanol is used as solvent. Moreover, *cis*-fused to *trans*-fused selectivity is also influenced by change of solvent. The experimental observations were further supported by computational studies.

A novel methodology for the synthesis of bicyclic isoxazolidines **110** via 1,3-dipolar nitrone-alkene cycloaddition using achiral **111** and chiral ionic liquids **112** has been reported by Hornyak et al. [36] (Scheme 25). The ω-unsaturated nitrone **109** was derived from D-galactose using a three step process. It was found in the results that ionic liquid catalyst **112** yielded only one isomer in comparison to ionic liquid catalyst **111** which gives two isomeric products.

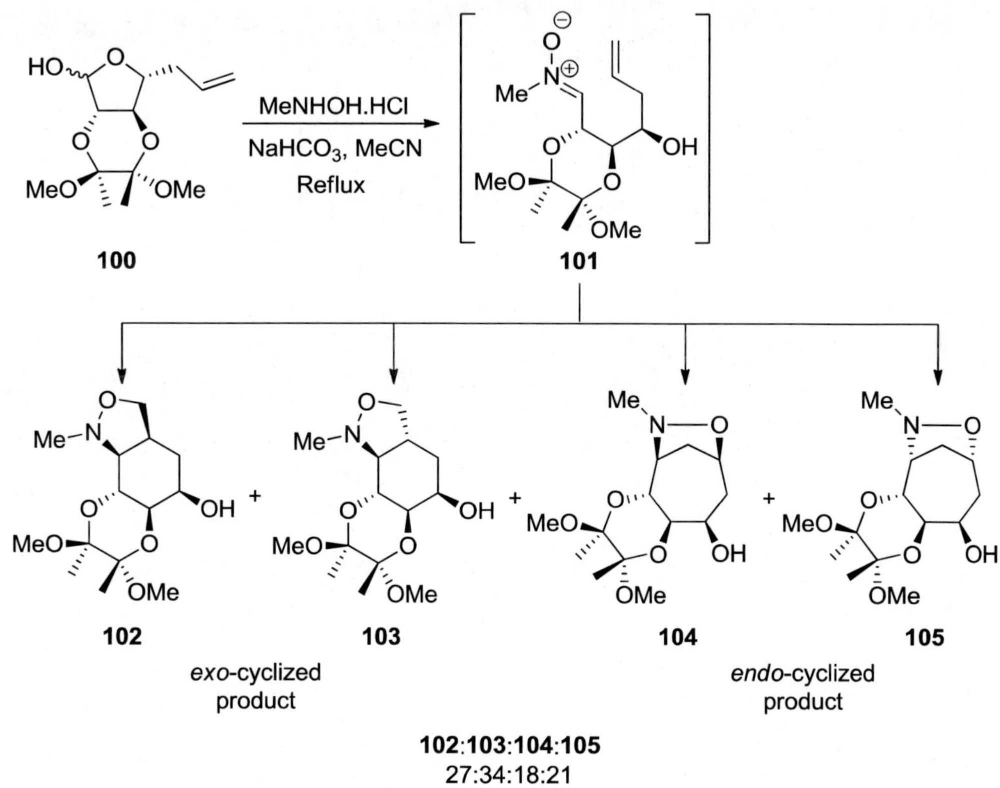

Scheme 24. Intramolecular alkene-nitrone cycloaddition of hept-6-enoses.

A novel series of naphtho[1,2-*b*]azepine derivatives **116** has been prepared via a mild and efficient protocol by Yepez et al. [37]. The reaction involves *in-situ* generated nitrones **114** from *N*-allyl-*N*-benzyl substituted α-naphthylamines **113** which subsequently undergo amino-Claisen rearrangement reaction, intramolecular Friedel-craft alkylation and nitrone-alkene cycloaddition to afford novel heterocyclic molecules **116** (Scheme 26).

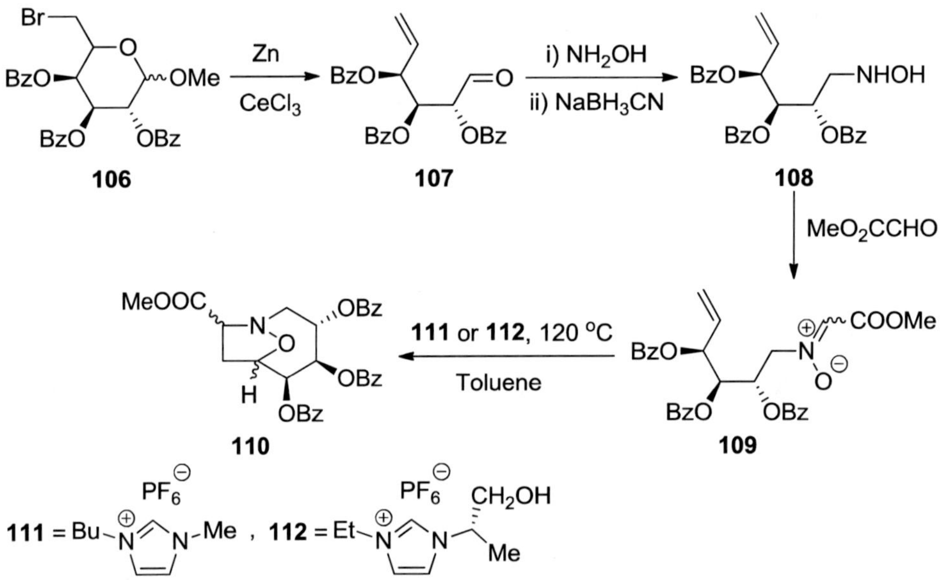

Scheme 25. Ionic liquid mediated synthesis of bicyclic isoxazolidine derivatives.

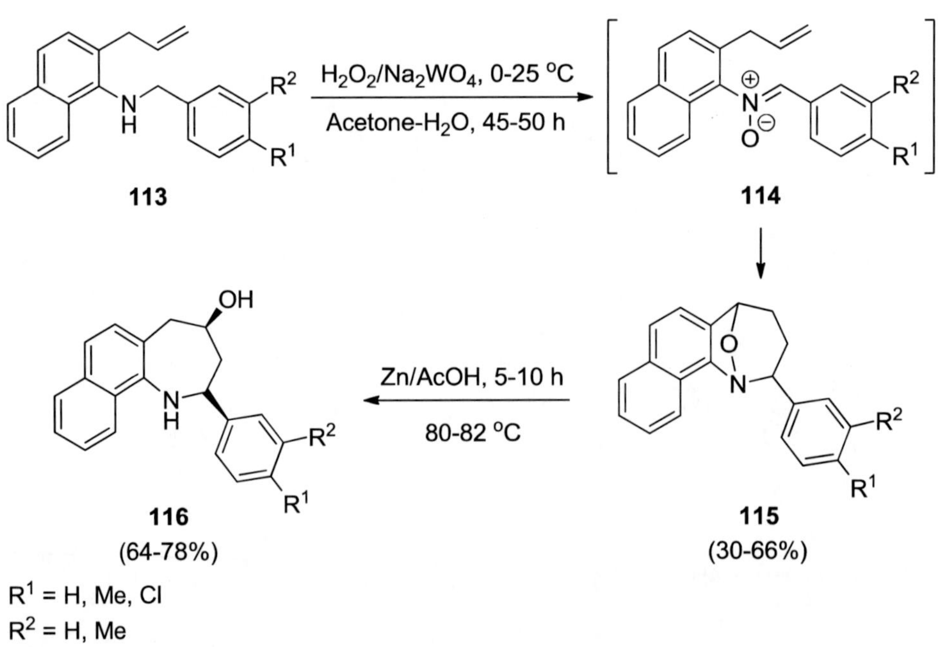

Scheme 26. Efficient synthesis of 2-aryl-4-hydroxy-2,3,4,5-tetrahydronaphtho[1,2-*b*]azepines.

Yamada and co-workers [38] have reported a highly efficient and selective intramolecular [3+2] nitrone-alkene cycloaddition (INAC) to furnish bridged bicyclo[4.2.1]isoxazolidines **118** in excellent yield (Scheme 27). The reaction resulted in the formation of *endo-* selective products which is controlled by presence of *trans*-acetonide group in hept-6-enoses **117**.

Chakraborty et al. [39] have listed the synthesis of functionalized isoxazolidines **121** via a regio and diastereoselective intramolecular 1,3-dipolar nitrone-olefin cycloaddition reaction (Scheme 28). The *N*-benzyl nitrones were generated *in-situ* from D-glucose derived substrate **120** by the treatment with *N*-benzyl hydroxylamine hydrochloride. The stereochemistry of the isoxazolidine derivatives was mainly influenced by *C*-5 OBn group. The isoxazolidine derivatives **121** were further transformed into various aminocyclohexitols **122** via reductive cleavage and removal of N–O bond.

A novel approach for the synthesis of pseudo-acarviosin **126** (potent inhibitor of α-glucosidases) has been reported by Cheng et al. [40] (Scheme 29). The strategy involves palladium-catalyzed coupling of substrates which in-turn were synthesized from D-glucose and L-arabinose. The *trans*-diacetal **123** was prepared from L-arabinose and transformed into nitrone **124** which undergo intramolecular [3+2] nitrone-alkene cycloaddition to afford functionalized isoxazolidine derivatives **125**. The cycloaddition reaction was highly selective for *trans*-fused isoxazolidine derivatives.

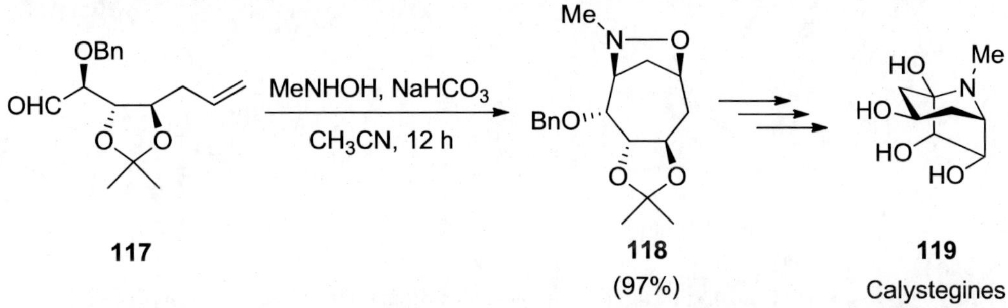

Scheme 27. Efficient and selective intramolecular nitrone-olefin cycloaddition.

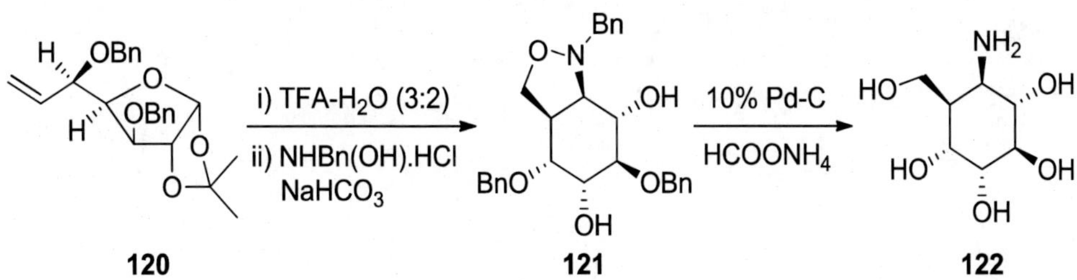

Scheme 28. Regio- and diastereoselective synthesis of functionalized isoxazolidines and amino cyclohexitols.

Stathakis and co-workers [41] have carried out synthesis of *ent*-gabosine E **130** from commercially available methyl α-D-mannopyranoside using a practical multistep approach with 12% overall yield (Scheme 30). This strategy involves an important step i.e., intramolecular 1,3-dipolar nitrone-alkene cycloaddition of mannose intermediate **127**. The cycloaddition resulted in the formation of products **128-129** in excellent yields with good selectivities (major:minor - 2:1).

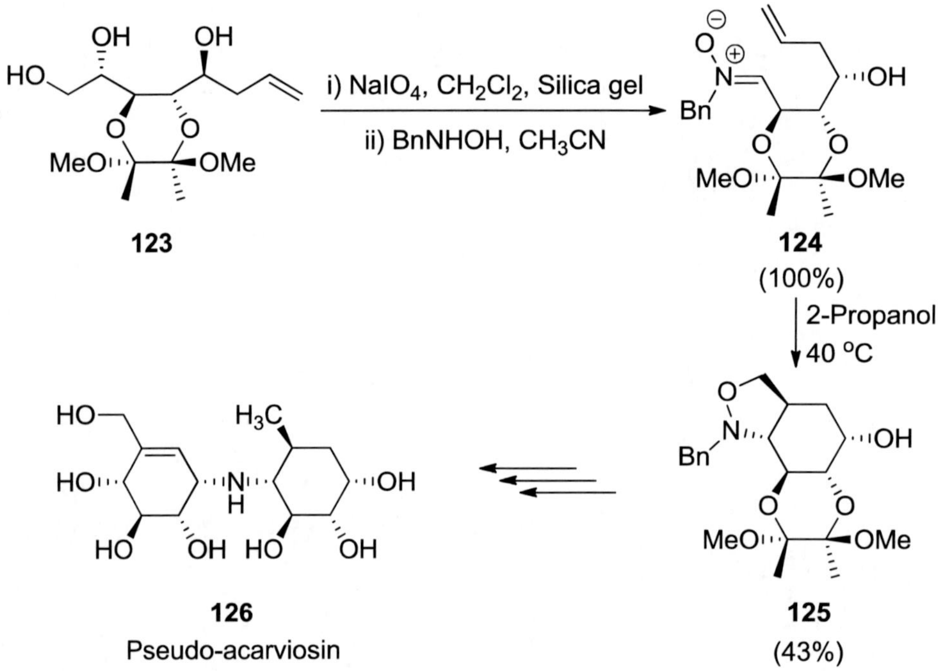

Scheme 29. *trans*-fused Isoxazolidines via intramolecular nitrone-alkene cycloaddition (INAC).

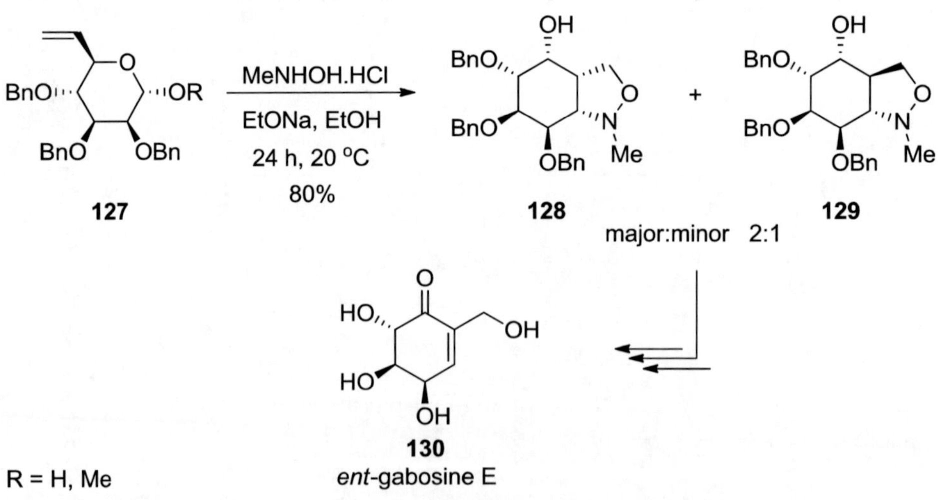

Scheme 30. Diastereoselective intramolecular nitrone-alkene cycloaddition.

A very efficient approach for the synthesis of enantiopure cocaine **135** analogues starting from D-(-)-ribose has been reported by Shing and So [42] (Scheme 31). The key step in this protocol is an *endo*-selective intramolecular nitrogen-olefin cycloaddition which resulted in the formation of a inseparable mixture of *endo*- (seven membered cycloadduct) and *exo*- (six membered cycloadduct) cyclized adduct **133-134**. The *endo*-selectivity in the cycloaddition was directed by the presence of *trans*-acetonide group presence in aldehyde intermediate **131**. Finally, *endo*- cyclized cycloadduct via several steps furnished the cocaine analogues in good overall yields.

Rajesh et al. [43] have synthesized a novel series of functionalized β-lactams anchored with polycyclic chromenoisoxazolidine and naphthopyranoisoxazolidine moieties (Scheme 32). The strategy involves intramolecular [3+2] nitrone-alkene cycloaddition of *O*-allylated salicylaldehyde substituted β-lactam derivatives **136** to furnish the polycyclic adducts **138** in very good yields. The substrate **136** in-turn were prepared from β-lactam aldehyde via Baylis-Hillman reaction. All the products were fully characterized using various spectroscopic techniques.

Stockman and co-workers [44] have a novel series of sp^3-rich 3-hydroxylactam and related analogues (drug-like molecules) via an efficient multistep protocol (Scheme 33). The key intermediate isoxazolidine derivative **141** was prepared from highly diastereo and regioselective 1,3-dipolar nitrone-alkene cycloaddition reaction. The nitrone was generated *in-situ* by the reaction of piperidinone **139** with *N*-substituted hydroxylamine hydrochloride to afford isoxazolidine derivatives in moderate to good yields.

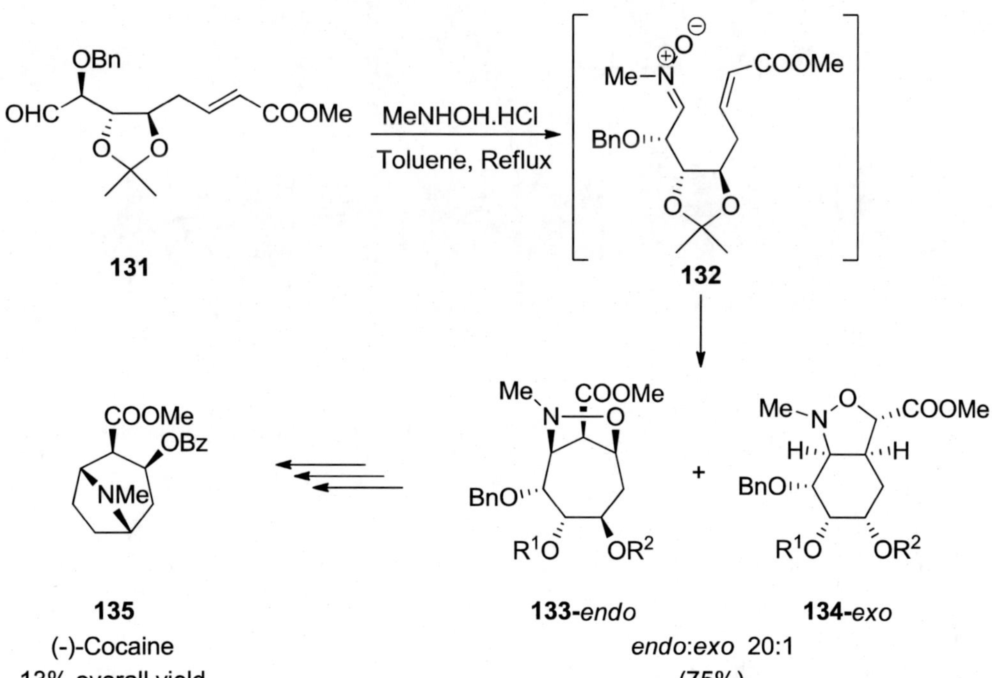

Scheme 31. Facile and efficient *endo*-selective intramolecular nitrone-olefin cycloaddition.

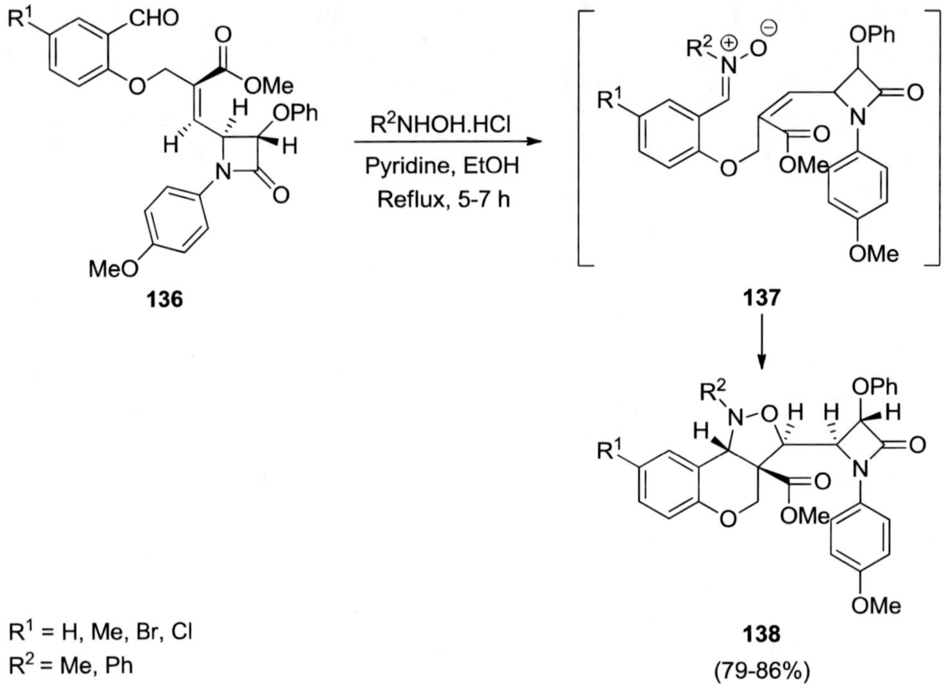

Scheme 32. Synthesis of β-lactam derivatives substituted with chromenoisoxazolidines.

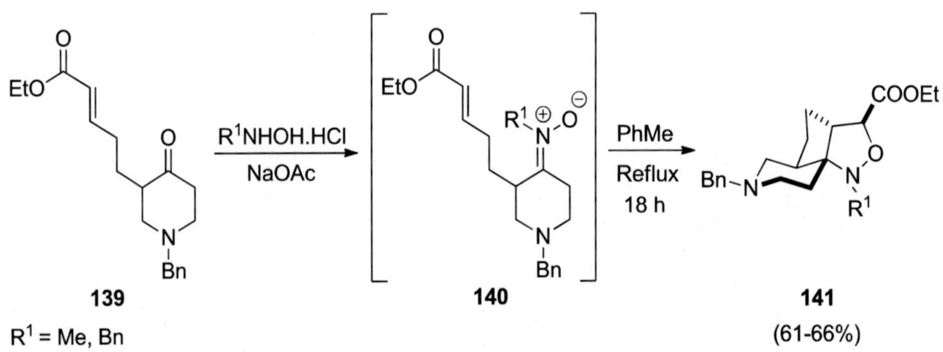

Scheme 33. Diastereoselective and regioselective synthesis of sp^3-rich isoxazolidine derivatives.

A new synthesis of 1,2-isoxazolidine based adducts **145** which acts as potential synthons for Zelekitoxin AB (a potent sodium channel blocker) has been reported Paladugu and Looper [45] (Scheme 34). The key intermediate i.e., *cis*-1,2-isoxazolidine adduct **144** has been prepared via intramolecular [3+2] nitrone-alkene cycloaddition reaction of aldehyde **142**. The reaction involves *in-situ* generation of nitrone followed by intramolecular cycloaddition which proceeds via transition state with all acetate in equatorial position, to afford *cis*-fused bicyclic isoxazolidine adduct **144** as single isomer in moderate yields.

Bai et al. [46] have reported the synthesis of novel isoxazolidine fused tricyclic molecules **149-150** in moderate to good yields (Scheme 35). The reaction involves *in-situ* generation of nitrone **148** which subsequently underwent intramolecular nitrone-olefin 1,3-dipolar cycloaddition to furnish the tricyclic isoxazolidine derivatives **149-150**. The reaction resulted in the formation of a mixture of *cis*- and *trans*- fused adducts which further depends upon type of substitutions present on aromatic carbo/heterocyclic ring. The electron withdrawing groups (EWGs) favors the formation of *cis*- or *exo*- adduct by decreasing the electron density on the ring whereas the effect is reversed when electron releasing groups (ERGs) are present.

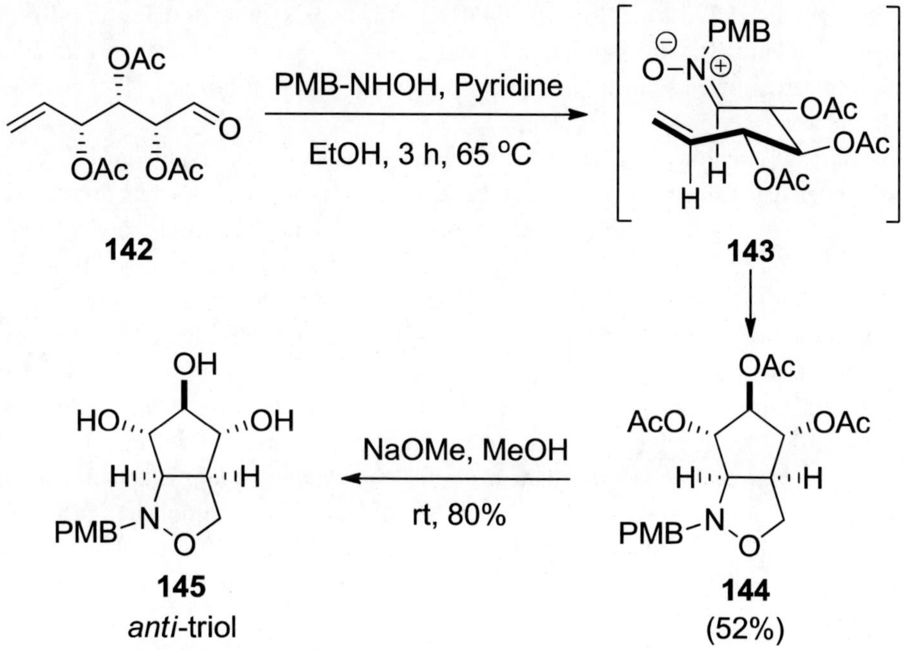

Scheme 34. Synthesis of bicyclic isoxazolidine derivatives via INAC.

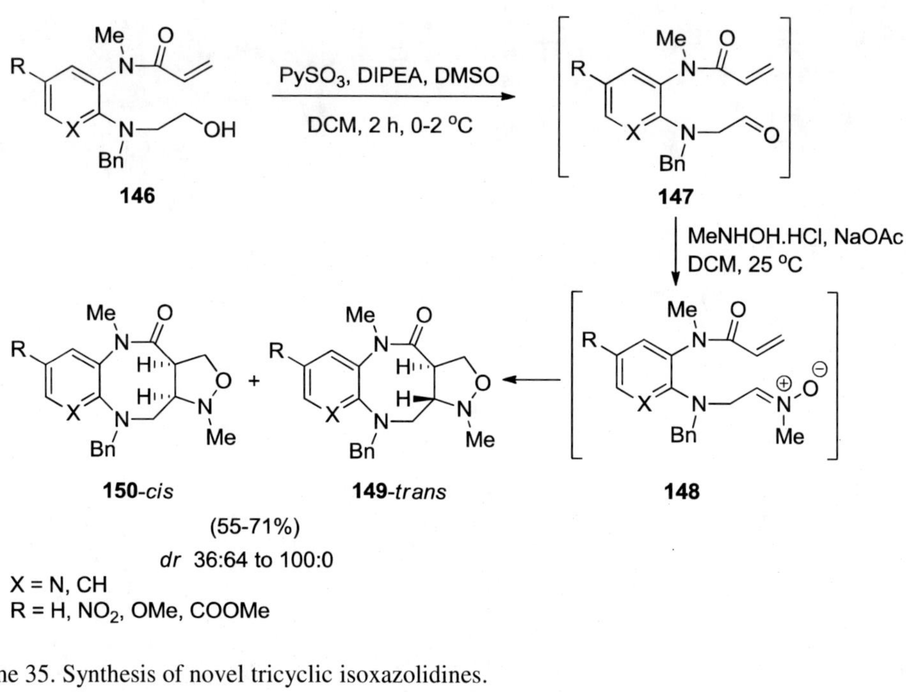

Scheme 35. Synthesis of novel tricyclic isoxazolidines.

2.3. Kinugasa Cycloaddition

Kinugasa reaction, a very important category of 1,3-dipolar nitrone-olefin cycloaddition reactions used for the synthesis of β-lactam or 2-azetidinone derivative which are known for their diverse biological activities [47]. This methodology was first reported by Kinugasa and Hashimoto [48] in 1972 as a direct method for β-lactam synthesis using copper derived dipolarophile (Figure 8). The significance and utility of Kinugasa reaction can be easily understood in terms of easy accessibility and high stability of substrates. This reaction also exhibits high atom economy and efficiency.

The Kinugasa reaction is different from classical nitrone-olefin cycloaddition reaction in using the copper species/catalyst which changes the overall outcome of the reaction. The copper free nitrone-alkyne cycloaddition reaction results in the formation of regioisomeric isoxazolines **152-153** whereas copper mediated nitrone-alkyne cycloaddition reaction (Kinugasa reaction) leads to *cis*- and/or *trans*-β-lactams **154** (Scheme 36) [49].

The reaction mechanism of copper free nitrone-olefin/alkyne has already been in discussed in this chapter. Therefore, it is also important to examine the reaction mechanism of copper mediated nitrone-alkyne cycloaddition reaction i.e., Kinugasa reaction. Many research groups have reported different mechanistic paths to explain the formation of β-lactams in Kinugasa reaction [50]. However, the exact mechanism could be much more complicated [51] but the fact of its similarity with CuAAC reaction can't be neglected. The simplified mechanism is presented in Scheme 37. Initial step involves the reaction of copper (I) acetylide with nitrone **156** to afford copper linked isoxazoline **157** which upon rearrangement furnished β-lactam ring **158**.

Pal and Basak [52] have reported a facile strategy for the synthesis of β-lactam fused macrocyclic enediynes **161-162** via intramolecular Kinugasa reaction (Scheme 38). The reaction was found to be more selective for the *trans*- isomers and products were obtained in moderate to good yields. The methodology was applicable to systems containing sensitive groups.

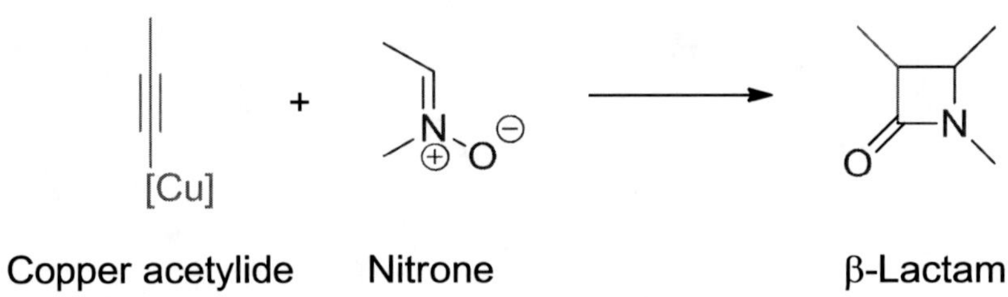

Figure 8. Kinugasa reaction.

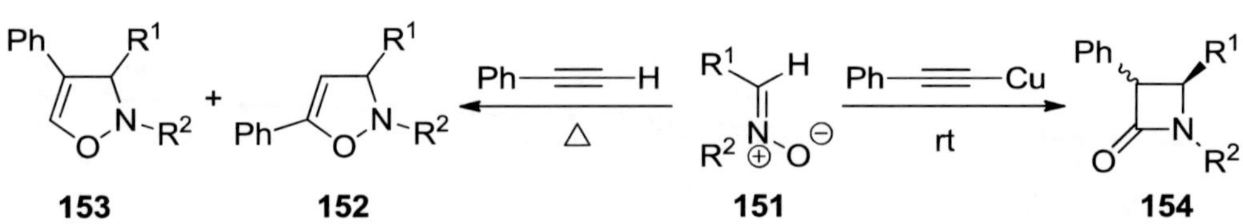

Scheme 36. Cu mediated and Cu free nitrone-olefin cycloaddition reaction.

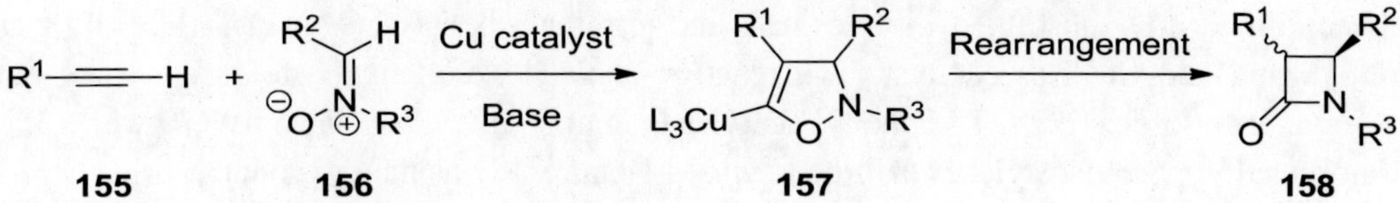

Scheme 37. General mechanism of Kinugasa reaction.

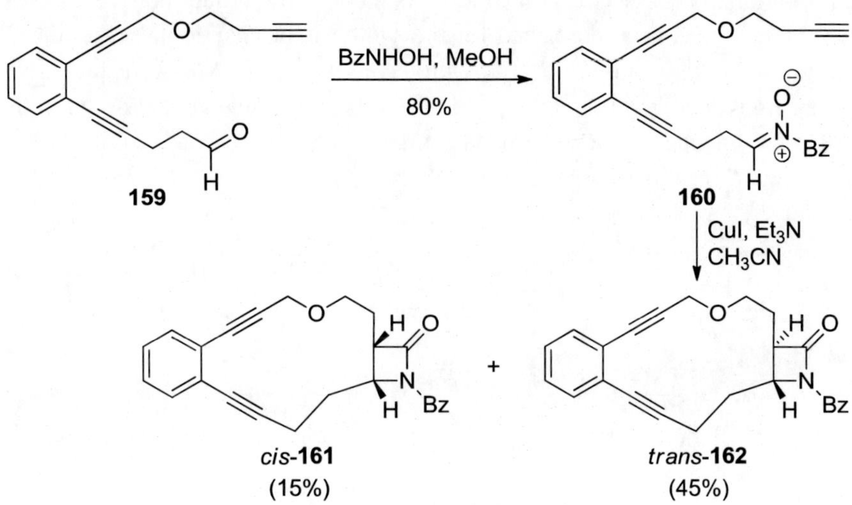

Scheme 38. Facile synthesis of β-lactam fused macrocyclic enediynes via intramolecular Kinugasa reaction.

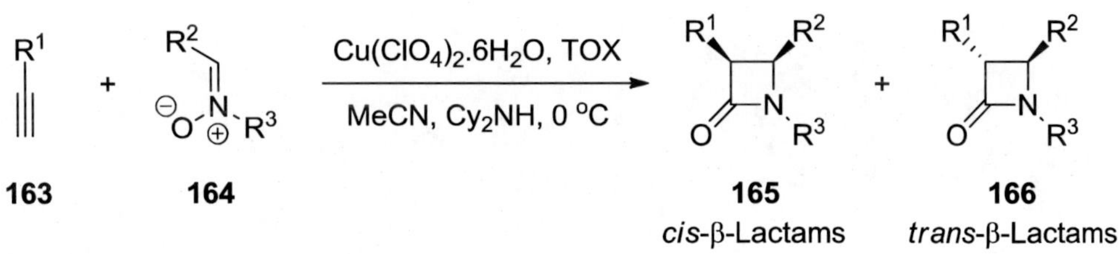

R^1 = Ph, 4-$CF_3C_6H_4$, $SiMe_3$, COOEt, 1-Cyclohexenyl
R^2 = Ph, 4-$CF_3C_6H_4$, 4-MeC_6H_4, 4-$OMeC_6H_4$, Cy, α-furyl
R^3 = Ph, Bz, 4-BrC_6H_4, 4-MeC_6H_4, 4-$OMeC_6H_4$, 4-$COOEtC_6H_4$

Scheme 39. Enantioselective synthesis of β-lactams via Kinugasa reaction.

An efficient methodology for the synthesis of enantiopure β-lactams **165-166** via copper catalyzed Kinugasa reaction has been reported by Tang et al. [53] (Scheme 39). The reaction was performed in the presence of $Cu(ClO_4)_2$/*i*-Pr-trisoxazoline as catalytic system and afforded the products in good yields with high enantioselectivities (up to 85%). It was also found in the results that secondary base exhibited better enantioselectivities. Furthermore, the diastereoselectivity was controlled by alkynes which mostly favor the formation of *cis*- isomer.

Coyne and co-workers [54] has reported asymmetric Kinugasa cycloaddition for the synthesis of 2-azetidinone derivatives **170-171** (Scheme 40). The reaction was carried out between differently substituted nitrones **167** and terminal alkynes **168** using HETPHOX as chiral ligand and CuCl as catalyst. The reaction showed excellent diastereoselectivity but moderate enantioselectivity. The phenyl acetylenes mainly gave *cis*-β-lactams **170** while 3,5-trifluoromethyl phenylacetylenes afforded *trans*-β-lactams **171** as major product.

Chmielewski et al. [55] have reported diastereoselective synthesis of novel carbapenam analogues **174-175** (Scheme 41). The Kinugasa reaction was carried out between various terminal alkynes **172** and cyclic nitrones **173** derived from malic/tartaric acid. The reaction favors the formation of *cis*-carbapenam which was induced by tertiary butoxy group.

A novel protocol has been described for the synthesis of enantiopure α-amino-β-lactams by Hsung and co-workers [56]. Initially, *N*-protected-β-lactams **178** were prepared in good yields via diastereoselective Kinugasa cycloaddition between ynamides **176** and nitrones **177** (Scheme 42). The β-lactams **178** were further transformed into chiral α-amino-β-lactams via hydrogenation and protection-deprotection strategy.

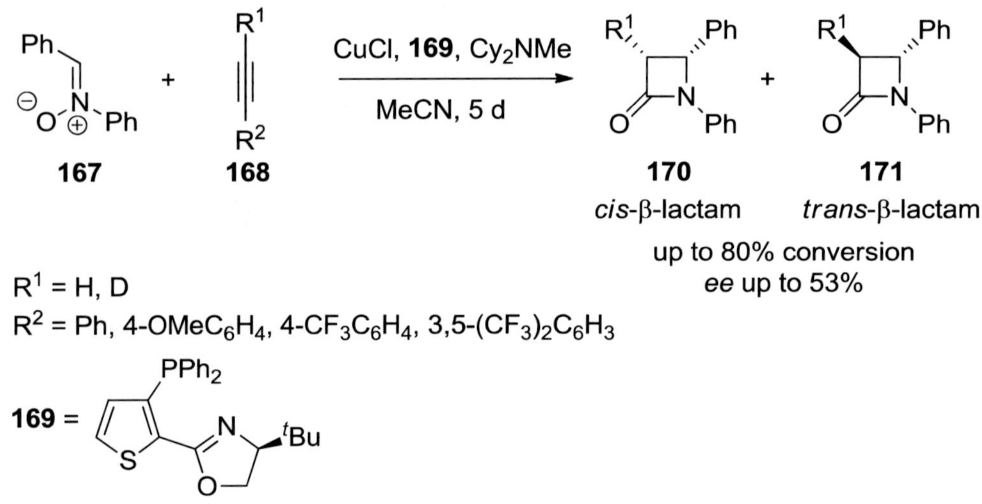

Scheme 40. Cu(I)/HETPHOX catalyzed asymmetric synthesis of β-lactams.

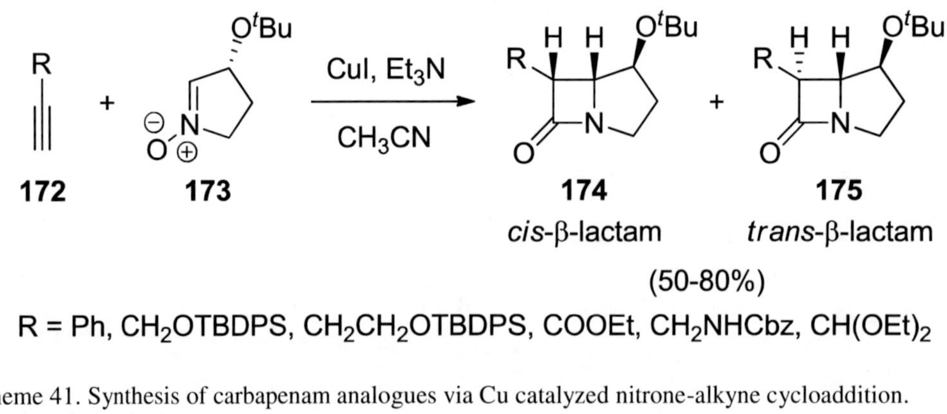

Scheme 41. Synthesis of carbapenam analogues via Cu catalyzed nitrone-alkyne cycloaddition.

A series of bicyclic β-lactams has been prepared by Stecko and coworkers [57] using copper catalyzed 1,3-dipolar cycloaddition between chiral acetylenes **179** and cyclic nitrones

180 (Scheme 43). The reaction was found to be highly selective and furnished a single isomer in most of the cases.

A mild and selective strategy for the synthesis of chiral β-lactams **185-186** in good yields and excellent selectivities has been reported by Saito and co-workers [58] (Scheme 44). The reaction was found to be general for a wide variety of differently substituted substrates **182-183**. The reaction was mainly selective for *cis*- isomer and significant enantioselectivities were achieved.

Mames et al. [59] have successfully carried out the synthesis of carbapenam analogues **189** via cycloaddition/rearrangement process (Scheme 45) and studied the reactivity of various dipolarophiles towards Kinugasa reaction. The reaction was highly diastereoselective and resulted in the formation of bicyclic β-lactams with *cis*- configuration. It was found in the results that glyceraldehyde and propargyl aldehyde derived acetylenes exhibited better reaction rates due to the interaction of oxygen atom/s with copper in rigid dinuclear Cu (I) intermediate and rigid structure of dioxolane rings in dipolarophile also stabilizes the intermediate. Furthermore, two geminal groups and dioxolane ring in dipolarophile is very essential for Kinugasa reaction and replacing dioxolane ring with OBz groups or increasing its size decreases the reactivity.

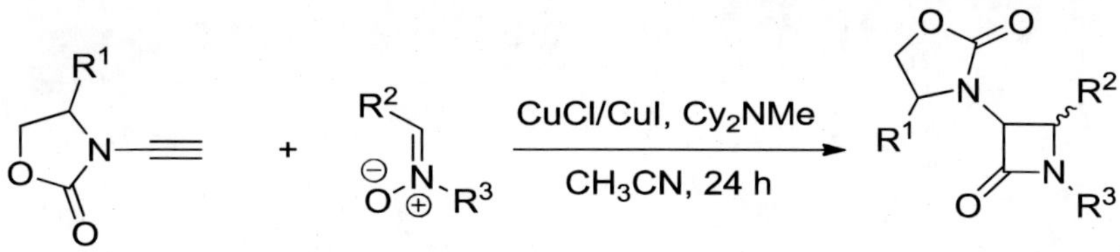

176 **177** **178**

(60-80%)
d.r. up to 95:5

R^1 = Ph, ^{i}Pr, Bn, $CHPh_2$
R^2 = Ph, 4-BrC_6H_4, styryl, 2-furyl, 1-naphthyl, 2-thiophenyl, c-hex
R^3 = Ph, 4-ClC_6H_4, 4-$COOEtC_6H_4$

Scheme 42. Synthesis of chiral 3-amino-2-azetidinones.

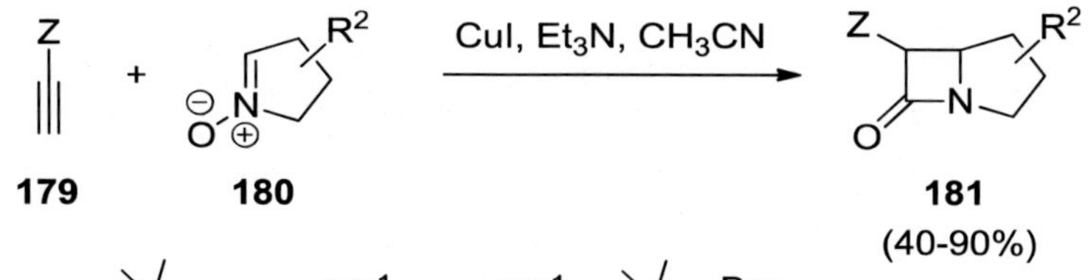

179 **180** **181**

(40-90%)

R^1 = Bn, $SiPh_2{}^tBu$
R^2 = H, 3-*t*-Bu, 4-*t*-Bu

Scheme 43. Asymmetric synthesis of bicyclic β-lactams via Kinugasa cycloaddition.

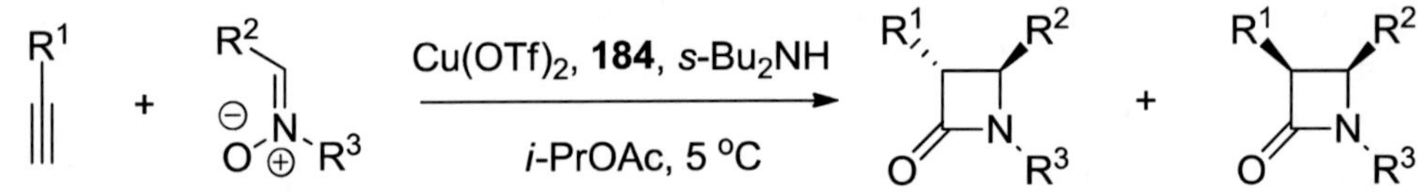

182 **183** *trans*-**185** *cis*-**186**

trans:cis 15-20:85-80

(40-70%)

R^1 = Ph, 4-MeC$_6$H$_4$, 4-OMeC$_6$H$_4$, 4-CF$_3$C$_6$H$_4$, 4-BrC$_6$H$_4$, 4-NO$_2$C$_6$H$_4$, COOEt, Cy

R^2 = Ph, 4-MeC$_6$H$_4$, 4-OMeC$_6$H$_4$, 4-BrC$_6$H$_4$, 4-CF$_3$C$_6$H$_4$, 4-NO$_2$C$_6$H$_4$

R^3 = Ph, 4-MeC$_6$H$_4$, 4-OMeC$_6$H$_4$, 4-ClC$_6$H$_4$, 4-COOEtC$_6$H$_4$

184 =

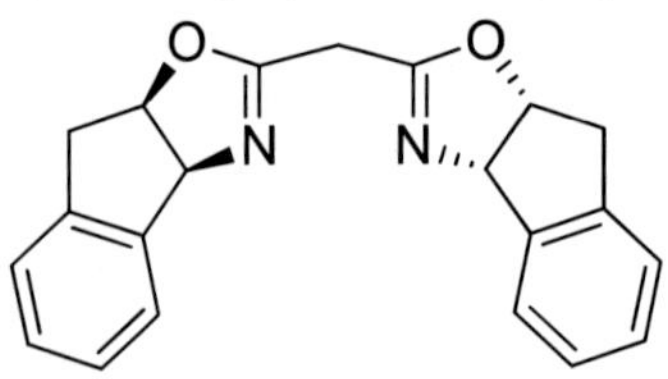

Scheme 44. Enantioselective synthesis of 2-azetidinones via IndaBox-Cu mediated Kinugasa reaction.

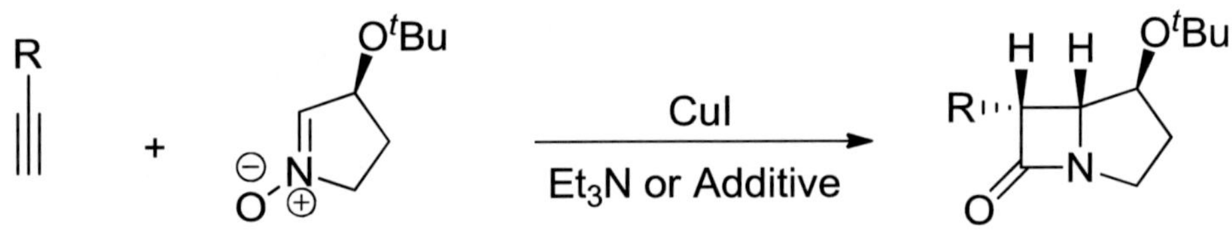

187 **188** **189**

(45-94%)

R =

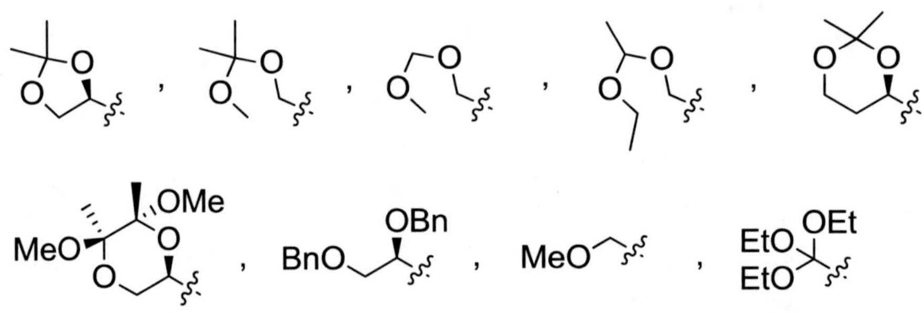

Scheme 45. Synthesis of carbapenam analogues via copper catalyzed alkyne-nitrone cycloaddition.

A novel strategy for the synthesis of bicyclic heterocycles with excellent diastereoselectivity and enantioselectivity has been reported by Worgull et al. [21]. The one pot strategy involves reaction between α,β-unsaturated aldehyde **190** and malononitrile derivative **191** in the presence of copper catalysts furnished bicyclic β-lactams **193** with high selectivity in moderate to good yields (Scheme 46). Moreover, different heterocyclic products can be obtained by slightly changing the conditions.

Khangarot and Kaliappan [60] has synthesized a novel series of sugar derived β-lactam derivatives **196** via asymmetric [3+2] cycloaddition reaction (Scheme 47). The alkynes **194** were derived from different sugars while cyclic nitrones **195** were obtained from various sugars and tartarate esters. The reaction between nitrone and alkyne exhibited high diastereoselectivities for *cis*- products with moderate to good yields.

Michalak and co-workers [61] have successfully utilized Kinugasa cycloaddition/rearrangement reaction for the first time in synthesis of (3*R*,4*S*)-β-lactam **199**, intermediate for ezetimibe (cholesterol absorption inhibitor). The Kinugasa reaction was performed between L-glyceraldehyde derived acetylenes **197** and differently substituted nitrones **198** (Scheme 48) and found to be highly efficient and selective with regard to target product.

Chen et al. [62] have carried out diastereo and enantioselective cycloaddition between alkynes **203** and nitrones **204** catalyzed by copper/tris(oxazoline) system (Scheme 49). The reaction resulted in the formation of β-lactam derivatives **205-206** with excellent enantioselectivities (up to 99%).

Grzeszczyk and co-workers [63] have studied asymmetric Kinugasa cycloaddition between cyclic nitrones **207** and different terminal alkynes (Scheme 50). The reaction resulted in the formation of β-lactams **208-211** with high diastereoselectivities but in moderate yields. The reaction was also analyzed for chiral substrates and other nitrones.

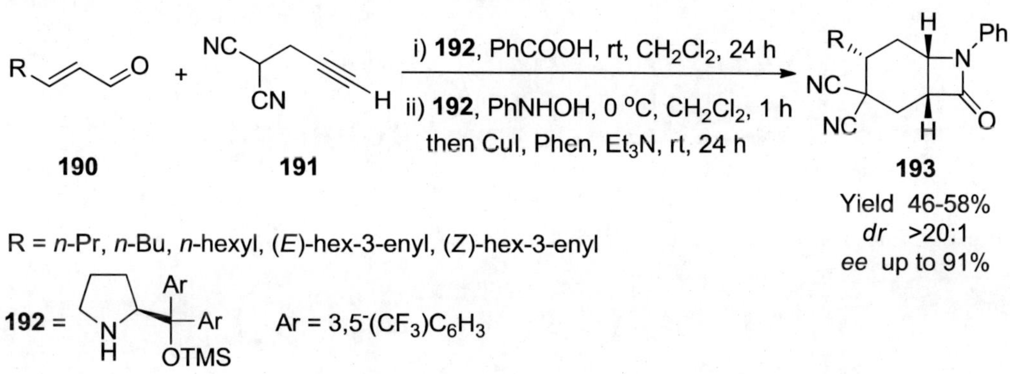

Scheme 46. Synthesis of bicyclic β-lactams via organocatalytic conjugate addition/cycloaddition sequence.

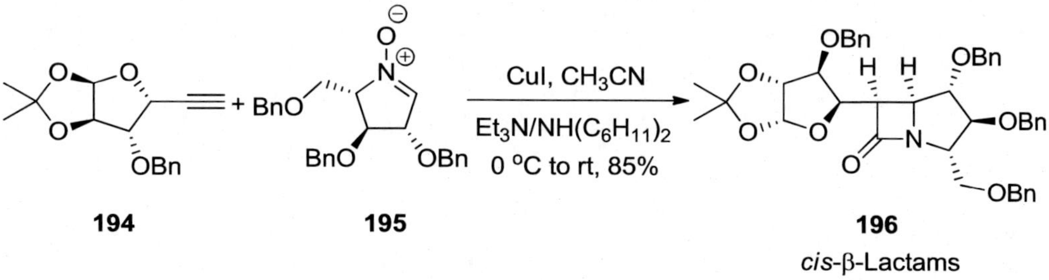

Scheme 47. Facile and efficient synthesis of bicyclic β-lactams.

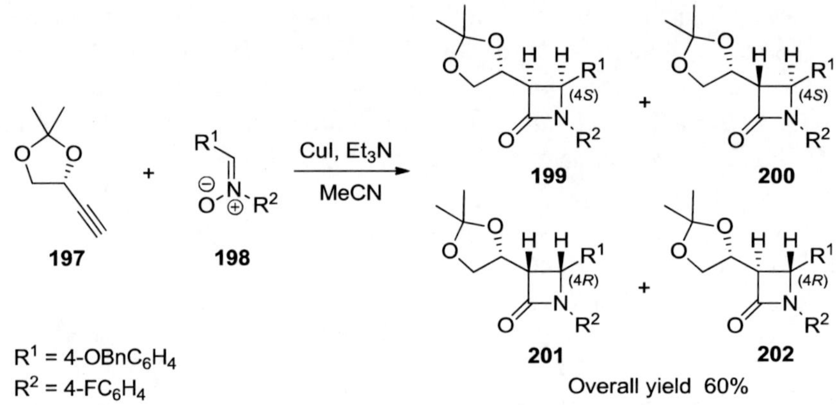

Scheme 48. *C,N*-Diaryl nitrones and acetylenes mediated synthesis of β-lactams.

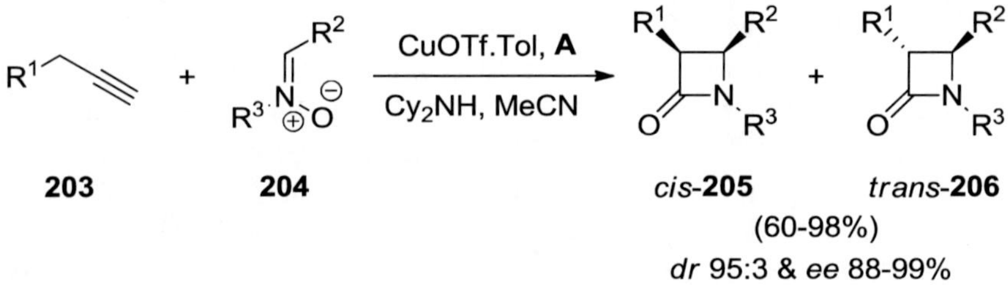

R^1 = Ph, 4-$OMeC_6H_4$, 4-BrC_6H_4, 4-MeC_6H_4, *n*-pentyl
R^2 = Ph, 4-$OMeC_6H_4$, 4-BrC_6H_4, 4-$COOEtC_6H_4$
R^3 = Ph, 4-$OMeC_6H_4$, 4-BrC_6H_4, 4-MeC_6H_4, 4-ClC_6H_4, 4-$CF_3C_6H_4$, 4-$NO_2C_6H_4$, 2-furyl

Scheme 49. Enantioselective synthesis of β-lactams.

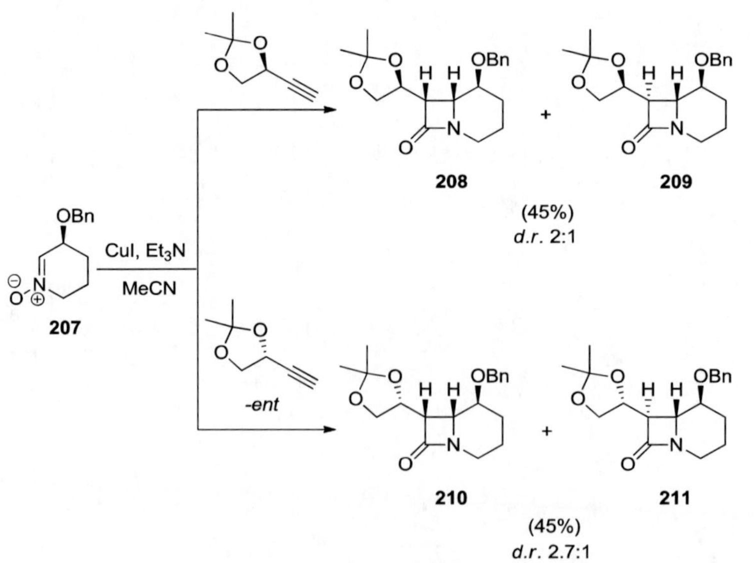

Scheme 50. Asymmetric and diastereoselective synthesis of bicyclic β-lactams.

Michalak et al. [64] have reported asymmetric synthesis of 1,4-diaryl-β-lactams **214-217** (Scheme 51). It was revealed in the results that nitrones containing electron withdrawing groups (EWGs) exhibited excellent diastereoselectivities. The authors have also utilized unprotected propargyl alcohols for the synthesis of β-lactams using same methodology.

Sierra and co-workers [65] have prepared a novel series of transition metal appended 2-azetidinone derivatives **220** via copper catalyzed 1,3-dipolar cycloaddition reaction (Scheme 52). The reaction was carried out between metal linked sandwich complexes **218** and *C,N*-aryl substituted nitrones **219** in the presence of CuCl/IndaBox system. However, the diastereo- and enantioselectivities of the reaction was very low.

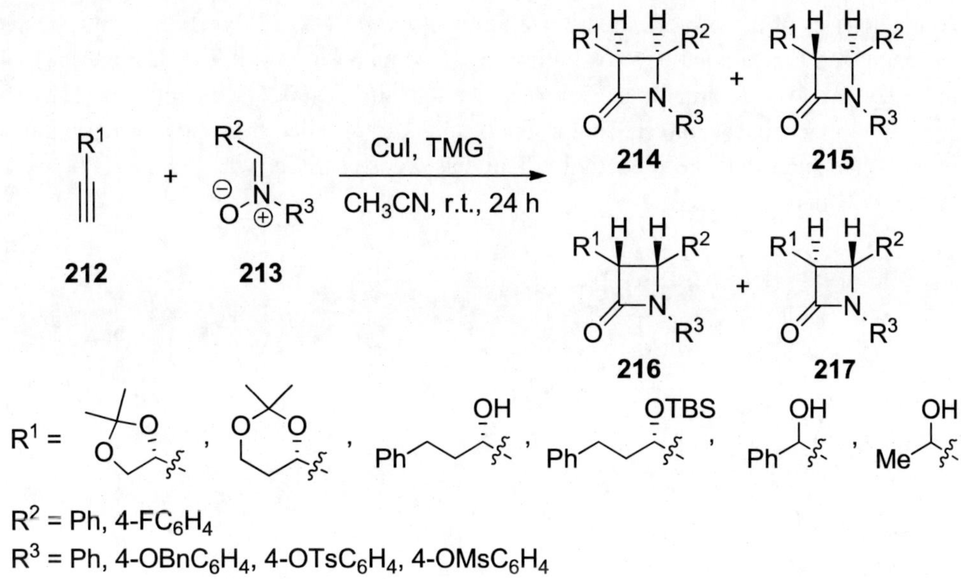

Scheme 51. Facile synthesis of *cis-* and *trans*-β-lactams.

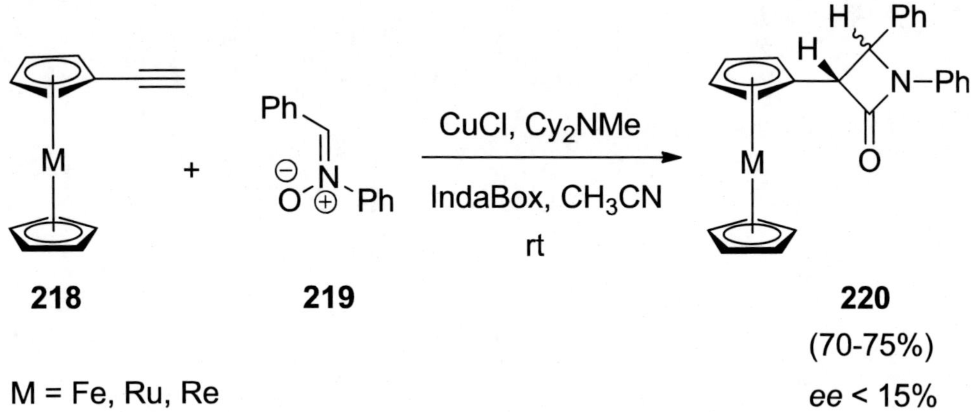

Scheme 52. Synthesis of transition metal substituted 2-azetidinones.

A novel synthesis of chiral β-lactams under aqueous conditions has been reported by Chen and co-workers [66] (Scheme 53). The reaction between various terminal alkynes **221** and differently substituted nitrones **222** resulted in the formation of products **224-225** in very good yields with excellent diastereo and enantioselectivities. The results revealed that electron rich phenyl acetylenes were more reactive in comparison to electron deficient phenyl acetylenes. The alkenyl substituted acetylenes exhibited low *dr* and high *ee* whereas aromatic alkynes showed good *dr* and *ee*. Further, different electron deficient and rich aryl group on nitrone carbon did not affect the diastereomeric ratio and yielded *trans*-β-lactams exclusively. On the other hand, electron deficient aryl groups on nitrone *N* furnished *trans-* isomers with high diastereoselectivity but electron rich aryl substituents gave low diastereoselectivities.

A novel series of radiofluorinated 2-azetidinone derivatives **228** has been synthesized via copper catalyzed 1,3-dipolar cycloaddition by Neumaier et al. [67] (Scheme 54). The methodology involves treatment of fluorine labelled nitrones **226** with alkynes **227** in the presence of copper salt as catalyst. The products i.e., [^{18}F]-labelled β-lactams were obtained in good yields and high diastereoselectivity. The diastereoselectivity was highly influenced by alkyne components.

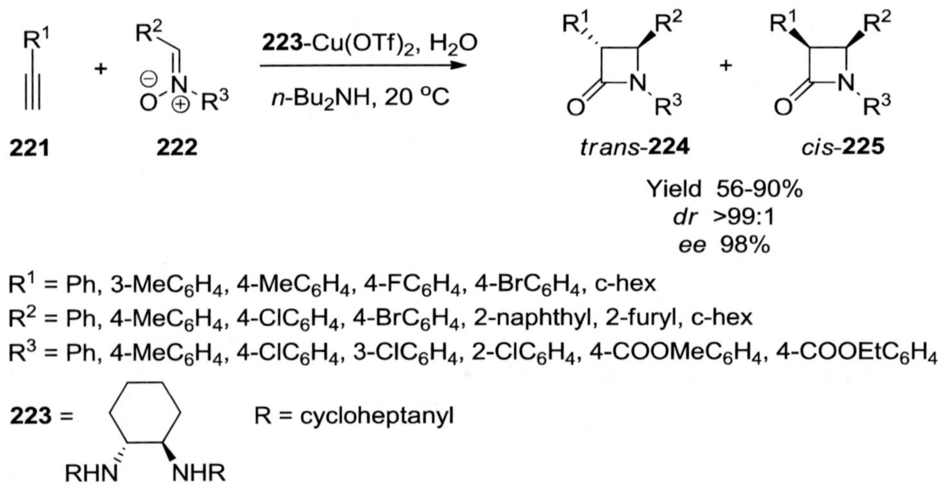

Scheme 53. Synthesis of β-lactams via asymmetric Kinugasa reaction.

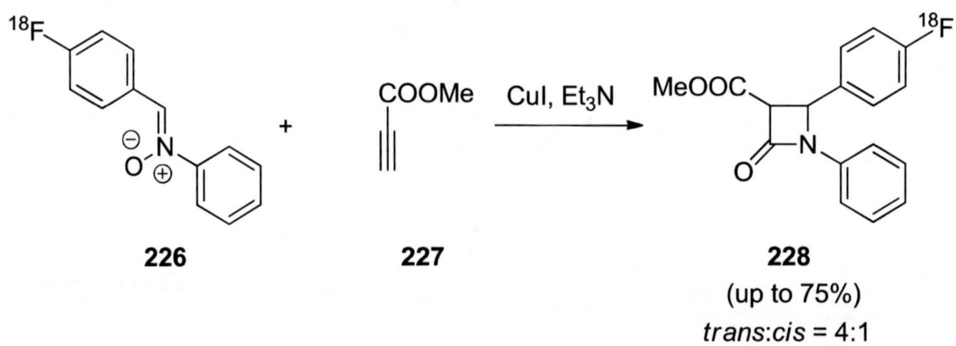

Scheme 54. Facile and efficient synthesis of [^{18}F]-labelled β-lactams.

3. Significance of Nitrone-Olefin Cycloaddition Reaction

Nitrone-olefin cycloaddition reaction is one of the most important reaction which has found applications in synthesis of many heterocyclic derivatives having potential applications in medicinal field viz. Pseudoacarviosin **126** (potent inhibitor of α-glucosidases), *ent*-gabosine E **130** (enzyme inhibition and anti-protozoal activity) and (-)-cocaine **135** [40-42].

The synthesis of variety of alkaloids and natural products involve this reaction as an important step. Total synthesis of hydroxycotinine (a metabolite of nicotine) **232** [68], lupinine **236** (a quinolizidine alkaloid found in plants of *Lupinus* species) [69] and porantheridine **240** (an alkaloid) [70] involves 1,3-dipolar nitrone-olefin cycloaddition reaction as an initial step to afford adducts which can be transformed into target products (Scheme 55-57).

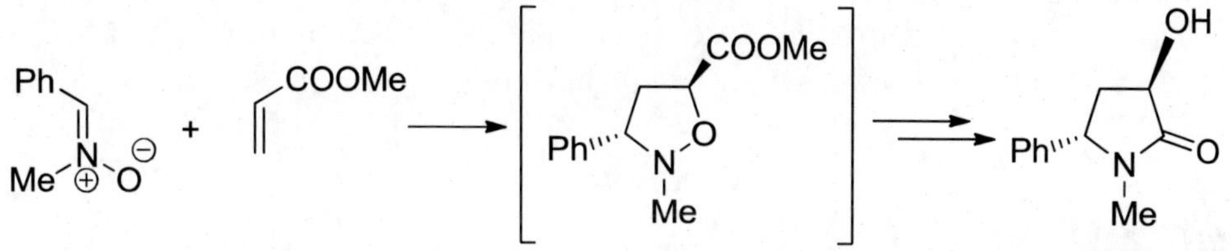

Scheme 55. Synthesis of Hydroxycotinine.

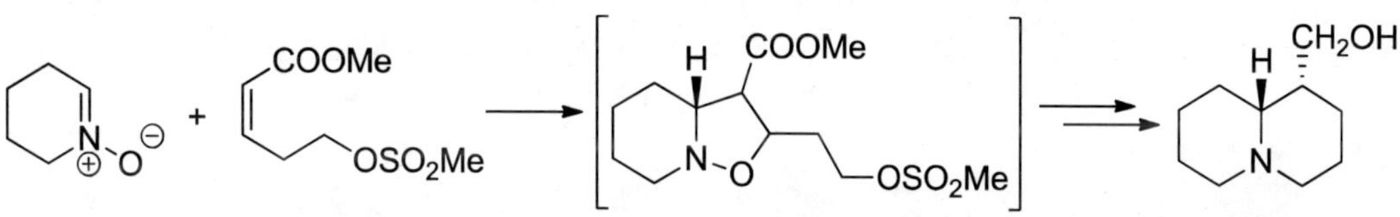

Scheme 56. Synthesis of Lupinine.

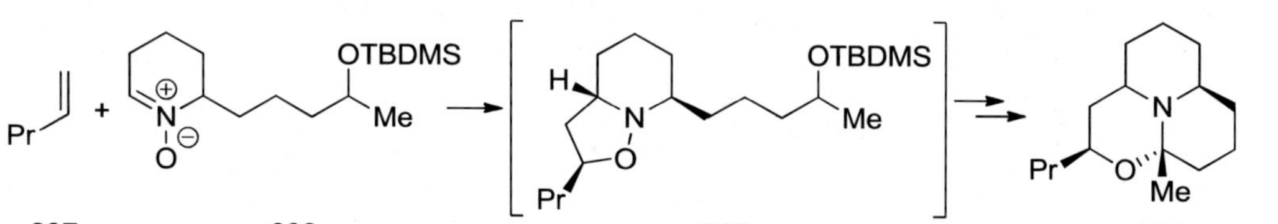

Scheme 57. Synthesis of Porantheridine.

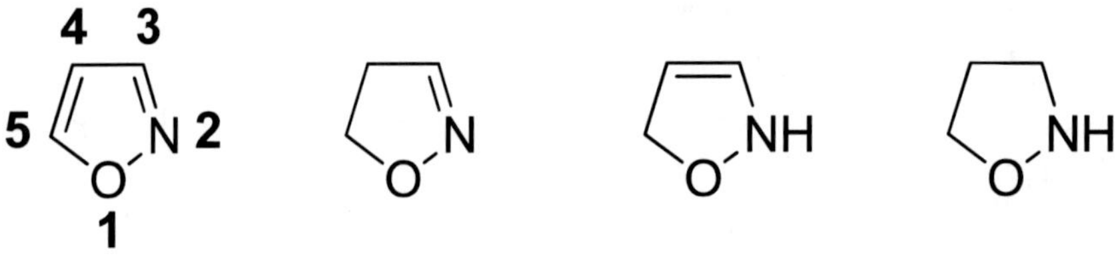

241 242 243 244

Figure 9. Basic structure of isoxazole, isoxazoline and isoxazolidine.

As we have already discussed that classical nitrone-olefin 1,3-dipolar cycloaddition reaction yields variety of *N* and *O* containing five membered heterocyclic products depending upon the nature of olefin viz. isoxazole **241**, isoxazoline **242-243** and isoxazolidine **244** (Figure 9). These heterocycles are considered very useful in bioorganic as well as medicinal chemistry.

3.1. Biologically Active Isoxazolines

Isoxazoline derivatives **242-243** are known to possess activities such as antimicrobial, anticancer, insecticidal and ulcerogenic activities. In addition to this, number of drugs containing isoxazoline moiety are available in the market which are used as anti-inflammatory and COX-2 inhibitory drugs.

A wide variety of diversely substituted 4,5-diphenyl-4-isoxazoline derivatives **245** has been designed and synthesized by Knaus and co-workers [71] (Figure 10). All the novel compounds were investigated for potential analgesic and selective COX-2 inhibitory anti-inflammatory agents. The 4,5-diphenyl substituted derivative lacking C-3 methyl group exhibited good analgesic and anti-inflammatory activity but with no significant selectivity. However, 2,3-dimethyl-5-(4-methylsulfonylphenyl)-4-phenylisoxazoline showed excellent analgesic and anti-inflammatory activities with high selectivity (IC_{50}; COX-1:COX-2 is 258:0.004 μM). Further, compound having 4-fluorophenyl group at *C*-4 of isoxazole analogue was also found to be selective but less potent inhibitor. The docking studies revealed that $MeSO_2$ and *C*-3 methyl groups are important for selective inhibition.

Sammelson and co-workers [72] have reported the synthesis of novel 3-(2-benzyloxyphenyl) isoxazolines **246** via nitrile oxide-alkene 1,3-dipolar cycloaddition (Figure 10). These isoxazoline derivatives were identified as activators of CFTR-dependent chloride transport. The best compound showed half-maximal stimulation at 20-50 μM in fluorescence cell-based halide transport assay.

Mishra et al. [73] have carried out diastereoselective synthesis of novel glycosyl isoxazolines **247** via 1,3-dipolar cycloaddition reaction and evaluated them for potential antifungal activity (Figure 10). The *in-vitro* antifungal screening was investigated against *C. albicans*, *C. parapsilosis*, *S. schenckii*, *A. fumigates*, *T. mentagrophytes* and *C. neoformans*. No activity was observed against *C. albicans*, *C. parapsilosis* and *A. fumigates* at tested concentrations. However, all the compounds showed good antifungal activity against other fungal strains (MIC 6.25-50 μg/mL).

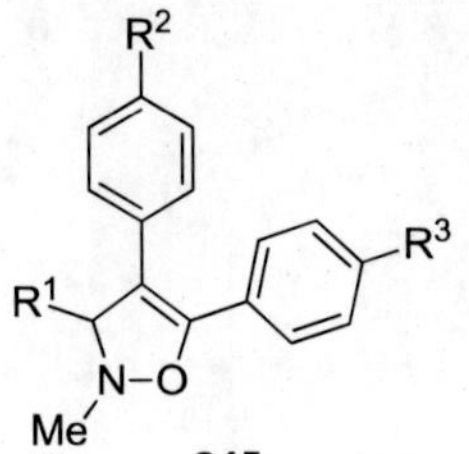

245

R^1 = H, Me; R^2 = H, F, SMe, SO_2Me; R^3 = H, F, SMe, SO_2Me

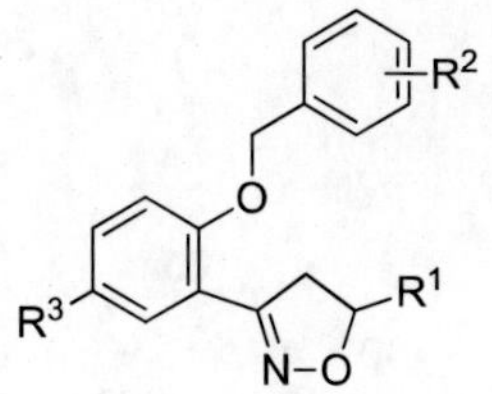

246

R^1 = CH_2Cl, CH_2OH, COOMe; R^2 = 2-OMe, 4-OMe, 4-Br; R^3 = H, Br

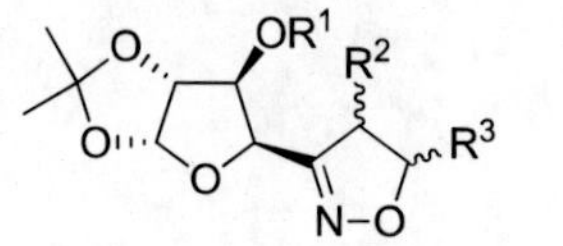

247

R^1 = Me, Bz; R^2 = H, COOH; R^3 = Me, CN, OEt, COOEt, OCOMe, COMe, CH_2OH, CH_2Br, CH_2PhOH, CH_2PhOCH_3

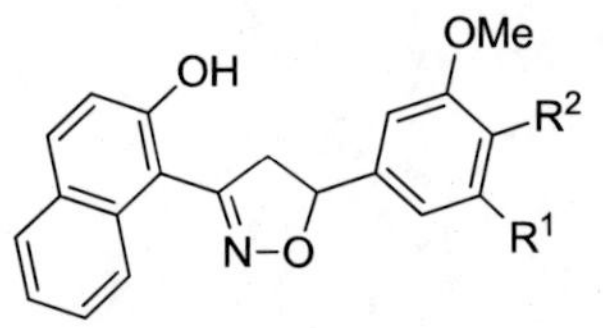

248

R^1 = H, I; R^2 = OC_2H_5, OC_3H_7, $OCH_2CH_2NEt_2$, OCH_2CH_2-morpholine, OCH_2CH_2-piperidine, OCH_2CH_2-pyrollidine

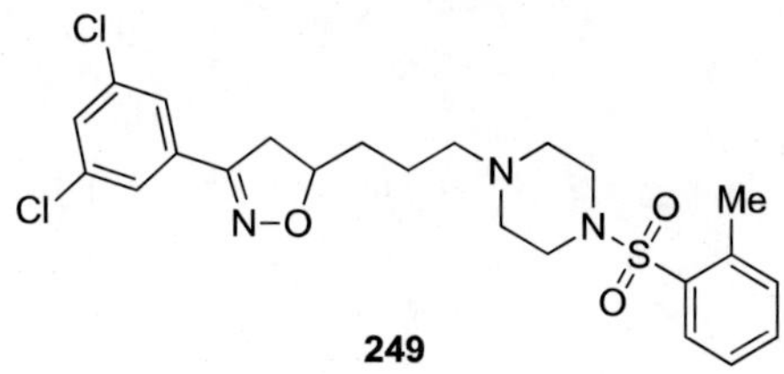

249

Figure 10. Biologically active isoxazoline derivatives 245-249.

Sailu and co-workers [74] have successfully synthesized a novel series of isoxazoline derivatives **248** from 2-hydroxyacetonaphthanone (Figure 10). All these compounds were evaluated for antibacterial activities against a panel of Gram-positive (*S. aureus*, *S. pyogenes*) and Gram-negative (*E. coli*, *P. aeruginosa*) bacterial strains and the results were compared with ampicillin as positive control. Most of the compounds were found to show significant antibacterial activities in comparison to standard drug. Further, compounds having R^1 = H and R^2 = piperidine and pyrrolidine group showed excellent activity against *E. coli* (Zone of inhibition is 22 mm).

Pae and co-workers [75] have reported the synthesis of novel isoxazoline derivatives in moderate to good yields (Figure 10). All the novel compounds were screened against 5-HT_{2A} and 5-HT_{2C} receptors. It was revealed in the results that molecules containing 2-MePh group showed better binding affinities towards 5-HT_{2A} and 5-HT_{2C} receptors. Also isoxazoline derivatives exhibited better results in comparison to isoxazole derivatives. In addition to this, most potent isoxazoline derivative **249** (IC_{50} 72-150 nM) was evaluated for antidepressant activity in animal model and results come out be highly promising.

Kariyappa et al. [76] have reported the synthesis and antimicrobial evaluation of a new series of isoxazoline nitrile derivatives **250** (Figure 11). The antibacterial activity was investigated against *S. aureus*, *B. subtilis*, *E. coli*, *S. typhimurium* whereas antifungal activity was checked against *C. albicans*, *A. flavus*, *A. niger* using paper disc diffusion assay. Some of the compounds were found to be significantly active. Furthermore, the novel compounds were also screened for potential antioxidant activity and showed good antioxidant activity in DPPH radical scavenging assay and ferric ion reducing power assay. It was also observed that electron withdrawing groups enhances the antioxidant activity.

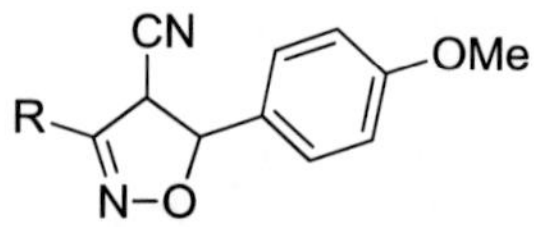

250

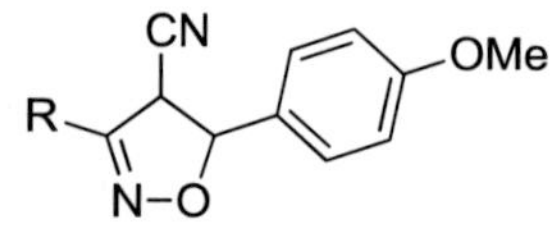

251

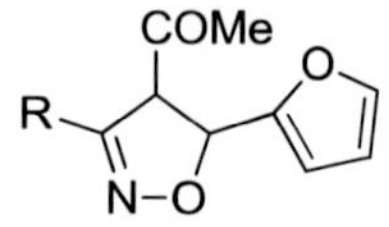

252

R = Ph, 2-ClC_6H_4, 4-ClC_6H_4, 4-FC_6H_4, 4-BrC_6H_4, 4-CNC_6H_4, 4-$OMeC_6H_4$, 2-NO_2-4-ClC_6H_3

R = Ph, 4-MeC_6H_4, 4-$NO_2C_6H_4$, 4-FC_6H_4, 4-ClC_6H_4, 4-$OMeC_6H_4$, 3,4-$(OMe)_2C_6H_3$, 3,4,5-$(OMe)_3C_6H_2$

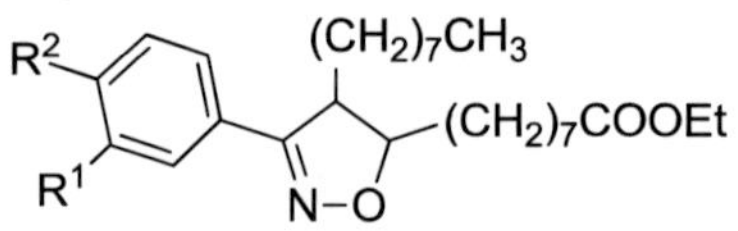

253

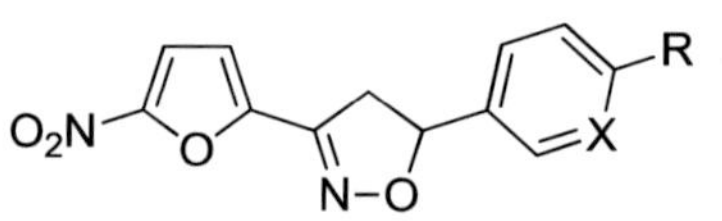

254

R^1 = H, OMe; R^2 = H, Me, OMe, F, Cl, NO_2

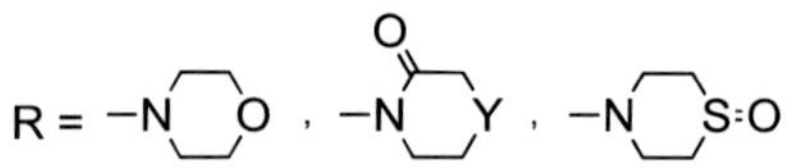

X = CH, N, CF; Y = O, NMe

Figure 11. Other biologically active isoxazolines 250-254.

A novel series of 3-aryl-5-(4-methoxyphenyl)-4,5-dihydroisoxazole-4-carbonitrile **251** and 4-acetyl-3-aryl-5-(furan-2-yl)-4,5-dihydroisoxazoles **252** has been screened for potential antibacterial, antifungal and antioxidant activities [77] (Figure 11). All the compounds showed promising results and presence of halogens (F, Cl) or NO_2 group exhibited high antimicrobial and antioxidant activities.

Kumar et al. [78] have performed the synthesis of novel isoxazoline derivatives **253** via 1,3-dipolar cycloaddition between nitrile oxide and ethyl oleate in moderate to very good yields (Figure 11). All the compounds were evaluated for potential antibacterial and antifungal activity and some of them exhibited promising activity. Compound having 4-chlorophenyl group is highly active and may act as future drug candidate.

A new series of tetracyclic nitrofuran isoxazoline derivatives **254** have been designed, synthesized and screened for potential antitubercular activity by Rakesh et al. [79] (Figure 11). In this study, initial lead compound was modified structurally in order to improve the solubility, decrease metabolic rate and affinity for protein binding. The best antitubercular activity against non-replicating *M. tuberculosis* was exhibited by compounds containing morpholine and pyridyl morpholine rings. The *in-vivo* screening of selected compounds exhibited longer half-life, good tolerance and high bioavailability.

3.2. Biologically Active Isoxazoles

Isoxazole derivatives **255** have been synthesized by reaction between nitrile oxide as 1,3-dipolar compound and various alkynes as dipolarophiles [72] (Figure 12). These compounds were found to be excellent activator of halide conductance CFTR proteins. The results revealed that the nature of substituent at *C*-5 of isoxazole derivatives play a crucial role towards their biological profile. Moreover, the activity was significantly reduced when R^3 is bromine.

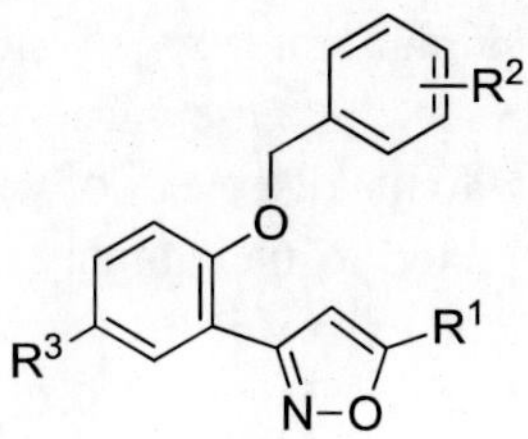

255

R^1 = CH_2Cl, CH_2OH, CH_2CH_2OH, CH_2OMe, COOMe, ${}^nBu$; R^2 = 2-OMe, 4-OMe, 4-Br; R^3 = H, Br

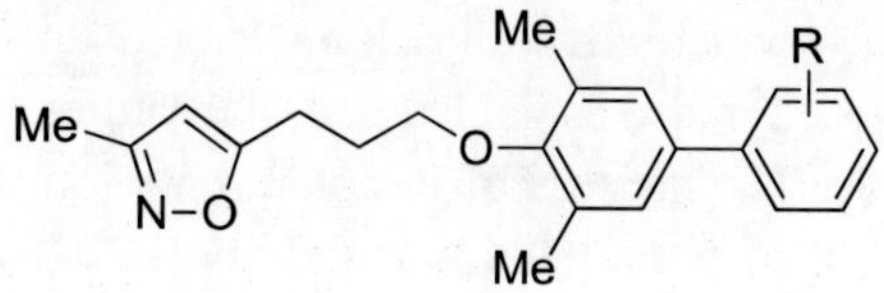

256

R = 3-F, 4-F, 3-NO_2, 3-CF_3, 4-CF_3, 4-Ph, 4-OCF_3, 2-Me, 3,5-$(CF_3)_2$, 3,4-Me_2, 3,4,5-$(OMe)_3$

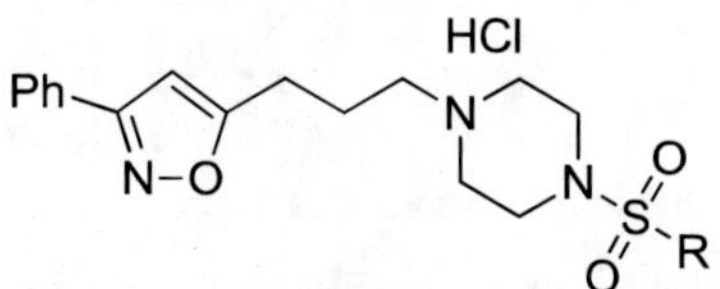

257

R = Ph, 2-FC_6H_4, 3-FC_6H_4, 4-FC_6H_4, 2-ClC_6H_4, 3-ClC_6H_4, 4-ClC_6H_4, 2-MeC_6H_4

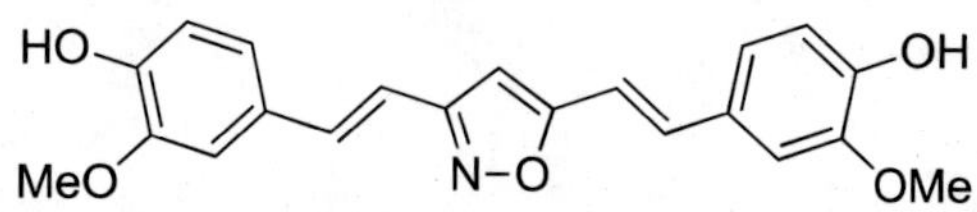

258

Figure 12. Biologically active isoxazoles 255-258.

Schmidtke and co-workers [80] have reported a novel methodology for the synthesis of [(biphenyloxy)propyl]isoxazole derivatives **256** in good yields (Figure 12) followed by their evaluation against human rhinovirus and coxsackievirus for potential antiviral activity. Some of the novel derivatives have showed excellent anti-HRV-2 activity (SI > 50-200) and good anti-CVB-3 activity (SI > 2-37.5). However, these derivatives showed negligible effect on HRV-14 induced cytopathic effect (CPE).

Pae et al. [75] have prepared isoxazole derivatives **257** in moderate yields (Figure 12). All the novel compounds were tested against 5-HT_{2A} and 5-HT_{2C} receptors. The results showed that molecules containing *ortho*-substituted phenyl groups showed better binding affinities towards 5-HT_{2A} and 5-HT_{2C} receptors.

D'Alessandro et al. [81] have carried out antitumor evaluation of curcumin-isoxazole derivative **258** and curcumin against MCF-7 and MCF-7R breast cancer cell line (Figure 12). The curcumin-isoxazole derivative exhibited promising antitumor activities as compared to curcumin. Moreover, growth inhibitory effects and cell death induction effects were not affected by MDR condition.

3.3. Biological Profile of β-Lactams

Moreover, 1,3-dipolar nitrone-olefin cycloaddition reaction results in the formation of diversely substituted four and five membered heterocyclic products viz. β-lactams, isoxazoline and isoxazolidine. All of these heterocycles are extremely important from the biological prospective. These heterocycles exhibit wide range of medicinal properties which will be briefly discussed in this section.

β-Lactams or 2-azetidinone derivatives constitute an important category of small ring heterocycles which are well acknowledged due to their antibiotic properties. The twentieth century was regarded as golden period of antibiotics mainly due to the discovery of penicillins (Figure 13). The success of β-lactam antibiotics may be attributed to their highly effective nature, broad spectrum and low toxicity [82]. The continuous and exhaustive research in this field has yielded many classical and non-classical β-lactam antibiotics (Figure 13).

The importance of β-lactam derivatives is not limited to their antibiotic properties. In the past few decades, it has been revealed that β-lactams exhibited wide range of other applications (Figure 14). The β-lactam derivatives have been found to possess diverse pharmacological activities such as antifungal [83], antitubercular [84], antiviral [85], anticancer [86], anti-inflammatory [87], antihyperglycemic [88], antihyperlipidemic [89] and CNS activity [90]. Furthermore, β-lactams also exhibited inhibitory action against many enzymes viz. chymase inhibitor [91], thrombin inhibitory [92], human leukocyte elastase inhibitor [93], cysteine protease inhibitor [94], cholesterol acyl transferase inhibitor [95], serine-dependent enzyme inhibitors [96], human cytomegalovirus inhibitors [97], apoptosis inductors [98] and vasopressin V1a antagonist [99].

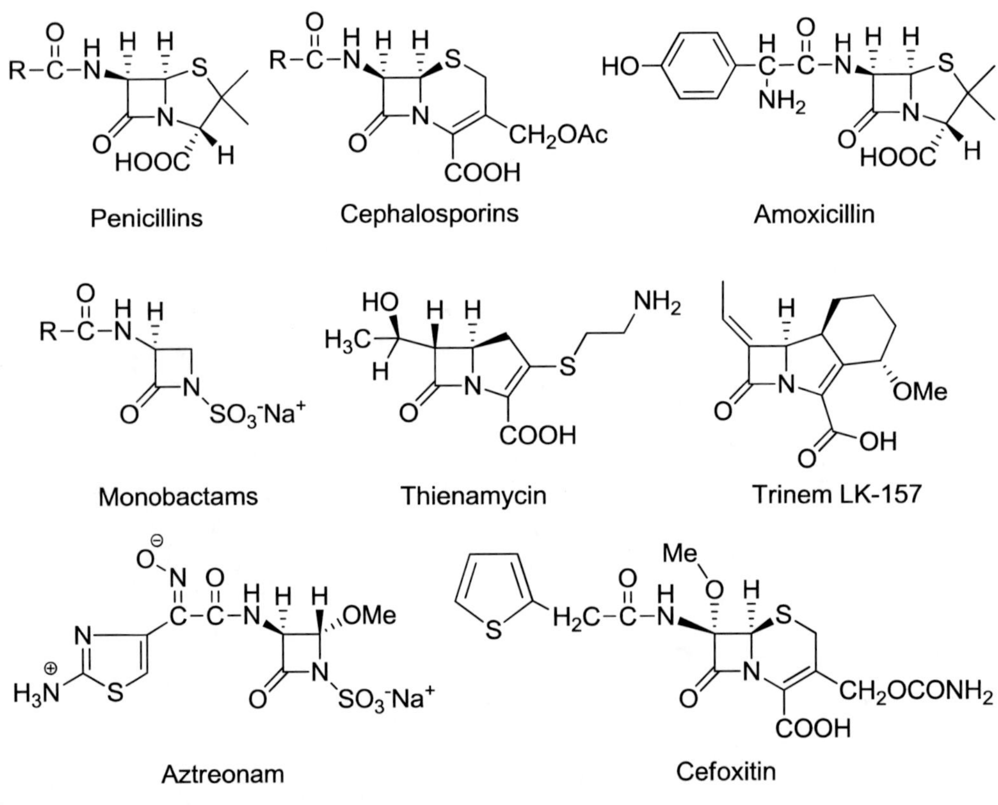

Figure 13. β-Lactam antibiotics.

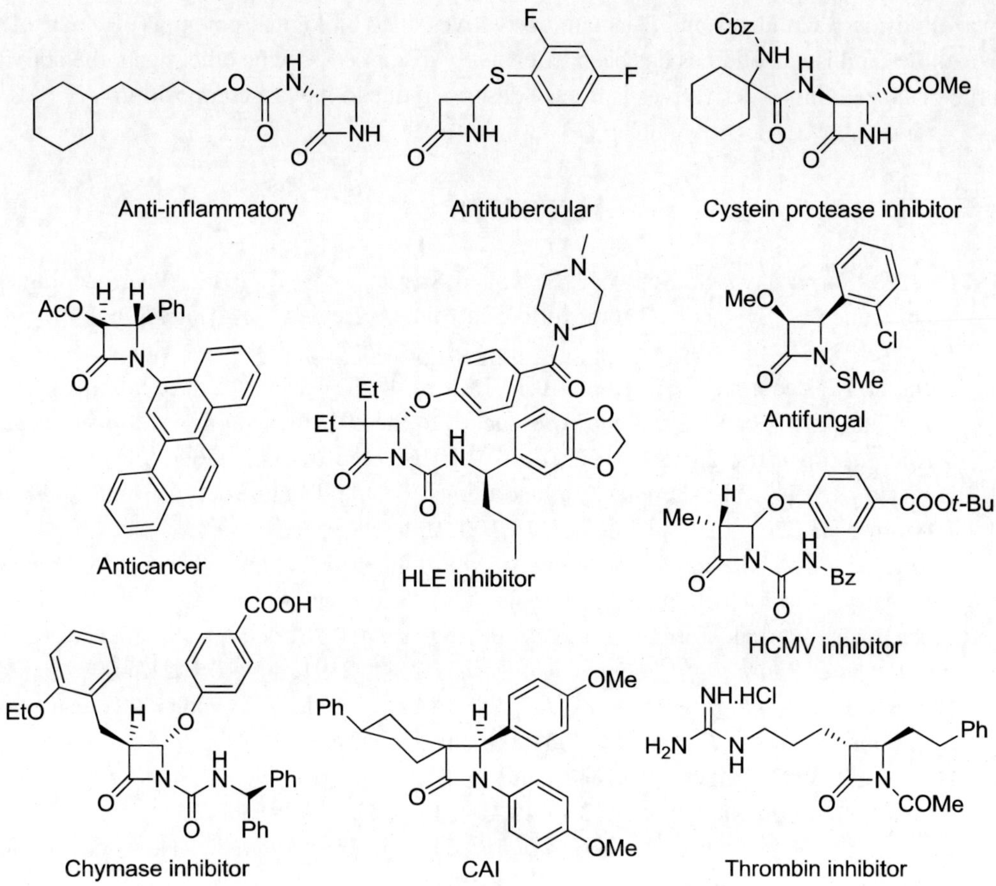

Figure 14. Biological activity profile of β-lactam derivatives.

ACKNOWLEDGMENTS

We would like to apologize to those scientists whose work may not have appeared in this review either due to the limited scope of the review or oversight. We gratefully acknowledge the financial support for our β-lactam research programme from Council of Scientific and Industrial Research, New Delhi, Department of Science and Technology (DST), New Delhi, Government of India (Project No. SR/FTP/CS-135/2006 and SR/FT/CS-037/2010) and University Grants Commission (UGC), India, vide sanction No. F.17-7(J)/2004 (SA-I) dated 03-10-2011.

CONCLUSION

In conclusion, different types of nitrone-olefin 1,3-dipolar cycloadditions, their synthetic applications for the synthesis of various heterocyclic adducts and biological application of different products has been described. The nitrone-olefin [3+2] cycloaddition is a mild, efficient and stereoselective approach for the synthesis of differently substituted isoxazoles, isoxazolines, isoxazolidines and β-lactam derivatives. The biological potential of isoxazoles, isoxazolines and isoxazolidines has been extensively discussed. On the other hand, the activity profile of 2-azetidinone derivatives is briefly described due to its vast composition.

REFERENCES

[1] Dua, R., Shrivastava, S., Sonwane, S. K. and Srivastava, S. K. 2011. "Pharmacological Significance of Synthetic Heterocyclic Scaffold: A Review." *Advances in Biological Research* 5:120–144. URL: https://idosi.org/abr/5(3)/2.pdf.

[2] Ding, H. X., Leverett, C. A., Kyne, R. E. Jr., Liu, K. K. C., Fink, S. J., Flick, A. C. and O'Donnell, C. J. 2015. "Synthetic Approaches to the 2013 New Drugs" *Bioorganic and Medicinal Chemistry* 23:1895–1922. doi: 10.1016/j.bmc.2015.02.056.

[3] Huisgen, R. 1963. "1,3-Dipolar Cycloadditions. Past and Future" *Angewandte Chemie International Edition* 2:565–598. doi: 10.1002/anie.196305651.

[4] Firestone, R. A. 1972. "Application of the Linnett Electronic Theory to Organic Chemistry. V. Orientation in 1,3-Dipolar Cycloadditions According to the Diradical Mechanism. Partial Formal Charges in the Linnett Structures of the Diradical Intermediate" *Journal of Organic Chemistry* 37:2181–2191. doi: 10.1021/jo00978a027.

[5] Gothelf, K. V. and Jorgensen, K. A. 1998. "Asymmetric 1,3-Dipolar Cycloaddition Reactions" *Chemical Reviews* 98:863–909. doi: 10.1021/cr970324e.

[6] Huisgen, R. 1963. "Kinetics and Mechanism of 1,3-Dipolar Cycloadditions" *Angewandte Chemie International Edition* 2:633–645. doi: 10.1002/anie.196306331.

[7] Padwa, A. 1984. *1,3-Dipolar Cycloaddition Chemistry, Volume 1*. New York: John Wiley & Sons. doi: 10.1002/jhet.5570230658.

[8] Houk, K. N. 1975. "Frontier Molecular Orbital Theory of Cycloaddition Reactions" *Accounts of Chemical Research* 8:361–369. doi: 10.1021/ar50095a001.

[9] Acharjee, N., Banerji, A. and Gayen, Biswajit. 2011. "Cycloaddition Approach to Five-membered Heterocycles – Some Recent Experimental and Theoretical Aspects" *Journal of Indian Chemical Society* 88:1857–1877. URL: http://indianchemicalsociety.com/.

[10] Kumar, K. A., Govindaraju, M., Renuka, N. and Kumar, G. V. 2015. "Isoxazolines: An Insight to their Synthesis and Diverse Application" *Journal of Chemical and Pharmaceutical Research* 7:250–257. URL: http://www.jocpr.com/articles/isoxazolines-an-insight-to-their-synthesis-and-diverse-applications.pdf.

[11] Kamath, A. and Ojima, I. 2012. "Advances in the Chemistry of β-Lactams and its Medicinal Applications" *Tetrahedron* 68:10640–10664. doi: 10.1016/j.tet.2012.07.090.

[12] Shindo, M., Ohtsuki, K. and Shishido, K. 2005. "Asymmetric Inverse Electron-demand 1,3-Dipolar Cycloaddition of Ynolates with a Chiral Nitrone Derived from L-serine Leading to β-Amino Acid Derivatives" *Tetrahedron: Asymmetry* 16:2821–2831. doi: 10.1016/j.tetasy.2005.07.023.

[13] Gonzalez-Cruz, D., Tejedor, D., de Armas, P., Morales, E. Q. and Garcia-Tellado, F. 2006. "Organocatalysis 'on Water'. Regioselective [3+2]-Cycloaddition of Nitrones and Allenolates" *Chemical Communications* 2798–2800. doi: 10.1039/b606096j.

[14] Canterbury, D. P., Frontier, A. J., Um, J. M., Cheong, P. H. Y., Goldfeld, D. A., Huhn, R. A. and Houk, K. N. 2008. "Origins of Stereoselectivity in the Oxido-Alkylidenation of Alkynes" *Organic Letters* 10:4597–4600. doi: 10.1021/ol8019154.

[15] Coutouli-Argyropoulou, E., Sarridis, P. and Gkizis, P. 2009. "Water as the Medium of Choice for the 1,3-Dipolar Cycloaddition Reactions of Hydrophobic Nitrones" *Green Chemistry* 11:1906–1914. doi: 10.1039/b916765j.

[16] Canterbury, D. P., Herrick, I. R., Um, J., Houk, K. N. and Frontier, A. J. 2009. "A Torquoselective Extrusion of Isoxazoline N-oxide. Application to the Synthesis of Aryl Vinyl and Divinyl Ketones for Nazarov Cyclization" *Tetrahedron* 65:3165–3179. doi: 10.1016/j.tet.2008.10.003.

[17] Ning, X., Temming, R. P., Dommerholt, J., Guo, J., Ania, D. B., Debets, M. F., Wolfert, M. A., Boons, G. J. and van Delft, F. L. 2010. "Protein Modification by Strain-Promoted Alkyne-Nitrone Cycloaddition" *Angewandte Chemie International Edition* 49:3065–3068. doi: 10.1002/anie.201000408.

[18] Lopes, S. M. M., Nunes, C. M. and Pinho e Melo, T. M. V. D. 2010. "4-Isoxazolines and Pyrroles from Allenoates" *Tetrahedron* 66:6078–6084. doi: 10.1016/j.tet.2010.06.010.

[19] Valizadeh, H. and Dinparast, L. 2010. "Rapid, Highly Efficient, and Room-temperature $TiCl_4$-Catalyzed Synthesis of Δ^4-Isoxazolines Under Solvent-free Conditions" *Synthetic Communications* 41:291–297. doi: 10.1080/00397910903537349.

[20] Sakakura, A., Hori, M., Fushimi, M. and Ishihara, K. 2010. "Catalytic, Enantioselective 1,3-Dipolar Cycloadditions of Nitrones with Propioloylpyrazoles and Acryloylpyrazoles Induced by Chiral π-Cation Catalysts" *Journal of American Chemical Society* 132:15550–15552. doi: 10.1021/ja1081603.

[21] Worgull, D., Dickmeiss, G., Jensen, K. L., Franke, P. T., Holub, N. and Jorgensen, K. A. 2011. "Optically Active Bicyclic N-heterocycles by Organocatalytic Asymmetric Michael Addition/Cyclization Sequences" *Chemistry - A European Journal* 17:4076–4080. doi: 10.1002/chem.201100233.

[22] Zlatopolskiy, B. D., Kandler, R., Mottaghy, F. M. and Neumaier, B. 2012. "C-(4-[^{18}F]fluorophenyl)-N-phenylnitrone: A Novel ^{18}F-Labeled Building Block for Metal Free [3+2] Cycloaddition" *Applied Radiation and Isotopes* 70:184–192. doi: 10.1016/j.apradiso.2011.09.002.

[23] Badoiu, A. and Kundig, P. E. 2012. "Electronic Effects in 1,3-Dipolar Cycloaddition Reactions of *N*-alkyl and *N*-benzyl Nitrones with dipolarophiles" *Organic and Biomolecular Chemistry* 10:114–121. doi: 10.1039/c1ob06144e.

[24] McKay, C. S., Chigrinova, M., Blake, J. A. and Pezacki, J. P. 2012. "Kinetics Studies of Rapid Strain-Promoted [3+2]-Cycloadditions of Nitrones with Biaryl-aza-cyclooctynone" *Organic and Biomolecular Chemistry* 10:3066–3070. doi: 10.1039/c2ob07165g.

[25] Chakraborty, B. and Luitel, G. P. 2013. "Synthesis of Some Novel Fluoro Isoxazolidine and Isoxazoline Derivatives using *N*-benzyl Fluoro Nitrone via Cycloaddition Reaction in Ionic Liquid" *Journal of Chemical Sciences* 125:1071–1077. doi: 10.1007/s12039-013-0495-6.

[26] Tran, T. Q., Savinkov, R. S., Diev, V. V., Starova, G. L. and Molchanov, A. P. 2013. "An Efficient Method of Synthesis of Isoxazolidine-Fused β-Lactams via Base-Promoted

Cyclisation-Ring opening of Carbamoylspirocyclopropane Isoxazolidines" *Tetrahedron* 69:5173–5177. doi: 10.1016/j.tet.2013.04.054.

[27] Chiroli, V., Benaglia, M., Puglisi, A., Porta, R., Jumde, R. P. and Mandoli, A. 2014. "A Chiral Organocatalytic Polymer-Based Monolithic Reactor" *Green Chemistry* 16:2798–2806. doi: 10.1039/c4gc00031e.

[28] Aouadi, K., Vidal, S., Msaddek, M. and Praly, J. P. 2013. "Stereoselective Synthesis of 1,2,3-Triazolyl-Functionalized Isoxazolidines via Two Consecutive 1,3-Dipolar Cycloadditions, as Precursors of Unnatural Amino Acids" *Tetrahedron Letters* 54:1967–1971. doi: 10.1016/j.tetlet.2013.01.125.

[29] Aouadi, K., Abdoul-Zabar, J., Msaddek, M. and Praly, J. P. 2014. "A Cycloaddition-Cyclization Combined Approach to Enantiopure 3-Glycinyl-4-hydroxypyrrolidines and 3-Substituted 4-Hydroxyprolines" *European Journal of Organic Chemistry* 4107–4114. doi: 10.1002/ejoc.201402215.

[30] Siriwardena, A., Sonawane, D. P., Bande, O. P., Markad, P. R., Yonekawa, S., Tropak, M. B., Ghosh, S., Chopade, B. A., Mahuran, D. J. and Dhavale, D. D. 2014. "Synthesis of 1,5-Dideoxy-1,5-iminoribitol C-Glycosides through a Nitrone-Olefin Cycloaddition Domino Strategy: Identification of Pharmacological Chaperones of Mutant Human Lysosomal β-Galactosidase" *Journal of Organic Chemistry* 79:4398–4404. doi: 10.1021/jo500328u.

[31] Brahmi, J., Aouadi, K., Msaddek, M., Praly, J. P. and Vidal, S. 2016. "A Stereoselective Method for the Synthesis of Enantiopure 3-Substituted 4-Hydroxyproline Derivatives via 1,3-Dipolar Cycloadditions" *Comptes Rendus Chimie* 19:933–935. doi: 10.1016/j.crci.2016.03.003.

[32] Hoogenboom, J., Lutz, M., Zuilhof, H. and Wennekes, T. 2016. "Exploring the Chemistry of Bicyclic Isoxazolidines for the Multicomponent Synthesis of Glycomimetic Building Blocks" *Journal of Organic Chemistry* 81:8826–8836. doi: 10.1021/acs.joc.6b01515.

[33] Abda, H., Aouadi, K., Brahmi, J., Msaddek, M. and Vidal, S. 2016. "Unexpected Synthesis of Aziridines under Cu(I) Catalyzed Kinugasa Conditions Assisted by Microwave Irradiation" *Comptes Rendus Chimie* 19:275–278. doi: 10.1016/j.crci.2015.09.009.

[34] Chatterjee, A. and Bhattacharya, P. K. 2006. "Stereoselective Synthesis of Chiral Oxepanes and Pyrans through Intramolecular Nitrone Cycloaddition in Organized Aqueous Media" *Journal of Organic Chemistry* 71:345–348. doi: 10.1021/jo051414j.

[35] Shing, T. K. M., Wong, A. W. F., Ikeno, T. and Yamada, T. 2006. "Experimental and Theoretical Studies on Stereo- and Regioselectivity in Intramolecular Nitrone-Alkene Cycloaddition of Hept-6-enoses Derived from Carbohydrates" *Journal of Organic Chemistry* 71:3253–3263. doi: 10.1021/jo060348y.

[36] Hornyak, M., Kele, Z., Kovacs, L., Forgo, P. and Howarth, N. M. 2006. "Investigation of 1,3-Dipolar Cycloaddition Reactions of Unsaturated Nitrones in Ionic Liquids" *Nucleosides, Nucleotides and Nucleic Acids* 25:867–870. doi: 10.1080/15257770600793851.

[37] Yepez, A. F., Palma, A., Stashenko, E., Bahsas, A. and Amaro-Luis, J. M. 2006. "2-Allyl-N-benzyl Substituted α-Naphthylamines as Building Blocks in Heterocyclic Syntheses of Benz[*e*]naphtha[1,2-*b*]azepine and Naphtha[1,2-*b*]azepine Derivatives" *Tetrahedron Letters* 47:5825–5828. doi: 10.1016/j.tetlet.2006.04.160.

[38] Shing, T. K. M., Wong, W. F., Ikeno, T. and Yamada, T. 2007. "*trans*-Acetonide Controlled *endo*-Selective Intramolecular Nitrone-Alkene Cycloaddition of Hept-6-enoses: A Facile Entry to Calystegines, Tropanes and Hydroxylated Aminocycloheptanes" *Organic Letters* 9:207–209. doi: 10.1021/ol062621o.

[39] Chakraborty, C., Vyavahare, V. P. and Dhavale, D. D. 2007. "Intra-molecular Nitrone–Olefin Cycloaddition of D-Glucose Derived Allylic Alcohol: Synthesis of New Aminocyclohexitols" *Tetrahedron* 63:11984–11990. doi: 10.1016/j.tet.2007.09.011.

[40] Shing, T. K. M., Cheng, H. M., Wong, W. F., Kwong, C. S. K., Li, J., Lau, C. B. S., Leung, P. S. and Cheng, C. H. K. 2008. "Enantiospecific Synthesis of Pseudoacarviosin as a Potential Antidiabetic Agent" *Organic Letters* 10:3145–3148. doi: 10.1021/ol8010503.

[41] Stathakis, C. I., Athanatou, M. N. and Gallos J. K. 2009. "Synthesis of *ent*-Gabosine E from D-Mannose by Intramolecular Nitrone-Olefin Cycloaddition" *Tetrahedron Letters* 50:6916–6918. doi: 10.1016/j.tetlet.2009.09.154.

[42] Shing, T. K. M. and So, K. H. 2011. "Facile and Enantiospecific Syntheses of (6*S*,7*R*)-6-Chloro-7-benzyloxy-, (7*S*)-Halo- and (7*S*)-Hydroxy-cocaine and Natural (–)-Cocaine from D-(–)-Ribose" *Organic Letters* 13:2916–2919. doi: 10.1021/ol2009686.

[43] Rajesh, R., Suresh, M. and Raghunathan, R. 2014. "A Tactical Approach for the Synthesis of Novel β-Lactam-substituted, Polycyclic-fused Isoxazolidine Derivatives via an Intramolecular [3+2] Cycloaddition Reaction" *Tetrahedron Letters* 55:699–705. doi: 10.1016/j.tetlet.2013.11.117.

[44] Rawling, M. J., Storr, T. E., Bawazir, W. A., Cully, S. J., Lewis, W., Makki, M. S. I. T., Strutt, I. R., Hamza, J. D. and Stockman, R. A. 2015. "Facile Access to a Heterocyclic, sp^3-Rich Chemical Scaffold via a Tandem Condensation/Intramolecular Nitrone-Alkene [3+2] Cycloaddition Strategy" *Chemical Communications* 51:12867–12870. doi: 10.1039/c5cc05070g.

[45] Paladugu, S. R. and Looper, R. E. 2015. "Preparation of a 1,2-Isoxazolidine Synthon for the Synthesis of Zetekitoxin AB" *Tetrahedron Letters* 56:6332–6334. doi: 10.1016/j.tetlet.2015.09.070.

[46] Xiang, J., Zhu, T., Dang, Q. and Bai, X. 2016. "Electronic Effects on the *cis/trans* Selectivity in Formation of Isoxazolidine-fused Eight-membered Ring via an Intramolecular Nitrone-Alkene Cycloaddition" *Chemistry of Heterocyclic Compounds* 52:601–608. doi: 10.1007/s10593-016-1938-z.

[47] Stecko, S., Furman, B. and Chmielewski, M. 2014. "Kinugasa Reaction: An 'Ugly Duckling' of β-Lactam Chemistry" *Tetrahedron* 70:7817–7844. doi: 10.1016/j.tet.2014.06.024.

[48] Kinugasa, M. and Hashimoto, S. 1972. "The Reaction of Copper(I) Phenylacetylide with Nitrones" *Journal of Chemical Society, Chemical Communications* 466–467. doi: 10.1039/C39720000466.

[49] Grigor'ev, I. A. 2007. "Nitrones: Novel Strategies in Synthesis" In *Nitrile Oxide, Nitrones, and Nitronates in Organic Synthesis*, edited by Feuer, H., 129–134. Hoboken, NJ, USA: John Wiley & Sons, Inc. doi: 10.1002/9780470191552.ch2.

[50] Ding, L. K. and Irwin, W. J. 1976. "*cis*- and *trans*-Azetidin-2-ones from Nitrones and Copper Acetylide" *Journal of Chemical Society, Perkin Transactions 1* 2382–2386. doi: 10.1039/P19760002382.

[51] Fokin, V. V. 2011. "New Reactions of Copper Acetylides: Catalytic Dipolar Cycloadditions and Beyond" In *Catalyzed Carbon–Heteroatom Bond Formation*, edited by Yudin, A. K., 199–225. Weinheim, Germany: Wiley VCH. doi: 10.1002/9783527633388.

[52] Pal, R. and Basak, A. 2006. "A Novel Synthesis of β-Lactam Fused Cyclic Enediynes by Intramolecular Kinugasa Reaction" *Chemical Communications* 2992–2994. doi: 10.1039/b605743h.

[53] Ye, M. C., Zhou, J. and Tang, Y. 2006. "Trisoxazoline/Cu(II)-Promoted Kinugasa Reaction. Enantioselective Synthesis of β-Lactams" *Journal of Organic Chemistry* 71:3576–3582. doi: 10.1021/jo0602874.

[54] Coyne, A. G., Muller–Bunz, H. and Guiry, P. J. 2007. "The Asymmetric Synthesis of β-Lactams: HETPHOX/Cu(I) Mediated Synthesis via the Kinugasa Reaction" *Tetrahedron: Asymmetry* 18:199–207. doi: 10.1016/j.tetasy.2007.01.006.

[55] Stecko, S., Mames, A., Furman, B. and Chmielewski, M. 2008. "Diastereoselective Synthesis of Carbapenams via Kinugasa Reaction" *Journal of Organic Chemistry* 73:7402–7404. doi: 10.1021/jo801212q.

[56] Zhang, X., Hsung, R. P., Li, H., Zhang, Y., Johnson, W. L. and Figueroa, R. 2008. "A Highly Stereoselective Synthesis of Chiral α-Amino-β-lactams via the Kinugasa Reaction Employing Ynamides" *Organic Letters* 10:3477–3479. doi: 10.1021/ol801257j.

[57] Stecko, S., Mames, A., Furman, B. and Chmielewski, M. 2009. "Asymmetric Kinugasa Reaction of Cyclic Nitrones and Nonracemic Acetylenes" *Journal of Organic Chemistry* 74:3094–3100. doi: 10.1021/jo900121x.

[58] Saito, T., Kikuchi, T., Tanabe, H., Yahiro, J. and Otani, T. 2009. "Enantioselective Synthesis of β-Lactams via the IndaBox-Cu(II)-catalyzed Kinugasa Reaction" *Tetrahedron Letters* 50:4969–4972. doi: 10.1016/j.tetlet.2009.06.050.

[59] Mames, A., Stecko, S., Mikozajczyk, P., Soluch, M., Furman, B. and Chmielewski, M. 2010. "Direct, Catalytic Synthesis of Carbapenams via Cycloaddition/Rearrangement Cascade Reaction: Unexpected Acetylene's Structure Effect" *Journal of Organic Chemistry* 75:7580–7587. doi: 10.1021/jo101355h.

[60] Khangarot, R. K. and Kaliappan, K. P. 2011. "A Stereoselective Synthesis of Sugar-Derived Chiral β-Lactams" *European Journal of Organic Chemistry* 6117–6127. doi: 10.1002/ejoc.201100953.

[61] Michalak, M., Stodulski, M., Stecko, S., Mames, A., Panfil, I., Soluch, M., Furman, B. and Chmielewski, M. 2011. "A Formal Synthesis of Ezetimibe via Cycloaddition/Rearrangement Cascade Reaction" *Journal of Organic Chemistry* 76:6931–6936. doi: 10.1021/jo2010846.

[62] Chen, J. H., Liao, S. H., Sun, X. L., Shen, Q. and Tang, Y. 2012. "Tris(oxazoline)/copper-catalyzed Coupling of Alkynes with Nitrones: A Highly Enantioselective Access to β-Lactams" *Tetrahedron* 68:5042–5045. doi: 10.1016/j.tet.2012.04.049.

[63] Grzeszczyk, B., Polawska, K., Shaker, Y. M., Stecko, S., Mames, A., Woznica, M., Chmielewski, M. and Furman, B. 2012. "Asymmetric Kinugasa Reaction Involving Six-membered Cyclic Nitrones" *Tetrahedron* 68:10633–10639. doi: 10.1016/j.tet.2012.09.031.

[64] Michalak, M., Stodulski, M., Stecko, S., Woznica, M., Staszewska-Krajewska, O., Kalicki, P., Furman, B., Frelek, J. and Chmielewski, M. 2012. "Synthesis of N,4-Diaryl Substituted β-Lactams via Kinugasa Cycloaddition/rearrangement Reaction" *Tetrahedron* 68:10806–10817. doi: 10.1016/j.tet.2011.11.007.

[65] Baeza, B., Casarrubios, L. and Sierra, M. A. 2013. "Towards a General Synthesis of 3-Metal-substituted β-lactams" *Chemistry - A European Journal* 19:11536–11540. doi: 10.1002/chem.201301114.

[66] Chen, Z., Lin, L., Wang, M., Liu, X. and Feng, X. 2013. "Asymmetric Synthesis of *trans*-β-Lactams by a Kinugasa Reaction on Water" *Chemistry - A European Journal* 19:7561–7567. doi: 10.1002/chem.201204373.

[67] Zlatopolskiy, B. D., Krapf, P., Richarz, R., Frauendorf, H., Mottaghy, F. M. and Neumaier, B. 2014. "Synthesis of ^{18}F-Labelled β-Lactams by Using the Kinugasa Reaction" *Chemistry - A European Journal* 20:4697–4703. doi: 10.1002/chem.201304056.

[68] Dagne, E. and Castagnoli, N. Jr. 1972. "Structure of Hydroxycotinine, a Nicotine Methobolite" *Journal of Medicinal Chemistry* 15:356–360. doi: 10.1021/jm00274a005.

[69] Tufariello, J. J. and Tegeler, J. J. 1976. "Nitrone Cycloaddition. A Route to the Lupin Class of Alkaloids" *Tetrahedron Letters* 17:4037–4040. doi: 10.1016/S0040-4039(00)92569-3.

[70] Gossinger, E. 1980. "Stereoselektive Synthesis Von (±)-Porantheridin" *Tetrahedron Letters* 21:2229–2232. doi: 10.1016/0040-4039(80)80010-4.

[71] Habeeb, A. G., Rao, P. N. P. and Knaus, E. E. 2001. "Design and Synthesis of 4,5-Diphenyl-4-isoxzolines: Novel Inhibitors of Cyclooxygenase-2 with Analgesic and Antiinflammatory Activity" *Journal of Medicinal Chemistry* 44:2921–2927. doi: 10.1021/jm0101287.

[72] Sammelson, R. E., Ma, T., Galietta, L. J. V., Verkman, A. S. and Kurth, M. J. 2003. "3-(2-Benzyloxyphenyl)isoxazoles and Isoxazolines: Synthesis and Evaluation as CFTR Activators" *Bioorganic and Medicinal Chemistry Letters* 13:2509–2512. doi: 10.1016/S0960-894X(03)00482-7.

[73] Mishra, R. C., Tewari, N., Verma, S. S., Tripathi, R. P., Kumar, M. and Shukla, P. K. 2004. "Diastereoselective Synthesis and Antifungal Activity of Glycosyl Isoxazolines" *Journal of Carbohydrate Chemistry* 23:353–374. doi: 10.1081/CAR-200037331.

[74] Sailu, B., Srinivas, M., Srinivas, S., Balram, B., Ram, B., Taara, B. and Vasudha B. 2012. "Synthesis and Antibacterial Activity of Novel Isoxazoline Derivatives" *Der Pharma Chemica* 4:2036–2041. URL: https://www.derpharmachemica.com/pharma-chemica/synthesis-and-antibacterial-activity-of-novel-isoxazoline-derivatives.pdf.

[75] Youn, H. S., Lee, E. J., Lee, J. E., Park, W. K., Baek, D. J., Cho, Y. S., Koh, H. Y., Choo, H. and Pae, A. N. 2009. "Synthesis and Biological Evaluation of Isoxazoline and Isoxazole Derivatives as 5-HT2A and 5-HT2C Receptor Ligands" *Bulletin of Korean Chemical Society* 30:1873–1876. URL: http://citeseerx.ist.psu.edu/viewdoc/download?doi=10.1.1.1013.5601&rep=rep1&type=pdf.

[76] Prabhashankar, J., Govindappa, V. K., Renuka, N., Akshatha, K. N., Mahadevamurthy, S. and Kariyappa, A. K. 2012. "Evaluation and Studies on the Structural Impact of 3-Aryl-5-(4-methoxyphenyl)-4,5-dihydroisoxazole-4-carbonitriles on their Biological Activities" *Der Pharmacia Lettre* 4:1685–1691. URL: http://eprints.uni-mysore.ac.in/12866/1/DPL-2012-4-6-1685-1691.pdf.

[77] Kumar, K. A., Lokeshwari, D. M. and Kumar G. V. 2012. "Evaluation and Studies on the Structural Impact of Substituted 4,5-Dihydroisoxazoles on Their Biological Activities" *International Journal of Pharmaceutical Sciences and Drug Research* 4:236–239. URL: http://ijpsdr.com/pdf/vol4-issue4/3.pdf.

[78] Govindaraju, M., Kumar, G. V. and Kumar, K. A. 2014. "Synthesis of Antimicrobial activity of Novel Isoxazolines by 1,3-Dipolar Cycloaddition Reactions" *International Journal of ChemTech Research* 6:886–890. URL: http://sphinxsai.com/2014/CTVOL6/CT=02(886-890)AJ14.pdf.

[79] Rakesh, Bruhn, D., Madhura, D. B., Maddox, M., Lee, R. B., Trivedi, A., Yang, L., Scherman, M. S., Gilliland, J. C., Gruppo, V., McNeil, M. R., Lenaerts, A. J., Meibohm, B. and Lee, R. E. 2012. "Antitubercular Nitrofuran Isoxazolines with Improved Pharmacokinetic Properties" *Bioorganic and Medicinal Chemistry* 20:6063–6072. doi: 10.1016/j.bmc.2012.08.023.

[80] Makarov, V. A., Riabova, O. B., Granik, V. G., Wutzler, P. and Schmidtke, M. 2005. "Novel [(Biphenyloxy)propyl]isoxazole Derivatives for Inhibition of Human Rhinovirus 2 and Coxsackievirus B3 Replication" *Journal of Antimicrobial Chemotherapy* 55:483–488. doi: 10.1093/jac/dki055.

[81] Poma, P., Notarbartolo, M., Labbozzetta, M., Maurici, A., Carina, V., Alaimo, A., Rizzi, M., Simoni, D. and D'Alessandro, N. 2007. "The Antitumor Activities of Curcumin and of its Isoxazole Analogue are not Affected by Multiple Gene Expression Changes in an MDR Model of the MCF-7 Breast Cancer Cell Line: Analysis of the possible molecular basis" *International Journal of Molecular Medicine* 20:329–335. doi: 10.3892/ijmm.20.3.329.

[82] Arya, N., Jagdale, A. Y., Patil, T. A., Yeramwar, S. S., Holikatti, S. S., Dwivedi, J., Shishoo, C. J. and Jain, K. S. 2014. "The Chemistry and Biological Potential of Azetidin-2-ones" *European Journal of Medicinal Chemistry* 74:619–656. doi: 10.1016/j.ejmech.2014.01.002.

[83] O'Driscoll, M., Greenhalgh, K., Young, A., Turos, E., Dickey, S. and Lim, D. V. 2008. "Studies on the Antifungal Properties of *N*-thiolated β-Lactams" *Bioorganic and Medicinal Chemistry* 16:7832–7837. doi: 10.1016/j.bmc.2008.06.035.

[84] Beck, T. N., Lloyd, D., Kuskovsky, R., Minah, J., Arora, K., Plotkin, B. J., Green, J. M., Boshoff, H. I., Barry, C. III., Deschamps, J. and Konaklieva, M. I. 2015. "Non-transpeptidase Binding Arylthioether β-Lactams Active against *Mycobacterium tuberculosis* and *Moraxella Catarrhalis*" *Bioorganic and Medicinal Chemistry* 23:632–647. doi: 10.1016/j.bmc.2014.11.025.

[85] Skiles, J. W. and McNeil, D. 1990. "Spiro Indolinone Beta-lactams, Inhibitors of Poliovirus and Rhinovirus 3*C*-proteinases" *Tetrahedron Letters* 31:7277–7280. doi: 10.1016/S0040-4039(00)88543-3.

[86] Banik, B. K., Banik, I. and Becker, F. F. 2010. "Asymmetric Synthesis of Anticancer β-Lactams via Staudinger Reaction: Utilization of chiral ketene from carbohydrate" *European Journal of Medicinal Chemistry* 45:846–848. doi: 10.1016/j.ejmech.2009.11.024.

[87] Nuzzi, A., Fiasella, A., Ortega, J. A., Pagliuca, C., Ponzano, S., Pizzirani, D., Bertozi, S. M., Ottonello, G., Tarozzo, G., Reggiani, A., Bandiera, T., Bertozi, F. and Piomelli, D. 2016. “Potent α-Amino-β-lactam Carbamic Acid Ester as NAAA Inhibitors. Synthesis and Structure-activity Relationship (SAR) studies” *European Journal of Medicinal Chemistry* 111:138–159. doi: 10.1016/j.ejmech.2016.01.046.

[88] Goel, R. K., Mahajan, M. P. and Kulkarni S. K. 2004. “Evaluation of Anti-hyperglycemic Activity of Some Novel Monocyclic Beta Lactams” *Journal of Pharmacy and Pharmaceutical Sciences* 7:80–83. URL: https://sites.ualberta.ca/~csps/JPPS7(1)/R.Goel/betalactams.pdf.

[89] Leach, C. A., Hickey, D. M. B., Lfe, R. J., Macphee, C. H., Smith, S. A. and Tew, D. G. 2001. “Lipoprotein-associated PLA_2 Inhibition – A Novel, Non-lipid Lowering Strategy for Atherosclerosis Therapy” *Il Farmaco* 56:45–50. doi: 10.1016/S0014-827X(01)01011-4.

[90] Goel, R. K., Singh, A., Naidu, P. S., Mahajan, M. P. and Kulkarni, S. K. 2005. “PASS Assisted Search and Evaluation of Some Azetidin-2-ones as CNS Active Agents” *Journal of Pharmacy and Pharmaceutical Sciences* 8:182–189. URL: https://sites.ualberta.ca/~csps/JPPS8(2)/R.Goel/cns.htm.

[91] Aoyama, Y., Uenaka, M., Kii, M., Tanaka, M., Konoike, T., Hayasaki-Kajiwara, Y., Naya, N. and Nakajima, M. 2001. “Design, Synthesis and Pharmacological Evaluation of 3-Benzylazetidine-2-one-based Human Chymase Inhibitors” *Bioorganic and Medicinal Chemistry* 9:3065–3075. doi: 10.1016/S0968-0896(01)00209-7.

[92] Han, W. T., Trehan, A. K., Wright, J. J. K., Federici, M. E., Seiler, S. M. and Meanwell, N. A. 1995. “Azetidin-2-one Derivatives as Inhibitors of Thrombin” *Bioorganic and Medicinal Chemistry* 3:1123–1143. doi: 10.1016/0968-0896(95)00101-L.

[93] Cvetovich, R. J., Chartrain, M., Hartner, F. W., Roberge, C., Amato, J. S. and Grabowski, E. J. 1996. “An Asymmetric Synthesis of L-694,458, a Human Leukocyte Elastase Inhibitor, via Novel Enzyme Resolution of β-Lactam Esters” *Journal of Organic Chemistry* 61:6575–6580. doi: 10.1021/jo960618k.

[94] Setti, E. L., Davis, D., Chung, T. and McCarter, J. 2003. “3,4-Disubstituted Azetidinones as Selective Inhibitors of the Cysteine Protease Cathepsin K. Exploring P2 Elements for Selectivity” *Bioorganic and Medicinal Chemistry Letters* 13:2051–2053. doi: 10.1016/S0960-894X(03)00304-4.

[95] Bhalla, A., Bari, S. S., Bhalla, J., Khullar, S. and Mandal, S. 2016. “Facile Synthesis of Novel Halogenated 4-Pyrazolylspirocyclic-β-lactams: Versatile Heterocyclic Synthons” *Tetrahedron Letters* 57:2822–2828 and reference cited therein. doi: 10.1016/j.tetlet.2016.05.047.

[96] Konaklieva, M. I. 2002. “β-Lactams as Inhibitors of Serine Enzymes” *Current Medicinal Chemistry - Anti-Infective Agents* 1:215–238. doi: 10.2174/15680120 23354910.

[97] Borthwick, A. D., Weingarte, G., Haley, T. M., Tomaszewski, M., Wang, W., Hu, Z., Bedard, J., Jin, H., Yuen, L. and Mansour, T. S. 1998. “Design and Synthesis of Monocyclic β-Lactams as Mechanism-based Inhibitors of Human Cytomegalovirus Protease” *Bioorganic and Medicinal Chemistry Letters* 8:365–370. doi: 10.1016/S0960-894X(98)00032-8.

[98] Kazi, A., Hill, R., Long, T. E., Kuhn, D. J., Turos, E. and Dou, Q. P. 2004. "Novel N-Thiolated β-Lactam Antibiotics Selectively Induce Apoptosis in Human Tumor and Transformed, but not Normal or Nontransformed, Cell" *Biochemical Pharmacology* 67:365–374. doi: 10.1016/j.bcp.2003.09.017.

[99] Guillon, C. D., Koppel, G. A., Brownstein, M. J., Chaney, M. O., Ferris, C. F., Lu, S. F., Fabio, K. M., Miller, M. J., Heindel, N. D., Hunden, D. C., Cooper, R. D. G., Kaldox, S. W., Skelton, J. J., Dressman, B. A., Clay, M. P., Steinberg, M. I., Bruns, R. F. and Simon, N. G. 2007. "Azetidinones as Vasopressin V1a Antagonists" *Bioorganic and Medicinal Chemistry* 15:2054–2080. doi: 10.1016/j.bmc.2006.12.031.

In: Organic and Medicinal Chemistry, Volume 1
Editor: Bimal Krishna Banik ISBN: 978-1-53614-453-6

Chapter 2

NOVEL ASPECTS IN OXIDATIVE ORGANIC TRANSFORMATIONS USING 2,3-DICHLORO-5,6-DICYANO-1,4-BENZOQUINONE (DDQ): SYNTHESIS OF MEDICINALLY IMPORTANT FRAMEWORKS

Dipika Narula and Aman Bhalla*
Department of Chemistry and Centre of Advanced Studies in Chemistry,
Panjab University, Chandigarh, India

ABSTRACT

Amongst the various quinonoid oxidizing agents, 2,3-dichloro-5,6-dicyano-1,4-benzoquinone (DDQ) serves as a versatile strong oxidant for carrying out a number of oxidative organic transformations. Efficient and atom-economic synthesis of complex molecules is a great endeavor in organic synthesis. The use of DDQ as an oxidant in the Cross Dehydrogenative Coupling (CDC) reactions is a new method for the generation of C-C bonds from C-H bonds in more atom-economic and environmental friendly manner. Many patents from the chemical and pharmaceutical industries mention several uses of DDQ. The goal of this chapter, however, is to take a comprehensive look at DDQ mediated organic transformations, particularly in the last four years, with prime focus on its use for oxidative C-C bond forming reactions, oxidative cyclizations/annulations, aromatizations as well as for oxidation reactions. A variety of medicinally important frameworks could be accessed by this oxidation protocol.

Keywords: 2,3-Dichloro-5,6-dicyano-1,4-benzoquinone, oxidation, cross dehydrogenative coupling, annulation, aromatization, cyclization

* Corresponding Author E-mail: amanbhalla@pu.ac.in.

ABBREVIATIONS

Å	Angstrom
Ac	Acetyl
Ar	Aryl
Boc	*N-tert*-Butoxycarbonyl
Bu	Butyl
i-Bu	*iso*-Butyl
n-Bu	*normal*-Butyl
t-Bu	*tert*-Butyl
Bn/Bz	Benzyl
°C	Degree centigrade
cat.	Catalytic
Cbz	*N*-Benzyloxycarbonyl
CDDO	2-Cyano-3,12-dioxooleana-1,9(11)-dien-28-oic acid
DBU	1,8-Diazabicyclo(5.4.0)undec-7-ene
DCE	1,2-Dichloroethane
DCM	Dichloromethane
$DDQH_2$	2,3-Dichloro-5,6-dicyano-1,4-hydroquinone
DKP	Diketopiperazines
DMC	Dimethylcarbonate
DMB	3,4-Dimethoxybenzyl
DMF	*N*,*N*-Dimethylformamide
DMP	3,4-Dimethoxyphenyl
DMSO	Dimethylsulfoxide
dppe	1,2-Bis(diphenylphosphino)ethane
Dr	Diastereomeric ratio
Ee	Enantiomeric excess
equiv.	Equivalent
Et	Ethyl
EWG	Electron-withdrawing group
FG	Functional group
G	Gram
H	Hour
c-hex	Cyclohexyl
n-hex	*normal*-Hexyl
hERG	Human Ether-a-go-go Related Gene
HIV	Human Immunodeficiency Virus
HCV	Hepatitis C Virus
HPLC	High Performance Liquid Chromatography
Hz	Hertz

IC_{50}	Half maximal Inhibitory Concentration
MeOH	Methyl alcohol
M	Micro
M	*meta*
min	Minute
Mg	Milligram
mL	Millilitre
mol	Mole
Me	Methyl
MsOH	Methanesulfonic acid
MS	Molecular Sieves
NS5A	Nonstructural protein 5A
O	*ortho*
P	*para*
Pa	Pascal
n-Pent	*normal*-Pentyl
Ph	Phenyl
Pin	Pinacol
Piv	Pivalate
PMB	*p*-Methoxybenzyl
PMP	*p*-Methoxyphenyl
PPA	Polyphosphoric acid
Pr	Propyl
i-Pr	*iso*-Propyl
n-Pr	*normal*-Propyl
Py	Pyridinyl
Rt	Room temperature
TBDPS	*tert*-Butyldiphenylsilyl
TBHP	*tert*-Butylhydroperoxide
TBS	*tert*-Butyldimethylsilyl
TES	Triethylsilyl
Tf	Trifluoromethanesulfonyl
TFA	Trifluoroacetic acid
THF	Tetrahydrofuran
TMS	Trimethylsilyl
TMSCN	Trimethylsilylcyanide
TMSOTf	Trimethylsilyltrifluoromethanesulfonate
Trp	Tryptophan
Ts	Tosyl
v/v	Volume/Volume

1. INTRODUCTION

Oxidation reactions are of fundamental importance in nature and play a crucial role in organic synthesis. Whereas inorganic chemists define oxidation in terms of increase in oxidation number or loss of electrons, the perspective of an organic chemist is removal of hydrogen, addition of oxygen or removal of an electron. These definitions in organic chemistry, though technically correct, are not easy to apply. The task is simplified by a qualitative categorization and arrangement of a series of functional groups in an order of increasing oxidation state. Then, it is easier to define oxidation as the conversion of a functional group in a molecule from one category to a higher one. The reverse process is therefore a reduction. Conversions within a category are neither oxidations nor reductions. Evidently, no oxidation is observed without a concurrent reduction. Such redox reactions (portmanteau of oxidation and reduction reactions) are common and vital to some of the basic functions of life, including photosynthesis, combustion and cellular respiration.

In synthetic organic chemistry, a variety of oxidants/oxidizing agents are available for carrying out the oxidative organic transformations. These transformations refer to cleavage of carbon-carbon bonds, replacement of hydrogen by oxygen, oxidative coupling, elimination of hydrogen, oxidation to inorganic esters, and reactions in which an unshared pair becomes bonded to oxygen. Therefore, oxidizing agents that attack one or a few types of substrate or oxidize one group selectively in a molecule without affecting the other oxidizable groups are needed. For this purpose, there is currently a demand for more sustainable and selective oxidation methods [1].

Quinones are an important class of organic compounds. Quinones are pervasive in nature [2], they are present in flowering plants, bacteria and arthropods. Their role in biological functions including oxidative phosphorylation and electron transfer is well documented in the literature [3]. Derivatized 1,4-benzoquinones (Figure 1) such as plastoquinones, phylloquinones serve as electron carriers in photosynthesis. Ubiquinone, on the other hand, has been recognized as a component of electron-transport chain in aerobic cellular respiration. Furthermore, the 1,4-benzoquinone ring is the core structure in pharmacologically active compounds (Figure 1) such as streptonigrin (STN) (antitumor, antibiotic) [4], bis-prenylated quinone (antiproliferative) [5], heliotropinone A (antifungal) [6], xylariaquinone (antimalarial) [7] and alopecuquinone (estereogenic active) [8]. In addition, antineoplastic [9], antitermite [10], anticoagulant [11], insect antifeedant [12] and herbicidal [13] activities of 1,4-benzoquinone derivatives have been reported.

In addition to exhibiting a vast pharmacological profile, the halogenated derivatives of benzoquinone *viz* 3,4,5,6-tetrachloro-1,2-benzoquinone (*o*-chloranil), 2,3,5,6-tetrachloro-1,4-benzoquinone (*p*-chloroanil) and 2,3-dichloro-5,6-dicyano-1,4-benzoquinone (DDQ) have been utilized as dehydrogenating/oxidizing agents in synthetic organic chemistry (Figure 2) [14-16]. Oxidative transformations facilitated by DDQ are well documented in the literature [17-21].

DDQ has multitude of uses as an electrophile [22], as a maleonitrile building block [23], as an efficient catalyst for the hydrolysis of acetals [24] and in the alcoholysis of epoxides [25]. Its role in oxidation [14-21] remains an essential part of current synthetic and mechanistic chemistry. The aim of this chapter is to provide aspirant synthetic organic chemists with a survey and flavor of the oxidative transformations promoted by DDQ in the last four years (2014-2017). The scope of the chapter is limited to DDQ-mediated dehydrogenative coupling reactions (both inter- and intramolecular), cyclizations/annulations, dehydrogenative aromatizations, and oxidation reactions. We present an overview of the evolution of methods which utilize the oxidizing capability of DDQ for the synthesis of biologically active compounds/novel synthetic assemblies/heterocyclic ring frameworks of medicinal interest.

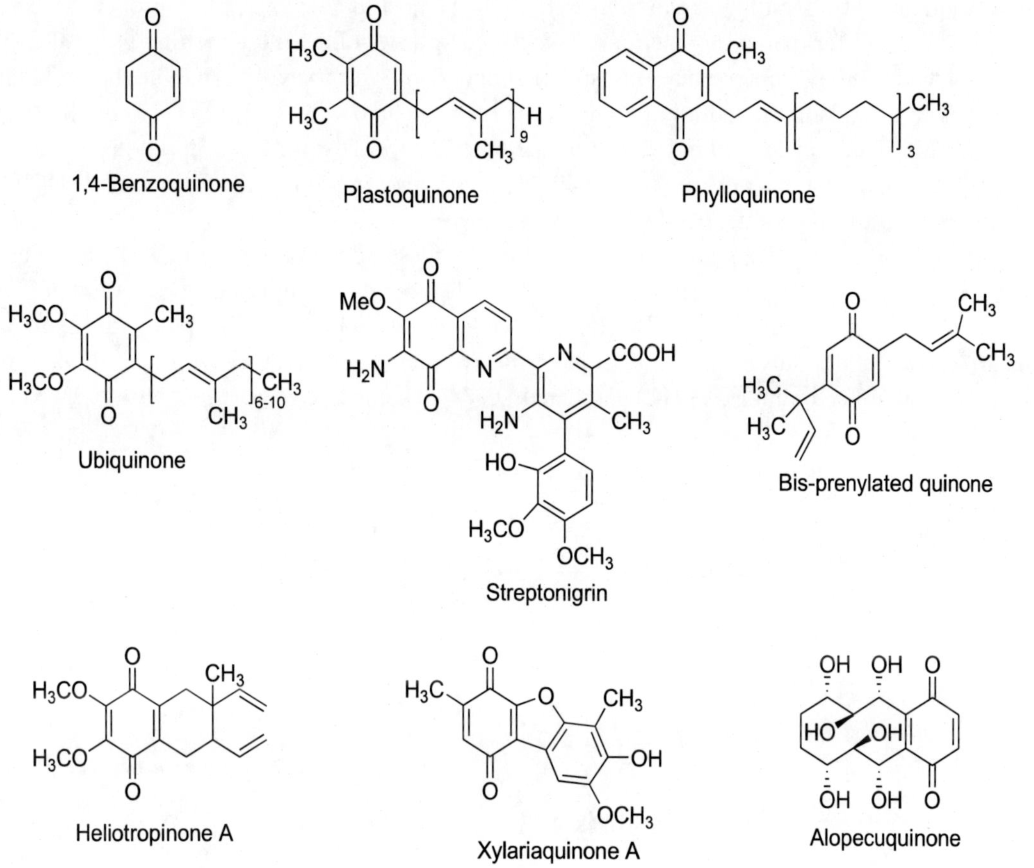

Figure 1. 1,4-Benzoquinone and its pharmacologically active derivatives.

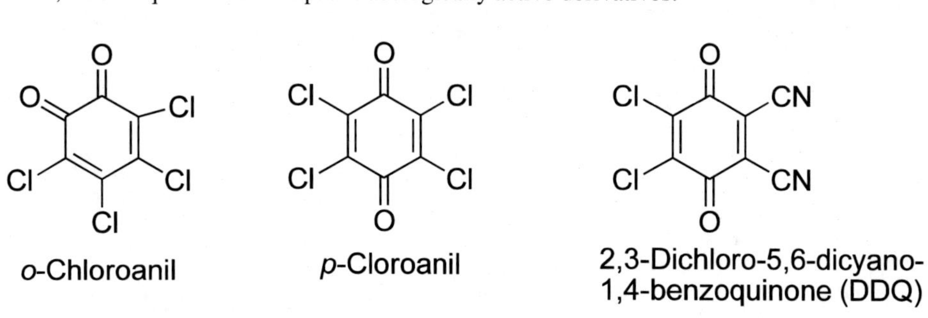

Figure 2. Halogenated benzoquinones.

2. 2,3-DICHLORO-5,6-DICYANO-1,4-BENZOQUINONE (DDQ): A VERSATILE REAGENT FOR OXIDATIVE ORGANIC TRANSFORMATIONS

2.1. Synthesis

A quinone with multiple electron-withdrawing substituents, DDQ has proven to be a versatile reagent for oxidation reactions. This bright yellow solid, with a melting point of 213-215°C is indefinitely stable in dry atmosphere, but decomposes to evolve HCN in the presence of water. The oxidation potential of DDQ is *ca.* 1.0 compared with 0.71 v for chloroanil [26].

In 1906, Thiele and Gunther [27] formulated the first synthesis of DDQ in six-steps starting from 1,4-benzoquinone (Scheme 1). A sequence of steps involving cyanation (with HCN), oxidation (with HNO_3 fumes) and chlorination (with HCl) led to DDQ.

In 1952, Brook [28] suggested mixed nitrogen oxides to be superior than nitric acid fumes in the oxidation steps. Eleven years later, Mitchell [29] found that an initial chlorination of 2,3-dicyanohydroquinone in refluxing acetic acid yielded 2,3-dichloro-5,6-dicyanohydroquinone in 40 - 50% yields. Subsequent oxidation of the 2,3-dichloro derivative with lead dioxide in aqueous ethanol containing 5% HCl generated DDQ (Scheme 2). No reaction was observed in the absence of acid.

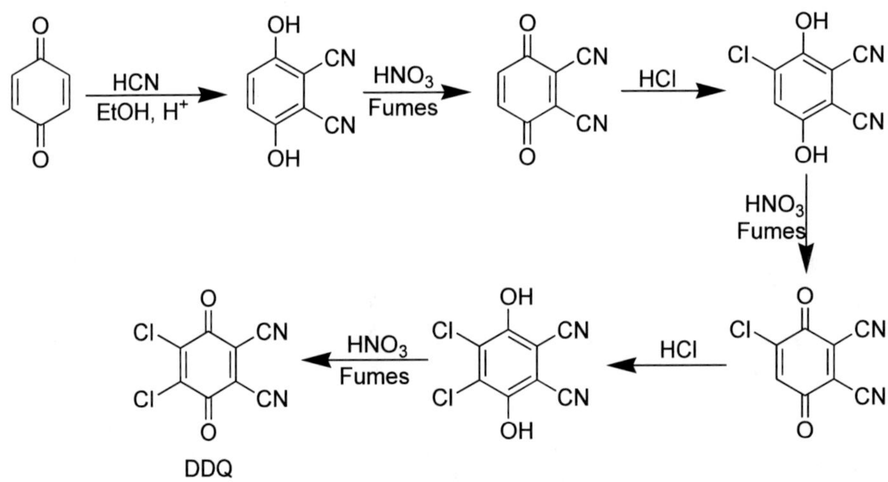

Scheme 1. Synthesis of DDQ from 1,4-benzoquinone.

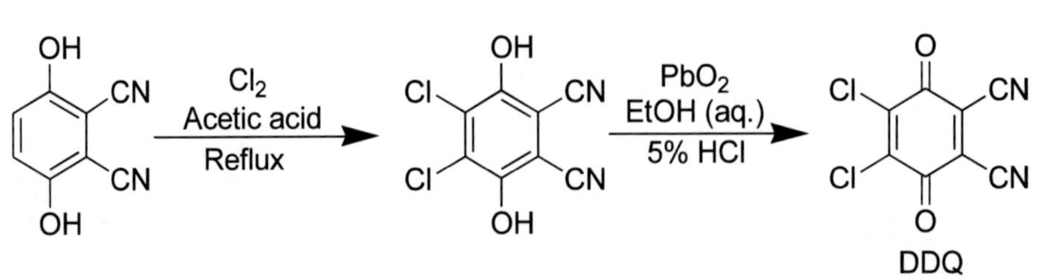

Scheme 2. Synthesis of DDQ from 2,3-dicyanohydroquinone.

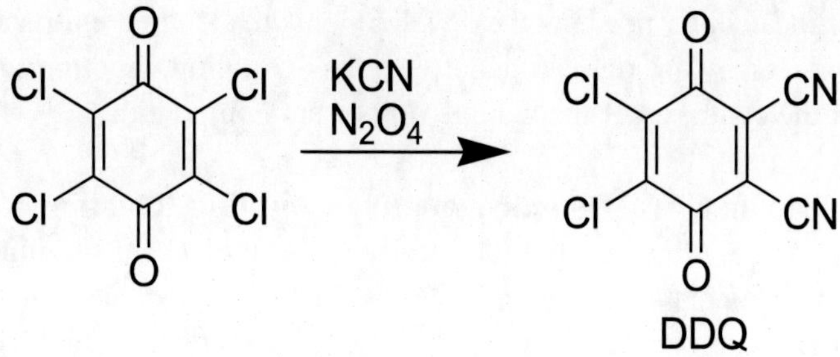

Scheme 3. Synthesis of DDQ from *p*-chloroanil.

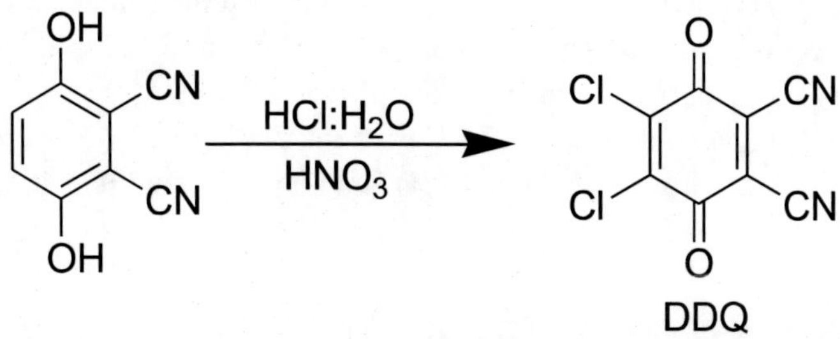

Scheme 4. Synthesis of DDQ from 2,3-dicyanohydroquinone.

Starting from *p*-chloroanil, Wallenfels and co-workers [30] obtained DDQ *via* potassium cyanide mediated cyanation and nitrogen tetraoxide mediated oxidation process (Scheme 3).

In a report by Walker and Waugh [31], stirring of 2,3-dicyanohydroquinone with 1 : 1 hydrochloric acid-water, is followed by addition of 70% nitric acid at 35°C to generate DDQ in 90% yield (Scheme 4).

2.2. Oxidative Transformations Promoted by DDQ

2.2.1. Cross Dehydrogenative Coupling (CDC) Reactions

The dehydrogenative cross coupling reactions for the C-C bond formation through selective C-H bond activation have attracted great attention in the past decade [17, 32-34]. It provides the opportunity to directly access the target from two readily accessible C-H reagents in minimum steps with minimum reactivity tuning, high atom economy, and avoids any pre-functionalization of the substrates.

Quinone-mediated oxidative couplings of the substrate with a nucleophile (Nu-H) has been generally associated with three mechanistic pathways (Figure 3):

(a) Single electron transfer (SET): It involves one electron oxidation of the neutral substrate into a radical cation which reacts with the nucleophile (Figure 3 (a)) [35, 36]
(b) α-Deprotonation (Figure 3 (b))
(c) Direct-hydride transfer (Figure 3 (c))

Both (b) and (c) lead to the formation of good electrophiles which couple with the nucleophile [37, 38] to generate the cross-coupled products.

This section of the chapter presents the developments made via the utilization of DDQ for carrying out a wide array of dehydrogenative cross-coupling reactions (both inter- and intramolecular) for the synthesis of medicinally important compounds.

2.2.1.1. Inter-Molecular Cross Dehydrogenative Coupling Reactions

Given the importance of indenes in biologically active molecules and functional materials [39, 40], Liu and co-workers [41] have described the metal-free synthesis of these carbocyclic compounds *via* a CDC approach employing DDQ as an oxidant. The stoichiometric ratios of the coupling partners and the oxidant was found to have a profound effect on the outcome of the CDC reaction. The coupling reaction between ketene dithioacetals **1** (1 equiv.) and benzylic compounds **2** (5 equiv.) promoted by DDQ (2 equiv.) furnished the coupled products **3** (Scheme 5). A range of cyclic/acyclic ketene dithioacetals **1** (with α-substituted cyano, acetyl, cinnamoyl, benzoyl groups) and benzylic compounds **2** (including diphenylmethanes, benzylmethyl ethers) underwent a facile CDC reaction. The products **3** are formed by DDQ-mediated CDC reaction of vinylic bond (sp^2 C-H) and benzylic bond (sp^3 C-H).

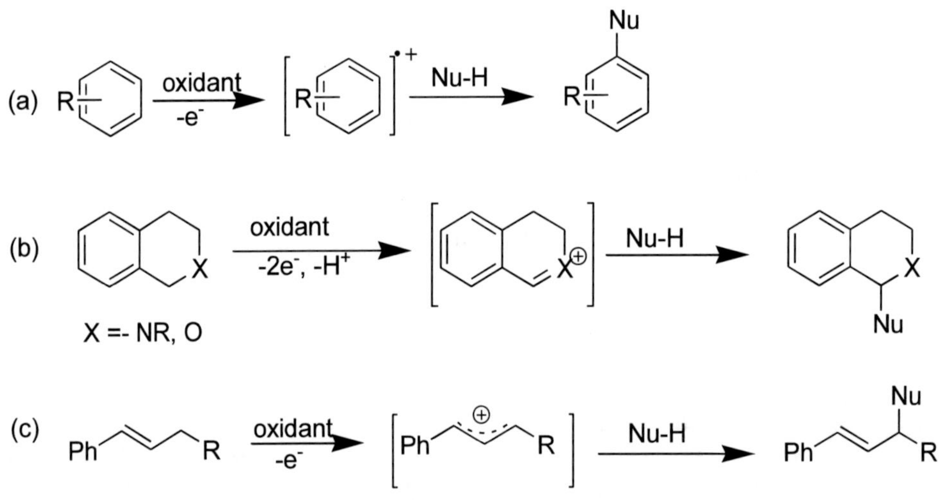

Figure 3. Mechanistic pathways for CDC reactions.

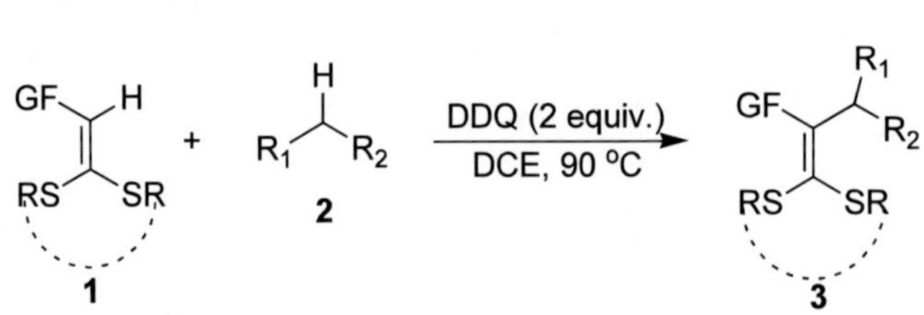

FG = CN, CH_3CO, $COC_6H_4.Cl(4)$, COC_6H_5, $COCH{=}CHC_6H_4.Cl(4)$
R = $(CH_2)_2$, CH_2CH_3, CH_3
R_1 = C_6H_5, $C_6H_4.CH_3(4)$, $C_6H_4.OCH_3(4)$, $C_6H_4.Br(4)$, $C_6H_4.F(4)$, $C_6H_4.Cl(4)$
R_2 = C_6H_5, $C_6H_4.CH_3(4)$, $C_6H_4.OCH_3(4)$, $C_6H_4.Br(4)$, $C_6H_4.F(4)$, $C_6H_4.Cl(4)$, $CH{=}CH_2$, OCH_3

Scheme 5. Synthesis of cross-coupled ketene thioacetals.

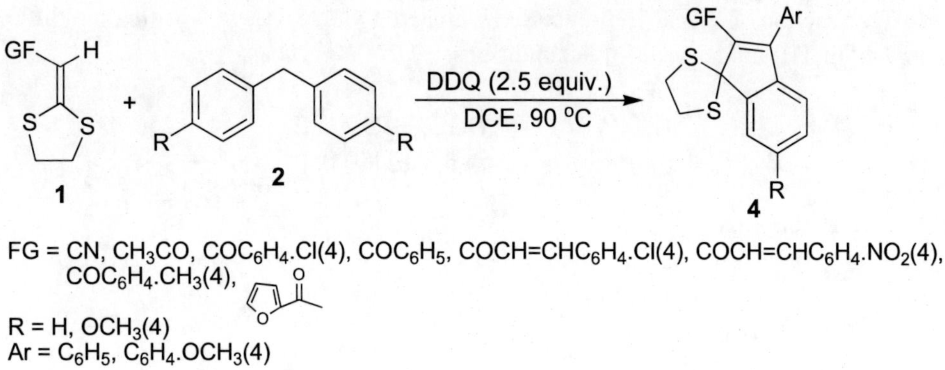

Scheme 6. Synthesis of polysubstituted indenes.

With a change in the stoichiometric ratios of cyclic ketene dithioacetals **1** (1 equiv.), benzylic compounds **2** (2 equiv.) and DDQ (2.5 equiv.), the polysubstituted indene compounds **4** were obtained in good to high yields (Scheme 6) [41]. An initial CDC reaction followed by DDQ-mediated intramolecular cyclization is expected to generate the indenes **4** by two $C(sp^2)$-$C(sp^2)$ and $C(sp^3)$-$C(sp^2)$ bond formation process.

An enantioselective cross dehydrogenative coupling reaction between cyclic benzylic ethers **5** and aldehydes **6** has been developed by Liu et al. [42]. Employing DDQ as an oxidant, the selective oxidation of C-H bond adjacent to an ethereal oxygen atom resulted in α-substituted oxygen/sulfur heterocycles **7** with excellent enantioselectivity (85 - 96% *ee*) and good to excellent yields (55 - 92%) (Scheme 7). Protic additive (H_2O), metal-salt additive (LiOTf), and phenylalanine-derived catalyst **A** provided the highest enantioselectivity. An efficient coupling reaction was observed when an initial oxidation of ethers **5** with DDQ/H_2O system in DCE was followed by nucleophilic addition with aldehydes **6** in nitromethane. The synthetic approach to heterocyclic motifs **7** has the advantages of broad scope for both cyclic ethers **5** (isochromans (X = O) **5a** and isothiochroman (X = S) **5b**) and aldehydes **6** (bearing olefins, aryls, substituted aryls, silyl ethers, benzyl ethers). On the contrary, no cross-coupled products were detected upon reaction of aldehydes **6** with acyclic benzyl ethers.

The DDQ-mediated CDC methodology [42] proved successful in the Bronsted acid **B** catalyzed coupling of isochroman **5a** with boronate esters **8** to form products **9** by alkenylation of isochroman **5a** and subsequent reduction (Scheme 8).

The existence of the chromene framework in pharmaceutical molecules and natural products has been explored extensively [43]. In 2014, Huang and co-workers [44] carried out L-proline catalyzed oxidative CDC reaction between chromene **10** and straight-chain aldehydes **11** using DDQ as an oxidant (Scheme 9). $LiClO_4$ and molecular sieves (4Å) served as additives. DDQ-mediated α-C-H bond functionalization of chromenes **10** generated the cross-coupled products **12** which were reduced with $NaBH_4$ to form **13** in good yields.

N-Aryltetrahydroisoquinolines (THIQs) have been recognized as potential anti-HIV agents *in vitro* [45] and neuroprotective agents [46]. A DDQ-mediated oxidative coupling reaction of *N*-phenyl THIQ **14** with 5-benzyl-2-phenyloxazole-4(5*H*)-one **15** was developed by Wang et al. [47] (Scheme 10). Employing CuBr as the metal catalyst in DCM and DDQ as the oxidant, the oxidative functionalization of sp^3 C-H bond adjacent to nitrogen was accomplished to furnish the dehydrogenatively cross-coupled products **16** in 23% yields. It is worth mentioning that the CDC reaction of **14** and **15** progressed with better yields (72 - 90%) utilizing $PhI(OAc)_2$ as an oxidant in THF under metal-free conditions.

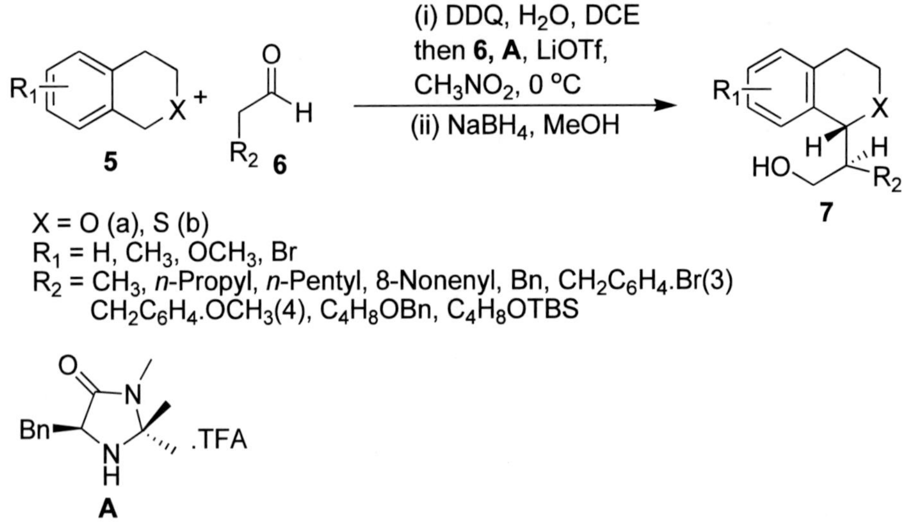

Scheme 7. Synthesis of α-substituted isochromans/isothiochromans.

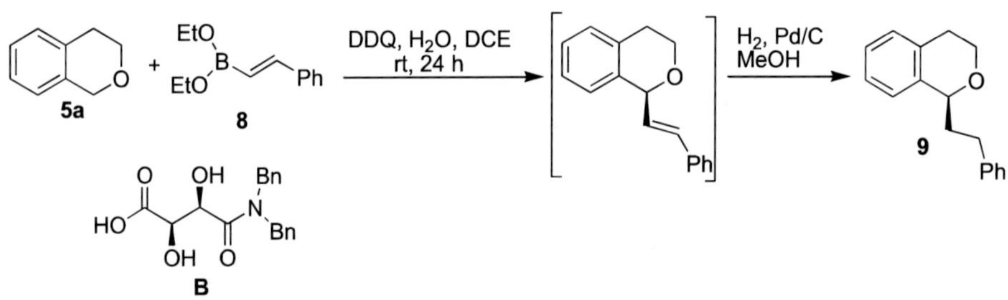

Scheme 8. Synthesis of α-substituted isochromans.

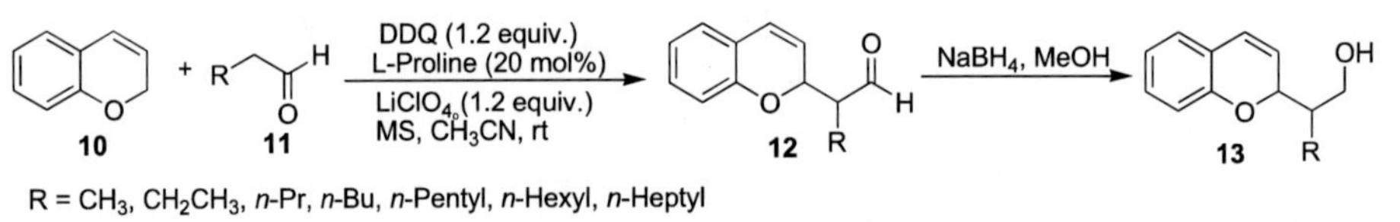

Scheme 9. Synthesis of α-substituted chromenes.

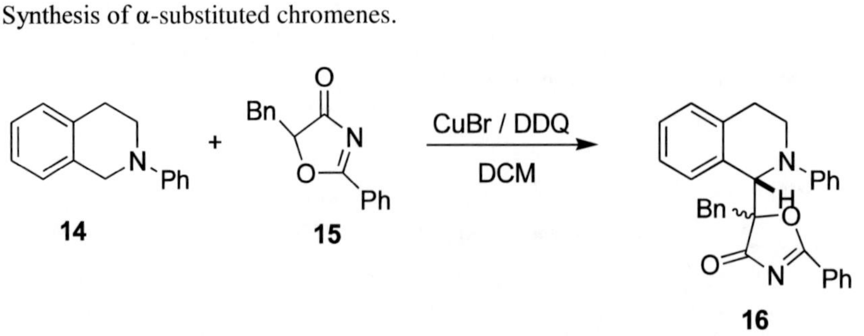

Scheme 10. Synthesis of α-substituted *N*-phenyltetrahydroisoquinolines (THIQs).

A novel and metal-free asymmetric synthesis of optically active α-heterocyclic tetrahydroisoquinoline scaffolds **19** by DDQ-mediated oxidative CDC reaction of *N*-aryltetrahydroisoquinolines (tertiary amine) **17** with α,β-unsaturated γ-butyrolactams **18** catalyzed by a bifunctional thiourea organocatalyst **C** was performed by Wang et al. (Scheme 11) [48]. The Morita-Baylis-Hillman-type products **19** were obtained with good to excellent enantioselectivities (up to 93% *ee*) and good yields (61 - 89%) *via* selective α-C(sp^3)-H bond functionalization adjacent to nitrogen atom of the tertiary amine **17**. Acyclic tertiary amines failed to generate the cross-coupled products **19**.

In 2014, Zhang's group [49] published the transition metal-free synthesis of substituted acrylamides **24** by the reaction of arylpropynes **20** and hydroxyl-amine hydrochloride **21** making use of the applicability of DDQ to serve as an efficient oxidizing agent in the cross-coupling reactions (Scheme 12). The transformation was compatible with electron-withdrawing substituents on the phenyl ring. On the contrary, electron-donating groups led to unwanted by products. An initial DDQ-promoted hetero-CDC reaction of aryl-propynes **20** with hydroxylamine hydrochloride **21** leads to C-N bond coupled product **22** which undergoes cascade aza-Meyer-Schuster rearrangement and Beckmann rearrangement to furnish acrylamides **24** by one C(sp^3)-H, one N-H and one C-C bond cleavages.

Yang et al. [50] carried out the cross-coupling reaction of hydrophosphine oxide **25** with a variety of nucleophiles (Nu-H) utilizing oxidizing ability of DDQ and catalytic efficiency of $CuBr_2$ in DMF at room temperature (Scheme 13). A series of cross-coupled products **26** can be obtained with methanol, benzyl alcohol, 4-methoxyphenol, and thiophenol as the nucleophilic coupling partners.

In 2014, Huang et al. [51] developed a mild and atom-economic synthesis of γ,δ-unsaturated ketones **29** and aldehydes **31** by cross-dehydrogenating coupling reaction between allylic C-H bonds of 1,3-diphenylpropene **27** and α-C-H bond of ketones **28** or aldehydes **30** employing DDQ as an oxidant and $Cu(OTf)_2$ as a Lewis acid catalyst in acetonitrile at room temperature (Scheme 14, 15). A range of aromatic/aliphatic ketones **28** (both cyclic and straight chain) and cyclic/acyclic aldehydes **30** furnished the coupled products in good to excellent yields.

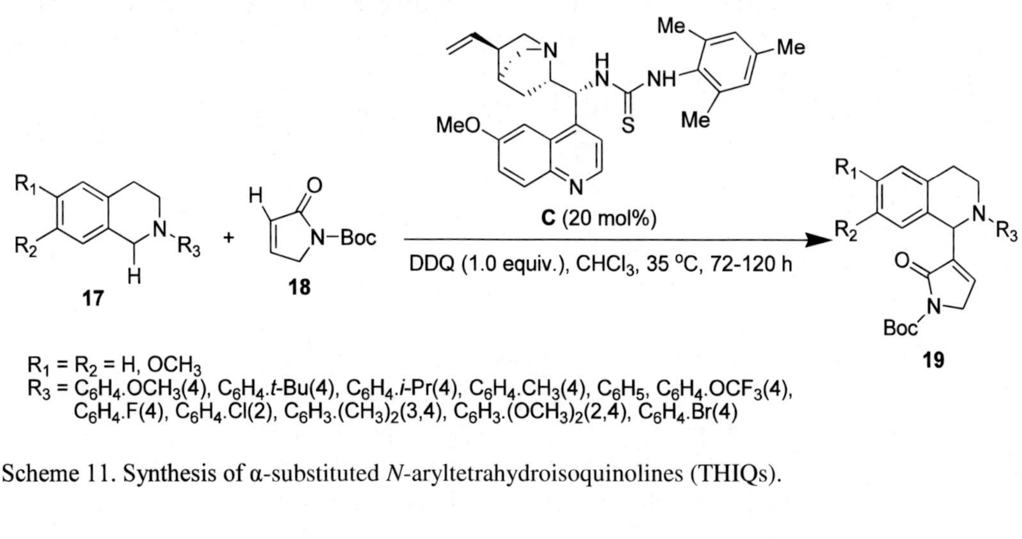

Scheme 11. Synthesis of α-substituted *N*-aryltetrahydroisoquinolines (THIQs).

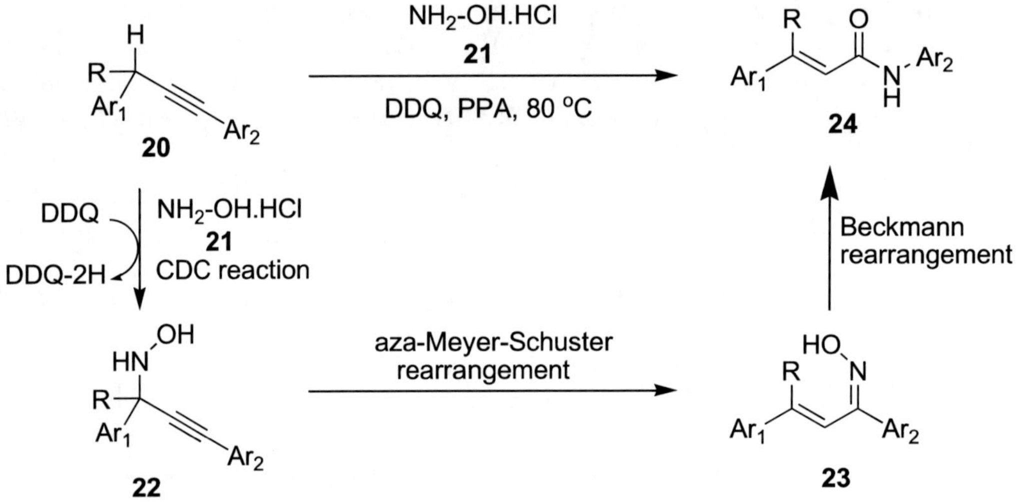

R = C_6H_5, C_6H_4.Cl(4), C_6H_4.F(4)
Ar_1 = C_6H_5, C_6H_4.Cl(4), C_6H_4.F(4), C_6H_4.CH_3(4), C_6H_4.OCH_3(4)
Ar_2 = C_6H_5, C_6H_4.Cl(4), C_6H_4.Br(4), C_6H_4.CH_3(4)

Scheme 12. Synthesis of substituted acrylamides.

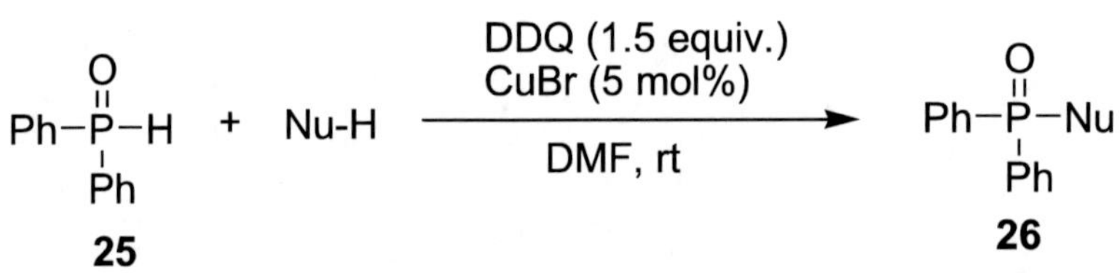

Nu-H = CH_3OH, $C_6H_5CH_2OH$, OH.C_6H_4.OCH_3(4), C_6H_5SH

Scheme 13. Synthesis of nucleophile substituted phosphine oxides.

In 2015, the group of Yan and co-workers [52] reported that the oxidative CDC reaction of 1,3-dicarbonyl compounds **32** with benzylic compounds **33** could be completed by using catalytic amounts of DDQ (1 mol%) (Scheme 16). The regeneration of DDQ was achieved using a combination of sodium nitrite and molecular oxygen (co-oxidant) in formic acid. A range of 1,3-dicarbonyl compounds **32** (including β-ketoesters, β-diketones) and symmetrical/unsymmetrical 1,3-diarylpropenes **33** furnished the cross coupled products **34** in good to excellent yields. With unsymmetrical 1,3-dicarbonyl compounds **32** ($R_1 \neq R_2$), **34** was obtained as a mixture of diastereomers. On the other hand, with unsymmetrical 1,3-diarylpropenes **33** ($R_3 \neq R_4$), a mixture of α- and γ- products were detected due to the intermediacy of the allyl cation.

Derivatized indoles exhibit cardiovascular [53] and anti-inflammatory [54] activities. Employing DDQ as an oxidant, a direct synthesis of 3-alkylated indoles **37** *via* cross-coupling of *N,N*-dimethyl-1*H*-indole-1-carboxamides **35** with diphenylmethane derivatives **36** was achieved by Chen et al. (Scheme 17) [55]. The $FeCl_2$-catalyzed and DDQ mediated oxidative activation of benzylic C-H bonds of **36** generates the benzyl cation (electrophile) which upon nucleophilic reaction with **35** yields the derivatized indoles **37**. Electron-withdrawing groups on indoles **35** and electron-donating groups on diphenylmethane derivatives **36** favors the cross-coupling reactions. A series of *N*-substituted indoles **35** bearing auxiliary group (R_3) other than *N,N*-dimethylcarbonyl moiety failed to yield the cross-coupled products.

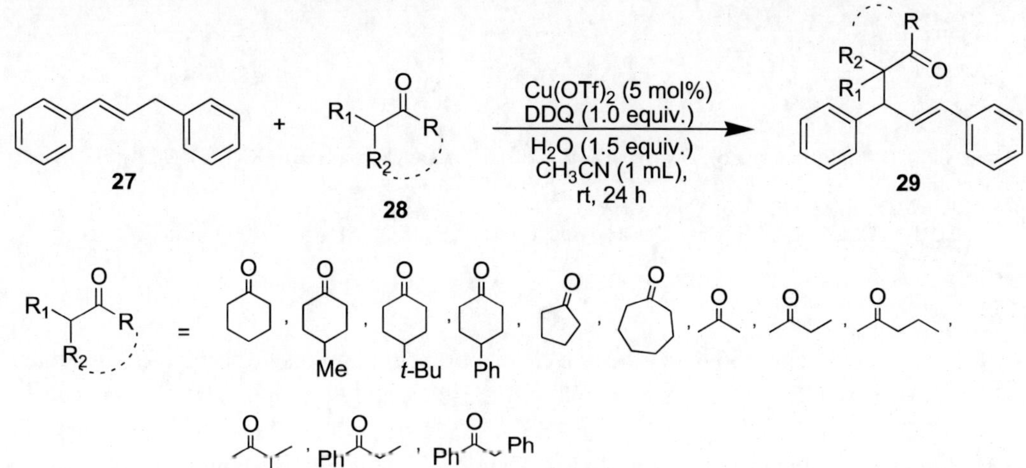

Scheme 14. Synthesis of γ,δ-unsaturated ketones.

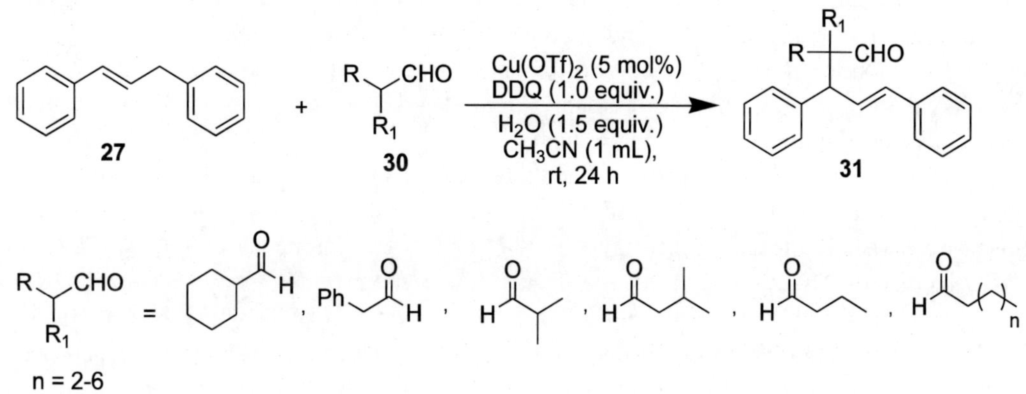

Scheme 15. Synthesis of γ,δ-unsaturated aldehydes.

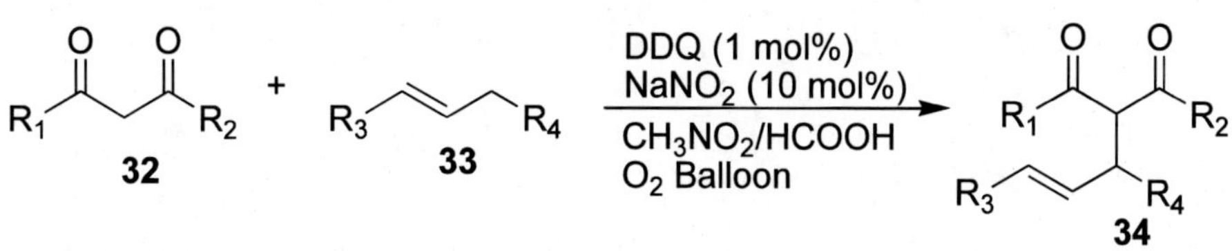

$R_1 = C_6H_5, CH_3$
$R_2 = C_6H_5, C_6H_4.CH_3(2), C_6H_4.OCH_3(3), C_6H_4.Br(3), C_6H_4.OCH_3(4), C_6H_4.F(4)$, Thienyl, Furyl, CH_3, OCH_2CH_3
$R_3 = C_6H_5, C_6H_4.OCH_3(4)$
$R_4 = C_6H_4.OCH_3(4), C_6H_4.CH_3(4), C_6H_5, C_6H_4.Br(4)$

Scheme 16. Synthesis of α-substituted 1,3-dicarbonyl compounds.

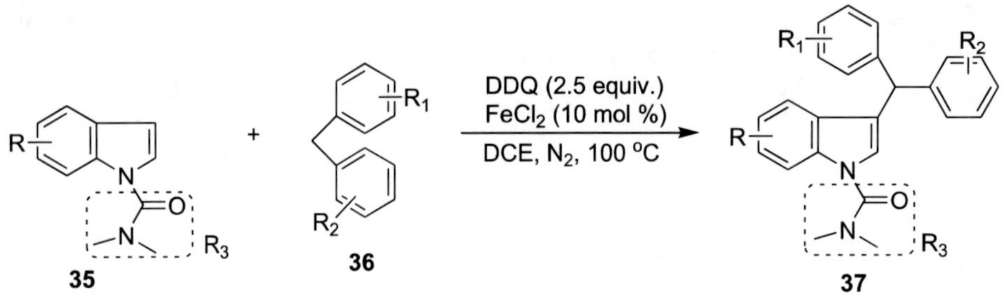

R = H, Cl(5), Br(5), F(5), CHO(5), CN(5), NO_2(5), Cl(6), F(6), $COOCH_3$(4), $COOCH_3$(6), OCH_3(5), CH_3(6)
R_1 = H, F(4)
R_2 = CH_3(4), OCH_3(4), OCH_3(3), Cl(4), CH_3(2), CH_3(3), $(CH_3)_2$(3,5), F(4)
R_3 = H, CH_3, COC_6H_5, $COCH{=}CH_2$, $COCH_3$, Boc, $CONHC_6H_5$, $SO_2C_7H_8$

Scheme 17. Synthesis of derivatized indoles.

Song and co-workers [56] described an efficient tandem approach for the DDQ-mediated double oxidative CDC of arylmethanes **38** and 1,3-dicarbonyl compounds **39** catalyzed by $FeCl_2$ in DCE at 90°C under air (Scheme 18). A range of arylmethanes **38** *viz m*-xylene, *p*-xylene, *p-tert*-butyltoluene, mesitylene and *o*-xylene and 1,3-dicarbonyl compounds **39** *viz* 1,3-diesters, β-ketoesters and 1,3-diketones furnished the coupled products **40** *via* a CDC protocol. An initial formation of diarylmethane derivatives ($PhCH_2Ph$) from two arylmethane molecules in one cycle *via* DDQ mediated, $FeCl_2$-catalyzed oxidation process and its cross-coupling with iron-chelated 1,3-dicarbonyl compounds in the second cycle give the couple products **40**. The coupling reaction leads to the formation of one new $C(sp^3)$-$C(sp^3)$ bond and one new $C(sp^3)$-$C(sp^2)$ bond in a one-pot cascade protocol. Amides and ether esters as the 1,3-dicarbonyl compounds failed to give the coupled products.

Wang and co-workers [57] demonstrated that DDQ acts as an oxidative dehydrogenating agent for the synthesis of nitriles **42** and **44** *via* CDC reaction (Scheme 19, 20). Metal-free and solvent-free cyanation reaction of benzyl ethers **41** or 1,3-diarylpropenes **43** with TMSCN *via* a DDQ-mediated CDC approach leads to selective functionalization of $C(sp^3)$-H benzylic bond leading to corresponding benzylic nitriles **42**, **44** in good to excellent yields. $K_3Fe(CN)_6$, $CH_2(CN)_2$, NH_4HCO_3/DMSO as cyanide sources failed to give the target cyano products. Electron-donating groups on the isochroman ring system **41** favored the cyanation reaction.

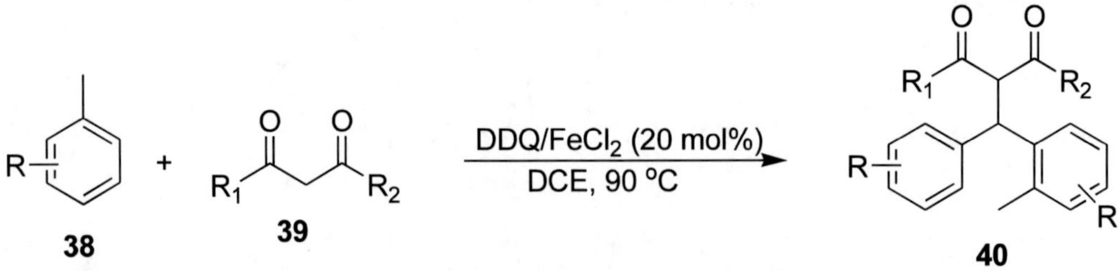

R = CH_3, *t*-Bu
R_1 = OCH_2CH_3, OCH_3, O*i*-Pr, CH_3, CH_2CH_3, C_6H_5, C_6H_4.Cl(4), CH_2OCH_3
R_2 = OCH_2CH_3, OCH_3, O*i*-Pr, CH_3, CH_2CH_3, $N(CH_3)_2$

Scheme 18. Synthesis of α-substituted 1,3-dicarbonyl compounds.

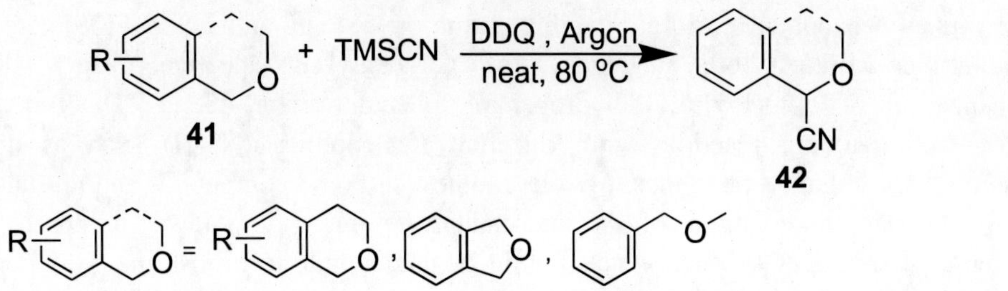

R = H, F(7), Cl(7), Br(7), *t*-Bu(7), CH_3(7), $(OCH_3)_2$(6,7), Br(5), CF_3(5), CH_3(3), $COOCH_3$(3)

Scheme 19. Synthesis of nitriles from benzyl ethers.

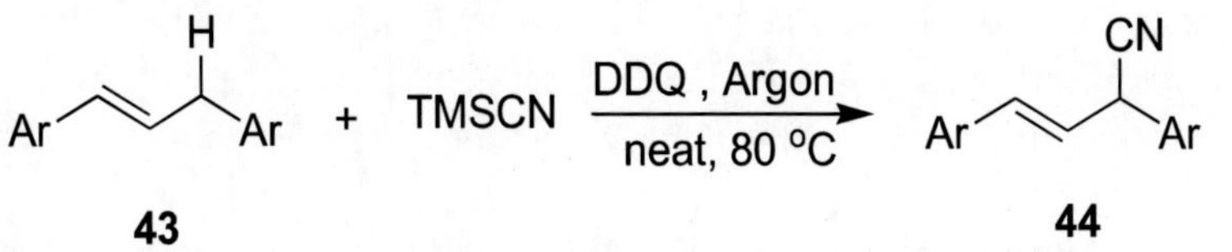

Ar = C_6H_5, C_6H_4.F(4), C_6H_4.Cl(4), C_6H_4.Br(4), C_6H_4.CH_3(4), C_6H_4.*t*-Bu(4)

Scheme 20. Synthesis of nitriles from 1,3-diaryl propenes.

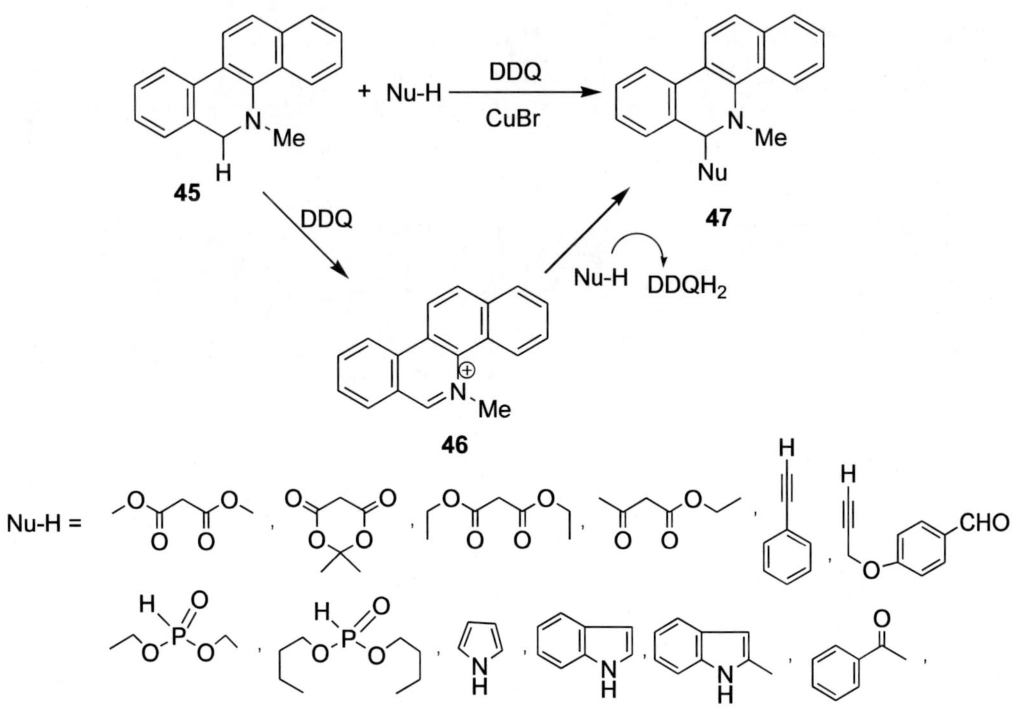

CH_3NO_2, $CH_3CH_2NO_2$, $CH_3CH_2CH_2NO_2$, $(CH_3)_3SiCN$

Scheme 21. Synthesis of α-substituted *N*-methyl-5,6-dihydrobenzo[*c*]phenanthridines.

Derivatized benzophenanthridines exhibit antitumor [58] and antibacterial [59] activities. Garcia and co-workers [60] developed the Cu-catalyzed selective C(sp^3)-H bond functionalization of *N*-methyl-5,6-dihydrobenzo[*c*]phenanthridine **45** *via* DDQ-mediated oxidative cross-coupling reactions with different nucleophiles (Nu-H) such as dialkyl malonates, alkynes, dialkyl phophites, pyrrole, indoles, carbonyl compounds and nitroalkanes (Scheme 21). The intervention of an intermediate iminium ion **46** is proposed upon interaction of **45** and DDQ, which is acted upon by nucleophile to furnish the cross-coupled products **47**.

An expeditious route involving Fe-catalyzed and DDQ mediated oxidative cross-coupling reaction between electron-rich arenes *viz* 1,3,5-trimethoxybenzene **48** and alkenes **49/51** was developed by Lei et al. (Scheme 22) [61]. Triaryl-ethylenes **50** were obtained as the cross-coupled products upon direct arylation of **48** with diaryl-ethylenes **49**. On the contrary, styrene derivatives **51** generated double arylation products **52** upon DDQ-mediated cross-coupling reactions catalyzed by Lewis acid $FeCl_3$. Diaryl-ethylenes **49** with both electron-donating and electron-withdrawing groups were well tolerated in the oxidative transformation with DDQ.

Helaja et al. [62] worked upon the DDQ/H$^+$ mediated intermolecular oxidative dehydrogenative coupling (ODH) reactions under metal-free conditions (Scheme 23). 2-Arylbenzo[*b*]furans **53** and 2-arylbenzo[*b*]thiophenes **55** underwent facile 3,3'-couplings on treatment with DDQ (oxidant), MsOH (acid additive) in DCE to furnish the 3,3'- homocoupled products *viz* 2,2'-bis(aryl)-3,3'-bibenzo[*b*]furans **54** and 2,2'-bis(aryl)-3,3'-bi benzo[*b*]thiophenes **56** in good to excellent yields. The oxidation potential of the reactants **53,55** and their frontier molecular orbital morphology served as the deciding factors in predicting the feasibility of the ODH reactions. The authors employed the 3,3'-homocoupling methodology to synthesize natural product Shandougenine B **58** from its monomer sophora **57** *via* DDQ mediated oxidative 3,3'-homocoupling and subsequent debenzylation (Scheme 24).

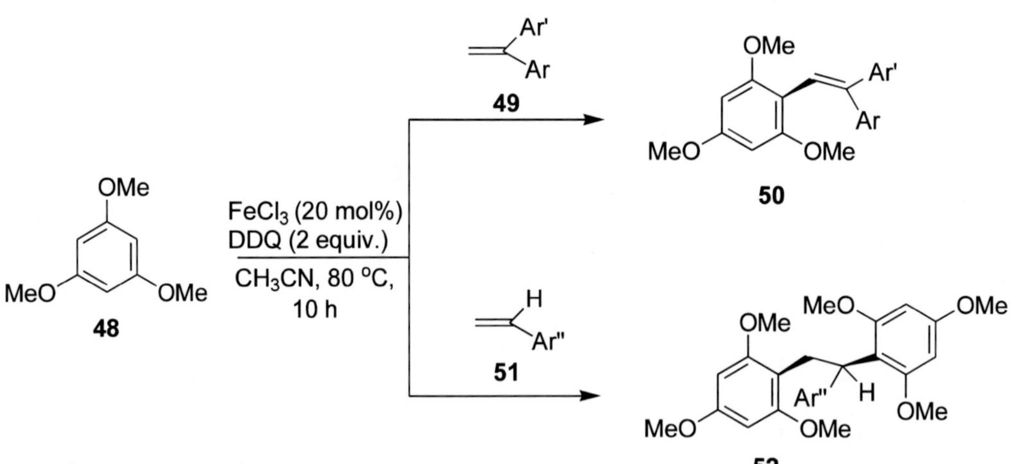

Ar = C_6H_5, $C_{10}H_7$, $C_6H_4.CF_3(4)$, $C_6H_4.Cl(4)$, $C_6H_4.Br(4)$
Ar' = C_6H_5, $C_{10}H_7$, $C_6H_4.CF_3(4)$, $C_6H_4.Cl(4)$, $C_6H_4.CH_3(4)$
Ar'' = $C_6H_4.Br(4)$, $C_6H_4.Br(3)$, $C_6H_4.Cl(4)$, $C_6H_4.F(4)$

Scheme 22. Synthesis of triarylethylenes and double arylation products.

An efficient DDQ-mediated oxidative allylic C-H bond functionalization of bulk hydrogenated graphene HG (graphane) **59** has been reported by Pumera et al. (Scheme 25) [63]. They subjected HG **59** to coupling reaction with tetrahydrothiophen-3-one (THT) **60** employing DDQ as an oxidant and $Cu(OTf)_2$ as the metal catalyst. The reaction afforded THT-functionalized hydrogenated graphene (S-HG) **61** by coupling of allylic C-H bond of **59** with α-C-H bond of THT **60**. Adsorption of THT **60** onto the HG **59** surface rather than its chemical functionalization was observed in the absence of oxidant DDQ and $Cu(OTf)_2$ catalyst. The polar carbonyl functionality in S-HG **61** promotes its hydrogen-bonding interactions with water, thereby exhibiting good dispersion solubility in comparison to HG **59** (water insoluble).

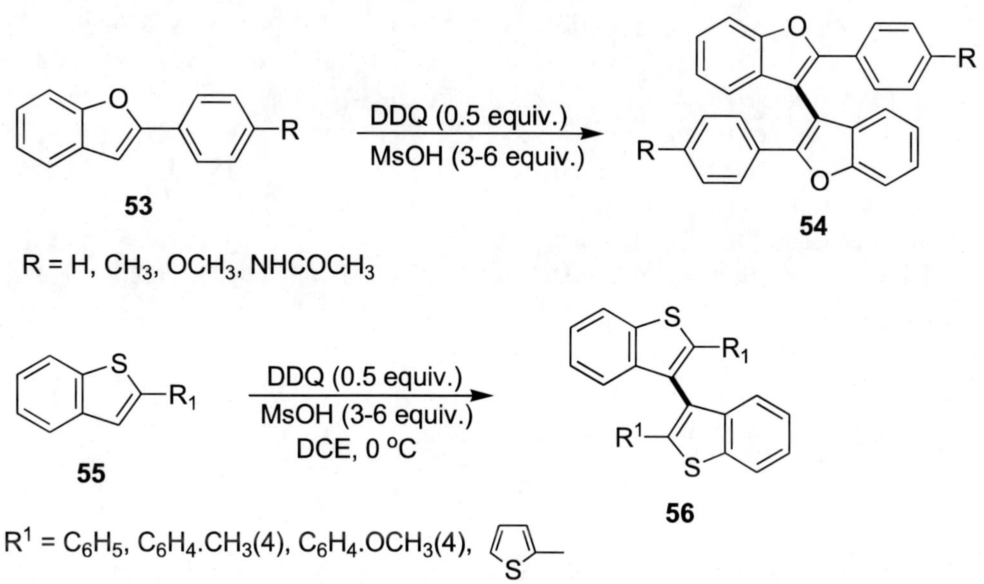

Scheme 23. Synthesis of 2,2'-bis(aryl)-3,3'-bibenzo[*b*]furans and 2,2'-bis(aryl)-3,3'-bibenzo[*b*]thiophenes.

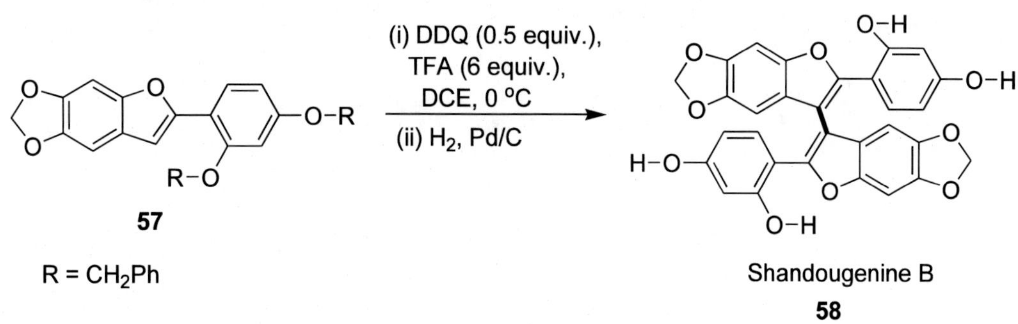

Scheme 24. Synthesis of Shandougenine B.

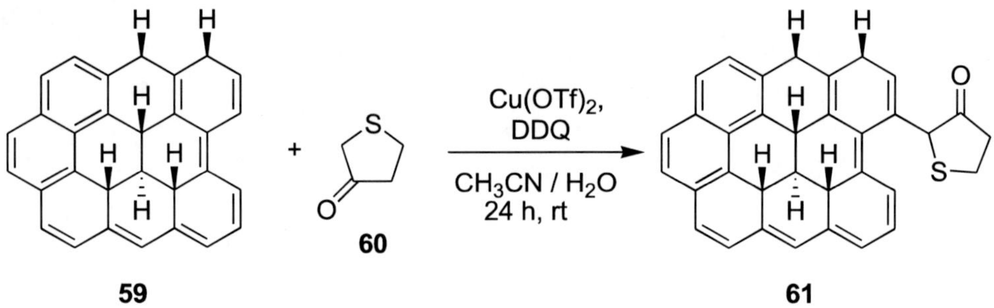

Scheme 25. Synthesis of THT-functionalized hydrogenated graphene (S-HG).

Su and co-workers [64] studied the mechanochemical DDQ-mediated oxidative CDC reaction of 3-benzylic indoles **62** and compounds with acidic methylene groups **63** *viz* diethylmalonates, dibenzylmalonates, malononitrile, ethyl-2-nitroacetate, ethyl acetoacetate, ethyl 4-chloroacetoacetate and ethyl-3-oxo-3-phenylpropanoate under solvent-free iron-catalysis (Fe^{3+}) (Scheme 26). The High Speed Ball Milling (HSBM) CDC reaction worked best with addition of the oxidant DDQ in three portions to the reaction vessel working at milling frequency (*υ*): 25 Hz, time (*t*): 21 min and grinding auxiliary: silica gel at room temperature. Both electron-donating and electron-withdrawing groups were well tolerated. The coupled products, 3-arylmethylindole derivatives **64** were obtained in moderate to high yields.

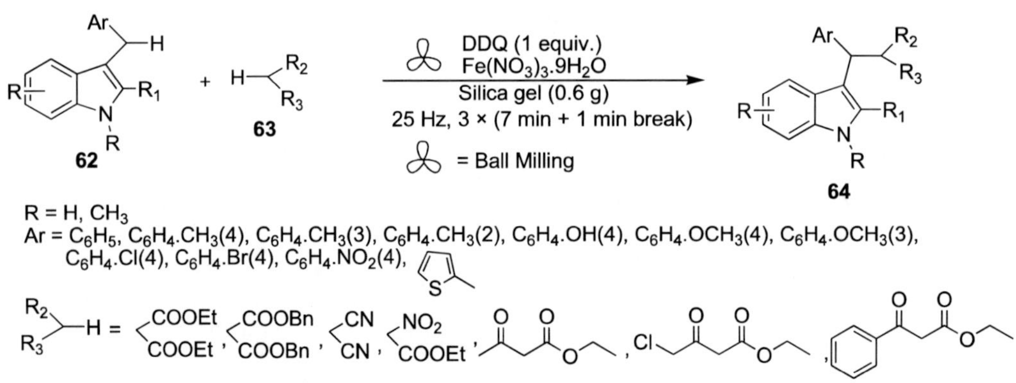

Scheme 26. Synthesis of 3-arylmethylindole derivatives.

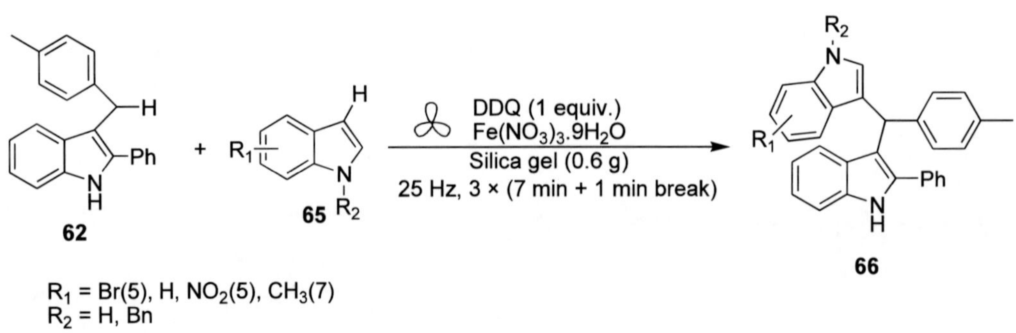

Scheme 27. Synthesis of bisindoles.

In the same report [64], utilization of indoles **65** as the nucleophilic coupling partner with 3-benzylic indoles **62** in the DDQ-mediated ball-milling reaction was successful yielding the corresponding bisindoles **66** (Scheme 27). Both *N*-substituted and *N*-free indoles **65** underwent a smooth transformation to the coupled products.

DDQ-mediated solvent-free oxidative Mannich reaction of *N-tert*-butoxycarbonyl (Boc)/*N*-benzyloxycarbonyl (Cbz)/*N*-aryl tetrahydroisoquinolines (THIQs) **67** with ketones **68** was achieved under high-speed ball-milling conditions by Su et al. (Scheme 28) [65]. The mechanochemical cross coupling reaction proceeded chemoselectively to afford the coupled products **69** in 40 - 89% yields. The optimized mechanochemical milling parameters were identified as milling frequency (υ): 20 Hz, time (t): 45 min, milling ball size (d_{MB}): 14 mm, milling ball filling degree (Φ_{MB}): 0.115 and grinding auxiliary: silica gel (450 mg) to facilitate the oxidative coupling reaction without any significant deprotection of the *N*-substituent as the side reaction. Cyclic/acyclic ketones as well as aralkyl ketones underwent the DDQ-mediated mechanochemical CDC reaction smoothly under mild oxidative conditions.

Liu and co-workers [66] disclosed the first DDQ-promoted intermolecular metal-free organocatalytic enantioselective coupling reaction of cyclic amines *viz N*-carbamoyl THIQs **70** and functionalized aldehydes **71** (Scheme 29). The cross-coupled products **72** generated with excellent enantiocontrol (upto 97% *ee*) are C_1-alkylated nitrogen heterocycles possessing two contiguous stereocentres. Chiral imidazolidinone in combination with TFA (**A**) served as the organocatalyst. The usage of 20 equiv. of water as an additive led to an enhancement in the enantiomeric excess. Methyl ester groups (R_2 = OMe) gave superior results than bulkier ester moieties (R_2 = OBn, O*t*-Bu). A range of aldehydes **71** bearing olefin, acetate, aryl, halide, benzyl ether and silyl ethers were well tolerated. *N*-Carbamoyl THIQs **70** with both electron-donating and electron-withdrawing substituents gave smooth transformations to the coupled products **72**.

The scope of the reaction was extended to the coupling reaction of *N*-carbamoyl tetrahydro-*β*-carbolines **73** with aldehyde **71** using DDQ as an oxidant (Scheme 30) [66]. The cross-coupled products **74** were formed with moderate *dr* but excellent *ee*.

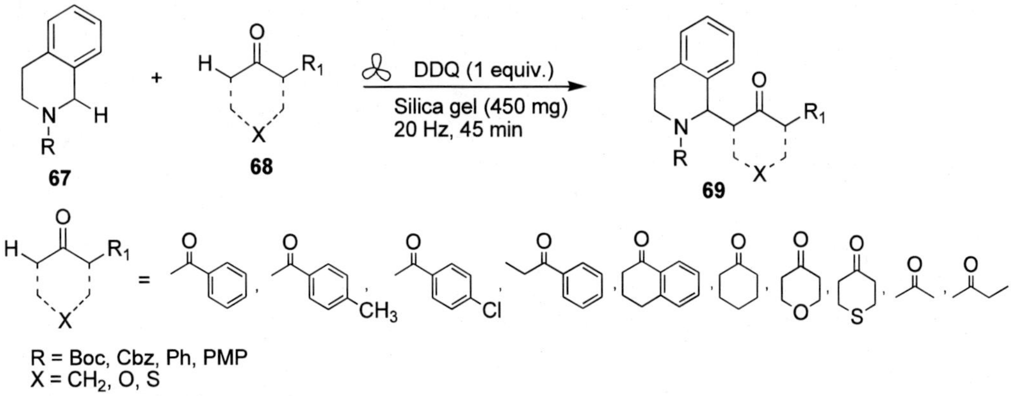

Scheme 28. Synthesis of α-substituted tetrahydroisoquinolines.

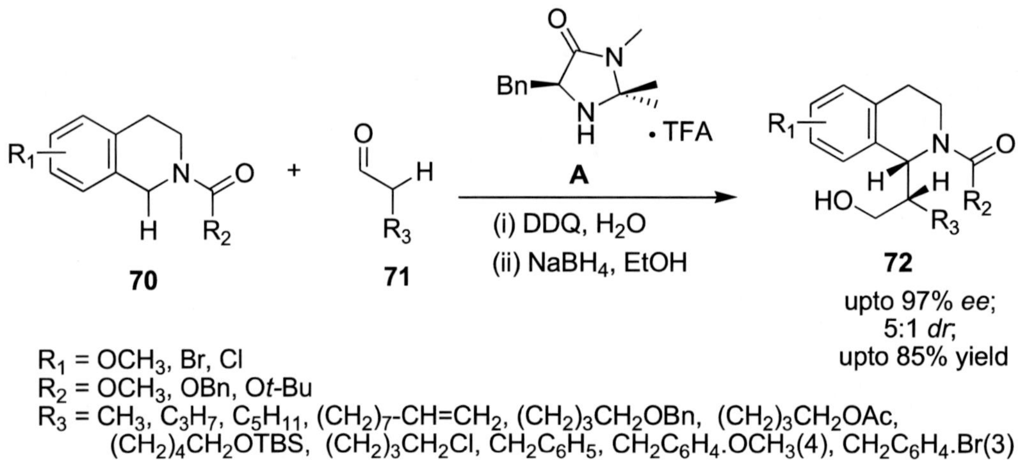

Scheme 29. Synthesis of α-substituted tetrahydroisoquinolines.

In an attempt to synthesize 2-hydroxy-3-alkyl-1,4-naphthoquinones **77**, Yan and co-workers [67] employed a transition metal-free CDC approach (Scheme 31). The coupling reaction between 2-hydroxy-1,4-naphthoquinone (lawsone) **75** and 1,3-diarylpropenes **76** with DDQ (1.2 equiv.) as an oxidant in 1,2-dichloroethane furnished the target cross-coupled products **77** as a mixture of α- and γ- isomers in good to excellent yields. The intervention of an allylic cation from 1,3-diarylpropenes is indicated that leads to both α- and γ- isomers of **77**.

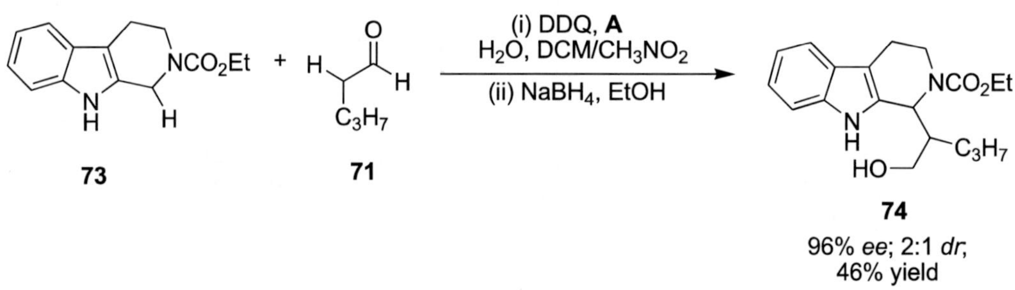

Scheme 30. Synthesis of α-substituted tetrahydro-*β*-carbolines.

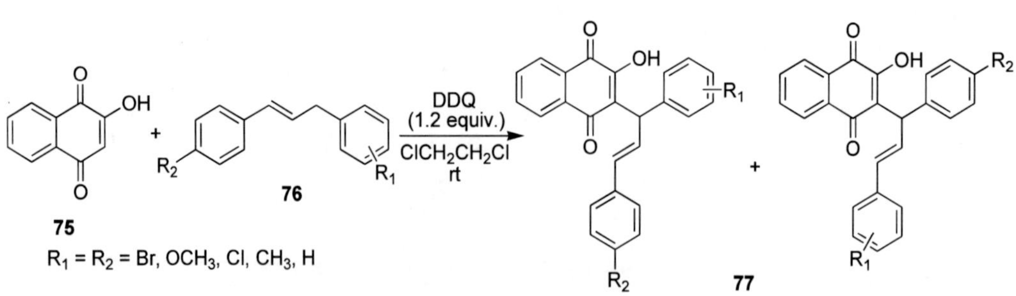

Scheme 31. Synthesis of 2-hydroxy-3-alkyl-1,4-naphthoquinones.

The authors [67] extended the coupling reaction methodology to the cross-coupling reactions of 1,3-diarylpropenes **76** with 4-hydroxycoumarins **78**/2-isopropylamin-onaphthoquinones **80**/2-pyrrolidene-naphthoquinone **82** employing oxidant DDQ (1.2 equiv) in 1,2-dichloroethane at room temperature (Scheme 32).

2.2.1.2. Intra-Molecular Cross Dehydrogenative Coupling Reactions

Qingmin and co-workers [68] have reported a novel conversion of 1,2-diarylethylenes **84** to phenanthrenes **85** under mild conditions using DDQ as an intramolecular oxidative coupling reagent in combination with TFA in dichloromethane (Scheme 33). Both *E*- and *Z*- isomer of **84** with electron-donating groups on phenyls and electron-withdrawing groups on double bonds furnished the products in excellent yields. A cation–radial mechanism promoting isomerization of substrates is suggested.

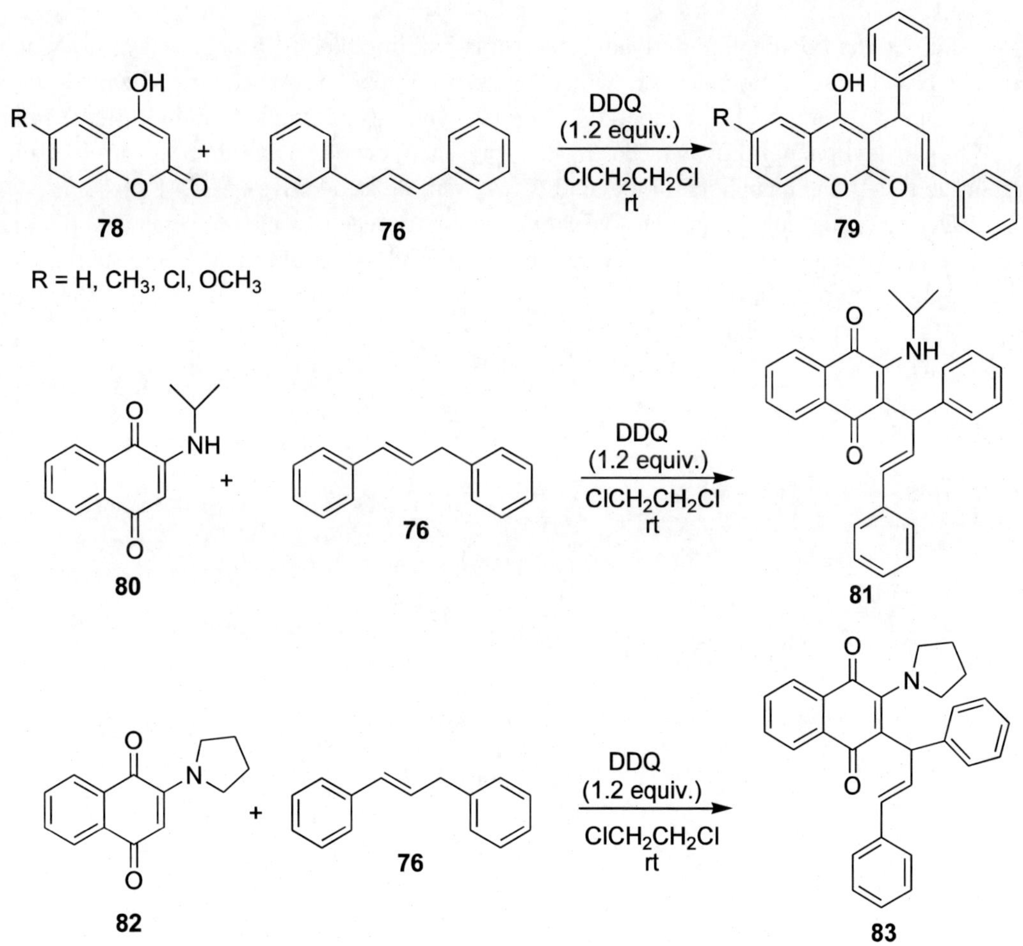

Scheme 32. Synthesis of 3-alkyl-4-hydroxycoumarins and 3-alkyl-1,4-naphthoquinones.

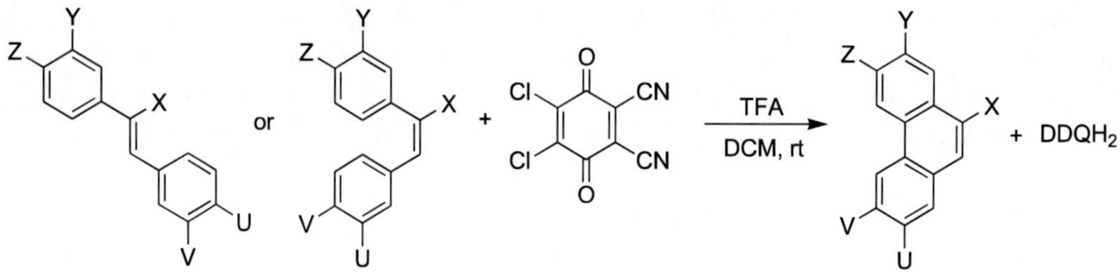

E/Z-**84** **85**

X = $COOCH_3$, COOH, CN
U,V,Y,Z = OCH_3
U-V, Y-Z = $-O-CH_2-O-$

Scheme 33. Synthesis of phenanthrenes.

Chiral tetrahydroquinoline derivatives serve as building blocks in organic synthesis and exhibit a broad array of biological activities thereby are important in medicinal chemistry [69, 70]. Kim and co-workers [71] reported the coupling synthesis of tetrahydroquinolines **87** from 3-(2-(azepan-1-yl)phenyl)propanal **86** in the presence of DDQ serving as an oxidant, organocatalyst **D** and camphorsulfonic acid (CSA) serving as additive in DCM (Scheme 34). The DDQ-mediated intramolecular oxidative coupling reaction of **86** generates **87** *via* enantioselective oxidative enamine catalysis followed by 1,5-hydride transfer and subsequent 6-endo ring closure/cyclization in a tandem sequence. The product **87** was obtained in high yield (75%) but with low enantioselectivity (19% *ee*).

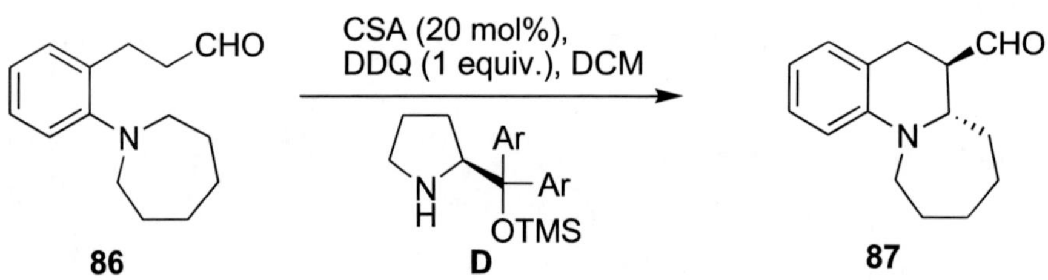

Ar = $C_6H_3(CF_3)_2(3,5)$

19% *ee*;
4:1 *dr*

Scheme 34. Synthesis of chiral tetrahydroquinoline derivatives.

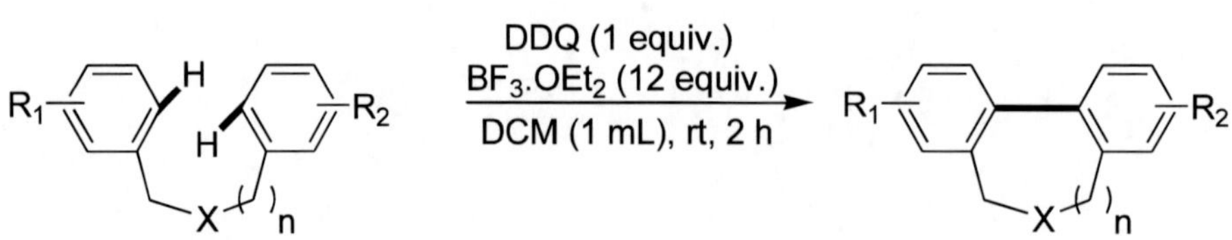

88 **89**

X = N($COCF_3$), S(=O), O, CH_2
n = 1, 2
$R_1 = R_2 = OCH_3$, $-O-CH_2-O-$

Scheme 35. Synthesis of biaryl compounds.

In 2014, Liu et al. [72] developed a novel DDQ and $BF_3.OEt_2$ promoted metal-free intramolecular CDC reaction to generate seven- and eight-membered heterocyclic biaryl compounds **89** (Scheme 35). The system was effective for the oxidative biaryl coupling of **88** *viz* symmetric and unsymmetric *N,N*-dibenzylamides (X = N-$COCF_3$), *N*-benzyl-*N*-phenethyl amides derivatives (X = N-$COCF_3$, n = 2), sulfone containing phenol ethers (X = S=O), dibenzyl ethers (X = O) and diarylbutanes (X = CH_2, n = 2). Individual DDQ or $BF_3.OEt_2$ failed to generate the coupled biaryl products **89**. The authors [72] succeeded in generating a range of six-membered tricyclic phenanthrenes **91** by the DDQ-mediated intramolecular oxidative biaryl coupling of stilbenes **90** in dichloromethane at room temperature (Scheme 36). Ester and carboxylic acid functionalities in the stilbenes **90** were well tolerated in the coupling reactions.

Zhang et al. [73] carried out synthesis of a novel series of diversely substituted olean-28,13β-lactams **93** from their corresponding amides **92** utilizing the oxidizing tendency of DDQ for the intramolecular dehydrogenative C-N coupling reactions in toluene (Scheme 37). A radical-ion mechanism for the dehydrogenative coupling is suggested.

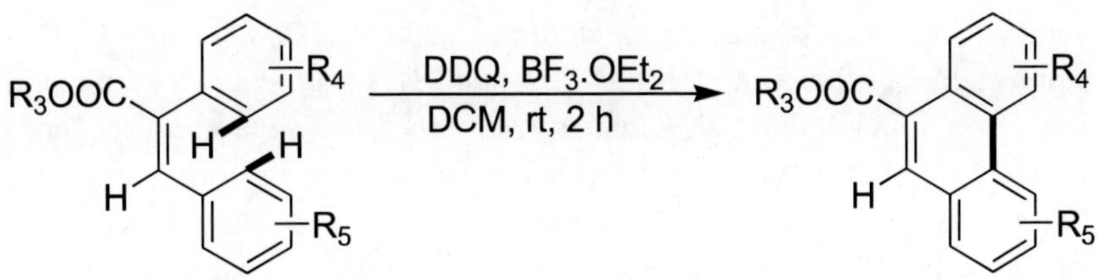

90 **91**

R_3 = H, OCH_3
R_4 = R_5 = OCH_3

Scheme 36. Synthesis of phenanthrenes.

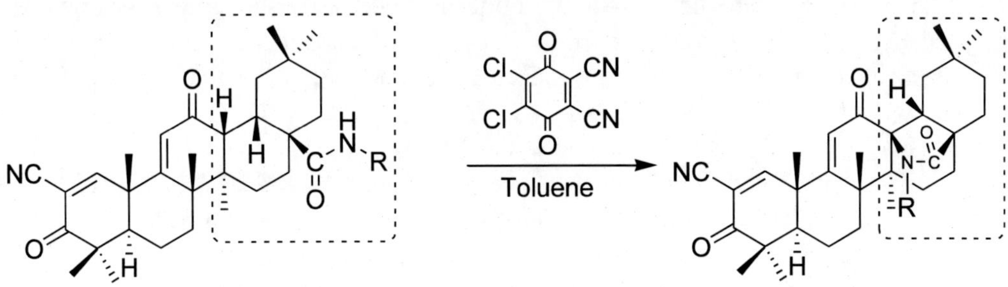

92 **93**

R = CH_3, CH_2CH_3, $CH_2CH_2CH_3$, $CH_2(CH_2)_2CH_3$, $CH_2(CH_2)_4CH_3$, $C(CH_3)_3$, CH_2CH_2OH,

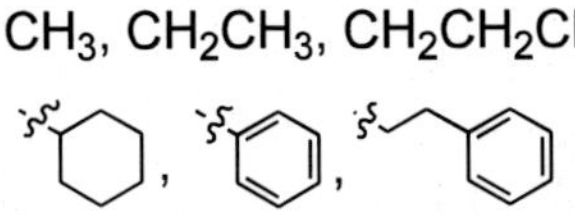

Scheme 37. Synthesis of olean-28,13β-lactams.

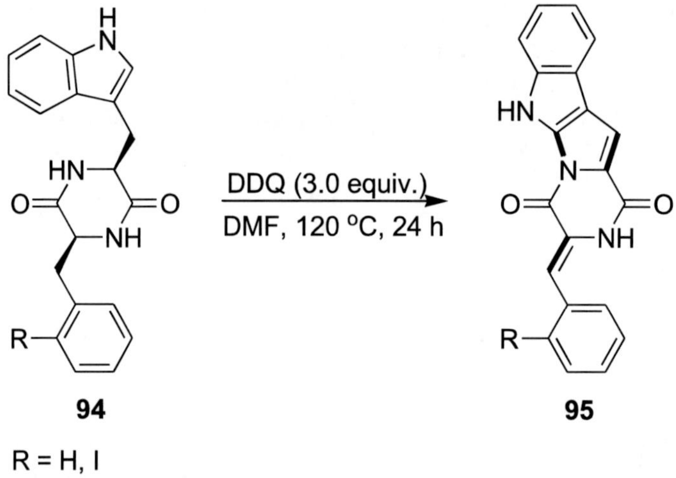

Scheme 38. Synthesis of pyrroloindole diketopiperazines.

Lavilla and co-workers [74] studied the intramolecular oxidative couplings of tryptophan-based diketopiperazines (c(Phe-Trp)DKP) **94** (Scheme 38). Treatment of **94** with DDQ in DMF at 120°C yields pyrroloindole DKP **95** *via* a cascade process. The domino sequence involves an initial double C_{α}-C_{β} dehydrogenation (Trp and Ph), followed by formation of a C-N bond between N^{α} group of Trp and C-2 of indole moiety. Polycyclic coupled products **95** incorporate a highly conjugated diacylated pyrroloindole core.

2.2.2. Oxidative Cyclizations Mediated by DDQ

2.2.2.1. DDQ Serving as a Sole Oxidant

In an attempt to synthesize heteroaromatic benzothiadiazine product **97**, the group of Lebegue and co-workers [75] subjected the *N*-alkylated benzothiadiazine **96** to an oxidative aromatization reaction with DDQ (2 equiv.) in toluene under reflux conditions (Scheme 39). Unexpectedly, the reaction generated the heteroannulated and chlorinated imidazobenzothiadiazine **98** as the sole product rather than the aromatized product **97** upon treatment with DDQ. The formation of **98** was explained by DDQ-induced oxidative intramolecular cyclization/heteroannulation and subsequent chlorination sequence. The reaction to form **98** proceeds with high diastereoselection to generate *trans* products (as supported by molecular mechanics calculation). Presence of methoxy substituent on the aryl ring was found essential for the DDQ-induced oxidative cyclization process.

Deng and co-workers [76] have carried out the direct transformation of (*Z*)-1,2,3-triaryl substituted propylenes **99** to the corresponding 1,2-diarylindene derivatives **100** utilizing DDQ as an oxidizing agent in nitromethane (Scheme 40). The $Cu(OAc)_2.H_2O$ catalyzed process involves DDQ-mediated benzylic/allylic sp^3 C-H bond activation of **99** and consequent intramolecular dehydrogenative cyclization to form the products **100** in moderate to excellent yields. Both electron-withdrawing and electron-donating groups on the aryl ring system **B** were tolerant in the transformation. The isomeric (*E*)-1,2,3-triaryl substituted propylenes failed to undergo the intramolecular cyclization under similar oxidative conditions.

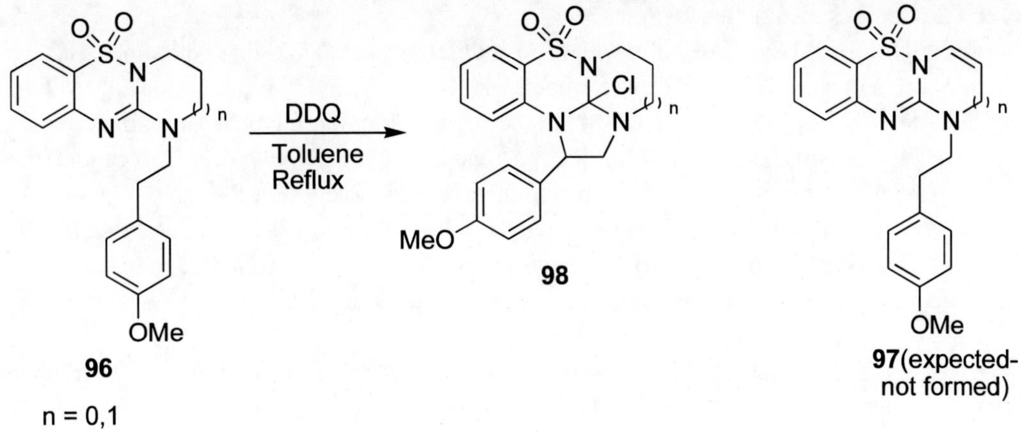

Scheme 39. Synthesis of chlorinated imidazobenzothiadiazine.

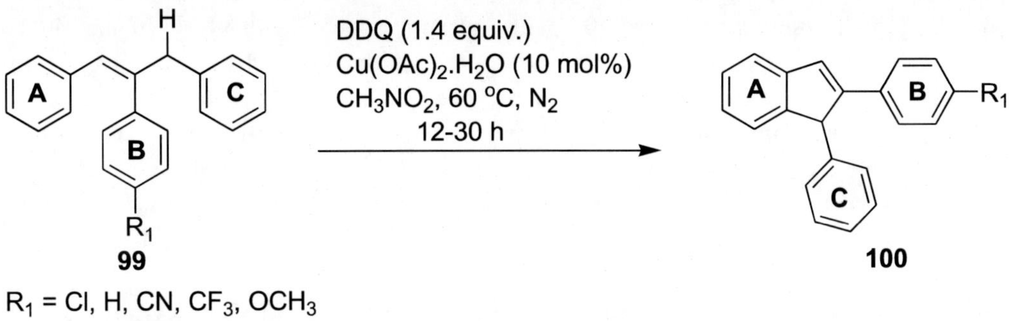

Scheme 40. Synthesis of 1,2-diarylindene derivatives.

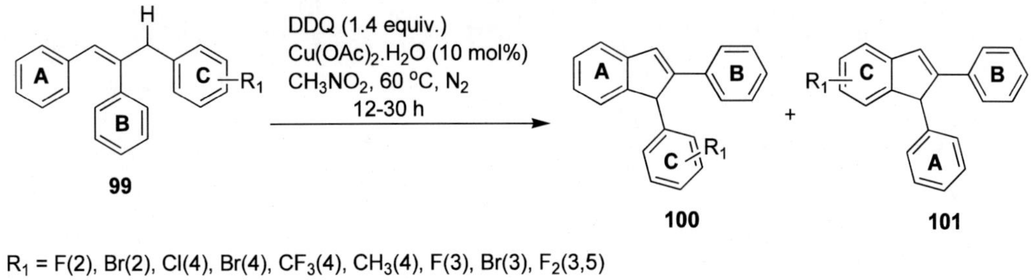

Scheme 41. Synthesis of 1,2-diarylindene derivatives.

In the same report [76], for substrates **99** with either *o*/*m*/*p* substituents on the aryl ring **C**, similar DDQ-mediated oxidative cyclizations generated a mixture of regioisomers **100** and **101** (Scheme 41). A preferential cyclization at the non-substituted ring **A** was observed with *o/p* electron-withdrawing groups on ring **C** to form 1,2-diarylindene derivative **100** preferentially. However, with electron-donating *p*- substituents on the **C**-aryl ring, opposite selectivity towards 1,2-diarylindene derivative **101** was observed. Poor regioselectivity was observed with *m*- substituted substrates **99**.

A straightforward synthesis of derivatized indoles from *N*-Ts-2-alkenylanilines **102** was reported by Youn and co-workers (Scheme 42) [77]. The C-H amination of *N*-Ts-2-styrylanilines **102** was carried in 1,1,1-trichloroethane employing DDQ as an oxidant to yield indoles **103** in good to excellent yields *via* an oxidative cyclization process. Both electron-donating and electron-withdrawing groups on the aromatic moiety (R_1, R_2) were tolerated. On the other hand, *N*-Ts-*β,β*-disubstituted 2-alkenylanilines **104** upon C-H amination with DDQ gave 2,3-disubstituted indoles **105**,**106** *via* an initial SET to form radical cation and an aryl migratorial process *via* a phenonium ion intermediate (Scheme 43) [77]. A preferential migration of an electron-rich aryl group was observed.

Diversely substituted 1,3,4-oxadiazoles occupy a prominent place in medicinal chemistry. These serve as potential therapeutic agents for treatment of HIV infections [78] and cancer [79]. A novel and direct approach to synthesize 5-aryl-2-(2-arylethenyl)-1,3,4,-oxadiazoles **110** *via* DDQ mediated oxidative cyclization of *N*-aroylhydrazones **109** has been reported by Kudelko et al. (Scheme 44) [80]. The starting *N*-aroylhydrazones **109** were synthesized by heating hydrazides **107** with aromatic aldehydes **108** in ethanol in the presence of catalytic HCl. Aldehydes **108** bearing electron-donating substituents gave superior yields of 1,3,4-oxadiazoles **110**. Simultaneously, one-pot reaction of a mixture of hydrazide **107**, aromatic aldehyde **108** and DDQ (serving as the potential oxidant) in toluene at reflux temperature afforded the 1,3,4-oxadiazoles **110** in comparable yields (Scheme 44).

Due to the poor availability of 10,12-peroxycalamenene **114** from natural sources, Woerpel et al. [81] reported the synthesis of 1,2-dioxepane ring system **114** from silyl peroxide **111** with catalytic amount of DDQ in the presence of water in THF (Scheme 45). Removal of the silyl groups of **111** to form intermediate hydroperoxide **112** is followed by intramolecular S_N1-type ring closure of the resulting benzylic cation **113** to form the seven-membered ring peroxide 10,12-peroxycalamenene **114**. The role of DDQ as an oxidant (single-electron acceptor) and a weak acid has been amply demonstrated.

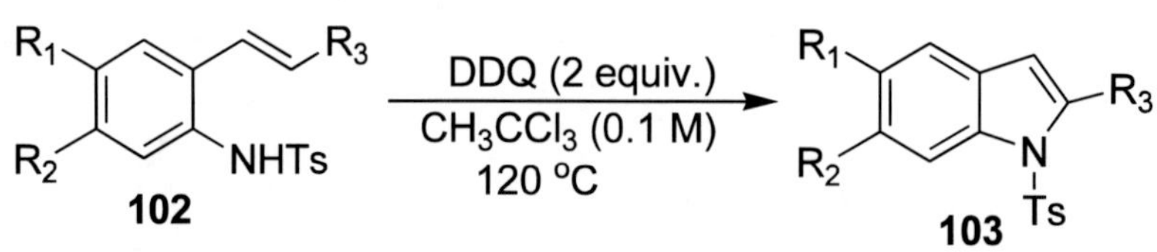

R_1 = H, OMe, Me, Cl, CF_3, NO_2
R_2 = H, OMe, Me, Cl, NO_2
R_3 = C_6H_5, $C_6H_4.CH_3(4)$, $C_6H_4.CH_3(3)$, $C_6H_4.CH_3(2)$, $C_6H_4.OCH_3(4)$, $C_6H_4.OCH_3(3)$, $C_6H_4.NO_2(4)$, $C_6H_4.CF_3(4)$, $C_6H_4.Cl(4)$, H, 1-naphthyl, 1-thienyl

Scheme 42. Synthesis of derivatized indoles.

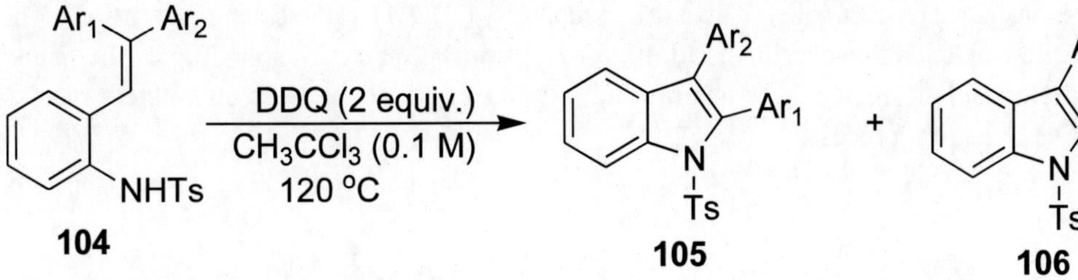

$Ar_1 = C_6H_5, C_6H_4.OCH_3(4)$
$Ar_2 = C_6H_5, C_6H_4.CH_3(4), C_6H_4.OCH_3(4), C_6H_4.Cl(4), C_6H_4.CF_3(4)$

Scheme 43. Synthesis of derivatized indoles.

Spiroacetal scaffolds occupy a prominent position in medicinal chemistry for microbial disruption [82], apoptosis induction [83] and B-cell chronic lymphocytic leukemia treatment [84]. Toflogliflozin possessing a spiroacetal subunit has been approved for treatment of type 2 diabetes [85]. Floreancig and co-workers [86] have reported one-pot synthesis of spiroacetals **119** from dihydropyrans **116** *via* [n+1]-DDQ-promoted bimolecular oxidative annulation process (Scheme 46). A Lewis acid (TMSOTf) catalyzed Ferrier fragment-coupling reaction of nucleophilic allylsilanes **115** and electrophilic dihydropyrans **116** in DCM generates the allylic ether **117** which was cleaved to alcohol **118** upon treatment with aq. $NaHCO_3$. Subsequent DDQ promoted oxidative carbon-hydrogen bond cleavage and cyclization of **118** leads to spiroacetals **119**. Dihydropyrans **116** possessing tertiary allylic/benzylic alcohols proved to be excellent substrates for the spiroacetals synthesis.

Uludag et al. [87] published a short and efficient total synthesis of 20-diethyl-4-demethyldasycarpidone **125** possessing a tetracyclic ring system *via* DDQ mediated penultimate-intramolecular cyclization process (Scheme 47). The synthesis involved the alkylation of 1-oxo-tetrahydrocarbazole **120** with chloroacetonitrile to form **121**. The protection of the ketone **121** as the thioketal **122** is followed by reduction with BH_3.THF to afford the aliphatic amine **123**. The cyclization to azocino[4,3-*b*]indole skeleton **124** was achieved on treatment of **123** with DDQ. Final deprotection of **122** with *bis*(trifluoroacetoxy)iodobenzene gave the tetracyclic product **125** in 31% yields.

DDQ-mediated oxidative cyclization of thiohydroximic acids **126** to form 1,2,4-oxathiazoles 1**27** was reported by Pierce et al. (Scheme 48) [88]. A variety of alkyl, aryl, and heteroaryl substituents on the thiohydroximic acid and the activating fragments were tolerated in the cyclization process. The oxidative cyclization method serves as a useful tool for medicinal chemists to access heterocyclic scaffolds for structure-activity relationship (SAR) studies.

Scott and co-workers [89] demonstrated that the metal-free Scholl type oxidation reactions using a combination of oxidant DDQ and trifluoromethanesulfonic acid (triflic acid, TfOH) could furnish triply-fused benzene ring systems from alkenes (Scheme 49). In this protocol, 4,7-di-tert-butylacenaphthylene 128 was converted into the corresponding 2,5,8,11,14,17-hexa-tert-butyldecacyclene 129 by intermolecular oxidative cyclotrimerization process using DDQ/TfOH in DCM at rt.

In addition [89], intermolecular DDQ/TfOH oxidative reactions of a range of o-difunctionalized benzenes 130 converted them into the corresponding trimers 131 which possess a triphenylene carbon skeleton (Scheme 50). In continuation to the approach [89], the authors subjected benzothiophene 132 to metal-free DDQ/TfOH oxidation reaction. The intermolecular oxidative cyclotrimerization process furnished the corresponding cyclic trimer 133 regioselectively (Scheme 51). Benzofuran, indole and thiophene were completely inert to DDQ/TfOH oxidation process.

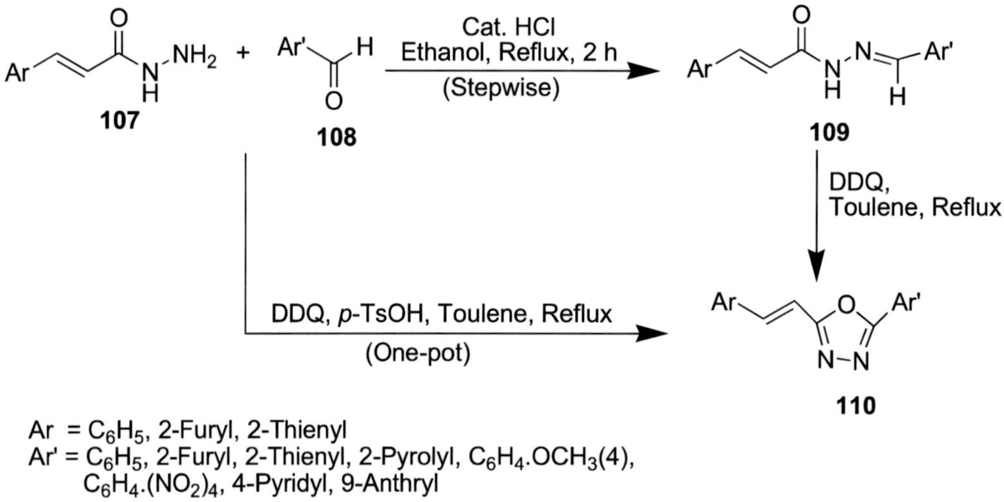

Scheme 44. Synthesis of 1,3,4-oxadiazoles.

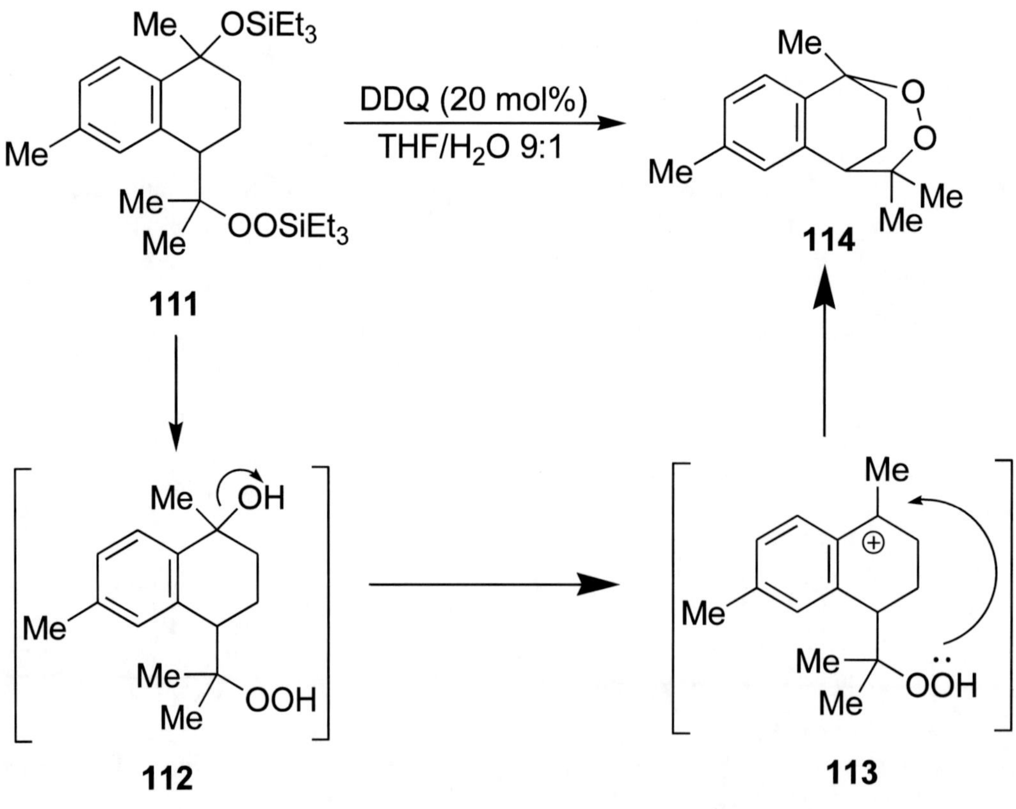

Scheme 45. Synthesis of 10,12-peroxycalamenene.

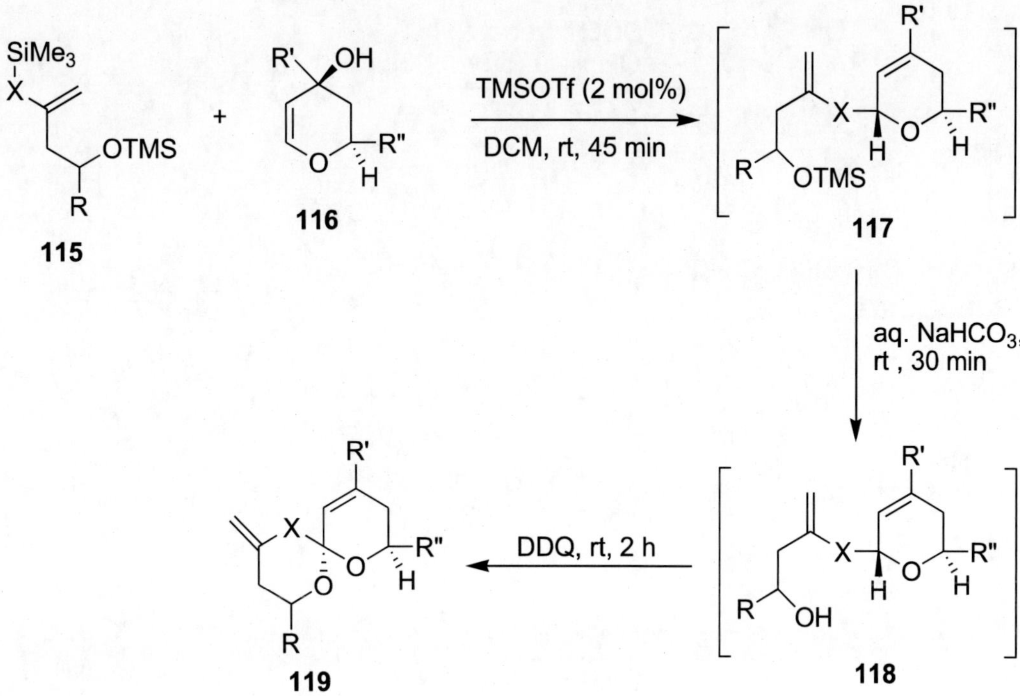

X = CH_2
R = $(CH_2)_2C_6H_5$, H, -C≡C−CH_3, CH_3
R' = CH_2OBn, CH_3, C_6H_5
R" = CH_2OBn, H, C_6H_5

Scheme 46. Synthesis of spiroacetals.

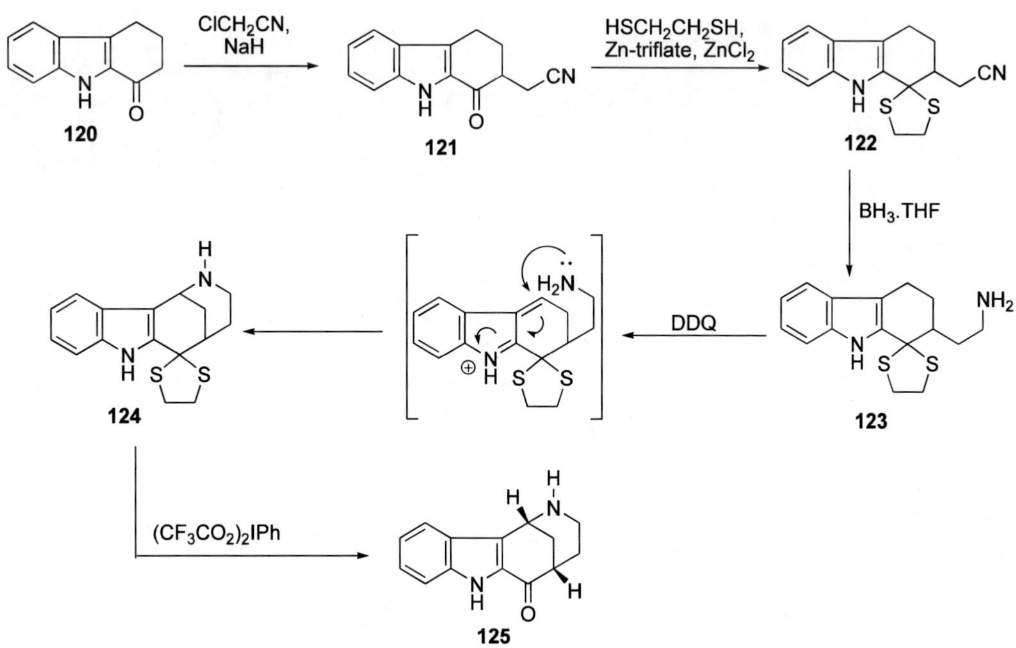

Scheme 47. Synthesis of 20-diethyl-4-demethyldasycarpidone.

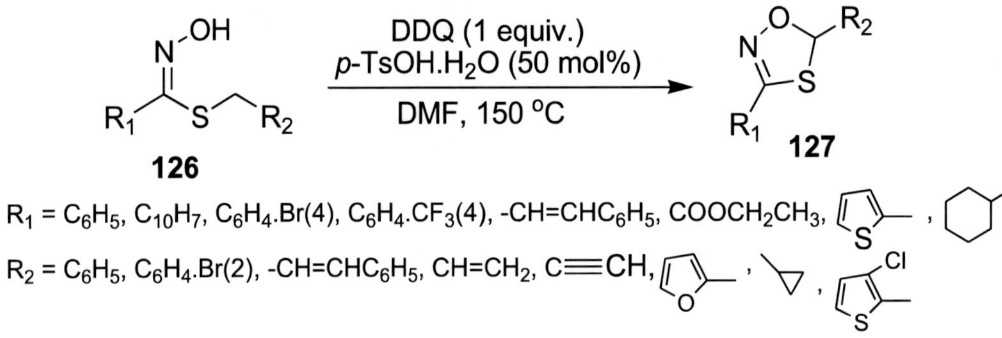

Scheme 48. Synthesis of 1,2,4-oxathiazoles.

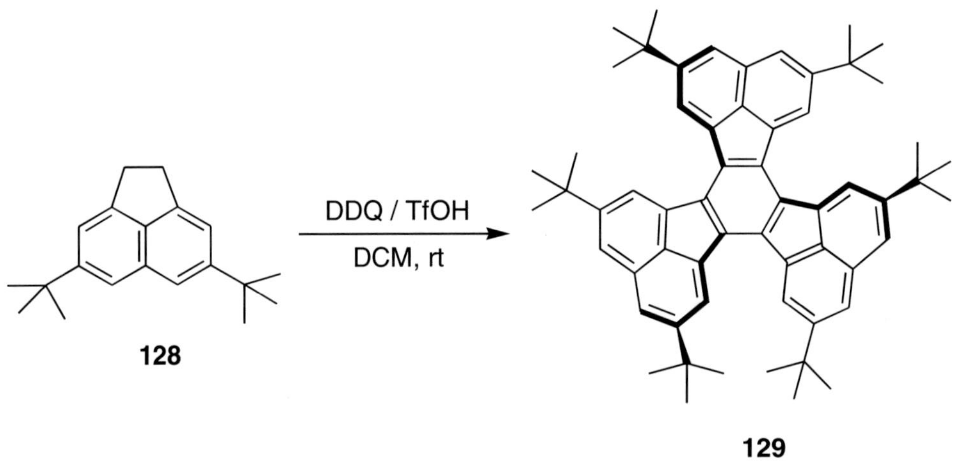

Scheme 49. Synthesis of 2,5,8,11,14,17-hexa-*tert*-butyldecacyclenes.

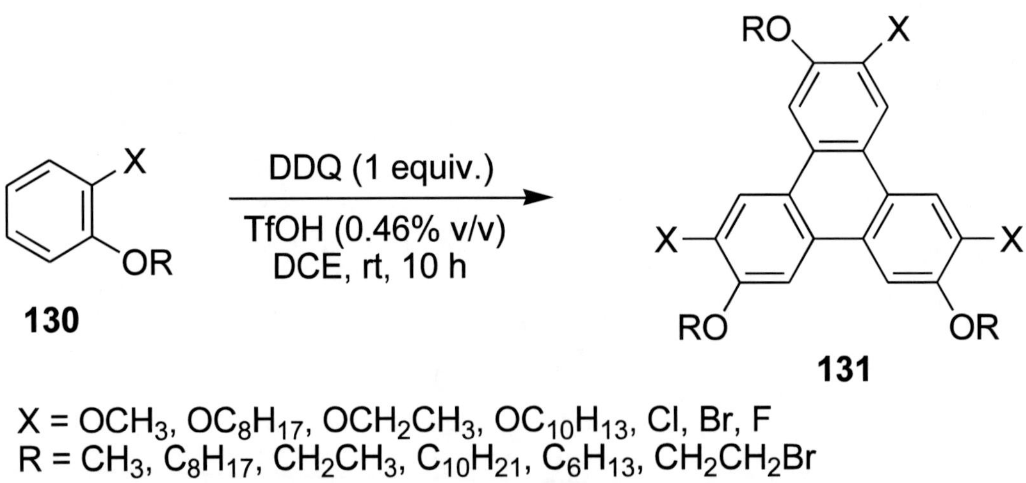

Scheme 50. Synthesis of triphenylenes.

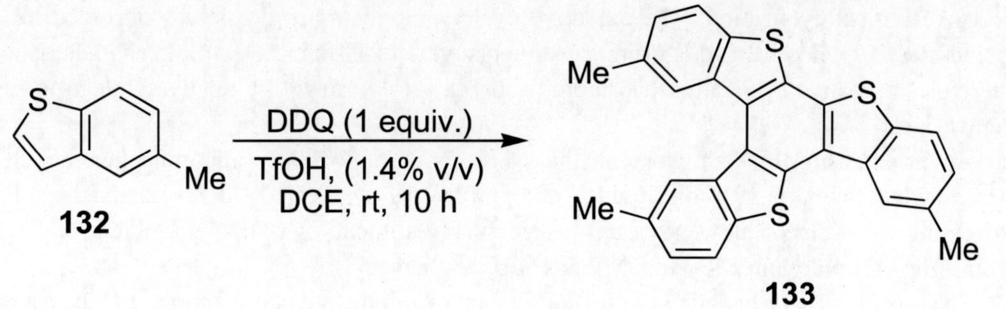

Scheme 51. Synthesis of cyclic trimers from benzothiophene.

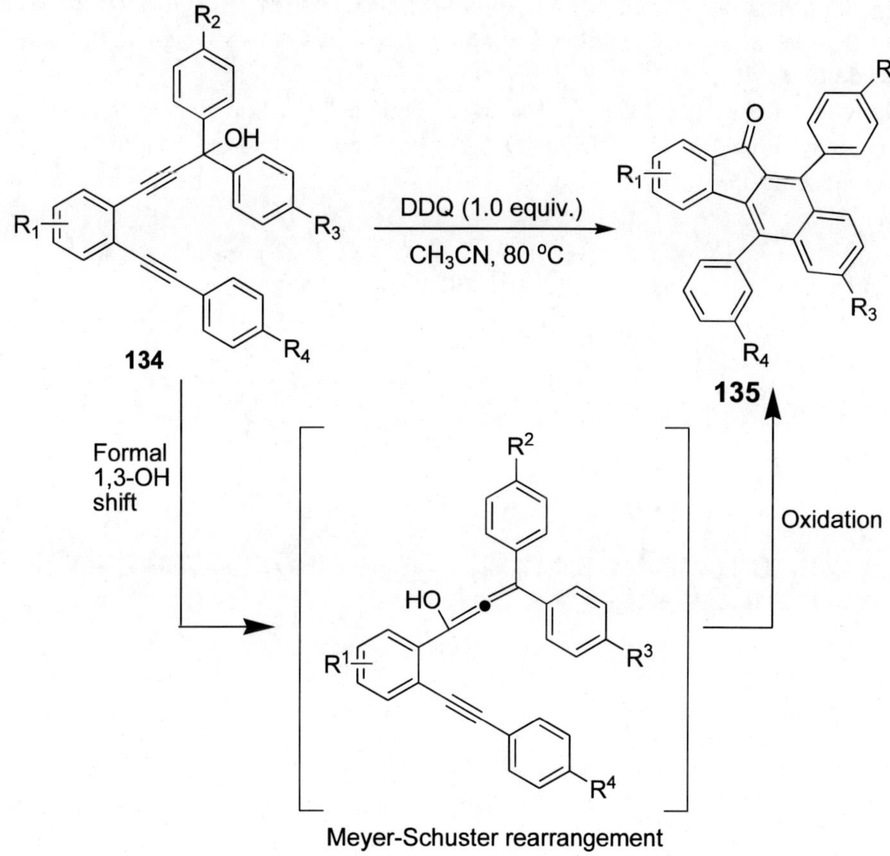

R_1 = H, OCH_3(4), CH_3(5), F(4), CF_3O(4)
R_2 = R_3 = H, OCH_3, Me, Cl, Br, F
R_4 = CH_3

Scheme 52. Synthesis of benzo[*b*]fluorenones.

Chen and co-workers [90] developed a facile approach for the DDQ-mediated oxidative synthesis of benzo[b]fluorenones 135 from 1,5-diynols 134 under metal-free conditions in acetonitrile at 80°C (Scheme 52). The oxidative radical cycloisomerization of 134 to 135 is believed to proceed by a formal 1,3-OH shift, subsequent Meyer-Schuster rearrangement and oxidative radical cyclization. The oxidative cycloisomerization process was observed to be chemoselective when R2 and R3 ring substituents were of different electronic characteristics. Therefore, oxidative cyclization had more probability to occur in the relatively electron-rich phenyl ring.

As an extension of their work on the synthesis of biologically important heterocycles, Pierce and co-workers [91] disclosed a straightforward synthesis of 1,2,4-oxadiazoles 137 possessing a variety of alkyl, aryl and heteroaryl substituents via DDQ-promoted oxidative cyclization of amidoximes 136 in DMF at 150°C (Scheme 53).

The group of Hayashi and co-workers [92] carried out the novel utilization of DDQ for the formation of chiral THF ring systems (Scheme 54). They treated the substrates **138** with DDQ in the presence of catalytic amount of additive $BF_3.Et_2O$ (15 mol%), leading to products **139** possessing a THF ring system *via* an intramolecular oxidative cyclization process. No cyclization was achieved in the absence of additive $BF_3.Et_2O$ or by usage of Bronsted acids (CSA) as additives.

In the same report [92], acetals **141** were obtained by oxidative cyclization of 2,4-dimethoxybenzyl ether **140** with DDQ in DCM at 0°C under anhydrous conditions (molecular sieves 4Å) (Scheme 55).

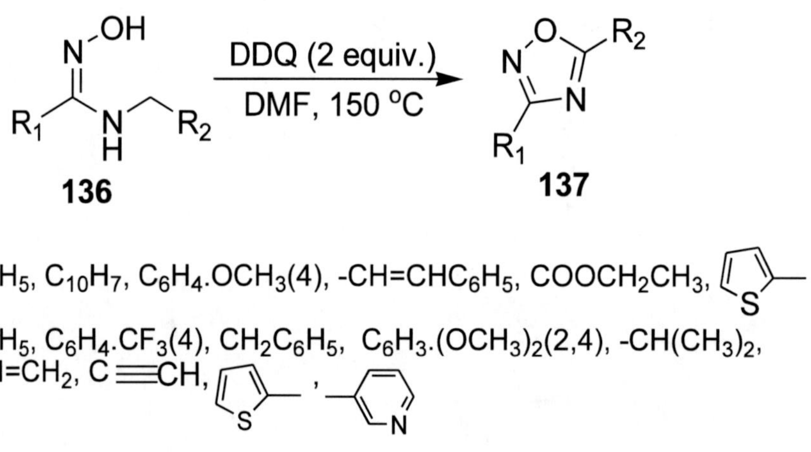

Scheme 53. Synthesis of 1,2,4-oxadiazoles.

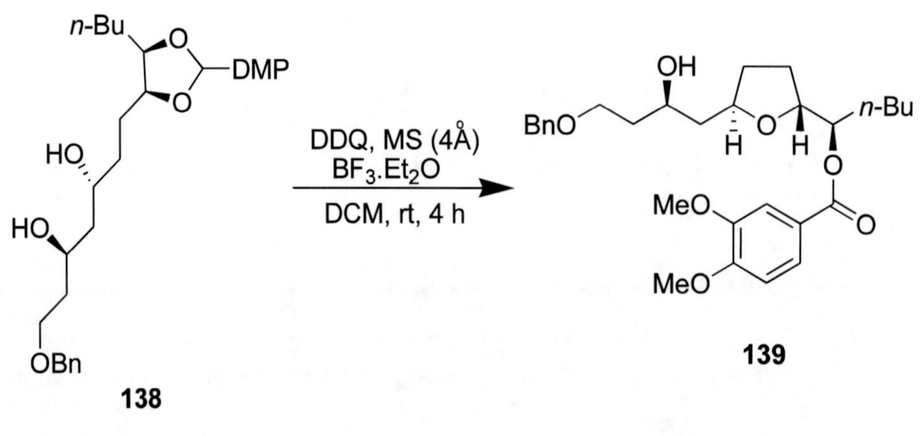

Scheme 54. Synthesis of chiral THF ring systems.

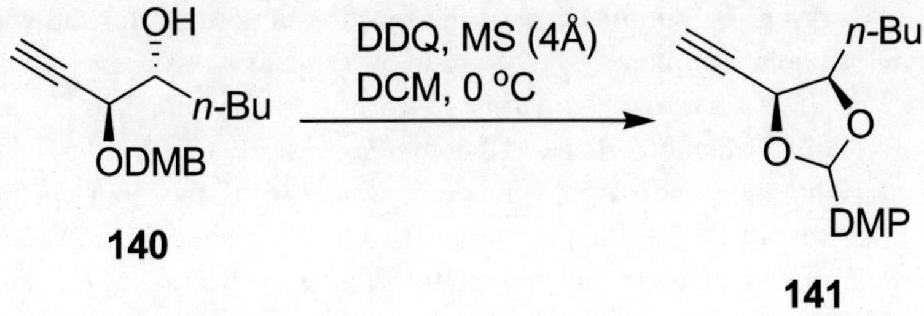

Scheme 55. Synthesis of acetals.

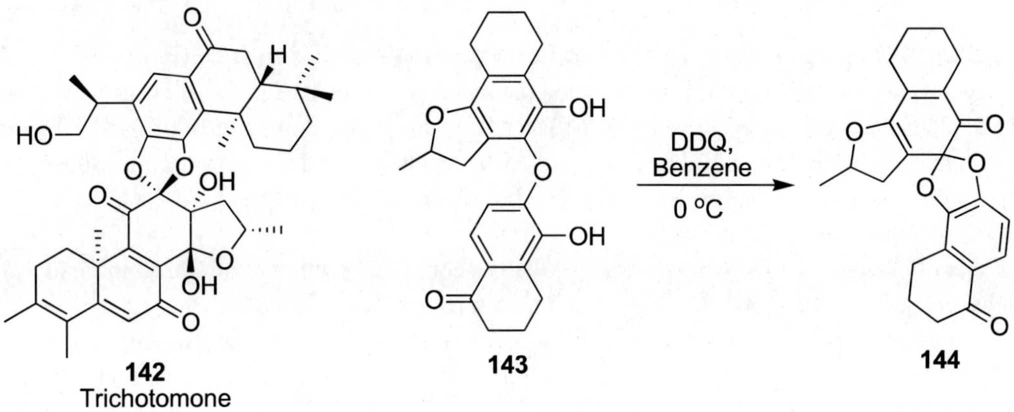

Scheme 56. Synthesis of ketal core sub-structure of trichotomone.

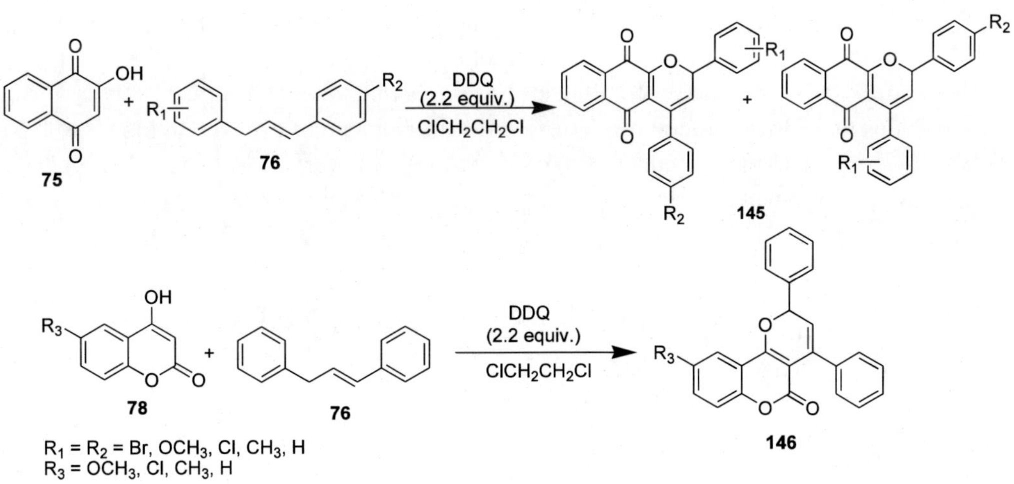

Scheme 57. Synthesis of pyranonaphthoquinones and pyranocoumarins.

The synthesis of the catecholic ketal core sub-structure of trichotomone **142** *viz* **144** was accomplished by the group of Liu and co-workers (Scheme 56) [93]. Subjecting substrate **143** to DDQ-mediated oxidation in benzene at 0°C, the target core structure of trichotomone **144** was formed as a 1:1 mixture of diastereomers by an oxidative dearomatization/ketalization process.

During their work on the optimization of the reaction conditions for the synthesis of 2-hydroxy-3-alkyl-1,4-naphthoquinones **77**, the group of Yan and co-workers [67] attempted the reaction of lawsone **75** / 4-hydroxycoumarins **78** with 1,3-diarylpropenes **76** using oxidant DDQ (2.2 equiv.) in 1,2-dichloroethane at room temperature (Scheme 57). The reaction generated pyranonaphthoquinones **145**/pyranocoumarins **146** as the only cyclized products (Scheme 57), rather than the cross-coupled products **77**/**79** (Scheme 31, 32). These products are expected to be formed by DDQ-mediated intramolecular cyclization of the initially formed cross-coupled products **77** / **79**. The coupled products react with DDQ to generate an allylic cation which undergoes nucleophilic attack of hydroxyl group to form the cyclized products **145**/**146**.

2.2.2.2. DDQ as an Oxidant as Well as Substrate in Annulation Reactions

Deng and co-workers [94] have reported the metal-free synthesis of 2,3-dicyanofurans **148** *via* oxidative radical coupling reaction between 1,3-dicarbonyl compounds **147** and DDQ in acetonitrile at room temperature (Scheme 58). Both aliphatic and aromatic substrates underwent the transformation to afford the furan-containing scaffolds **148** in moderate to excellent yields.

Simple ketones **149** undergo similar oxidative coupling reactions with DDQ to yield 2,3-dicyanofuran products **150** only at reflux temperatures (Scheme 59) [94].

A range of 2,3-dicyanothiophenes **152** were obtained in excellent yields by employing β-ketothioamides **151** as substrates for DDQ-mediated oxidative radical coupling reaction (Scheme 60) [94].

Li and co-workers [95] carried out the direct one-step transformation of electron-rich aryl ketones **153** to 5-aryl-2,3-dicyanofurans **154** by oxidative annulation reaction with DDQ (Scheme 61). The metal-free α-C-H functionalization of **153** and subsequent C-C and C-O bond formation in **154** was effective employing DDQ in 1,2-dichloroethane (DCE) in a sealed tube at 100°C. The presence of an electron-donating alkoxy group(s) on the aryl ring at *o*/*p* positions of **153** was crucial for the oxidative intermolecular cyclization to proceed. No products were observed with *m*-alkoxy substituted aryl ketones **153**. These 5-aryl-2,3-dicyanofurans **154** exhibited strong photonic fluorescence both in the solid state and in solution.

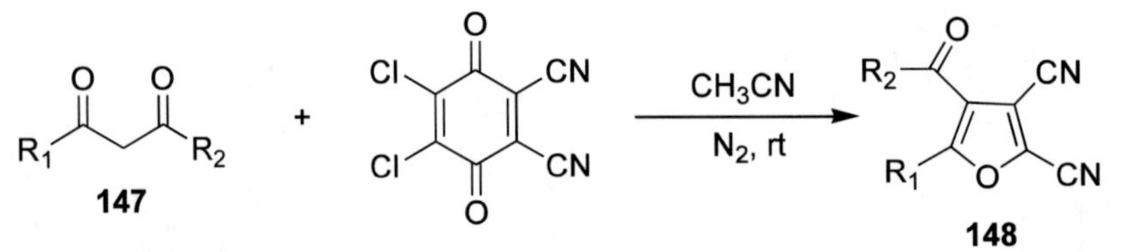

R_1 = CH_3, CH_2Cl, CF_3, CH_2CH_3, $CH(CH_3)_2$, $C(CH_3)_3$, CH_2OBn, C_6H_5, C_6H_4.Br(2), C_6H_4.Br(3), C_6H_4.Br(4), C_6H_4.Cl(4), $C_6H_4.CH_3$(4), $C_6H_4.OCH_3$(4), $C_6H_4OCH_2C_6H_4.CH_3$(4), Napthyl,

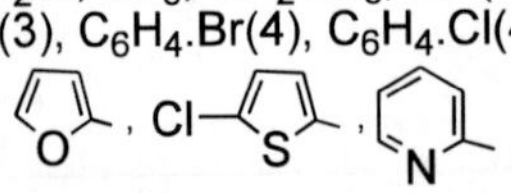

R_2 = OCH_3, OCH_2CH_3, O*i*-Pr, O*t*-Bu, OBn, CH_3, C_6H_5

Scheme 58. Synthesis of 2,3-dicyanofurans from 1,3-dicarbonyl compounds.

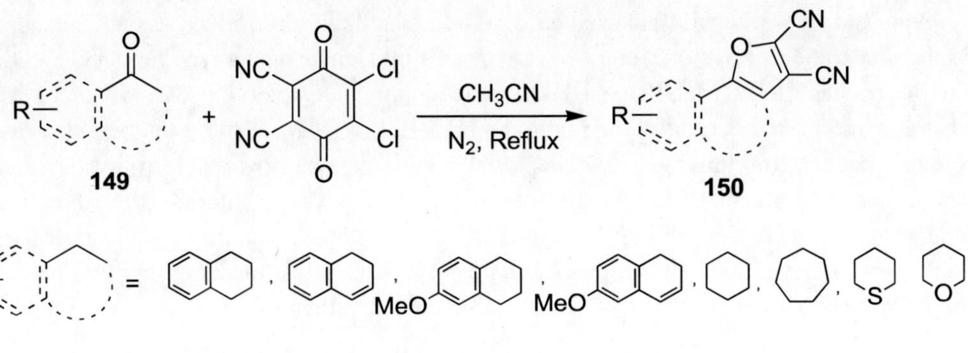

Scheme 59. Synthesis of 2,3-dicyanofurans from cyclic ketones.

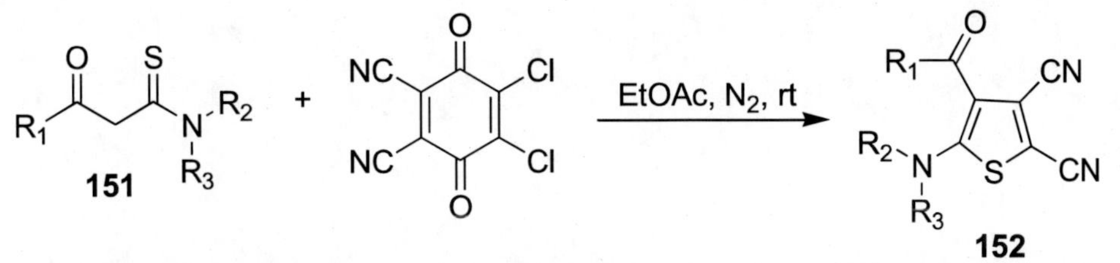

$R_1 = OCH_3, OCH_2CH_3, Ot\text{-}Bu, OBn, CH_3, CH_2CH_3, C_6H_5, C_6H_4.Br(4)$
$R_2 = R_3 = CH_3, CH_2CH_3$

Scheme 60. Synthesis of 2,3-dicyanothiophenes.

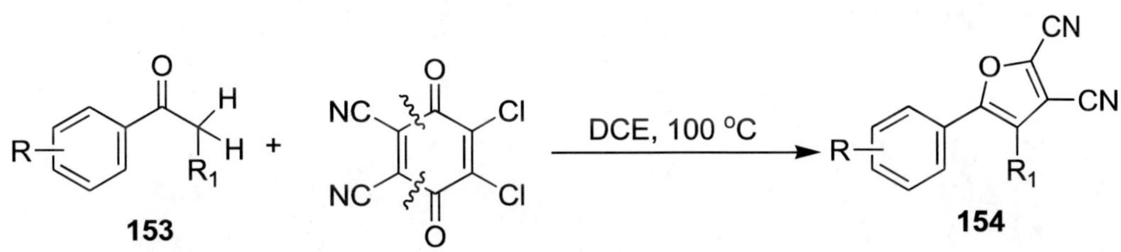

$R = OCH_3(4), OCH_3(2), OCH_3(3), (OCH_3)_2(3,5), OCH_2CH_3(2), OBn(2), (OCH_3)_2(2,4), (OCH_2CH_3)_2(2,4)$
$R_1 = H, CH_3$

Scheme 61. Synthesis of 5-aryl-2,3-dicyanofurans.

Furo[3,2-*c*]coumarins have been known to exhibit potent pharmacological and biological activities, thereby serving as lead structures for drug discovery [96, 97]. Li and co-workers [98] succeeded in establishing a mild and efficient protocol for the synthesis of furo[3,2-*c*]coumarin derivatives **156** from 4-hydroxycoumarins **155** *via* DDQ mediated oxidative annulation process in acetonitrile at rt (Scheme 62). Electron-withdrawing groups (Cl, Br) on **155** favour the oxidative annulation in comparison to electron-donating substituents. The authors [98] extended the methodology to the synthesis of furo[2,3-*c*]quinolones **158** from *N*-substituted-4-hydroxyquinolin-2(1*H*)-ones **157** and DDQ through the oxidative cyclization process (Scheme 63).

Zhou and co-workers [99] carried out an efficient and diastereoselective synthetic approach for the preparation of cyclohexene derivatives **160** by Dehydrogenative Diels-Alder (DHDA) reaction (Scheme 64). Metal-free DHDA reaction of DDQ (serving as oxidant as well as substrate) and prenyl derivatives *viz* prenyl ethers **159** in DCM at room temperature gave the cyclized products **160** in good to excellent yields with excellent *trans* selectivity. The reaction was proposed to proceed through the formation of an isoprene intermediate (diene) which undergoes DA reactions with DDQ (dienophile) to give the cycloaddition products **160**. Better yields were achieved with substrates bearing electron-rich substituents at the *para* position of the phenyl ring than those bearing electron-deficient substituents.

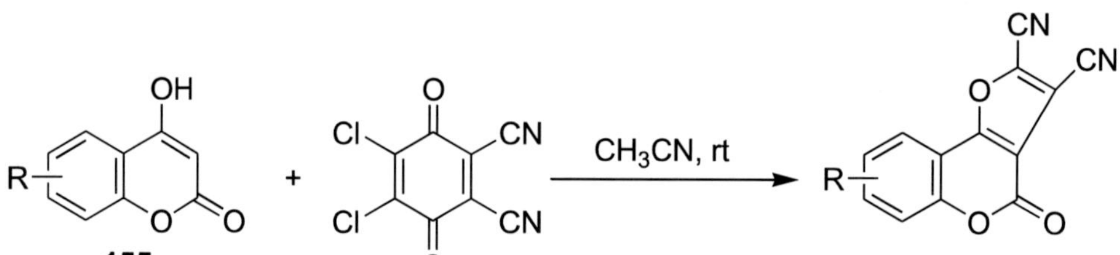

155 156

R = CH_3(1), CH_3(2), CH_3(3), *t*-Bu(3), *i*-Pr(1), OCH_3(2), Cl(3), Br(3), Cl(2)

Scheme 62. Synthesis of furo[3,2-*c*]coumarins.

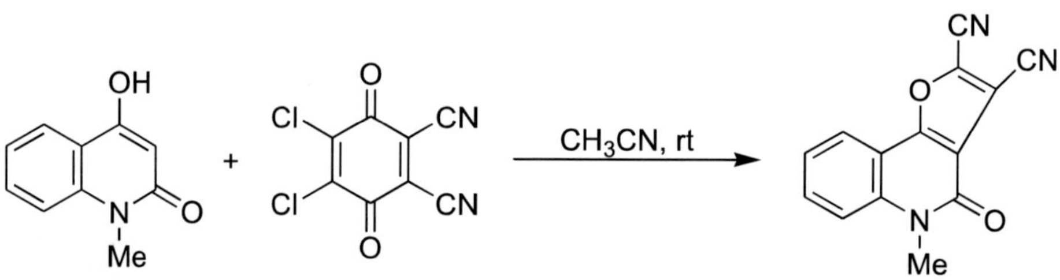

157 158

Scheme 63. Synthesis of furo[2,3-*c*]quinolones.

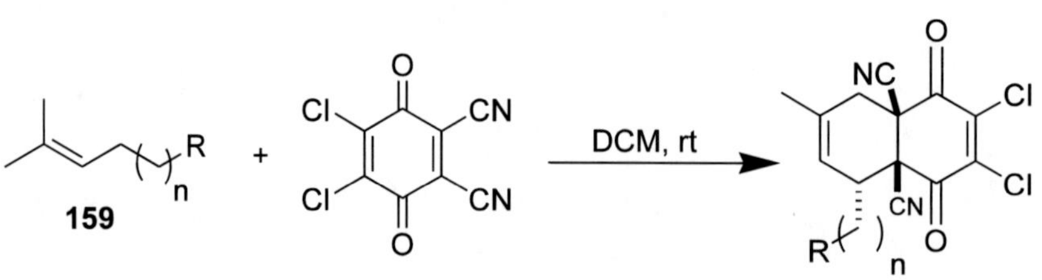

159 160

n = 0, 1, 2
R = OC_6H_5, $OC_6H_4.CH_3$(4), $OC_6H_4.OCH_3$(4), $OC_6H_4.Cl$(4), $OC_6H_4.Br$(4), $OC_6H_4.NO_2$(4), $OC_6H_4.CH_3$(3), $OC_6H_4.Cl$(3), $OC_6H_4.Br$(3), $OC_6H_4.CH_3$(2), $OC_6H_4.Cl$(2), $OC_6H_4.Br$(2), C_6H_5, OTs, OBn, 2,3-dimethyl-2-butene

Scheme 64. Synthesis of cyclohexene derivatives.

Exclusive *trans* selectivity for the cyclized products **162** was also obtained in DHDA reaction of DDQ and prenyl derivatives **161** containing nitrogen functional groups (Scheme 65) [99].

A novel and direct domino synthesis of 5,6-dicyanobenzo[*c*]fluorenes **164a-c** *via* DDQ-mediated oxidative approach using tetraaryl[3]cumulenes **163a-c** as the substrates was reported by Diederich and co-workers (Scheme 66) [100]. The one-step cascade reaction of **163** and DDQ (2 equiv.) proceeds in 1,2-dichloroethane under simple stirring conditions to yield benzo[*c*]flourenes **164** with remarkable optoelectronic properties. The mechanistic details proposed indicate both electron transfer/dehydrogenation and cycloaddition reactivities of DDQ. The corresponding tetraaryl[3]cumulenes **163** with electron-donating groups (R = OCH_3) and electron-withdrawing groups (R = F, Cl) on the phenyl ring did not show any conversion to benzo[*c*]fluorenes **164**.

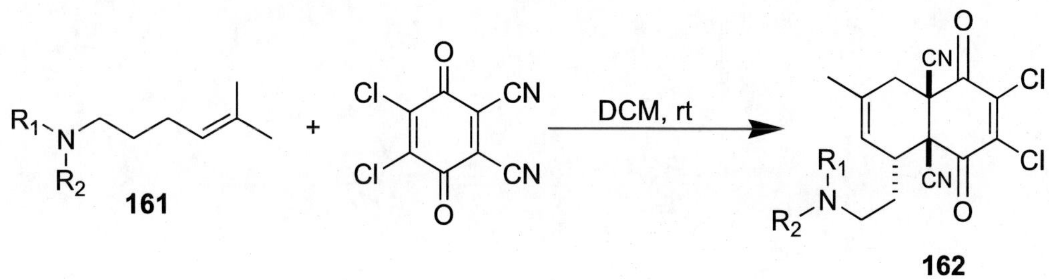

$R_1 = SO_2C_6H_4.CH_3(4), SO_2C_6H_4.NO_2(4), SO_2C_6H_4.OCH_3(4), SO_2C_6H_4.Br(4), C_6H_5$
R_2 = Bn, PMB, C_6H_5

Scheme 65. Synthesis of cyclohexene derivatives.

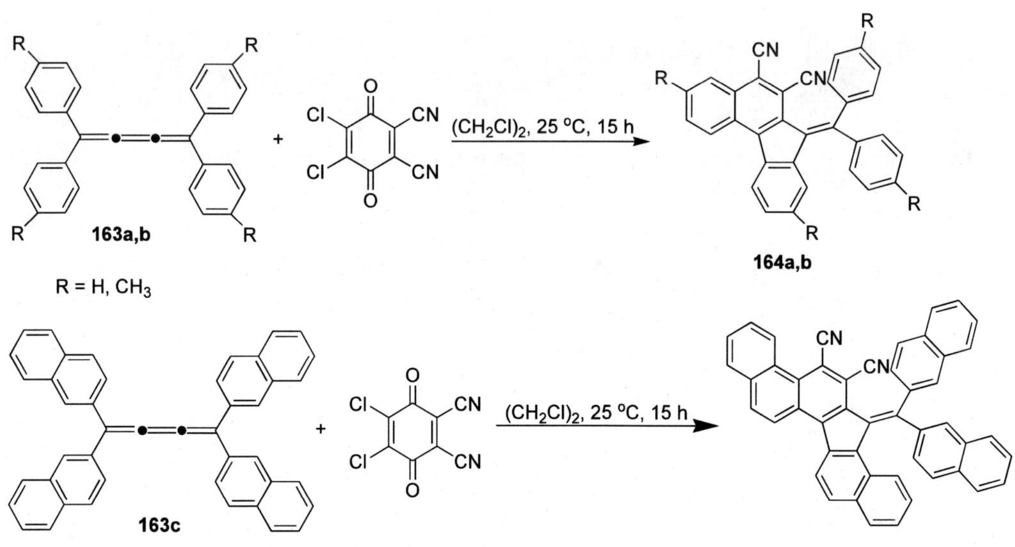

Scheme 66. Synthesis of 5,6-dicyanobenzo[*c*]fluorenes derivatives.

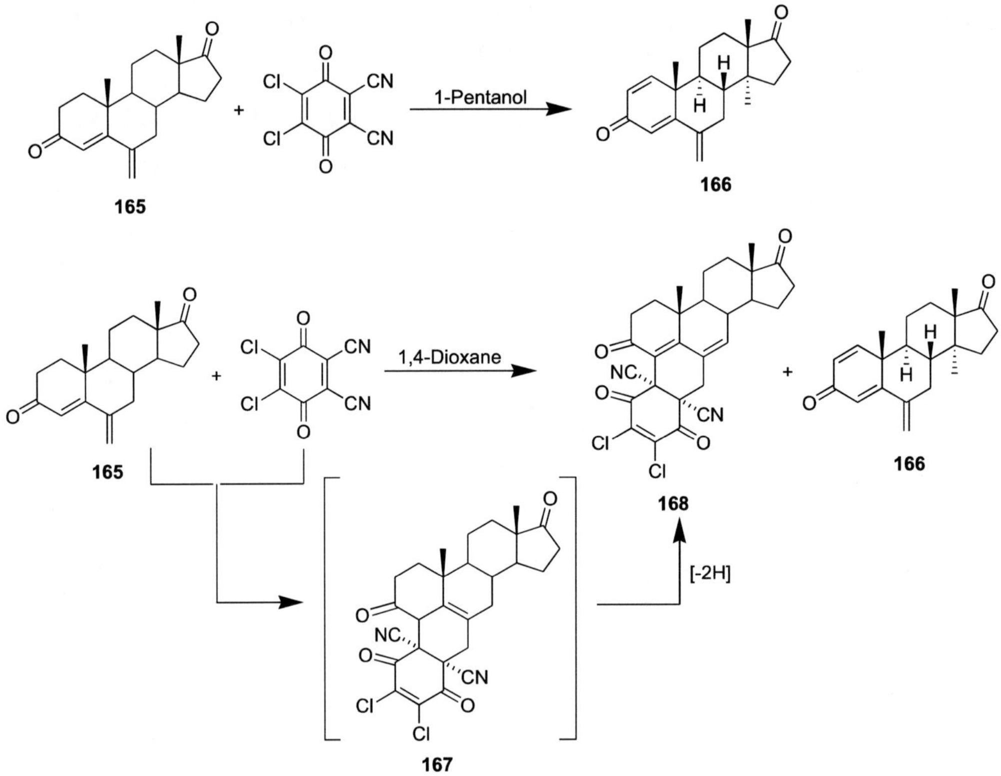

Scheme 67. Synthesis of exemestane and hexacyclic quinoid steroidal byproduct.

In a report by Giovenzana et al. [101], the substrate 6-methyleneandrost-4-ene-3,17-dione **165** on oxidative dehydrogenation with DDQ in polar protic solvents (AcOH, 1-pentanol, 1-butanol, *tert*-butanol) led to exemestane **166** by selective introduction of Δ^1-unsaturation (Scheme 67). Carrying out the oxidative transformation of **165** with DDQ in a polar aprotic solvent (1,4-dioxane) furnished a hexacyclic quinoid steroidal byproduct **168** with a *cis*-ring junction in addition to exemestane **166**. An initial Diels-Alder cycloaddition of **165** (serves as the diene) with DDQ (serves as the dienophile) yields adduct **167** possessing a hexacyclic skeletal system. Oxidation of **167** with another molecule of DDQ leads to selective Δ^6-unsaturation generating 4,6-diene-3-one **168** as a by product in 27% yields.

2.2.3. DDQ-Promoted Oxidative Aromatization Reactions

Aromatization reactions involve the elimination of hydrogen. In alicyclic systems, aromatization is easily accomplished if the ring is fused to an aromatic ring or if the ring previously possesses one or two double bonds. The mechanism of aromatization with DDQ involves hydride transfer from the unaromatized substrate **E** to the quinone oxygen leading to phenolate ion **F**. Transfer of a second proton from **F** to $DDQH^-$ forms the aromatized product **G** and $DDQH_2$ (Figure 4).

Arylnaphthalene lactones isolated from medicinal or dietary plants have been known to exhibit anticancer [102], antiviral [103] and ion-channel blocking activities [104]. Shin and co-workers [105] employed DDQ-mediated oxidative aromatization of dihydronaphthalenes **169**, **171** and **173** in benzene at 80°C in the synthesis of the arylnaphthalene lactones *viz* Taiwanin C (Type I) **170**, Justicidin E (Type II) **172** and Daurinol (Type I) **174** (Scheme 68).

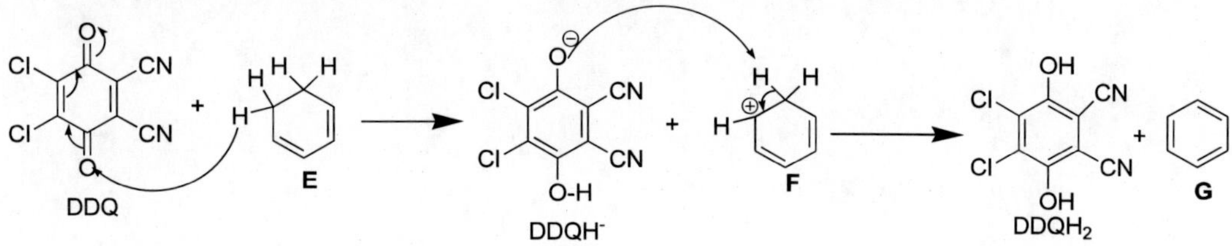

Figure 4. Mechanism of DDQ-mediated aromatizations.

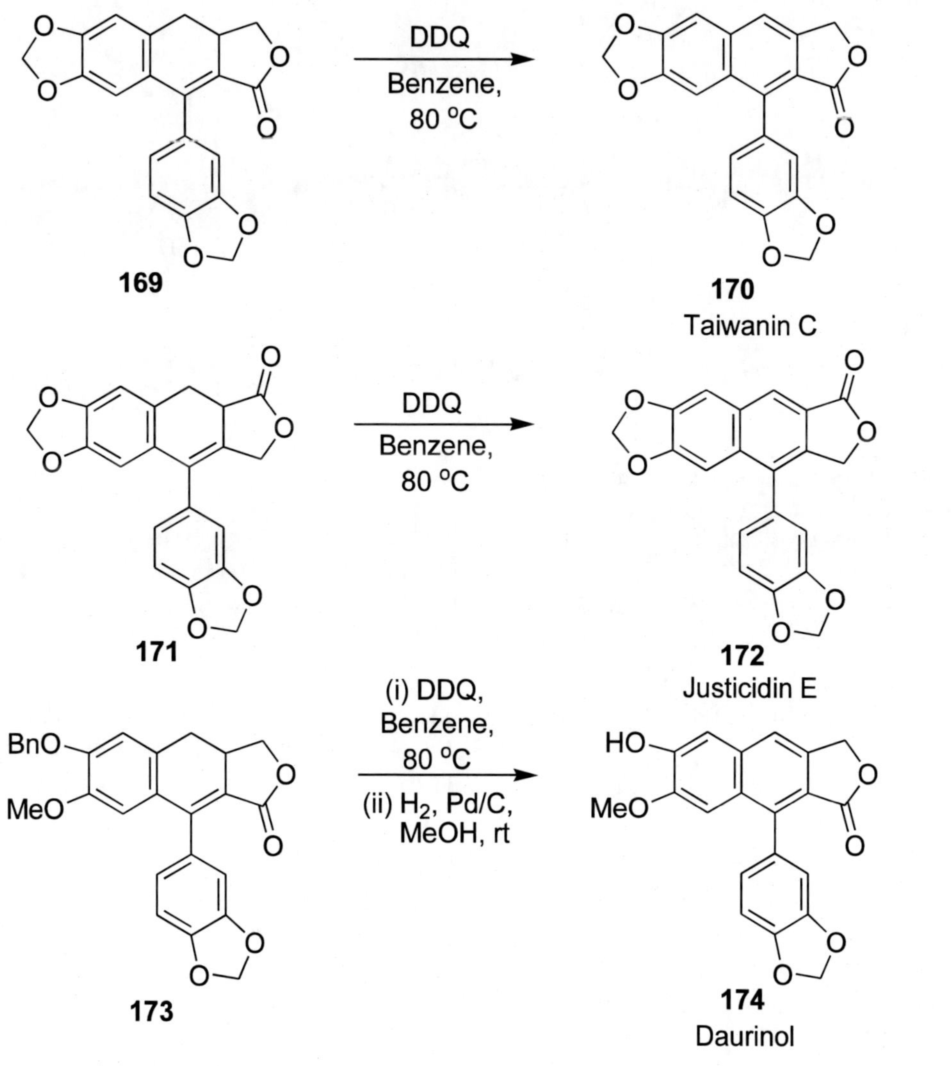

Scheme 68. Synthesis of arylnaphthalene lactones.

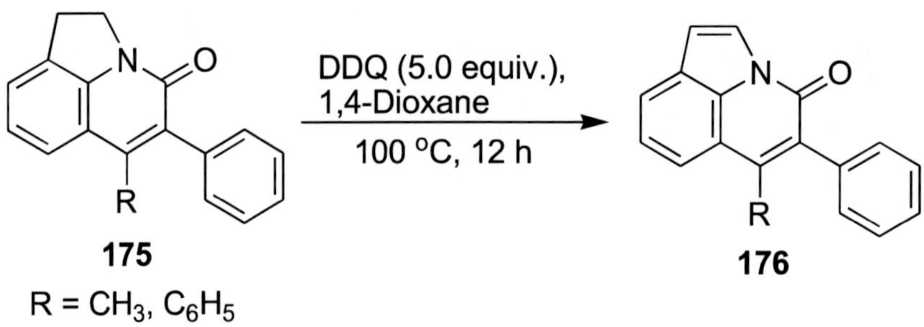

Scheme 69. Synthesis of indole derivatives.

A convenient route to synthesize indole derivatives **176** *via* oxidative aromatization of pyrroloquinolinone derivatives **175** with DDQ was reported by Jeganmohan et al. (Scheme 69) [106].

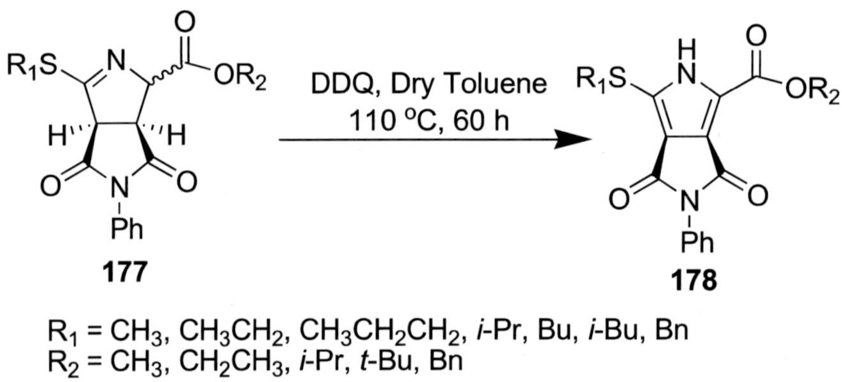

Scheme 70. Synthesis of polysubstituted pyrroles.

Oikonomou et al. [107] reported the aromatization *via* oxidation of dihydropyrroles (*syn*/*anti* mixture) **177** to the corresponding polysubstituted pyrroles **178** using DDQ as an oxidant in dry toluene under an inert atmosphere in a sealed tube heated at 110°C for 60 h (Scheme 70). The *syn* diastereomers reacted two times faster than the *anti* isomers.

Xanthones incorporating 1,4-disubstituted-1,2,3-triazole moiety have been reported to be antitumor in nature [108]. An efficient DDQ-mediated oxidative aromatization strategy was reported by Silva et al. [109] for the synthesis of xanthone-1,2,3-triazole dyads **181**, **183** (Scheme 71). Starting from substrates **179** *viz* 4-(arylethynyl)-2-methyl-3a,4,5,11b-tetrahydrochromeno[3,2-*e*]isoindole-1,3,11(2*H*)-triones, an initial DDQ-mediative oxidative aromatization in toluene at 100°C leads to aromatized derivative **180**. Subsequent treatment of **180** with sodium azide in refluxing DMF furnishes xanthone-1,2,3-triazole dyads **181** in excellent yields. A second approach towards synthesizing xanthone-1,2,3-triazole dyads **183** was directed towards direct oxidative aromatization of 4-(4-aryl-2-methyl-2*H*-1,2,3-triazol-5-yl)-2-methyl-3a,4,5,11b-tetrahydrochromeno[3,2-*e*]isoindole-1,3,11(2*H*)-triones **182** with DDQ in toluene at 100°C.

Kimura and co-workers [110] reported the synthesis of isoquinoline analogues **189** using DDQ as a debenzylation oxidant (Scheme 72). The initial palladium catalyzed ampiphilic allylation of 2,3-bismethylenebutane-1,4-diol derivatives **184** [alcohol (R = H), acetate (R =

Ac), benzylether (R = Bn)] with aldimines **185** using triethylborane or diethylzinc as promoters generates 3,4-bismethylene piperidine ring **186** *via* a formal [4+2] cycloaddition reaction. Diels-Alder reaction of piperidines **186** and acetylenedicarboxylates **187** afforded bicyclic hexahydroisoquinolines **188**. Oxidative debenzylation/aromatization of **188** using DDQ in xylene affords the target isoquinoline compounds **189**.

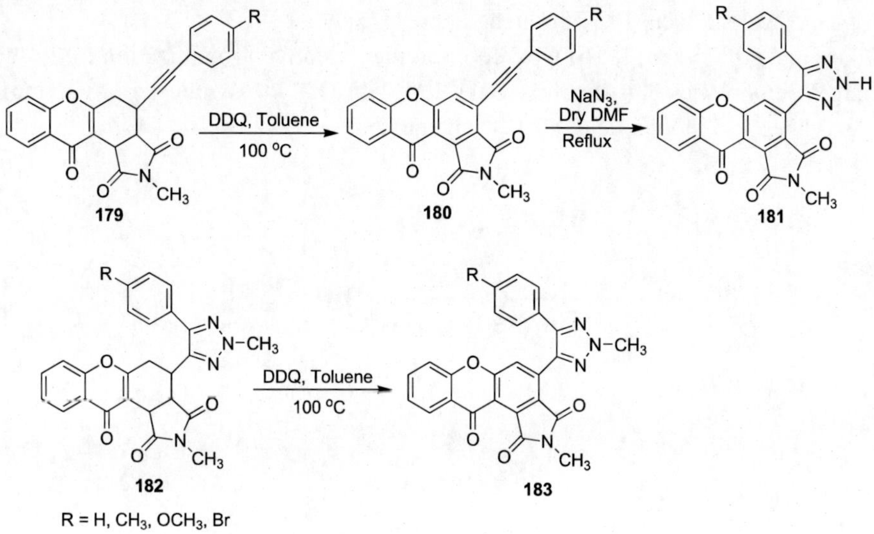

Scheme 71. Synthesis of xanthone-1,2,3-triazole dyads.

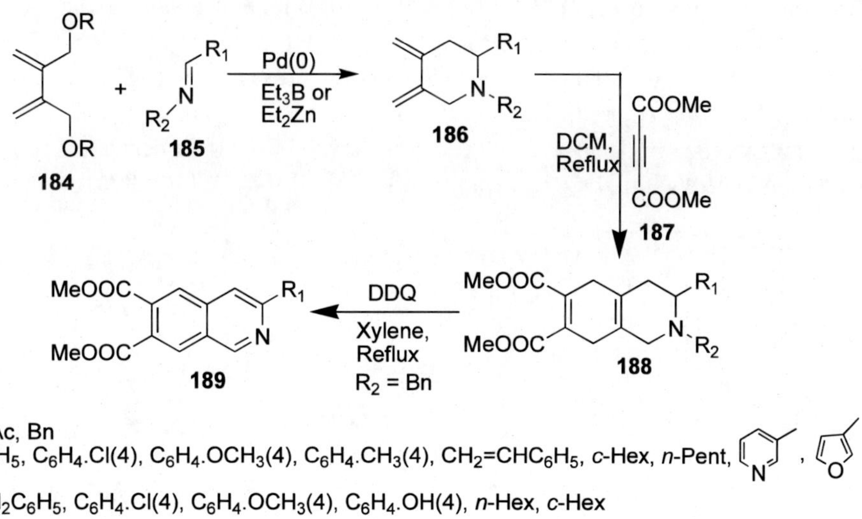

Scheme 72. Synthesis of isoquinolines.

Several compounds containing the quinazolinone skeleton are known to display anti-depressant [111] and antihypertensive [112] activities. Furthermore, quinazolinone derivatives possessing long Π-electron conjugated system have been utilized as organic dyes [113]. Oh et al. [114] developed a series of novel 5-substituted-12H-quinolino[2,1-b]quinazolin-12-ones 191 with fully conjugated Π-system by 2,3-dichloro-5,6-dicyanobenzoquinone (DDQ) promoted aromatization of 5-substituted-5,6-dihydro-12H-quinolino[2,1-b]quinazolin-12-ones 190 in benzene at 100°C (Scheme 73). The authors utilized derivatized quinazolinones 191 as substrates for cross coupling reactions viz Heck and Suzuki reactions to synthesize products with longer Π-conjugation.

In 2016, the group of Minuti and co-workers [115] synthesized a series of novel 4-aryl[2,2]paracyclophanes **193** by DDQ-mediated oxidative aromatization of cyclohexadienyl-paracyclophane cycloadducts **192** in toluene at 80°C (Scheme 74).

Kostikov and co-workers [116] worked upon the oxidative aromatization of 2,4-dihydro-1*H*-azeto[1,2-*a*]quinolones-3,4-dicarboxylates **194** with DDQ to synthesize 2-vinylquinoline-3,4-dicarboxylates **195** in toluene at 80°C (Scheme 75).

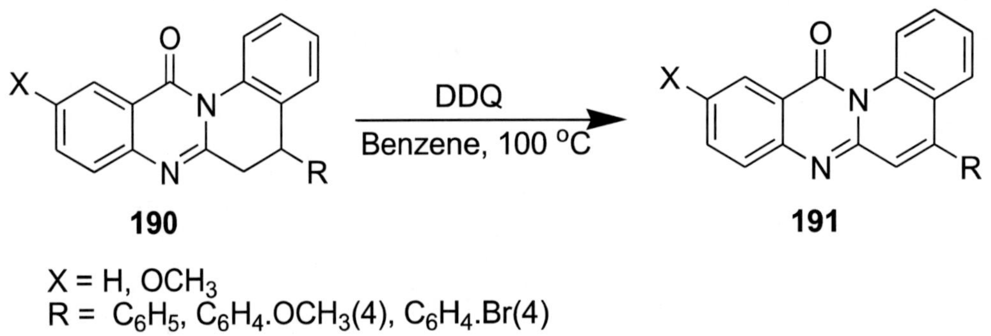

Scheme 73. Synthesis of quinazolinone derivatives.

Basavaiah and co-workers [117] utilized Baylis-Hillman acetates **196** in the synthesis of pyrrolo[2,1-*a*]isoquinolines **200** employing Cu(I)/TBHP-promoted oxidative intramolecular cyclization and subsequent DDQ-mediated oxidative aromatization (Scheme 76). An initial alkylation of **196** with tetrahydroisoquinoline **197** leads to allyl amine **198** *in situ*, which upon intramolecular cyclization in *m*-xylene generates the dihydropyrrole derivatives **199**. Oxidative aromatization of **199** with DDQ furnishes pyrrolo[2,1-*a*]isoquinolines **200**.

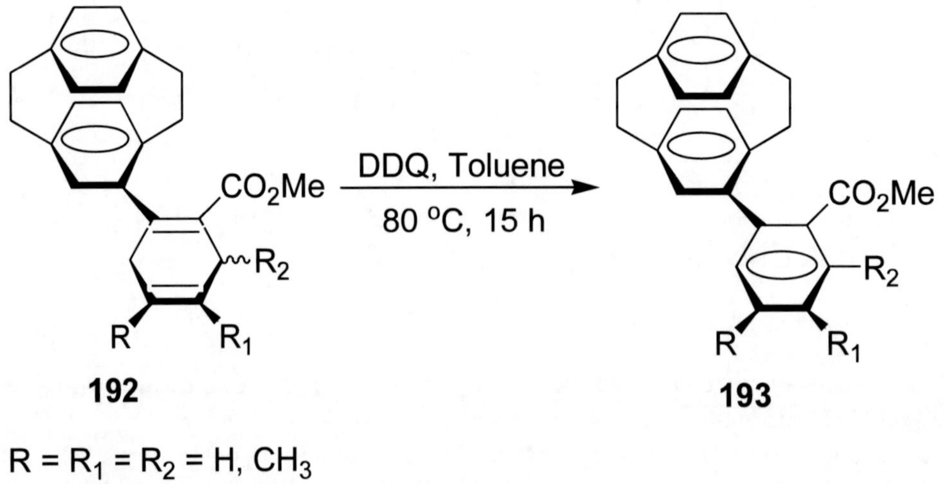

Scheme 74. Synthesis of 4-aryl[2,2]paracyclophanes.

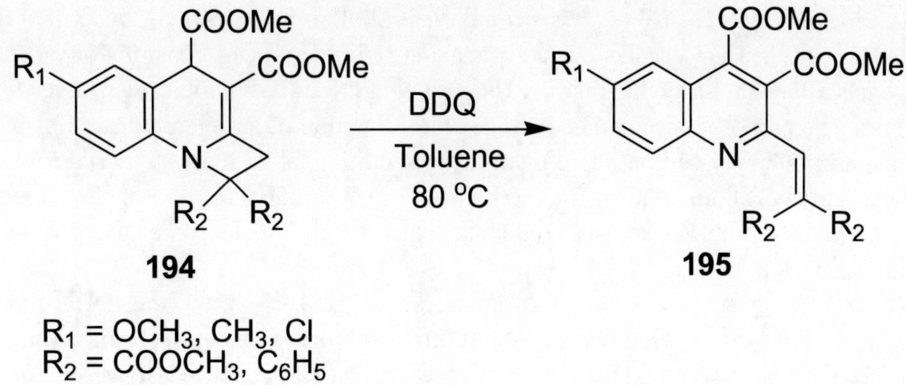

Scheme 75. Synthesis of 2-vinylquinoline derivatives.

Yu and co-workers [118] synthesized a series of tetracyclic indole-based chromane containing HCV NS5A inhibitors **204**. The central core indole-based skeleton **203** was generated from indolines **201** by an oxidative transformation process with DDQ (Scheme 77). Treatment of the indoline compound **201** with substituted aldehydes **202** under acidic conditions generates the tetracyclic indoline intermediate by ring cyclization. Oxidative aromatization of the intermediate to the tetracyclic indole-based scaffold **203** was accomplished upon treatment with DDQ in toluene at 115°C. A series of transformations on **203** could lead to the chromane containing HCV NS5A inhibitors **204**.

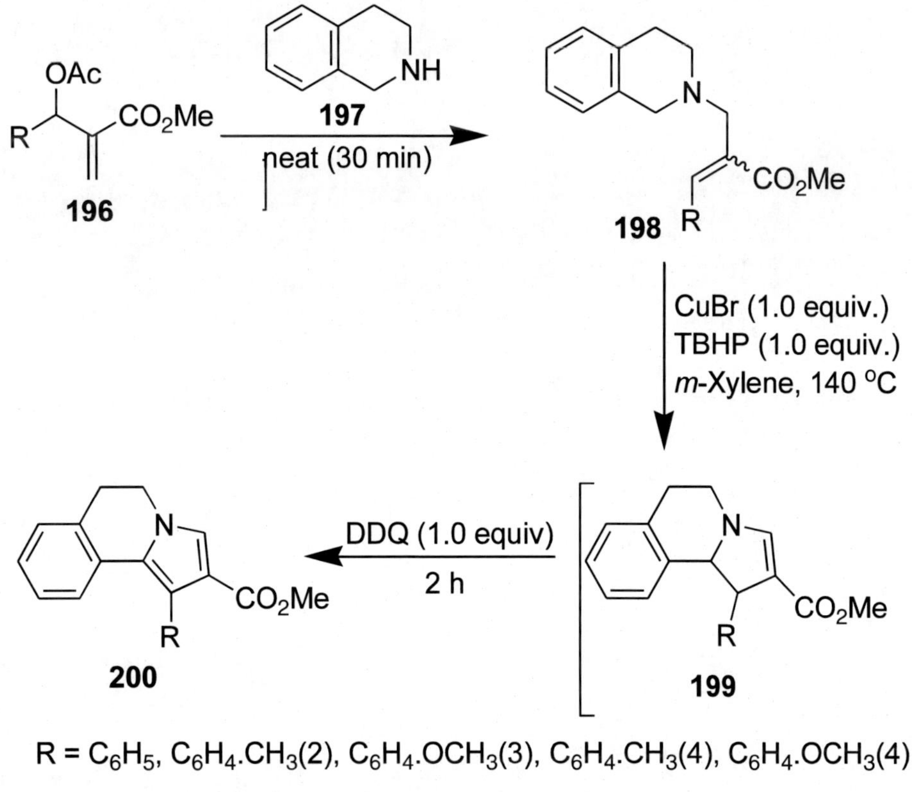

Scheme 76. Synthesis of pyrrolo[2,1-*a*]isoquinolines.

An efficient approach to benzocoumarins **208** was reported by Chi et al. (Scheme 78) [119]. They carried out a formal [4+2] benzene ring construction of **208** by treating enals **205** with 3-acetylcoumarins **206**, utilizing DBU (base), molecular sieves (4Å) in chloroform. The proposed mechanism includes the formation of an intermediate dihydride of benzocoumarin **207** by γ-carbon deprotonation, 1,4-addition, aldol reaction, acetal formation, and the acetate elimination. The dihydride **207** undergoes a facile oxidative aromatization using DDQ as an oxidant in refluxing dichloromethane to furnish benzocoumarins **208**.

Recently, the group of Chen and co-workers [120] explored the DDQ-mediated aromatization of substrates **209** under xylene reflux conditions (Scheme 79). The synthesized aromatic 1,16-diaryl-substituted benzo[5]helicene derivatives *viz rac*-**210** were efficiently resolved through HPLC into individual enantiomers. The resolved enantiomers exhibited high specific optical rotations, thermostability, and racemization barriers. The X-ray analysis of **210** indicates that they adopt a twisted helical conformation due to the presence of two aryl substituents on the inner part of the helix.

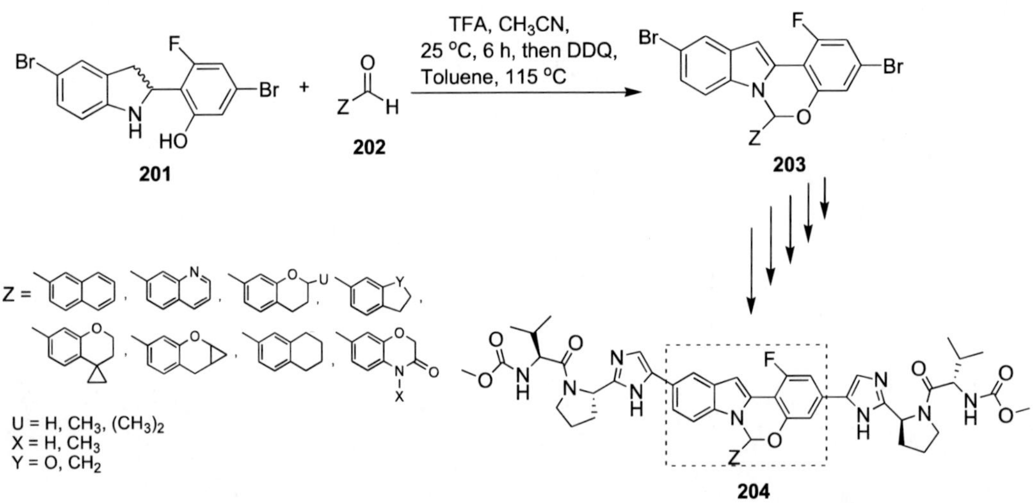

Scheme 77. Synthesis of indole-based chromane containing HCV NS5A inhibitors.

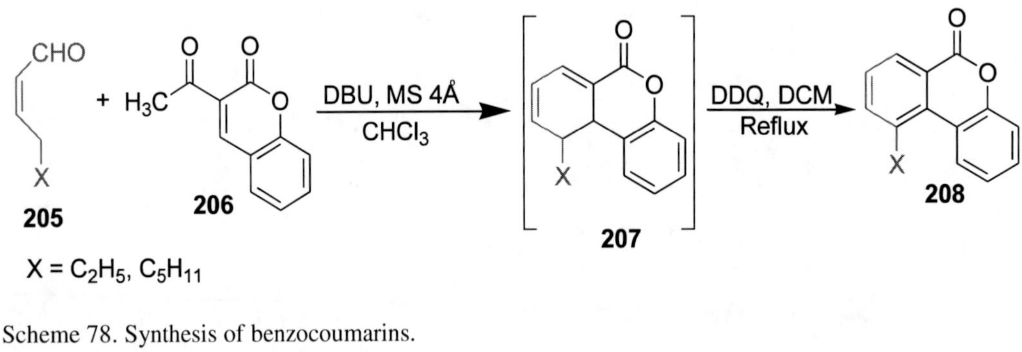

Scheme 78. Synthesis of benzocoumarins.

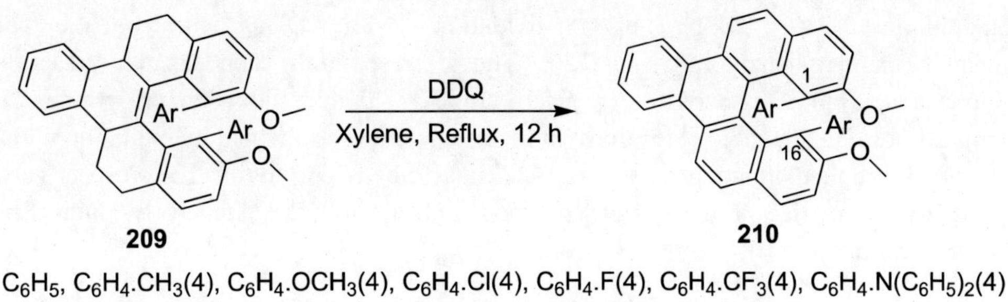

Scheme 79. Synthesis of 1,16-diaryl-substituted benzo[5]helicene derivatives.

2.2.4. DDQ-Promoted Oxidative Transformations

Moody et al. [121] described the usefulness of DDQ as an organo-oxidant, sodium nitrite as co-catalyst and oxygen as the terminal oxidant for the photochemical oxidation of benzylic alcohols **211**, allylic alcohols **213**, and β-O-4 lignin models **215** under visible-light irradiation conditions (Scheme 80). Benzylic alcohols **211** with electron-withdrawing substituents on the aromatic ring were conveniently oxidized under visible light irradiation. The oxidation reactions inferred well even in relatively poor sunlight. For the β-O-4 lignin models **215**, selective solar photochemical oxidation of the secondary benzylic alcohols was observed to generate the corresponding benzylic ketones **216**.

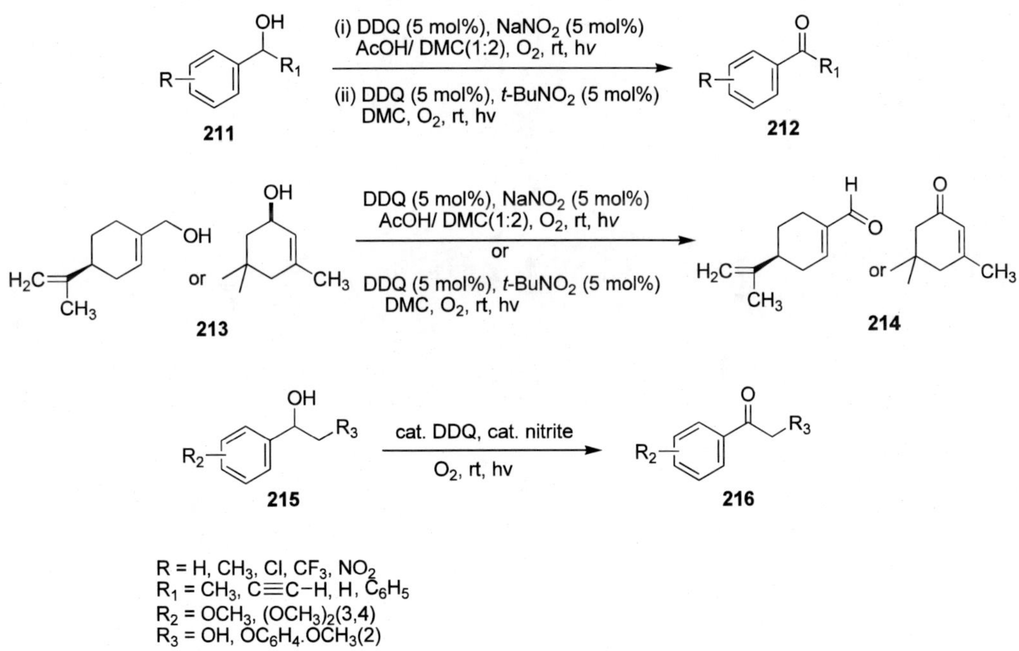

Scheme 80. Synthesis of ketones.

Benziporphyrins and dibenziporphyrins represent the porphyrin analogues in which one or two of the pyrrole rings is replaced with a 1,3-phenylene group. An efficient synthesis of these derivatives was accomplished by Lash et al. [122] utilizing the tendency of DDQ to facilitate the oxidative organic transformations (Scheme 81). The condensation reaction of dimethoxybenzitripyrranes **217** with pyrrole dialdehyde **218** and TFA in chloroform leads to benziphlorins **219**, which upon oxidation with DDQ furnish the desired diphenyldimethoxybenziporphyrins **220**. The corresponding condensation reactions of thiophene dicarbinol **221** and benzene dicarbinol **223** with dimethoxybenzitripyrranes **217** and boron trifluoride etherate in chloroform at room temperature followed by oxidation with DDQ generated tetraphenylthiabenziporphyrins **222** and dibenziporphyrins **224**, respectively. The metallation properties of the synthesized benziporphyrins was extensively studied by this research group.

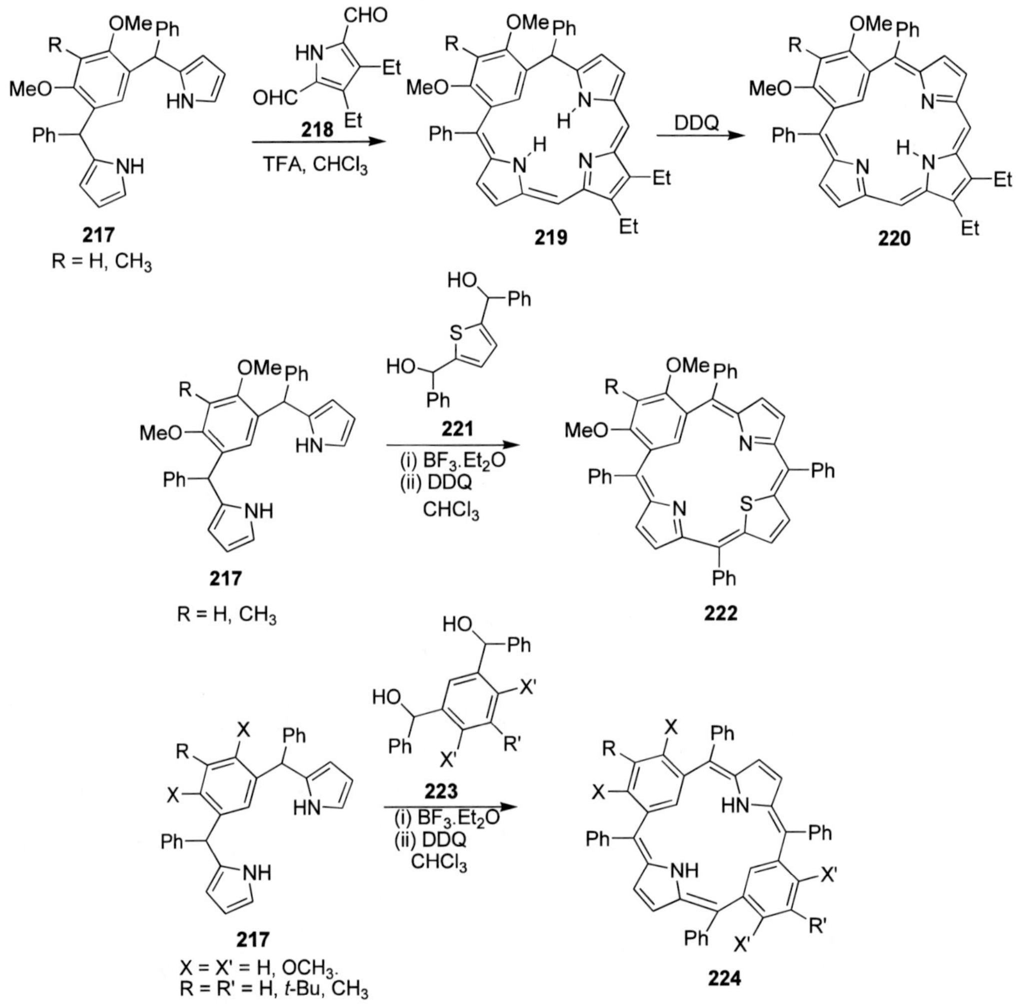

Scheme 81. Synthesis of benziporphyrins and dibenziporphyrins.

In a report by Hilt and co-workers [123], a three-step reaction sequence involving enyne metathesis, regiodivergent Diels-Alder reaction, and oxidation was employed to synthesize bifunctional terphenyl derivatives **228** and **229** (Scheme 82). Initially, bromo-substituted phenylacetylenes **225** underwent cross-metathesis with ethene under ruthenium catalysis to generate an intermediate diene **226**. The cobalt-complex **H/I** catalyzed Diels-Alder reaction of **226** (diene) with aryl-substituted alkynyl boronic esters **227** (dienophile) and subsequent oxidation with DDQ gave the bifunctional terphenyl derivatives **228** and **229**. The regioselectivity of the

Diels-Alder reaction to generate **228** and **229** was controlled by the choice of the ligands on the cobalt catalyst precursor **H/I**. This methodology provides access to architecturally different monomers **228** and **229** from the same starting materials by a regioselective approach. The bi-functional building blocks **228** and **229** were successfully polymerized *via* Suzuki poly-condensation approach to generate poly-phenylenes.

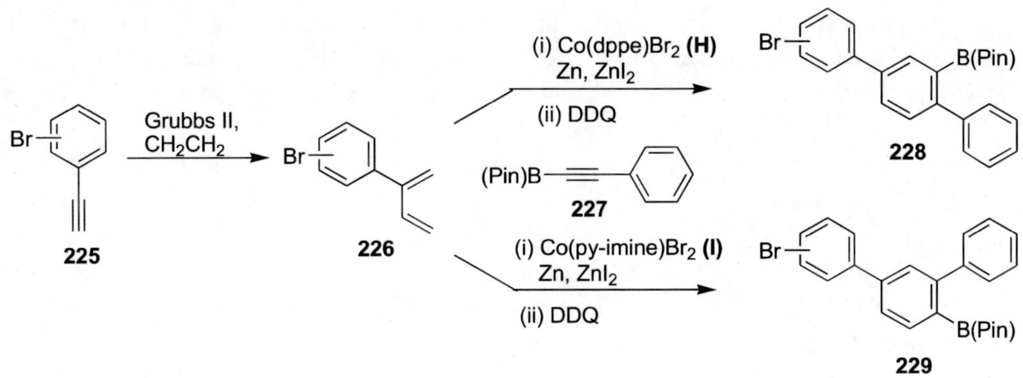

Scheme 82. Synthesis of bi-functional terphenyl derivatives.

Reyns et al. [124] described the multi-residue determination of fourteen dyes (triarylmethanes **230**, xanthenes **231**, phenothiazines **232**, phenoxazines **233** and rhodamines **234**) in fish tissues by using an oxidation protocol (Scheme 83). Accordingly, a liquid-liquid dye extraction with acetonitrile in 2g of fish sample is followed by oxidation with DDQ. DDQ-mediated oxidation serves as an *in situ* transformation step for detection of dyes in their chronic form. Removal of excess DDQ is accomplished by tandem solid phase extraction process. The quantification of dyes in fish tissues was accomplished by combination of ultra performance liquid chromatography and electrospray ionization tandem mass spectrometry analysis. The validity of this method was proved by tracing the presence of synthetic dyes in caught wildlife European eel (Belgium).

An efficient and green approach towards the deprotection of *p*-methoxybenzyl (PMB) ethers **235** (1 M solution in acetic acid) into the corresponding alcohols **236** was devised by Moody et al. (Scheme 84) [125]. The reaction was catalyzed by a combination of DDQ and sodium nitrite with oxygen serving as the terminal oxidant. Selective deprotection of PMB ether in contrast to other orthogonal protecting groups (silyl ether, ethoxy methoxy (EOM), benzyl ether and acetate ester) was ensured by DDQ/$NaNO_2$/O_2 catalytic system.

In a report by Osuka and co-workers [126], DDQ was employed as an oxidant for the condensation reaction of 5,10-bis-(pentafluorophenyl)tripyrrane **237** with 3,5-di-*tert*-butyl-4-hydroxybenzaldehyde **238** in the presence of methanesulfonic acid (MSA) (catalyst) at 0°C (Scheme 85). The condensation product *viz* cross-conjugated quinonoidal hexaphyrin **239** bearing two *meso*-oxacyclohexadienylidenyl (OCH) groups was obtained in 17% yield. In the same report [126], the oxidation reactions of bis-rhodium[28]hexaphyrin complex **240** with DDQ was studied (Scheme 86). Upon careful oxidation of **240** with 1 equiv. of DDQ, the oxidized cross-conjugated complex **241** possessing one OCH and one 3,5-di-*tert*-butyl-4-hydroxyphenyl groups was generated. However, with excess DDQ, the quinonoidal rhodium hexaphyrin complex **242** with two *meso*-OCH groups was formed.

The efficiency of DDQ as an oxidizing agent as well as a chlorinating agent was studied by Wu et al. (Scheme 87) [127]. They carried out the regioselective ligand-directed *ortho*-chlorination of 2-arylpyridines **243** using a palladium salt as a catalyst and DDQ. 2-Arylpyridines **243** possessing a *meta*-substituent in the aryl ring produced mono-chlorinated products **244** regioselectively by chlorination at the less sterically hindered *ortho* C-H position. With *para*-substituted 2-arylpyridines **243**, *ortho*-di-chlorinated products **245** were obtained. Electron-donating substituents on the aromatic ring furnished products with better yields. The scope of the reaction was extended to the synthesis of *ortho*-chlorinated benzo[*h*] quinolines **247** and *ortho*-chlorinated 2-thienylpyridines **249** *via* regioselective *ortho*-chlorination of benzo[*h*]quinoline **246** and 2-thienylpyridines **248** with DDQ/$PdCl_2$ system.

Laha and co-workers [128] utilized the oxidizing ability of DDQ to synthesize phenazines **251** from 10,11-dihydro-5*H*-dibenzo[*b*,*e*][1,4]diazepines **250** in acetonitrile at 90°C (Scheme 88). The tandem C-C and C-N bond cleavage and subsequent new aryl C-N bond formation leads to the oxidative removal of the benzylic methylene group under transition-metal-free conditions transforming diazepines **250** to phenazines **251**. The application of DDQ in the secondary benzylamine oxidation under radical conditions is described.

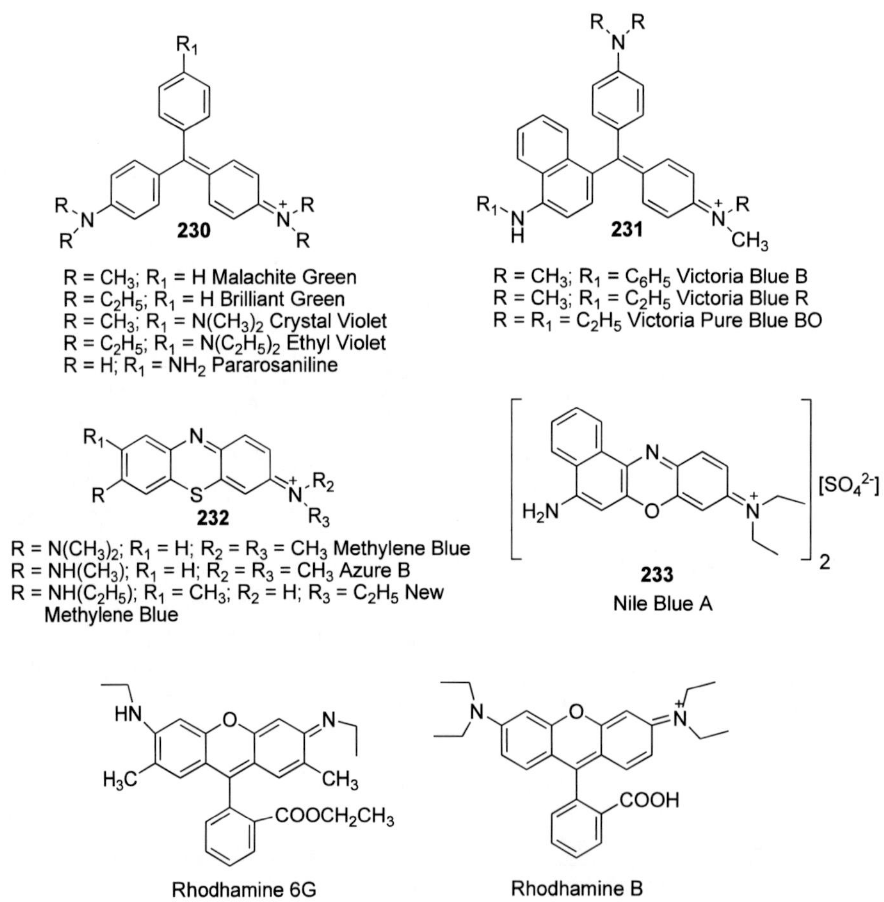

234

Scheme 83. Dyes extracted from fish tissues.

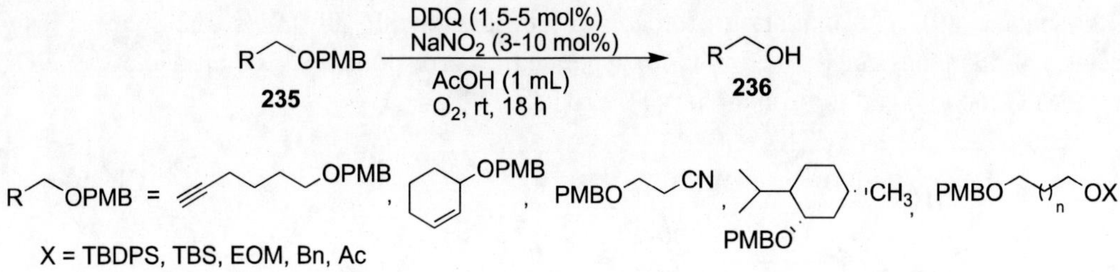

Scheme 84. Synthesis of alcohols.

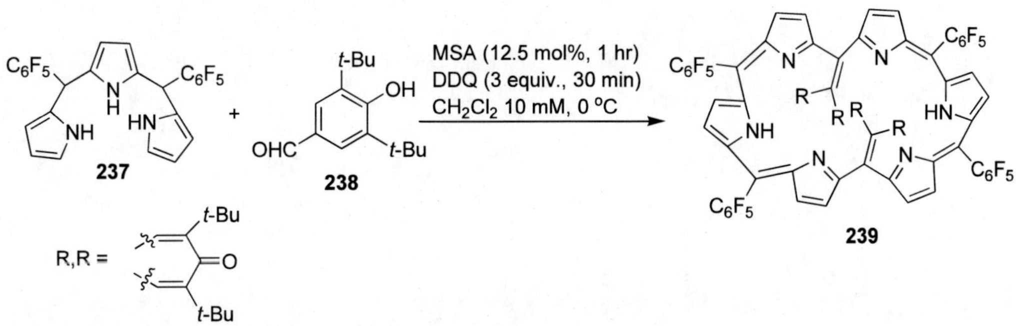

Scheme 85. Synthesis of cross-conjugated quinonoidal hexaphyrins.

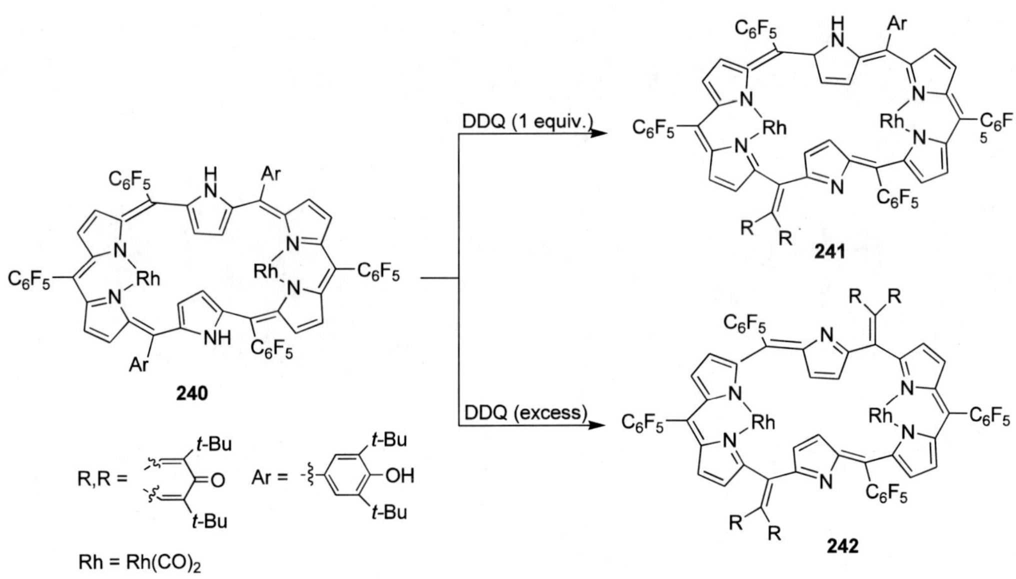

Scheme 86. Synthesis of cross-conjugated and quinonoidal hexaphyrins.

Vicinal tricarbonyl (VTC) compounds have been utilized in the synthesis of various biologically active antibiotics [129], heterocyclic compounds [130], and potent inhibitors of hydrolytic enzymes [131]. These compounds serve as building blocks in synthetic organic chemistry. Deng and co-workers [132] utilized a combination of oxidizing agent DDQ and TEMPO ((2,2,6,6-tetramethylpiperidin-1-yl)oxyl) for the synthesis of vicinal tricarbonyl compounds **253** *via* straightforward oxidation of sp^3 C-H bond of 1,3-dicarbonyl compounds

252 (Scheme 89). The oxidation protocol was efficient with substrates **252** such as aroyl acetates, β-diketones and β-ketoamides to generate the corresponding VTC products **253**. DDQ or TEMPO alone failed to furnish the VTC products.

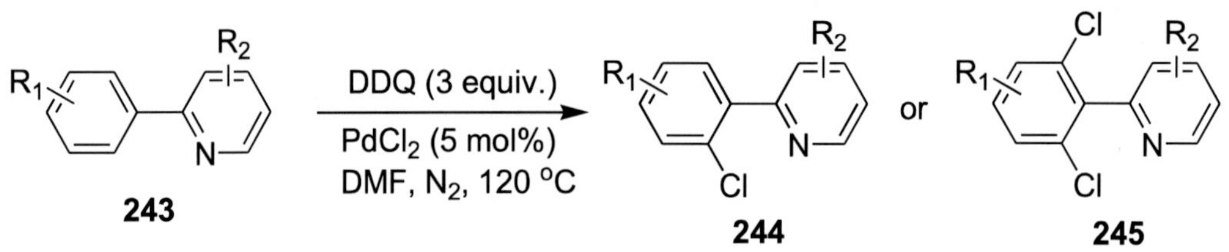

$R_1 = OCH_2C_6H_5(2), O(CH_2)_3CH_3(2), OCH_3(3), CH_3(3), OCH_2CH(CH_3)_2(3), OCH_3(4), C(CH_3)_3(4), CH_3(4), Br(4), F(4), COOCH_2CH_3(4)$
$R_2 = H, CH_3(4)$

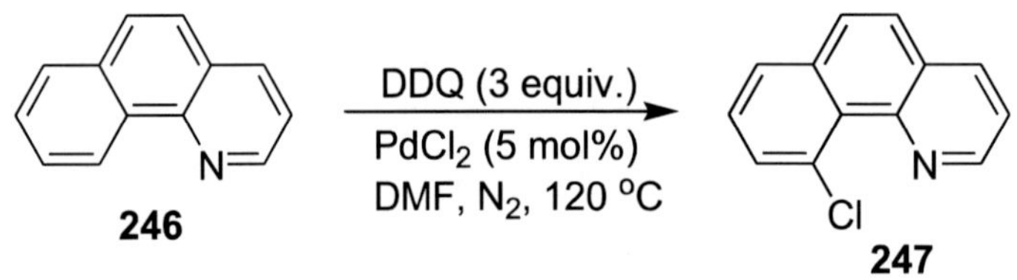

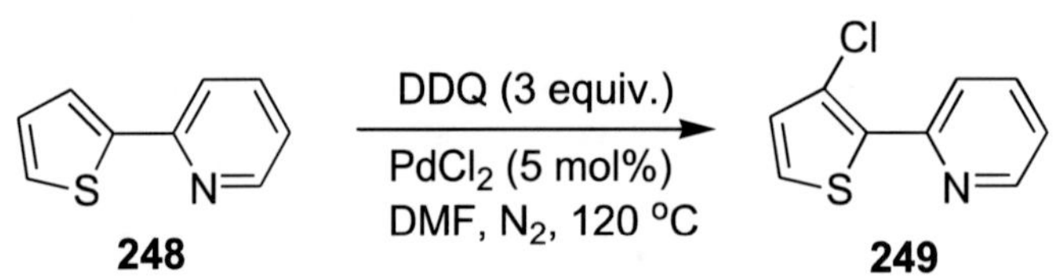

Scheme 87. Synthesis of *ortho*-chlorinated 2-arylpyridines, benzo[*h*]quinolines and 2-thienylpyridines.

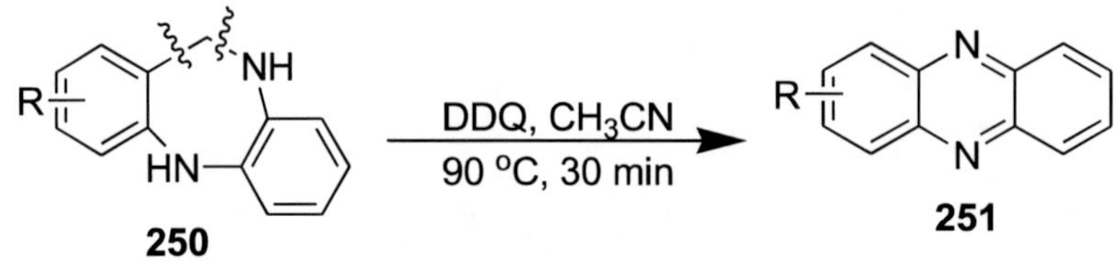

$R = H, COC_6H_5, COOCH_3, Cl, CH_3, CN$

Scheme 88. Synthesis of phenazines.

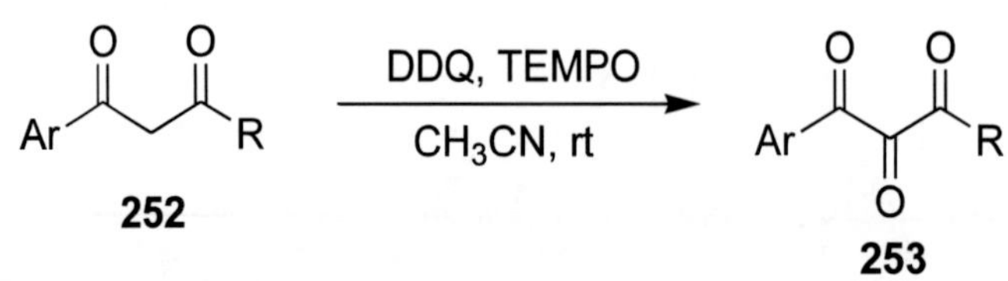

$Ar = C_6H_5, C_6H_4.CH_3(4), C_6H_4.Cl(4), C_6H_4.Br(4)$, 2-Naphthyl, 2-Thienyl, 5-Cl-2-Thienyl
$R = OCH_2CH_3, OCH_3, C_6H_5, N(CH_3)_2, NHCH_2CH_3$

Scheme 89. Synthesis of vicinal tricarbonyl compounds.

The reaction methodology [132] was employed in the synthesis of quinoxaline derivatives **257**, expected to be formed by reaction of *in situ* generated VTC product **255** (obtained by reaction of aliphatic 1,3-dicarbonyl compound **254** with DDQ/TEMPO) with *o*-phenylene diamine **256** (Scheme 90).

For the methylsulfonyl- and nitro-ketone compounds **258**, oxidation with DDQ/TEMPO produced the stable TEMPO-substrate adducts **259** which were transformed to VTC products **260** on treatment with 1.1 equiv. of DDQ (Scheme 91) [132].

Yang et al. [50] employed a DDQ-mediated oxidation-NaF assisted fluorination strategy to synthesize organophosphorus-fluorine compounds (Scheme 92). The reaction of diaryl/ dialkyl phosphine oxides **261** with DDQ (oxidant/hydrogen acceptor), sodium fluoride (fluoride source) and CuBr (catalyst/Lewis acid) in DMF proceeds smoothly at room temperature to give phosphoric fluorides **262** in good to excellent yields. The reaction of **261** with NaF proceeded with either electron-donating or electron-withdrawing groups.

The oxidation of saturated 3β-hydroxysteroids **263** to the corresponding 3-ketosteroids **264** using DDQ in combination with TEMPO in DCM was developed by Shen et al. (Scheme 93) [133].

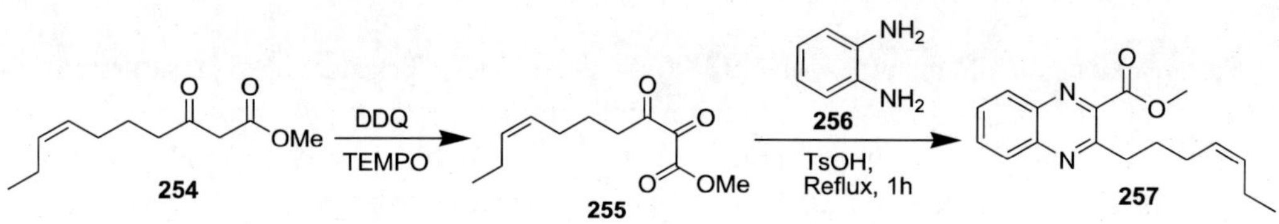

Scheme 90. Synthesis of quinoxalines.

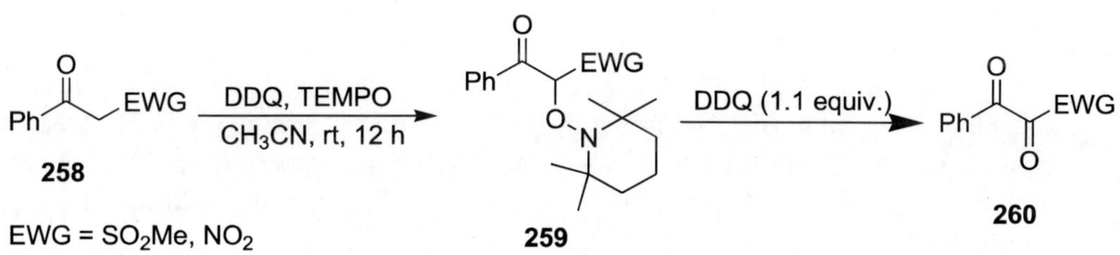

Scheme 91. Synthesis of vicinal tricarbonyl compounds.

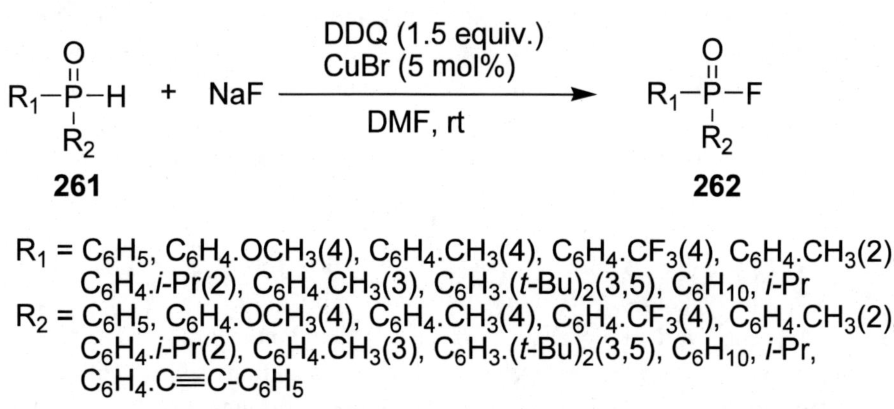

Scheme 92. Synthesis of phosphoric fluorides.

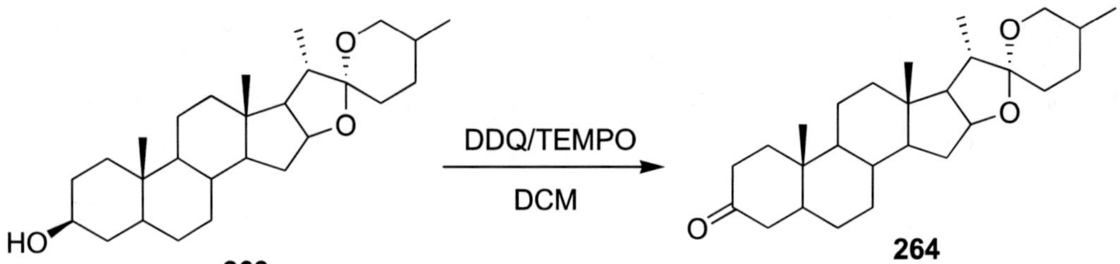

Scheme 93. Synthesis of 3-ketosteroids.

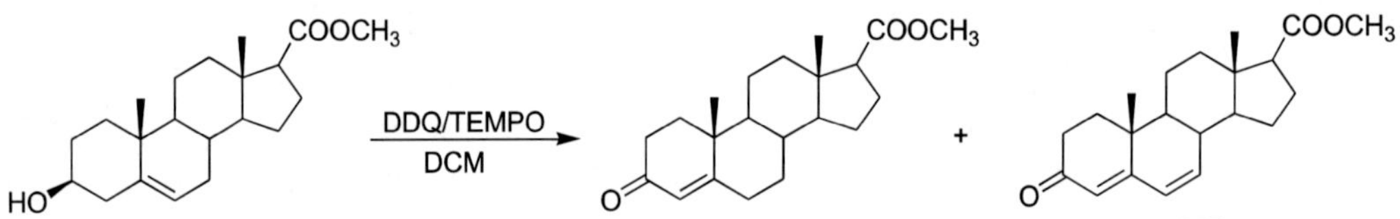

Scheme 94. Synthesis of 4-en-3-one steroids and 4,6-diene-3-one steroids.

As an extension of the protocol [133] for unsaturated substrates *viz* 5-unsaturated 3β-hydroxysteroids **265**, DDQ/TEMPO oxidation major products were 4-en-3-one derivatives **266** (with 1.2 equiv. DDQ) or 4,6-diene-3-one derivatives **267** (with 2.2 equiv. DDQ) (Scheme 94). No 1,4-diene-3-one steroidal products were detected in the oxidation process.

Benzimidizole derivatives have been used as anti-inflammatory [134], antifungal [135] agents and angiotensin II inhibitors [136]. Naeimi and co-workers [137] developed a mild synthesis of 2-substituted benzimidizole derivatives **269** by reaction of *o*-phenylenediamine **256** with aromatic aldehydes **268** using DDQ as an oxidant in acetonitrile at room temperature (Scheme 95). Aromatic aldehydes **268** with electron-withdrawing groups furnished best results.

The oxidizing capability of DDQ in the methylene extrusion reactions was explored by Lacour and co-workers [138] for the synthesis of highly configurationally stable Troger Bases (Scheme 96). The enantiospecific transformation of ethano-Troger Bases **270** through CH_2 extrusion was carried out under mild metal-free conditions employing DDQ as an oxidant in wet nitromethane (1.3% H_2O). The presence of an electron-withdrawing group ($COOR_1$) on the ethano-bridge was determinant for the positive methylene extrusion. Enantiospecificities of 98% or higher were observed suggesting an efficient chirality transfer in the ring contraction process. The configuration of **271** was established by vibrational circular dichroism (VCD) as ($5S_N$,$11S_N$) indicating complete retention of configuration after methylene extrusion. The extruded CH_2 group was converted to formaldehyde.

A direct aerobic oxidation of diarylmethane sp^3 C-H bonds of **272** to synthesize diarylketones **273** was reported by Shen et al. (Scheme 97) [139]. The treatment of diarylmethanes **272** with DDQ/*tert*-butyl nitrite (TBN)/O_2 catalytic systems was carried out in 1,1,2,2-tetrachloroethane (TeCA) in the presence of AcOH at 130°C. Herein, DDQ serves as the catalyst, molecular oxygen as the terminal oxidant and TBN as the NO equivalent to activate the molecular oxygen. The carbonylation products *viz* diarylketones **273** were produced in excellent yields with substrates **272** bearing electron-donating substituents. However, comparatively lower yields resulted from substrates with electron-withdrawing substituents. In addition, the heterocyclic substrates underwent a transformation to the corresponding carbonylation products as well.

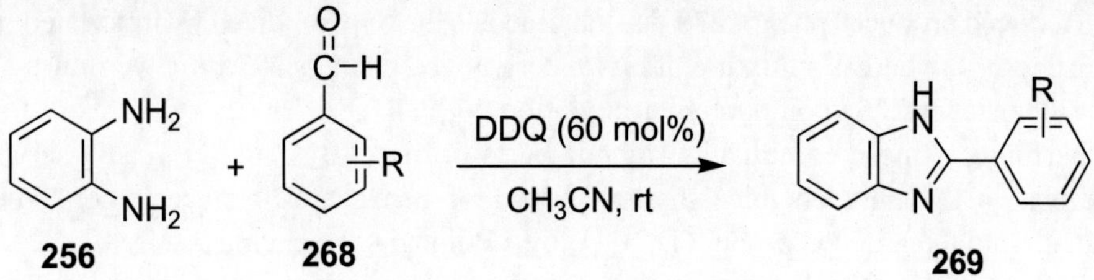

R = CH_3(4), OCH_3(3), $(OCH_3)_2$(3,4), OH(4), $(OH)_2$(2,3), Cl(2), Cl(4), $(Cl)_2$(2,4), NO_2(2), NO_2(4), NO_2(3).Cl(4)

Scheme 95. Synthesis of 2-substituted benzimidazole derivatives.

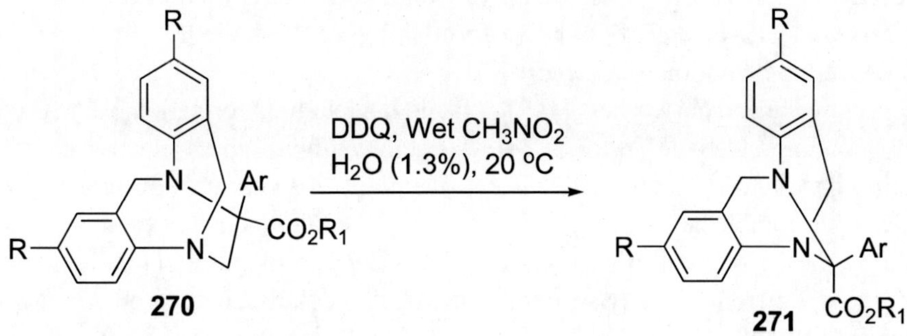

R = CH_3, H
R_1 = CH_3, Bn, CH_2=CH-CH_2-
Ar = C_6H_5, $C_6H_4.CH_3$(4), C_6H_4.Br(4) $C_6H_4.CF_3$(4), $C_6H_4.OCH_3$(4), Napthyl

Scheme 96. Synthesis of Troger Bases.

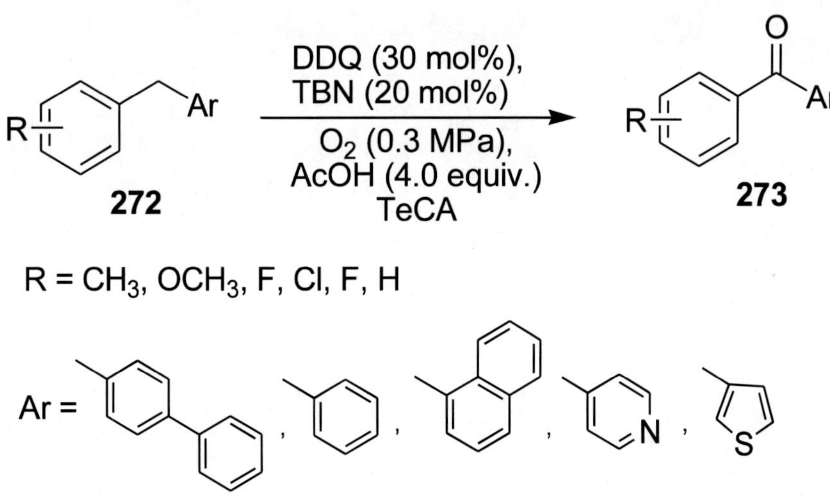

R = CH_3, OCH_3, F, Cl, F, H

Scheme 97. Synthesis of diarylketones.

In 2015, Fukuzumi et al. [140] carried out the photo-induced solvent-free hydroxylation of deareated benzene **274** and derivatized benzenes **276** employing DDQ as a super photo-oxidant in the presence of water under visible-light irradiation conditions (Scheme 98). In addition to $DDQH_2$ (reduced product), phenol **275** was obtained as the only oxidized hydroxylated product from benzene **274**, whereas a mixture of *o*/*m*/*p* derivatized phenols **277a/b/c** were furnished by derivatized benzenes **276** upon photohydroxylation with DDQ.

Furthermore, using a combination of *tert*-butyl nitrite (TBN) and O_2, DDQ served as a photocatalyst in the photochemical hydroxylation of benzene **274** since it could be back regenerated from the reduced product $DDQH_2$ by TBN and O_2 (Scheme 99) [140].

4,4-Difluoro-4-bora-3a,4a-diaza-*s*-indacene dyes (also called borondipyrromethenes or BODIPYs) with high extinction coefficients, photochemical and thermal stabilities [141] are useful in drug delivery [142], protein and DNA-labelling [143], medical imaging and theranostics [144]. Particularly, formyl-BODIPYs have been recognized as fluorescent sensors [145]. Smith et al. [146] have made efforts towards regioselective synthesis of 5-formyl-3,8-diphenyl-6-ethyl-1,2,7-trimethyl-BODIPY **279** from 3,8-diphenyl-6-ethyl-1,2,5,7-tetramethyl-BODIPY **278** *via* DDQ-mediated oxidative formylation of the 5-methyl substituent of **278** in toluene under reflux conditions (Scheme 100).

In a report by Hu and co-workers [147], a facile tandem transformation of PMB ethers **280** to the corresponding aldehydes/ketones **281** was achieved with good selectivity under metal-free oxidation conditions (Scheme 101). The sequential PMB deprotection/selective oxidation process was accomplished by employing a three-component catalyst system of DDQ/TEMPO/TBN along with molecular oxygen serving as the terminal oxidant. A range of PMB ethers derived from primary/secondary benzylic alcohols, heteroaromatic alcohols and primary/secondary aliphatic alcohols underwent a convenient transformation to the products with excellent functional group compatibility.

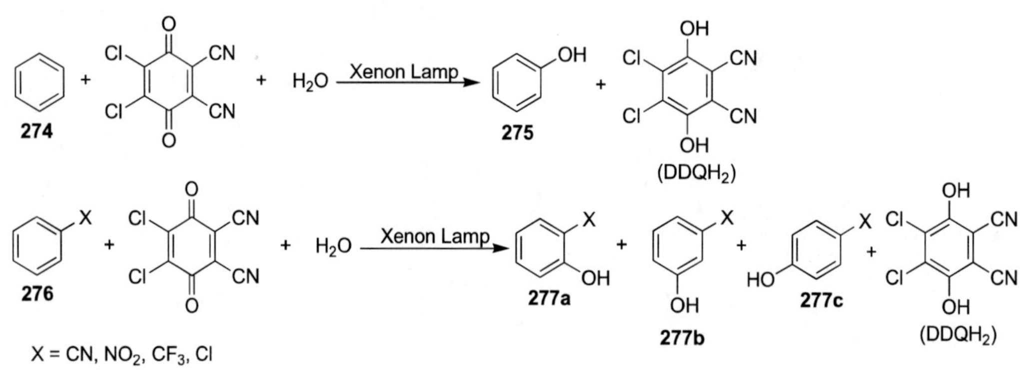

Scheme 98. Synthesis of hydroxylated benzenes/phenols.

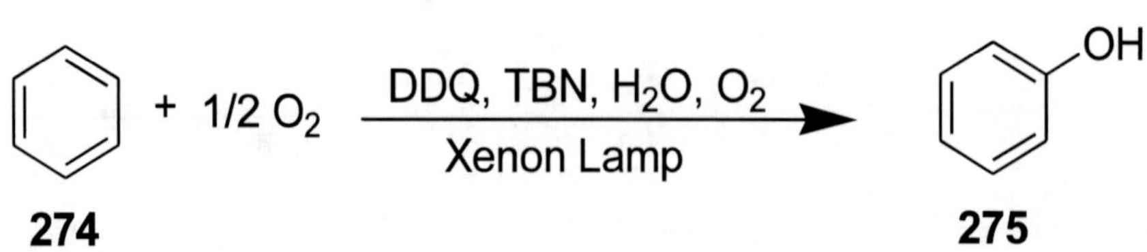

Scheme 99. Synthesis of phenol.

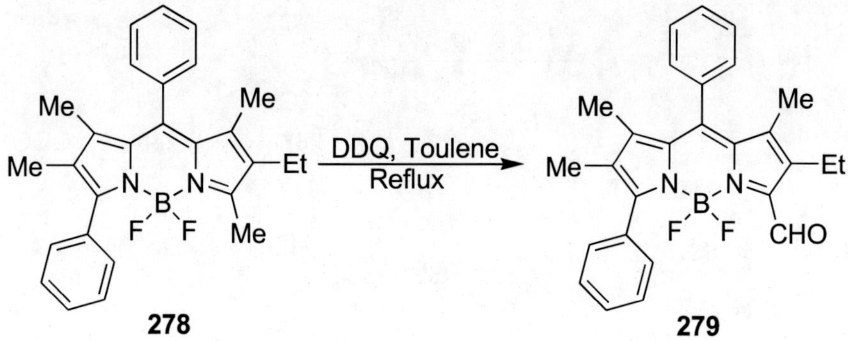

Scheme 100. Synthesis of 5-formyl-3,8-diphenyl-6-ethyl-1,2,7-trimethyl-BODIPY.

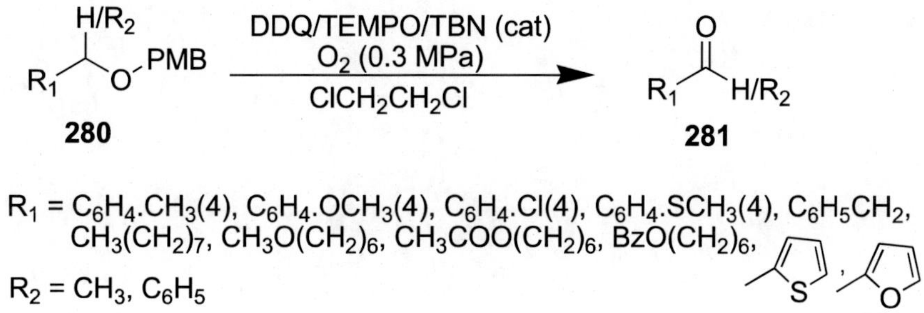

Scheme 101. Synthesis of aldehydes/ketones.

Imahori and co-workers [148] employed DDQ as an effective oxidant for the synthesis of phosphole-fused porphyrin dimer **284** that exhibits excellent optoelectronic properties (Scheme 102). The methanesulfonic acid (MSA) catalyzed condensation reaction of bis(pyrrolo)phosphole **282** and tripyrranecarbinol **283** in 1,2-dichloroethane for 1 hour at 60°C and subsequent oxidation with DDQ at rt for 3 hours furnished the novel phosphorus containing porphyrins **284**. These porphyrin dimers with high electron-accepting ability have the potential to serve as n-type semiconductors.

Chai et al. [149] have worked upon the oxidation of four catecholamines *viz N*-acetyl dopamine (NADA) **285**, *N*-acetyl 3,4-dihydroxy-phenylalanine methyl ester (NADOPAME) **286** and their dehydroderivatives NAΔDA **289** and NAΔDOPAME **290** to the corresponding oxidized quinones **287/288** and **291/292** employing DDQ as an oxidizing agent in THF at -40°C (Scheme 103). The regioselectivities and reactivities of quinones **287/288/291/292** towards nucleophilic trapping with ethanethiol was studied for comparison.

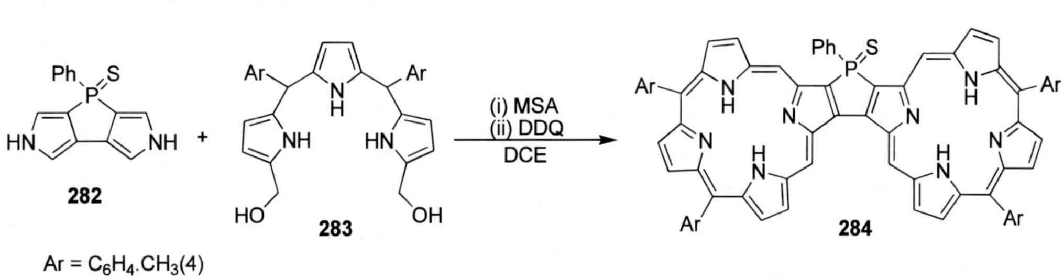

Scheme 102. Synthesis of phosphole-fused porphyrin dimer.

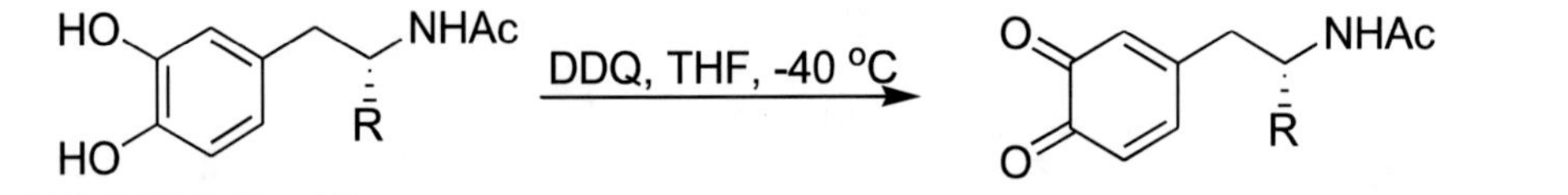

NADA **285** (R = H)
NADOPAME **286** (R = COOMe)

287 (R = H)/**288** (R = COOMe)

NA△DA **289** (R = H)
NA△DOPAME **290** (R = COOMe)

291 (R = H)/**292** (R = COOMe)

Scheme 103. Synthesis of quinones.

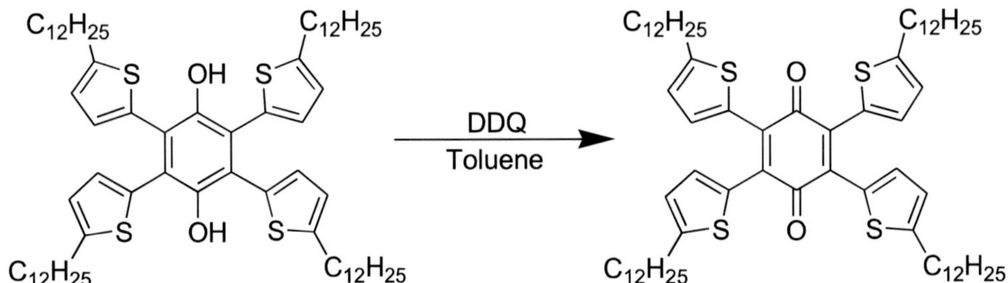

293 **294**

Scheme 104. Synthesis of derivatized benzoquinones.

In 2016, Xu and co-workers [150] reported the synthesis of 2,3,5,6-tetrakis(5'-dodecylthiophenyl-2-yl)benzoquinone **294** from its reduced hydroxy derivative **293** by an oxidation reaction with DDQ in toluene (Scheme 104).

Vanderwal et al. [151] reported a two-step sequence involving Cs-F mediated deprotection of bis(silylether) **295** and subsequent DDQ-mediated selective oxidation to synthesize the corresponding ketone **296** (Scheme 105). A series of transformations on ketone **296** led to exiguaquinol dessulfate **297** (desulfated analogue of natural product exiguaquinol which inhibits *H. pylori* Murl [152]).

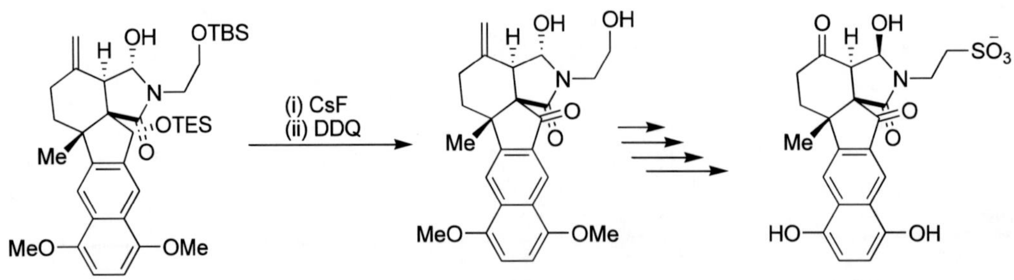

295 **296** **297**

Scheme 105. Synthesis of bis ketone.

Non-aromatic Siamese twin porphyrins **298** (with 28 Π electrons) incorporate two porphyrin-like binding pockets (each bearing 14 Π electrons) separated by pyrazole moiety. Meyer and co-workers [153] attempted to aromatize **298** by oxidation with 1 equiv. of DDQ in dichloromethane (Scheme 106). However, no aromatization was achieved, but rather an oxidatively-fused product **299** bearing fused N^{pz}-$C^{o\text{-}Ph}$ linkage was obtained by two-hydrogen atom oxidation process. Consecutively, upon oxidation of **298** with 2 equiv. of DDQ, four-hydrogen atom oxidation product **300** was generated with two successively fused N^{pz}-$C^{o\text{-}Ph}$ linkages. Product **300** was also accessed by oxidizing **299** with another equiv. of DDQ. The non-aromatized products **299**, **300** incorporate pyrazolo[1,5-*a*]indole sub-units within the macrocyclic framework accessed by the fusion of pyrazole *N*H and neighboring inactivated *meso*-phenyl-CH groups. These oxidatively-fused products **299**, **300** had more twisted and folded over-topologies in comparison to the substrates as indicated by the X-ray diffraction analysis.

Hayashi et al. [92] employed DDQ as an oxidant in the deprotection of PMB ethers **301**, **303** to form the corresponding alcohols **302**, **304** (Scheme 107). For substrate **301**, the selective deprotection of the PMB ether of secondary alcohol was observed with DDQ in contrast to primary alcohol PMB ether. This was ascribed to greater steric crowding around the primary PMB ether moiety. In addition, the driving force is the relief of steric hindrance on the PMB ether from the secondary alcohol.

Ravikanth and co-workers [154] synthesized mono-*meso*-thienyl/pyrrolyl/furyl substituted oxasmaragdyrins **307/309/311** by an oxidative approach (Scheme 108). The [3+2] TFA-catalyzed oxidative coupling reaction of 16-oxatripyrrane **305** (1 equiv.) with *meso*-thienyl/pyrrolyl/furyl dipyrromethanes **306/308/310** (1 equiv.) in DCM under an inert atmosphere is followed by an oxidation reaction with DDQ (3 equiv.) in open air to furnish the products **307/309/311** in 20 - 25% yields.

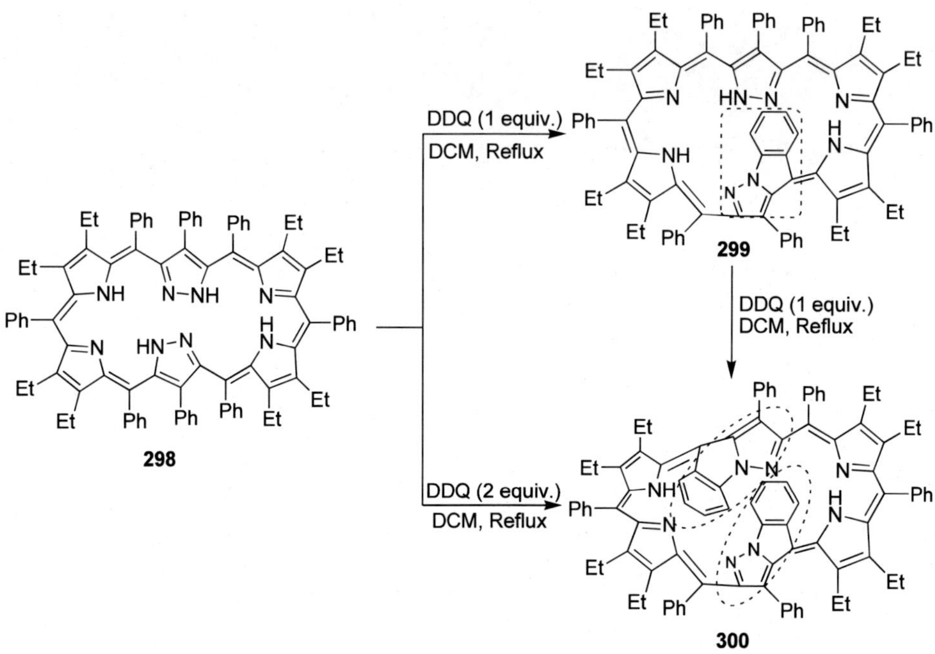

Scheme 106. Synthesis of non-aromatic porphyrins.

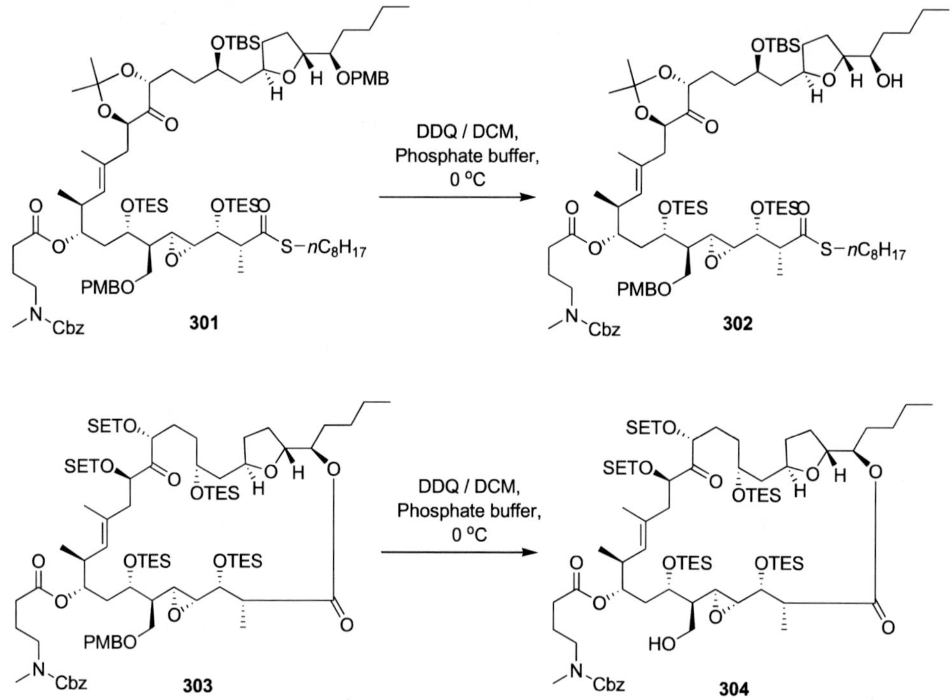

Scheme 107. Synthesis of alcohols from PMB ethers.

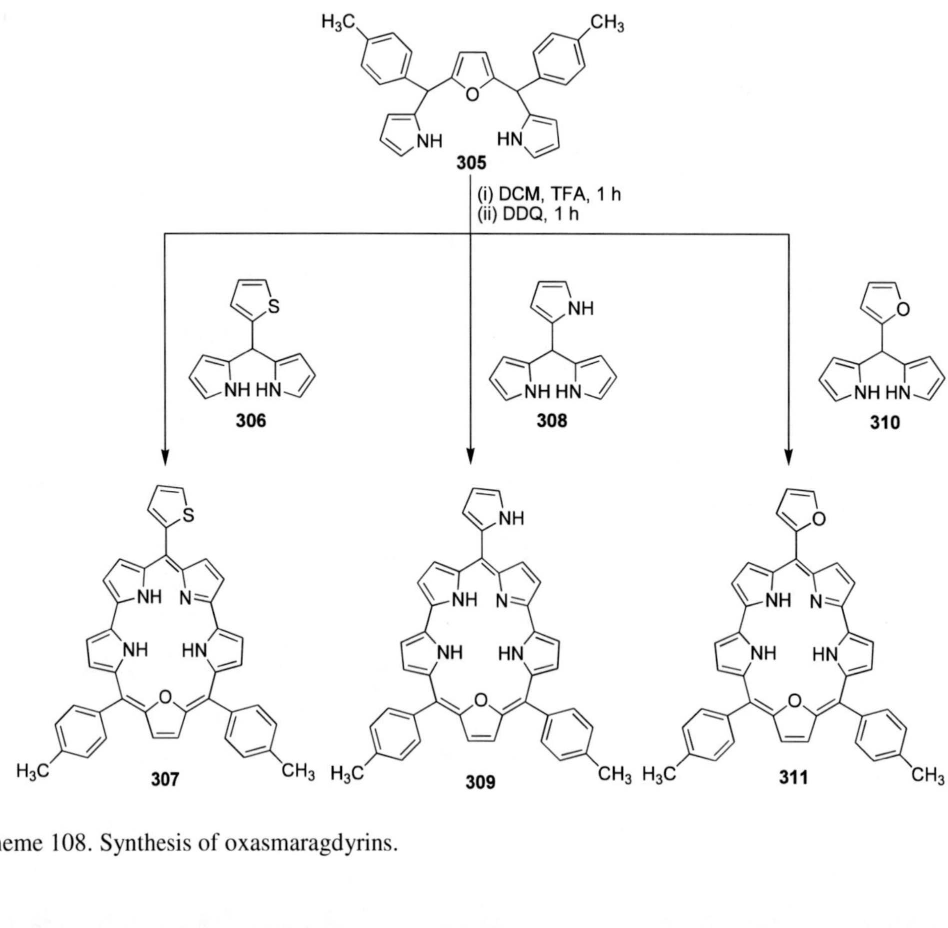

Scheme 108. Synthesis of oxasmaragdyrins.

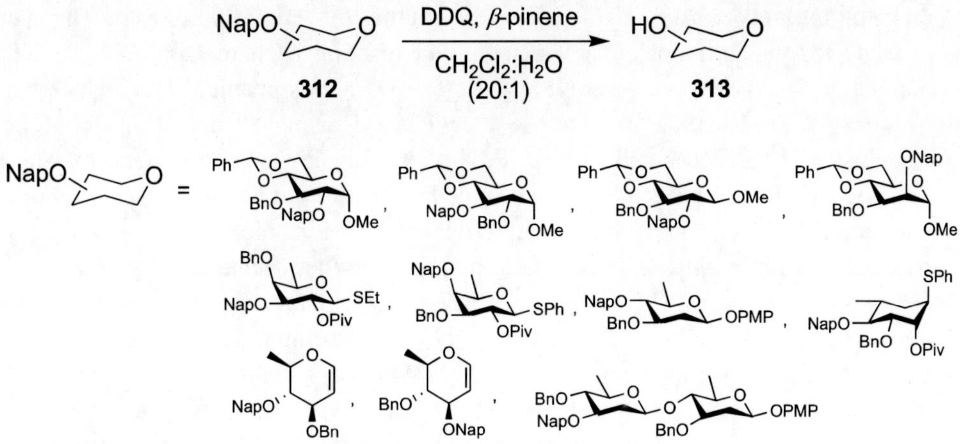

Scheme 109. Synthesis of Nap deprotected products.

The group of Bennett and co-workers [155] utilized a combination of DDQ (oxidant) and a strained alkene *β*-pinene (additive) for the removal of 2-naphthylmethyl (Nap) ether protecting group in methylene chloride: water (20:1) solvent system (Scheme 109). A series of substrates **312** *viz.* disaccharides, benzylidene acetals, thioglycosides of fucose and rhamnose, glycal structures with Nap ether protecting group were transformed to the deprotected products in 68 - 96% yields. Selective Nap ether deprotection was observed in the presence of other protecting groups (PMP, Me, Bn) on the substrate **312**. The use of tri-*tert*-butylpyrimidine (TTBP), 2,6-di-*tert*-butylpyridine (DTBP), or 2,6-di-*tert*-butyl-4-methylpyridine (DTBMP) as additives led to lower yields of the deprotected products.

3. Biological Evaluation Reports

This section of the chapter highlights the applications of DDQ-mediated oxidative organic transformations in the synthesis of biologically active compounds of medicinal interest.

The group of Liu and co-workers [42] gained success in synthesizing a series of isochroman derivatives **7** by DDQ-mediated CDC protocol (Scheme 7). Three of the randomly selected isochroman-derived cross-coupled products **7a-c** were efficient in inhibiting human prostate cancer (PC3) cells (IC_{50} = 18.1-29.7 μM) (Figure 5). *cis*-Platin was used as the positive control.

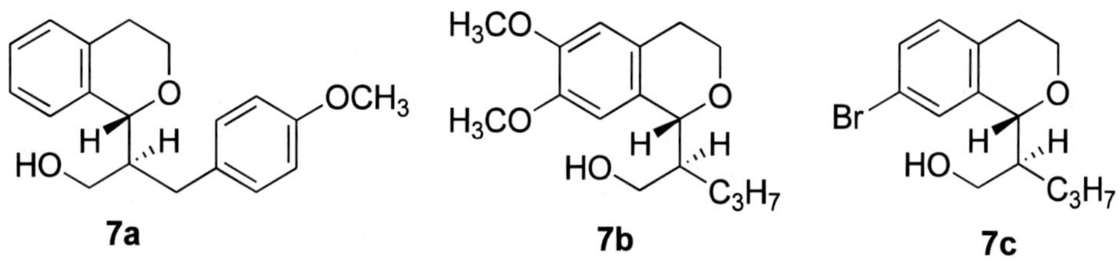

Figure 5. Biologically active α-Substituted isochromans.

A range of diversely substituted olean-28,13*β*-lactams **93** were synthesized by Zhang et al. utilizing DDQ for the intramolecular C-N coupling reaction (Scheme 37) [73]. Biological evaluation studies showed these olean-28,13*β*-lactams **93** act as potential antiprostate cancer agents (Figure 6). Product **93a** in which R = cyclohexyl displayed potent antiproliferative activity against human cancer cells while 13.84-16.92-fold less inhibitory activity on non-cancer cells *in vitro*. **93a** induces prostate cancer cell cycle arrest at G0/G1 phase and apoptosis *in vitro*. It had little hERG channel inhibitory activity and exhibits high stability in human liver microsomes and rat plasma than CDDO-Me. It had relatively less acute toxicity in mice.

Woerpel and co-workers [81] synthesized 10,12-peroxycalamenene **114** by DDQ-mediated intramolecular cyclization process (Scheme 45). **114** is a sesquiterpene endoperoxide natural product (Figure 7) derived from the roots of nut-grass (*Cyperus rotundus*) and has been reported as an active component in traditional medicines [156]. Blood disorder and leprosy were traditionally treated with the extracts of this plant [157].

Exemestane **166** was synthesized by Giovenzana et al. by DDQ-mediated oxidative dehydrogenative process (Scheme 67) [101]. The clinical efficiency of exemestane (6-methyleneandrosta-1,4-diene-3,17-dione) **166** (Figure 8) for the treatment of breast cancer in postmenopausal women has been reported previously, wherein it acts as an irreversible aromatase inhibitor [158].

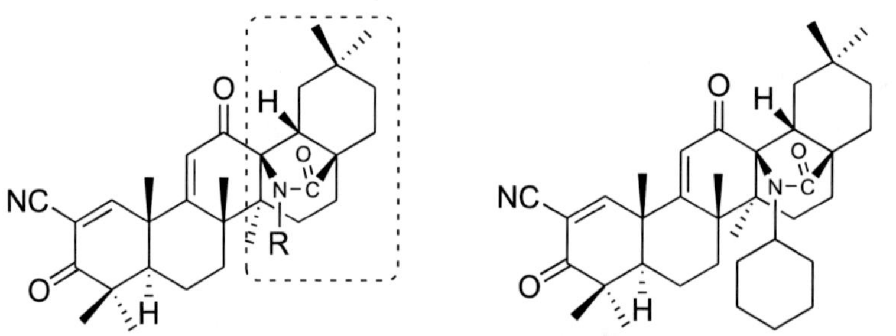

93 **93a**

R = CH_3, CH_2CH_3, $CH_2CH_2CH_3$, $CH_2(CH_2)_2CH_3$, $CH_2(CH_2)_4CH_3$, $C(CH_3)_3$, CH_2CH_2OH,

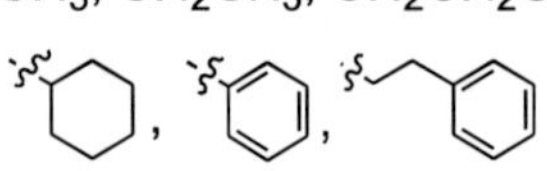

Figure 6. Biologically active olean-28,13*β*-lactams.

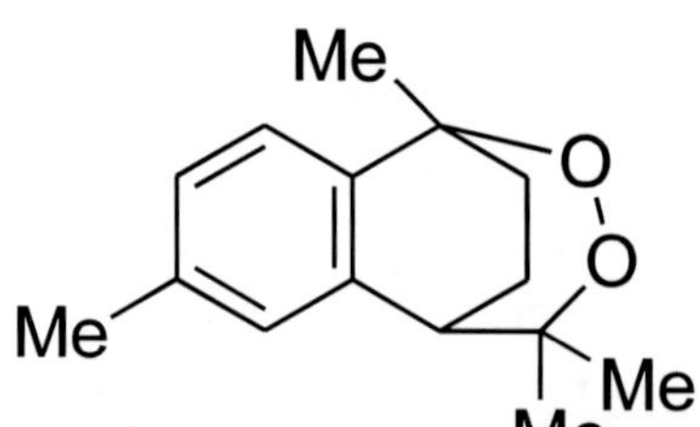

114

Figure 7. Medicinally important 10,12-peroxycalamenene.

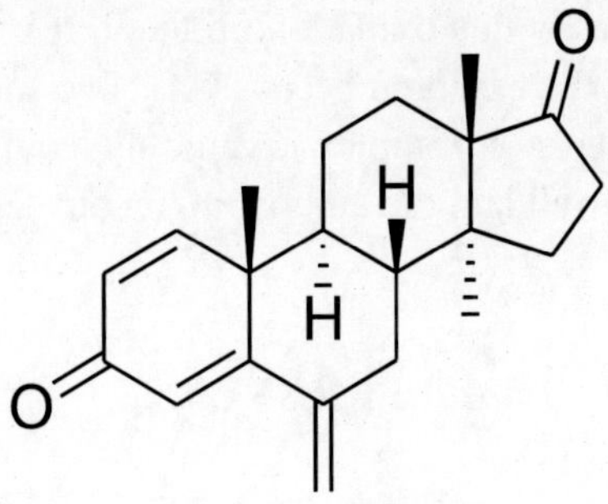

166

Figure 8. Exemestane: An aromatase inhibitor.

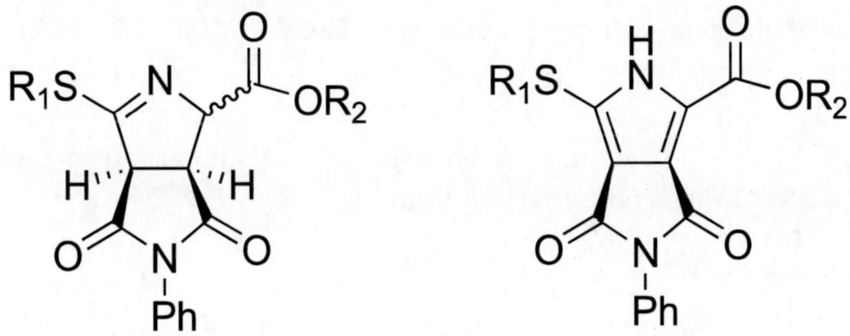

177 **178**

$R_1 = CH_3, CH_2CH_3, CH_3CH_2CH_2$, *i*-Pr, Bu, *i*-Bu, Bn
$R_2 = CH_3, CH_2CH_3$, *i*-Pr, *t*-Bu, Bn

Figure 9. Biologically active dihydropyrroles and polysubstituted pyrroles.

The DDQ-mediated oxidative aromatization strategy was adopted by Oikonomou et al. [107] for the synthesis of polysubstituted pyrroles **178** from dihydropyrroles **177** (Scheme 70). The potential of these compounds as antioxidants and as inhibitors of soybean lipoxygenase (LOX) was studied by this group (Figure 9). Nordihydroguaiaretic acid (NDGA) and Trolox were used as the standard for the evaluation of antioxidant activity. Both **177** and **178** had lower anti-lipid peroxidation activity than Trolox. But, pyrroles **178** with R_1 = Pr, Bu and R_2 = Me exhibited significantly high antioxidant activities. Also, polysubstituted pyrroles **178** were found to be more potent lipoxyginase inhibitors than dihydro derivatives **177** due to the presence of pyrrole group in them.

CONCLUSION

This chapter summarizes the recent advances made towards the utilization of DDQ for carrying out oxidative organic transformations *viz* cross dehydrogenative couplings, cyclizations/annulations, aromatizations and oxidations. Diverse types of heterocyclic frameworks have been accessed by this oxidation protocol. It is expected that DDQ-mediated transformations will continue to be explored because of its versatility, molecular economy and exciting potential for the synthesis of complex organic molecules. In a chapter of this scope, some omissions may have occurred and the authors regret any such oversights.

REFERENCES

[1] Backvall, J. E. Ed. (2004). *Modern Oxidation Methods*, VCH-Wiley, Weinheim.

[2] Thomson, R. H. (1997). *Naturally Occurring Quinones IV. Recent Advances*, 4th Edition, Blackie: London, UK pp. 746.

[3] Morton, R. A. (1965). *Biochemistry of Quinones*, Academic Press: New York, pp. 585.

[4] Rao, K. V. & Cullen, W. P. (1959-1960). "Streptonigrin, an antitumor substance. I. Isolation and characterization" *Antibiotics Annual*, *7*, 950–953. URL:https://www.ncbi.nlm.nih.gov/labs/articles/14436228/ .

[5] Sansom, C. E., Larsen, L., Perry, N. B., Berridge, M. V., Chia, E. W., Harper, J. L. & Webb, V. L. (2007). "An Antiproliferative Bis-prenylated Quinone from the New Zealand Brown Alga *Perithalia capillaris*" *Journal of Natural Products*, *70*, 2042–2044. doi: 10.1021/np070436t.

[6] Guntern, A., Ioset, J. R., Queiroz, E. F., Foggin, C. M. & Hostettmann, K. (2001). "Quinones from *Heliotropium ovalifolium*" *Phytochemistry*, *58*, 631–635. URL: http://doi.org/10.1016/S0031-9422(01)00302-8.

[7] Tansuwan, S., Pronpakakul, S., Roengsumran, S., Petsom, A., Muangsin, N., Sihanonta, P. & Chaichit, N. (2007). "Antimalarial Benzoquinones from an Endophytic Fungus, *Xylaria* sp." *Journal of Natural Products*, *70*, 1620–1623. doi: 10.1021/np0701069.

[8] Nassar, M. I., Abdel-Razik, A. F., El-Din, E., El-Khrisy, A. M., Dawidar, A. A. M., Bystorm, A. & Mabry, T. J. (2002). "A benzoquinone and flavonoids from *Cyperus alopecuroides*" *Phytochemistry*, *60*, 385–387. URL: http://doi.org/10.1016/S0031-9422(02)00060-2.

[9] Lin, A. J., Lillis, B. J. & Sartorelli, A. C. (1975). "Potential bioreductive alkylating agents. Antineoplastic activity of quinolone-5,8-diones, naphthazarins, and naphthoquinones" *Journal of Medicinal Chemistry*, *18*, 917–921. doi: 10.1021/ jm00243a012.

[10] Mozaina, K., Cantrell, C. L., Mims, A. B., Lax, A. R., Tellez, M. R. & Osbrink, W. L. A. (2008). "Activity of 1,4-Benzoquinones Against Formosan Subterranean Termites (*Coptotermes formosanus*)" *Journal of Agriculture and Food Chemistry*, *56*, 4021–4026. doi: 10.1021/jf800331r.

[11] Dowd, P. & Zheng, Z. B. (1995). "On the mechanism of the anticlotting action of vitamin E quinone" *Proceedings of the National Academy of Sciences of the United States of America*, *92*, 8171–8175. URL:http://www.pnas.org/content/92/ 18/8171.full.pdf.

[12] Bernays, E., Lupi, A., Bettolo, R. M., Matrofrancesco, C. & Tagaliatesta, P. (1984). "Antifeedant nature of the quinone primin and its quinol miconidin from *Miconia* spp." *Experientia*, *40*, 1010–1011. doi: 10.1007/BF01946484.

[13] Gonzalez-Ibarra, M., Farfan, N., Trejo, C., Uribe, S. & Lotina-Hennsen, B. (2005). "Selective Herbicide Activity of 2,5-Di(benzylamine)-*p*-benzoquinone against the

Monocot Weed *Echinochloa crusgalli*. An *in Vivo* Analysis of Photosynthesis and Growth" *Journal of Agriculture and Food Chemistry, 53*, 3415–3420. doi:10.1021/jf047883o.

[14]Fu, P. P. & Harvey, R. G. (1978). "Dehydrogenation of polycyclic hydroaromatic compounds" *Chemical Reviews, 78*, 317–361. doi: 10.1021/cr60314a001.

[15]Li, C. J. (2009). "Cross-Dehydrogenative Coupling (CDC): Exploring C-C Bond Formations beyond Functional Group Transformations" *Accounts of Chemical Research, 42*, 335–344. doi: 10.1021/ar800164n.

[16]Walker, D. & Hiebert, J. D. (1967). "2,3-Dichloro-5,6-dicyanobenzoquinone and its reactions" *Chemical Reviews, 67*, 153–195. doi: 10.1021/cr60246a002.

[17]Kun, Y., Xin-yi, Y., Dong-ping, C. & Ji-zhong, Y. (2014). "DDQ in the application of oxidative coupling reaction" *Zhejiang Huagong, 45*, 21–28.

[18]Kiyoshi, T., Yoko, N. & Tsuneo, S. (2011). "2,3-Dichloro-5,6-dicyano-*p*-benzoquinone (DDQ) as the useful synthetic reagent" *Nippon Shika Daigaku Kiyo, Ippan Kyoiku-kei., 40*, 31–34.

[19]Liqun, S., Fuhou, L. & Huang, S. (2011). "Progress in the applications of DDQ to organic reactions" *Huaxue Tongbao, 74*, 497–507.

[20]Utley, J. H. P. & Rozenberg, G. G. (2003). "DDQ mediated anodic oxidation of 2-methyl- and 2-benzylnaphthalenes" *Journal of Applied Electrochemistry, 33*, 525–532. doi: 10.1023/A:1024474620525.

[21]Yao-Chang, X. (2000). "DDQ-induced, oxidative formation of C-O, and C-C bonds for isochromans and isothiochromans, application to the synthesis of dihydropyran containing anticancer agents and natural products" *Recent Research Developments in Organic Chemistry, 4*, 423–441.

[22]Guo, X. & Mayr, H. (2013). "Manifestation of Polar Reaction Pathways of 2,3-Dichloro-5,6-dicyano-*p*-benzoquinone" *Journal of the American Chemical Society, 35*, 12377–12387. doi: 10.1021/ja405890d.

[23]Li, J. S., Xue, Y., Li, J. W., Liu, W. D., Lu, C. H. & Zhao, P. X. (2013). "An Efficient Access to Fluorescent 2,3,4-Tricyanofurans from α-Cyano Ketones Using DDQ as Maleonitrile Building Block" *Synlett, 24*, 2003–2005. doi: 10.1055/s-0033-1339501.

[24]Tanemura, K., Suzuki, T. & Horaguchi, T. (1992). "2,3-Dichloro-5,6-dicyano-*p*-benzoquinone as a mild and efficient catalyst for the deprotection of acetals" *Journal of the Chemical Society, Chemical Communications*, 979–980. doi: 10.1039/ C39920000979.

[25]Iranpoor, N. & Baltork, I. M. (1990). "2,3-Dichloro-5,6-Dicyano-*p*-benzoquinone, an efficient, mild, neutral and highly regioselective catalyst for alcoholysis of epoxides" *Tetrahedron Letters, 31*, 735–738. URL: https://doi.org/10.1016/S0040-4039(00)94616-1.

[26]Jackman, L. M. (1960). "Hydrogenation-dehydrogenation reactions" *Advances in Organic Chemistry. Methods and Results*, 329–366. Vol. II, Interscience Publishers. Inc., New York, N.Y.

[27]Thiele, J. & Gunther, F. (1906). "Descendants of the dicyanhydroquinones" *Justus Liebigs Annalen der Chemie, 349*, 45–66. doi: 10.1002/jlac.19063490103.

[28]Brook, A. G. (1952). "The preparation of substituted quinones by a new oxidizing agent" *Journal of the Chemical Society*, 5040–5041.

[29]Mitchell, P. W. D. (1963). "The preparation of 2,3-Dicyano-5,6-Dichloroquinone (DDQ)" *Canadian Journal of Chemistry, 41*, 550–553. doi: 10.1139/v63-081.

[30]Wallenfels, K., Bachman, G., Hoffman, D. & Kern, R. (1965). "Cyano substituted quinone: 2,3-, 2,5-, 2,6-dicyanoquinone and tetracyanobenzoquinone" *Tetrahedron*, *21*, 2239–2256. doi: https://doi.org/10.1016/S0040-4020(01)93879-7.

[31]Walker, D. & Waugh, T. D. (1965). "2,3-Dichloro-5,6-Dicyanobenzoquinone (DDQ). A New Preparation" *The Journal of Organic Chemistry*, *30*, 3240–3240. doi: 10.1021/jo01020a529.

[32]Li, B. J. & Shi, Z. J. (2012). "From C(sp^2)-H to C(sp^3)-H: Systematic studies on transition metal-catalyzed oxidative C-C formation" *Chemical Society Reviews*, *41*, 5588–5598. doi: 10.1039/C2CS35096C.

[33]Girard, S. A., Knauber, T. & Li, C. J. (2014). "The Cross-dehydrogenative coupling of Csp3-H bonds: A versatile strategy for C-C bond formations" *Angewandte Chemie International Edition*, *53*, 74–100. doi: 10.1002/anie.201304268.

[34]Bergman, R. G. (2007). "C–H activation" *Nature* 446:391–393. doi:10.1038/446391a.

[35]Grzybowski, M., Skonieczny, K., Butenschön, H. & Gryko, D. T. (2013). "Comparison of Oxidative Aromatic Coupling and the Scholl Reaction" *Angewandte Chemie International Edition*, *52*, 9900–9930. doi: 10.1002/anie.201210238.

[36]Ohkubo, K., Fujimoto, A. & Fukuzumi, S. (2013). "Visible-Light-Induced Oxygenation of Benzene by the Triplet Excited State of 2,3-Dichloro-5,6-dicyano-*p*-benzoquinone" *Journal of the American Chemical Society*, *135*, 5368–5371. doi: 10.1021/ja402303k.

[37]Narayan, R., Matcha, K. & Antonchick, A. P. (2015). "Metal-Free Oxidative C-C Bond Formation through C-H Bond Functionalization" *Chemistry A European Journal*, *21*, 14678–14693. doi: 10.1002/chem.201502005.

[38]Yeung, C. S. & Dong, V. M. (2011). "Catalytic dehydrogenative cross-coupling: forming carbon-carbon bonds by oxidizing two carbon-hydrogen bonds" *Chemical Reviews*, *111*, 1215–1292. doi: 10.1021/cr100280d.

[39]Zhu, X., Tsuji, H., Nakabayashi, K., Ohkoshi, S. I. & Nakamura, E. (2011). "Air- and Heat-Stable Planar Tri-p-quinodimethane with distinct Biradical Characteristics" *Journal of the American Chemical Society*, *133*, 16342–16345. doi: 10.1021/ja206060n.

[40]Stefano, A. D., Sozio, P., Cacciatore, I., Cocco, A., Giorgioni, G., Costa, B., Lucacchini, M. A., Martini, C., Spoto, G., Pietrantonio, D. F., Matteo, D. E. & Pinnen, F. (2005). "Preparation and Pharmacological Characterization of *trans*-2-Amino-5(6)-fluoro-6(5)-hydroxy-1-phenyl-2,3-dihydro-1H-indenes as D_2-like Dopamine Receptor Agonists" *Journal of Medicinal Chemistry*, *48*, 2646–2654. doi: 10.1021/jm040889k.

[41]Wang, H., Zhao, Y., Li, L., Li, S. & Liu, Q. (2014). "Metal-Free 2,3-Dichloro-5,6-dicyano-1,4-benzoquinone (DDQ)-Mediated Cross-Dehydrogenative-Coupling (CDC) of Benzylic C(sp^3)-H Bonds and Vinylic C (sp^2)-H Bonds: Efficient One-Pot Synthesis of 1*H*-Indenes" *Advanced Synthesis & Catalysis*, *356*, 3157–3163. doi: 10.1002/adsc. 201400357.

[42]Meng, Z., Sun, S., Yuan, H., Lou, H. & Liu, L. (2014). "Catalytic Enantioselective Oxidative Cross-Coupling of Benzylic Ethers with Aldehydes" *Angewandte Chemie International Edition*, *53*, 543–547. doi: 10.1002/anie.201308701.

[43]Sperry, J., Wilson, Z. E., Rathwell, D. C. K. & Brimble, M. A. (2010). "Isolation, biological activity and synthesis of benzannulated spiroketal natural products" *Natural Products Reports*, *27*, 1117–1137. doi: 10.1039/B911514P.

[44]Wu, J., Qian, J., Wang, T. & Huang, Z. (2014). "Proline-catalyzed dehydrogenative cross-coupling reaction between chromene and aldehydes" *Synlett*, *25*, 2072–2074. doi: 10.1055/s-0034-1378496.

[45]Cheng, P., Huang, N., Jiang, Z. Y., Zhang, Q., Zheng, Y. T., Chen, J. J., Zhang, X. M. & Ma, Y. B. (2008). "1-Aryl-tetrahydroisoquinoline analogs as active anti-HIV agents *in vitro*" *Bioorganic & Medicinal Chemistry Letters*, *18*, 2475–2478. doi:10.1016/j.bmcl.2008.02.040.

[46]Gao, M., Kong, D., Clearfield, A. & Zheng, Q. H. (2006). "Synthesis of carbon-11 and fluorine-18 labeled N-acetyl-1-aryl-6,7-dimethoxy-1,2,3,4-tetrahydroisoquinoline derivatives as new potential PET AMPA receptor ligands" *Bioorganic & Medicinal Chemistry Letters*, *16*, 2229–2233. oi:10.1016/j.bmcl.2006.01.042.

[47]Liu, X., Zhang, J., Ma, S., Ma, Y. & Wang, R. (2014). "Oxidative cross-dehydrogenative coupling between *N*-aryl tetrahydroisoquinolins and 5H-oxazol-4-ones through two methodologies: copper catalysis or a metal-free strategy" *Chemical Communications*, 15714–15717. doi: 10.1039/c4cc04508d.

[48]Ma, Y., Zhang, G., Zhang, J., Yang, D. & Wang, R. (2014). "Organocatalyzed Asymmetric Oxidative Coupling of α-Csp^3-H of Tertiary Amines to α,β-Unsaturated γ-Butyrolactam: Synthesis of MBH-Type Products" *Organic Letters*, *16*, 5358–5261. URL: dx.doi.org/10.1021/ol5025597.

[49]Qiu, J. & Zhang, R. (2014). "Direct transformation of arylpropynes to acrylamides *via* a three-step tandem reaction" *Organic & Biomolecular Chemistry*, *12*, 1556–1560. doi: 10.1039/c3ob42444h.

[50]Liu, N., Mao, L., Yang, B. & Yang, S. (2014). "Copper-promoted oxidative-fluorination of arylphosphine under mild conditions" *Chemical Communications*, *50*, 10879–10882. doi: 10.1039/c4cc04830j.

[51]Huang, X., Salman, M. & Huang, Z. (2014). "Copper-Catalyzed Dehydrogenative Cross-Coupling Reaction between Allylic C-H Bonds and α-C-H Bonds of Ketones or Aldehydes" *Chemistry A European Journal*, *20*, 6618–6621. doi: 10.1002/chem.201402114.

[52]Cheng, D., Yuan, K., Xu, X. & Yan, J. (2015). "The oxidative coupling of benzylic compounds catalyzed by 2,3-dichloro-5,6-dicyano-benzoquinone and sodium nitrite using molecular oxygen as a co-oxidant" *Tetrahedron Letters*, *56*, 1641–1644. URL: http://dx.doi.org/10.1016/j.tetlet.2015.02.018.

[53]Singh, N., Agarwal, R. C. & Singh, C. P. (2013). "Synthesis and Evaluation of Some Substituted Indole Derivatives for Cardiovascular Activity" *International Journal of Pharmaceutical Sciences and Drug Research*, *5*, 14–17. doi: http://ijpsdr.com/pdf/vol5-issue1/2.pdf.

[54]Rani, P., Srivastava, V. K. & Kumar, A. (2004). "Synthesis and antiinflammatory activity of heterocyclic indole derivatives" *European Journal of Medicinal Chemistry*, *39*, 449–452. doi:10.1016/j.ejmech.2003.11.002.

[55]Guo, S., Li, Y., Wang, Y., Guo, X., Meng, X. & Chen, B. (2015). "Iron-Catalyzed Cross Dehydrogenative Coupling (CDC) of Indoles and Benzylic C-H Bonds" *Advanced Synthesis & Catalysis*, *357*, 950–954. doi: 10.1002/adsc.201400938.

[56]Yang, K. & Song, Q. (2015). "Fe-Catalyzed Double Cross-Dehydrogenative Coupling of 1,3-Dicarbonyl Compounds and Arylmethanes" *Organic Letters*, *17*, 548–551. doi: 10.1021/ol503556x.

[57]Kong, S., Zhang, L., Dai, X., Tao, L., Xie, C., Shi, L. & Wang, M. (2015). "DDQ-mediated Direct C(sp^3)-H Cyanation of Benzyl Ethers and 1,3-Diarylpropenes under Solvent- and

Metal-free Conditions" *Advanced Synthesis & Catalysis*, *357*, 2453–2456. doi: 10.1002/adsc.201500096.

[58]Kock, I., Heber, D., Weide, M., Wolschendorf, U. & Clement, B. (2005). "Synthesis and Biological Evaluation of 11-Substituted 6-Aminobenzo[*c*]phenanthridine Derivatives, a New Class of Antitumor Agents" *Journal of Medicinal Chemistry*, *48*, 2772–2777. doi: 10.1021/jm0490888.

[59]Parhi, A., Kelley, C., Kaul, M., Pilch, D. S. & LaVoie, E. J. (2012). "Antibacterial activity of substituted 5-methylbenzo[c]phenanthridinium derivatives" *Bioorganic & Medicinal Chemistry Letters*, *22*, 7080–7083. URL:http://dx.doi.org/10.1016/j.bmcl. 2012.09.097 .

[60]Romo-Pérez, A., Miranda, L. D. & García, A. (2015). "Synthesis of *N*-methyl-5,6-dihydrobenzo[*c*]phenanthridine and its sp^3 C(6)–H bond functionalization via oxidative cross-dehydrogenative coupling reactions" *Tetrahedron Letters*, *56*, 6669–6673. doi: http://dx.doi.org/10.1016/j.tetlet.2015.10.018.

[61]Ma, Y., Zhang, D., Yan, Z., Wang, M., Bian, C., Gao, X., Bunel, E. E. & Lei, A. (2015). "Iron-Catalyzed Oxidative C−H/C−H Cross-Coupling between Electron-Rich Arenes and Alkenes" *Organic Letters*, *17*, 2174–2177. doi: 10.1021/acs.orglett.5b 00775.

[62]Wirtanen, T., Muuronen, M., Hurmalainen, J., Tuononen, H. M., Nieger, M. & Helaja, J. (2016). "Intermolecular oxidative dehydrogenative 3,3'-coupling of benzo[*b*]furans and benzo[*b*]thiophenes promoted by DDQ/H+: total synthesis of shandougenine B" *Organic Chemistry Frontiers*, *3*, 1738–1745. doi: 10.1039/c6qo00331a.

[63]Chua, C. K., Sofer, Z. & Pumera, M. (2016). "Functionalization of Hydrogenated Graphene: Transition-Metal-Catalyzed Cross-Coupling Reactions of Allylic C-H Bonds" *Angewandte Chemie International Edition*, *55*, 10751–10754. doi: 10.1002/ anie.201605457.

[64]Yu, J., Zhang, Y., Jiang, Z. & Su, W. (2016). "Mechanically Induced Fe(III) Catalysis at Room Temperature: Solvent-Free Cross-Dehydrogenative Coupling of 3-Benzylic Indoles with Methylenes/Indoles" *The Journal of Organic Chemistry*, *81*, 11514–11520. doi: 10.1021/acs.joc.6b02197.

[65]Yu, J., Peng, G., Jiang, Z., Hong, Z. & Su, W. (2016). "Mechanochemical Oxidative Mannich Reaction: Evaluation of Chemical and Mechanical Parameters for the Mild andChemoselective Coupling of *N-tert*-butoxycarbonyltetrahydroquinolines and Ketones" *European Journal of Organic Chemistry*, 5340–5344. doi:10.1002/ejoc. 201600987.

[66]Xie, Z., Zan, X., Sun, S., Pan, X. & Liu, L. (2016). "Organocatalytic Enantioselective Cross-Dehydrogenative Coupling of *N*-Carbamoyl Cyclic Amines with Aldehydes" *Organic Letters*, *18*, 3944–3947. doi: 10.1021/acs.orglett.6b01625.

[67]Cheng, D., Wu, L., Lv, H., Xu, X. & Yan, J. (2017). "CDC Reaction and Subsequent Cyclization for the Synthesis of 2-Hydroxy-3-alkyl-1,4-naphthoquinones and Pyranonaphthoquinones" *The Journal of Organic Chemistry*, *82*, 1610-1617. doi: 10. 1021/acs.joc.6b02787.

[68]Ling, L., Yanna, H., Kailiang, W. & Qingmin, W. (2014). "DDQ/TFA: an Efficient Nonmetallic Reagent System for the Oxidative Coupling to Construct Phenanthrene Rings" *Chemical Research in Chinese Universities*, *30*, 619–623. doi: 10.1007/s40242-014-3561-9.

[69]Katritzky, A. R., Rachwal, S. & Rachwal, B. (1996). "Recent progress in the synthesis of 1,2,3,4-tetrahydroquinolines" *Tetrahedron*, *52*, 15031–15070. URL: https://doi.org/10.1016/S0040-4020(96)00911-8.

[70]Zhou, Y. G. (2007). "Asymmetric Hydrogenation of Heteroaromatic Compounds" *Accounts of Chemical Research*, *40*, 1357–1366. doi: 10.1021/ar700094b.

[71]Kang, Y. K. & Kim, D. Y. (2014). "Enantioselective organocatalytic oxidative enamine catalysis–1,5-hydride transfer–cyclization sequences: asymmetric synthesis of tetrahydroquinolines" *Chemical Communications*, *50*, 222–224. doi: 10.1039/ c3cc46710d.

[72]Sun, S., Yang, J., Li, F., Lv, Z., Li, W., Lou, H. & Liu, L. (2014). "Seven- and eight-membered heterocyclic biaryl synthesis through a metal-free oxidative coupling reaction" *Tetrahedron Letters*, *55*, 6899–6902. URL:http://dx.doi.org/ 10.1016/j.tetlet.2014.10.103 .

[73]Ai, Y., Hu, Y., Kang, F., Lai, Y., Jia, Y., Huang, Z., Peng, S., Ji, H., Tian, J. & Zhang, Y. (2015). "Synthesis and Biological Evaluation of Novel Olean-28,13β-lactams as Potential Antiprostate Cancer Agents" *Journal of Medicinal Chemistry*, *58*, 4506–4520. doi: 10.1021/jm5020023.

[74]Mendive-Tapia, L., Albornoz-Grados, A., Bertran, A., Albericio, F. & Lavilla, R. (2017). "Oxidative couplings on tryptophan-based diketopiperazines leading to fused and bridged chemotypes" *Chemical Communications*, *53*, 2740-2743. doi: 10.1039/ c7cc00555e.

[75]Verones, V., Flouquet, N., Farce, A., Berthelot, P. & Lebegue, N. (2014). "Unexpected Heteroannulation and Chlorination of Benzothiadiazine Derivatives Mediated by DDQ" *Synthesis*, *46*, 235–241. doi: 10.1055/s-0033-1340249.

[76]Li, Y., Cao, L., Luo, X. & Deng, W. (2014). "The synthesis of 1,2-diarylindenes *via* DDQ-mediated dehydrogenative intramolecular cyclization" *Tetrahedron*, *70*, 5974–5979. URL: http://dx.doi.org/10.1016/j.tet.2014.05.088.

[77]Jang, Y. H. & Youn, S. W. (2014). "Metal-Free C-H Amination for Indole Synthesis" *Organic Letters*, *16*, 3720–3723. URL: dx.doi.org/10.1021/ol5015398.

[78]El-Emam, A. A., Al-Deeb, O. A., Al-Omar, M. & Lehmann, J. (2004). "Synthesis, antimicrobial, and anti-HIV-1 activity of certain 5-(1-adamantyl)-2-substituted thio-1,3,4-oxadiazoles and 5(1-adamantyl)-3-substituted aminomethyl-1,3,4-oxadiazoline-2-thiones" *Bioorganic & Medicinal Chemistry*, *12*, 5107–5113. doi: doi:10.1016/j.bmc. 2004.07.033.

[79]Kumar, D., Sundaree, S., Johnson, E. O. & Shah, K. (2009). "An efficient synthesis and biological study of novel indolyl-1,3,4-oxadiazoles as potent anticancer agents" *Bioorganic & Medicinal Chemistry Letters*, *19*, 4492–4494. doi: 10.1016/j.bmcl. 2009.03.172.

[80]Jasiak, K. & Kudelko, A. (2015). "Oxidative cyclization of *N*-aroylhydrazones to 2-(2-arylethenyl)-1,3,4-oxadiazoles using DDQ as an efficient oxidant" *Tetrahedron Letters*, *56*, 5878–5881. URL: https://www.sciencedirect.com/science/article/pii/ S0040403915300605?via%3Dihub.

[81]Hilf, J. A., Witthoft, L. W. & Woerpel, K. A. (2015). "An S_N1-type Reaction To Form the 1,2-Dioxepane Ring: Synthesis of 10,12-Peroxycalamenene" *The Journal of Organic Chemistry*, *80*, 8262–8267. doi: 10.1021/acs.joc.5b01326.

[82]Uckun, F. M., Mao, C., Vassilev, A. O., Huang, H. & Jan, S. T. (2000). "Structure-based design of a novel synthetic spiroketal pyran as a pharmacophore for the marine natural product spongistatin 1" *Bioorganic & Medicinal Chemistry Letters*, *10*, 541–545. doi: http://doi.org/10.1016/S0960-894X(00)00044-5.

[83]Mitsuhashi, S., Shima, H., Kawamura, T., Kikuchi, K., Oikawa, M., Ichihara, A. & Oikawa, H. (1999). "The spiroketals containing a benzyloxymethyl moiety at C8 position showed the most potent apoptosis-inducing activity" *Bioorganic & Medicinal Chemistry Letters*, *9*, 2007–2012. URL: https://doi.org/10.1016/S0960-894X(99)00309-1.

[84]Milroy, L. G., Zinzalla, G., Loiseau, F., Qian, Z., Prencipe, G., Pepper, C., Fegan, C. & Ley, S. V. (2008). "Natural-product-like spiroketals and fused bicyclic acetals as potential therapeutic agents for B-cell chronic lymphocytic leukaemia" *ChemMed Chem*, *3*, 1922–1935. doi: 10.1002/cmdc.200800265.

[85]Ohtake, Y., Sato, T., Kobayashi, T., Nishimoto, M., Taka, N., Takano, K., Yamamoto, K., Ohmori, M., Yamaguchi, M., Takami, K., Yeu, S. Y., Ahn, K. H., Matsuoka, H., Morikawa, K., Suzuki, M., Hagita, H., Ozawa, K., Yamaguchi, K., Kato, M. & Ikeda, S. (2012). "Discovery of Tofogliflozin, a Novel *C*-Arylglucoside with an *O*-Spiroketal Ring System, as a Highly Selective Sodium Glucose Cotransporter 2 (SGLT2) Inhibitor for the Treatment of Type 2 Diabetes" *Journal of Medicinal Chemistry*, *55*, 7828–7840. doi: 10.1021/jm300884k.

[86]Peh, G. & Floreancig, P. E. (2015). "Convergent One-Pot Oxidative [n+1] Approaches to Spiroacetal Synthesis" *Organic Letters*, *17*, 3750–3753. doi: 10.1021/acs.orglett. 5b01736.

[87]Uludag, N. & Yakup, M. (2015). "Concise Total Synthesis of 20-Deethyl-4-demethyldasycarpidone" *Organic Preparations and Procedures International*, *47*, 454–460. doi: 10.1080/00304948.2015.1088756.

[88]Lemercier, B. C. & Pierce, J. G. (2015). "Synthesis of 1,4,2-Oxathiazoles *via* Oxidative Cyclization of Thiohydroximic Acids" *Organic Letters*, *17*, 4542–4545. doi: 10.1021/acs.orglett.5b02256.

[89]Cho, H. V. & Scott, L. T. (2015). "Oxidative cyclotrimerization of unsaturated compounds with DDQ and triflic acid: An efficient synthetic route to triply-fused benzene rings" *Tetrahedron Letters*, *56*, 3458-3462. URL:http://dx.doi.org/10.1016/j. tetlet.2014.12.141 .

[90]Zhu, H. & Chen, Z. (2016). "DDQ-Mediated Oxidative Radical Cycloisomerization of 1,5-Diynols: Regioselective Synthesis of Benzo[*b*]fluorenones under Metal-Free Conditions" *Organic Letters*, *18*, 488–491. doi: 10.1021/acs.orglett.5b03533.

[91]Parker, P. D. & Pierce, J. G. (2016). "Synthesis of 1,2,4-Oxadiazoles via DDQ-Mediated Oxidative Cyclization of Amidoximes" *Synthesis*, *48*, 1902–1909. doi: 10.1055/s-0035-1561597.

[92]Ochiai, K., Kuppusamy, S., Yasui, Y., Harada, K., Gupta, N. R., Takahashi, Y., Kubota, T., Kobayashi, J. & Hayashi, Y. (2016). "Total Synthesis of the 7,10-Epimer of the Proposed Structure of Amphidinolide N, Part II: Synthesis of C17–C29 Subunit and Completion of the Synthesis" *Chemistry A European Journal*, *22*, 3287–3291. doi : 10.1002/chem.201504675.

[93]Yin, H., Meng, Z. & Liu, B. (2016). "Construction of the Core Structure of Trichotomone" *Synthesis*, *48*, 3951–3956. doi: 10.1055/s-0035-1562491.

[94]Wang, Z. L., Li, H. L., Ge, L. S., An, X. L., Zhang, Z. G., Luo, X., Fossey, J. S. & Deng, W. P. (2014). "DDQ-Mediated Oxidative Coupling: An Approach to 2,3-Dicyanofuran (Thiophene)" *The Journal of Organic Chemistry*, *79*, 1156–1165. URL: dx.doi.org/10.1021/jo4026034.

[95]Li, J. S., Cai, F. F., Li, Z. W., Liu, W. D., Simpson, J., Xue, Y., Pang, H. L., Huang, P. M., Cao, Z. & Li, D. L. (2014). "One-step metal-free construction of fluorescent 5-aryl-2,3-

dicyanofurans from simple aryl ketones with DDQ" *Royal Society of Chemistry Advances*, *4*, 474–478. doi: 10.1039/c3ra45462b.

[96]Conforti, F., Marrelli, M., Menichini, F., Bonesi, M., Statti, G., Provenzano, E. & Menichini, F. (2009). "Natural and Synthetic Furanocoumarins as Treatment for Vitiligo and Psoriasis" *Current Drug Therapy*, *4*, 38–58. URL: https://doi.org/10.2174/157488509787081886.

[97]Wang, X., Bastow, K. F., Sun, C. M., Lin, Y. L., Yu, H. J., Don, M. J., Wu, T. S., Nakamura, S. & Lee, K. H. (2004). "Antitumor agents. Isolation, Structure Elucidation, Total Synthesis. And Anti-Breast Cancer Activity of Neo-tanshinlactone from Salvia miltiorrhiza" *Journal of Medicinal Chemistry*, *47*, 5816–5819. doi: 10.1021/jm040112r.

[98]Li, J. S., Fu, D. M., Xue, Y., Li, Z. W., Li, D. L., Da, Y. D., Yang, F., Zhang, L., Lu, C. H. & Li, G. (2015). "One-step synthesis of furocoumarins *via* oxidative annulation of 4-hydroxycoumarins with DDQ" *Tetrahedron*, *71*, 2748–2752. doi: http://dx.doi.org/10.1016/j.tet.2015.03.025.

[99]Feng, H. X., Wang, Y. Y., Chen, J. & Zhou, L. (2015). "A Dehydrogenative Diels–Alder Reaction of Prenyl Derivatives with 2,3-Dichloro-5,6-dicyanobenzoquinone" *Advanced Synthesis & Catalysis*, *357*, 940–944. doi: 10.1002/adsc.201401038.

[100] Gawel, P., Halabi, E. A., Schweinfurth, D., Trapp, N., Ruhlmann, L., Boudon, C. & Diederich, F. (2016). "Synthesis of Dicyano-Substituted Benzo[*c*]fluorenes from Tetraaryl[3]cumulenes" *European Journal of Organic Chemistry*, *17*, 2919–2924. doi: 10.1002/ejoc.201600470.

[101] Giovenzana, G. B., Masciocchi, N., Negri, R., Palmisano, G., Penoni, A. & Toma, L. (2017). "Quinone-related hexacyclic by-products in the production process of exemestane" *Steroids*, *120*, 26–31. URL: http://dx.doi.org/10.1016/j.steroids .2017.01.007.

[102] Yu, Z., Dan, W., Jie, H. & Li, Z. (2010). "Synthesis and bioevaluation of novel analogues of Justicidin A" *Medicinal Chemistry Research*, *19*, 71–76. doi: 10.1007/s00044-009-9172-1.

[103] Janmanchi, D., Tseng, Y. P., Wang, K. C., Huang, R. L., Lin, C. H. & Yeh, S. F. (2010). "Synthesis and the biological evaluation of arylnaphthalene lignans as anti-hepatitis B virus agents" *Bioorganic & Medicinal Chemistry*, *18*, 1213–1226. doi:10. 1016/j.bmc.2009.12.038.

[104] Leung, Y. M., Tsou, Y. H., Kuo, C. S., Lin, S. Y., Wu, P. Y., Hour, M. J. & Kuo, Y. H. (2010). "Arylnaphthalene lignans from *Taiwania cryptomerioides* as novel blockers of voltage-gated K^+ channels" *Phytomedicine*, *18*, 46–51. doi:10.1016/j.phymed.2010.04. 001.

[105] Park, J. E., Lee, J., Seo, S. Y. & Shin, D. (2014). "Regioselective route for arylnaphthalene lactones: convenient synthesis of taiwanin C, justicidin E, and daurinol" *Tetrahedron Letters*, *55*, 818–820. URL: http://dx.doi.org/10.1016/j.tetlet.%202013.12.014.

[106] Manoharan, R. & Jeganmohan, M. (2015). "Ruthenium-catalyzed cyclization of *N*-carbamoyl indolines with alkynes: an efficient route to pyrroloquinolinones" *Organic & Biomolecular Chemistry*, *13*, 9276–9284. doi: 10.1039/c5ob01146a.

[107] Oikonomou, K., Georgiou, D., Katsamakas, S., Hadjipavlou-Litina, D. & Elemes, Y. (2015). "Sulfanyl 5*H*-dihydropyrrole derivatives *via* 1,3-dipolar cycloaddition, their further chemical manipulation and antioxidant activity" *Arkivoc*, (iii), 214–231. URL: http://dx.doi.org/10.3998/ark.5550190.0016.315.

[108] Zou, Y., Zhao, Q., Hu, H., Hu, L., Yu, S., Xu, M. & Wu, Q. (2012). "Synthesis and *in vitro* antitumor activities of xanthone derivatives containing 1,4-disubstituted-1,2,3-triazole moiety" *Archives of Pharmacal Research*, *35*, 2093–2104. doi: 10.1007/ s12272-012-1206-4.

[109] Albuquerque, H. M. T., Santos, C. M. M., Cavaleiro, J. A. S. & Silva, A. M. S. (2015). "(*E*)-2-(4-Arylbut-1-en-3-yn-1-yl)chromones as Synthons for the Synthesis of Xanthone-1,2,3-triazole Dyads" *European Journal of Organic Chemistry*, 4732–4743. doi: 10.1002/ejoc.201500448.

[110] Hirata, G., Yamada, N., Sanada, S., Onodera, G. & Kimura, M. (2015). "Palladium-Catalyzed [4+2] Cycloaddition of Aldimines and 1,4-Dipolar Equivalents via Amphiphilic Allylation" *Organic Letters*, *17*, 600–603. doi: 10.1021/ol503614d.

[111] Gökhan-Kelekçi, N., Koyunoglu, S., Yabanoglu, S., Yelekçi, K., Özgen, Ö., Uçar, G., Erol, K., Kendi, E. & Yesilada, A. (2009). "New pyrazoline bearing 4(3H)-quinazolinone inhibitors of monoamine oxidase: Synthesis, biological evaluation, and structural determinants of MAO-A and MAO-B selectivity" *Bioorganic & Medicinal Chemistry*, *17*, 675–689. doi:10.1016/j.bmc.2008.11.068.

[112] Alagarsamy, V. & Pathak, U. S. (2007). "Synthesis and antihypertensive activity of novel 3-benzyl-2-substituted-3*H*-[1,2,4]triazolo[5,1-b]quinazolin-9-ones" *Bioorganic & Medicinal Chemistry*, *15*, 3457–3462. doi:10.1016/j.bmc.2007.03.007.

[113] Parekh, N. M. & Maheria, K. C. (2014). "Colorimetric studies of heterocyclic monoazo reactive dyes and their dyeing applications on cotton, silk, and wool fibres" *Research on Chemical Intermediates*, *40*, 1003–1019. doi: 10.1007/s11164-012-1017-8.

[114] Oh, B. K., Ko, E. B., Han, J. W. & Oh, C. H. (2015). "Synthesis of *N*-heteropolycyclic compounds including quinazolinone skeleton using Friedel-Crafts alkylation" *Synthetic Communications*, *45*, 768–776. doi: 10.1080/00397911.2014.987353.

[115] Minuti, L., Barattucci, A., Bonaccorsi, P. M., Siciliano, C. & Temperini, A. (2016). "New access to 4-aryl[2,2]paracyclophanes by high-pressure Diels–Alder reaction" *Tetrahedron Letters*, *57*, 917–919. URL: http://dx.doi.org/10.1016/j.tetlet.2016.01.053.

[116] Molchanov, A. P., Tran, T. Q., Stepakov, A. V. & Kostikov, R. R. (2016). "Oxidation of Reaction Products of *p*-Aryl-*C,C* disubstituted Nitrones and Dimethyl 3-Methylenecyclopropane-1,2-dicarboxylate" *Russian Journal of Organic Chemistry*, *52*, 1603–1605. doi: 10.1134/S1070428016110099.

[117] Basavaiah, D., Lingaiah, B., Reddy, G. C. & Sahu, B. C. (2016). "Baylis–Hillman Acetates in Synthesis: Copper(I)/*tert*-Butyl Hydroperoxide Promoted One-Pot Oxidative Intramolecular Cyclization Protocol for the Preparation of Pyrrole-Fused Compounds and the Formal Synthesis of (±)-Crispine A" *European Journal of Organic Chemistry*, 2398–2403. doi: 10.1002/ejoc.201600384.

[118] Yu, W., Tong, L., Hu, B., Zhong, B., Hao, J., Ji, T., Zan, S., Coburn, C. A., Selyutin, O., Chen, L., Rokosz, L., Agrawal, S., Liu, R., Curry, S., McMonagle, P., Ingravallo, P., Asante-Appiah, E., Chen, S. & Kozlowski, J. A. (2016). "Discovery of Chromane Containing Hepatitis C Virus (HCV) NS5A Inhibitors with Improved Potency against Resistance-Associated Variants" *Journal of Medicinal Chemistry*, *59*, 10228–10243. doi: 10.1021/acs.jmedchem.6b01234.

[119] Mou, C., Zhu, T., Zheng, P., Yang, S., Song, B. & Chi, Y. R. (2016). "Green and Rapid Access to Benzocoumarins *via* Direct Benzene Construction through Base-Mediated

Formal [4+2] Reaction and Air Oxidation" *Advanced Synthesis & Catalysis*, *358*, 707–712. doi: 10.1002/adsc.201500771.

[120] Lin, W., Li, M., Fang, L., Shen, Y. & Chen, C. (2017). "Synthesis, Structures, Resolution, and Chiroptical Properties of 1,16-Diaryl-Substituted Benzo[5]helicene Derivatives" *Chemistry An Asian Journal*, *12*, 86–94. doi: 10.1002/asia.201601305.

[121] Walsh, K., Sneddon, H. F. & Moody, C. J. (2014). "Solar Photochemical Oxidations of Benzylic and Allylic Alcohols Using Catalytic Organo-oxidation with DDQ: Application to Lignin Models" *Organic Letters*, *16*, 5224–5227. URL: dx.doi.org/ 10.1021/ol502664f.

[122] Fosu, S. C., Ferrence, G. M. & Lash, T. D. (2014). "Synthesis and Metalation of Dimethoxybenziporphyrins, Thiabenziporphyrins, and Dibenziporphyrins" *The Journal of Organic Chemistry*, *79*, 11061–11074. URL: dx.doi.org/10.1021/jo502063w.

[123] Kuttner, J. R. & Hilt, G. (2014). "Regiodivergent Cobalt-Catalyzed Diels−Alder Reactions for the Synthesis of Bifunctional Building Blocks and Their Suzuki-Cross-Coupling Polymerizations" *Macromolecules*, *47*, 5532–5541. URL: dx.doi.org/10.1021/ ma5012446.

[124] Reyns, T., Belpaire, C., Geeraerts, C. & Loco, J. V. (2014). "Multi-dye residue analysis of triarylmethane, xanthene, phenothiazine and phenoxazine dyes in fish tissues by ultra-performance liquid chromatography–tandem mass spectrometry" *Journal of Chromatography B*, 92–101. URL: http://dx.doi.org/10.1016/j.jchromb.2014.02.002.

[125] Walsh, K., Sneddon, H. F. & Moody, C. J. (2014). "Sustainable, mild and efficient *p*-methoxybenzyl ether deprotections utilizing catalytic DDQ" *Tetrahedron*, *70*, 7380–7387. URL: http://dx.doi.org/10.1016/j.tet.2014.07.003.

[126] Naoda, K., Sung, Y. M., Lim, J. M., Kim, D. & Osuka, A. (2014). "Cross-Conjugated Hexaphyrins and Their Bis-Rhodium Complexes" *Chemistry A European Journal*, *20*, 7698–7705. doi: 10.1002/chem.201402328.

[127] Zhang, Q., Yang, F. & Wu, Y. (2014). "DDQ: the chlorinating reagent and oxidant for the ligand-directed ortho-chlorination of 2-arylpyridines" *Organic Chemistry Frontiers*, *1*, 694–697. doi: 10.1039/c4qo00076e.

[128] Laha, J. K., Tummalapalli, K. S. S. & Gupta, A. (2014). "Transition-Metal-Free Tandem Oxidative Removal of Benzylic Methylene Group by C−C and C−N Bond Cleavage Followed by Intramolecular New Aryl C−N Bond Formation under Radical Conditions" *Organic Letters*, *16*, 4392–4395. URL: dx.doi.org/10.1021/ol501766m.

[129] Wasserman, H. H., Rotello, V. M. & Krause, G. B. (1992). "Oxidation of α-Ylido, β-Keto Amides to Vicinal Tricarbonyls. Synthesis of a Diketopiperazine Precursor of Bicyclomycin" *Tetrahedron Letters*, *33*, 5419–5422. URL:https://doi.org/10.1016 /S0040-4039(00)79110-6 .

[130] Adlington, R. M., Baldwin, J. E., Catterickand, D. & Pritchard, G. J. (2001). "The efficient, enantioselective synthesis of quinoxaline, pyrazine and 1,2,4-triazine substituted α-amino acids from vicinal tricarbonyls" *Journal of the Chemical Society, Perkin Transactions*, *1*, 668–679. doi: 10.1039/B010134F.

[131] Wasserman, H. H., Chen, J. H. & Xia, M. (1999). "Total Syntheses of Depsipeptide Elastase Inhibitors YM-47141 and YM-47142 with use of Ylide Protection and Coupling Methods" *Journal of the American Chemical Society*, *121*, 1401–1402. doi: 10.1021/ja9840302.

[132] Wang, Z., An, X., Ge, L., Jin, J., Luo, X. & Deng, W. (2014). "DDQ-mediated oxidation of sp^3 C-H bond for the direct synthesis of vicinal tricarbonyl compounds" *Tetrahedron, 70*, 3788–3792. URL: http://dx.doi.org/10.1016/j.tet.2014.04.021.

[133] Zhang, W., Pan, D., Wu, A. & Shen, L. (2015). "Effective and mild method for converting 3β-hydroxysteroids to 3-ketosteroids *via* DDQ/TEMPO" *Steroids*, *96*, 16–20. URL: https://www.sciencedirect.com/science/article/pii/S0039128X1 5000215?via%3 Dihub.

[134] Rana, A., Siddiqui, N. & Khan, S. A. (2007). "Benzothiazoles: A New Profile of Biological Activities" *Indian Journal of Pharmaceutical Sciences*, *69*, 10–17. doi: 10.4103/0250-474X.32100.

[135] Lin, S. N. & Yang, L. H. (2005). "A Simple and Efficient Procedure for the Synthesis of Benzimidazoles Using Air as the Oxidant" *Tetrahedron Letters*, *46*, 4315–4319. URL: doi:10.1016/j.tetlet.2005.04.101.

[136] Kim, J. S., Gatto, B., Yu, C., Liu, A., Liu, L. F. & LaVoie, E. J. (1996). Substituted 2,5'-Bi-1*H*-benzimidazoles: Topoisomerase I Inhibition and Cytotoxicity. *Journal of Medicinal Chemistry*, *39*, 992–998. doi: 10.1021/jm950412w.

[137] Naeimi, H. & Babaei, Z. (2015). "A Mild and Simple One-pot Synthesis of 2-Substituted Benzimidazole Derivatives Using DDQ as an Efficient Oxidant at Room Temperature" *Journal of Chinese Chemical Society*, *62*, 41–46. doi: 10.1002/jccs. 201400293.

[138] Pujari, S. A., Besnard, C., Bîrgi, T. & Lacour, J. (2015). "A Mild and Efficient CH_2-Extrusion Reaction for the Enantiospecific Synthesis of Highly Configurationally Stable Troger Bases" *Angewandte Chemie International Edition*, *54*, 7520–7523. doi: 10.1002/anie.201500435.

[139] Ma, J., Hu, Z., Li, M., Zhao, W., Hu, X., Mo, W., Hu, B., Sun, N. & Shen, Z. (2015). "DDQ/*tert*-Butyl nitrite-catalyzed aerobic oxidation of diarylmethane sp^3 C-H bonds" *Tetrahedron*, *71*, 6733–6739. doi: http://dx.doi.org/10.1016/j.tet.2015.07.042.

[140] Ohkubo, K., Hirose, K. & Fukuzumi, S. (2015). "Solvent-Free One-Step Photochemical Hydroxylation of Benzene Derivatives by the Singlet Excited State of 2,3-Dichloro-5,6-dicyano-p-benzoquinone Acting as a Super Oxidant" *Chemistry A European Journal*, *21*, 2855–2861. doi: 10.1002/chem.201404810.

[141] Lu, H., Mack, J., Yang, Y. & Shen, Z. (2014). "Structural modification strategies for the rational design of red/NIR region BODIPYs" *Chemical Society Reviews*, *43*, 4778–4823. doi: 10.1039/C4CS00030G.

[142] Kobayashi, H., Ogawa, M., Alford, R., Choyke, P. L. & Urano, Y. (2010). "New Strategies for Fluorescent Probe Design in Medical Diagnostic Imaging" *Chemical Reviews*, *110*, 2620–2640. doi: 10.1021/cr900263j.

[143] Jang, H. G., Park, M., Wishnok, J. S., Tannenbaum, S. R. & Wogan, G. N. (2006). "Hydroxyl-specific fluorescence labeling of ABP-deoxyguanosine, PhIP-deoxyguanosine, and AFB1-formamidopyrimidine with BODIPY-FL" *Analytical Biochemistry*, *359*, 151–160. doi:10.1016/j.ab.2006.08.030.

[144] Liu, S., Lin, T., Li, D., Leamer, L., Shan, H., Li, Z., Gabbaı, F. P. & Conti, P. S. (2013). "Lewis acid-assisted isotopic 18F-19F exchange in BODIPY dyes: Facile generation of positron emission tomography/fluorescence dual modality agents for tumor imaging" *Theranostics*, *3*, 181–189. doi: 10.7150/thno.5984.

[145] Madhu, S., Rao, M. R., Shaikh, M. S. & Ravikanth, M. (2011). "3,5-Diformylboron Dipyrromethenes as Fluorescent pH Sensors" *Inorganic Chemistry*, *50*, 4392–4400. doi: 10.1021/ic102499h.

[146] Zhao, N., Vicente, M. C. H., Fronczek, F. R. & Smith, K. M. (2015). "Synthesis of 3,8-Dichloro-6-ethyl-1,2,5,7-tetramethyl–BODIPY from an Asymmetric Dipyrroketone and Reactivity Studies at the 3,5,8-Positions" *Chemistry A European Journal*, *21*, 6181–6192. doi: 10.1002/chem.201406550.

[147] Shen, Z., Chen, M., Fang, T., Li, M., Mo, W., Hu, B., Sun, N. & Hu, X. (2015). "Transformation of ethers into aldehydes or ketones: a catalytic aerobic deprotection/oxidation pathway" *Tetrahedron Letters*, *56*, 2768–2772. URL: https://www.sciencedirect.com/science/article/pii/S0040403915006607?via%3Dihub.

[148] Higashino, T., Yamada, T., Sakurai, T., Seki, S. & Imahori, H. (2016). "Fusing Porphyrins and Phospholes: Synthesis and Analysis of a Phosphorus-Containing Porphyrin" *Angewandte Chemie International Edition*, *55*, 12311–12315. doi: 10.1002/anie.201607417.

[149] Cheah, Y. S., Santhanakrishnan, S., Sullivan, M. B., Neoh, K. G. & Chai, C. L. L. (2016). "The chemical reactivities of DOPA and dopamine derivatives and their regioselectivities upon oxidative nucleophilic trapping" *Tetrahedron*, *72*, 6543–6550. URL: http://dx.doi.org/10.1016/j.tet.2016.08.068.

[150] Ye, Q., Zhang, Z., Png, Z. M., Neo, W. T., Lin, T., Zeng, H., Xu, H. & Xu, J. (2016). "Cyclization of Tetraaryl-Substituted Benzoquinones and Hydroquinones through the Scholl Reaction" *The Journal of Organic Chemistry*, *81*, 9219–9226. doi: 10.1021/acs.joc.6b01785.

[151] Schwarzwalder, G. M., Scott, D. R. & Vanderwal, C. D. (2016). "A Synthesis of Exiguaquinol Dessulfate" *Chemistry A European Journal*, *22*, 17953–17957. doi: 10.1002/chem.201604506.

[152] Leone, P. D. A., Carroll, A. R., Towerzey, L., King, G., McArdle, B. M., Kern, G., Fisher, S., Hooper, J. N. A. & Quinn, R. J. (2008). "Exiguaquinol: A Novel Pentacyclic Hydroquinone from *Neopetrosia exigua* that Inhibits *Helicobacter pylori* Murl" *Organic Letters*, *10*, 2585–2588. doi: 10.1021/ol800898z.

[153] Vogel, A., Dechert, S., John, M., Bruckner, C. & Meyer, F. (2016). "Siamese-Twin Porphyrin Origami: Oxidative Fusing and Folding" *Chemistry A European Journal* 22:2307–2316. doi: 10.1002/chem.201503699.

[154] Umasekhar, B., Samanta, P., Chatterjee, T. & Ravikanth, M. (2016). "Synthesis and properties of Oxasmaragdyrins containing one Five-membered Heterocycle at *Meso*-position" *Journal of Chemical Sciences*, *128*, 1709–1715. doi: 10.1007/s12039-016-1178-x.

[155] Lloyd, D., Bylsma, M., Bright, D. K., Chen, X. & Bennett, C. S. (2017). "Mild Method for 2-Naphthylmethyl Ether Protecting Group Removal Using a Combination of 2,3-Dichloro-5,6-dicyano-1,4-benzoquinone (DDQ) and β-Pinene" *The Journal of Organic Chemistry*, *82*, 3926-3934. doi: 10.1021/acs.joc.7b00065.

[156] Thebtaranonth, C., Thebtaranonth, Y., Wanauppathamkul, S. & Yuthavong, Y. (1995). "Antimalarial sesquiterpenes from tubers of *cyperus rotundus*: Structure of 10,12-peroxycalamenene, a sesquiterpene endoperoxide" *Phytochemistry*, *40*, 125–128. URL: https://doi.org/10.1016/0031-9422(95)00260-E.

[157] Srivastava, R. K., Singh, A. and Shukla, S. V. 2013. "Clinical investigation and Pharmaceutical action of *Cyperus rotundus* - A review" *Journal of Biologically Active Products from Nature, 3,* 166–172. URL: http://dx.doi.org/10.1080/22311866.2013.833381.

[158] Kaufmann, M., Bajetta, E., Dirix, L. Y., Fein, L. E., Jones, S. E., Zilembo, N., Dugardyn, J. L., Nasurdi, C., Mennel, R. G., Cervek, J., Fowst, C., Polli, A., Salle, E., Arkhipov, A., Piscitelli, G., Miller, L. L. & Massimini, G. (2000). "Exemestane Is superior to megestrol acetate after tamoxifen failure in postmenopausal women with advanced breast cancer: results of a phase III randomized double-blind trial" *Journal of Clinical Oncology, 18,* 1399–1411, doi: 10.1200/JCO.2000.18.7.1399.

In: Organic and Medicinal Chemistry, Volume 1
Editor: Bimal Krishna Banik
ISBN: 978-1-53614-453-6

Chapter 3

BENZOXAZOLE: PRIVILEGED SCAFFOLD FOR DRUG DESIGN IN THERAPEUTIC MEDICINE

Preety Saini and Aman Bhalla*
Department of Chemistry and Centre of Advanced Studies in Chemistry,
Panjab University, Chandigarh, India

ABSTRACT

Benzoxazole is an integral component of numerous valuable natural products and commercially marketed drugs. It represents an interesting template to annelate differently functionalized benzoxazole scaffolds of useful therapeutic value. Owing to enormous target based research and the vast literature on the benzoxazole scaffolds, a systematic compilation is highly in demand to understand the significance of benzoxazole nucleus in remedial research. Present chapter is designed with the main aim to give an appreciation for the wide-ranging importance of benzoxazole-based heterocycles in medicine in the light of new knowledge and to include recent work (2010-2017). The introductory paragraph outline the evolution of chemistry in medicinal field followed by an overview of natural and synthetic potential of benzoxazole scaffold. The other sections of this article deal with the documentation of various literature reports highlighting the significance of pharmacokinetic profile and pharmacological applications of benzoxazole nucleus.

Keywords: benzoxazole, small molecule, privileged scaffold, biological activity, drug design

ABBREVIATIONS

i-Bu	*iso*-Butyl
n-Bu	*n*-Butyl
t-Bu	*tert*-Butyl

* Corresponding Author Email: amanbhalla@pu.ac.in (Department of Chemistry and Centre of Advanced Studies in Chemistry, Panjab University, Chandigarh 160 014, India. Fax: +91-172-254-5074; Tel.: +91-172-253-4417/+91-9872659217.)

Bn	Benzyl
Bz	Benzoyl
C	Carbon
CA-4	Combretastatin A-4
CC_{50}	Half maximal cytotoxic concentration
COX	Cyclooxygenase
CYP	Cytochrome P450
3-D	Three dimensional
DENV	Dengue virus
DMPK	Drug metabolism and pharmacokinetics
DNA	Deoxyribonucleic acid
DPPH	2,2-Diphenyl-1-picrylhydrazyl
DORA	Dual orexin receptor antagonist
EC_{50}	Half maximal effective concentration
Et	Ethyl
ER	Estrogen receptor
GI_{50}	Half maximal growth concentration
H-bond	Hydrogen bond
HCV	Hepatitis C virus
HIV	Human immunodeficiency virus
HO	Heme oxygenase
HPSE	Heparanase
5-HT	5-Hydroxytryptamine
IBRD	Irritable bowel syndrome with diarrhea
IC_{50}	Half maximal inhibitory concentration
IC_{90}	Concentration for 90% inhibition
IRSE	Internal ribosome entry site
IVF	Influenza B virus
Kg	Kilogram
K_i	Affinity constant
M	Molar
μ	Micro
MDR-TB	Multidrug-resistant tuberculosis
Me	Methyl
MIC	Minimum inhibitory concentration
Mg	Milligram
mL	Millilitre
N	Nitrogen
n	Nano

NMT	*N*-myristoyltransferase
NNRTI	Non-nucleoside reverse transcriptase inhibitor
NSAID/NSAIA	Non-steroidal anti-inflammatory drug/agent
O	Oxygen
OX	Orexin
P	Probability
Ph	Phenyl
PI	Protection index
pK_a	Acidity constant
i-Pr	*iso*-Propyl
n-Pr	*n*-Propyl
QSAR	Quantitative structure-activity relationship
RNA	Ribonucleic acid
SAR	Structure-activity relationship
SI	Selectivity index
SRB	Sulphorhodamine-B
$t_{1/2}$	Half life
TB	Tuberculosis
XDR-TB	Extensively-drug resistant tuberculosis

1. INTRODUCTION

Natural products have been a major source of drugs for centuries. Before the development of modern chemistry, only inorganic materials or natural organic products were used to cure maladies. With the gradual decline of the fallacy and enchantment, scientists began to look for evidence of effectiveness of these folk remedies. Antoine Fourcroy, the son of an apothecary, from Paris applied chemistry to pharmacy and medicine for the first time. One of the first attempts to correlate chemical structure with biological activity was performed by Arthur Crum Brown, the professor of chemistry at the University of Edinburgh that thereby, led to the marketing of the first semisynthetic ethyl ether morphine derivative (Dionin) as a cough sedative in 1898.

Despite progress to date in drug discovery, the development of drug-resistant strains against existing medications is the major setback that led to an increased number of infections and extensive therapeutic failure in clinical settings [1,2]. Therefore, pressing the urgency to produce new remedial compounds to combat these drug-resistant strains.

Heterocyclic organic compounds containing nitrogen, oxygen and sulfur in the ring system are expected to be more active due to the presence of hetero atoms [3]. The majority of the drugs used in human medicine are either small molecules containing heterocycles or have heterocyclic structural components [4,5]. Because of the fundamental importance of benzoxazole-based heterocycles, this chapter explores the concept of using these benzoxazoles as a privileged scaffold to identify biologically active compounds for building chemical libraries.

2. Benzoxazole: Privileged Scaffold

After the universal prevalence of the concept that "the molecular structure does influence the biological activity of chemical compounds" there is an upsurge in research to find other organic functional groups or nucleus (rings).

The chemical scaffold present in several biologically active ligands which is responsible for influencing the molecule's specificity (Evans et al.) is known as privileged scaffold [6]. The benzoxazole is a privileged scaffold present in natural and synthetic compounds [7].

Benzoxazole is a biologically active skeleton present in numerous categories of therapeutic agents such as antimicrobial, antiviral, anticancer, anti-inflammatory, antioxidant, anticoagulant, antidepressant, anti-HIV, antimycobacterial, antileishmanial, antidiabetic, lipid level modulator, immunomodulatory, etc. and thereby has become an indispensable anchor for the development of new therapeutic agents [8].

2.1. Natural Occurrence

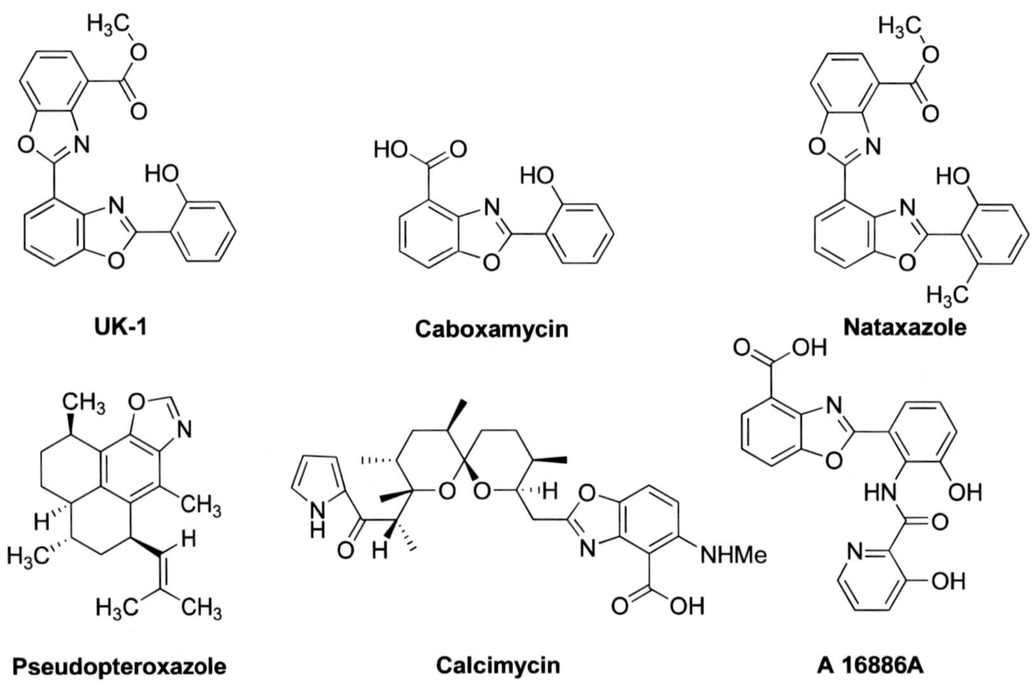

Figure 1. Benzoxazole "privileged scaffold" present in natural products.

Benzoxazoles are the structural isostere of natural nucleotides such as guanine and adenine and thereby is capable to intermingle easily with the biological targets (most commonly a protein or nucleic acid) [9,10]. Since, this ubiquitous pharmacophore motif is present in natural products as well as synthetic drug molecules, therefore, this 'Master Key" holds the potential to unlock a multitude of synthetic and biologically active compounds of pharmacological properties. Benzoxazole is an important heterocyclic system, which is present in many biologically active natural products such as UK-1 (antibiotic) [11], caboxamycin (antibiotic and antifungal) [12], nataxazole (antitumor) [13], pseudopteroxazole (anti-tubercular) [14], A 16886A (antibiotic) and calcimycin (antibiotic) [15] (Figure 1).

2.2. Synthetic Potential

Benzoxazoles are heterocyclic targets of substantial importance in the pharmaceutical sector. As discussed before, benzoxazole is a multifunctional nucleus susceptible to chemical modifications with varied functionalities or chemical entities to create a whole new series of organic molecules with complex architecture of multifaceted pharmacological applications. This heterocyclic scaffold is the core structure found within the chemical structures of several commercial drugs [15,16] including, flunoxaprofen (anti-inflammatory), benoxaprofen (anti-inflammatory), prinaberel (agonist) and zoxazolamine (skeletal muscle relaxant and uricosuric) (Figure 2).

Besides being a core of valuable framework, benzoxazoles found uses in medical field as fluorescent probes to study biological samples of patients [17,18]. In addition to this, due to natural property of being fluorescent, the benzoxazole derivatives found application for laundry purposes as optical brightener detergents [19,20] viz. 4,4'-bis(benzoxazolyl)-*cis*-stilbene and 2,5-bis(benzoxazol-2-yl)thiophene (Figure 3).

In order to get deeper insight into how organic functional groups or nuclei (rings) that comprise any given structure of a drug or drug like candidate would exert specific biologic response, a careful understanding of physical and chemical properties of this component is highly amenable.

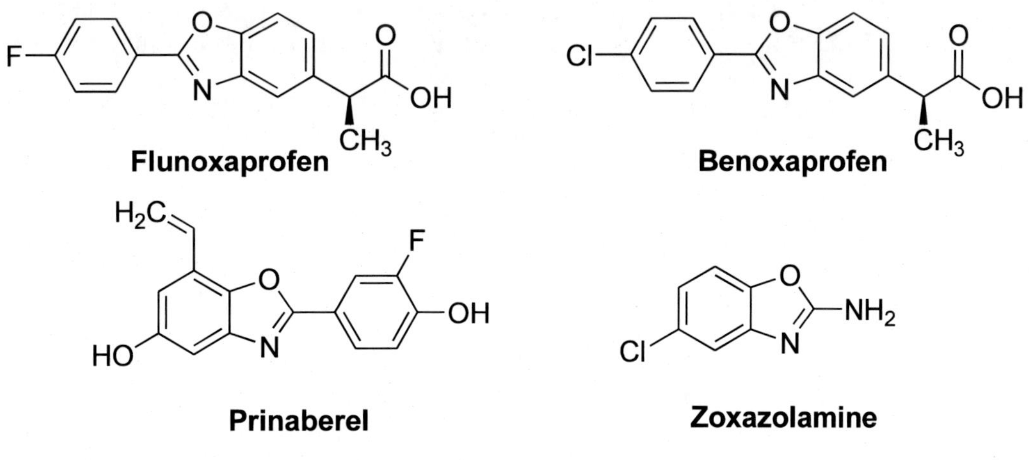

Figure 2. Commercially marketed benzoxazole-based drugs.

4,4'-Bis(benzoxazolyl)-*cis*-stilbene **2,5-Bis(benzoxazole-2-yl)thiophene**

Figure 3. Commonly marketed benzoxazole-based laundry detergents.

2.3. Physical Properties

To design a better medicinal agent or to improve the drug efficacy, it is important for a practicing pharmacist to better investigate the physical and chemical properties of the medicinal agents that he/she is dispensing. Benzoxazole is an aromatic heterocyclic organic compound (molecular formula C_7H_5NO and pK_a: 4.5) that contain benzene and the oxazole ring as a single unit. The oxygen (*O*-1) and nitrogen (*N*-3) atoms of benzoxazole can actively engage in hydrogen bonding to the host molecules. The aromatic ring aids in forming π-π stacking and π-cation interactions with the targeting host. Furthermore, the lipophilic character of the benzoxazole moiety adds favourable hydrophobic interactions to the biologic target [6].

Theoretically, it poses five possible directions (Figure 4) for molecular modification with a variety of chemical entities or groups. Accordingly, the new chemical structure may be *mono-*, *di-* or *tri-* or *tetra-* or *penta-* substituted derivatives of the benzoxazole-appended framework. The knowledge of physical and chemical properties of a molecule will allow the pharmacist to predict formulation problems, unpleasant interactions with other drugs as well as aid in studying their structure-activity relationships (SAR) to better understand what structural features of the molecule contribute to, or take away from, the desired biological activity of the molecule of interest.

2.4. Chemical Development for Library Synthesis

A good chemical synthesis, which is extremely reproducible, both in chemistry and product purity is the key concern for commercialising a drug chemical. Different protocols have been developed to access this interesting class of benzoxazole frameworks. The massive synthetic significance of benzoxazole-based derivatives can be ascertained from regular attempts that have been made very recently [21-29] to fashion libraries of this scaffold. Among the documented routes, the fundamental method is the condensation of 2-amino phenol with carboxylic acid derivatives under appropriate reaction conditions and reagents [30] (Figure 4).

3. Benzoxazole: Pharmacological Potential

Drug design is an ingenious blueprint of finding innovative prescriptions based on the understanding of a biological target associated with a specific pathology. Due to several advances on many fronts, including the ability of synthetic chemists to prepare libraries containing thousands of bioactive benzoxazole-based compounds, it is difficult for a medicinal chemist to understand the physicochemical properties of the benzoxazole-based molecules without careful compilation of research reports. Therefore, a systematic documentation of literature discussing useful pharmacological potential of benzoxazole-based heterocycles for effective medications is highly in demand.

For convenience, the following sections are grouped according to the different pharmacological activities of benzoxazole-based heterocycle.

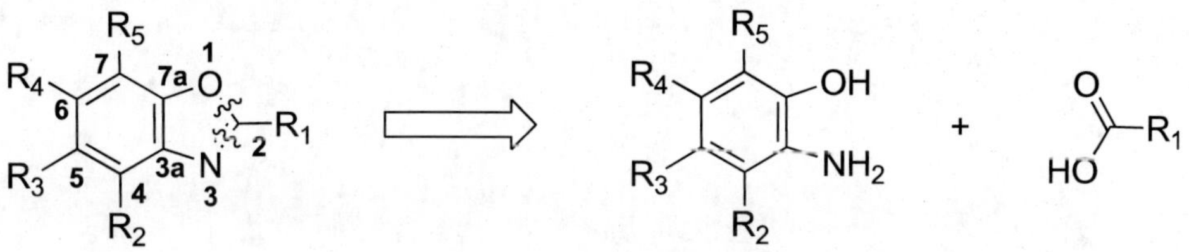

Figure 4. Pictorial representation of retro synthesis of benzoxazole nucleus.

3.1. Antimicrobial

Benzoxazole-based heterocyclic contribution to the antibiotics is well-established by the presence of a benzoxazole nucleus in natural antibiotic products such as UK-1, caboxamycin, A 16886A and calcimycin (Figure 1). Most of the pathogens have developed a novel mechanism of resistance against old existing drugs that thereby causes therapeutic collapse and an ever-increasing public health hazards.

Benzoxazoles are clinically approved therapeutic scaffold for treatment of bacterial and microbial ills and infections. In a need of new compounds to fight against microbial pathogens, various research reports have been listed in recent years.

Y. Sheena Mary and co-workers [31] reported the novel methodology for the synthesis of 5-[(4-methylphenyl)acetamido]-2-(4-*tert*-butylphenyl)benzoxazole **1** (Figure 5) followed by their broad spectrum antibacterial assessment against Gram-positive and Gram-negative bacteria and yeasts with minimum inhibitory concentrations, MIC ranging between 8 to 128 μg/mL. Whereas, its antifungal activity against *Candida krusei* was observed to be of the same potency (MIC: 64 μg/mL) as that of standard fluconazole drug.

In another study, Tugba Ertan-Bolelli and co-workers [32] reported the synthesis and antimicrobial evaluation of 5-amino-2-(4-substitutedphenyl/benzyl)benzoxazole **2**, 2-substituted-5-(4-nitrophenylsulfonamido)benzoxazole **3** and 2-substituted-5-(4-aminophenyl-sulfonamido)benzoxazole **4** derivatives (Figure 6). Further, the derivatives **2** with X = -, R = H; X = -, R = F and X = -, R = CH_3 exhibited significant antimicrobial activity against *Pseudomonas aeruginosa* ATCC 27853, *Staphylococcus aureus* ATCC 29213, *Escherichia coli* ATCC 25922, *Enterococcus faecalis* ATCC 29212 and *Mycobacterium tuberculosis*

H37RV ATCC 27294 and their drug-resistant isolates *Candida albicans* ATCC 10231 and *Candida krusei* ATCC 6258 with MIC ranging between 256 to 8 μg/mL against both *M. tuberculosis* and its drug-resistant isolate.

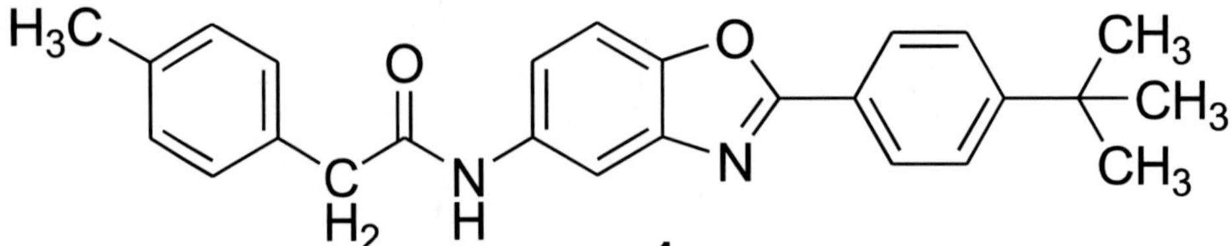

1

Figure 5. 5-[(4-Methylphenyl)acetamido]-2-(4-*tert*-butylphenyl)benzoxazole **1**.

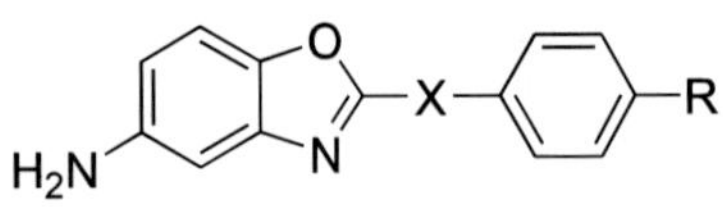

2

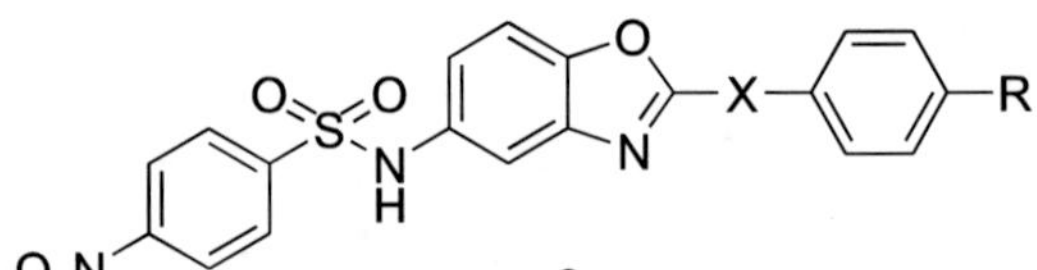

3

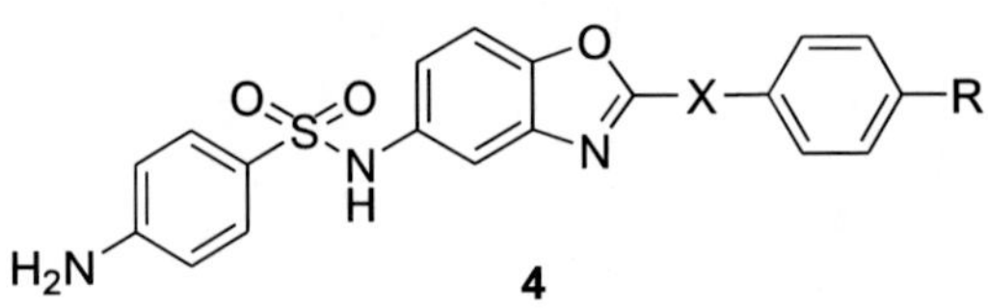

4

X = -, CH_2
R = H, Cl, F, Br, C_2H_5, CH_3, OCH_3

Figure 6. Benzoxazole derivatives **2-4**.

Shana Parveen S. and group [33] performed the synthesis and antimicrobial evaluation of 5-ethylsulphonyl-2-(4-aminophenyl)benzoxazole **5** (Figure 7) against *Staphylococcus aureus* ATCC 29213, *Staphylococcus aureus* isolated (MRSA), *Enterococcus faecalis* ATCC 29212, *Enterococcus faecalis* isolated (VRE), *Escherichia coli* ATCC25922, *Escherichia coli* isolated (ESBL), *Pseudomonas aeruginosa* ATCC 27853, *Pseudomonas aeruginosa* isolated (gentamycin resistant), *Candida albicans* ATCC 10231 and *Candida krusei* ATCC 6258 and found that **5** possess lower activity against *Candida albicans* with MIC value of 64 μg/mL than the compared reference drug fluconazole and amphotericin B and the same activity as that of fluconazole against *Candida krusei* with MIC value of 64 μg/mL.

For the amino-liked benzoxazolyl quinazolines derivatives **6** (Figure 8), Lingaladinne Mallikarjuna Reddy L. et al. [34] reported that the Gram-positive bacteria found to be more susceptible than Gram-negative one.

In another study, carried by the same research group Padamvathi Venkatapuram et al. [35] reported the synthesis of a new class of sulfonylmethane linked bisheterocycle of benzoxazole **7** (Figure 8) and screened them for antibacterial and antifungal activity against the microorganisms *Staphylococcus aureus*, *Bacillus subtilis* (Gram-positive) and *Pseudomonas aeruginosa*, *Klebsiella pneumonia* (Gram-negative), *Aspergillus niger* and *Penicillium chrysogenum* at two 50 and 100 mg/mL concentrations and the results indicated that the tested derivatives found to be more susceptible towards the Gram-negative bacteria than Gram-positive bacteria.

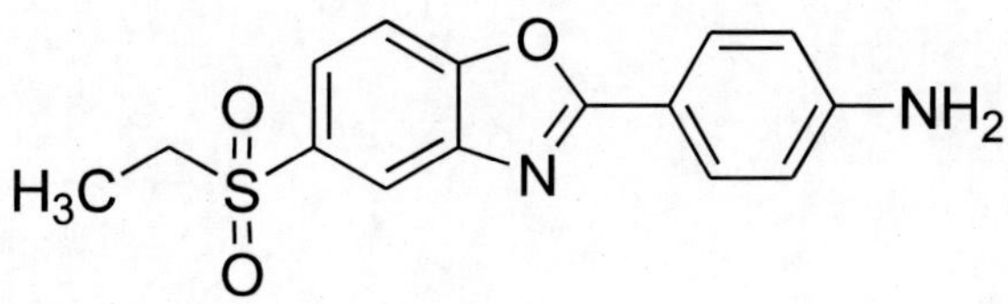

5

Figure 7. 5-Ethylsulphonyl-2-(4-aminophenyl)benzoxazole **5**.

In another study from the same research group, D. Seenaiah and co-workers [36], reported the synthesis and significant antibacterial activity of thio, methylthio and amino linked pyrimidinyl benzoxazoles **8** (Figure 8) against Gram-positive (*Staphylococcus aureus*), Gram-negative (*Escherichia coli*) and *Pseudomonas aeruginosa* and antifungal activity against *Aspergillus niger* and *Penicillium chrysogenum*.

Vikas S. Padalkar and group [37] reported the best broad-spectrum antimicrobial activity dipodal-benzoxazole derivative **9** (Figure 9) against bacterial strain (*Escherichia coli*, *Staphylococcus aureus*) and fungal strain (*Candida albicans*, *Aspergillus niger*) cultures with MIC value of 62.5 μg/mL.

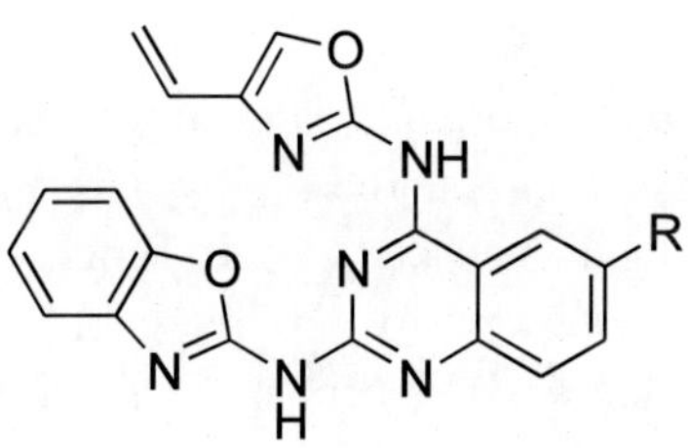

6

R = H, Br, Cl, CH_3, OCH_3, NO_2

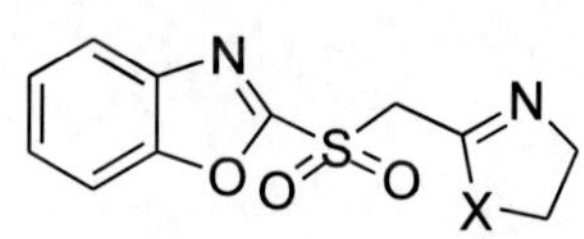

7

X = O, S, NH

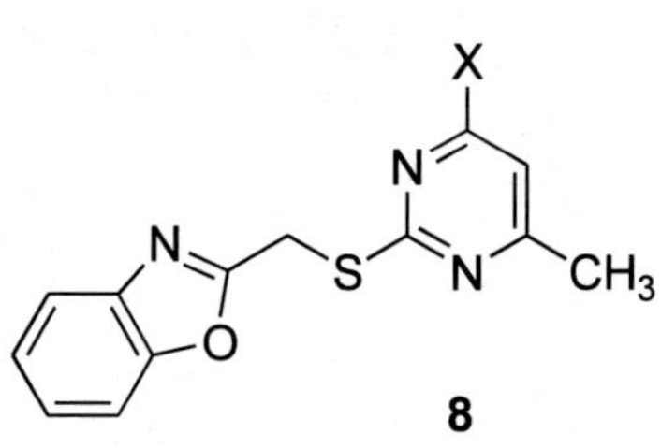

8

X = Cl ,

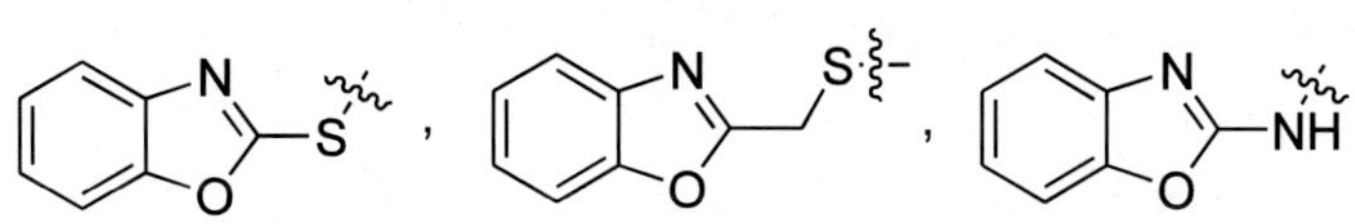

Figure 8. Benzoxazole-based quinazoline **6**, bisheterocycle **7** and pyrimidine **8**.

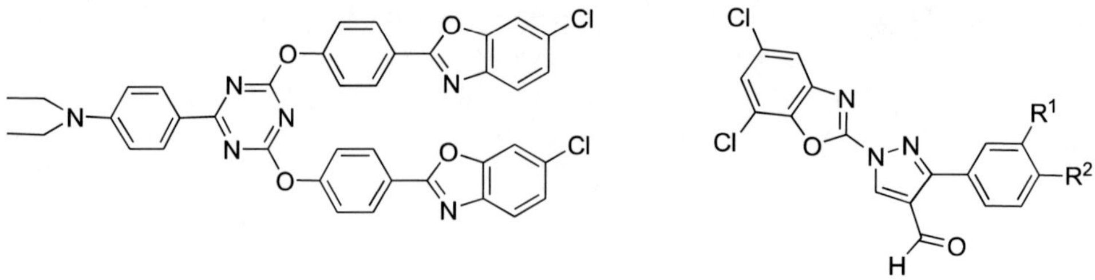

9 **10**

R^1 = H, Cl
R^2 = OCH_3, OH, CH_3, Br, Cl

Figure 9. Benzoxazole-based dipodal derivative **9** and pyrazolyl derivative **10**.

N. D. Jayanna et al. [38] reported the synthesis and *in-vitro* antimicrobial activity of the novel 1-(5,7-dichloro-1,3-benzoxazole-2-yl)-3-phenyl-1*H*-pyrazole-4-carbaldehyde derivative **10** (Figure 9) against four bacterial strains, namely *Staphylococcus aureus*, *Bacillus subtilis* (Gram positive), *Salmonella typhi*, *Escherichia coli* (Gram negative) and the two fungal strains, namely *Aspergillus niger* and *Candida albicans*. The results indicated that all the tested compounds were found to be equipotent with the compounds having R^1 = H, R^2 = OH and R^1 = H, R^2 = Br exhibited higher antimicrobial potency as compared with standard drugs ciprofloxacin and fluconazole.

C. H. Lakshmi Praveena et al. [39] performed the synthesis and antimicrobial evaluation of 6-nitro-1*H*-benzo[*d*]oxazole/thiazole-2-yl)methyl)-6-oxido-4,8-dihydro-1*H*-[1,3,2]dioxaphosphepine[5,6-c]pyrazole-6-yl)urea/carboximide **11/12** (Figure 10) against the *Staphylococcus aureus* (Gram-positive), *Bacillus Cereus*, *Escherichia coli* (Gram-negative), *Pseudomonas aeruginosa*, *Aspergillus niger*, *Candida albicans* organisms. The results from the study indicated that among the tested derivatives, the compound **11** (R = Cl) and **12** exhibited higher antibacterial and antifungal activity against bacteria and fungi, while compounds **11** having R = H, CH_3 and OCH_3 show lower or no activity under given experimental conditions when tested using amoxicillin, cefaclor and ketoconazole as reference.

Afaf Bouchoucha and group [40] reported the isolation of a new benzoxazole ligand namely 2-(5-methylpyridine-2-yl)benzoxazole **13** along with its iron (II) and nickel (II) metal complexes **14** (Figure 11) and further screened them for *in-vitro* antimicrobial activity, acute and subacute toxicity against different human bacteria and fungi. The outcome of the performed studies indicated that **13** and **14** exhibited an excellent bacteria growth inhibition on *Staphylococcus aureus* and *P. aeruginosa* strains.

A. M. Vijesh et al. [41] screened the new Schiff's bases **15** and **16** derived from benzoxazole analogues containing pyrazole moiety (Figure 12) for antimicrobial evaluation. The tested compounds **15** and **16** exhibited antibacterial activity against *S. aureus*, *B. subtilis* and *E. coli* with MIC ranging between 3.125 to 12 μg/mL and antifungal activity against *C. albicans* with MIC ranging between 6.25 to 15μg/mL.

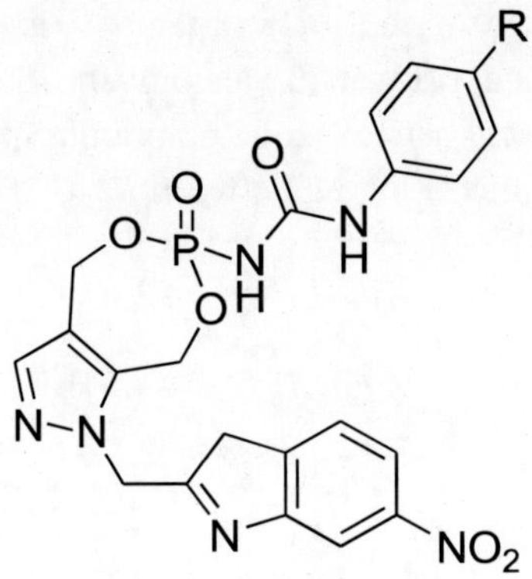

11

R = H, CH_3, OCH_3, Cl

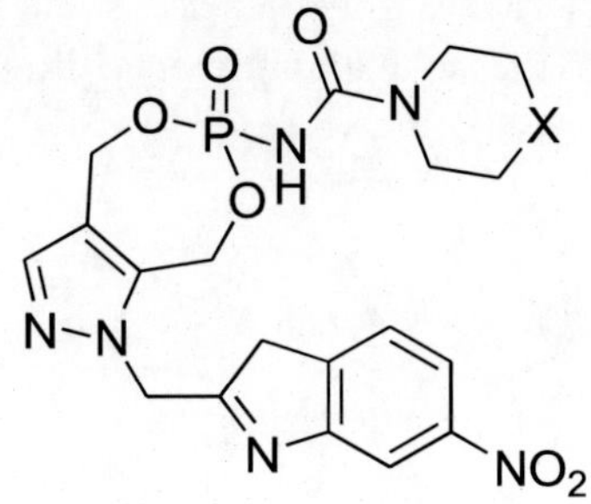

12

X = O, CH_2, NCH_3

Figure 10. 6-Nitro-1*H*-benzo[*d*]oxazole/thiazole-2-yl)methyl)-6-oxido-4,8-dihydro-1*H*-[1,3,2]dioxaphosphepine[5,6-c]pyrazole-6-yl)urea/carboximide **11/12**.

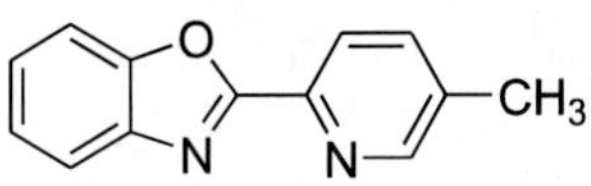

13

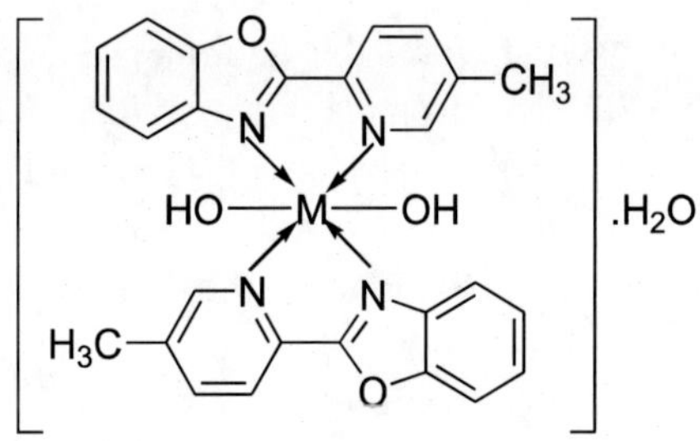

14

M = Fe, Ni

Figure 11. 2-(5-Methylpyridine-2-yl)benzoxazole **13** and its iron (II) and nickel (II) metal complexes **14**.

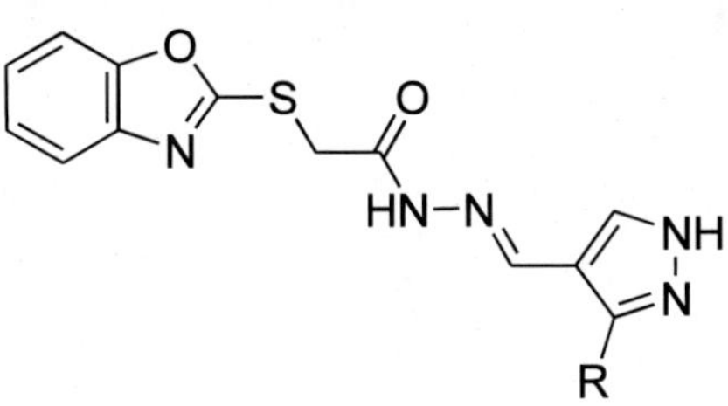

15

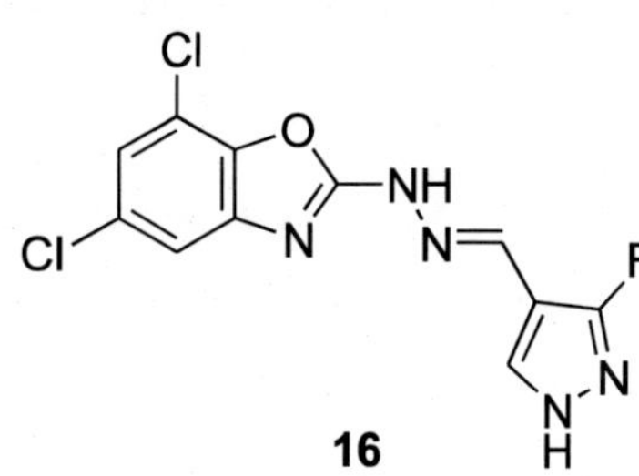

16

R = 2,4-$(Cl)_2C_6H_3$, 4-$SCH_3C_6H_4$, $C_{12}H_{10}$, 2,5-dichlorothiophene

Figure 12. Benzoxazole containing pyrazolyl Schiff's bases **15** and **16**.

Mustafa Arisoy et al. [42] reported the broad spectrum of antibacterial and antifungal activities of 2-(4-substituted-benzyl)-5-[[4-(4-chloro/fluoro-phenyl)piperazin-1-yl]acet-amido]-benzoxazole **17** and 2-(4-substituted-benzyl)-5-substituted-benzoxazole **18**, (Figure

13) against standard strains and drug resistant isolates of the microorganisms and further, compared the results with ampicillin, gentamicin sulfate, ofloxacin, vancomycin, fluconazole and amphotericin B trihydrate drugs. They found the tested benzoxazoles compounds to possess similar activities to that of reference drugs with MIC values ranging between 32 to 256 μg/mL.

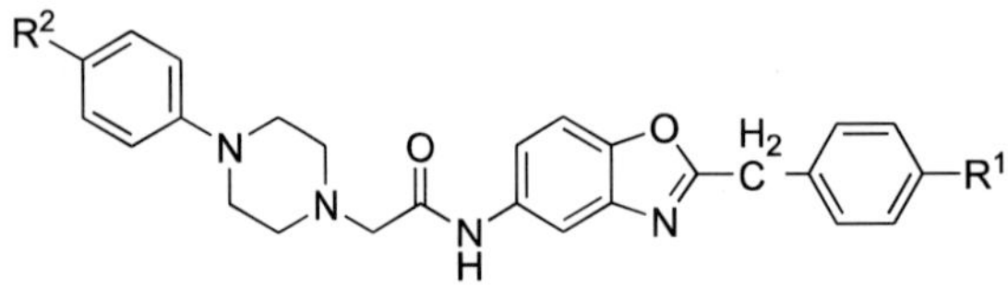

17

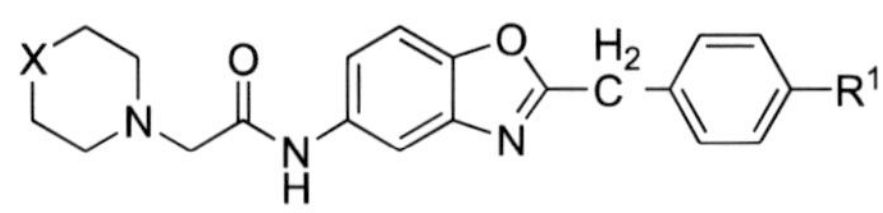

18

R^1 = Cl, F
R^2 = Cl, CH_3, H, Br, F

X = O, CH_2, CH_3-N, C_6H_5-N
R^1 = Cl, CH_3, H, Br, F

Figure 13. 5-Substituted benzoxazoles **17** and **18**.

As discussed before, the natural product pseudopteroxazole (Figure 1) is an effective antibiotic exhibiting strong activity against *Mycobacterium tuberculosis* with tenuous toxicity. Malcolm W. B. McCulloch et al. [43] reported the synthesis and potent antimicrobial evaluation of non-natural pseudopteroxazole **19** and *iso*-pseudoptreoxazole **20** (Figure 14) against a range of important Gram-positive pathogens including *Mycobacterium* spp. and vancomycin-resistant *Enterococcus faecium.* The results of evaluation indicated that all non-natural pseudoptreoxazoles **19** (except R = 2-$OCH_3C_6H_4$) exhibited broader spectrum of activity against methicillin-resistant *Staphylococcus aureus* compared to pseudopteroxazole.

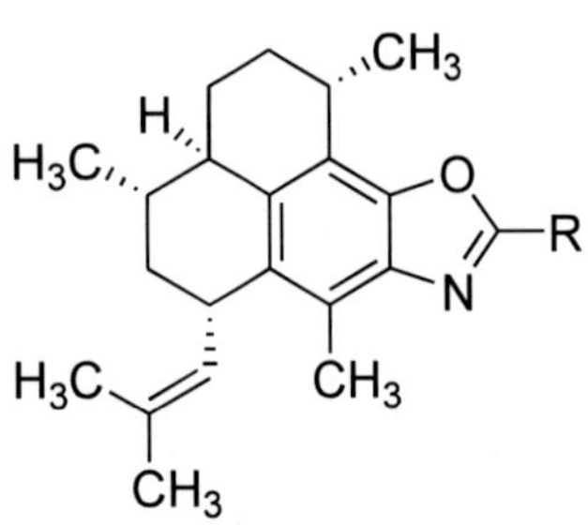

19

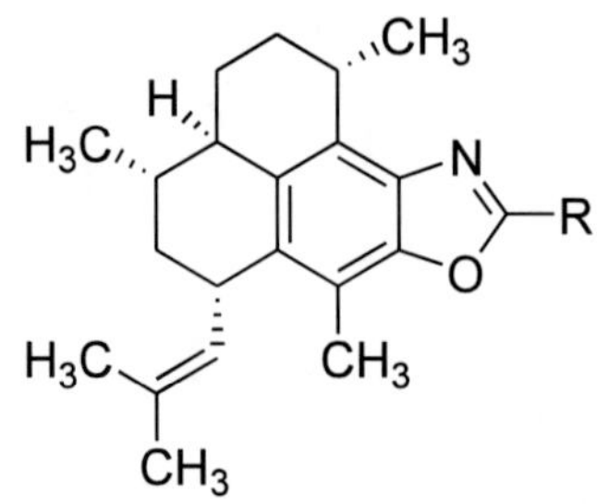

20

R = H (Natural product Pseudoptreoxazole or *iso*-Popseudoptreoxazole)
R = 2-$OCH_3C_6H_4$, CH_3, $CH(CH_3)CH_2CH_3$, $CH_2CH_2SCH_3$, $CH_2C_6H_5$, $CH_2C(O)NH_2$, $CH(OH)CH_3$, $CH_2CH_2C(O)OCH_3$, $CH_2CH_2C(O)OH$

Figure 14. Non-natural pseudopteroxazole **19** and *iso*-pseudoptreoxazole **20**.

Hadi Adibi and research group [44] reported the synthesis and *in-vitro* biological evaluation of a series of catecholthioethers containing benzoxazole **21** (Figure 15) against three Gram-positive bacteria including *Staphylococcus aureus* (clinical isolated), *Staphylococcus aureus* ATCC 25922, *Enterococcus faecium* (clinical isolated) and two Gram-negative bacteria including *Klebsiella pneumonia* (clinical isolated) and *Pseudomonas aeruginosa* 27853 and the yeast *Candida-albicans* in comparison with control drugs Sulfamethoxazole and Fluconazole. The synthesized compounds possessed a broad-spectrum of activity against the tested microorganisms with MIC: between 4 to 256 μg/mL range.

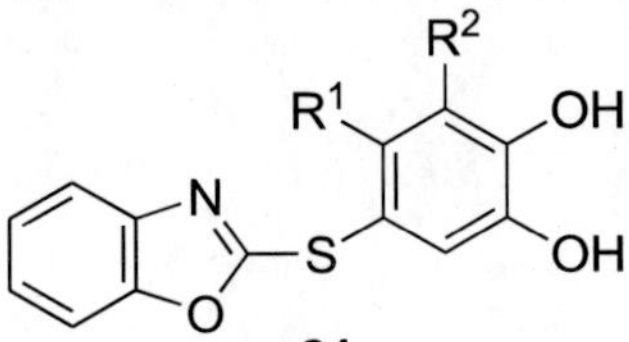

21

R^1 = H, $C(CH_3)_3$, CH_3
R^2 = H, OCH_3

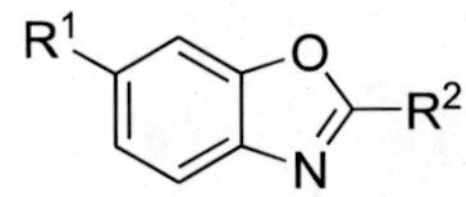

22

R^1 = H, Cl
R^2 = 2-thienyl, 5-methyl-2-thienyl, 2-thienyl, 5-methyl-2-thienyl, 3-pyridyl, C_6H_5, 4-$CH_3C_6H_4$, 4-$NO_2C_6H_4$, 3-OCH_3-4-OHC_6H_3

Figure 15. Benzoxazole-based heterocycles **21** and **22**.

Ravi Kumar and co-workers [45] screened the 2-aryl/hetarylbenzoxazole **22** (Figure 15) for antimicrobial evaluation against *Bacillus subtilis*, *Bacillus stearothermophilus*, *Escherichia coli*, *Pseudomonas putida*, *Aspergillus flavus* and *Aspergillus niger* strains. Among the screened benzoxazoles the compounds having R^1 = H, R^2 = 2-thienyl; R^1 = H, R^2 = 5-methyl-2-thienyl; R^1 = Cl, R^2 = 2-thienyl; R^1 = Cl, R^2 = 3-methyl-2-thienyl; R^1 = Cl, R^2 = 3-pyridyl; R^1 = Cl, R^2 = C_6H_5 and R^1 = Cl, R^2 = 4-$NO_2C_6H_4$ possessed equipotent activity with MIC 100 μg/mL.

Ananya Srivastava et al. [46] synthesized and evaluated the combined biological impact of benzoxazolyl-triazole heterocycle **23** (Figure 16). In particular the compounds **23** having R^1 = R^2 = R^3 = H; R^1 = R^2 = H, R^3 = 4-NO_2 and R^1 = R^3 = H, R^2 = 4-OCH_3 showed potent Gram-negative activity against clinical isolates of *Staphylococcus aureus* and *Escherichia coli.*

Mustafa Arisoy and co-workers [47] carried out the synthesis of 5-(4-substituted benzamido/phenylacetamido)-2-(4-*tert*-butylphenyl)benzoxazole **24** (Figure 17) and tested their antibacterial profile against *Staphylococcus aureus*, *Enterococcus faecalis*, *Escherichia coli*, *Klebsiella pneumonia*, antifungal profile against *Candida albicans* and *Candida krusei* and antimycobacterial profile against *Mycobacterium tuberculosis.* All the compounds displayed broad-spectrum activity against sensitive strains with MICs range between 16 to 128 μg/mL. Among the tested benzoxazoles the compound with X = -, R = H exhibited significant antibacterial activity against *Enterococcus faecalis* isolate with MIC value of 16 μg/mL and twice the potency of the compared vancomycin and gentamicin sulfate drugs. Further, it also showed moderate antitubercular activity against *Mycobacterium tuberculosis* and its clinical isolate with MIC ranging between 8-128 μg/mL.

From the biological evaluation study conducted by Liang Ouyang and co-workers [48], a novel benzoxazol-2-amine **25** (Figure 18) screened for antibacterial activity against ATCC bacterial strains (*S. aureus* 25923, *E faecalis* 29212 and *E. coli.* 29425) showed MIC ranging from 0.5 to > 64 μg/mL. The results of the study concluded that the benzoxazoles containing more substituted halogen groups and propanediamine linker were found to be more potent compared to non-halogenated ones.

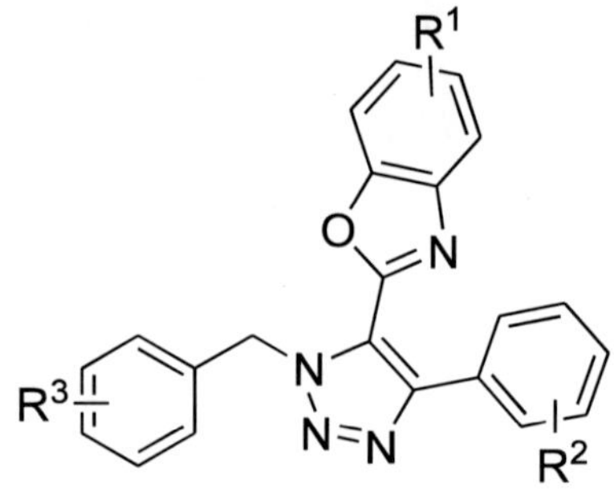

23

R^1 = H, 5-CH_3
R^2 = H, 4-OCH_3, 4-CH_3, 2-Br, 4-NO_2, 4-CF_3
R^3 = H, 2,5-$(F)_2$, 4-CN, 4-NO_2, 4-F, 4-CH_3

Figure 16. Benzoxazolyl-triazole heterocycle **23**.

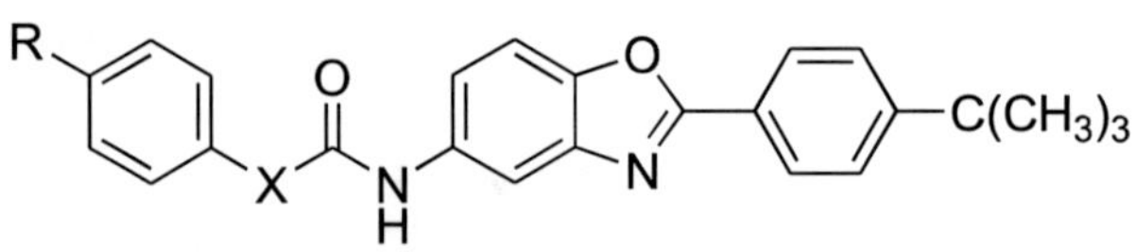

24

X = -, CH_2
R = H, Cl, F, Br, CH_3, NO_2, CN, OCH_3, C_2H_5, C_4H_9

Figure 17. 5-(4-Substituted benzamido/phenylacetamido)-2-(4-*tert*-butylphenyl)benzoxazole **24**.

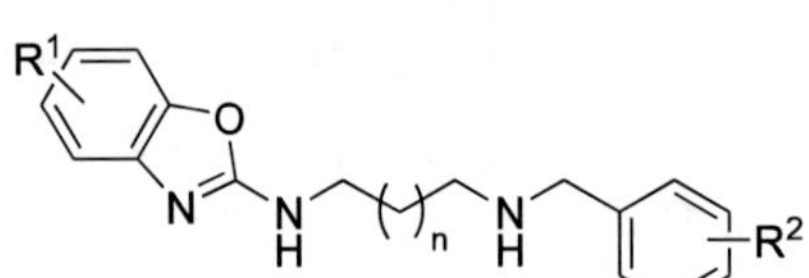

25

n = 0, 1
R^1 = H, 6-OCH_3, 6-Cl, 6-F, 5-F, 5,6-$(F)_2$
R^2 = H, 4-Cl, 3-Br, 4-OH, 4-OCH_3, 3,4-$(Cl)_2$, 2,4-$(Cl)_2$

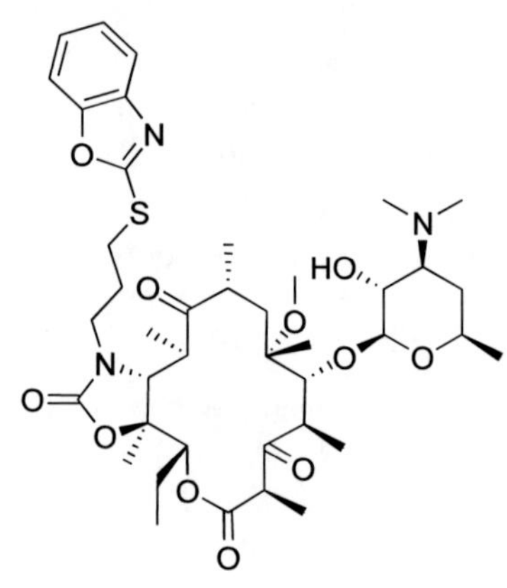

26

Figure 18. Biologically active benzoxazol-2-amine **25** and ketolide **26**.

Xiao-zhuo Chen and group [49] reported the newly synthesized ketolide with benzoxazole heterocycle **26** (Figure 18) as promising antibacterial activity against key macrolide sensitive and resistant respiratory pathogens.

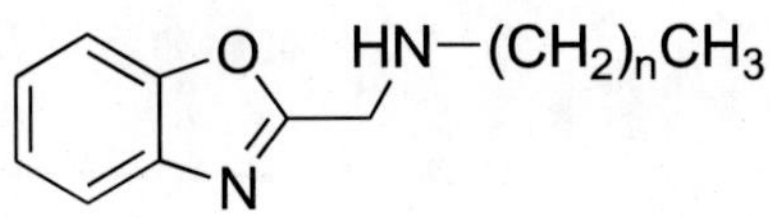

27

n = 5, 6, 7

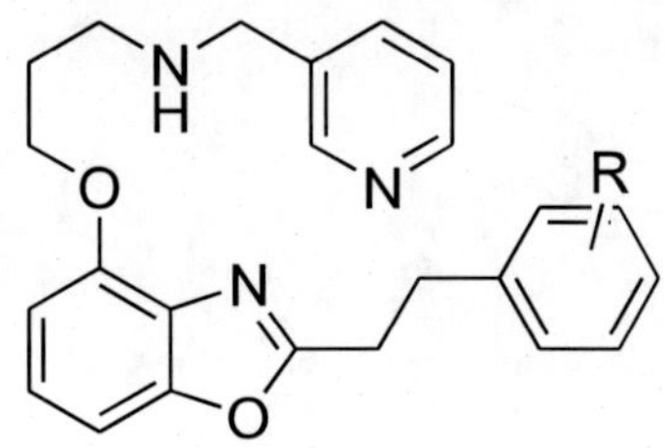

28

R = 2-F, 4-F, H, 2,4-$(F)_2$, 2,6-$(F)_2$, 2,3,4-$(F)_3$, 2,4,6-$(F)_3$, 3-NO_2

Figure 19. Biologically active benzoxazoles **27** and **28** as NMT inhibitors.

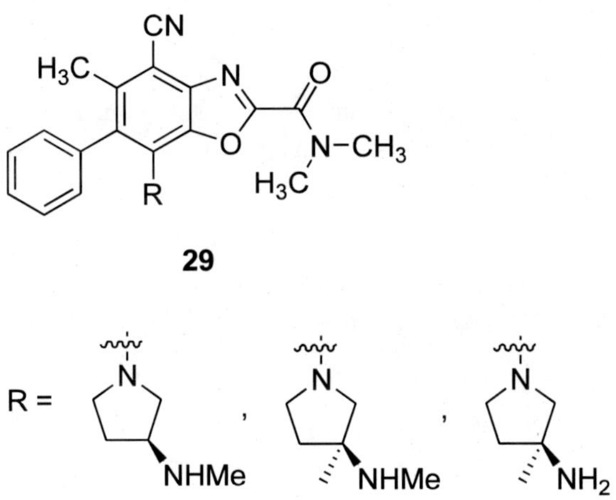

29

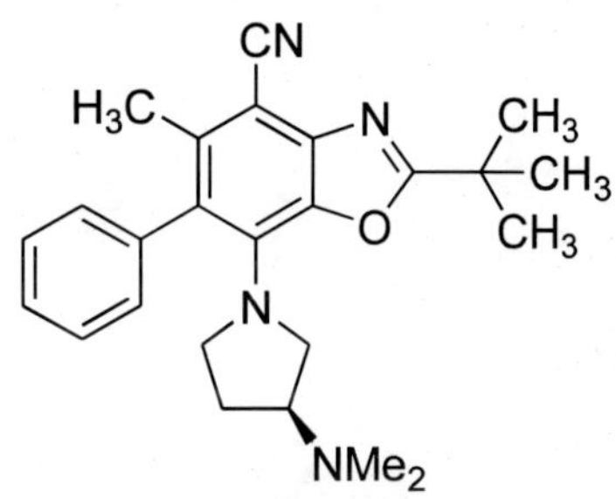

30

Figure 20. Benzoxazole-4-carbonitriles **29** and **30**.

Chunquan Sheng and co-workers [50], reported the benzoxazole scaffold **27** (Figure 19) for the antifungal activity against both standard and clinically resistant fungal pathogens.

In another study, by the same research group Chunquan Sheng et al. [51] reported *in-vitro* antifungal activity of a series of benzoxazoles **28** (Figure 19) as *N*-myristoyltransferase (NMT) inhibitors. Further the molecular docking studies revealed that the hydrogen bonding interaction between the benzoxazole core and NMT are essential for the orientation of the NMT-inhibitor. Among the tested compounds the antifungal activity of benzoxazole **28** having R = 2,3,4-$(F)_3$ was found to superior than fluconazole and therefore, can be a hopeful candidate for further structural diversity studies of the benzoxazole-based NMT inhibitors.

Jun-ichi Kuroyanagi and research group [52] described the synthesis of a series of 1,3-benzoxazole-4-carbonitriles **29** (Figure 20) and evaluated for antifungal activity, solubility and metabolic stability. Among these benzoxazoles the derivative with R = (3*S*)-3-methyl-3-(methylamino)pyrrolidin-1-yl was found to be a superior scaffold structure exhibiting potent *in-vitro* activity against *Candida* species, higher water solubility, and improved metabolic stability compared to lead compound **30** (Figure 20). It further showed potent *in-vivo* efficacy and bioavailability against mice *Candida* infection.

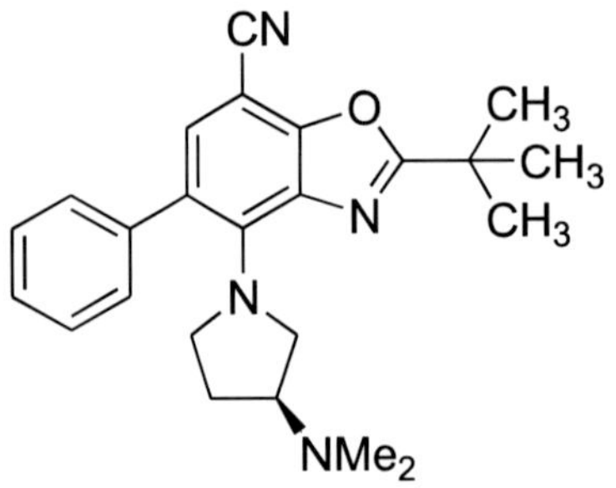

31

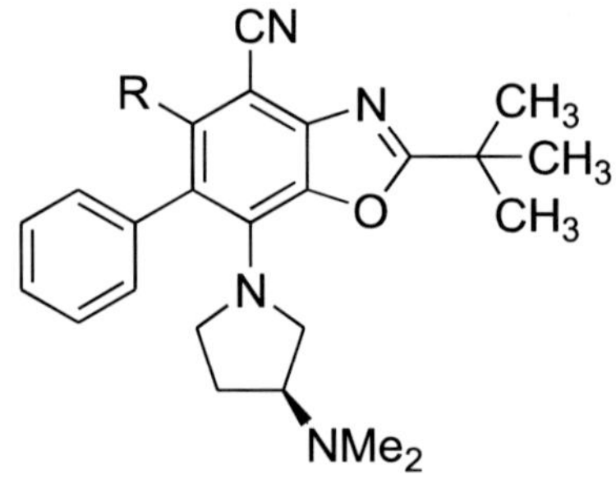

32

R = H, CH_3

Figure 21. Benzoxazole-4-carbonitriles **31** and **32**.

From another study, by the same group of researchers [53] *in-vitro* antifungal evaluations of 1,3-benzoxazole-7-carbonitrile **31** and 1,3-benzoxazole-4-carbonitrile **32** (Figure 21). Among them, 1,3-benzoxazole-4-carbonitrile **32** with R = CH_3 was found to the most potent with moderate growth inhibition against *Candida* species compared to 5-desmethyl derivative with R = H.

3.2. Analgesic and Anti-Inflammatory

Analgesic, anti-inflammatory and anesthetics all comes under the same drug class that is medicated to relieve pain perception. The success of this research is limited because the analgesic and the anti-inflammatory agents have a clear cut demarcation from anesthetics in a sense that analgesic refers specifically to pain killing drugs, whereas anti-inflammatory agents often have a non-inflammatory/swelling effect associated with their pain killing action. On the other hand, an anesthetic (general or local) inhibit the pain transmission by hindering voltage-regulated sodium and calcium channels.

Typically, analgesics are classified in many ways, such as cyclooxygenase COX-2 inhibitors, medical cannabis, nonsteroidal anti-inflammatory drugs/agents NSAID/NSAIAs, opiods, alcohols and paracetamol/acetaminophen, etc.

In addition to being antibiotic, the reported benzoxazole-pyrazole Schiff's bases **15** and **16** (Figure 12) have also been found to possess good analgesic properties as NSAID conjugates.

The Schiff's bases of methyl 2-amino-*N*-(substituted arylidene)benzoxazole-5-carbohydrazide **33** (Figure 22) showed good to moderate activity when compared with the standard refecoxib [54]. Among the tested benzoxazoles, the Schiff's bases **33** with R = 4-OHC_6H_4, 4-$N(CH_3)_2C_6H_4$, 4-ClC_6H_4 and 4-$NO_2C_6H_4$ presented moderate activity with IC_{50}:

12.69, 19.39, 20.13 and 23.85%, respectively compared to standard refecoxib drug (IC_{50}: 7.79%). Further, the exceptionally potent COX-2 inhibitory activity was observed with R = 4-$OCH_3C_6H_4$, 4-FC_6H_4, 2-OHC_6H_4 and C_6H_5 with IC_{50}: 6.40, 10.72, 8.01 and 7.61%, respectively.

Recently, Kapileshwar Seth and research group [55], evaluated the anti-inflammatory activity 2-(2-arylphenyl)benzoxazole **34** (Figure 23) for cyclooxygenase-2 (COX-2) enzyme. The benzoxazoles **34** with R^1 = H, R^2 = Cl, R^3 = OCH_3; R^1 = H, R^2 = Cl, R^3 = OH and R^1 = R^2 = H, R^3 = SO_2CH_3 exhibited selective inhibition for COX-2 only. The selectivity index, SI of **34** having R^1 = H, R^2 = Cl, R^3 = OH (112.80±5.62) was found to be much better compared to that of COX-2 selective NSAID celecoxib (92.87±6.47).

Further, the *in-vivo* anti-inflammatory potency (% inhibition of rat paw volume) of benzoxazoles **34** with R^1 = H, R^2 = Cl, R^3 = OCH_3 was 53.85% at 12.5 mg/Kg and 64.42% at 25 mg/Kg and R^1 = H, R^2 = Cl, R^3 = OH was 68.27% at 12.5 mg/Kg and 82.69% at 25 mg/Kg which is comparable to that of celecoxib (60.57% at 12.5 mg/Kg and 74.61% at 25 mg/Kg) and the nonselective NSAID diclofenac (67.31% at 12.5 mg/Kg and 84.62% at 25 mg/Kg).

The benzoxazole having R^1 = R^2 = H, R^3 = SO_2CH_3 displayed better potency compared to these clinically used NSAIDs with % inhibition of 62.50 and 75.00 at 12.5 mg/Kg and 25 mg/Kg doses, respectively.

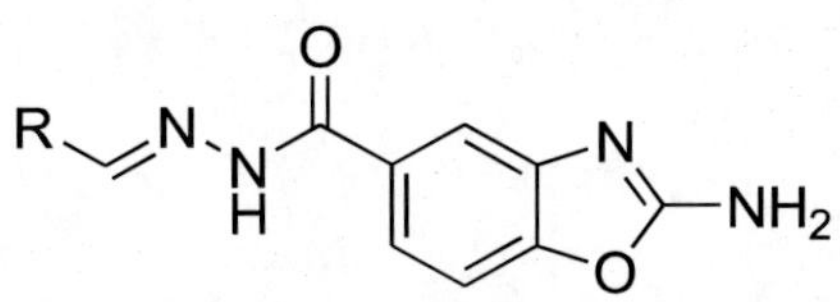

33

R = 4-OHC_6H_4, 4-$N(CH_3)_2C_6H_4$, 4-ClC_6H_4, 4-$NO_2C_6H_4$, 4-$OCH_3C_6H_4$, 4-FC_6H_4, 2-OHC_6H_4, C_6H_5

Figure 22. Benzoxazole-based Schiff's base **33** as COX-2 inhibitor.

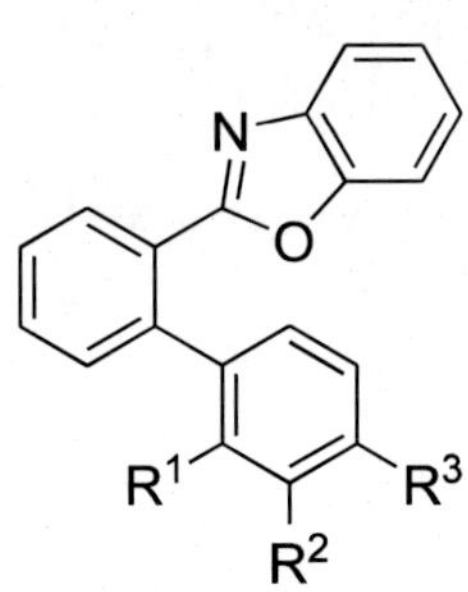

34

R^1 = H, OCH_3
R^2 = H, Cl, OCH_3, OC_5H_9
R^3 = H, OCH_3, SCH_3, $O(CH_3)_2CH$, $OCH_2C_6H_5$, F, CF_3, CHO, CN

Figure 23. Substituted benzoxazole **34** as COX-2 inhibitor.

Arunadevi Parlapalli and group [56] investigated anti-inflammatory activity of 2-amino-*N*-(substituted alkyl)benzoxazole-5-carboxamide **35** (Figure 24) and compared their results against standard diclofenac sodium. The benzoxazoles **35** having R = C_2H_5, C_3H_7 and $(CH_3)_2CH$ presented reduced the carrageenan induced inflammation with 34.60, 35.30 and 33.33% inhibitions, respectively.

Shivendra Pratap Singh and research group [57] conducted anti-inflammatory evaluation of benzoxazoles **36** and **37** (Figure 24). Further, the benzoxazoles **36** and **37** with X = CH, R^1 = CH_3, R^2= R^3 = H; X = NH, R^1 = R^3 = H, R^2 = CH_3; X = NH and R^1 =R^3 = H, R^2 = 4=FC_5H_4 displayed significant anti-inflammatory activity comparable to diclofenac sodium.

From another study, by P. Christina Ruby Stella and group [58], the anti-inflammatory activity of benzoxazole derivative **38** (Figure 25) was observed to be significant over carrageenan induced hind paw edema. Further, the benzoxazole **38** (X = N, Y = Z = CH) showed 44.05% inhibition potency compared to control indomethacin (63.80% inhibition).

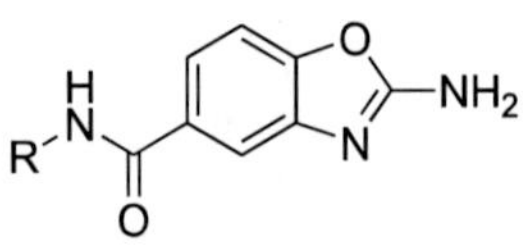

35

R = CH_3, $(C_2H_5)_2$, C_2H_5, $(CH_3)_2$, C_3H_7, $(CH_3)_2CH$, di-Pr

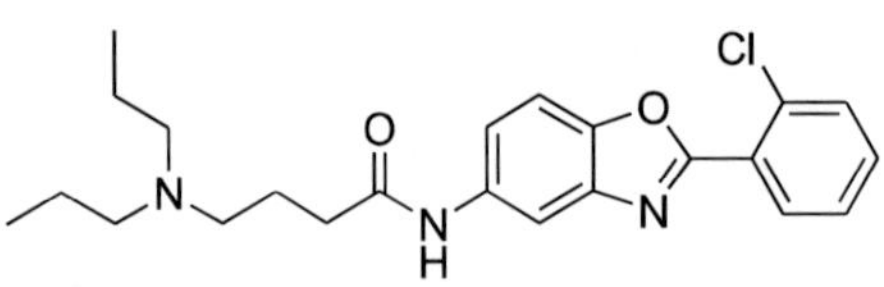

36

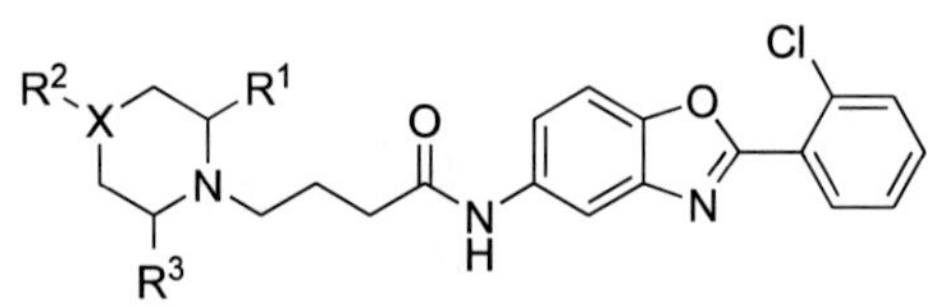

37

X = CH, NH
R^1 = H, CH_3
R^2 = H, CH_3, C_2H_5, C_6H_5, 4-FC_6H_4
R^3 = H, CH_3

Figure 24. Benzoxazoles **35**-**37** as anti-inflammatory agents.

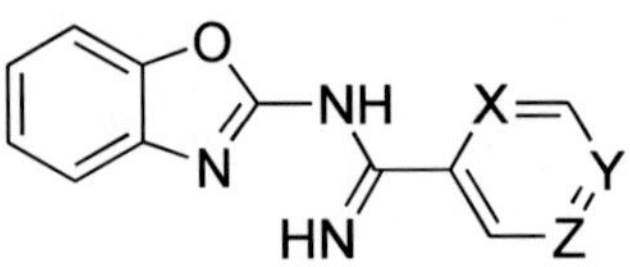

38

X, Y, Z = N, CH

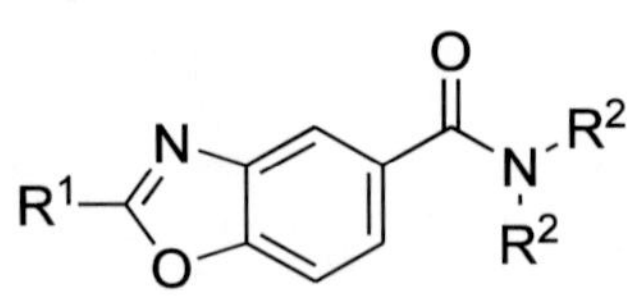

39

R^1 = H, CH_3, C_2H_5
R^2 = CH_3, C_2H_5, C_6H_5

Figure 25. Benzoxazoles **38** and **39** as anti-inflammatory agents.

In another study by Reena M and group [59] reported the anti-inflammatory screening of 2-substituted ((*N,N*-disubstituted)-1,3-benzoxazole)-5-carboxamide **39** (p: < 0.01) (Figure 25) over carrageenan induced rat paw edema. The benzoxazole **39** with $R^1 = H$, $R^2 = CH_3$ (61.69% inhibition); $R^1 = CH_3$, $R^2 = C_2H_5$ (77.47% inhibition), $R^1 = C_2H_5$, $R^2 = C_2H_5$ (75.34% inhibition) and $R^1 = H$, $R^2 = C_6H_5$ (73.97% inhibition) were found to be more potent comparable to standard diclofenac sodium (69.86% inhibition).

3.3. Antimalarial

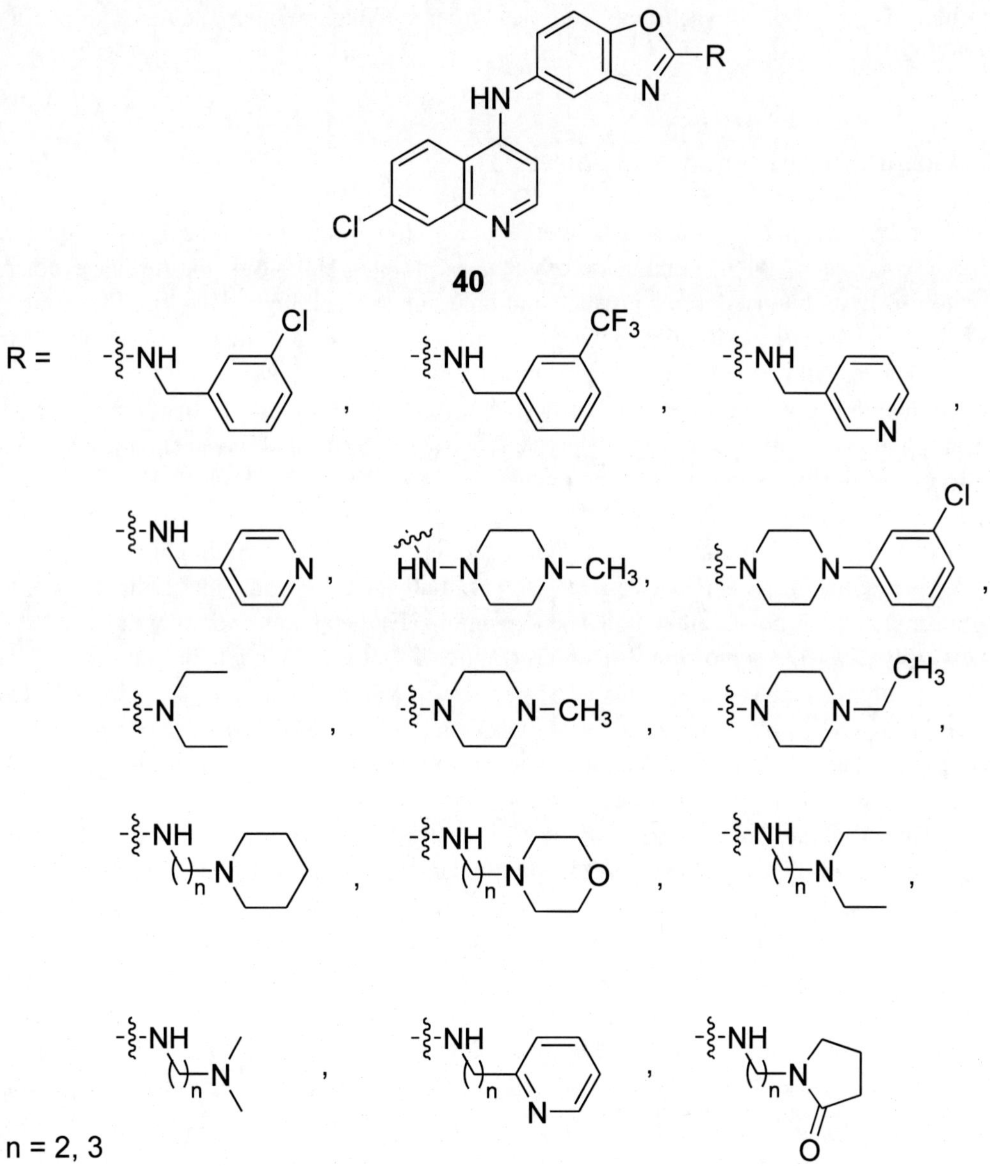

Figure 26. Benzoxazole scaffold **40** as antimalarial.

Dennis S. B. Ongarora and co-workers [60] reported the antimalarial evaluation of benzoheterocyclic analogues of amodiaquine **40** (Figure 26). The antiplasmodial activity of tested analogues was evaluated against chloroquine-sensitive NF54 and multidrug resistant K1 strain *Plasmodium falciparum*. The most potent compounds **40** having R = NH-1-ethylpyrollidine (0.015 μM), NH-1-ethylpiperidine (0.010 μM), NH-1-propylpiperidine (0.010 μM), $NHCH_2CH_2NEt_2$ (0.011 μM), $NHCH_2CH_2CH_2NEt_2$ (0.014 μM), $NHCH_2CH_2NMe_2$ (0.012 μM) had equal to or comparable IC_{50} to that of chloroquine (0.001 μM) against the chloroquine-sensitive NF54 strain. The structure-activity relationship studies of these highly promising leads suggested them for *in-vivo* antimalarial efficacy evaluation in *Plasmodium berghei* infected mice where they exhibited good activity following oral administration. In particular, **40** (R = 1-ethylpyrollidine) completely cured treated mice at a lower multiple dose (4 x 10 mg/Kg).

3.4. Antiproliferative and Anticancer

Recently, Loredana Salerno and co-workers [61] reported the benzoxazole **41** (Figure 27) as heme oxygenase 1 (HO-1) and heme oxygenase 2 (HO-2) inhibitors and further evaluated the cytotoxicity in human derived prostate and breast cancer cell lines (DU-145, PC3, LnCap, MDA-MB-231 and MCF-7). The authors concluded that imidazole ring connected to a hydrophobic benzoxazole moiety by means of alkyl or thioalkyl chains of different length is crucial for the activity these benzoxazoles. Further, the alkyl chain played a critical role to aid **41** in discriminating between the two isoforms and increasing the selectivity against one of the two isoforms of heme oxygenase. Furthermore, for both chain lengths i.e., n = 3 (IC_{50} HO-1: 16.9 μM and IC_{50} HO-2: 49 μM) and n = 4 (IC_{50} HO-1: 30.9 μM and IC_{50} HO-2: 74.1 μM) the **41** was found to possess significant HO-1 and HO-2 inhibition potency.

Werner Ginzinger and co-workers [62] initially carried out the structure-activity relationship study of novel ruthenium(II) and osmium (II) metal arene complexes with 3-(1*H*-benzoxazoly-2-yl)-1*H*-quinoxalin-2-one **42** (Figure 27) to evaluate their antiproliferative activity in three human cancer cell lines (A549, CH1, SW480). The best fit **42** ($R^1 = R^2 = R^3 =$ H) yielded IC_{50}s: 38 μM (A549), 7.5 μM (CH1) and 14μM (SW480) for cancer cell lines, respectively. Further, the complexation with osmium metal resulted in the increase in cytotoxicity compared to ruthenium metal.

Pu Xiang and co-workers [63] conducted a lead optimization study on anticancer agent lead compound **43** (Figure 28) to resolve its problematic poor solubility issue of *in-vitro* studies. The successful replacement of benzothiazole scaffold with benzoxazole moiety and introduction of *N*-methyl-piperazine group at the 2nd and 6th position of the heterocycle and led to new benzoxazole **44** (Figure 28) such that the activities were fundamentally retained. The *in-vitro* antiproliferative evaluation against human hepatocellular carcinoma (HepG2) and human colon carcinoma (HCT116) cell lines suggested that **44** led to cytosolic vacuolization at relatively lower micromolar concentrations (IC_{50} HepG2: 12.2 μM and IC_{50} HCT116: 2.61 μM) and further, constitute a novel lead for anticancer agents.

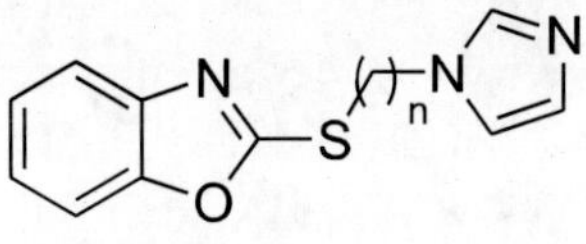

41

n = 3, 4

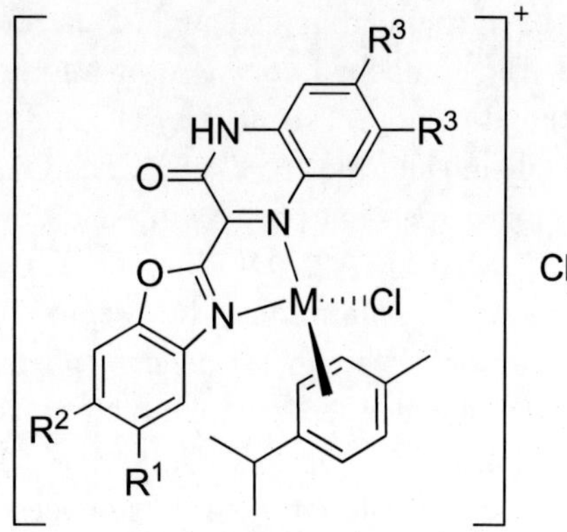

42

M = Ru, Os
R^1, R^2, R^3 = H, CH_3

Figure 27. Benzoxazole scaffolds **41** and **42** as antiproliferatives.

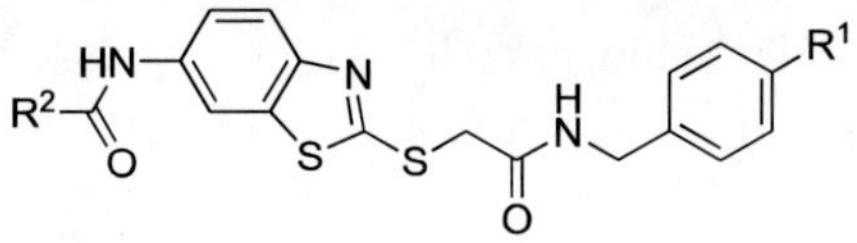

43

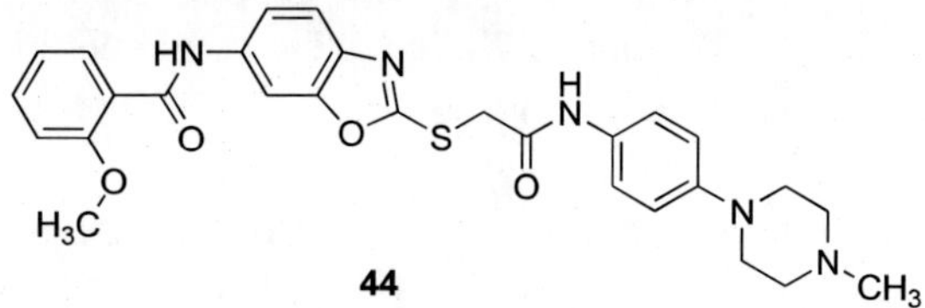

44

R^1 = H, Cl
R^2 = 2-$OCH_3C_6H_4$, $ClCH_2$

Figure 28. Benzoxazole scaffold **44** with more potent anticancer activity compared to **43**.

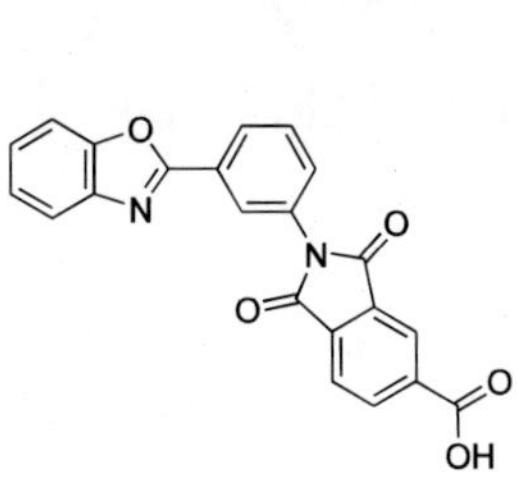

45

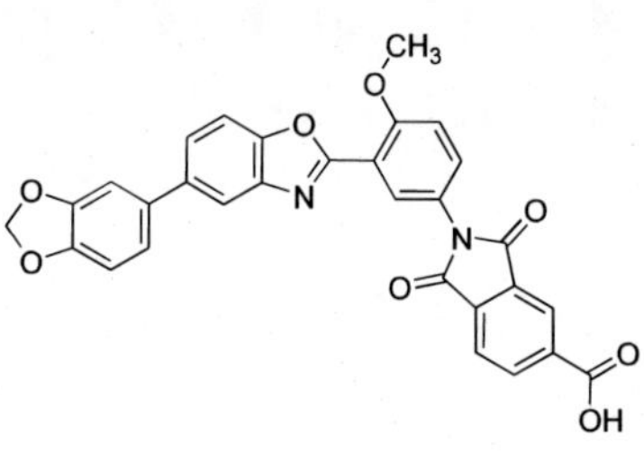

46

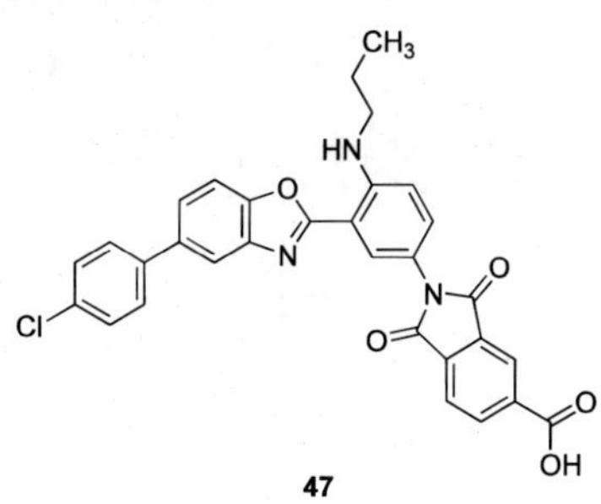

47

Figure 29. Benzoxazole-based 2,3-dihydro-1,3-dioxo-1*H*-isoindole-5-carboxylic acids **45-47** as HPSE inhibitors.

High throughput screening conducted by Li Jia and Shutao Ma [64] resulted **45** (Figure 29) as a lead inhibitor with anti-heparanase activity (IC_{50}: 8 μM) and anti-angiogenic activity (IC_{50}: 40 μM). Encouraged by results, they developed a new series of banzoxazole-based 2,3-dihydro-1,3-dioxo-1*H*-isoindole-5-carboxylic acids **46** and **47** (Figure 29) as active heparanase (HPSE) inhibitors. The investigations demonstrated that for the enhanced inhibitory activity, the carboxylic acid group, isoindole-5-carboxylic acid moiety and benzoxazole ring played crucial role. The compound **46** was shown to possess more potent anti-heparanase activity (IC_{50}: 0.2 μM) and angiogenesis activity (IC_{50}: 3 μM) compared to **47**. They further exploited **46** with solubilising substitution like $O(CH_2)_2OCH_3$, $O(CH_2)_2N(CH_3)_2$ and $-NH(CH_2)_2OCH_3$, etc. at 4-position of pendant phenyl ring and concluded the heparanase activity (IC_{50}) was enhanced from > 10 μM to 0.25 μM upon increase in size of substitution from $-NHCH_3$ to -NH-*n*-Pr. The results of the optimization, by keeping the substitution fixed with -NH-*n*-Pr and studying the substitutions effect in the pendant phenyl ring indicated that the 4'-F and 4'-Cl substituents resulted in better heparanase inhibitory activity with IC_{50}s: 0.2 μM and 0.5 μM, respectively.

In order to improve the DMPK (Drug metabolism and pharmacokinetics) profile of heparanase inhibition and angiogenesis inhibition in furanyl-1,3-thiazol-2-yl-acetic acid derivative **48** (Figure 30), authors [64] further reported the attempts to replace the furanythiazole unit with other bicyclic heteroaromatic moieties in **48** the same research report that thereby resulted in **49** (IC_{50} heparanase inhibition: 0.4 μM), **50** (IC_{50} heparanase inhibition: 0.2 μM) and **51** (Figure 30). In general benzoxazol-5-yl acetic acids **51** displayed superior pharmacokinetic (PK) profile and anti-angiogenesis activity. They further concluded that the substitution with 2-F substituent in the central phenyl ring presented IC_{50} of 0.2 μM for heparanase and of 2 μM for angiogenesis.

In an attempt to design new synthetic biomimetic analogues of natural CA-4, N. Bramhananda Reddy and co-workers [65], synthesized a series of benzoxazole fused analogues **52** (Figure 31). The *in-vitro* anticancer evaluation of **58** exhibited GI_{50} of < 0.1-27.30 μM range against human cancer cell lines Colo-205 (colon), A549 (lung), MCF-7 (breast) compared to control drug Adriamycin (< 0.1-0.13 μM). Among the tested those having R = 6-OCH_3, 5-OCH_3, 5,6-$(OCH_3)_2$ and 5,6-$(CH_3)_2$ showed better anticancer potency. In particular, **52** with 5,6-$(OCH_3)_2$ found to be most potent (GI_{50}: 2.30 μM and 2.10 μM) against A549 and MCF-7 cell lines compared to Adriamycin with GI_{50} of < 0.1 μM and 0.13 μM, respectively.

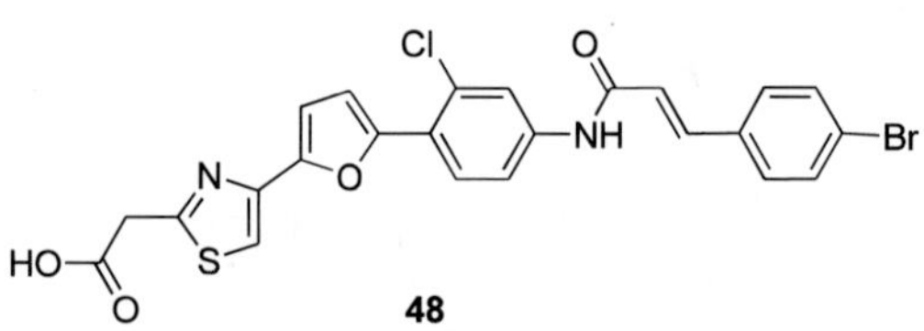

48

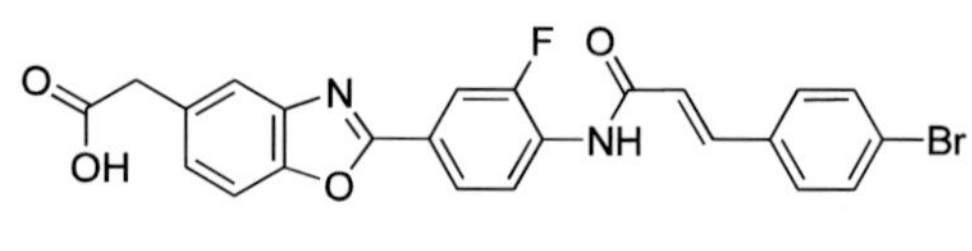

49

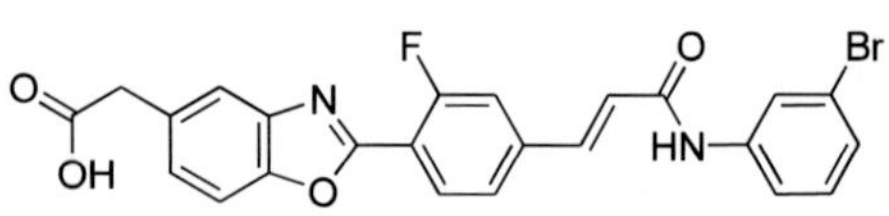

50

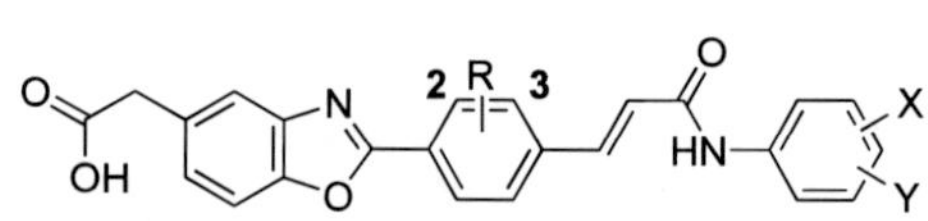

51

X, Y = 2,4-$(Cl)_2$, 3,4-$(Cl)_2$
R = 2-F

Figure 30. Furanyl-1,3-thiazol-2-yl-acetic acid **48** and benzoxazol-5-yl acetic acids **49-51**as heparanase inhibitors and angiogenetics.

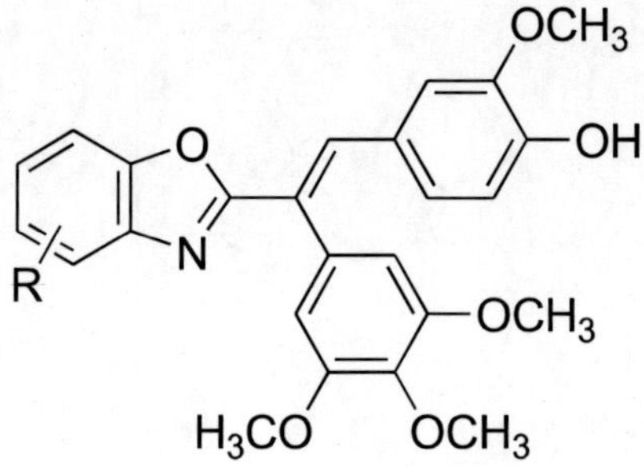

52

R = H, 6-OCH_3, 5-OCH_3, 5,6-$(OCH_3)_2$, 6-NO_2, 5-Cl-6-NO_2, 5,6-$(NO_2)_2$, 5-Cl, 5-CH_3, 5,6-$(CH_3)_2$

Figure 31. Benzoxazole-based analogues **52** of natural CA-4 with anticancer activity.

Ahmed A. Fadda et al. [66] screened benzoxazoles **53-57** (Figure 32) for the cytotoxic effects on the viability of various HepG2 (hepatocellular carcinoma), HTC-116 (colorectal carcinoma), PC-3 (human prostate cancer carcinoma), MCF-7 (mammary gland breast cancer) cancer cell lines and displayed moderate (**56** and **57**) to strong cytotoxic effects towards the tested cancer cell lines. Out of all the tested derivatives **53** and **54** derived from 2-hydrazinylbenzoxazole moiety showed the stronger cytotoxic effects against carcinoma and breast cancer cell lines with IC_{50}s: 5.5±0.22 μg/mL and 5.6±0.32 μg/mL, respectively close to the standard control anticancer drug 5-fluorouracil. The benzoxazole **55** exhibited very strong cytotoxicity (IC_{50}: 9.7±0.32 μg/mL) against HepG2 cells among the other three cancer cell lines.

Recently, Sunisha Kottukulam Subran et al. [67] evaluated the cytotoxicity of ruthenium (II) metal complex with *p*-cymene 2-pyridinylbenzoxazole **58** (Figure 33) against HeLa (human epitheloid cervix carcinoma), MCF-7 (human breast carcinoma), MRC-5 and A2780 (human ovarian carcinoma) cell lines using cisplatin as standard. The tested derivative **58** showed IC_{50}s: 13.66 ± 0.88, 12.02 ± 0.65, 9.02 ± 0.65 and 100.32 ± 2.62 μM against MCF-7, HeLa, A2780 and MRC-5 cell lines, respectively.

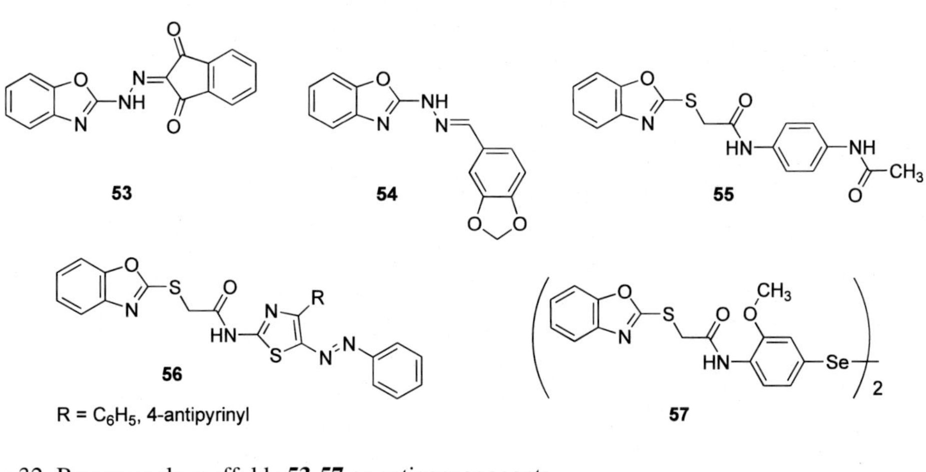

Figure 32. Benzoxazole scaffolds **53-57** as anticancer agents.

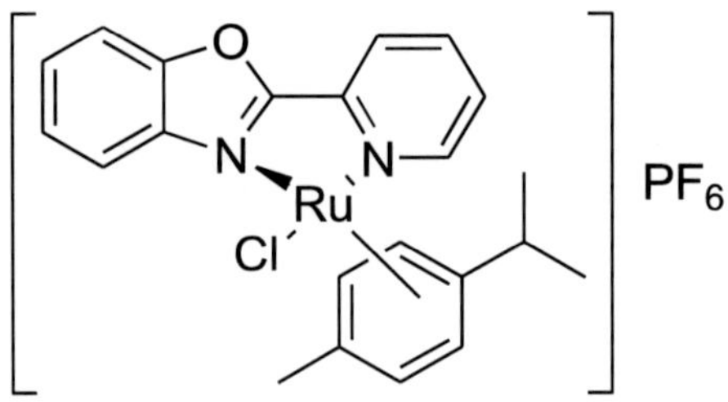

58

Figure 33. Ruthenium (II) metal complex with *p*-cymene 2-pyridinylbenzoxazole **58** as cytotoxic agent.

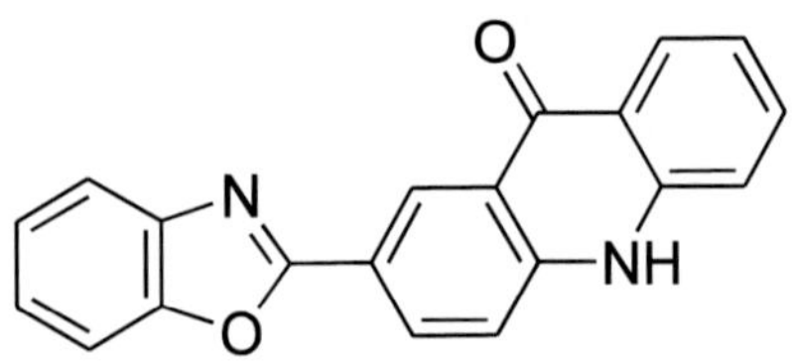

59

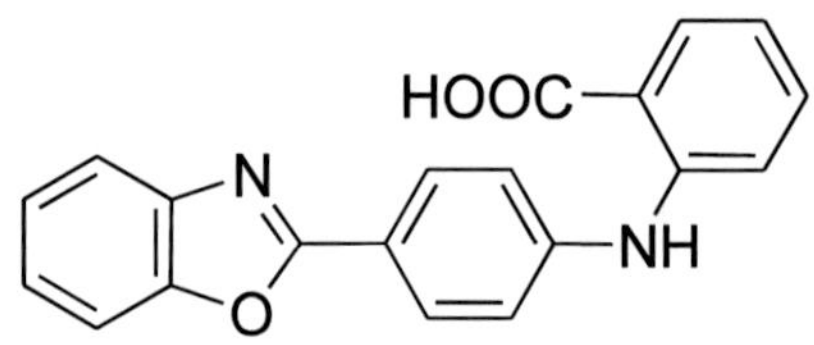

60

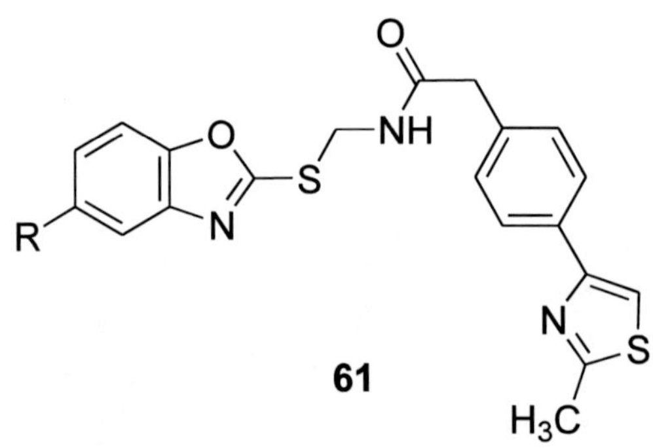

61

R = H, Cl, CH_3, NO_2

Figure 34. Benzoxazoles **59-61** as cytotoxic agents.

In another study Manal M. Kandeel et al. [68] reported the synthesis of 2-(benzo[*d*]oxazole-2-yl)phenyl)acridine-9(10*H*)-one **59** and 10-(2-((4-benzo[*d*]oxazol-2-yl)phenyl)amino)-2-oxo-9,10-dihydroacridine-4-carboxylic acid **60** (Figure 34) and their cytotoxic evaluation for antitumor activity against a panel of MCF-7 (human breast cancer), HepG-2 (hepatocellular carcinoma) and HCT-116 (colon) cell lines using SRB assay and doxorubicin as a reference positive control drug. The tested derivatives **59** and **60** displayed potent antitumor activity against colon cancer (HCT-116) line with IC_{50} range between 4.75 to 5.25 μM/mL. In general, **60** found to be more potent than **59** when compared their IC_{50}.

Leyla Yurttas et al. [69] reported the synthesis and anticancer activity evaluation of a new class of benzoxazole containing *N*-[4-(2-methylthiazol-4-yl)phenylacetamide **61** (Figure 34) against A549 and C6 tumor cell lines.

Ananya Srivastava et al. [46] reported the potential anticancer activity of novel benzoxazole-triazole scaffold **23** (Figure 16). Out of all tested derivatives **23** having R^1 = 5-CH_3, R^2 = 4-OCH_3, R^3 = 2,5-$(F)_2$ displayed moderate cytotoxic % viability of 45±0.86 against HeLa cells compared to standard anticancer drug daunomycin (22±0.86%). After being encouraged by the results scientists carried out further cytotoxicity evaluation on SKBr and Hep G2 cancer cells whereas the **23** having R^1 = 5-CH_3, R^2 = 4-OCH_3, R^3 = 2,5-$(F)_2$ showed IC_{50}s: 6.8 ± 0.34, 7.1 ± 0.41 and 11.2 ± 0.64 μg/mL against HeLa, SKBr3 and Hep G2 cells lines, respectively. Further a dose dependent cytotoxicity analysis of **23** (R^1 = 5-CH_3, R^2 = 4-OCH_3) indicated that decrease in the percent viability against the three cancer cell lines has been observed with an increase in the concentration from 5-30 μg/mL.

In order to enhance cytotoxicity and easy uptake by the cancer cell, Amarnath Velidandi and co-workers [70], developed a series of ammonium based cationic lipids of benzoxazole **62**-**64** (Figure 35). The results indicated that heterocycle based lipids with alkyl chain of *C*-14 possessed better cytotoxic effect lung and breast cancer cell lines than those with *C*-16 and *C*-18.

Yusuf Ozkay et al. [71] reported cytotoxic evaluation of novel imidazolyl-benzoxazole derivative **65** (Figure 36) against HT-29 (colon) and MCF-7 (breast carcinoma) cancer cell lines using standard cisplatin as the positive control drug. All the targeted derivatives **65** exhibited more cytotoxic sensitivity against HT-29 (colon) cell lines than the MCF-7 (breast carcinoma) cancer cell line.

M. S. R. Murty and co-workers reported [72] the cell growth inhibitory activity (GC_{50}) of piperazinyl benzoxazoles fused to 1,3,4-oxadizole-2-thiol pharmacophore **66** (Figure 37) against $HepG_2$ (liver), MCF-7 (breast), A549 (lung), A431 (skin), HeLa (cervical) cell lines.

The sensitivity against the targeted followed the order of A431 (skin) > MCF7 (breast) > A549 (lung) > $HepG_2$ (liver) > HeLa (cervical) cell lines. All the derivatives induced inhibitory activity against proliferation of these cell lines, but **66** having R = $CH_2C(O)$ NHC_6H_5 (2-*N* phenyl acetamide) exhibited maximum cytotoxicity due to the presence of O=C-N-H group. Further, among the tested derivatives for anticancer activity **66** having R = $CH_2C(O)NHC_6H_5$ and NO_2 displayed good anticancer activity.

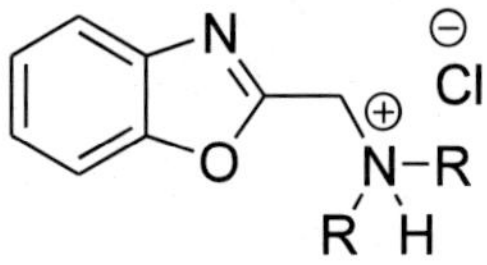

62

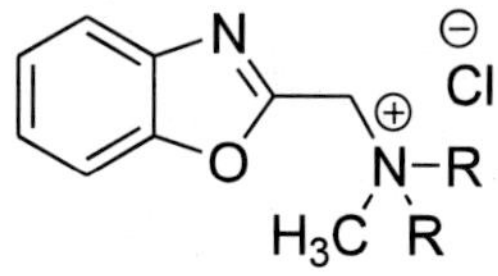

63

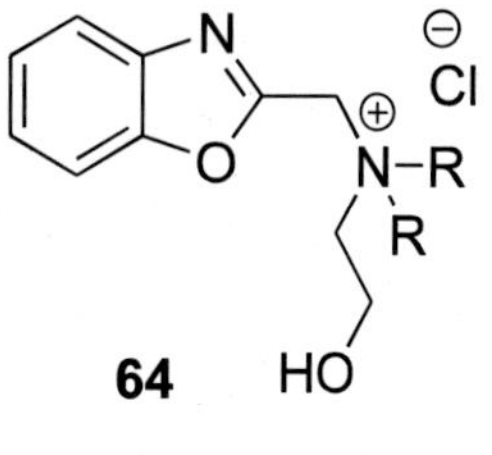

64

R = *n*-$C_{14}H_{29}$, *n*-$C_{16}H_{33}$, *n*-$C_{18}H_{37}$

Figure 35. Benzoxazole-based ammonium lipids **62-64**.

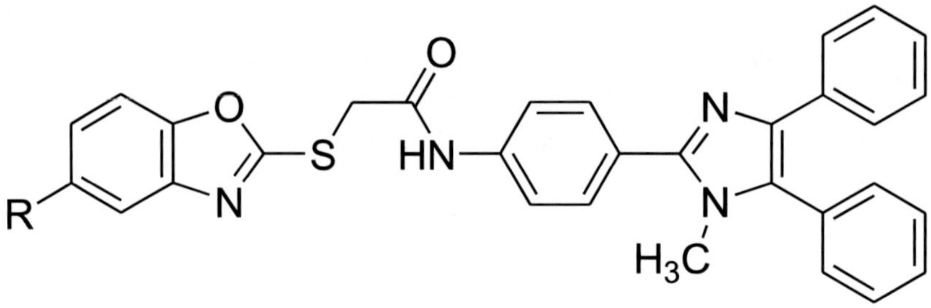

65

R = H, Cl, CH_3

Figure 36. Imidazolyl-based benzoxazole derivative **65** as cytotoxic agent.

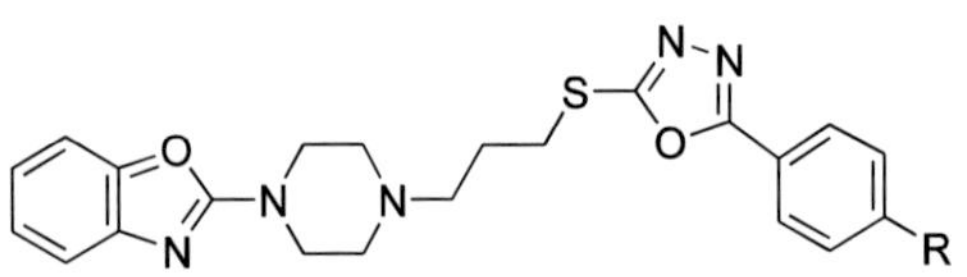

66

R = OCH_3, NO_2, CH_3, Cl, F, Br, $C(CH_3)_3$, OH, H, $CH_2NHC(O)C_6H_5$

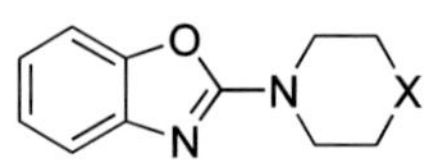

67

X = NCH_3, NC_2H_5, NC_6H_5, $NCH_2C_6H_5$, CH_2, O, $N(3\text{-}ClC_6H_4)$

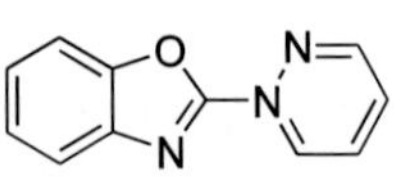

68

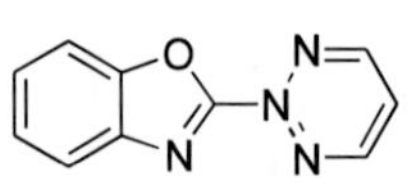

69

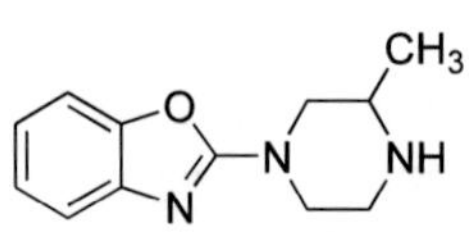

70

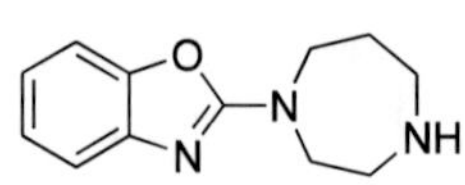

71

Figure 37. Benzoxazole scaffolds **66-71** as cytotoxic agents.

The cytotoxicity biological evaluation of 2-substituted-1,3-benzoxazoles **67-71** (Figure 37) against human breast cancer (MCF-7) and lung cancer cells (A549) than cervical cancer (HeLa) and colon cancer (SW-480) as reported by the same group of researches [73] were found to be of good to moderate anticancer activity. Among tested derivatives, the benzoxazole **67** (X = $NCH_2C_6H_5$) observed to be most potent against HeLa (cervical) cells and A549 (lung) cells with IC_{50}s: 17.31 and 32.35 μM, respectively. Whereas the benzoxazole **67** (X = NC_6H_5) showed equipotency (IC_{50}: 51-54 μM) to the growth of A549 (lung), MCF-7 (breast) and SW-480 (colon) cell lines and almost double the amount (IC_{50}: 102.02 μM) needed for inducing the same cytotoxicity in HeLa (cervical) cell lines. The benzoxazole **67** (X = NC_2H_5) showed almost equal cytotoxicity in MCF-7 (breast) (IC_{50}: 54.55 μM) as well as A549 (lung) cell (IC_{50}: 52.51 μM) and no observable effect in HeLa (cervical) and SW-480 (colon) cancer cell lines. The benzoxazole **68** displayed no significant cytotoxicity except only to MCF-7 (breast) cells with IC_{50}: 52.57 μM.

3.5. Antimycobacterial

To improve the antimycobacterial activity (MIC: 1.9 μM, intercellular IC_{90}: 0.20 μM), against *Mycobacterium tuberculosis* ($H_{37}Rv$), cytotoxicity for $HepG_2$ (> 19.95 μM) and metabolic blood stability ($t_{1/2}$: < 5 min) of 2-(6-methoxy-2-methylquinolin-4-yloxy)-*N*-(3,5-dimethylphenyl)acetamide, Eleni Pitta and co-workers [74] carried out lead optimization that resulted in 2-(((6-methoxy-2-methylquinolin-4-yl)oxy)-methyl)-5-methylbenzo[*d*]oxazole **72** (Figure 38). The lead benzoxazole **72** exhibited MIC: 5.5 μM, intercellular IC_{90}: 5 μM, cytotoxicity for $HepG_2$> 100 μM and blood stability $t_{1/2}$: > 240 min. Therefore, **72** displayed useful lead for new anti-tuberculosis agents.

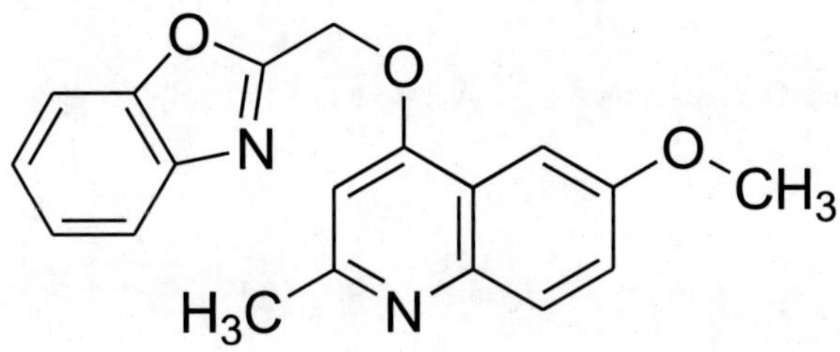

72

Figure 38. 2-(((6-Methoxy-2-methylquinolin-4-yl)oxy)methyl)-5-methylbenzo[*d*]oxazole **72** as anti-TB agent.

A novel series of substituted pyrazoline-based benzoxazole derivatives **73** and **74** (Figure 39) have been evaluated for *in-vitro* antituberculosis against *Mycobacterium tuberculosis* $H_{37}Rv$, multidrug-resistant TB (MDR-TB) and extensively-drug resistant TB (XDR-TB) strains [75] where they displayed good selectivity index (SI) with MIC range of 1.25 to 25 μg/mL. The results of the screening showed that the compounds **73** having $R^1 = R^2 = R^4 = R^5 = H$, $R^3 = Cl$; $R^1 = R^2 = R^3 = R^5 = H$, $R^4 = Cl$; $R^1 = OCH_3$, $R^2 = R^3 = R^5 = H$, $R^4 = Cl$; $R^1 = OCH_3$, $R^2 = R^3 = R^5 = H$, $R^4 = Br$ and $R^1 = OCH_3$, $R^2 = R^4 = R^5 = H$, $R^3 = Cl$ exhibited potent activity against *M.tuberculosis* $H_{37}Rv$.

The compounds **73** having $R^1 = R^2 = R^3 = R^5 = H$, $R^4 = Cl$; $R^1 = OCH_3$, $R^2 = R^4 = R^5 = H$, $R^3 = Cl$ and $R^1 = OCH_3$, $R^2 = R^4 = R^5 = H$, $R^3 = Br$ and the compounds **74** having $R^1 = R^2 = R^5 = H$, $R^3 = R^4 = OCH_3$; $R^1 = R^2 = R^3 = R^5 = H$, $R^4 = Cl$ and $R^1 = OCH_3$, $R^2 = R^3 = R^5 = H$, $R^4 = Cl$ were found to have potent activity against MDR-TB strain.

The compounds **73** having $R^1 = R^2 = R^4 = R^5 = H$, $R^3 = Cl$; $R^1 = OCH_3$, $R^2 = R^3 = R^5 = H$, $R^4 = Br$ and $R^1 = OCH_3$, $R^2 = R^4 = R^5 = H$, $R^3 = Cl$ exhibited potent activity against XDR-TB strain than isoniazid (MIC: 12.5 μg/mL).

The compounds **73** ($R^1 = R^2 = R^3 = R^5 = H$, $R^4 = Cl$ and $R^1 = OCH_3$, $R^2 = R^4 = R^5 = H$, $R^3 = Cl$) and the compounds **74** ($R^1 = R^2 = R^5 = H$, $R^3 = R^4 = OCH_3$) were found to be equipotent to isoniazid against MDR-TB strain (MIC: 6.25 μg/mL).

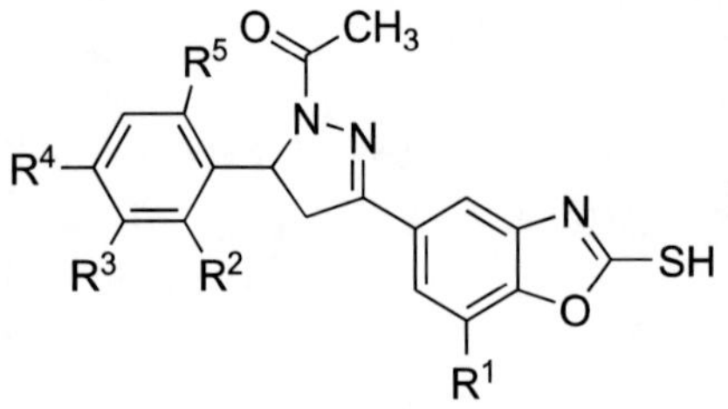

73

R^1, R^5 = H, OCH_3
R^2 = H, Cl, OCH_3
R^3, R^4 = H, Cl, Br, OCH_3

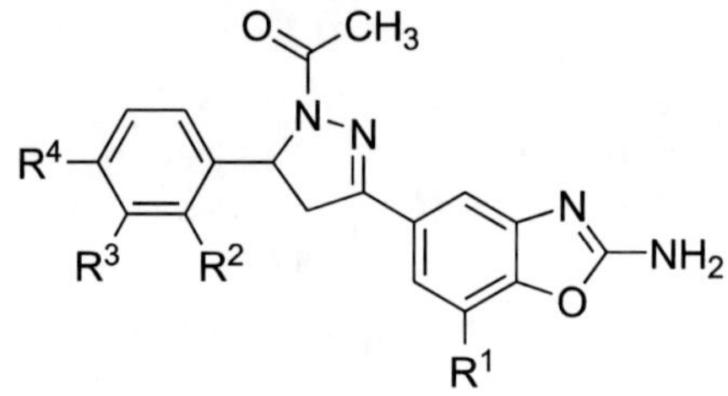

74

R^1 = H, OCH_3
R^2 = H, Cl
R^3 = H, Cl, OCH_3
R^3, R^4 = H, Cl, Br, OCH_3

Figure 39. Pyrazoline-based benzoxazole derivatives **73** and **74** as anti-TB agents.

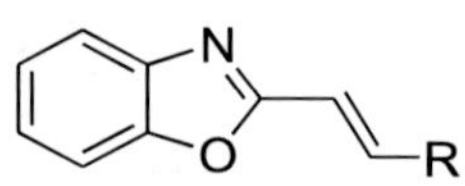

75

R = C_6H_5, 2-$OCH_3C_6H_4$, 4-$OCH_3C_6H_4$, 4-$SCH_3C_6H_4$, 4-$CH_3C_6H_4$, 4-ClC_6H_4, 4-$CF_3C_6H_4$

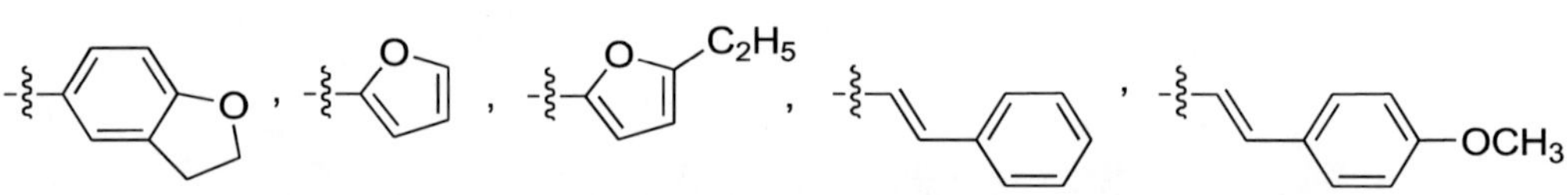

Figure 40. 2-[(*E*)-2-substituted-ethenyl]-1,3-benzoxazole **75** as antimycobacterial agent.

The compounds **73** (R^1 = OCH_3, R^2 = R^4 = R^5 = H, R^3 = Cl) and the compounds **75** (R^1 = R^2 = R^3 = R^5 = H, R^4 = Cl and R^1 = OCH_3, R^2 = R^3 = R^5 = H, R^4 = Cl) were found to be more potent than isoniazid (MIC: 3.25 μg/mL).

Ales Imramovsky and research group [76] reported the antimycobacterial evaluation of 2-[(*E*)-2-substituted-ethenyl]-1,3-benzoxazole **75** (Figure 40) against *Mycobacterium tuberculosis*, *Mycobacterium kansasii* and *Mycobacterium avium* strains. Among the tested benzoxazoles **75** those having R = 4-$OCH_3C_6H_4$, 4-$SCH_3C_6H_4$ and 2,3-dihydro-1-benzofuran-5-yl displayed highest potency against tested strains and significantly higher activity against *Mycobacterium kansasii* and *Mycobacterium avium* strains compared than reference positive control isoniazid drug.

3.6. Antiviral

The property of viruses to use host's cells to replicate, makes the fabrication of safe and effective antiviral drug a complicated task.

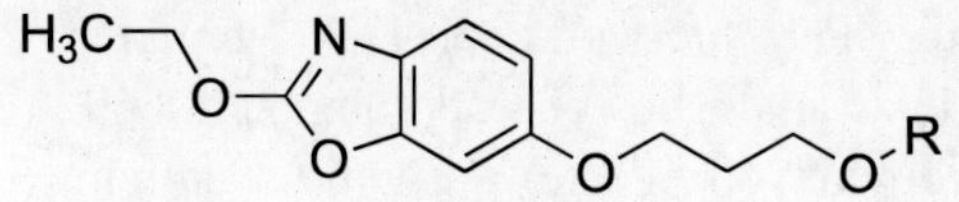

76

R = 4-CNC_6H_4, 4-$OCH_3C_6H_4$, 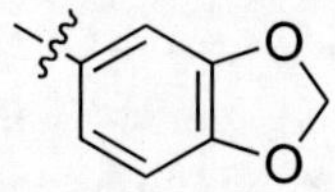

Figure 41. Benzoxazoles **76** with antiviral activity against group A and B rhinoviruses.

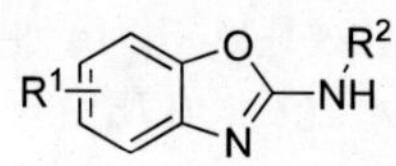

77

R^1 = H, 6-NO_2, 5-NH_2, 6-NH_2, 7-NH_2

R^2 = H, 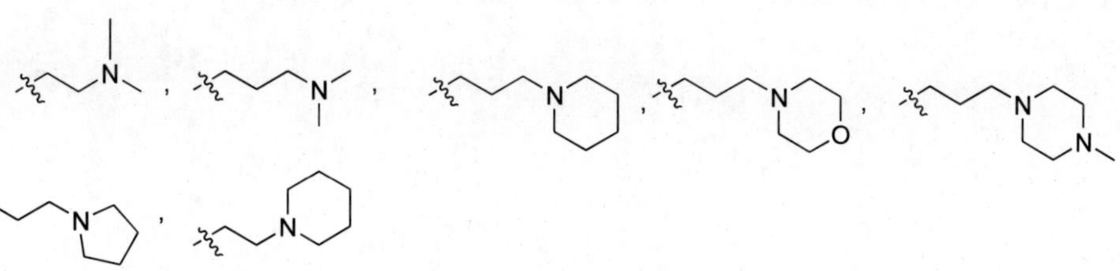

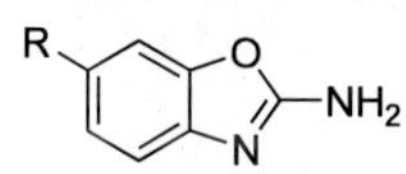

78

R = 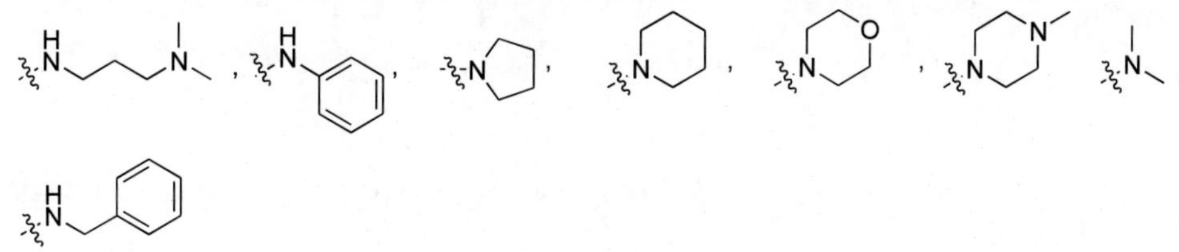

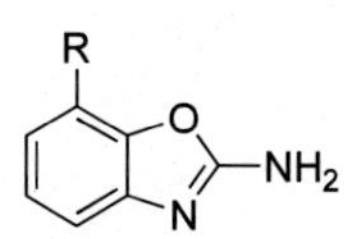

79

R = 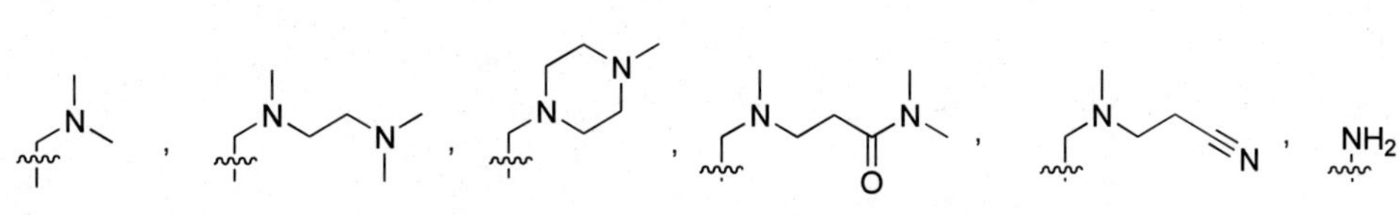

Figure 42. 2-Amino benzoxazoles **77**-**79** with antiviral activity against HCV.

Angela M. Bernard and co-workers [77] reported the new series of benzoxazoles **76** (Figure 41) as analogues of lead compound 6602 (ethyl 4-{3-[2-(3-methyl-1,2-isoxazol-5-yl)ethoxy]propyloxy}benzoate) with antiviral properties against group A and B rhinoviruses with improved protection index (PI) and lower toxicity than Pirodavir.

Interestingly, **76** (R = 4-$OCH_3C_6H_4$) presented highest improved activity and cytotoxicity on cultured cells. The **76** (R = 4-$OCH_3C_6H_4$) displayed less toxicity (30 times) compared to Pirodavir and more active (2 times) on the group A strain of rhinovirus HRV 14 with PI: 400 compared to pirodavir and 6602 with PI: 258 and 200, respectively. Furthermore, **76** (R = 4-$OCH_3C_6H_4$) exhibited more activity (4 times) on the group B strain with PI: 2000 compared to pirodavir and 6602 with PI: 516 and 714, respectively.

Kevin D. Rynearson and co-workers [78], explored the results of hydrogen bonding interactions of 2-aminobenzoxazole scaffolds **77-79** (Figure 42) as ligands of the hepatitis C virus (HCV) internal ribosome entry site (IRES) RNA by their structure-binding activity relationships. The studied benzoxazoles showed EC_{50}: 25 to 45 μM range. Further, authors concluded that these benzoxazoles have ability to impart superior drug like properties to inhibitors.

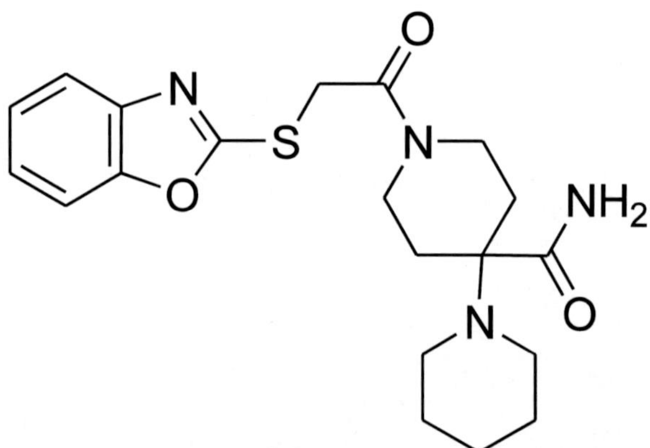

80

Figure 43. 1-(2-(Benzo[*d*]oxazol-2-ylthio)acetyl)-4-(piperidin-1-yl)piperidine-4-carboxamide (ST-610) **80** as DENV inhibitor.

Chelsea M. Byrd and co-workers [79] reported the 1-(2-(benzo[*d*]oxazol-2-ylthio)acetyl)-4-(piperidin-1-yl)piperidine-4-carboxamide (ST-610) **80** (Figure 43) as small molecule DENV (dengue virus) inhibitor. As compared to vehicle controls, this benzoxazole based DENV inhibitor **80** has been found to be highly potent (EC_{50}: 0.045 to 0.377 μM range), nontoxic (CC_{50}: > 100 μM both in 2 and 5 day assay) that significantly shrinks viremia and viral load. In general, **80** selectively inhibit the four serotypes of *in-vitro* dengue virus replication that targets NS3 helicase.

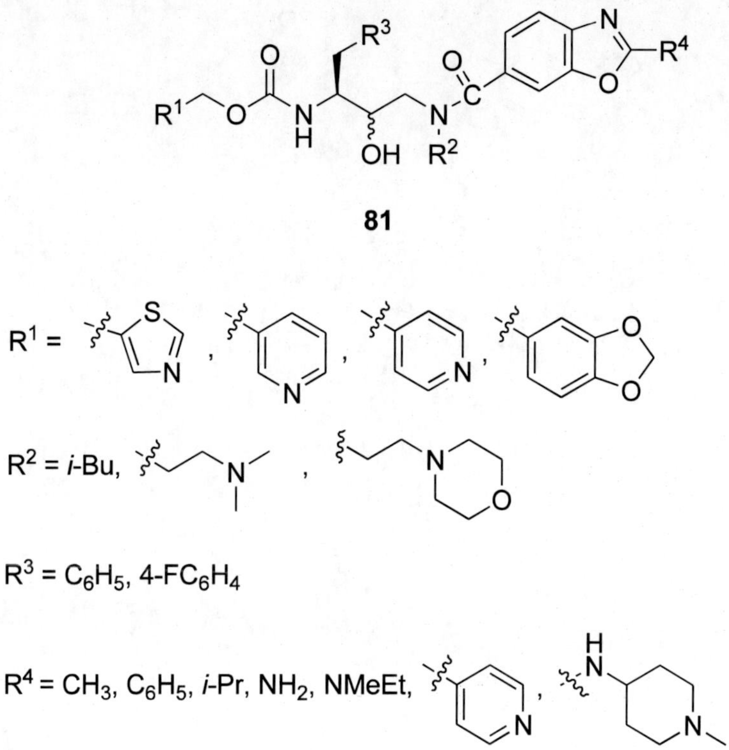

Figure 44. Benzoxazole amide **81** as HIV protease inhibitor.

Tim H. M. Jonckers et al. [80] have reported the benzoxazole amide **81** (Figure 44) potent as HIV protease inhibitors compared to ritonavir. Among other similar analogues of **81**, the benzoxazole amides namely, thiazol-5-ylmethyl (2*S*,3*R*)-4-(2-(ethyl(methyl)amino)-*N*-isobutylbenzo[*d*]oxazole-6-carboxamido)-3-hydroxy-1-phenylbutan-2-ylcarbamate (**81** having R^1 = 5-thiazolyl, R^2 = *i*-Bu, R^4 = Ph, R^5 = NMeEt) has been found to lead exhibiting the most significant CYP3A4 inhibiting properties. The results of *in-vivo* studies in dogs suggested that all these analogues can serve as better alternatives to Ritonavir standard.

Zhenyu Li and et al. [81], reported the *in-vitro* anti-influenza B virus (anti-IVF) inhibition of 5-substituted-2-(4-substituted phenyl)-1,3-benzoxazole **82** (Figure 45). Further, **92** having R = 3,4-$(OCH_3)_2C_6H_3$ exhibited antiviral activity of higher selectivity index (SI: > 5) against A/H3N2 with EC_{50}: 37.03 μM compared to oseltamivir (EC_{50}: 59.00 μM and SI: > 1) standard drug.

Jeremie Boyer and co-workers [82] reported the difluoromethylbenzoxazole pyrimidine thioether **83** (Figure 46) as the potent non-nucleoside HIV-1 reverse transcriptase inhibitors. In general, the targeted derivatives **83** possessed significant potency against drug-resistant strains with lower levels of toxicity. Among the tested derivatives the compound **83** having (X = S, R^1 = H and R^2 = C_2H_5 exhibited a significant EC_{50} (6.4 nM) against HIV-1 IIIB and moderate EC_{50} (54 μM) against an NNRTI resistant double mutant (K103N+Y181C). Furthermore, it exhibited an outstanding SI: > 15477 and CC_{50}: > 100μM.

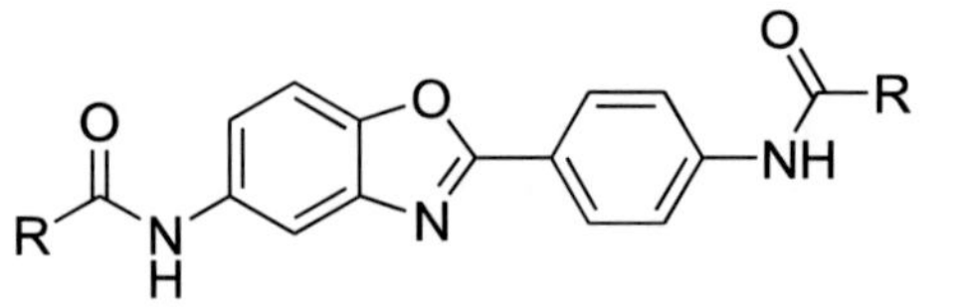

82

R^1 = 4-$NO_2C_6H_4$, 4-FC_6H_4, 4-ClC_6H_4, 4-BrC_6H_4, 4-IC_6H_4, 3,4-$Cl_2C_6H_3$, 3,4-$(CH_3)_2C_6H_3$, 4-Cl-2-$NO_2C_6H_3$, C_6H_5, 4-$NH_2C_6H_4$, 4-$CH_3C_6H_4$, 4-$^tBuC_6H_4$, 4-$^nPrC_6H_4$, 3-FC_6H_4, 3-ClC_6H_4, 3-$CH_3C_6H_4$, 2-$OCH_3C_6H_4$, CH_3, $ClCH_2CH_2$

Figure 45. 5-Substituted-2-(4-substituted phenyl)-1,3-benzoxazole **82** as anti-IVF agent.

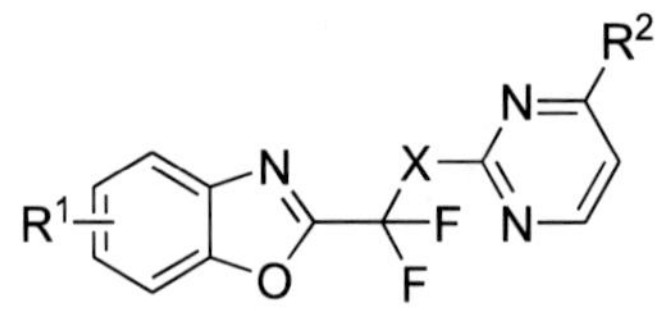

83

X = S, SO_2
R^1 = H, CH_3, Cl
R^2 = CH_3, C_2H_5, $CH(CH_3)_2$, CH_2OCH_3, NH_2, $NHC(O)CH_3$, 

Figure 46. Difluoromethylbenzoxazole pyrimidine thioether **83** as HIV-1 reverse transcriptase inhibitor.

3.7. Antioxidant

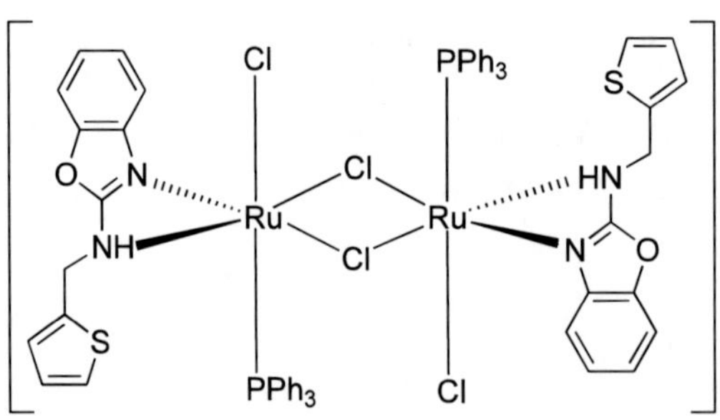

84

Figure 47. Ruthenium complex with *N*-(thiophene)methyl-benzoxazole-2-amine (L) ligand, $(\mu\text{-}L,Cl)_2[Ru^{II}ClPPh_3]_2$ **84** as DPPH radical scavenger.

Irvin Noel Booysen et al. [83] reported the synthesis of dinuclear ruthenium complex with *N*-(thiophene)methyl-benzoxazole-2-amine (L) ligand as $(\mu\text{-}L,Cl)_2[Ru^{II}ClPPh_3]_2$ **84** (Figure 47) and further for DPPH radical scavenging activities. The complex **84** exhibited superior DPPH radical scavenging activity (low IC_{50}) compared to corresponding free ligands and natural oxidant, vitamin C (IC_{50}: 147 μM) because the presence of two metal centres, unpaired *d*-electron configurations of ruthenium (III) metal centres and aromatic hydrocarbon ligands aid in proton donation to neutralize radicals.

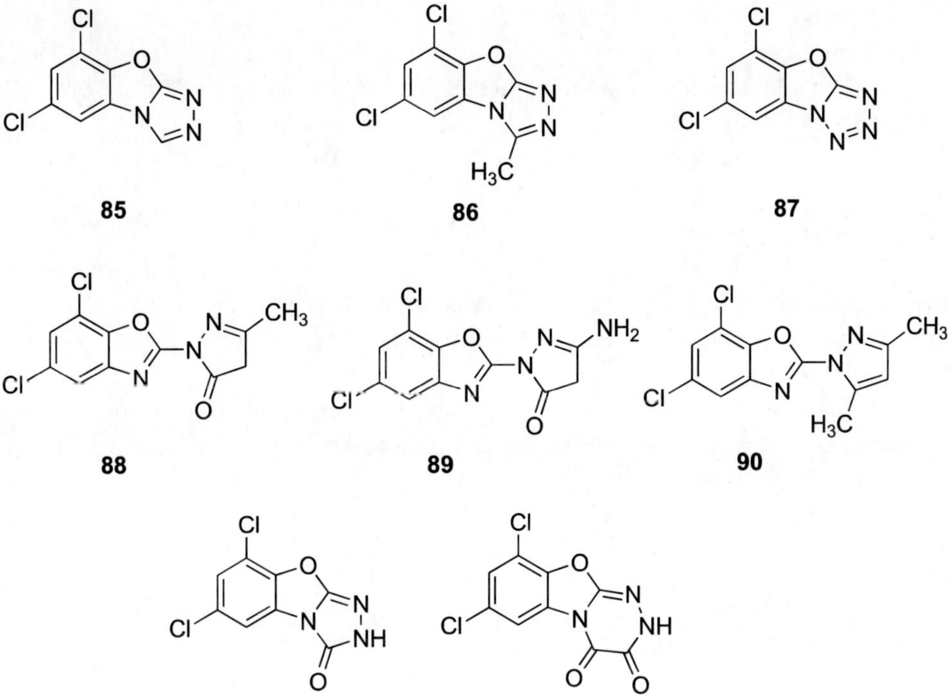

85 86 87 88 89 90 91 92

Figure 48. 5,7-Dichloro-1,3-benzoxazoles **85-92** as antioxidants.

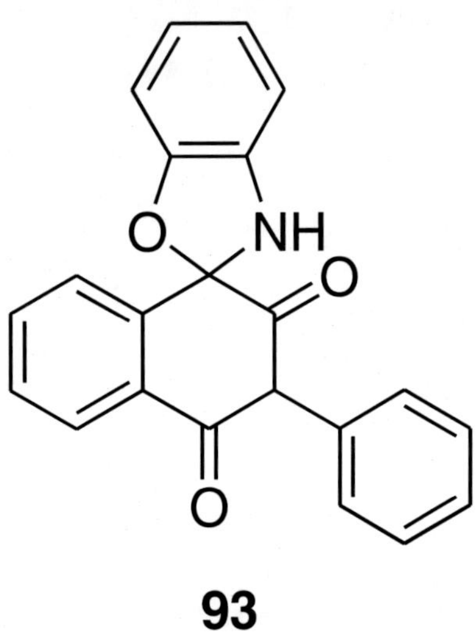

93

Figure 49. 2'-Phenyl-1'*H*,3*H*-spiro[1,3-benzoxazole-2,4'-isoquinoline]-1',3'(2'*H*)-dione **93** as lipid peroxidation inhibitor.

N. D. Jayanna et al. [84] reported synthesis of a novel series of 5,7-dichloro-1,3-benzoxazoles **85-92** (Figure 48) and further, evaluated them for DPPH radical scavenging. All test derivatives were found to be potent antioxidants. In particular, benzoxazoles **90-92** were found to possess better scavenging properties compared to other test derivatives.

In other study, Mohamed Mohamed Youssef et al. [85] reported the synthesis new benzoxazole spiro-derivatives **93** (Figure 49) and further evaluated its antioxidant activity. The result from *in-vivo* lipid peroxidation inhibition, indicated that the spiro derivative **93** exhibit protective activity towards DNA.

3.8. Antagonist

Benzoxazole scaffolds contributed significantly to antagonist's family. Benzoxazole heterocycle is a core structure of naturally occurring prinaberel, ERB-041 (Figure 1) an agonist (estrogen receptor ERβ ligand).

Dinthia Sampson et al. [86] reported the novel selective dopamine D_4 receptor benzoxazole ligand **94** (Figure 50) with good to moderate affinity for the D_4 receptor with K_iD_4 falling in the nanomolar (nM) range.

Sina Rezazadeh and co-workers [87] (Figure 51) reported the synthesis of 2-heteroarylbenzoxazol-5-diols **95-103**, non natural analogues of ERB-04. All these derivatives emerged as the potent ligands with high estrogen affinity.

Laxman M. Prajapati and co-workers [88] studied QSAR of phenyl benzoxazole ligand **115** (Figure 52) for evaluating estrogen receptor inhibition using Dragon software (version 1.11). The results of the study for IC_{50}s of Erα and ERβ (fold selectivity) will help in designing of more potent ERβ inhibitor ligands.

Zhicai Yang et al. [89] discovered the 2-substituted benzoxazole carboxamides **105** and **106** (Figure 53) as $5\text{-}HT_3$ receptor antagonists with nanomolar (nM) *in-vitro* activity. Further, these analogues secured themselves to cure malfunctioning of $5\text{-}HT_3$ receptor and gastrointestinal disorder IBS-D.

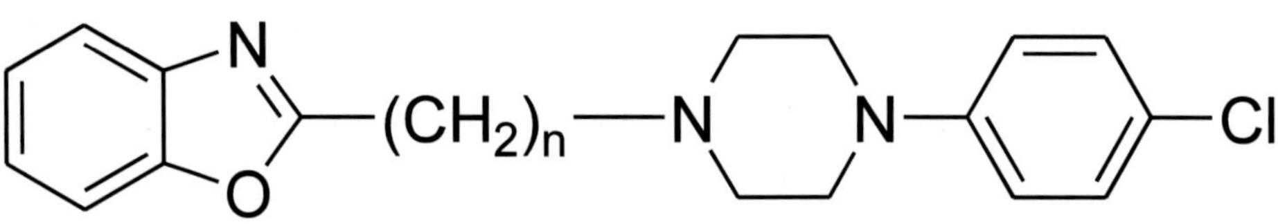

94

n = 2, 3, 4, 5

Figure 50. Benzoxazole scaffold **94** as D_4 dopamine receptor ligand.

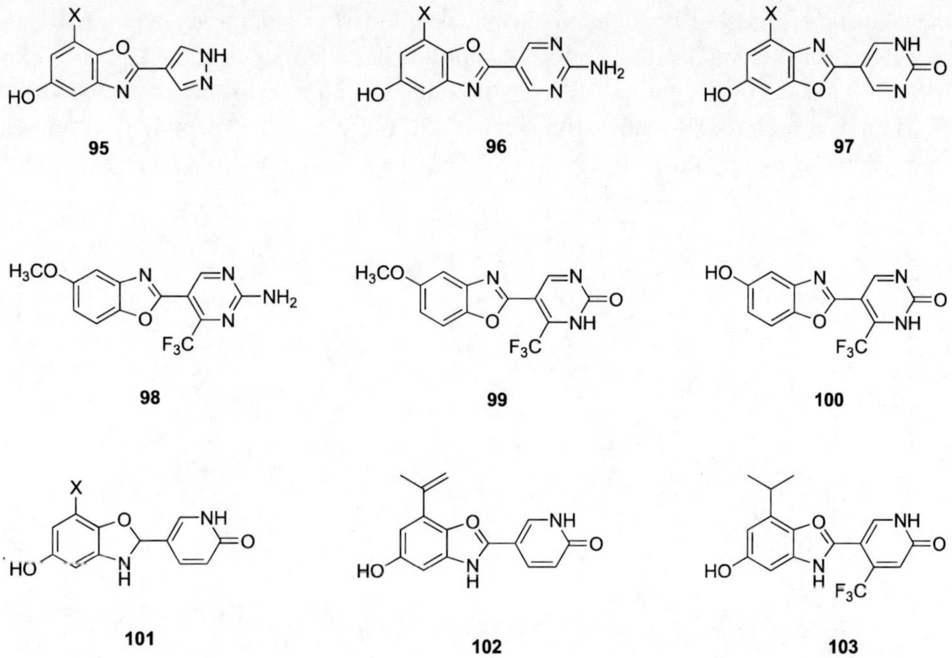

X = H, Br

Figure 51. 2-Heteroarylbenzoxazol-5-diols **95-103** as ER ligands.

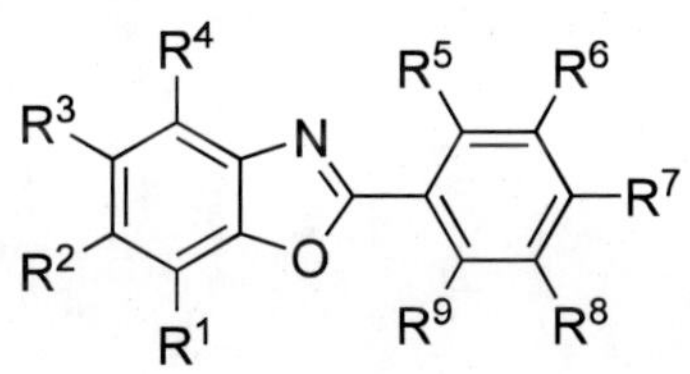

104

R^1 = H, OCH_3, Br, CN, CH_2Br, CHO, *n*-Bu, ethynyl, vinyl, 2-F-vinyl, 2-Br-vinyl, 3-CH_3-vinyl, C_6H_5, 2-furyl, 2-thiazole, cyclopentane, OCH_3

R^2 = H, Br, OH

R^3 = H, Cl, Br, OH

R^4 = H, Br

R^5 = H, F, OH

R^6 = H, OH, F, Cl, $C(CH_3)_2$, O-$^nC_4H_9$

R^7 = H, OH

R^8 = H, F

R^9 = H, CH_3

Figure 52. Substituted phenyl benzoxazole **104** as ER ligand.

Christopher D. Cox and co-workers [90] have reported the benzoxazole based diazepane (MK-4305) **107** (Figure 54) as dual orexin receptor antagonist (DORA) with improved pharmacokinetic profile. Further, it has to be noticed that **107** is undergoing in phase III clinical trials for the curing insomnia and therefore, can be a better alternate ligand.

Sarah L. Watson and group [91] reported the *N*-(2-methyl-6-benzoxazolyl)-*N*''-1,5-naphthyridin-4-yl-urea (SB-334867) **108** (Figure 55) as the first non-peptide OX_1 antagonist that significantly increased nociceptive responsiveness when administered alone in rats and has the potential for further clinical testings.

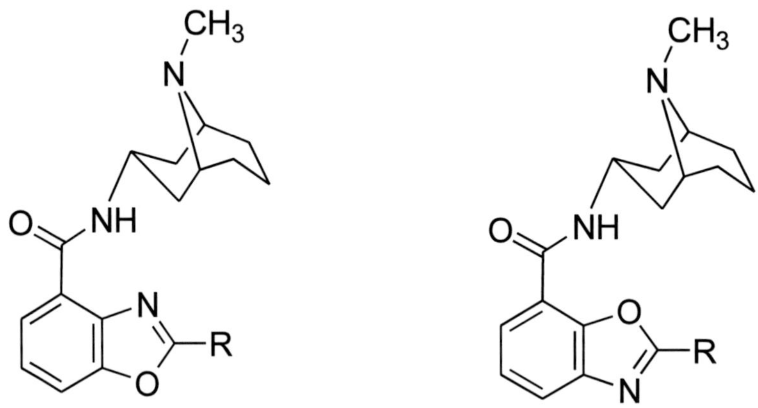

105 **106**

R = C_6H_5, 4-FC_6H_4, 4-ClC_6H_4, 4-$OCH_3C_6H_4$

Figure 53. 2-Substituted benzoxazole carboxamides **105** and **106** as $5\text{-}HT_3$ receptor ligands.

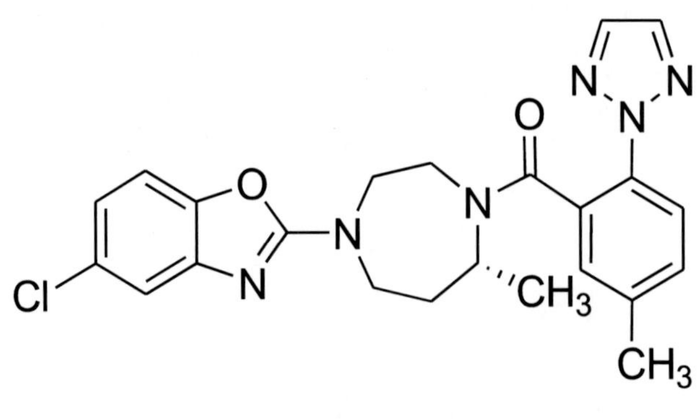

107

Figure 54. Benzoxazole based diazepane **107** as dual DORA.

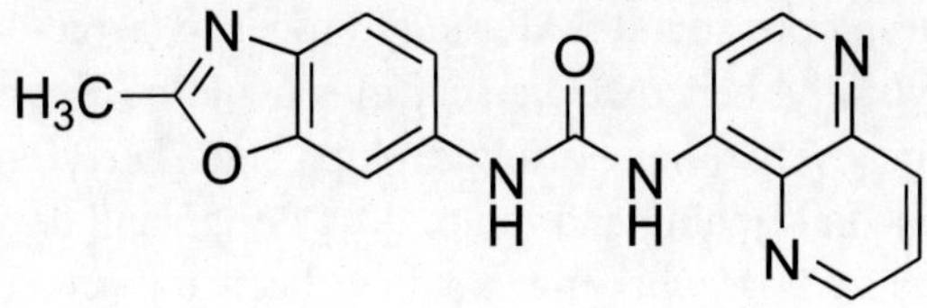

108

Figure 55. *N*-(2-Methyl-6-benzoxalyl)-*N*''-1,5-naphthyridin-4-yl-urea **108** as OX_1 antagonist.

3.9. Antidiabetic Activity

Jagbir Singh et al. [92] studied the 3-D QSAR model of benzoxazole benzenesulfonamide derivatives **109-111** (Figure 56) using PHASE3.4 (Schrodinger 9.0) to examining correlation of H-bond acceptor/donor groups and hydrophobic/nonpolar groups to anidiabetic activity of the derivatives. Most of the evaluated derivatives presented best docking score than parent *N*-(1,3-Benzoxazol-2-yl)benzene sulfonamide analogue and therefore, can be used as lead for other antidiabetic agents in drug discovery programme.

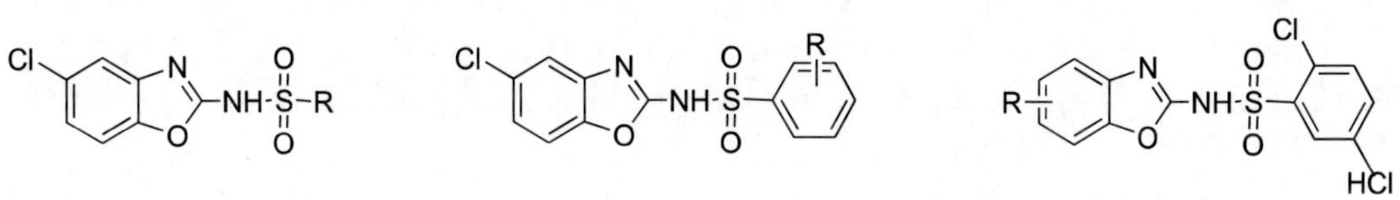

109 **110** **111**

R = 1,4-$Cl_2C_5H_3$, C_6H_5, CH_3, *n*-Bu, 2-Br, 2-CN, 2-C_2H_5, 3-Cl, 3-NO_2, 3-C_6H_5, 3-(4-^{t}BuBz), 4-F, 4-OCF_3, 4-CH_3, 4-^{t}BU, 4-C_6H_5, 4 (3-furanyl), 4-NHC(O)$CH_2CH_2CH_3$, 2,6-$(Cl)_2$, 2-Cl-6-CH_3, 4-I, 4-(4-pyridyl), 5-Br, 5-CH_3, 5-^{t}Bu, 5-(3-furanyl), 6-Cl, 6-CH_3, 6-OCH_3, 6-OCH_3-7-(4-OCH_3-3-pyridyl), 5-OCH_3-7 (3-OHC_6H_4), 5-OCH_3-7 (4-OHC_6H_4), 5-OCH_3-7 (3-$NH_2CH_2C_6H_4$), 5-OCH_3-7 (4-$NH_2C_6H_4$), 5-OCH_3-7 (3-$NH_2C_6H_4$), 2-thiophenyl, 2-(imidazol-1-yl), imidazol-4-yl, naphthalene, anthracene

Figure 56. Benzoxazole benzenesulfonamide derivatives **109-111** as lead for antidiabetics.

4. Conclusion

In summary, various benzoxazole derivatives with multiple pharmacological activities are reported on the basis of substitution differentiation around the core benzoxazole motif. The present study gives an insight into the pharmacological profiles of benzoxazole-based heterocyclic frameworks that purvey non-debatable avenues to novel useful bioactive molecules. This comprehensive compilation of various research reports in each activity will assist all those interested in benzoxazole heterocyclic moiety to carry out target oriented research for upshooting clinically viable agents and tasks for further investigations. This will aid medicinal chemists to bloom sound SAR study on benzoxazole-derived compounds with refined activity. Unequivocally, benzoxazole scaffolds could be asserted as highly privileged "sub-structures" for the headway of new drug developments. Therefore, benzoxazole scaffold is a pressing suitor in remedial finding and future drug design and development. We sincerely apologize to those scientists whose work may not have been appeared in this chapter either due to the limited scope of the review or oversight.

REFERENCES

[159] Neu, H. C. (1992). "The Crisis in Antibiotic Resistance" *Science*, *257*, 1064–1073. doi: 10.1126/science.257.5073.1064.

[160] Davies, J. (1994). "Inactivation of Antibiotics and the Dissemination of Resistance Genes" *Science*, *264*, 375–382. doi: 10.1126/science.815624.

[161] Ding, H. X., Leverett, C. A., Kyne, R. E. Jr., Liu, K. K., Fink, S. J., Flick, A. C. & O'Donnell, C. J. (2015). "Synthetic Approaches to the 2013 New Drugs" *Bioorganic and Medicinal Chemistry*, *23*, 1895–1922. doi: 10.1016/j.bmc.2015.02.056.

[162] Welsch, M. E., Snyder, S. A. & Stockwell, B. R. (2010). "Privileged Scaffolds for Library Design and Drug Discovery" *Current Opinion in Chemical Biology*, *14*, 347–361. doi: 10.1016/j.cbpa.2010.02.0218.

[163] Thompson, L. A. & Ellman, J. A. (1996). "Synthesis and Application of Small molecule Libraries" *Chemical Reviews*, *96*, 555–600. doi: 10.1021/cr9402081.

[164] Evans, B. E., Rittle, K. E., Bock, M. G., DiPardo, R. M., Friedinger, R. M., Whitter, W. L., Lundell, G. F., Veber, D. F., Anderson, P. S., Chang, R. S. L., Lotti, V. J., Cerino, D. J., Chen, T. B., Kling, P. J., Kunkel, K. A., Springer, J. P. & Hirshfield, J. (1988). "Methods for Drug Discovery: Development of Potent, Selective, Orally Effective Cholecystokinin Antagonists" *Journal of Medicinal Chemistry*, *31*, 2235–2246. doi: 10.1021/jm00120a002.

[165] Demmer, C. S. & Bunch, L. (2015). "Benzoxazoles and oxazolopyridines in medicinal studies" *European Journal of Medicinal Chemistry*, *97*, 778–785. doi: 10.1016/j.ejmech.2014.11.064.

[166] Singh, S., Veeraswamy, G., Bhattarai, D., Goo, J. L., Lee, K. & Choi, Y. (2015). "Recent Advances in the Development of Pharmacologically Active Compounds that Contain a Benzoxazole Scaffolds" *Asian Journal of Organic Chemistry*, *4*, 1338–1361. doi: 10.1002/ajoc.201500235.

[167] Vinsova, J., Cermakova, K., Tomeckova, A., Ceckova, M., Jampilek, J., Cemak, P., Kunes, J., Dolezal, M. & Staud, F. (2006). "Synthesis and Antimicrobial Evaluation of New 2-Substituted 5,7-*tert*-butylbenzoxazoles." *Bioorganic and Medicinal Chemistry*, *14*, 5850–5865. doi: 10.1016/j.bmc.2006.05.030.

[168] Oehlers, L., Mazzitelli, C. L., Brodbelt, J. S., Rodriguez, M. & Kerwin, S. (2004). "Evaluation of Complexes of DNA Duplexes and Novel Benzoxazoles or Benzimidazoles by Electrospray Ionization Mass Spectrometry" *Journal of the American Society for Mass Spectrometry*, *15*, 1593–1603. doi: 10.1016/j.jasms.2004.07.015.

[169] Ueki, M., Ueno, K., Miyadoh, S., Abe, K., Shibata, K., Taniguchi, M. & Oi, S. M. (1993). "UK-1, a Novel Cytotoxic Metabolite from *Streptomyces* sp. 517-02. I. Taxonomy,

Fermentation, Isolation, Physico-chemical and Biological Properties" *The Journal of Antibiotics*, *46*, 1089–1094. doi: 10.7164/antibiotics.46.1089.

[170] Hohmann, C., Schneider, K., Bruntner, C., Irran, E., Nicholson, G., Bull, A. T., Jones, A. L., Brown, R., Stach, J. E., Goodfellow, M., Beil, W., Kramer, M., Imhoff, J. F., Sussmuth, R. D. & Fiedler, H. P. (2009). "Caboxamycine, a New Antibiotic of the Benzoxazole Family Produced By the Deep-Sea Strain *Streptomyces* sp. NTK 937" *The Journal of Antibiotics*, *62*, 99–104. doi: 10.1038/ja.2008.24.

[171] Sommer, P. S. M., Almeida, R. C., Schneider, K., Beli, W., Sussmuth, R. D. & Fiedler, H. P. (2008). "Nataxazole, a New Benzoxazole Derivative with Antitumor Activity Produced by *Streptomyces* sp. Tu 6176" *The Journal of Antibiotics*, *61*, 683–686. doi: 10.1038/ja.2008.97.

[172] Rodriguez, A., Ramirez, C., Rodriguez, I. I. & Gonzalez, E. (1999). "Novel Antimycobacterial Benzoxazole Alkaloids, from the West Indian Sea Whip *Pseudopterogorgia elisabethae*" *Organic Letters*, *1*, 527–530. doi: 10.1021/ol9907116.

[173] Wu, Q., Liang, J., Lin, S., Zhou, X., Bai, L., Deng, Z. & Wang, Z. (2011). "Characterization of Biosynthesis Gene Cluster for the Pyrrole Polyether Antibiotic (A23187) in *Streptomyces Chartreusis* NRRL 3882" *Antimicrobial Agents and Chemotherapy*, *55*, 974–982. doi: 10.1128/AAC.01130-10.

[174] Shao, J., Cui, Y., Zhang, M., Wang, T. W. D. & Wang, C. (2017). "Synergistic *in Vitro* Activity of Sodium Houttuyfonate with Fluconazole Against Clinical Candida Albicans Strains Under Planktonic Growing Conditions" *Pharmaceutical Biology*, *55*, 355–359. doi: 10.1080/13880209.2016.1237977.

[175] Roman-Blas, J. A., Castaneda, S., Cutolo, M. & Herrero-Beaumont, G. (2010). "Efficacy and Safety of a Selective Estrogen Receptor β Agonist, ERB-041, in Patients With Rheunmatoid Arthritis: A 12-Week, Randomized Placebo-Controlled, Phase II Study" *Arthritis Care and Research*, *62*, 1588–1593. doi: 10.1002/acr.20275.

[176] Holler, M. G., Campo, L. F., Brandelli, A. & Stefani, V. (2002). "Synthesis and Characterisation of 2-(2'-Hydroxyphenyl)benzazole isothiocyanates as New Fluorescent Probes for Proteins" *Journal Photochemistry and Photobiology A: Chemistry*, *149*, 217–225. doi: 10.1016/S1010-6030(02)00008-4.

[177] Kozich, V., Dreyer, J., Vodchits, A. & Wernere, W. (2005). "Time-resolved Resonance Raman Scarttering of the Excited Singlet of 2(2'-Hydroxyphenyl)benzoxazole After Excited State Intramolecular Proton Transfer" *Chemical Physics Lettres*, *415*, 121–125. doi: 10.1016/j.cplett.2005.08.148.

[178] Smulders, E. & Sung, E. (2012). Laundry Detergent, 2. Ingredients and Products in *Ullmann's Encyclopedia of Industrial Chemistry* Wiley-VCH, Weinheim. doi: 10.1002/14356007.o15_013.

[179] Fourati, M. A., Maris, T., Skene, W. G., Bazuin, C. G. & Prud'homme, R. E. (2011). "Photophysical, Electrochemical and Crystallographic Investigations of the Fluorophore 2,5-Bis(5-*tert*-butyl-benzoxazole-2yl)thiophene" *The Journal of Physical Chemistry B*, *115*, 12362–12369. doi: 10.1021/jp207136k.

[180] Neumann, K. T., Lindhardt, A. T., Bang-Andersen, B. & Skrydstrup, T. (2015). "Access to 2-(Het)aryl and 2-Styryl Benzoxazoles via Palladium-catalyzed Aminocarbonylation of Aryl and Vinyl Bromides" *Organic Letters*, *17*, 2094–2097. doi: 10.1021/acsorglett.5b00642.

[181] Nimual, P., Tummatorn, J., Thongsornkleeb, C. & Ruchirawat, S. (2015). "Utility of Nitrogen Extrusion of Azido Complexes for the Synthesis of Nitriles, Benzoxazoles, and Benzisoxazoles" *Journal of Organic Chemistry*, *80*, 8657–8667. doi: 10.1021/acs.joc.5b01305.

[182] Mayo, M. S., Yu, X., Zhou, X., Feng, X., Yamamoto, Y. & Bao, M. (2014). "Synthesis of Benzoxazoles from 2-Aminophenols and β-Diketones Using a Combined Catalyst of Bronsted Acid and Copper Iodide" *Journal of Organic Chemistry*, *79*, 6310–6314. doi: 10.1021/jo500604x.

[183] Yang, D., Zhu, X., Wei, W., Jiang, M., Zhang, N., Ren, D., You, J. & Wang, H. (2014). "Magnetic Copper Ferrite Nanoparticles: An Inexpensive, Efficient, Recyclable Catalyst for the Synthesis of Substituted Benzoxazoles via Ullmann-type Coupling Under Ligand-Free Conditions" *Synlett*, *25*, 729–735. doi: 10.1055/s-0033-1340599.

[184] Gorepatil, P. B., Mane, Y. D. & Ingle, V. S. (2013). "Samarium (III) Triflate as an Efficient and Reusable Catalyst for Synthesis of Benzoxazoles and Benzothiazoles in Aqueous Medium" *Synlett*, *24*, 2241–2244. doi: 10.1055/s-0033-1339758.

[185] Lassalas, P., Marsais, F. & Hoarau, C. (2013). "DMAP-Catalyzed Regel-Type Direct C-2 (Heero)Aroylation of Oxazoles and Thiazoles Derivatives with Acid Chlorides" *Synlett*, *24*, 2233–2240. doi: 10.1055/s-0033-1339858.

[186] Bastug, G., Eviolitte, I. E. & Marko, I. E. (2012). "Functionalized Orthoesters as Powerful Building Blocks for the Efficient Preparation of Heteroaromatic Bicycles" *Organic Letters*, *14*, 3502–3505. doi: 10.1021/ol301472a.

[187] Kasthuri, M., Babu, H. S., Kumar, K. S., Sudhakar, C. & Kumar, P. V. N. (2015). "A Facile Synthesis of 2-Aminobenzoxazoles and 2-Aminobenimidazoles Using *N*-cyano-*N*-phenyl-p-toluenesulfonamide (NCTS) as an Efficient Electrophillic Cyanating Agent" *Synlett*, 26, 857–900. doi: 10.1055/s-0034-1380166.

[188] Patil, S. S. & Bobade, V. D. (2010). "Simple and Efficient One-pot Synthesis of 2-Substituted Benzoxazole and Benzothiazole" *Synthetic Communications*, *40*, 206–212. doi: 10.1080/00397910902906602.

[189] Mary, Y. S., Alzoman, N. Z., Menon, V. V., Al-Abdullah, E. S., El-Emam, A. A., Panicker, C. Y., Temiz-Arpaci, O., Armakovi, S., Armakovi, S. J. & Alsenoy, C. V. (2017). "Reactive, Spectroscopic and Antimicrobial Assessments of 5-[(4-Methyl)acetamido]-2-(4-*tert*-butylphenyl)benzoxazole: Combined Experimental and Computational Study" *Journal of Molecular Structure*, *1128*, 694–706. doi: 10.1016/j.molstruc. 2016.09.024.

[190] Ertan-Bolelli, T., Yildizi, I. & Ozgen-Ozgacar, S. (2016). "Synthesis, Molecular Docking and Antimicrobial Evaluation of Novel Benzoxazole Derivatives" *Medicinal Chemical Research*, *25*, 553–567. doi: 10.1007/s0004-015-1499-1.

[191] Parveen, S. S., Al-Alshaikh, M. A., Panicker, C. Y., El-Emam, A. A., Arisoy, M., Temiz-Arpaci, O. & Alsenoy, C. V. (2016). "Synthesis, Vibrational Spectroscopic investigations, Molecular Docking, Antibacterial and Antimicrobial Studies of 5-Ethylsuphonyl-2-(p-aminophenyl)benzoxazole" *Journal of Molecular Structure*, *1115*, 94–104. doi: 10.1016/j.molstruc.2016.02.057.

[192] Reddy, L. M., Reddy, G. D., Padmaja, A. & Padmavathi, V. (2015). "Synthesis and Antimicrobial Activity of Amino Linked Heterocycles" *Chemical and Pharmaceutical Bulletin*, *63*, 75–80. doi: 10.1248/cpb.c14-00521.

[193] Venkatapuram, P. Dandu, S., Chokkappagari, P. & Adivireddy, P. (2014). "Synthesis and Antimicrobial Activity of a New Class of Sulfonylmethane Linked Bisheterocycles" *Journal of Heterocyclic Chemistry*, *51*, 1757–1763. doi: 10.1002/jhet.1874.

[194] Seenaiah, D., Reddy, P. R., Reddy, G. M., Padmaja, A., Padmavathi, V. & Krishna, N. S. (2014). "Synthesis, Antimicrobial and Cytotoxic Activities of Pyrimidinyl Benzoxazole, Benzothiazole and Benzimidazole" *European Journal of Medicinal Chemistry*, *77*, 1–7. doi: 10.1016/j.ejmech.2014.02.050.

[195] Padalkar, V. S., Gupta, V. D., Phatangare, K. R., Patil, V. S., Umape, P. F. & Sekar, N. (2014). "Synthesis of Novel Dipodal-benzimidazole, Benzoxazole and Benzothiazole from Cyanuric Chloride: Structural, Photophysical and Antimicrobial Studies" *Journal of Saudi Chemical Society*, *18*, 262–268. doi: 10.1016/j.jscs.2011.07.001.

[196] Jayanna, N. D., Vagdevi, H. M., Dharshan, J. C., Raghavendra, R. & Telkar, S. B. (2013). "Synthesis, Antimicrobial, Analgesic Activity, and Molecular Docking Studies of Novel 1-(5,7-Dichloro-1,3-dichloro-1,3-benzoxazole-2-yl)-3-phenyl-1*H*-pyrazole-4-carbaldehyde Derivatives" *Medicinal Chemistry Research*, *22*, 5814–5822. doi: 10.1007/s00044-013-0565-9.

[197] Praveena, C. H. L., Rani, V. E.,Spoorthy, Y. N. & Ravindranath, L. K. (2013). "Synthesis Characterization and Antimicrobial Activity of 6-Nitro-1*H*-benzo[d]oxazole/thiazol-2-yl)methyl)-6-oxido-4,8-dihydro-1*H*-[1,3,2]dioxaphosphephino[5,6-c]pyrazole-6-yl) ureas/carboxamides" *Der Pharma Chemica*, *5*, 58–70. URL: https://www.derpharmachemica.com/pharma-chemica/synthesis-characterization-and-antimicrobial-activity-of-6nitro1hbenzod-oxazol-thiazol2ylmethyl6oxido48dihydro1h132dioxa.pdf.

[198] Bouchoucha. A., Terbouche, A., Zaouani, M., Derridj, F. & Djebbar, S. (2013). "Iron and nickel complexes with heterocyclic ligands: Stability, Synthesis, Spectral Characterization, Antimicrobial Activity, Acute and Subacute Toxicity" *Journal of Trace Elements in Medicine and Biology*, *27*, 191–202. doi: 10.1016/j.jtemb.2012.12.001.

[199] Vijesh, A. M., Isloor, A. M., Shetty, P., Sundershan, S. & Fun, H. K. (2013). "New Pyrazole Derivatives Containing 1,2,4-triazoles and Benzoxazoles as Potent Antimicrobial and Analgesic Agents" *European Journal of Medicinal Chemistry*, *62*, 410–415. doi: 10.1016/j.ejmech.2012.12.057.

[200] Arisoy, M., Temiz-Arpaci, O., Kaynak-Onurdag, F. & Ozgen, S. (2012). "Synthesis and Antimicrobial Activity of Novel Benzoxazoles" *Zeitschrift fur Naturforschung*, *67c*, 466–472. URL: http://www.znaturforsch.com/s67c/s67c0466.pdf.

[201] McCulloch, M. W. B., Berrue, F., Haltli, B. & Kerr, R. G. (2011). "One-pot Synthesis of Pseudopteroxazoles from Pseudopterosins: A Rapid Route to Non-natural Congeners with Improved Antimicrobial Activity" *Journal of Natural Products*, *74*, 2250–2256. doi: 10.1021/np2006555.

[202] Hadi, A., Rashid, A., Khodael, M. M., Alizadeh, A., Majnooni, M. B., Pakravan, N., Abiri, R. & Nematollahi, D. (2011). "Catecholthioether Derivatives: Preliminary Study of *in-Vitro* Antimicrobial and Antioxidant Activities" *Chemical Pharmaceutical Bulletin*, *59*, 1149–1152. doi: 10.1248/cpb.59.1149.

[203] Kumar, R., Nair, R. R., Dhiman, S. S., Sharma, J. & Prakesh, O. (2010). "Iodine (III)-Mediated Synthesis of Some 2-Aryl/hetarylbenzoxazoles as Antibacterial/antifungal Agents" *Medicinal Chemistry Research*, *19*, 541–550. doi: 10.1007/s00044-009-9211-y.

[204] Srivastava, A., Aggarwal, L. & Jain, N. (2015). "One-pot Sequential Alkylation and Cycloaddition: Regioselective Construction and Biological Evaluation of Novel

Benzoxazole-Triazole Derivatives" *ACS Combinatorial Science*, *17*, 39–48. doi: 10.1021/co500135z.

[205] Arisoy, M., Temiz-Arpaci, O., Kaynak-Onurdag, F. & Ozgen, S. (2013). "Novel Benzoxazoles: Synthesis and Antibacterial, Antifungal, and Antitubercular Activity against Antibiotic-Resistant and -Sensitive Microbes" *Zeitschrift fur Naturforschung*, *68c*, 453–460. URL: http://www.znaturforsch.com/s68c/s68c0003.pdf.

[206] Ouyang, L., Huang, Y., Zhao, Y., He, G., Xie, Y., Liu, J., He, J., Liu, B. & Wei, Y. (2012). "Preparation, Antibacterial Evaluation and Prelimnary Structure-activity Realationship (SAR) Study of Benzothiazole- and Benzoxazol-2-amine Derivatives" *Bioorganic and Medicinal Chemistry Letters*, *22*, 3044–3049. doi: 10.1016/j.bmcl.2012.03.079.

[207] Chen, X.-z., Xu, P., Liu, L., Zheng, D. & Lie, P. S. (2011). "Synthesis and Antibacterial Activity of Novel Ketolides with 11,12-Sulfur Contained Aryl Alkyl Side Chains" *European Journal of Medicinal Chemistry*, *46*, 208–217. doi: 10.1016/j.ejmech.2010.11.004.

[208] Sheng, C., Che, X., Wang, W., Wang, S., Cao, Y., Yao, J., Miao, Z. & Zhang, W. (2011). "Design and Synthesis of Antifungal Benzoheterocyclic Derivatives by Scaffold Hopping" *European Journal of Medicinal Chemistry*, *46*, 1706–1712. doi: 10.1016/j.ejmech.2011.01.075.

[209] Sheng, C., Xu, H., Wang, W., Cao, Y., Dong, G., Wang, S., Che, X., Ji, H., Miao, Z., Yao, J. & Zhang, W. (2010). "Design, Synthesis and Antifungal Activity of Isosteric Analogues of Benzoheterocyclic N-myristoyltransferase Inhibitors" *European Journal of Medicinal Chemistry*, *45*, 3531–3540. doi: 10.1016/j.ejmech.2010.03.007.

[210] Kuroyanagi, J.-i., Kanai, K., Horiuchi, T., Takeshita, H., Kobayashi, S., Achiwa, I., Yoshida, K., Nakamura, K. & Kawakami, K. (2011). "Structure-Activity Relationships of 1,3-Benzoxazole-4-carbonitriles as Novel Antifungal Agents with Potent *in Vitro* Efficacy" *Chemical Pharmaceutical Bulletin*, *59*, 341–352. doi: 10.1248/cpb.59.341.

[211] Kuroyanagi, J.-i., Kanai, K., Sugimoto, Y., Horiuchi, T., Achiwa, I., Takeshita, H. & Kawakami, K. (2010). "1,3-Benzoxazole-4-carbonitrile as a Novel Antifungal Scaffold of β-1,6-Glucan Synthesis Inhibitors" *Bioorganic and Medicinal Chemistry*, *18*, 7593–7606. doi: 10.1016/j.bmc.2010.08.044.

[212] Garrepalli, S., Sarangapani, M., Garrepally, P. & Chilukala, S. (2011). "Design, Synthesis and Biological Evaluation of Benzoxazole Derivatives as Cyclogenase-2-Inhibitors" *Der Pharmacia Lettre*, *3*, 427–432. URL: https://www.scholarsresearchlibrary.com/articles/design-synthesis-and-biological-evaluation-of-benzoxazole-derivatives-as-cyclooxygensase2-inhibitors.pdf.

[213] Seth, K., Garg, S. K., Kumar, R., Purohit, P., Meena, V. S., Goyal, R., Banerjee, U. C. & Chakraborti, A. K. (2014). "2-(2-Arylphenyl)benzoxazole As a Novel Anti-Inflammatory Scaffolds: Synthesis and Biological Evaluation" *ACS Medicinal Chemistry Letters*, *5*, 512–516. doi: 10.1021/ml400500e.

[214] Parlapalli, A., K., Cidda, M. & Manda, S. (2014). "Antiinflammatory and Antioxidant Activities of 2-Amino-N-(substituted alkyl)benzoxazole-5-carboxamide Derivatives" *International Journal of Pharmacy and Pharmaceutical Sciences*, *6*, 311–314. URL:https://innovareareacademics.in/journal/index/index.php/ijpps/article/view/1787 .

[215] Singh, S. P., Verma, N. K., Rai, P. K. & Tripathi, A. K. (2014). "Synthesis and Pharmacological Screening of Some Noval Benzoxazole" *Der Pharmacia Lettre*, *6*,

283–288. URL: https://www.scholarsresearchlibrary.com/articles/synthesis-and-pharmacological-screening-of-some-noval-benzoxazole-derivatives.pdf.

[216] Stella, P. C. R., Rajam, S. & Venkatraman, B. R. (2012). "Synthesis, Characterization and Biological Evaluation of Benzoxazole Derivatives" *Journal of Chemical and Pharmaceutical Research*, *4*, 2988–2993. URL: http://www.jocpr.com/articles/synthesis-characterization-and-biological-evaluation-of-benzoxazole-derivatives.pdf.

[217] M. R., G. K., G. R., J. V. R. & M. S. (2010). "Synthesis and Anti-inflammatory Activity of 2-Substituted-((*N*,*N*-disubstituted)-1,3-benzoxazole)-5-carboxamides" *Acta Pharmaceutica Sinica*, *45*, 730–734. URL: http://www.academia.edu/1585638/Synthesis_and_anti-inflammatory_activity_of_2-substituted-_N_Ndisubstituted_-_1_3-benzoxazole_-5-carboxamide.

[218] Ongarora, D. S. B., Strydom, N., Wicht, K., Njoroge, M., Wiesner, L., Egan, T. J., Wittlin, S., Jurva, U., Masimirembwa, C. M. & Chibale, K. (2015). "Antimalarial Benzoheterocyclic 4-Aminoquinolines: Structure-activity Relationship, in Vivo Evaluation, Mechanistic and Bioactivation Studies" *Bioorganic and Medicinal Chemistry*, *23*, 5419–5432. doi: 10.1016/j.bmc.2015.07.051.

[219] Salerno, L., Pittal, V., Romeo, G., Modica, M. N., Marrazzo, A., Siracusa, M. A., Sorrenti, V., Giacomo, C. D., Vanella, L., Parayath, N. N. & Greish, K. (2015). "Novel Imidazole Derivatives as Heme Oxygenase-1 (HO-1) and Heme Oxygenase-2 (HO-2) Inhibitors and Their Cytotoxic Activity in Human Derived Cancer Cell Lines" *European Journal of Medicinal Chemistry*, *96*, 162–172. doi: 10.1016/j.ejmech.2015.04.003.

[220] Ginzinger, W., Muhlgassner, G., Arion, V. B., Jakupec, M. A. & Roller, A. (2012). "A SAR Study of Novel Antiproliferative Ruthenium and Osmium Complexes with Quinoxalinone Ligands in Human Cancer Cell Lines" *Journal of Medicinal Chemistry*, *55*, 3398–3413. doi: 10.1021/jm3000906.

[221] Xiang, P., Zhou, T., Wang, L., Sun, C. Y., Hu, J., Zhao, Y. L. & Yang, L. (2012). "Novel Benzothiazole, Benzimidazole Derivatives as Potential Antitumor Agents: Synthesis and Prelimnary *in Vitro* Biological Evaluation" *Molecules*, *17*, 873–883 and references cited therein. doi: 10.3390/molecules17010873.

[222] Jia, L. & Ma, S. (2016). "Recent Advances in the Discovery of Heparanase Inhibitors as Anti-cancer Agents" *European Journal of Medicinal Chemistry*, *121*, 209–220 and references cited therein. doi: 10.1016/j.ejmech.2016.05.052.

[223] Reddy, N. B., Burra, V. R., Ravindranath, L. K., Sreenivasulu, R. & Kuma, V. N. (2016). "Synthesis and biological evaluation of Benzoxazole Fused Combretastatin Derivatives as Anticancer Agents" *Monatshefte fur Chemie*, *147*, 593–598. doi: 10.1007/s00706-016-1685-y.

[224] Fadda, A. A., Abdel-Latif, E., Tawfik, E. H. & Mohammed, R. M. (2016). "Synthesis Some New Benzoxazole Derivatives and Their Cytotoxicity to Human Cancer Cell Lines" *Research Journal of Pharmaceutical, Biological and Chemical Sciences*, *7*, 1820–1826. URL: http://www.rjpbcs.com/pdf/2016_7(1)/[256].pdf.

[225] Subran, S. K., Banerjee, S., Mondal, A. & Paira, P. (2016). "Amberlite IR-120(H)-Mediated "On water" Synthesis of Novel Anticancer Ruthenium(II)-p-cymene 2-pyridinylbenzothiazole (BTZ), 2-Pyridinylbenzoxazole (BOZ) & 2-Pyridinylbenzimidazole (BIZ) Scaffolds" *New Journal of Chemistry*, *40*, 10333–10343. doi: 10.1039/C6NJ02049F.

[226] Kandeel, M. M., Ali, S. M., Abdelgawad, M. A., Abdel-Bakky, S. M. & Mohamed, F. E. A. (2016). "Synthesis and Cytotoxic Activity of Acridine Derivatives Substituted with Benzimidazole, Benzoxazole and Benzothiazole" Der *Pharma Chemica*, *8*, 117–123. URL: https://www.derpharmachemica.com/abstract/synthesis-and-cytotoxic-activity-of-acridine-derivatives-sustituted-withbenzimidazole-benothiazole-4706.html.

[227] Yurttas, L., Ozkay, Y., Akalim-Ciftci, G. & Ulusoylar-Yildirim, S. (2014). "Synthesis and Anticancer Activity Evaluation of N-[4-(2-methylthiazol-4-yl)phenyl]acetamide Derivatives Containing (Benz)azole Moiety" *Journal of Enzyme Inhibition and Medicinal Chemistry*, *29*, 175–184. doi: 10.3109/14756366.2013.763253.

[228] Velidandi, A., Gadidasu, K. K. & Patri, S. V. (2015). "Design, Synthesis and Evaluation of Novel Benzoxazole, Benzimidazole and Bezthiazole Based Cationic Lipids as Anticancer Agents" *Research Journal of Pharmaceutical, Biological and Chemical Sciences*, *6*, 404–418. URL: http://www.rjpbcs.com/pdf/2015_6(4)/[63].pdf.

[229] Ozkay, Y., Isikdag, I., Incesu, Z. & Akalin, G. (2010). "Synthesis of 2-Substituted-*N*-[4-(1-methyl-4,5-diphenyl-1*H*-imidazole-2-yl)phenyl]acetamide Derivatives and Evaluation of Their Anticancer Activity" *European Journal of Medicinal Chemistry*, *45*, 3320–3328. doi: 10.1016/j.ejmech.2010.04.015.

[230] Murty, M. S. R., Rao, B. R., Katiki, M. R., Nath, L. R. & Anto, R. J. (2013). "Synthesis of Piperazinyl Benzothiazol/benzoxazole Derivatives Coupled with 1,3,4-oxadizole-2-thiol: Novel Heterocycles as Anticancer Agents" *Medicinal Chemistry Research*, *22*, 4980–4991. doi: 10.1007/s00044-013-0510-y.

[231] Murty, M. S. R., Ram, K. R., Rao, R. V., Yadav, J. S., Rao, J. V., Cheriyan, V. T. & Anto, R. J. (2011). "Synthesis and Prelimnary Evaluation of 2-Substituted-1,3-benzoxazole and 3-[(3-substituted)propyl]-1,3-benzoxazol-2(3*H*)-one Derivatives as Potent Anticancer Agents" *Medicinal Chemistry Research*, *20*, 576–586. doi: 10.1007/s00044-010-9353-y.

[232] Pitta, E., Rogacki, M. K., Balabon, O., Huss, S., Cunningham, F., Lopez-Roman, E. M., Joossens, J., Augustyns, K., Ballel, L., Bates, R. H. & Vaken, P. V. D. (2016). "Searching for New Leads for Tuberculosis: Design, Synthesis, and Biological Evaluation of Novel 2-Quinolin-4-yloxyacetamides" *Journal of Medicinal Chemistry*, *59*, 6709–6728 and references cited therein. doi: 10.1021/acs.jmedchem.6b00245.

[233] Rana, D. N., Chhabria, M. T., Shah, N. K. & Brahmkshatriya, P. S. (2014). "Pharmacophore Combination as a Useful Strategy to Discover New Antitubercular Agents" *Medicinal Chemistry Research*, *23*, 370–381. doi: 10.1007/s00044-013-0645-x.

[234] Imramovsky, A., Kozic, J., Pesko, M., Stolarikova, J., Vinsova, J., Kralova, K. & Jampilek, J. (2014). "Synthesis and Antimycobacterial and Photosynthesis-Inhibiting Evaluation of 2-[(E)-2-Substituted-ethryl]-1,3-benzoxazoles" *The Scientific World Journal*, *705973*, 1–11. doi: 10.1155/2014/705973.

[235] Bernard, A. M., Cabiddu, M. G., Montis, S. D., Mura, R. & Pompei, R. (2014). "Synthesis of New Compounds with Promising Antiviral Properties Against Group A and B Human Rhinoviruses" *Bioorganic and Medicinal Chemistry*, *22*, 4061–4066 and references cited therein. doi: 10.1016/j.bmc.2014.05.066.

[236] Rynearson, K. D., Charrette, B., Gabriel, C., Moremo, J., Boerneke, M. A., Dibrov, S. M. & Hermann, T. (2014). "2-Aminobenzoxazole Ligands of the Hepatitus C Virus Ribosome Entry Site" *Bioorganic and Medicinal Chemistry Letters*, *24*, 3521–3525. doi: 10.1016/j.bmcl.2014.05.088.

[237] Byrd, C. M., Grosenbach, D. W., Berhanu, A., Dai, D., Jones, K. F., Cardwell, K. B., Schneider, C., Yang, G., Tyavanagimatt, S., Harver, C., Wineinger, K. A., Page, J., Stavale, E., Stone, M. A., Fuller, K. P., Lovejoy, C., Leeds, J. M., Hruby, D. E. & Jordan, R. (2013). "Novel Benzoxazole Inhibitors of Dengue Virus Replication That Targets the MS3Helicase" *Antimicrobial Agents and Chemotherapy*, *57*, 1902–1912. doi: 10.1128/AAC.02251-12.

[238] Jonckers, T. H. M., Rouan, M. C., Hache, G., Schepens, W., Hallenberger, S., Baumeister, J. & Sasaki, J. (2012). "Benzoxazole and Benzothiazole Amides as Novel Pharmacokinetic Enhancers of HIV Protease Inhibitors" *Bioorganic and Medicinal Chemistry Letters*, *22*, 4998–5002. doi: 10.1016/j.bmcl.2012.06.022.

[239] Li, Z., Zhan, P., Naesens, L., Vanderlinden, E., Liu, A., Du, G., Clercq, E. D. & Liu, X. (2012). "Synthesis and Preliminary Biologic Evaluation of 5-Substituted-2-(4-substituted phenyl)-1,3-Benzoxazoles as A Novel Class of Influenza Virus A Inhibitors" *Chemical Biology* and *Drug Design*, *79*, 1018–1024. doi: 10.1111/j.1747-0285.2012.01344.x.

[240] Boyer, J., Arnoult, E., Medebielle, M., Guillemount, J., Unge, J. & Jochmans, D. (2011). "Difluoromethylbenzoxazole Pyrimidine Thioether Derivatives: A Novel Class of Potent Non-Nucleoside HIV-1 Reverse Transcriptase Inhibitors" *Journal of Medicinal Chemistry*, *54*, 7974–7985. doi: 10.1021/jm200766b.

[241] Booysen, I. N., Adebisi, A. & Akerman, M. P. (2015). "Formation, Electrochemical and Scavenging Properties of Novel Ruthenium Compounds with N, X-donor (X = O, N) Heterocyclic Chelators" *Inorganic Chemica Acta*, *433*, 13–20. doi: 10.1016/j.ica.2015.04.031.

[242] Jayanna, N. D., Vagdevi, H. M., Dharshan, J. C., Kekuda, T. R. P. & Hanumanthappa, B. C. (2013). "Synthesis and Biological Evaluation of Novel 5,7-Dichloro-1,3-Benzoxazole Derivatives" *Journal of Chemistry*, *864385*, 1–9. doi: 10.1155/2013/864385.

[243] Youssef, M. M. & Amin, M. A. (2010). "Microwave Assisted Synthesis of Some New Heterocyclic Spiro-Derivatives with Potential Antimicrobial and Antioxidant Activity" *Molecules*, *15*, 8827–8840. doi: 10.3390/molecules15128827.

[244] Sampson, D., Zhu, X. Y., Eyunni, S. V. K., Etukala, J. R., Ofori, E., Bricker, B., Lamango, N. S., Setola, V., Roth, B. L. & Ablordeppey, S. Y. (2014). "Identification of a New Selective Dopamine D_4 Receptor Ligand" *Bioorganic and Medicinal Chemistry*, *22*, 3105–3114. doi: 10.1016/j.bmc.2014.04.026.

[245] Rezazadeh, S., Navidpour, L. & Shafiee, A. (2013). "Synthesis of Substituted 2-Heteroarylbenzazol-5-ol Derivatives as Potential Ligands for Estrogen Receptors" *Tetrahedron*, *69*, 6076–6082. doi: 10.1016/j.tet.2013.05.084.

[246] Prajapati, L. M., Parmar, V. K., Patel, M. J. & Patel, J. R. (2012). "Development of QSAR Model for Prediction of Fold Selectivity of Phenyl Benzoxazole as Estrogen Receptor Inhibitors" *Medicinal Chemistry Research*, *21*, 3681–3686. doi: 10.1007/s00044-011-9920-x.

[247] Yang, Z., Fairfax, D. J., Maeng, J. H., Masih, L., Usyatinsky, A., Hassler, C., Isaacson, S., Fitzpatrick, K., DeOrazio, R. J., Chen. J., Harding, J. P., Isherwood, M., Dobritsa, S., Christensen, K. L., Wierschke, J. D., Bliss, B. I., Peterson, L. H., Beer, C. M., Cioffi, C., Lynch, M., Rennells, M., Richard, J. J., Rust, T., Khmelnitsky, Y. L., Cohen, M. L. & Manning, D. D. (2010). "Discovery of 2-Substituted Benzoxazole Carboxamides as 5-HT_3 Receptor Antagonists" *Bioorganic and Medicinal Chemistry Letters*, *20*, 6538–6541. doi: 10.1016/j.bmcl.2010.09.038.

[248] Cox, C. D. Breslin, M. J., Whitman, D. B., Schreier, J. D., McGaughey, G. B., Bogusky, M. J., Roecker, A. J., Mercer, S. P., Bednar, R. A., Lamaire, W., Bruno, J. G., Reiss, D. R., Harrell, C. M., Murphy, K. L., Garson, S. L., Doran, S. M., Prueksartanont, T., Anderson, W. B., Tang, C., Roller, S., Cabalu, T. D., Cui, D., Hartman, G. D., Yang, S. D., Koblan, K. S., Winrow, C. J., Renger, J. J. & Coleman, P. J. (2010). "Discovery of the Dual Orexin Receptor Antagonist [(7R)-4-(5-Chloro-1,3-benzoxazole-2-yl)-7-methyl-1,4-diazepan-1-yl-2-(2H-1,2,3-triazol-2-yl)phenyl]methanone (MK-4305) for the Treatment of Insomnia" *Journal of Medicinal Chemistry*, *53*, 5320–5332. doi: 10.1021/jml100541c.

[249] Watson, S. L., Waston, C. J., Baghdoyan, H. A. & Lydic, R. (2010). "Thermal Nociception is Decreased by Hypocretin-1 and an Adenosine A_1 Receptor Agonists Microinjected into the Pontine Reticular Formation of Sprague Dawley Rat" *Journal of Pain*, *11*, 535–544. doi: 10.1016/j.jpain.2009.09.010.

[250] Singh, J., Mansuri, R., Sahoo, C. G. & Kumar, M. (2016). "*3-D* QSAR Modeling of Benzenesulfonamides Substituted Derivatives and Feature Based Inhibitor Identification as Antidiadetic Agent" *International Journal of Chemical Studies*, *4*, 24–32. URL: http://www.chemijournal.com/archives/2016/vol4issue4/PartA/4-3-1-894.pdf.

In: Organic and Medicinal Chemistry, Volume 1
Editor: Bimal Krishna Banik
ISBN: 978-1-53614-453-6

Chapter 4

THERAPEUTIC POTENTIALS OF PYRIMIDINE DERIVATIVE: A SCAFFOLD FOR NEW DRUG DEVELOPMENT

Biswa Mohan Sahoo[1,*] and Bimal K. Banik[2,†]

[1]Department of Pharmacy, Vikas Group of Institutions, Nunna, Vijayawada, India
[2]Community Health System of South Texas, Edinburg, Texas, US

ABSTRACT

A wide range of research activities were made to synthesize a large number of heterocyclic compounds and their derivatives. These were also screened for their pharmacological activities against various diseases. Heterocyclic compounds having nitrogen in their skeleton are the most fascinated compound by scientists due to their diverse biological activities. Among the various heterocyclic compounds, pyrimidine derivatives play an important role in the medicinal chemistry because they posses promising activities such as antifungal, antibacterial, anthelmintic, anticancer, antioxidant, vasodilator, neurotropic, anesthetic, antineoplastic, diuretics, antihypertensive, antipsychotic, anti-anxiety, anticonvulsant, antidiabetic, antimalarial, antihistaminic, antitubercular, antiparkinsonian, analgesic, anti-inflammatory, anti-HIV, expectorant and mucolytic, etc. This review is focused on the therapeutic potentials of pyrimidine derivatives that are now in development.

Keywords: heterocyclic compound, pyrimidine, SAR, therapeutic potentials

ABBREVIATIONS

ATP	Adenosine triphosphate
cAMP	Cyclic adenosine monophosphate

* Corresponding Author Email: drbiswamohansahoo@gmail.com.
† bimalbanik10@gmail.com.

COX	Cyclooxygenase
dGTP	Deoxyguanosine triphosphate
DHF	Dihydrofolic acid
DHFR	Dihydro-folate reductase
DHPS	Dihydropteroate synthetase
DNA	Deoxyribonucleic acid
FU	Fluorouracil
GABA	Gamma-aminobutyric acid
H1	Histamine
HAD	Hydrogen donor/acceptor
HIV	Human immuno deficiency virus
H_2O_2	Hydrogen peroxide
5-HT	5-hydroxytryptamine
I. R.	Infrared Spectroscopy
INH	Isoniazid
$KMnO_4$	Potassium permanganate
MABA	Microplate Alamar Blue Assay
MWI	Microwave irradiation
nACh	Nicotinic acetylcholine
NMR	Nuclear magnetic resonance
NRTI	Nucleoside reverse-transcriptase inhibitor
pKa	Acid dissociation constant
PZA	Pyrazinamide
RNA	Ribonucleic acid
RNR	Ribonucleotide reductase
SAR	Structure activity relationships
THF	Tetrahydrofolic acid
ThMP	Thymidine monophosphate
TMPKmt	Thymidine monophosphate kinase *Mycobacterium tuberculosis*
U. V.	Ultraviolet.

1. INTRODUCTION

Heterocyclic compounds impart considerable attention in the field of medicinal chemistry and serves as a key template for the development of various therapeutic agents. It provides scaffolds on which pharmacophores can be built to yield potent and selective drug candidates. The heterocycle often confers stability on a ligand which would otherwise be rapidly degraded *in-vivo* and enables the ligand to be absorbed from the gut and/or to penetrate the blood brain barrier. However a heterocycle is being introduced into a ligand as a bioisostere, the key considerations in choice of heterocycle are thus to maintain or improve the activity of the ligand while improving its absorption, brain penetration and metabolic stability characteristics.

Heterocyclic compounds containing two nitrogen atoms represent an important group of organic compounds because many of them exhibit significant pharmacological activity. Among these, pyrimidine and its derivatives offer enormous scope due to their unique structures and association with a wide spectrum of medicinal activities including anticonvulsant, antibacterial, antifungal, antiviral and anticancer properties. This broad spectrum of biological targets has been facilitated by the synthetic versatility of pyrimidine, which has allowed the generation of a large number of structurally diverse derivatives derived from substitution of the aryl ring, and/or derivatization of the pyrimidine nitrogen and C-2/C-4/C-5/C-6 carbon positions. Pyrimidines are synthetically versatile substrates. The structure and chemical reactivity of pyrimidine can be used for the synthesis of a large variety of heterocyclic compounds and as raw material for drug synthesis [1, 2].

2. Pyrimidine: Privileged Scaffold

Pyrimidines mediate their biological effects by interactions with distinct cell-surface receptors. Pyrimidine is a six membered heterocyclic aromatic organic compound containing two nitrogen atoms at positions 1 and 3 of the ring with molecular formula $C_4H_4N_2$ and molecular weight is 80 dalton. Pyrimidine was first isolated by Gabriel & Colman in 1899. It is the most important heterocyclic organic compound under member of diazines. This ring system is present in numerous living organisms, for examples purines, uric acid, xanthine, caffeine, folic acid, nucleosides, etc [3, 4]. It is isomeric with two other forms of diazine i.e., Pyridazine, with the nitrogen atoms in positions 1 and 2 and Pyrazine, with the nitrogen atoms in positions 1 and 4.

2.1. Structure of Pyrimidine

The replacement of 'C-H' units meta to each other in a benzene ring by two nitrogen atoms gives pyrimidine, generally results in a reduced symmetry with bonds of unequal length. However, it retains its symmetry about 2, 5 axis so that three differing pairs of equal bond density centers for the π-electrons which are otherwise equally distributed about the ring system in benzene (Figure 1). Therefore, the reactivity of 2, 4, 5 and 6 carbon atoms of pyrimidine, as well as, the substituent attached to them vary individually [5].

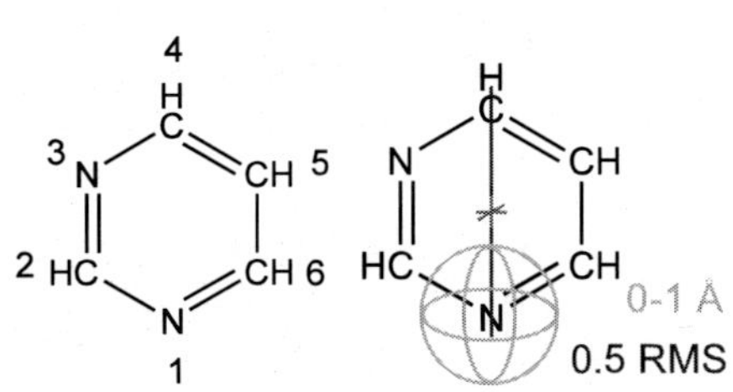

Figure 1. Molecular scaffold of pyrimidine and it's structure showing plane, centroid and exclusion sphere.

2.2. Geometry and Electronic Structure of the Pyrimidine

The pyrimidine ring is virtually flat. The depletion of electron density at position *ortho* and *para* to the electronegative nitrogen atoms is more marked in the pyrimidine than the other two diazines (pyrazine and pyridazine). This is because the two 'N' atoms of pyrimidine is so positioned that their individual effects reinforce each other. That is why the resultant electronic effect is greater in the pyrimidine than its isomeric diazines. The presence of hydroxyl or amino group at the 2, 4 and 6 position makes tautomeric oxo and imino compounds respectively. Tautomerism occurs with 2- and 4-hydroxypyrimidines and these are normally present in keto form (Figure 2) [6].

Thus, the marked positions at C-2, C-4 and C-6 in the pyrimidine ring become strong electropositive centers, while the position C-5 retains some electronegativity (Figure 3).

At the same time, the localization of π-electrons at two 'N' atoms results in decreased aromaticity of the pyrimidine ring. Thus, pyrimidine is not truly aromatic: the partial aromatic character of this molecule is due to C-5 position of the ring system. This leads to a sharp fall in its stability reflected by low resonance energy of 26 kcal/mole (Figure 4). As pyrimidine has close relationship with benzene, it is considered as aromatic but is less aromatic as compared to pyridine. Pyrimidine is the resonance hybrid of following resonating structures and these canonical structures provide evidences for aromatic property of pyrimidine.

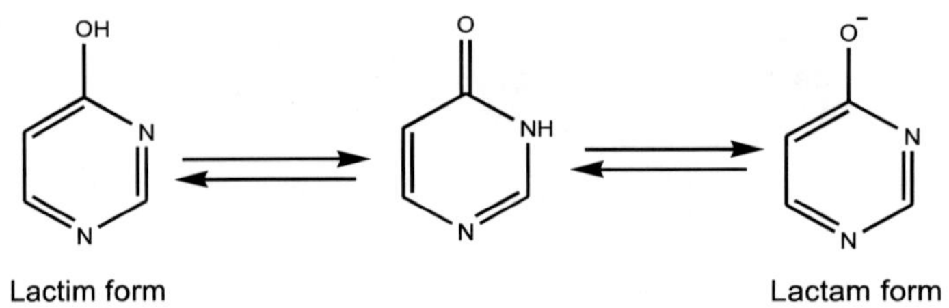

Figure 2. The most probable way of tautomerism shown by pyrimidine.

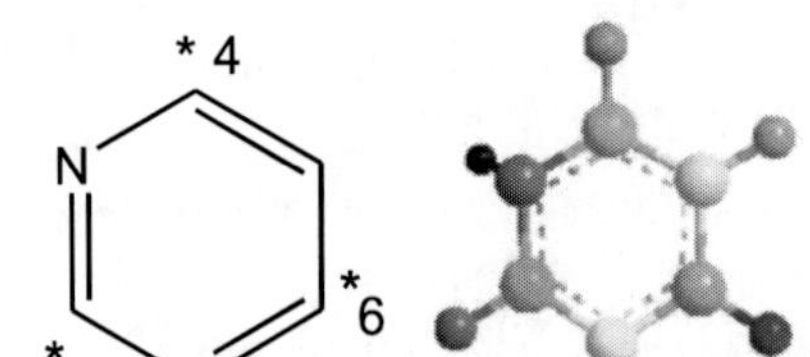

Figure 3. Electropositive centers in the pyrimidine ring and it's energy minimized 3D structure.

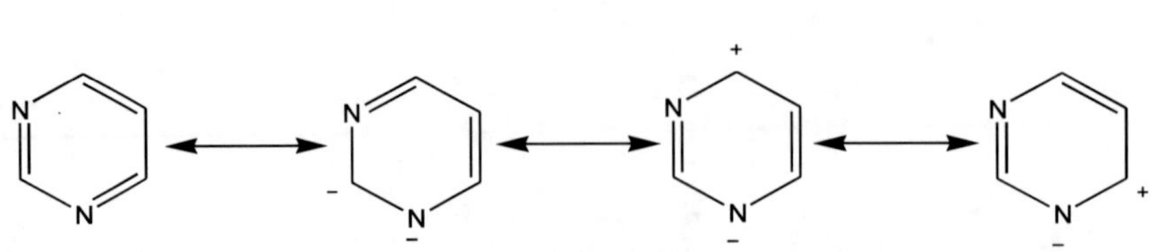

Figure 4. Resonance structure of pyrimidine.

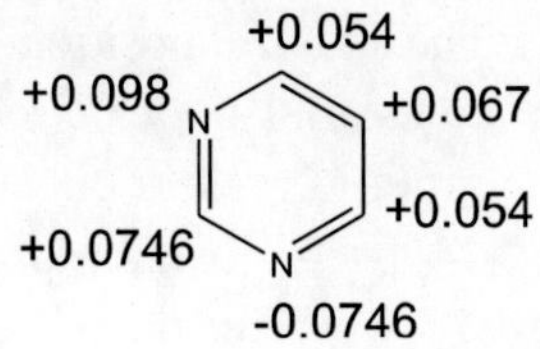

Figure 5. Distribution of electrical charges on the pyrimidine ring.

The following electron density diagram shows the distribution of electrical charges i.e., the gain or loss of the π-electrons at each atom of the molecule has been calculated for the pyrimidine (Figure 5).

2.3. Physical Properties of Pyrimidine

Pyrimidine is a colorless compound with melting point 22.5^{0}C and boiling point 124^{0}C. It has dipole moment of 2.42 D. Its dimensions have been determined by an X-ray diffraction study of a crystal at -2^{0}C and it closely resembles those of pyridine (six membered hetrocyclics with one nitrogen atom in their ring). Pyrimidine is a weaker base (pKa 1.31) than pyridine (pKa 5.2). The decrease in basicity is due to the electron withdrawing effect of the second nitrogen atom present in the ring. Presence of alkyl group enhances the basicity, thus 4-methylpyrimidine has pKa 2.0 while 4,6-dimethylpyrimidine has pKa value of 2.8. The 2- and 4-aminopyrimidines are more basic with pKa 3.54 and 5.71 respectively [7].

2.4. Chemical Properties [8, 9]

2.4.1. Reaction with Acids

Though pyrimidine is a weak base, it can be protonated in presence of acids. Diprotonation also takes place in presence of strong acids because the nitrogen atoms are not present in adjacent position as in pyridazine. In presence of strongly acidic solution, pyrimidine is converted to the pyrimidinium cation (Figure 6).

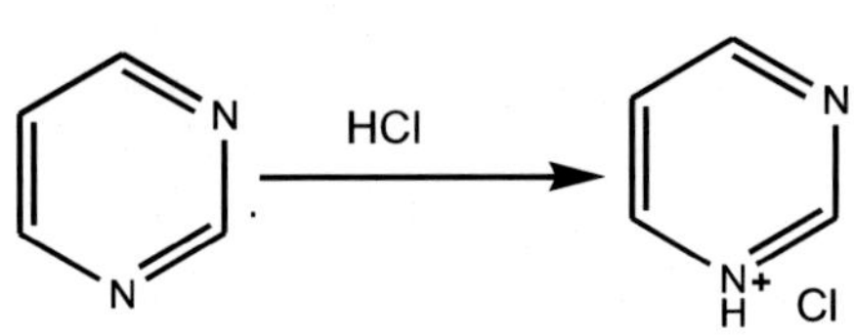

Figure 6. Reaction of pyrimidine with acids.

2.4.2. Reaction with Electrophilic Reagent

Like pyridines, in pyrimidines the π-electron density is decreased to a greater extent. Therefore electrophilic aromatic substitution is more difficult whereas nucleophilic aromatic substitution is facilitated. The attack of electrophile at position 2, 4, 6 is retarded because of electron deficiency at these positions. The attack at position-5 is also difficult as it is influenced by inductive effect of two nitrogen atoms. But the electrophilic substitution at position-5 is possible only when electron releasing group is present on the pyrimidine ring. e.g., presence of hydroxyl or amino group in the 2, 4 or 6 position of pyrimidine enables electrophilic substitution reaction at the 5-position. Two electron releasing groups at the ring make the molecule more reactive for electrophilic substitution. Thus 2- or 4-hydroxy pyrimidine does not nitrate easily but 2,4-dihydroxy pyrimidine undergoes nitration in boiling fuming nitric acid to give 2,4-dihydroxy-5-nitropyrimidine (Figure 7).

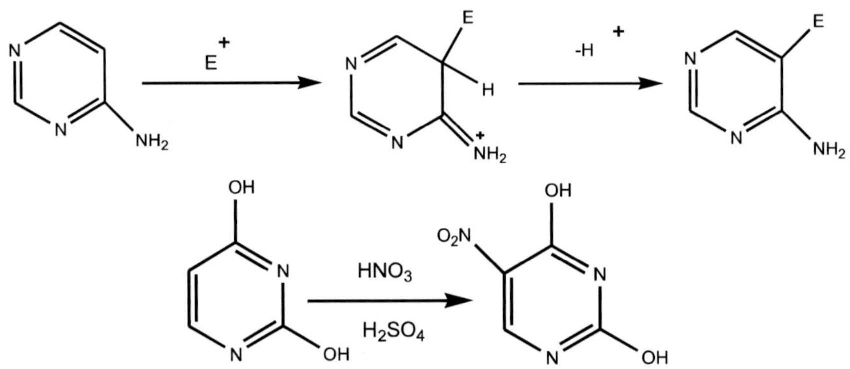

Figure 7. Reaction of pyrimidine with electrophilic reagent.

2.4.3. Reaction with Nucleophilic Reagents

The attack of a nucleophile takes place easily on the pyrimidine ring similar to pyridine, quinoline and isoquinoline. The position-2, 4 and 6 are susceptible to the nucleophilic attack. In general, the position-4 is the most reactive, for example, Grignard reagents readily add to the 3,4-bond of pyrimidines at room temperature as illustrated by the reaction of phenylmagnesium bromide on pyrimidine. The intermediate on hydrolysis and subsequent oxidation of the dihydropyrimidine with $KMnO_4$ in acetone yields 4-phenylpyrimidine (Figure 8).

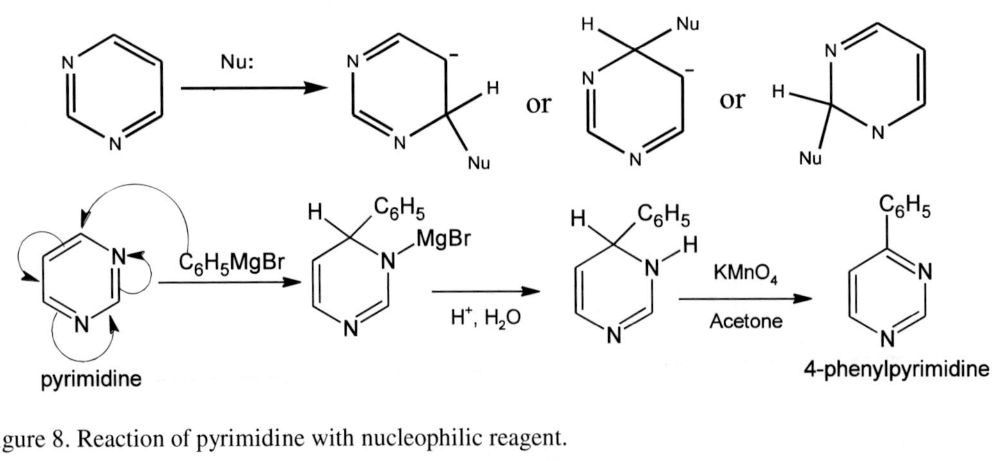

Figure 8. Reaction of pyrimidine with nucleophilic reagent.

2.4.4. Reaction with Oxidizing Reagent

Pyrimidine gives low yield of N-oxide on oxidation with peracids. The pyrimidine ring is greatly destroyed during N-oxide formation. But the alkyl pyrimidine gives a satisfactory yield of N-oxides. 4-Methylpyrimidine reacts with H_2O_2 and produces 4-methylpyrimidine N-oxide (Figure 9).

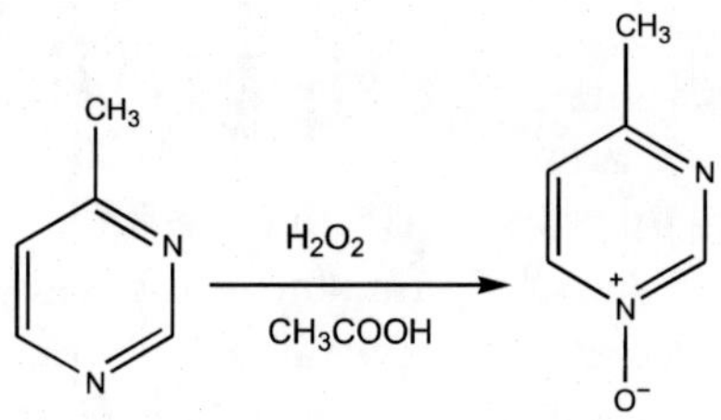

Figure 9. Reaction of pyrimidine with oxidizing reagent.

2.4.5. Rearrangement Reaction of Pyrimidine Derivatives

2,4 or 6-alkoxypyrimidines undergo rearrangement to N-alkylpyrimidinones on strong heating but allylethers of pyrimidines give both N-allylpyrimidinones and C-allyl pyrimidinones by Claisen rearrangement reaction (Figure 10).

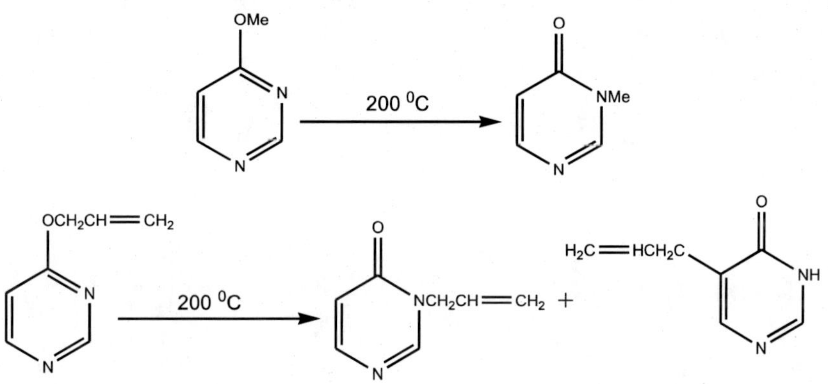

Figure 10. Rearrangement reaction of pyrimidine derivatives.

2.4.6. Spectral Properties of Pyrimidine

The U. V. spectrum of pyrimidine depends on the state of the nucleus i.e., aromatic (partly reduced or fully reduced), position, number and nature of any chromophoric substituent attached to it and also on the ionic species present in the solution being measured. The U. V. absorption of pyrimidine occurs in two bands centered at 243 and 298 nm in cyclohexane. The electronic transition in U. V. spectra is observed to be hypsochromic (blue shift) when electron donor groups are attached to the heterocyclic system. The I. R. spectrum of pyrimidine is observed basically due to vibrational transition. It is well established that N-H stretching vibration takes place in the range of 3360-3420 cm^{-1} and the C-H stretching is observed between 3070-3060 cm^{-1}. The ^1H NMR spectra of pyrimidine present three different types of protons such as H-2 (most deshielded), H-4, H-6 (both are equivalent proton) and H-5 (most shielded proton) (Table 1). On the basis of fragmentation pattern of pyrimidine ring, the mass spectrum is obtained [10-12].

Table 1. ¹H NMR and ¹³C NMR spectral data (δ, ppm; $CDCl_3$ as solvent) of pyrimidine

¹H NMR						¹³C NMR					
H-1	H-2	H-3	H-4	H-5	H-6	C-1	C-2	C-3	C-4	C-5	C-6
---	9.26	---	8.78	7.46	8.78	---	158.4	---	156.9	121.9	156.9

2.5. General Methods for Synthesis of Pyrimidine [13-17]

In 1818, Gasfare B. isolated the first pyrimidine derivative, alloxan, by the oxidation of uric acid with nitric acid. In 1878, the first pyrimidine was synthesized i.e., barbituric acid from malonic acid and urea. The various synthetic methods of pyrimidines are categorized on the basis of components employed in the cyclization of pyrimidine such as a) one component synthesis; b) two component synthesis; c) three component synthesis.

2.5.1. One Component Synthesis

This involves the intramolecular cyclization of certain open chain intermediates to produce the pyrimidine nucleus. One of the examples of this type is the synthesis of pyrimidine by condensation of 1,3-dicarbonyl compounds with amidines which proceeds *via* vinylamidine as intermediate. Intramolecular cyclization of the vinylamidine leads to the formation of 3,4(1,6)-bond. Then vinylamidine is cyclized under acidic and basic conditions to give corresponding functionalized pyrimidines (Figure 11).

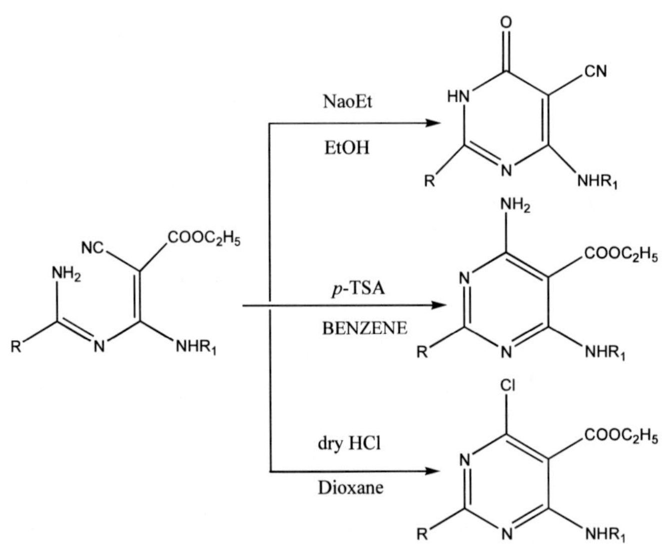

Figure 11. One component synthesis of pyrimidines.

2.5.2. Two Components Synthesis

It is the most widely used method of pyrimidine synthesis. It involves the condensation of two reactants. One of the components may contribute three, four or five atoms of the pyrimidine ring system, while the other contributes three, two or one atom, respectively. The different types of two component synthesis are summarized as follows (Table 2).

Table 2. Two components synthesis of pyrimidine

1.	C-C-C	+	N-C-N
2.	C-C-N	+	C-N-C
3.	C-C-C-N	+	C-N
4.	C-N-C-N	+	C-C
5.	C-C-N-C	+	N-C
6.	N-C-C-C-N	+	C
7.	C-C-C-N-C	+	N
8.	C-C-N-C-N	+	C

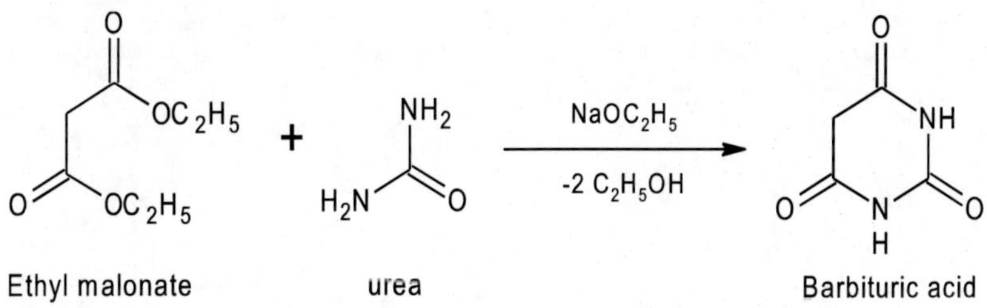

Figure 12. Synthesis of barbituric acid.

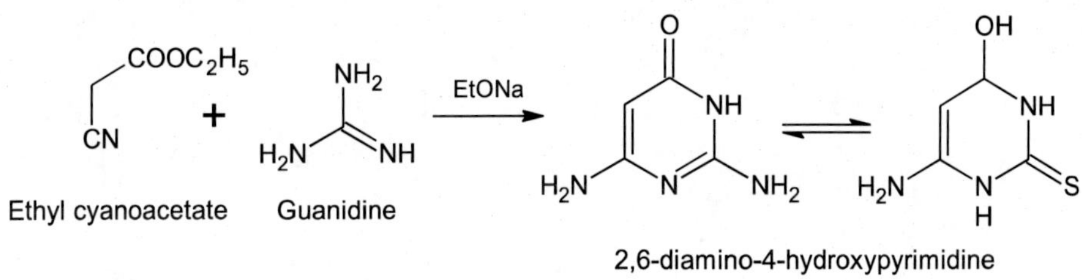

Figure 13. Synthesis of 2,6-diamino-4-hydroxypyrimidine.

Out of the above eight methods, method-1 is most widely used and also known as "principal synthesis" of pyrimidine. Synthesis of pyrimidine ring involves a condensation between a malonic ester and urea in presence of a base to yield barbituric acid. Gabriel was first to prepare pyrimidine from barbituric acid (Figure 12).

Ethyl cyanoacetate reacts with guanidine in the presence of sodium ethoxide to give 2,6-diamino-4-hydroxypyrimidine (Figure 13).

2.5.3. Three Components Synthesis

Very few reports are available for three component synthesis. Generally, each of the three components contributes two atoms each for cyclization to produce pyrimidine ring (Figure 14).

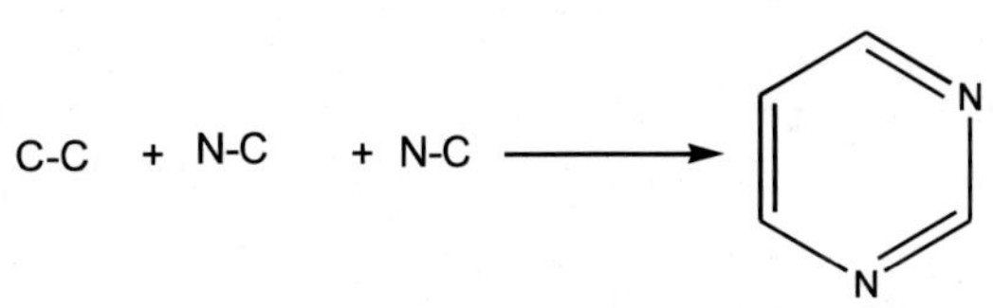

Figure 14. Three components synthesis of pyrimidines.

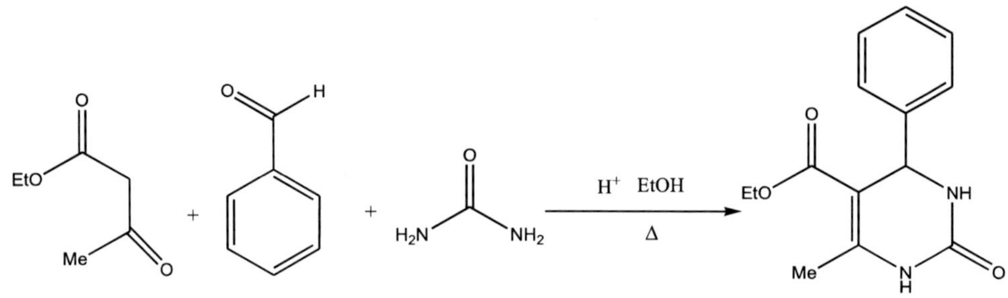

Figure 15. Biginelli dihydropyrimidine synthesis.

In 1893, P. Biginelli reported the acid-catalyzed cyclocondensation reaction of ethyl acetoacetate, benzaldehyde, and urea for the synthesis of dihydropyrimidines (Figure 15).

3. Structure Activity Relationships (SAR) Study of Pyrimidine

SAR study gives insights into the molecular properties causing receptor affinity and selectivity. The promising nature of the pyrimidine compounds may be attributed to the substitutions at the hydrophobic domain. These compounds had electron withdrawing and donating groups at the *ortho, meta* and *para* position of the hydrophobic aryl ring (Figure 16). In general it was observed that the substituted derivatives were more active than the other derivatives. This may be because of the fact that the substituted derivatives are better fitted into the receptor site. In all of the revolutionary experiments, important core fragments are defined by the presence of hydrogen donor/acceptor unit (HAD), hydrophobic domain (A) aryl ring substituted/unsubstituted, and electron donor atom (D). These common features are found in the structures of well established pyrimidine drugs [18].

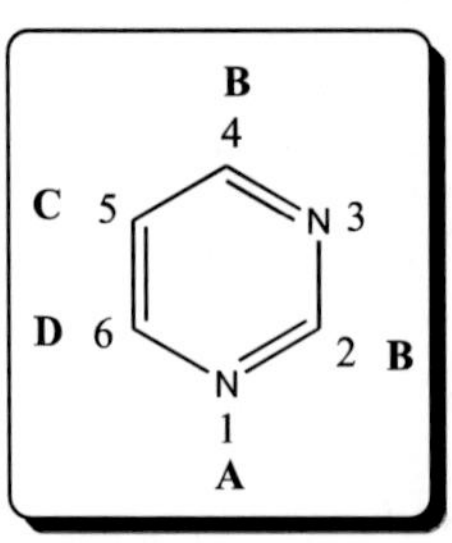

Figure 16. General Structure Activity Relationships (SAR) study of pyrimidine.

A: Five membered saturated heterocyclic ring substitution leads to anticancer and antiviral activities. B: 2nd position six or five membered saturated heterocyclic ring substitution leads to anthelmintic, antiparkinsonism expectorant and treatment of GI disturbance, peripheral neuropathies. B: 2nd and 4th position keto group substitution or amino substitution or mixed keto, amino groups substitution leads to anticancer, antiviral, antibacterial, antifungal, and treatment of respiratory tract infection and liver disorder. C: 5th position with halogen or substituted amine or saturated distal heterocyclic ring substitution leads to antibacterial and anticancer activities. D: 5th and 6th position fused with heterocyclic ring and *o*, *m*, *p* substituted distal aryl ring substitution leads to anticancer, antiviral, antibacterial, vasodilatation and treatment of urinary tract infection.

4. PYRIMIDINE: THERAPEUTIC POTENTIALS

Diverse pyrimidines are found to be biologically active and exhibit considerable therapeutic potentials. In the view of wide spectrum biological activities such as antibacterial, antifungal, antiviral, antitumor, anthelmintic, anti-inflammatory, anesthetic, anti-HIV, antihypertensive, antidiabetic, antiparkinsonian, anticonvulsant, antimalarial, anti-thyroid, antipsychotic, antigout and antianxiety activities, exhibited by synthetic pyrimidine-based scaffolds, a large number of analogues have gained considerable attention in the field of pharmaceutical research (Table 3). The very useful and convincing structural variability of pyrimidine nucleus has been achieved through chemical functionalization of all the six positions around the core (Figure 17).

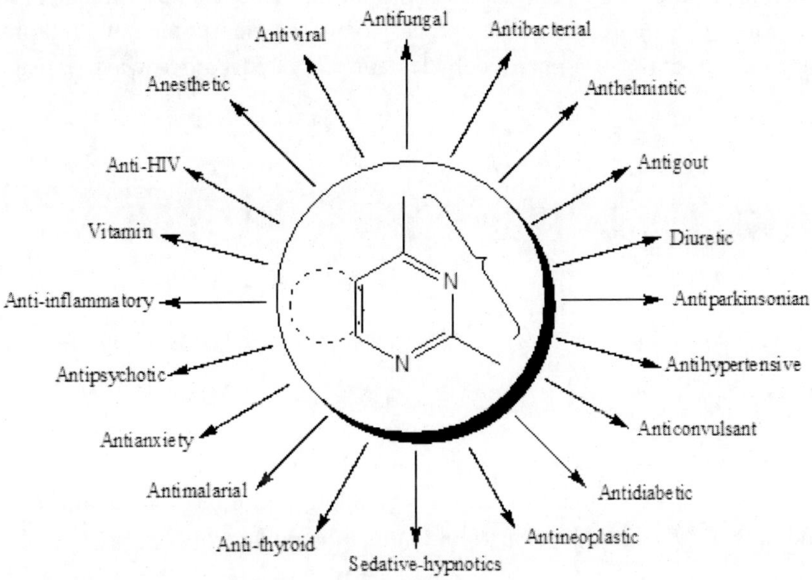

Figure 17. Therapeutic potentials of pyrimidines and their derivatives.

Bahekar et al. reported the synthesis of some [2-amino-6-(4-substitutedaryl)-4-(4-substituted phenyl)-1,6-dihydropyrimidin-5-yl]acetic acid derivatives and evaluated for antiinflammatory activity by carageenan induced rat paw edema method using diclofenac sodium as standard drug [19].

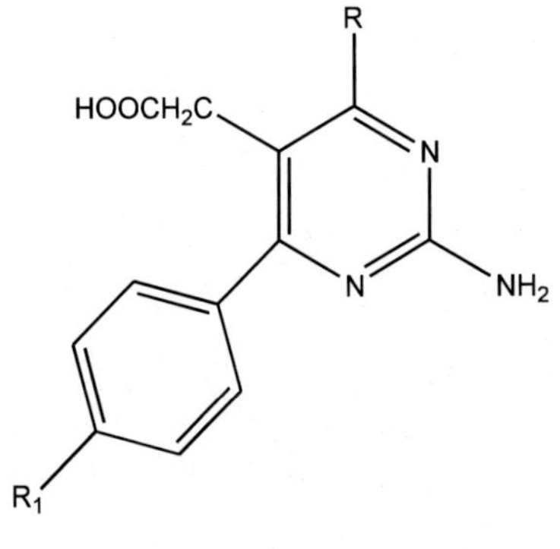

R= H,Cl,CH_3

R_1= phenyl, 4-chloro phenyl, furfural, 4-methoxy phenyl, 2-thiophene, 3-nicotine

Alagarsamy et al. synthesized a variety of novel 2-methylthio-3-substituted-5,6,7,8-tetrahydrobenzothieno[2,3-d]pyrimidin-4(3H)-ones by reacting (2-methylthio-4-oxo-3H-5,6,7,8-tetrahydrobenzothieno[2,3-*d*]pyrimidin-3yl)dithiocarbamic acid methyl ester with a variety of amines [20].

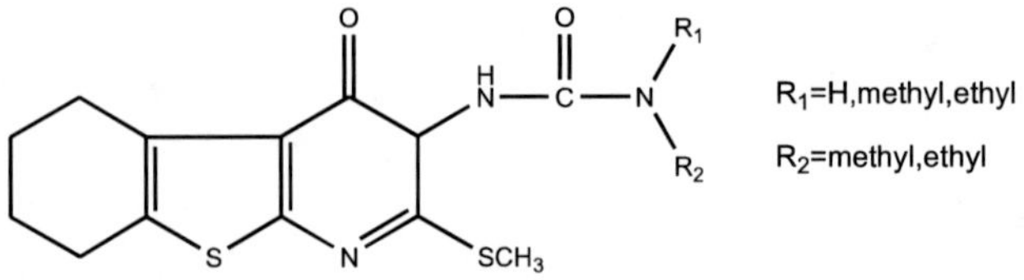

Dabiri et al. synthesized pyrimido[4,5-d]pyrimidine-2,5-dione derivatives in high yields in a novel, one pot, and efficient process by condensation of 6-amino-3-methyl-2(methylthio)pyrimidin-4(3H)-one, aldehydes and urea under microwave-assisted conditions [21].

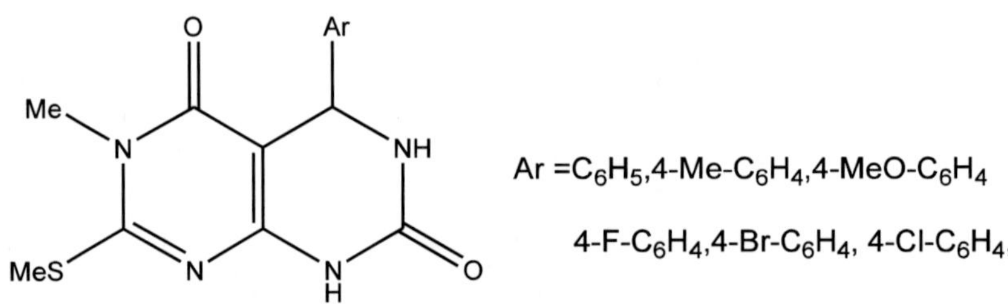

Agrawal, S. K. et al. reported pyrimidines derivative as COX-1, COX-2 and COX-3 inhibitors. Some of the compounds showed 64% inhibition of rat paw edema at 10mg/kg body weight and 52% COX-2 inhibition in whole human blood assay [22].

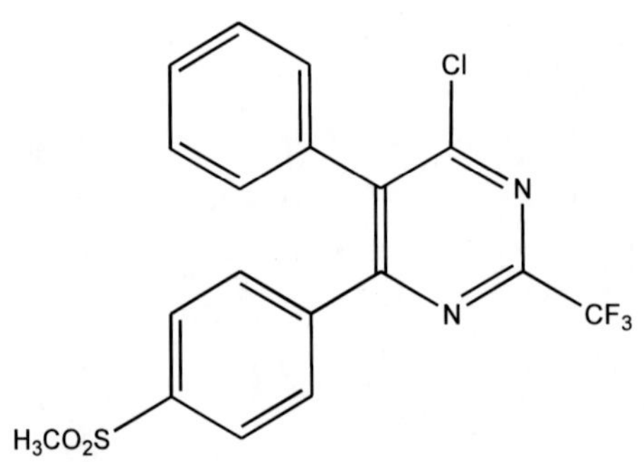

Chhabria, M. T. et al. synthesized 5-carbethoxy-4-chloro-6-(substituted amino) pyrimidines via cyclization of N-[2-ethoxycarbonyl-2-cyano-1-(isopropylamino)-vinyl] formamidine in presence of dry HCl in dioxane followed by nucleophilic substitution of 4-chloro group with substituted aromatic amine or phenoxide. Target compounds were evaluated for their analgesic and anti-inflammatory potential by known experimental models [23].

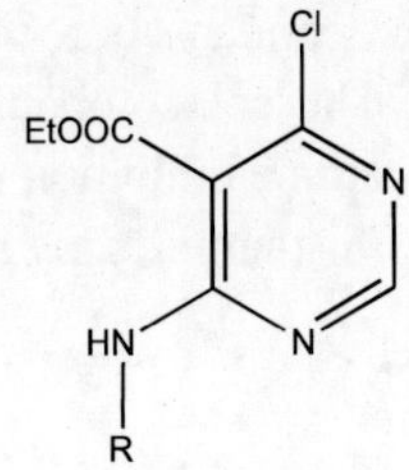

R = $(CH_3)_2CH$, 4-Cl-C_6H_4,4-CH_3-C_6H_4,
4-F-C_6H_4, 4-OCH_3-C_6H_4

Venkatesan et al. synthesized 4,6-diarylsubstituted-4,5-dihydro-2-aminopyrimidines by the reaction of chalcones with guanidine nitrate. The compounds were screened for antibacterial activity against 16 gram-positive bacterial strains [24].

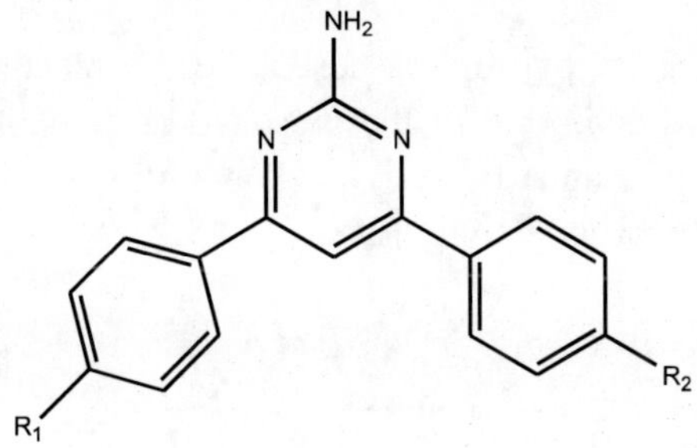

R_1=H, 4-OH,4-NO_2
R_2=H, 2-Cl,4-Cl

Sridhar et al. synthesized a number of 2-amino-4-(2,'5'-dimethyl-3'-furyl)-6-(aryl)-pyrimidines by treating the 1-(2,' 5'-dimethyl-3'-furyl)-3-(aryl)-2-propen-1-one with guanidine hydrochloride in presence of potassium hydroxide and ethanol. All these compounds were evaluated for anticancer activity and some of them were found to possess significant activity [25].

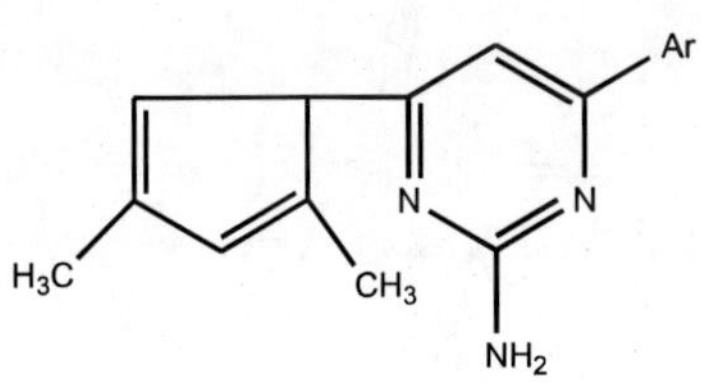

Ar=2-thienyl,2-pyridinyl,3-pyridinyl, 4-fluorophenyl,
4-nitrophenyl,4-methylphenyl,4-chlorophenyl

Xie et al. synthesized novel 2,4,5-substitutedpyrimidine derivatives and evaluated *in-vitro* for inhibition against human hepatocellular carcinoma BEL-74502 Cell proliferation. Several compounds showed potent anticancer activity with an IC_{50} less than 0.10 micromolar [26].

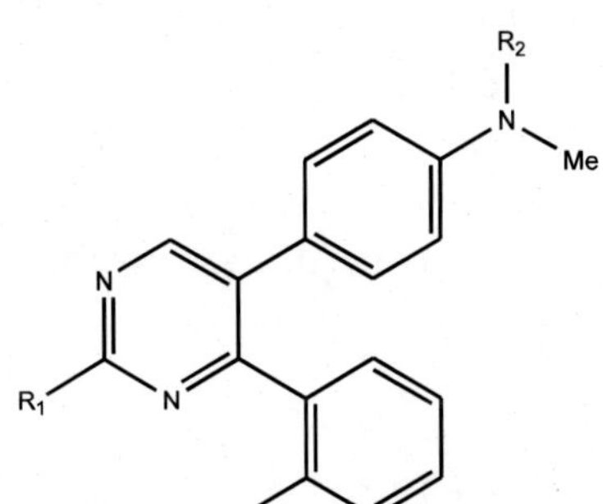

R_1=CH_3,NH_2
R_2=H,CH_3,CH_3CH_2

Rahaman et al. synthesized novel pyrimidines by the condensation of chalcones of 4'-piperazine acetophenone with guanidine HCl. These compounds were also screened for antihistaminic activity. The recorded % of histamine inhibition showed significant antihistaminic activity when compared to the reference antihistaminic drug mepiramine [27].

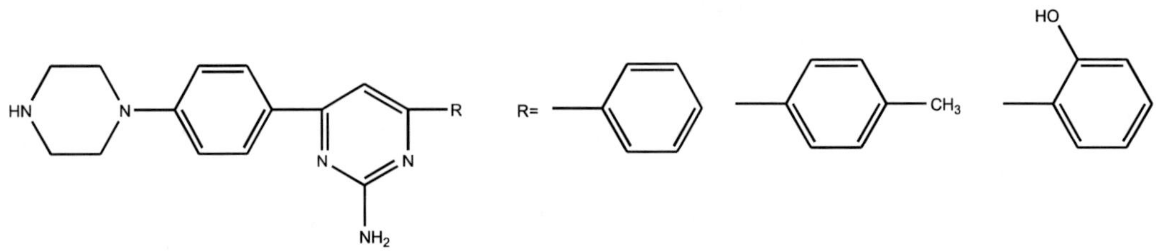

Ramesh et al. synthesized some new pyrimidine derivatives by reacting chalcones of 2-acetyl thiophene with guanidine hydrochloride in the presence of alcohol. The synthesized compounds were screened for anti-inflammatory activity. Some of these compounds showed moderate to considerable anti-inflammatory activity [28].

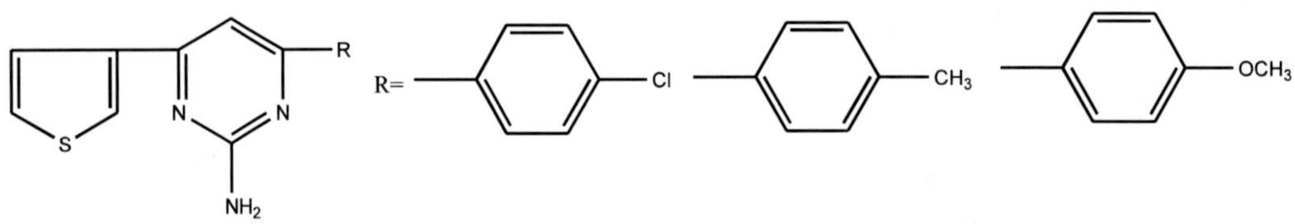

Rao et al. synthesized some new pyrimidine derivatives by reacting chalcones of 2-acetyl furan with guanidine hydrochloride in presence of alcohol. The synthesized compounds were screened for anticancer activity. Some of these compounds showed moderate to considerable anticancer activity [29].

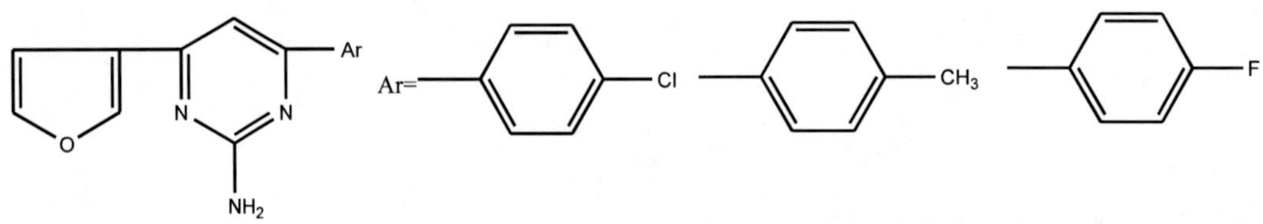

Chhabria et al. designed novel imidazo[1,2-c]pyrimidines. The designed molecules were synthesized by nucleophilic displacement of chloro group of various substituted 4-chloro pyrimidines by ethanolamine followed by cyclization of these 4-(2-hydroxyethyl)amino pyrimidines to imidazo[1,2-c]pyrimidines in good yield. All the compounds were screened for their antimycobacterial activity on Mycobacterium tuberculosis H37Rv strain by 1% proportion method. Some of the synthesized compounds exhibited potent antimycobacterial activity with MIC values in the range of 2–20 mg/mL [30].

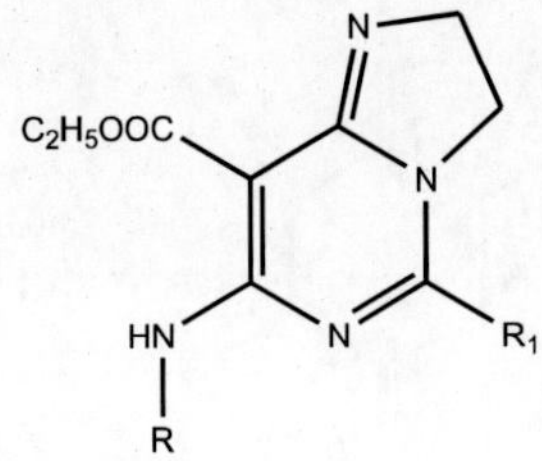

R=alkyl/aryl
R_1=H, Ph

Jain et al. synthesized 2-amino-5-cyano-4,6-disubstituted pyrimidines under microwave at 80W for 25-50 minutes. The reaction was completed in just few minutes instead of the 18 hrs required under conventional heating. The products were obtained in high yield (76%-90%) and purity (>95%) [31].

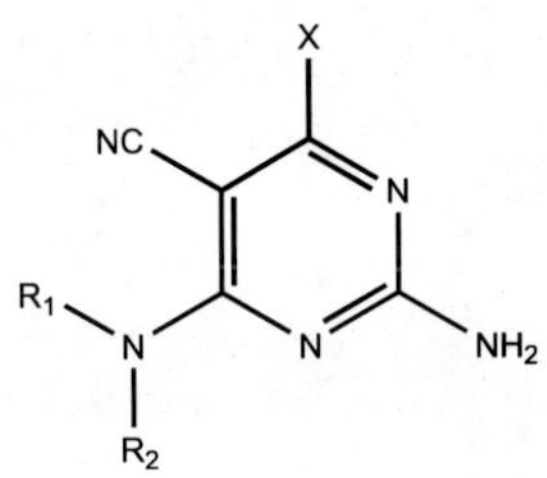

X=NH_2, OH

R_1R_2NH = 4-morpholinyl, 4-$CH_3OC_6H_4$NH, 4-$CH_3C_6H_4$NH,4-Br C_6H_5NH

Trivedi et al. synthesized a small library of 30 dihydropyrimidines and evaluated for their *in vitro* antitubercular activity against Mycobacterium tuberculosis H_{37}Rv. Two compounds such as ethyl4-[3-(4-fluorophenyl)-1-phenyl-1H-pyrazol-4-yl]-6-methyl-2-oxo-1,2,3,4-tetrahydro-pyrimidine-5carboxylate and ethyl4-[3-(4-nitrophenyl)-1-phenyl-1H-pyrazol-4-yl]-6-methyl-2-oxo-1,2,3,4-tetrahydropyrimidine-5-carboxylate were found to be the most active compounds *in vitro* with MIC of 0.02 lg/mL against MTB and were more potent than isoniazid [32].

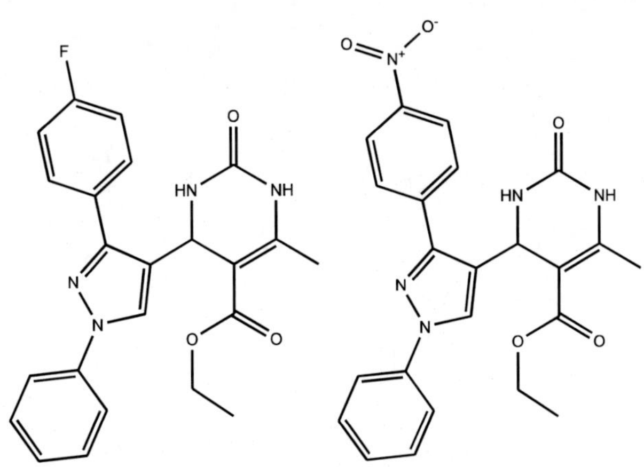

Shanmugasundaram et al. synthesized pyrido(2,3-d)pyrimidine-carboxylate derivatives through nucleophilic substitution reactions with the use of amidines, followed by 4-haloanilines and malonic acid. The synthesized novel derivatives were screened for anticancer activity [33].

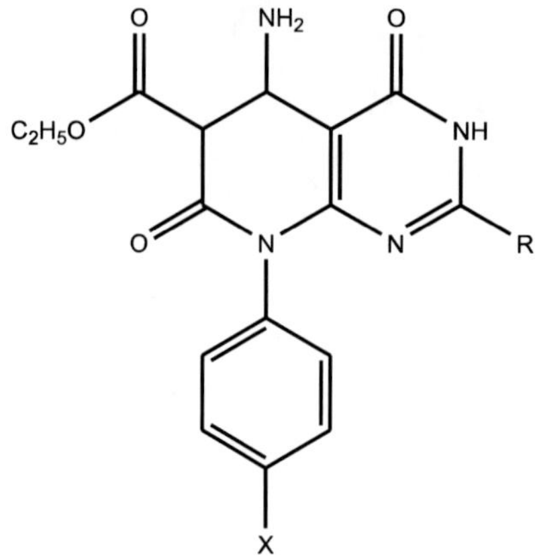

R=$COOCH_3$, C_6H_5
X=F,Cl,Br

Virsodia et al. reported the synthesis and 3D-QSAR analysis of a set of substituted N-phenyl-6-methyl-2-oxo-4-phenyl-1,2,3,4-tetrahydropyrimidine-5-carboxamides as antitubercular agents. Substituted acetoacetanilides were reacted with various aromatic aldehydes and urea which yielded the tetrahydropyrimidine derivatives with a phenyl carbamoyl group at C-5 position, and with various substitutions on the 4-phenyl and the N-phenyl aromatic rings. All compounds were screened for antitubercular activity against Mycobacterium tuberculosis $H_{37}Rv$ strain [34].

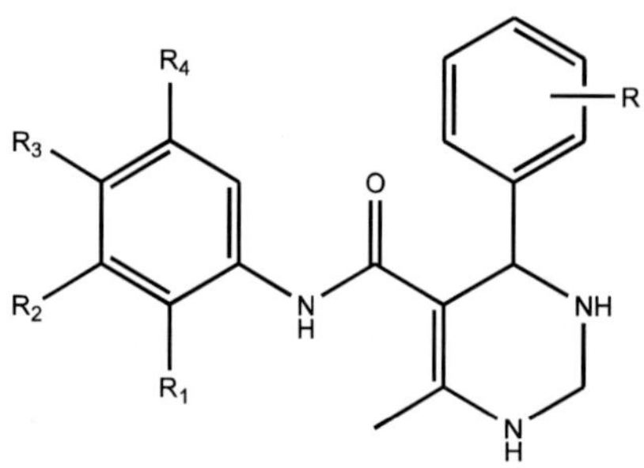

R= 4-OCH_3, 2-NO_2, 3-NO_2, 4-NO_2
R_1 = H, CH_3, Cl, F
R_3= H, CH_3, Cl, F, NO_2
R_4= H, CH_3, Cl

Thore et al. synthesized a series of triazolo[4,3-a]tetrahydrobenzo(b)thieno[3,2-e]pyrimidine-5(4H)-ones and evaluated for CNS depressant, skeletal muscle relaxant and anticonvulsant activities by photoactometer, rotarod and pentylenetetrazole induced the convulsions method respectively in Swiss albino mice. Diazepam was used as standard drug [35].

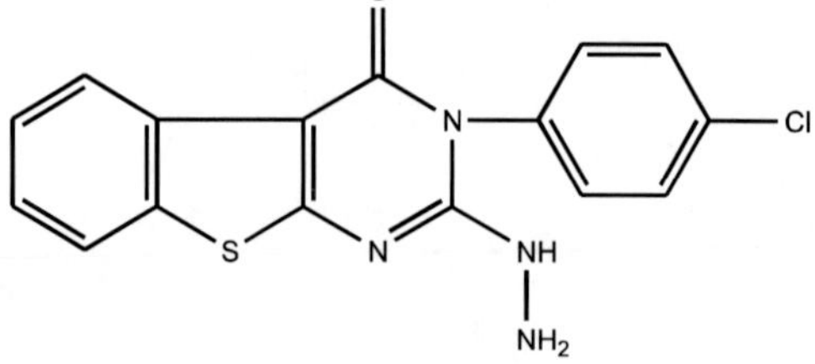

Chitre et al. synthesized pyrimidinedione and performed Pharmacophore Optimization of Selective Thymidine Monophosphate Kinase inhibitors using group QSAR Studies as potential antitubercular agents. G-QSAR was specifically done for knowing the structure activity relationship for carrying out variations in substitution at specific substitution sites. Also mathematical model for the prediction of activities of the new molecules was developed. G-QSAR studies were carried out using VLife molecular design suite (V Life MDS) software [36].

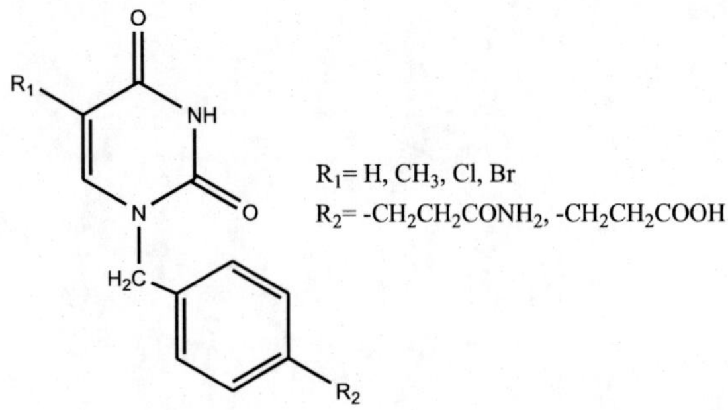

Read et al. have synthesized pyrimidine analogs and screened for antibacterial activity against *Mycobacterium tuberculosis* $H_{37}Rv$ *in vitro*. Several active compounds were identified and the best results were observed for 5-formamidopyrimidines. These compounds generally displayed IC_{90} values ≤ 1 μg/mL, and they exhibited low toxicity towards mammalian cells. Imidazolyl pyrimidines, which may be regarded as fleximer analogs of the parent purines, were also synthesized and one of them was found to be quite a potent inhibitor of Mycobacterium tuberculosis (IC_{90} 14 μg/mL) [37].

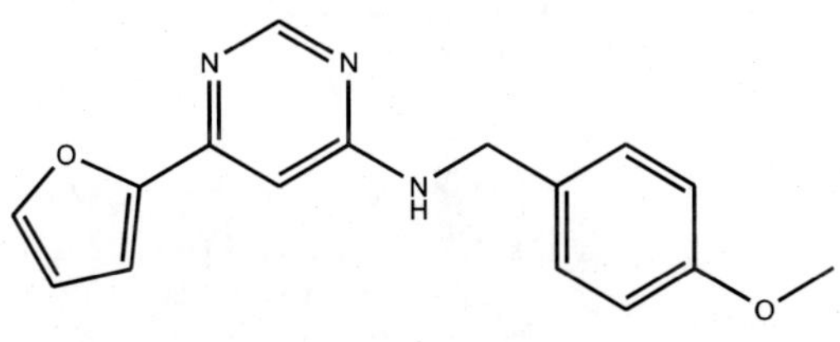

Dongarwar et al. synthesized ethyl-2-amino-4-chloro-6-(2,4,6 trichloro phenyl amino)pyrimidine-5-carboxylate and ethyl-2-amino 4-chloro-1,3-dimethyl-6-(phenyl amino)-1,2,3,4 tetrahydro pyrimidine-5-carboxylate derivatives through cyclocondensation and Michael type of addition reaction with the aromatic anilines, guanidine and benzamidines. These novel derivatives have been screened for antibacterial, antifungal and anti-inflammatory activity [38].

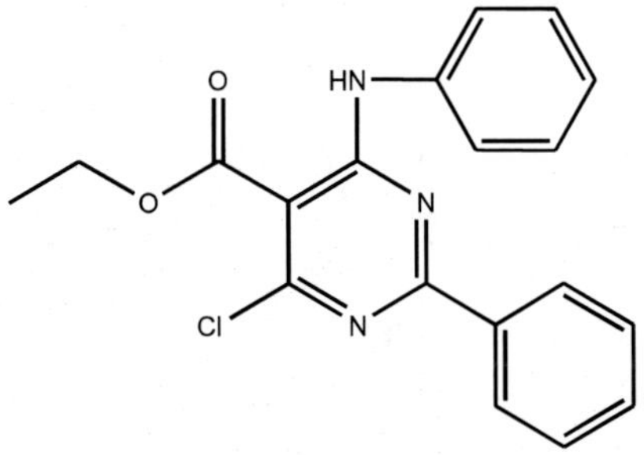

Shah et al. have been synthesized four series of *N*-Mannich bases of 3,4-dihydropyrimidine-2(1*H*)-thiones. In this reaction, four different derivatives of 3,4-dihydropyrimidine-2(1*H*)-thiones have been used as hydrogen active compound. These precursors have been derived by a Biginelli reaction of four aromatic aldehydes, namely, benzaldehyde, salicylaldehyde, anisaldehyde and vanillin respectively with ethyl acetoacetate and thiourea [39].

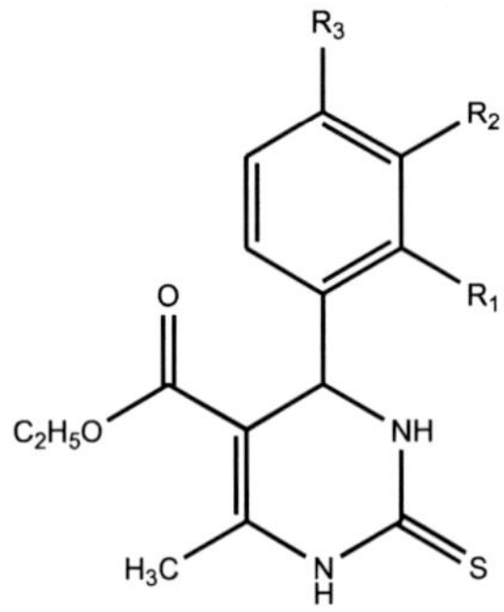

R_1=H, R_2=OMe, R_3=OH

Johar et al. carried out synthesis and *in vitro* anti-mycobacterial activity of 5-substituted-pyrimidine nucleosides against M. bovis and M. avium. 5-(C-1 Substituted)-2'-deoxyuridine derivatives emerged as potent inhibitors of *M. avium* (MIC_{90} = 1-5 μg/mL range). The nature of C-5 substituents in the 2'-deoxyuridine series appeared to be a determinant of anti-mycobacterial activity [40].

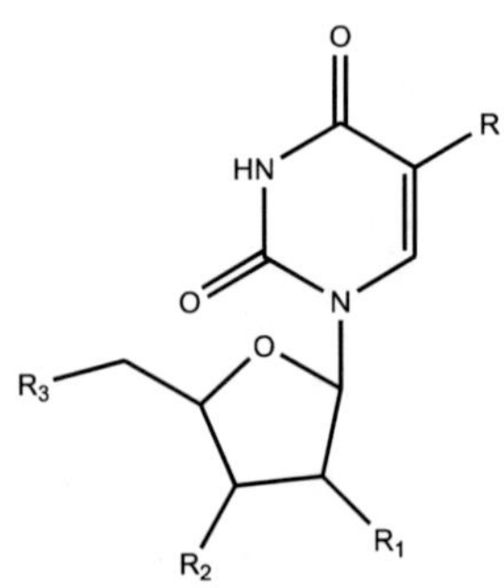

R= F, Cl, Br, I, OH
R_1= H, OH, OCH_3
R_2= H, F, OH, OAc
R_3 = OH, OAc

Sawant et al. synthesized a series of 2-methylthio-1,4-dihydropyrimidine derivatives in good yield. The compounds were tested for analgesic activity by acetic acid-induced writhing method. The observed analgesic activity is mainly because of inhibition of the peripheral pain mechanism by the title compounds [41].

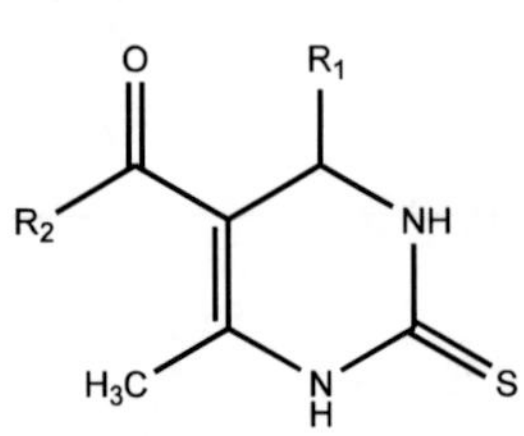

R_1= OC_2H_5, OCH_3
R_2= C_6H_5, p-Cl-C_6H_4,
p-$OCH_3C_6H_4$, p-$NO_2C_6H_4$

Kathiravan et al. described synthesis and antihyperlipidemic activity of a series of novel condensed 2-chloroalkyl-4-chloro/hydroxy-5,6-di-substituted pyrimidines. All the newly synthesized compounds have been evaluated by the Triton WR 1339 induced hyperlipidemia in albino Wistar rats model for antihyperlipidemic activity, and their activity was superior to that exhibited by the standard gemfibrozil used in the study. 3D QSAR study has also been performed to delineate the effect of the substituents at 5 and 6 positions on the antihyperlipidemic activity of 2-chloromethyl-5,6-substituted thieno(2,3-d)pyrimidin-4(3H)-ones [42].

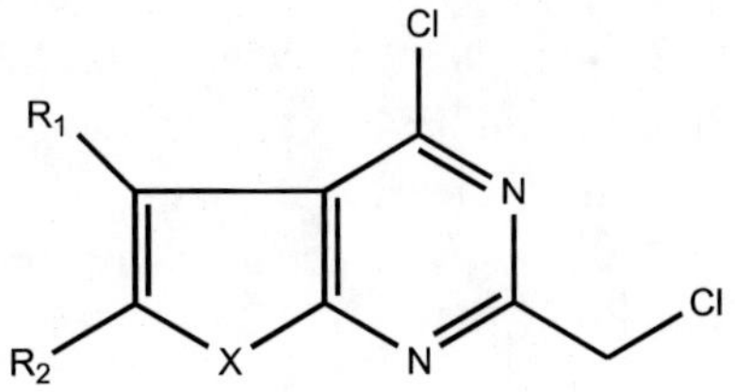

X= S, CH=CH

R_1= R_2= H, Alkyl, Aryl, Cycloalkyl, Heteroalkyl

Jain et al. synthesized a series of novel 2-unsubstituted 4-(substituted) anilinothieno [2,3-*d*]pyrimidines through the chlorination of the corresponding 2-unsubstituted-thieno[2,3-*d*]-pyrimidin-4-ones, followed by the nucleophilic displacement of the 4-Cl group, with a variety of anilines. Microwave irradiation method (MWI) was used to perform four steps of this synthesis and the entire process requires only 2h [43].

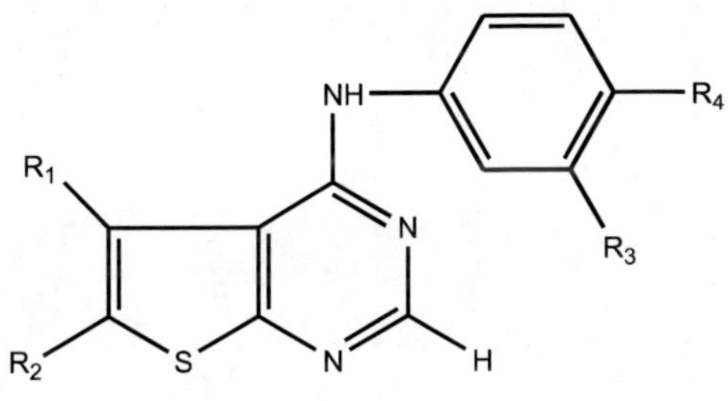

R_1, R_2 = H, alkyl, cycloalkyl

R_3, R_4 = H, halo, alkyl, alkoxy

Shishoo et al. reported a novel series of condensed 3-amino-2-(substituted) methylpyrimidin-4(3*H*)-ones with potential H1 receptor antagonistic activity. The IC_{50} values for 23 compounds were found to be in the micromolar range. When evaluated by the in vivo method, some compounds were found to protect guinea-pigs from the histamine induced asphyxia and antagonized histamine in a competitive and reversible manner. With a pA2 value of 8.7 and protection time of 9.5 min (in vivo test), one of the compounds was found to be the most active amongst other compounds. The isosteric replacement of the side chain -NH- series by oxygen and -$NHSO_2$- functional groups, was undertaken to investigate the role of two amino functionalities in the receptor binding. This isosteric replacement with -O- does not affect the antihistaminic activity and the sedative potential of the series. Preliminary molecular modeling studies indicate that the compounds with -$NHSO_2$- in the side chain exhibit a closer fit with temelastine than their -O- isosteres [44].

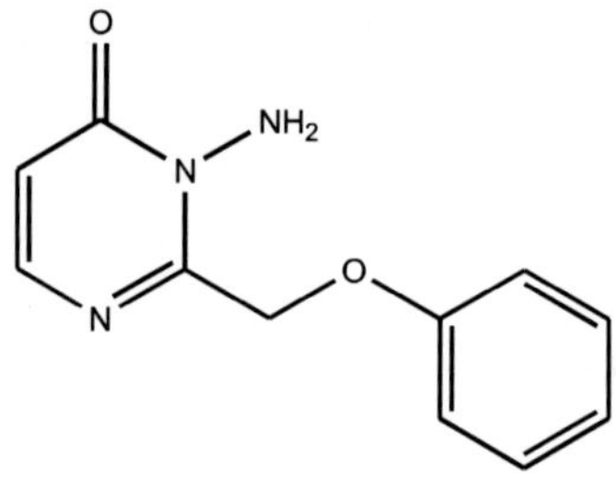

Rathod et al. described synthesis and diuretic activity of some novel 2,4-diamino-6-aryl-7-arylaminopyrimido[4,5-*d*]pyrimidin-5(6H)-ones. The series was designed on the basis of structural similarity between pteridines and pyridopyrimidines. Designed molecules were synthesized by cyclocondensation of 5-cyano-6-methylmercapto-3-aryl-2-arylaminopyrimidin-4(3H)-one with guanidine. All the compounds were screened for their diuretic activity using triamterene and furosemide as standard drugs [45].

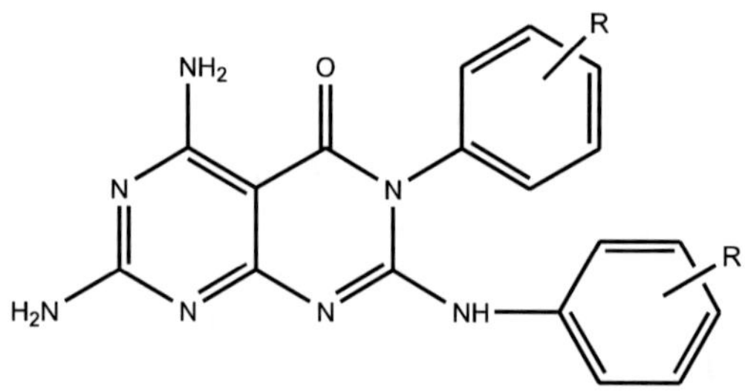

R=o-CH_3,m-CH_3, p-CH_3,
o-OCH_3,m-OCH_3,p-OCH_3,o-OC2H_5

Choudhari et al. synthesized different series of 3,6- and 4,5-disubstituted 1,2,3,4-tetrahydropyrimidine derivatives and screened for their *in vitro* COX-1 and COX-2 inhibition activity. The selectivity index was determined from the results of *in vitro* COX-1 and COX-2 activities. The data set was employed for QSAR analysis on Vlife Sciences MDS 3.5. The molecules were divided into training set and test set based on activity. The activity of the molecules of test set was predicted according to the QSAR equation fit. The outcomes of QSAR studies indicate that hydrophilic nature of compounds lead to selective COX-2 inhibition *in vitro* [46].

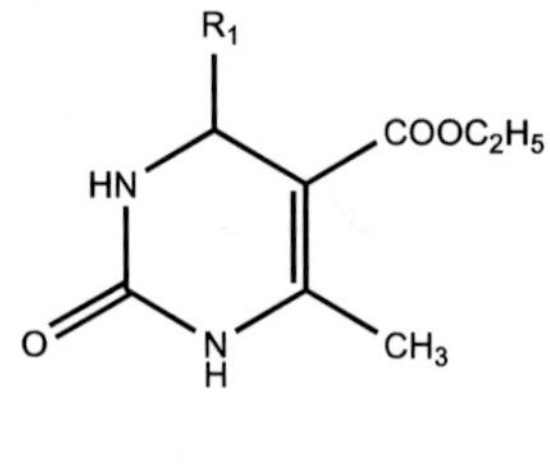

R_1=4-$OCH_3C_6H_4$, C_6H_5

Mohamed et al. synthesized a series of new indolyl-pyrimidine-5-carbonitriles. Antibacterial and antifungal activities of the synthesized derivatives were done in comparison with penicillin and fluconazole as standard to reveal the potency of synthesized derivatives [47].

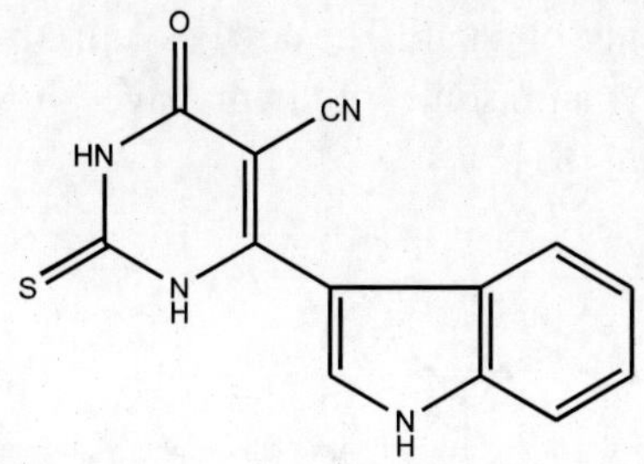

Chikhalia et al. synthesized pyrimidine derivatives and screened for antimicrobial activity. In comparison with standard drugs, some compounds showed maximum zone of inhibition against *E-coli., S.aureus, S.typhi and B.subtitlis* [48].

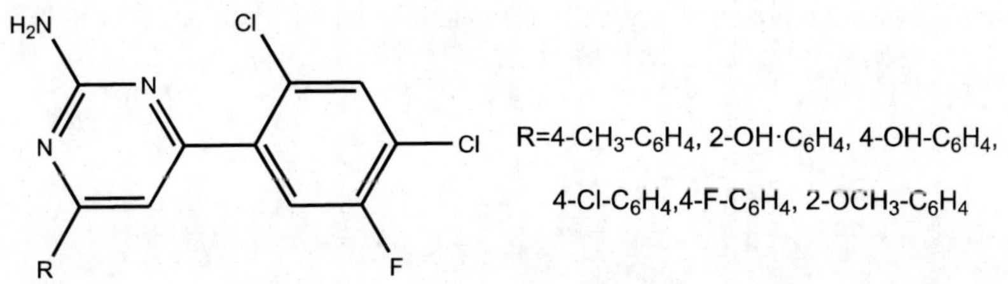

Fathalla et al. synthesized a series of some new pyrimidine derivatives *via* the reaction of ethyl cyanoacetate with thiourea and the appropriate aldehydes, namely 2-methyl-benzaldehyde and 2-methoxy-benzaldehyde. The antibacterial activity and anticancer activity of some of these compounds were studied [49].

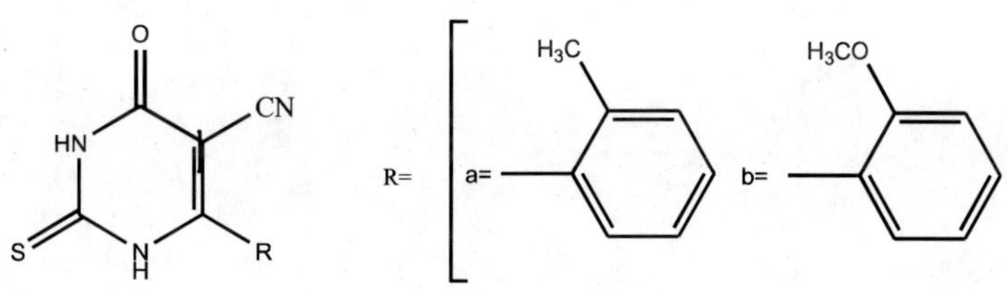

Ebtehal et al. synthesized new series of 6-phenyl-2,4-disubstituted pyrimidine-5-carbonitriles and tested for *in vitro* activities against a panel of Gram-positive and Gram negative bacteria and the yeast-like pathogenic fungus Candida albicans [50].

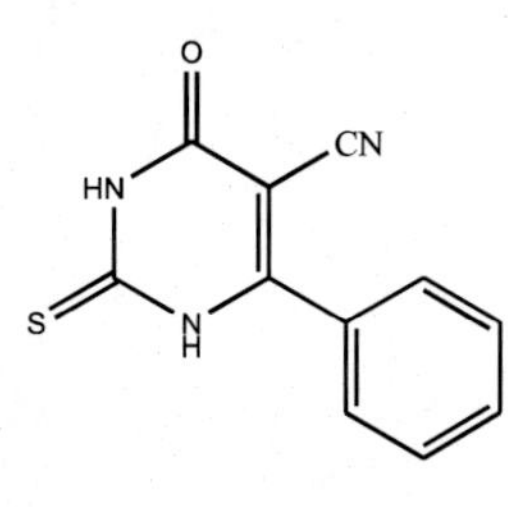

Anupama et al. synthesized a variety of 2-amino-4-(substituted)-6-(3″-thienyl) pyrimidines by reacting various chalcones with guanidine hydrochloride. The required chalcones were prepared by condensation of 3-acetylthiophene with various substituted aromatic /heteroaromatic aldehydes in the presence of alkali. The newly synthesized compounds were evaluated for their antimicrobial activity and some of them have shown significant activity when compared with the standard drug [51].

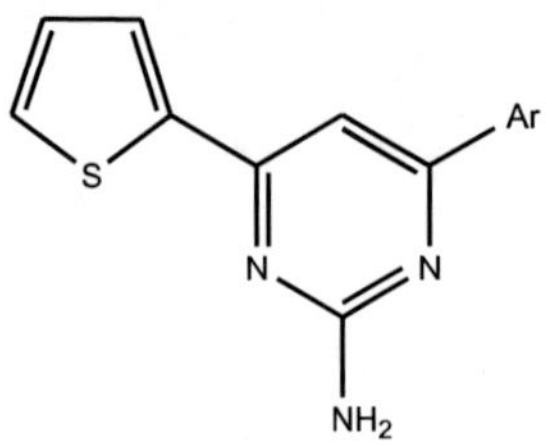

Ar=4'-chlorophenyl,4'-methoxyphenyl
4'-fluorophenyl,3'-nitrophenyl

Sawant et al. synthesized a series of 2-methylthio-1,4-dihydropyrimidine derivatives in good yields and were tested for analgesic activity by acetic acid-induced writhing method [52].

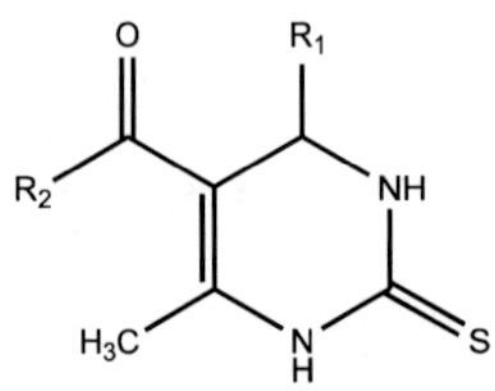

R1=C_6H_5,p-Cl-C_6H_5,p-NO_2-C_6H_5
R2=OC_2H_5, OCH_3

Agarwal et al. synthesized 2-alkylsulfanyl-6-aryl-4-chloropyrimidine-5-carbonitrile and evaluated for in vitro antibacterial activity against six pathogenic bacteria including virulent and non-virulent strains of Mycobacterium tuberculosis. All the synthesized compounds were also evaluated for their antimycotic activity [53].

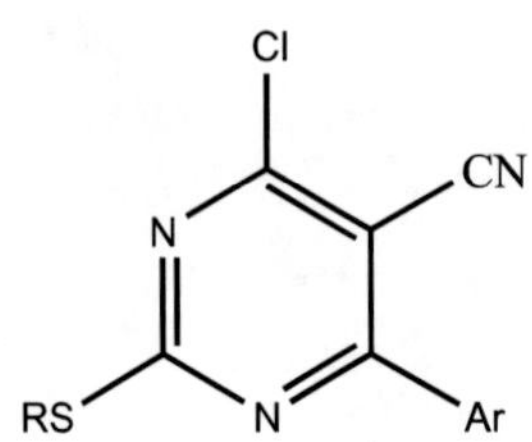

R=CH_3,C_2H_5,C_3H_7

Ar=2-Thienyl,2-Furyl

Singh et al. synthesized various 1,2,3,4-tetrahydropyrimidine derivatives prepared from urea and substituted aldehydes using microwave irradiation technique. These compounds were evaluated for their calcium channel inhibition using nifedipine as an analog [54].

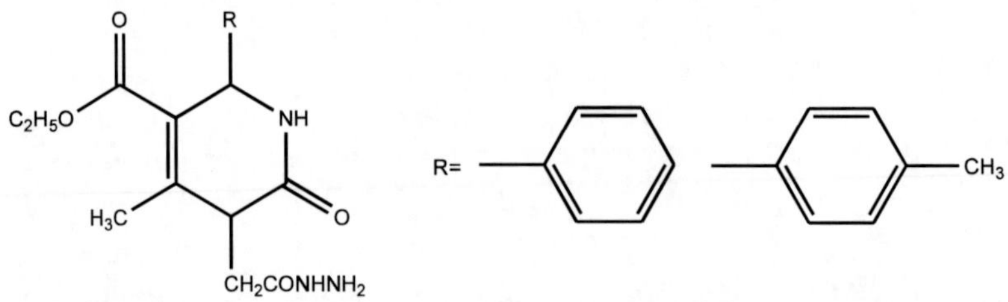

Borisagar et al. synthesized a series of triazolopyrimidines from different acetoacetamides, 4-(phenoxymethyl)benzaldehyde and 5-amino-1,2,4-triazole using one-pot microwave irradiation within 20-30 minutes with high yield [55].

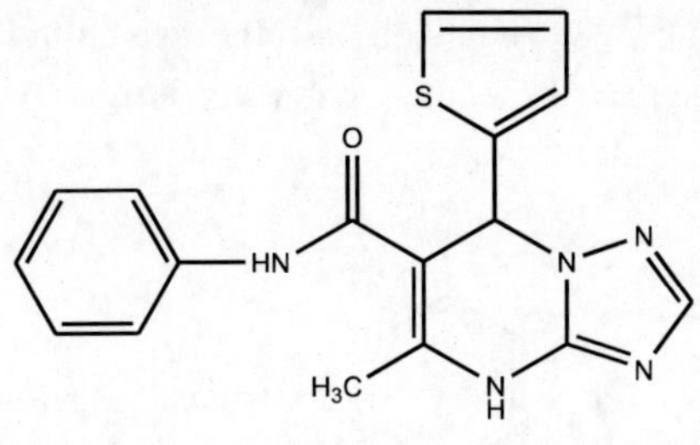

Chitre et al. designed 2-amino-4,6-dimethylpyrimidine-5-carbonitrile as inhibitors of thymidine monophosphate kinase of *Mycobacterium tuberculosis* (TMPKmt). The newly synthesized substituted pyrimidine derivatives exhibited promising antitubercular activities against *Mycobacterium tuberculosis* $H_{37}Rv$ using microplate alamar blue assay (MABA). Isoniazid (INH) and Pyrazinamide (PZA) were employed as the reference antimycobacterial agents [56].

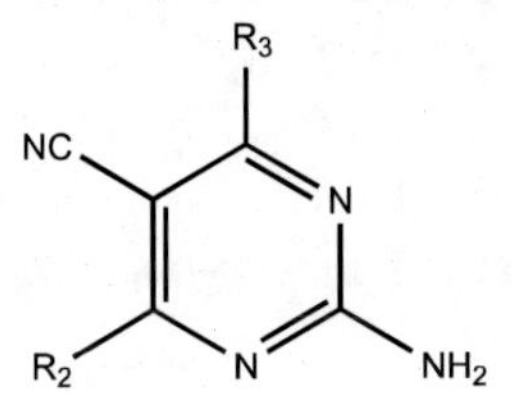

R_2= Amines

R_3=NH_2, OH

Rajesh et al. carried out green expedient Multi-component reactions (MCRs) to synthesize pyridopyrimidine-2-thiones and screened for their *in vitro* activity against Mycobacterium tuberculosis H37Rv. Among them, (E)-6-benzyl-8-(2,4-dichlorobenzylidene)-4-(2,4-dichlorophenyl)-3,4,5,6,7,8-hexahydropyrido[4,3-d]pyrimidine-2(1H)-thione (MIC 2.8 μM) displays the maximum activity, being 2.7 and 1.7 times more active than the first line antitubercular drugs ethambutol and ciprofloxacin, respectively, and less active than rifampicin and isoniazid, by 28 and 7 times, respectively [57].

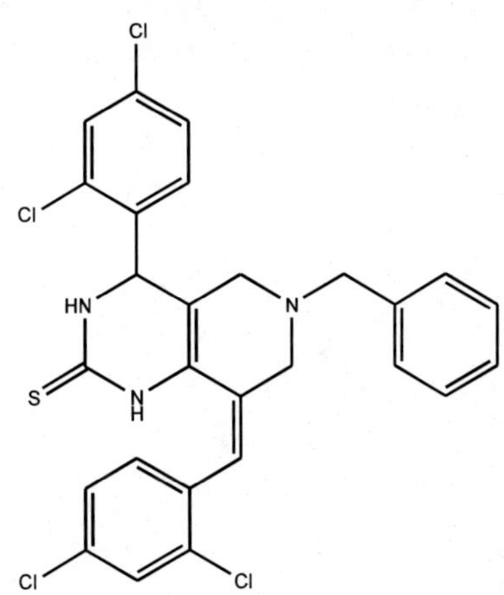

Chitre et al. also carried out synthesis and docking study of a novel series of 5-cyano 4,6-disubstituted, 2-thiouracil derivatives. Docking study was performed to rationalize the possible interactions between test compounds and active site of TMPKmt. The test compounds were screened for antimycobacterial activity using Micro plate Alamar Blue assay (MABA) against M. tuberculosis H37Rv (ATCC 27294) as well as INH resistant clinical strain. The SAR study reveals the importance of presence of electronegative group for better activity [58].

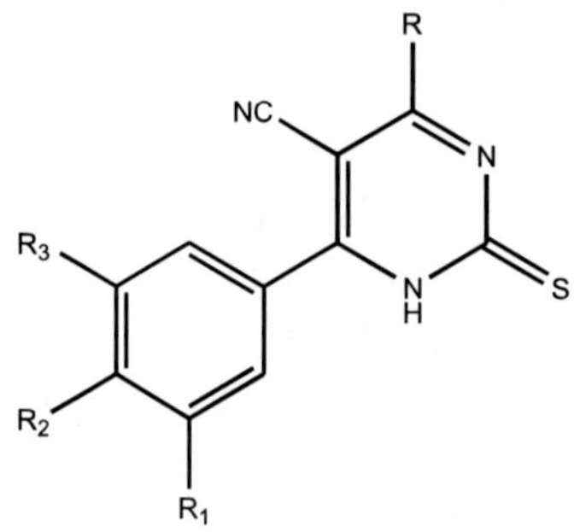

R= O,Cl, R_1= R_3= H, R_2= Cl, F

Pore et al. carried Quantitative structure activity relationship (QSAR) studies on 5-cyano, N1, 6-disubstituted, 2-thiouracil derivatives as central nervous system depressants (locomotor) activity using VlifeMDS3.5 software. In this context, a series of twenty four 5-cyano-N1,6-disustituted 2-thiouracil derivatives were synthesized by the tertiary condensation of aryl-substituted thiourea, appropriate aldehyde and ethylcyanoacetate catalyzed by potassium carbonate in presence of small amount of ethanol using microwave irradiation technique [59].

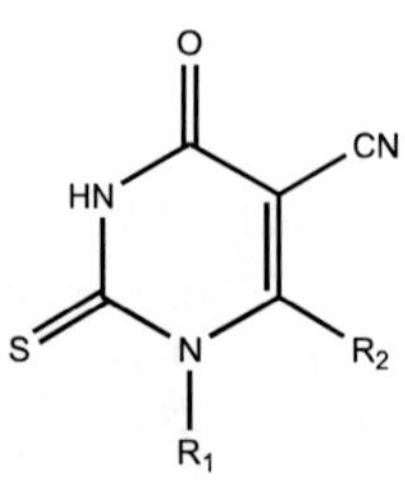

R_1=Phenyl,4-Chlorophenyl,4-Methylphenyl,4-Fluorophenyl

R_2= Phenyl,4-hydroxy phenyl, 4-methoxy phenyl

Basavaraja et al. synthesized several 6-(substituted)-5-cyano-2-thiouracils by the condensation of different aldehydes with ethylcyanoacetate and thiourea through a single step reaction by conventional and microwave irradiation technique. All the synthesized compounds were screened for their antimicrobial activity [60].

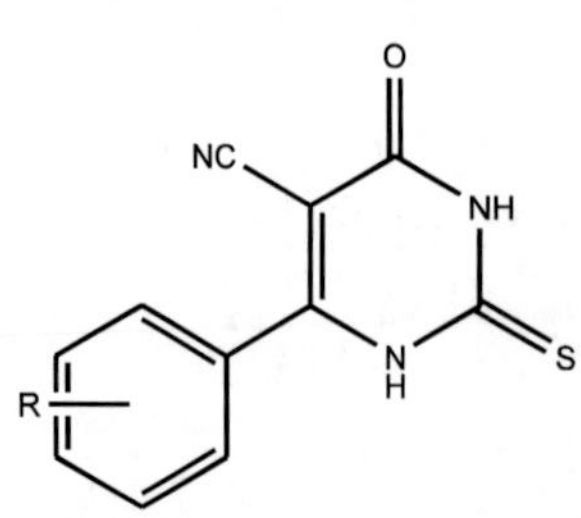

R=4-CH_3O-C_6H_4,3-NO_2-C_6H_4

Mishra et al. synthesized 4-phenyl-5-carboxyethyl-6-methyl-1,2,3,4-tetrahydropyrimdin-2-ones using the principle of Biginelli condensation and the compound are tested for antimicrobial activity against gram positive and gram negative bacteria using serial dilution technique [61].

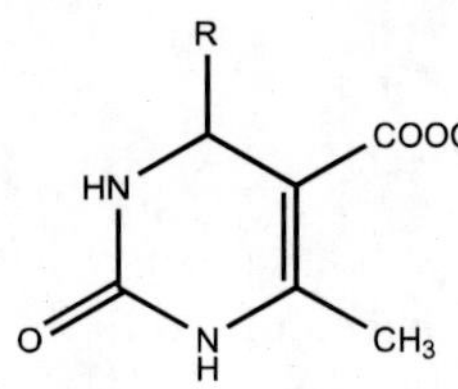

R= $-CH_3$, C_6H_5

Abdolmohammadi et al. prepared a series of pyrido[2,3-*d*]pyrimidines derivatives by one-pot three-component reaction of 4(6)-aminouracil, malononitrile and aromatic aldehydes. This efficient synthesis was done under microwave irradiation. This procedure has the advantages of good yields, easy work-up, and benign environmentally friendly character. Reaction could proceed via domino Knoevenagel-Michael-cyclization reactions [62].

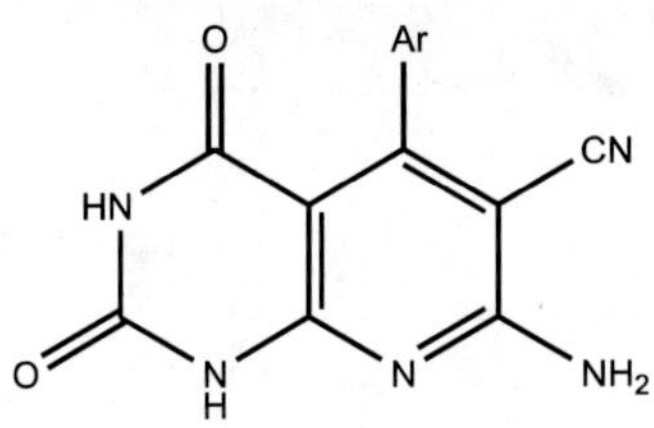

Ar=C_6H_5, 2-Cl-C_6H_4, 2,4-Cl_2-C_6H_3, 3-OH-C_6H_4,

4-OCH_3-C_6H_4, 4-CH_3-C_6H_4, 3-NO_2-C_6H_4, 4-NO_2-C_6H_4

Rajasekaran et al. synthesized a series of some dihydropyrimidine-2(1*H*)-thione derivatives. Antitubercular and antibacterial activities were performed by microbroth dilution and cup-plate method respectively [63].

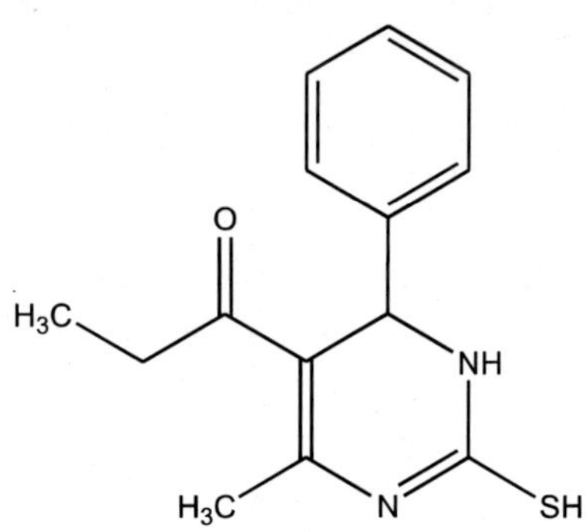

Morgan et al. carried out synthesis of novel anilinopyrimidine analogues which have been found to have micromolar activity against Mycobacteriumn tuberculosis. This could potentially generate new lead compounds in the fight against multi-drug resistant tuberculosis [64].

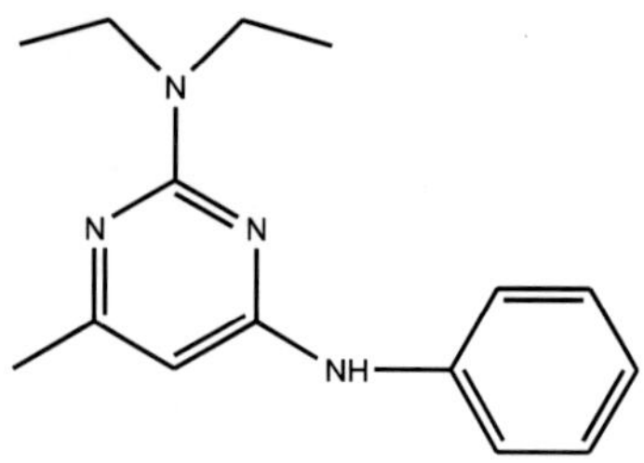

Bhatewara et al. have been developed a simple and efficient one-pot multi component synthesis of aryl and heteroaryl substituted pyrimidines via initial Knoevenagal, subsequent addition and final cyclization of aldehyde, ethylcyanoacetate and guanidine nitrate [65].

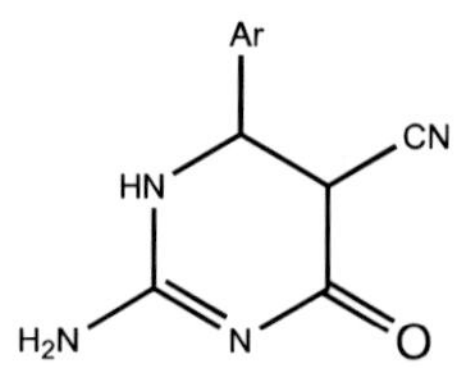

Ar=C_6H_5, 4-(CH_3O)-C_6H_4, 3,4-(CH_3O)-C_6H_3, 4-NO_2-C_6H_4

Deshmukh et al. has carried out a green, simple and environmentally friendly approach towards one-step synthesis of 2,6-diamino-4-phenyl pyrimidine-5-carbonitrile by three component condensation of aromatic aldehydes, malononitrile and guanidine hydrochloride in aqueous medium using potassium carbonate and in the presence of tetrabutyl ammonium bromide [66].

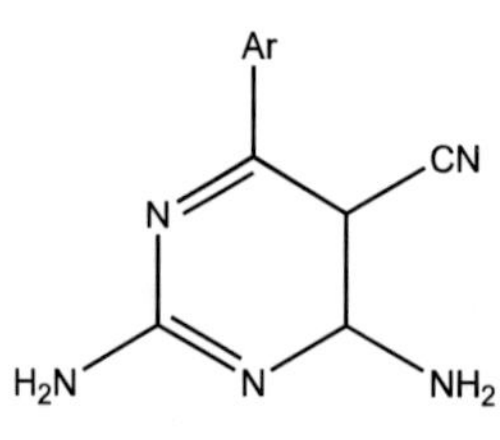

Ar=2-OH-C_6H_4, 4-OH-C_6H_4, 3-Cl-C_6H_4, 4-OCH_3-C_6H_4

Hamouly et al. carried out synthesis of new 4-aryl-isoxazolo[5,4-d]pyrimidin-6-one(thione) and 4-aryl-pyrazolo[3,4-*d*]-pyrimidin-6-one derivatives of potential antihypertensive activity. Some aromatic aldehydes are subjected to react with urea (or thiourea) and acetyl acetone in a one-pot Biginelli-type cyclocondensation reaction to give 5-acetyl-4-aryl-6-methyl-1,2,3,4-tetrahydro-pyrimidines. Aldehydes with ortho-hydroxy substituent namely salicylaldehyde and 2-hydroxy-3-methoxybenzaldehyde undergo Michael-type addition of the hydroxyl-proton to the C5-C6 double bond of the pyrimidine ring to form the tricyclic derivatives [67].

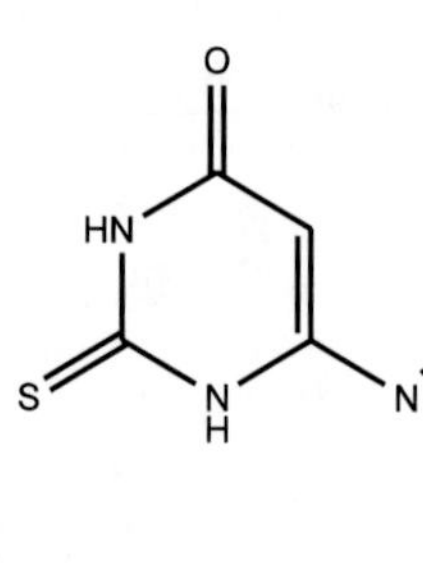

Ar=-C_6H_5, 3-indolyl

Mohamed et al. carried out synthesis of certain pyrimidine derivatives as antimicrobial agents and anti-Inflammatory agents. Some compounds showed good anti-inflammatory activity as compared to the standard Ibuprofen [68].

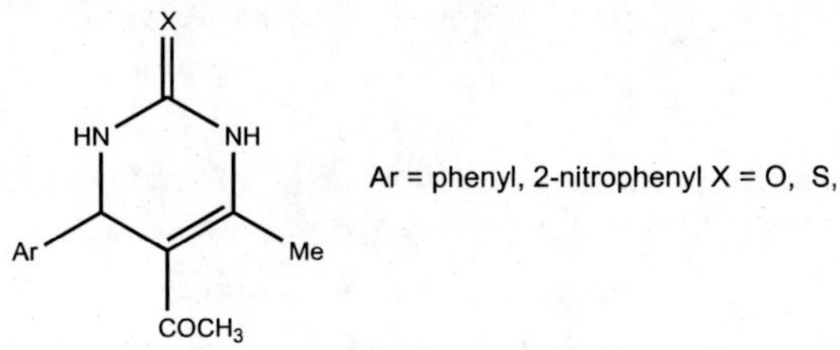

Ramesh et al. carried out design, synthesis and anticancer activity of some new pyrimidine derivatives. Some of these compounds showed moderate to considerable anticancer activity with IC_{50} values ranges from 5.87 to 13.86. Synthesized compounds were tested for their anticancer activity (prostate cancer) by MTT based cytotoxicity assay. Du-145 cell lines were used for the experiment [69].

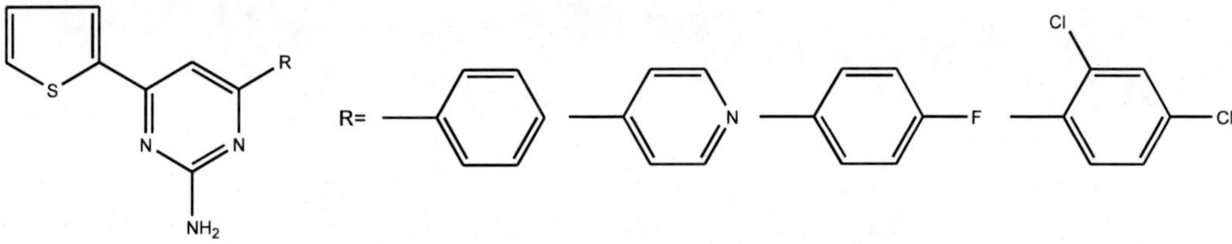

Patil et al. carried out synthesis of 3,4-Dihydropyrimidine-2(1H)-one derivatives using Microwave. The compounds were tested for antihypertensive activity by non-invasive tail-cuff, and evaluated by carotid artery cannulation method for determining the diastolic blood pressure. Hypertension was induced by DOCA-salt. Anti-inflammatory activity was carried out by carrageenan induced rat-paw oedema method. Test compounds exerted comparative anti-hypertensive activity at 10 mg/kg dose level compared to nifedipine [70].

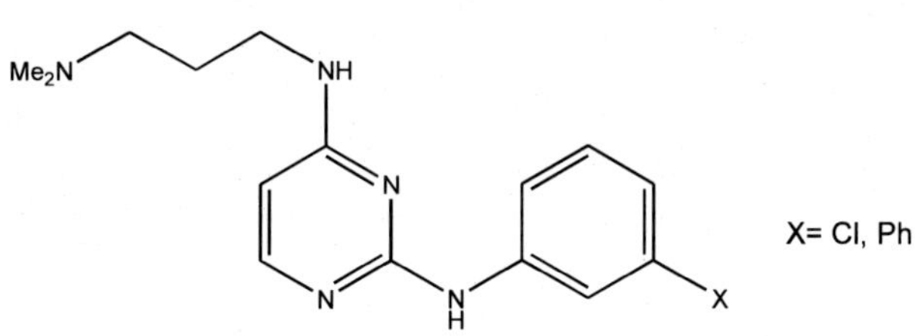

Raju et al. used $CaCl_2$ as an efficient, inexpensive and readily available catalyst for the three component, one-pot condensation reaction of an aldehyde, β-ketoester and urea in refluxing ethanol to afford the corresponding dihydropyrimidinones in high yield. This method provides an environmentally friendly, easy workup and isolation process [71].

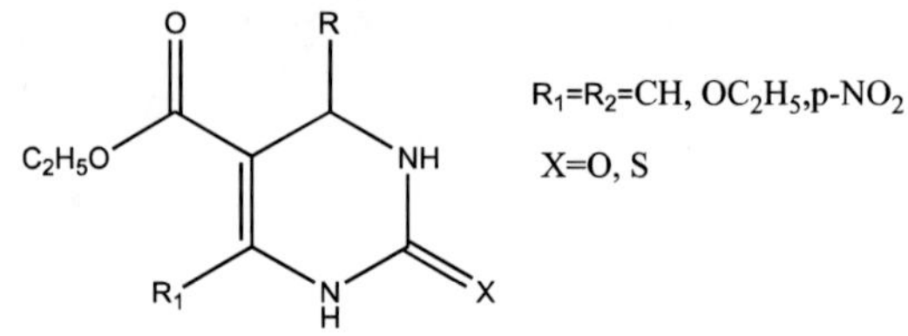

Sandhu et al. carried out microwave enhanced, solvent-free green protocol for the production of 3,4-dihydropyrimidine-2-(1*H*)-ones. These types of compounds are synthesized via a multicomponent reaction of aldehyde, urea or thiourea and 1,3-dicarbonyl compounds using microwave irradiation catalyzed by $AlCl_3.6H_2O$ under solvent-free conditions. The catalyst being reported here is cheap, safe to handle and the whole procedure is eco-friendly [72].

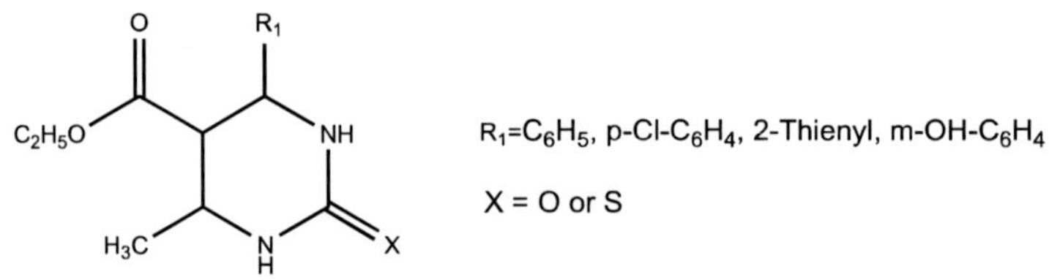

Nongkhlaw et al. synthesized 3,4-Dihydropyrimidin-2(*1H*)-ones and 3,4,5,6,7,8-hexahydro-quinazolin-2(*1H*)-ones from the condensation of ketones, aldehydes and urea/thiourea in the presence of sodium tertiarybutoxide by the solventless microwave irradiation (MWI) process as well as the conventional process [73].

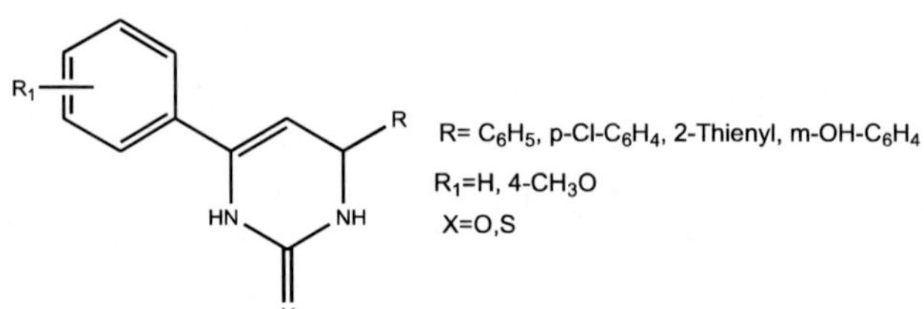

Kumar et al. carried out an environmentally benign protocol for the synthesis of 3,4-dihydropyrimidin-2(1*H*)-ones using solid acid catalysts under solvent-free conditions. The low cost catalysts exhibited remarkable reactivity and reusability for the aforementioned synthesis [74].

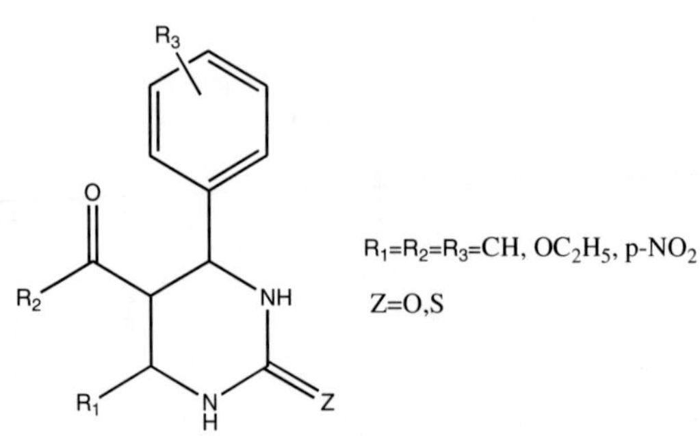

Mohamad, Y. et al. synthesized pyrido[2,3-*f*]quinoxaline-3-carboxylic acid as antitumor drug. This drug shows IC_{50} value of 0.05 μM [75].

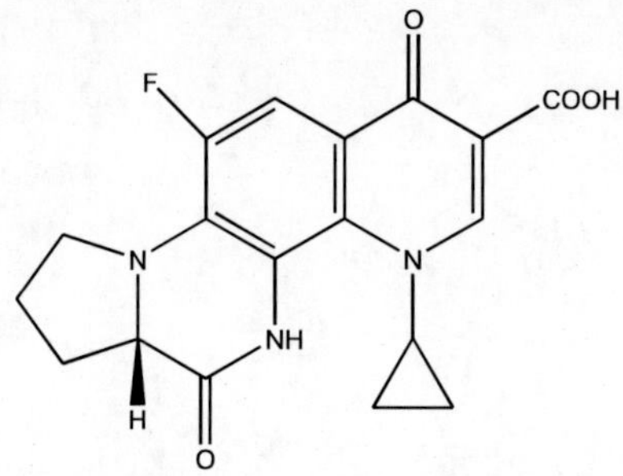

Hood, J. et al. have reported preparation of biaryl-2,4-pyrimidinediamines as inhibitor of kinase for treatment of cancer [76].

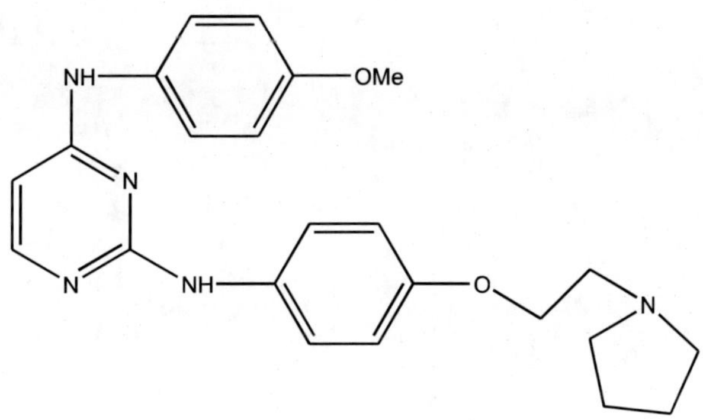

Stephen, T. et al. reported preparation of phenyl substituted pyrimidines as kinase inhibitors for treating cancer [77].

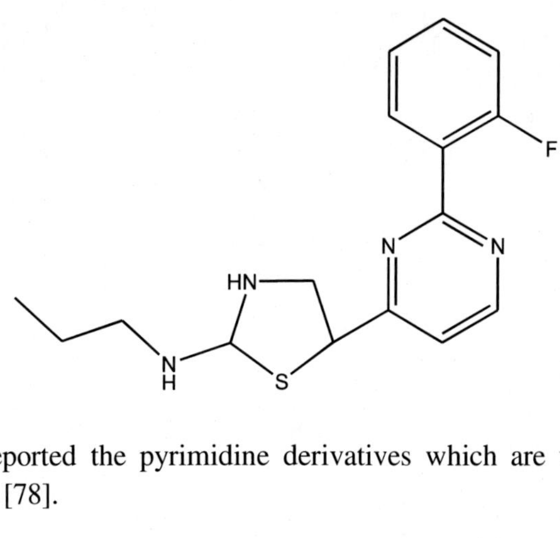

Pass, M. et al. reported the pyrimidine derivatives which are useful in producing an antiproliferative effect [78].

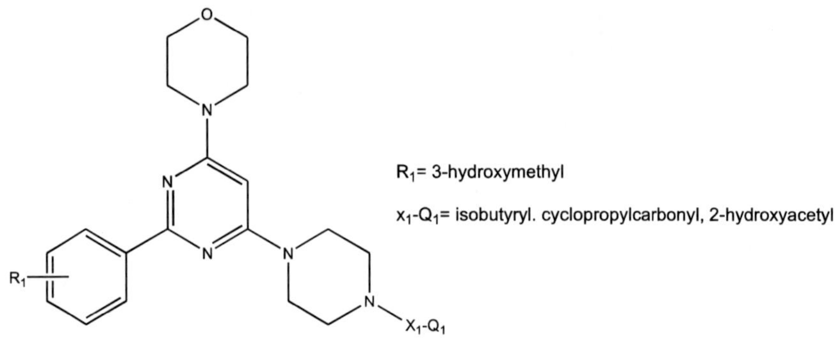

Kozina, E. S. et al. prepared 4-indol-3-yl-N-phenyl pyrimidin-2 amines as tyrosine kinase inhibitors for treatment of cancer [79].

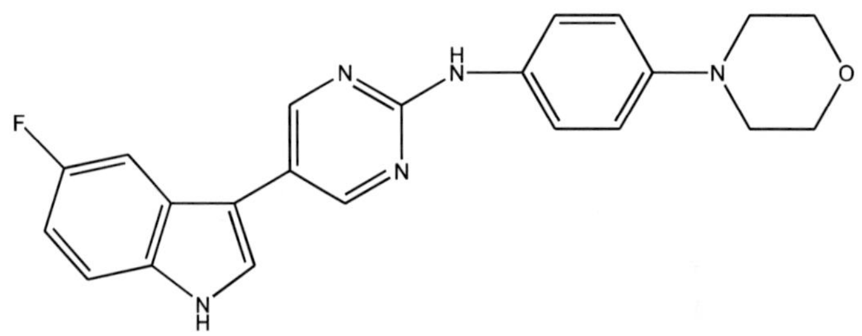

Djung, J. et al. reported the preparation of 2-anilino-4-aminoalkyl amino pyrimidines as protein kinase C-alpha (PKC-α) inhibitors. These drugs showed IC_{50} value of 0.8 nM and 0.4 nM respectively [80].

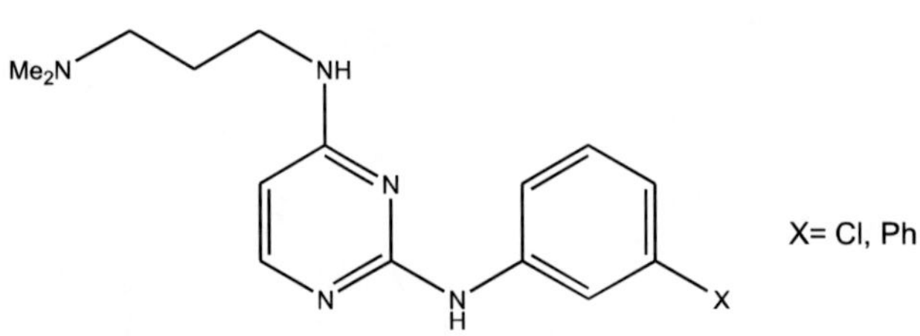

Fraley, et al. synthesized 4-(thiazolyl)-2-pyrimidinamine which exhibited excellent anticancer activity with an IC_{50} value 0.01 μM [81].

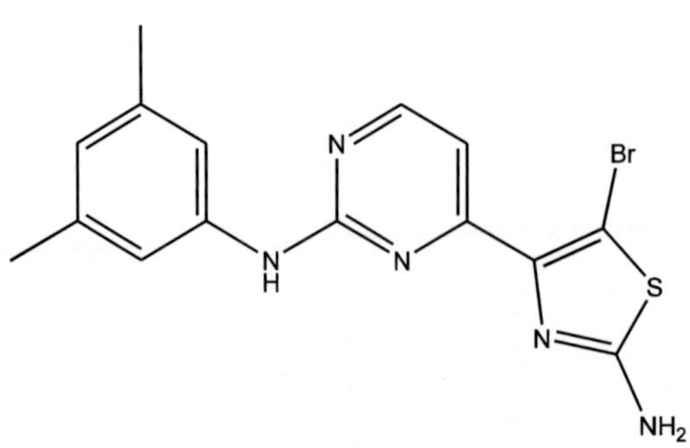

Mitchell et al. reported preparation of dihydrothieno[3,4-*d*]pyrimidine derivatives as AKt kinase inhibitors. These compounds inhibited human recombinant AKt-1 with IC_{50}<10 μM for treatment of hyperproliferative disease [82].

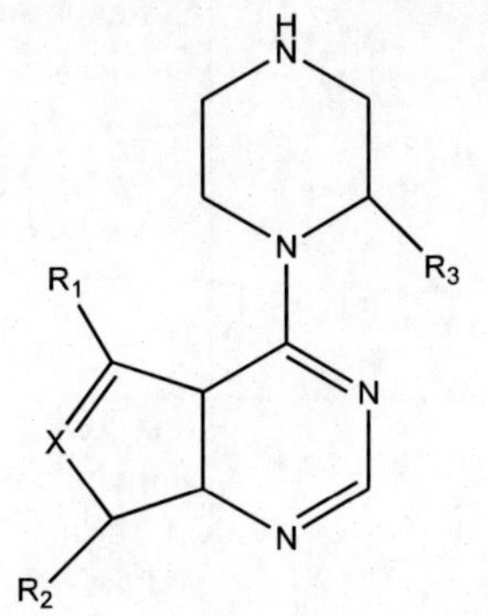

R_1= H, Me, Et, CF_3, CH_2F

R_2= H, Me

R_3= H, Me, Et, CF_3

X= S, SO, SO_2

Nasser et al. synthesized pyrazolo[3,4-*d*]pyrimidine based scaffold derivatives as anticancer agents. These derivatives have been explored for their inhibitory activity towards various protein kinase enzymes [83].

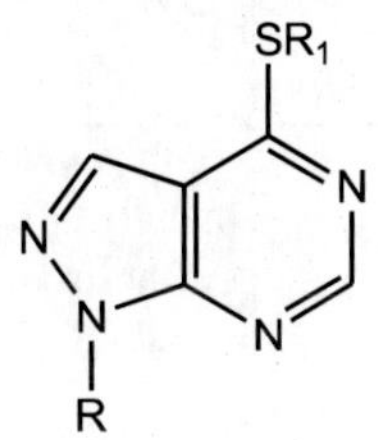

R= I, Br

R_1= $-CH_2CH_3$, $-CH_2CH_2OCH_3$,

Valarmathy et al. synthesized a series of novel 2-aminopyrimidine derivatives and screened them for their anticonvulsant activity. In this series, compounds and showed the excellent anticonvulsant activity [84].

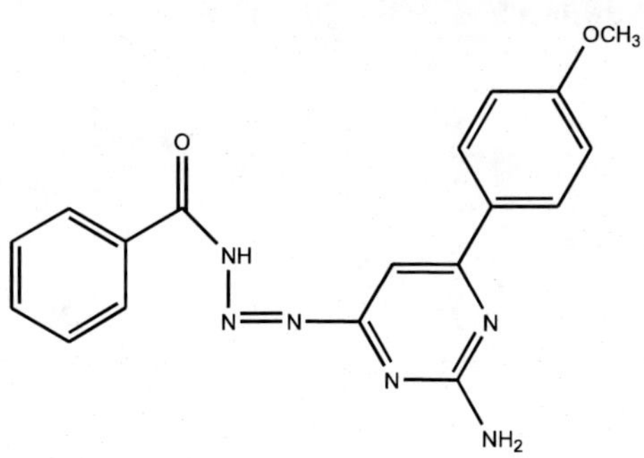

Pandeya et al. synthesized a series novel of pyrimidine hydrazones. All the synthesized compounds were evaluated for their anticonvulsant activity using Isoniazid (INH) as standard drug. Among, all the screened derivatives, *N*'-(2-isopropyl-5-methylcyclohexylidene)-6-methyl-2-oxo-4-phenyl-1,2,3,4-tetrahydropyrimidine-5-carbohydrazide was found to be active compound [85].

Table 3. Clinically useful drugs containing pyrimidine derivatives

Antifungal agent [87]		
Drug	Structure	Mechanism of Action
Cyprodinil	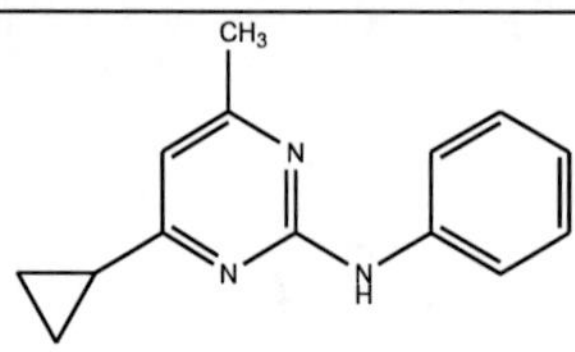 4-Cyclopropyl-6-methyl-N-phenyl-pyrimidin-2-amine	Systemic and translocated fungicide.
Flucytosine	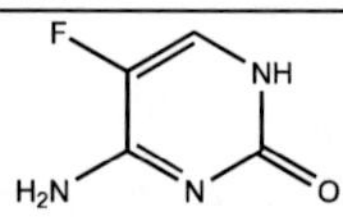 4-Amino-5-fluoro-1,2-dihydro pyrimidin-2-one	It is converted to the antimetabolite 5-fluorouracil (5-FU) which inhibits the action of thymidylate synthetase and DNA synthesis in fungal cells.
Mepanipyrim	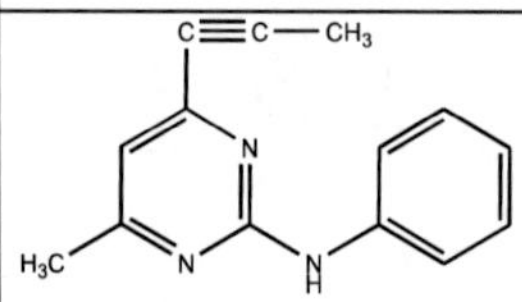 4-Methyl-N-phenyl-6-(1-propynyl)-2-pyrimidinamine	Non-systemic with preventative action and inhibits secretion of host-cell wall degrading enzymes.
Pyrimethanil	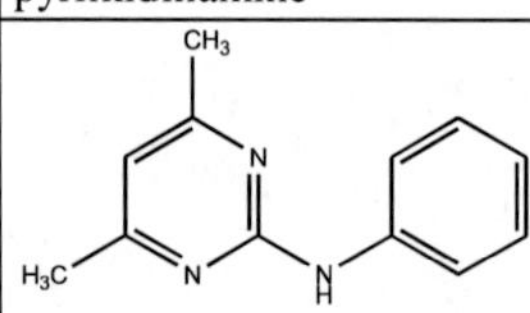 4,6-Dimethyl-N-phenylpyrimidin-2-amine	Inhibits methionine biosynthesis, there by affecting protein synthesis and subsequently cell division.
Antibacterial agent [88]		
Sulphamethiazine	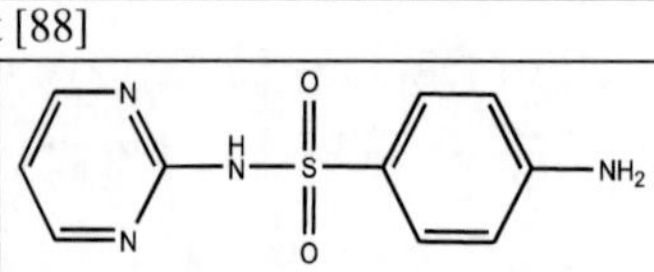 4-Amino-N-pyrimidin-2-yl-benzene- sulfonamide	Act as competitive inhibitors of the enzyme dihydropteroate synthetase (DHPS), an enzyme involved in folate synthesis.
Trimethoprim	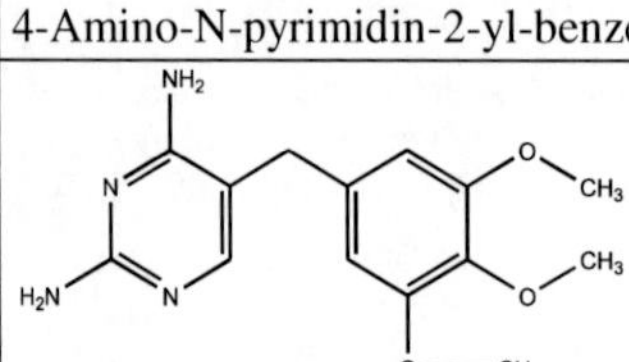 5-(3,4,5-Trimethoxybenzyl) pyrimidine-2,4-diamine	Binds to dihydrofolate reductase and inhibits the reduction of dihydrofolic acid (DHF) to tetrahydrofolic acid (THF). THF is the precursor for thymidine synthesis and interfere bacterial DNA synthesis.
Anthelmintic drug [89]		
Pyrantel	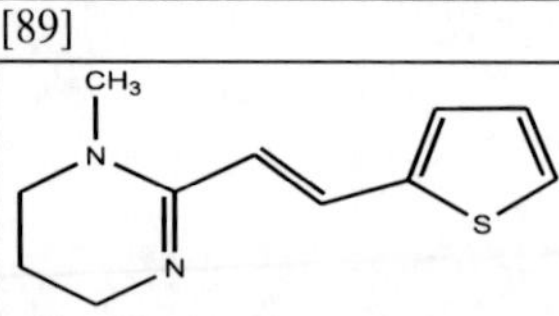 1-Methyl-2-[(E)-2-thiophen-2-ylethenyl]-5,6-dihydro-4H-pyrimidine	Elicit spastic muscle paralysis in parasitic worms due to prolonged activation of the excitatory nicotinic acetylcholine (nACh) receptors.

Morantel	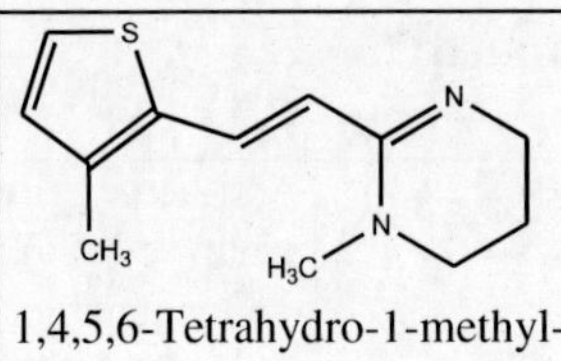 1,4,5,6-Tetrahydro-1-methyl-2-[(E)-2-(3-methyl-2-thienyl)ethenyl]pyrimidine	Elicit spastic muscle paralysis by allosteric modulation of neuronal nicotinic acetylcholine receptors.
Vasodilators [90, 91]		
Dipyridamole	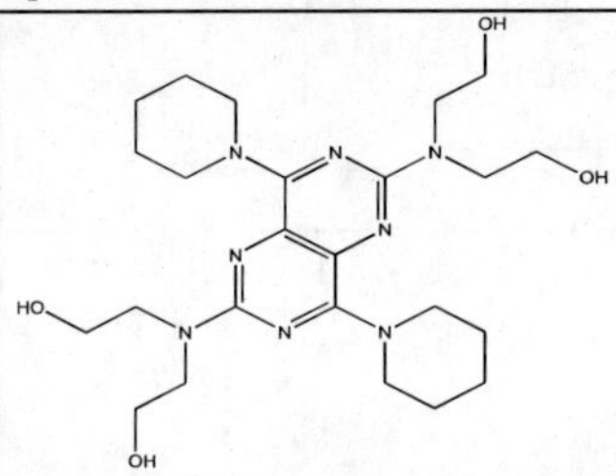 2,2,'2','2'''-(4,8-Di(piperidin-1-yl)pyrimido[5,4-d] pyrimidine-2,6-diyl)bis(azanetriyl)tetraethanol	It inhibits thrombus formation that leads to vasodilation. This action is due to inhibition of platelet cAMP-phosphodiesterase along with potentiation of adenosine inhibition of platelet function by blocking reuptake by vascular and blood cells.
Trapidil	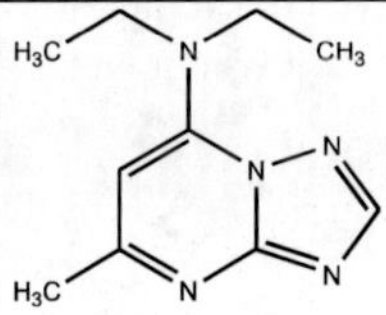 N,N-Diethyl-5-methyl-[1,2,4]triazolo[1,5-α]pyrimidin-7-amine	Trapidil has antithrombotic activity which may be mediated by the inhibition of platelet function which is characterized by the inhibition of both thromboxane synthetase and phosphodiesterase activities.
Treatment of respiratory tract and ear infections [92]		
Brodimoprim	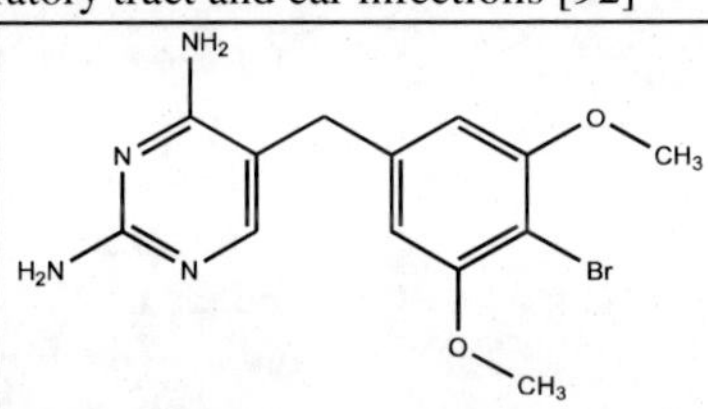 5-[(4-Bromo-3,5-dimethoxyphenyl)methyl]pyrimidine -2,4-diamine	Selective inhibitor of bacterial dihydrofolate reductase.
Antiviral drug [93, 94]		
Broxuridine	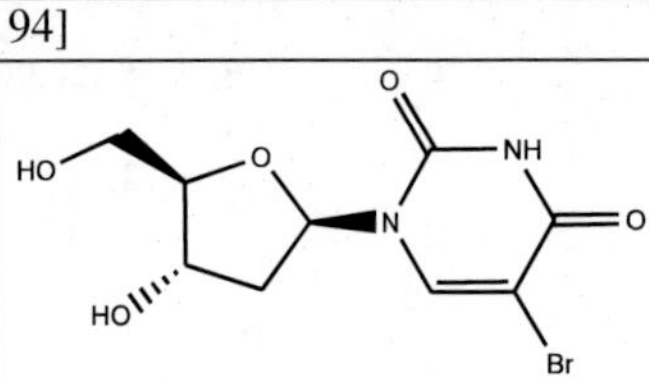 5-Bromo-1-(2-deoxy-β-d-ribofuranosyl)pyrimidine-2,4(1H,3H)-dione	Substitute for thymidine in DNA and thus acts as an anti-metabolite.
Idoxuridine	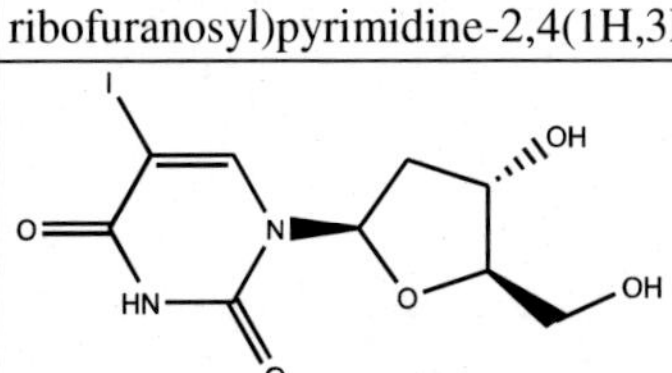 1-[(2R,4S,5R)-4-Hydroxy-5-(hydroxyl-methyl) oxolan-2-yl]-5-iodo-1,2,3,4-tetrahydro pyrimidine-2,4-dione	Anti-herpes virus and the iodine atom added to the uracil component blocks the base pairing in DNA.

Table 3. (Continued)

Triflouridine	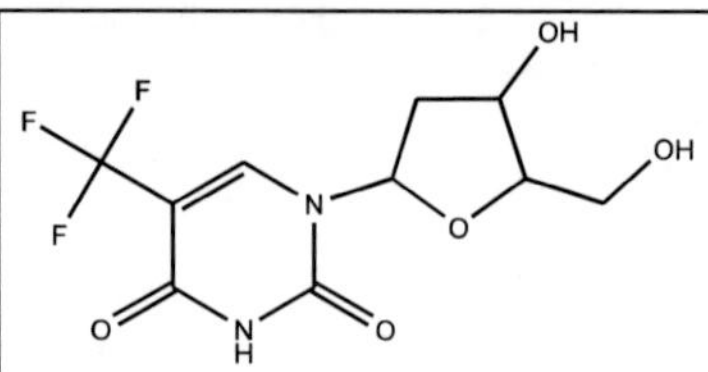 1-[4-Hydroxy-5-(hydroxymethyl) oxolan-2-yl]-5-(trifluoromethyl) pyrimidine-2,4-dione	Anti-herpes virus and the -CF3 group added to the uracil component blocks the base pairing in DNA.
Neurotropic agent [95]		
Isaxonine	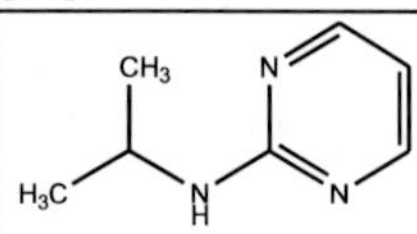 2-(Isopropylamino)pyrimidine	It has favorable effect on nerve regeneration and promotes neurite outgrowth.
Anesthetics [96]		
Thiopental Sodium	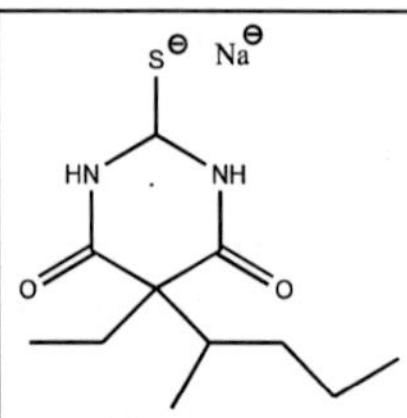 (RS)-[5-Ethyl-4,6-dioxo-5-(pentan-2-yl)-1,4,5,6-tetra hydropyrimidin-2-yl]sulfanide sodium	It is a rapid-onset short-acting barbiturate general anesthetic.
Thimylal	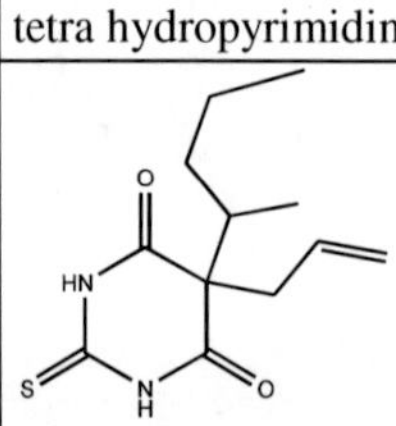 5-Allyl-5-(1-methylbutyl)-2-thioxodihydro pyrimidine-4,6(1H,5H)-dione	It is used as surgical anaesthesia. It produces its action by binding to the site associated with a Cl- ionopore at the GABAA receptor and thereby increasing the duration of time for which the Cl- ionopore is opened.
Anti-inflammatory agents [97]		
Octotiamine	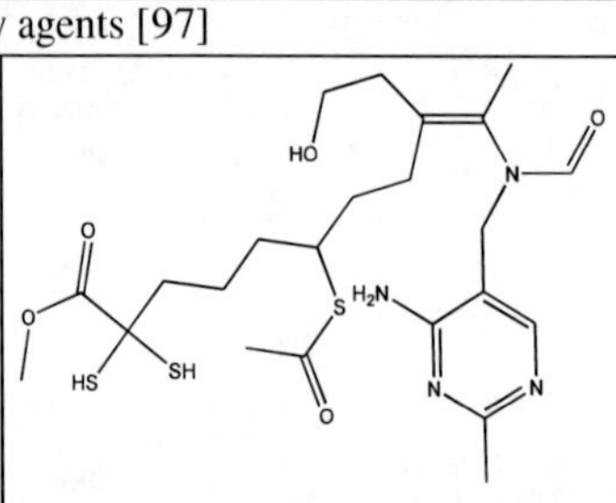 Methyl-6-acetylthio-8-[[(E)-2-[(4-amino-2-methylpyrimidin-5-yl)methyl-formylamino]-5-hydroxypent-2-en-3-yl]dithio]octanoate	It inhibits the enzyme COX-2 which is responsible for producing inflammation.
Anti-HIV drug [98]		

Cidofovir	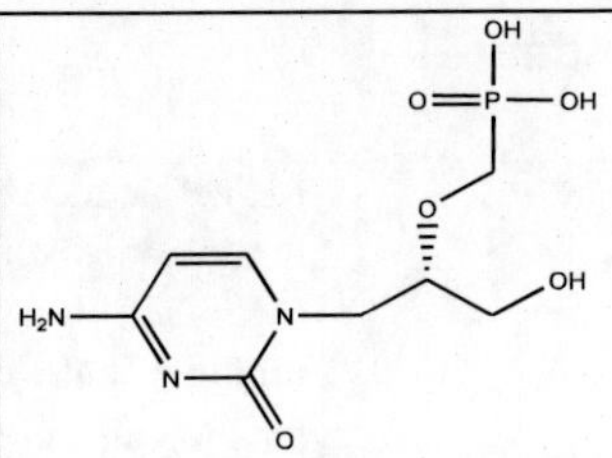 ({[(S)-1-(4-Amino-2-oxo-1,2-dihydropyrimidin-1-yl)-3-hydroxypropan-2yl]oxy}methyl)phosphonic acid	It suppresses cytomegalovirus replication by selective inhibition of viral DNA polymerase.
Ganciclovir	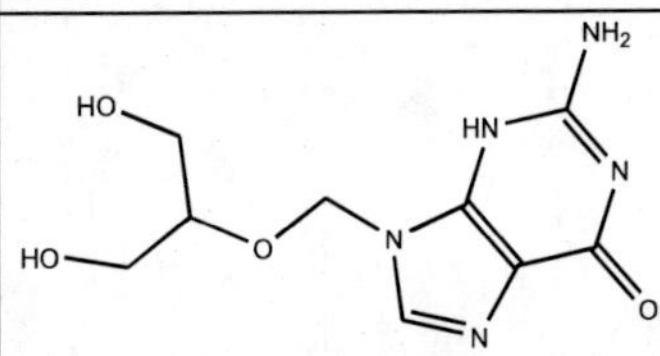 2-Amino-9-{[(1,3-dihydroxypropan-2-yl)oxy] methyl}-6,9-dihydro-3H-purin-6-one	It is a synthetic analogue of 2'-deoxy-guanosine. It is first phosphorylated to deoxyguanosine triphosphate (dGTP) which competitively inhibits the incorporation of dGTP by viral DNA polymerase and results in the termination of elongation of viral DNA.
Lamivudine	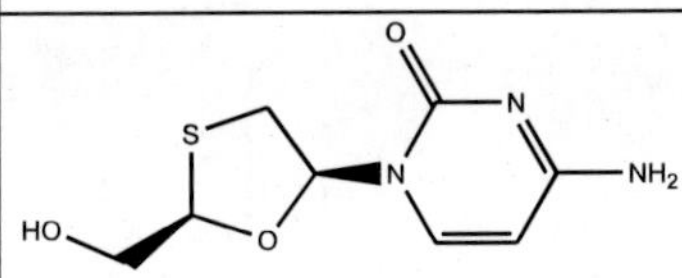 4-Amino-1-[(2R,5S)-2-(hydroxymethyl)-1,3-oxathiolan-5-yl]-1,2-dihydropyrimidin-2-one.	It is an analogue of cytidine which inhibits both types (1 and 2) of HIV reverse transcriptase and act as chain terminator of DNA synthesis.
Stavudine	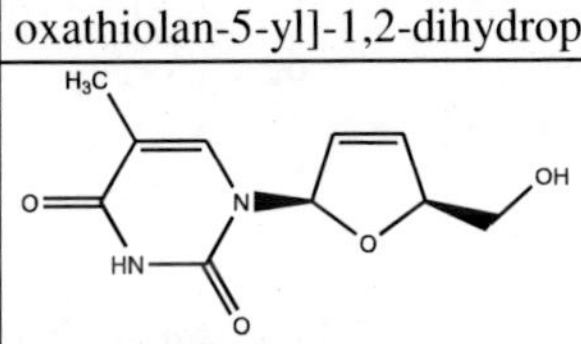 1-((2R,5S)-5-(Hydroxymethyl)-2,5-dihydro furan-2-yl)-5-methylpyrimidine-2,4(1H, 3H)-dione	It is an analog of thymidine which on phosphorylation produces Stavudine triphosphate by cellular kinases. This triphosphate inhibits the HIV reverse transcriptase and causes termination of DNA replication.
Zidovudine	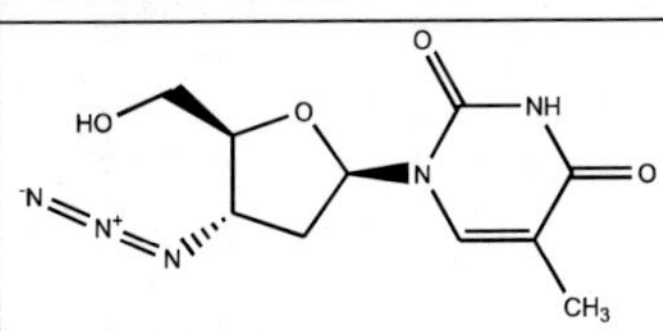 1-[(2R,4S,5S)-4-Azido-5-(hydroxymethyl) oxolan-2-yl]-5-methylpyrimidine-2,4-dione	It is a nucleoside analog and acts as reverse-transcriptase inhibitor (NRTI).
Antineoplastic agent [99,100]		
Azacitidine	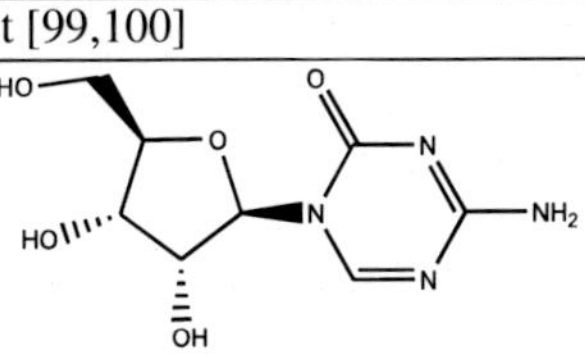 4-Amino-1-β-D-ribofuranosyl-1,3,5-triazin-2(1H)-one	It induces antineoplastic activity via two mechanisms such as inhibition of DNA methyl-transferase at low doses, causing hypomethylation of DNA, and direct cytotoxicity in abnormal hematopoietic cells in the bone marrow through its incorporation into DNA and RNA at high doses, resulting in cell death.

Table 3. (Continued)

Azathioprine	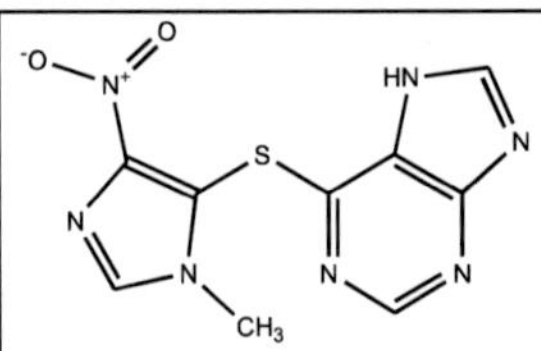 6-[(1-Methyl-4-nitro-1H-imidazol-5-yl)sulfanyl]-7H-purine	It's active metabolite methyl-thioinosine monophosphate (MeTIMP) is a purine synthesis inhibitor that works by blocking the enzyme amidophosphoribosyl transferase.
Cytarabine	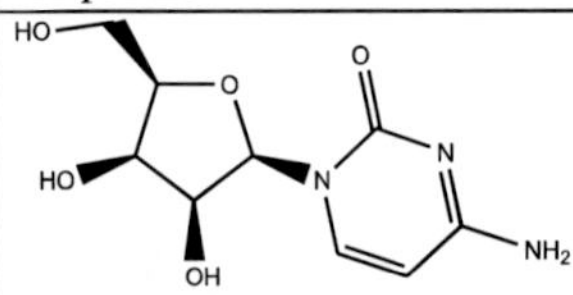 4-Amino-1-[(2R,3S,4R,5R)-3,4-dihydroxy-5-(hydroxymethyl)oxolan-2-yl]pyrimidin-2-one	Its mode of action is due to its rapid conversion into cytosine arabinoside triphosphate, which damages DNA during S phase of cell cycle.
Floxuridine	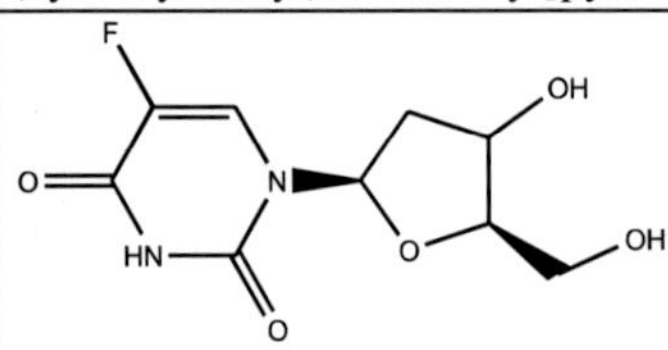 5-Fluoro-1-[4-hydroxy-5-(hydroxymethyl) tetrahydrofuran-2-yl]-1H-pyrimidine-2,4-dione	Floxuridine is rapidly catabolized to 5-fluorouracil which interferes with the synthesis of DNA and RNA.
Fluorouracil	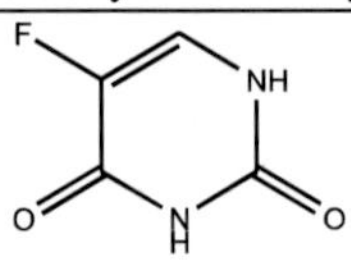 5-Fluoro-1H,3H-pyrimidine-2,4-dione	It works through irreversible inhibition of thymidylate synthase.
Gemcitabine	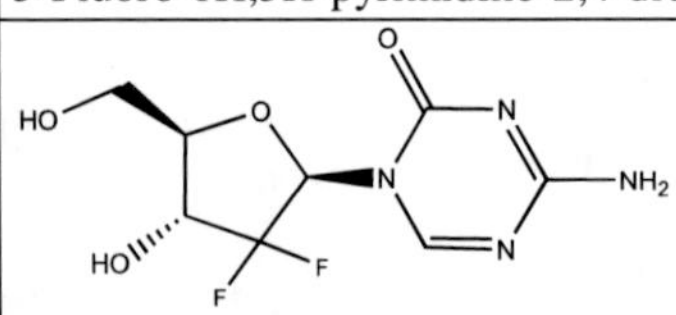 4-Amino-1-(2-deoxy-2,2-difluoro-β-D-erythro-pentofuranosyl)pyrimidin-2(1H)-one	It acts as an ribonucleotide reductase (RNR) inhibitor.
Tegafur	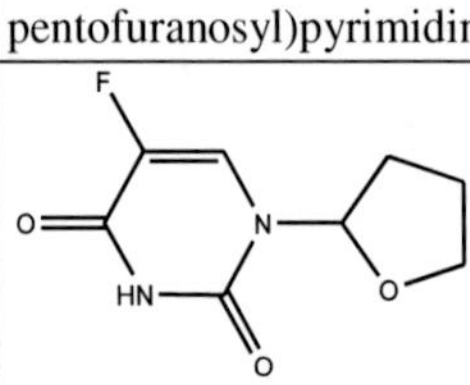 (RS)-5-Fluoro-1-(tetrahydrofuran-2-yl)pyrimidine-2,4(1H,3H)-dione	It is a chemotherapeutic fluorouracil prodrug and acts as antimetabolite.
Tetroxoprim	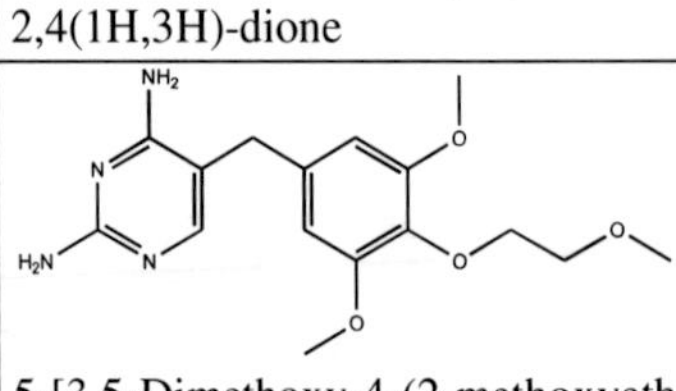 5-[3,5-Dimethoxy-4-(2-methoxyethoxy)benzyl] pyrimidine-2,4-diamine	Dihydrofolate reductase inhibitor

Thioguanine	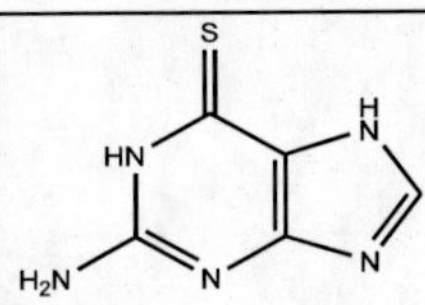 2-Amino-1H-purine-6(7H)-thione	It is an antimetabolite because it competes with hypoxanthine and guanine for the enzyme hypoxanthine-guanine phospho-ribosyltransferase which results inhibition of purine synthesis.
Uramustine	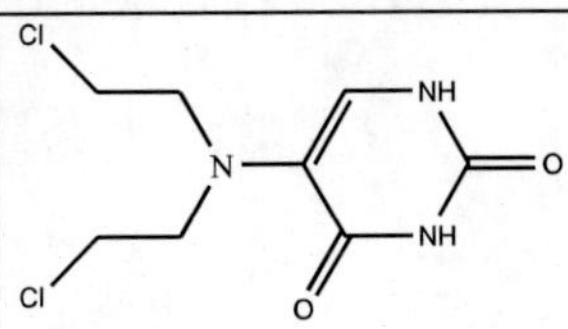 5-(Bis(2-chloroethyl) amino)pyrimidine-2,4 (1H,3H)-dione	Alkylating agents as it binds preferentially to the guanine and cytosine moieties of DNA, leading to cross-linking of DNA, there by inhibiting DNA synthesis and function.
Capecitabine	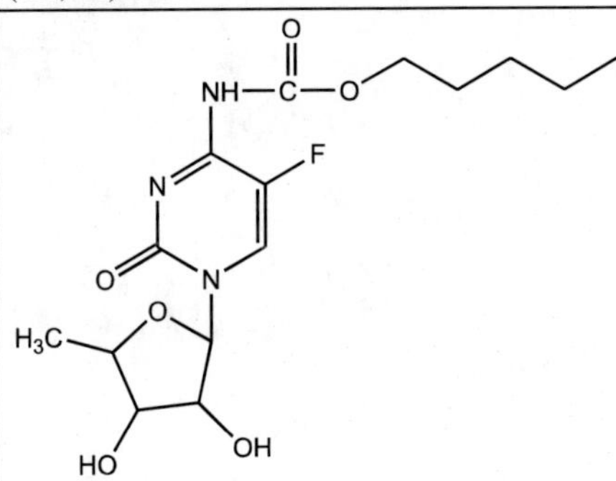 Pentyl 1-(3,4-dihydroxy-5-methyl-tetrahydrofuran-2-yl)-5-fluoro-2-oxo-1,2-dihydropyrimidin-4-ylcarbamate	Capecitabine is metabolised to 5-FU which is a thymidylate synthase inhibitor. So it inhibits synthesis of thymidine monophosphate (ThMP), the active form of thymidine which is required for the de novo synthesis of DNA
Diuretics [101]		
Orotic Acid	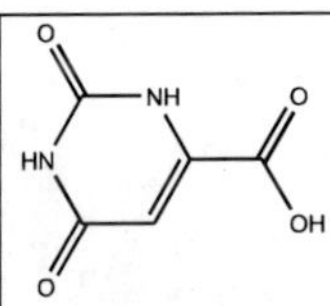 1,2,3,6-Tetrahydro-2,6-dioxo-4-pyrimidine carboxylic acid	It regulates water-salt exchange, by increasing diuresis and reducing the volume of extracellular fluid.
Antihypertensive drug [102]		
Minoxidil	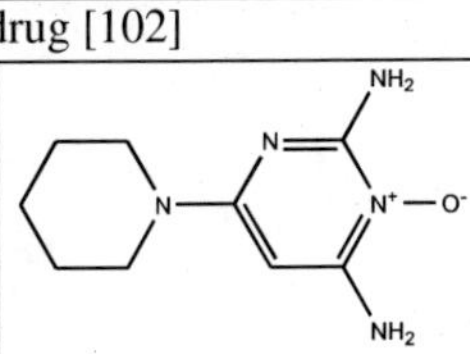 6-Piperidin-1-ylpyrimidine-2,4-diamine3-oxide	It is a potassium channel opener that leads to hyperpolarization of cell membranes.
Antipsychotic agent [103]		
Mezilamine	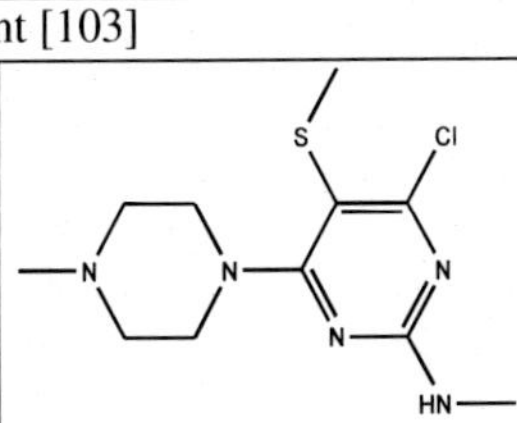 2-Methylamino-4-N-methylpiperazino-5-methylthio-6-chloropyrimidine	Dopamine antagonists

Table 3. (Continued)

CI-943	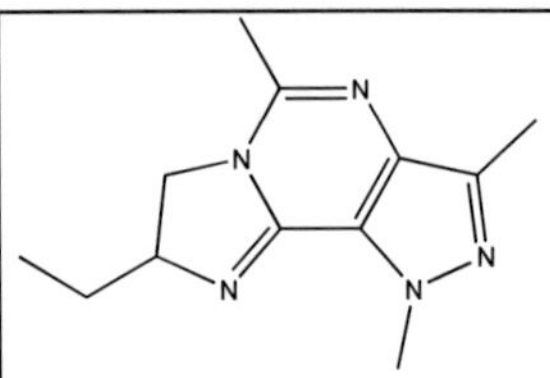 8-Ethyl-7,8-dihydro-1,3,5-trimethyl-1H-imidazo[1,2-c]pyrazolo[3,4-e]-pyrimidine	It reduces spontaneous locomotion
Management of anxiety disorders [104, 105]		
Buspirone	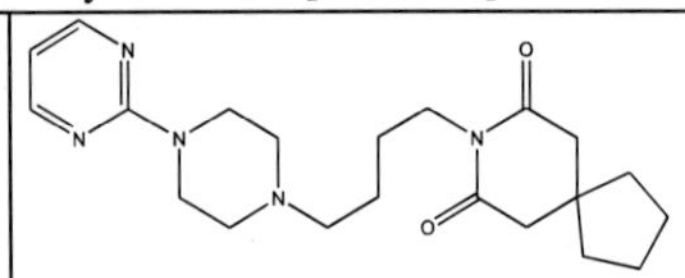 8-[4-(4-Pyrimidin-2-ylpiperazin-1-yl)butyl]-8-azaspiro[4.5]decane-7,9-dione	It is a serotonin 5-HT1A receptor partial agonist.
Ritanserin	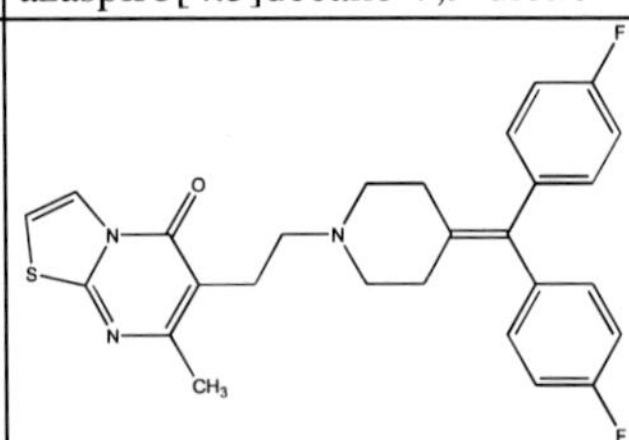 6-[2-[4-[Bis(4-fluorophenyl)methylidene piperidin-1-yl]ethyl]-7-methyl-[1,3] thiazolo [2,3-b]pyrimidi-5-one	It acts as a 5-HT2A and 5-HT2C receptor antagonist.
Treatment of urinary tract infections [106]		
Pipemidic Acid	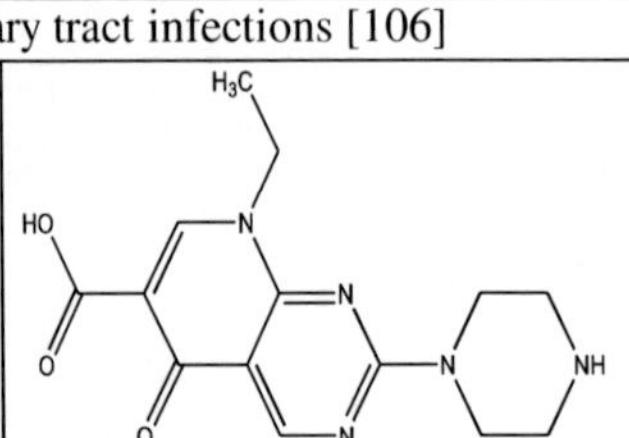 8-Ethyl-5-oxo-2-piperazin-1-yl-5,8-dihydro pyrido [2,3-d]pyrimidine-6-carboxylic acid	It targets bacterial DNA gyrase, an enzyme which reduces DNA strain during replication. This enzyme is essential for DNA replication and synthesis and finally cell division is inhibited.
Sedative-hypnotics [107]		
Allobarbital	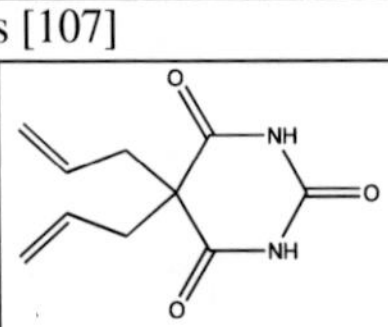 5,5-Diprop-2-enyl-1,3-diazinane-2,4,6-trione	It works by binding to the GABAA receptor at either the α or β-sub unit.
Pentobarbital	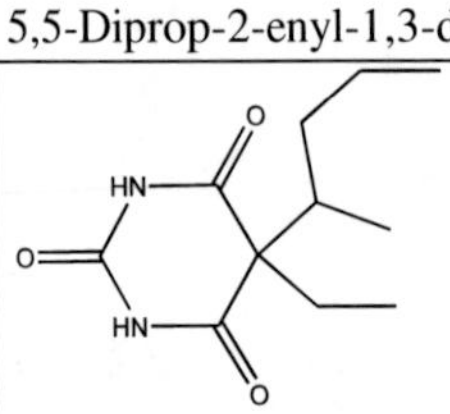 5-Ethyl-5-(1-methylbutyl)-2,4,6(1H,3H,5H)-pyrimidinetrione	It acts as GABA Modulators by binding to the site associated with a Cl- ionopore at the GABAA receptor thereby increasing the duration of time for which the Cl- ionopore is opened.

Phenobarbitone	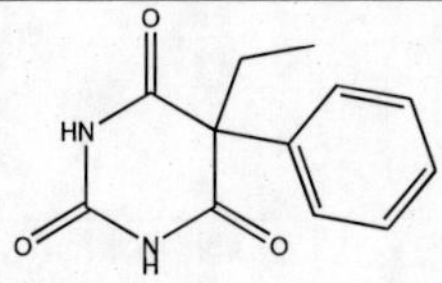 5-Ethyl-5-phenylpyrimidine-2,4,6 (1H,3H,5H)-trione	It acts on GABAA receptors and increases synaptic inhibition.
Anticonvulsant [108]		
Primidone	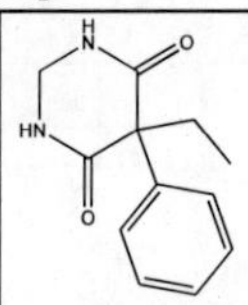 5-Ethyl-5-phenyl-hexahydro pyrimidine-4,6-dione	It is GABA receptor agonist. It is also believed to work via interactions with voltage-gated sodium channels which inhibit high-frequency repetitive firing of action potentials.
Vitamins [109]		
Thiamine	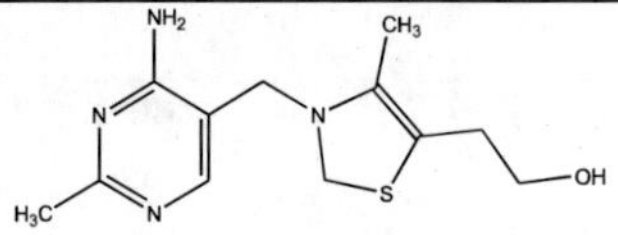 2-[3-[(4-Amino-2-methyl-pyrimidin-5-yl) methyl]-4-methyl-thiazol-5-yl]ethanol	Thiamine is a water-soluble vitamin. It combines with adenosine triphosphate (ATP) to form a coenzyme, thiamine pyrophosphate which is required for carbohydrate metabolism.
Antidiabetic [110]		
Alloxan	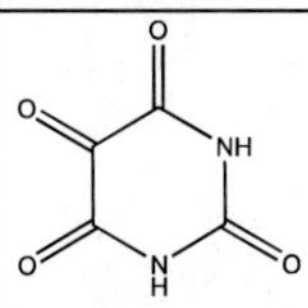 1,3-Diazinane-2,4,5,6-tetrone	It is selectively toxic to pancreatic β cells as it preferentially accumulates in the β cells as glucose analogues.
Antiparkinsonian drug [111]		
Piribedil	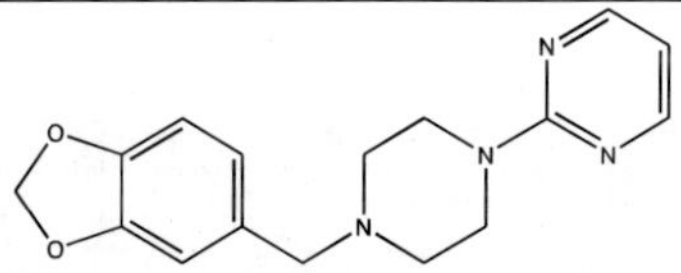 2-[4-(Benzo[1,3]dioxol-5-ylmethyl) piperazin-1-yl] pyrimidine	It acts as both D2 and D3 receptor agonist.
Antigout agent [112]		
Tisopurine	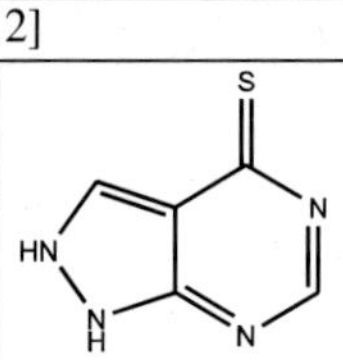 1,2-Dihydro-4H-pyrazolo[3,4-d-pyrimidine-4-thione	It has uricosuric activity.
Antimalarial drug [113]		
Pyrimethamine	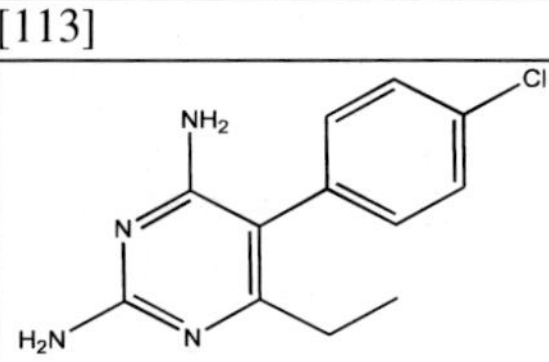 5-(4-Chlorophenyl)-6-ethyl-2,4-pyrimidine diamine	Interferes with tetrahydrofolic acid synthesis from folic acid by inhibiting the enzyme dihydro-folate reductase (DHFR).
Expectorant and mucolytic agent [114]		

Table 3. (Continued)

Tasuldine	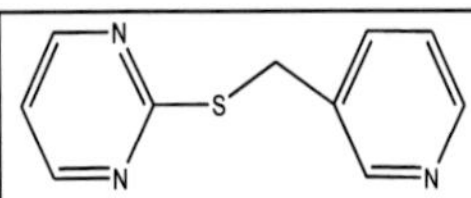 2-[(3-Pyridinylmethyl)thio]pyrimidine	It is a broncho secretolytic agent. It decreases the sialomucin content of the mucus which results decrease in viscosity of the mucus.
Antihistaminic drugs [115]		
Pemirolast	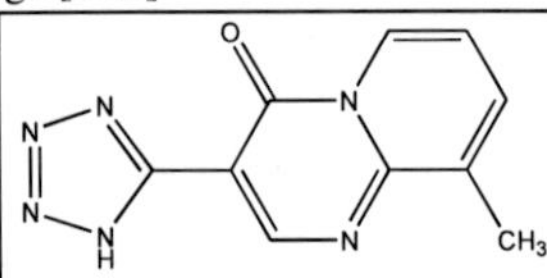 9-Methyl-3-(1H-tetrazol-5-yl)-4H-pyrido[1,2-a]pyrimidin-4-one	It binds to the histamine H1 receptor. This block the action of endogenous histamine, which subsequently leads to temporary relief of the negative symptoms brought on by histamine.
Thonzylamine	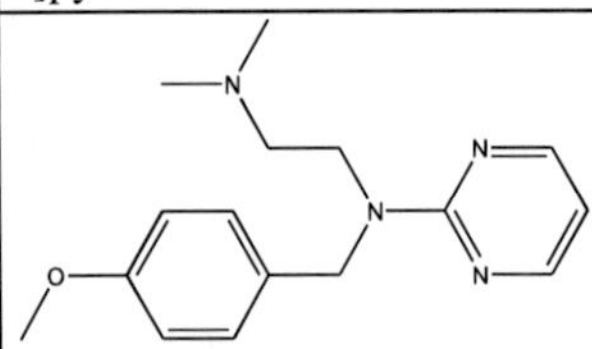 N-(4-Methoxybenzyl)-N,'N'-dimethyl-N-pyrimidin-2-ylethane-1,2-diamine	It is histamine H1-receptor antagonist.
Anti-thyroid drug [116]		
2-Thiouracil	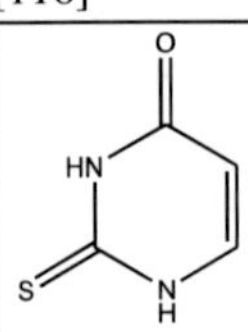 2-Thioxo-1H-pyrimidin-4-one	It inhibits thyroid activity by blocking the enzyme thyroid peroxidase.
Propylthiouracil	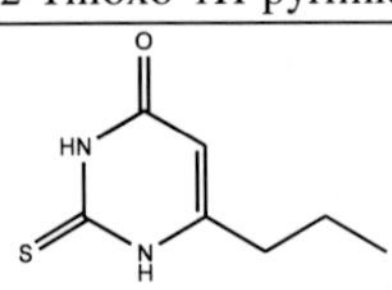 6-Propyl-2-sulfanylpyrimidin-4-one	It inhibits the enzyme thyroperoxidase, which normally acts in thyroid hormone synthesis by oxidizing the anion iodide

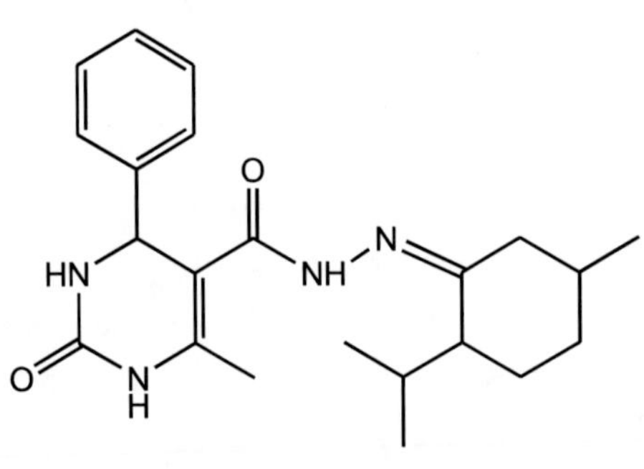

Sindhu et al. synthesized a series of novel pyrimidine derivatives and screened for their anti-depressant activity using forced swim test and tail suspension test in mice. In this study,

4,6-bis(4-chlorophenyl)pyrimidin-2-yloxy)-*N,N*-dimethylmethanamine was found to be effective in significant reduction of immobility period as compared to standard drug fluoxetine [86].

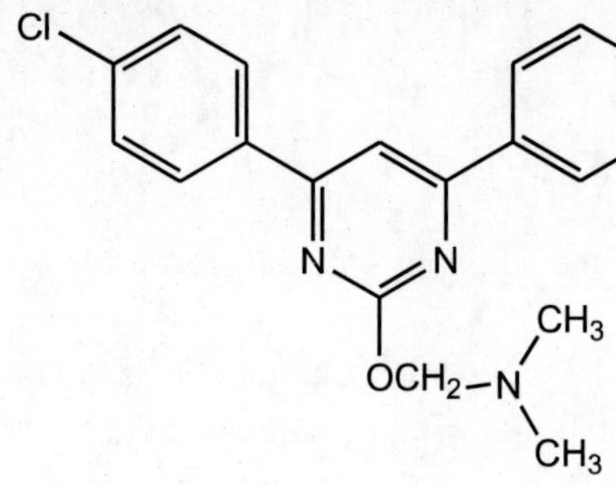

5. Natural Occurrence of Pyrimidines

The pyrimidine nucleus is involved widely in nature such as in nucleic acids, nucleotides, alkaloids and in uric acid etc. Most of the natural occurring pyrimidines derivatives consist of hydroxyl and amino substituents. Vicine was first pyrimidine derivative found in nature (seeds of Vicia sativa) and was discovered by Ritthausen in 1870 [117-120].

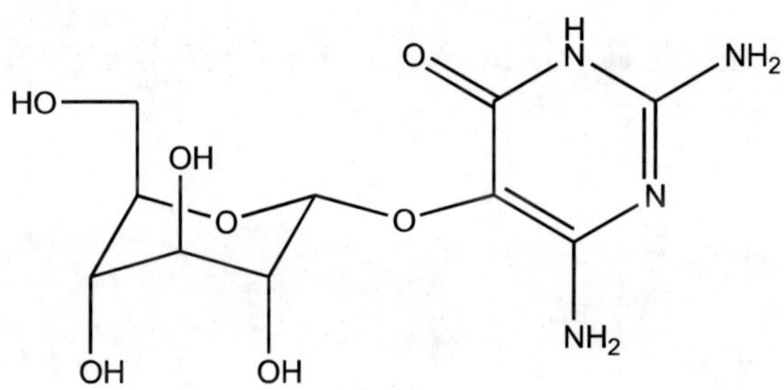

Vicine

Xanthine was first discovered by Marget in 1817 in urinary bladder stone and also occurs in tea coffee, cocoa along with hypoxanthine.

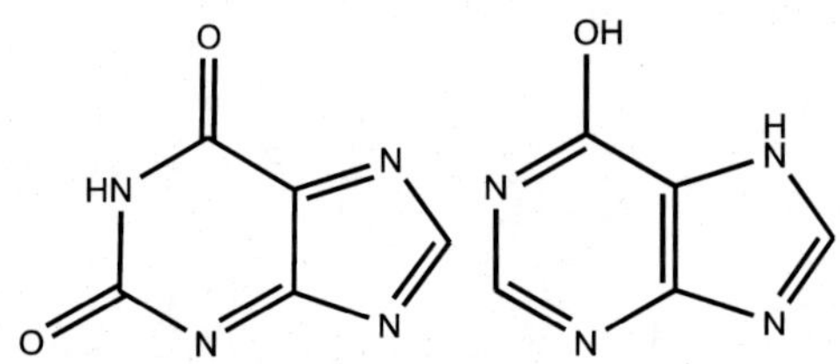

Xanthine Hypoxanthine

The coffee bean contains about 1.5% caffeine, tea leaves contain about 1.5% caffeine and some theophylline, cocoa bean contains about 1.3% theobromine. These methylated derivatives of xanthines have stimulating effect on the central nervous system.

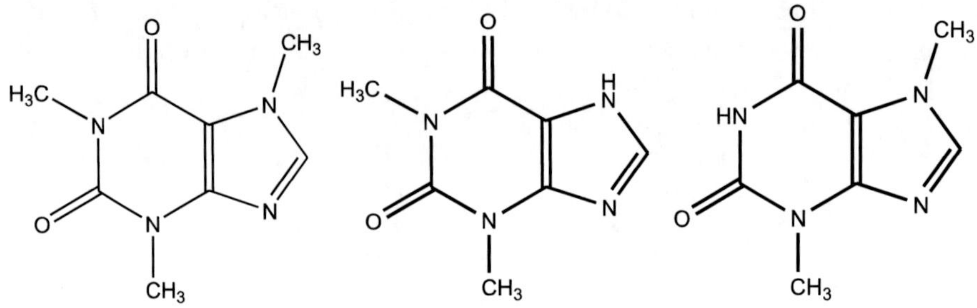

Caffeine Theophylline Theobromine

The uric acid was discovered by Swedish chemist Scheele in 1776 as a constituent of human urinary calculi and is a metabolite of purine nucleoside.

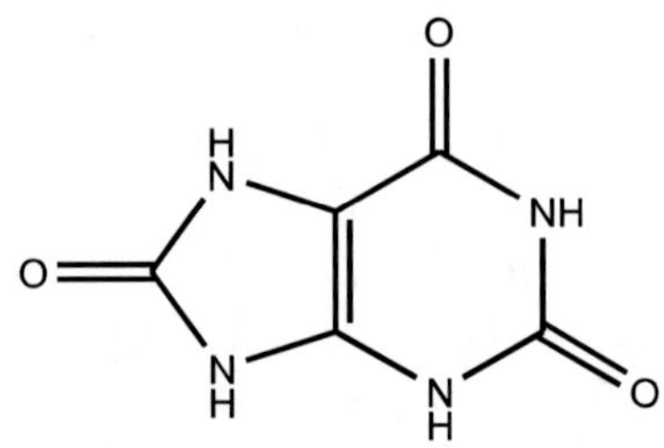

Uric acid

Folic acid which is chemically pteroyl glutamic acid was isolated from liver and yeast. It was considered as an important growth factor for several organisms.

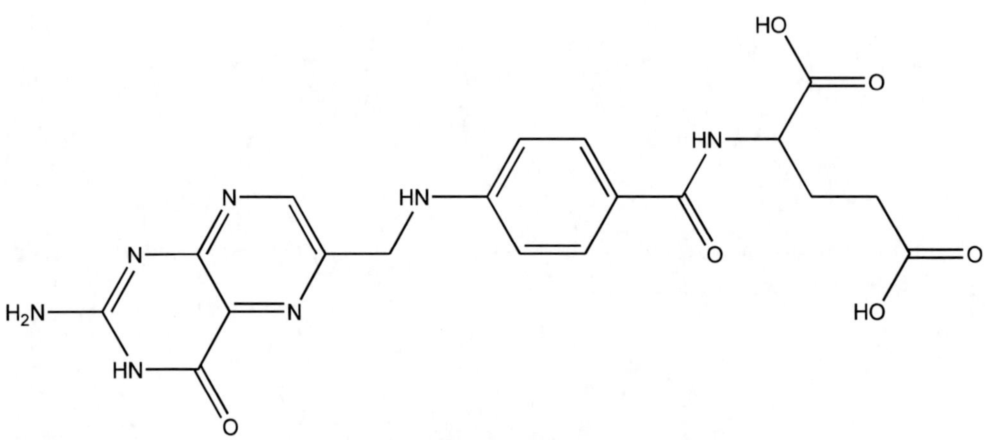

Folic acid

Riboflavin or vitamin B_2 was discovered in 1920, isolated in 1933, and first made in 1935. The active forms of riboflavin are flavin mononucleotide (FMN) and flavin adenine dinucleotide (FAD) which function as cofactors for a variety of enzyme reactions.

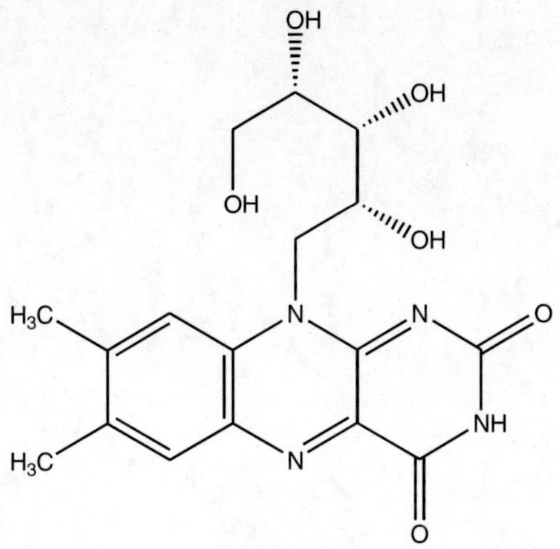

Riboflavin

Similarly Biopterin is pterin derivatives and act as endogenous enzyme cofactors in many species of animals, bacteria and fungi.

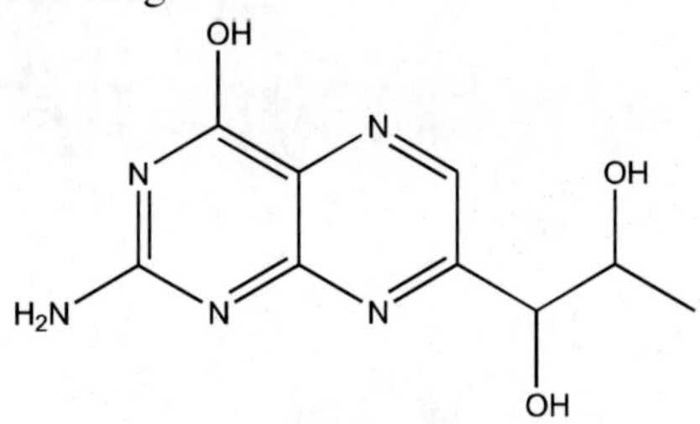

Biopterin

The nucleic acids such as Deoxyribonucleic acid (DNA) and Ribonucleic acid (RNA) possess pyrimidine nucleus in their structures. Both possess nitrogenous base (purine or pyrimidine), pentose (2-deoxyribose or ribose) and phosphoric acid. Both DNA and RNA contain the purine nitrogenous bases such as adenine (A), guanine (G) and the pyrimidine cytosine (C), but in DNA a second pyrimidine is thymine (T), whereas in RNA it is uracil (U).

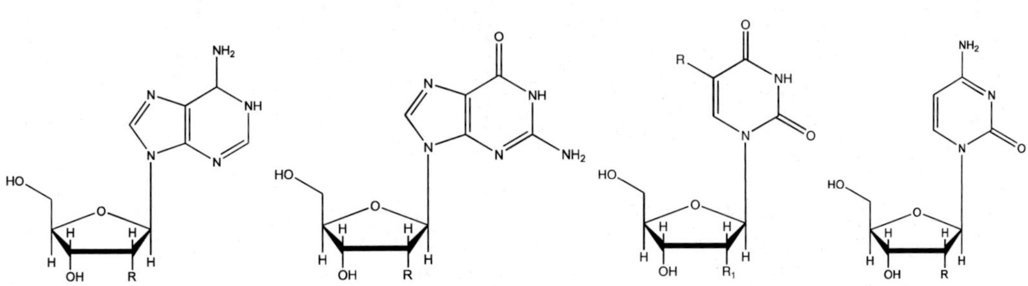

Adenosine (R = OH) Guanosine (R= OH Uridine (R = H, R_1 = OH) Cytidine (R = OH)
Deoxyadenosine (R = H) Deoxyguanosine (R = H) Thymidine (R = CH_3, R_1 = H)
Deoxycytidine (R = H)

Adenosine triphosphate (ATP) is a nucleotide and is considered as the "molecular unit of currency" of intracellular energy transfer. ATP is used as a substrate in signal transduction pathways by kinases that phosphorylate proteins and lipids.

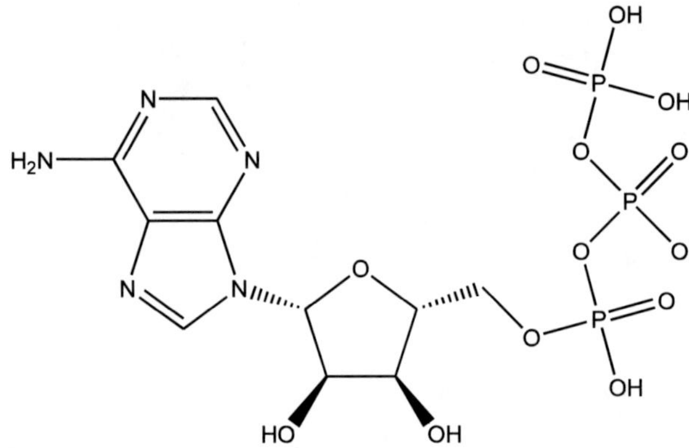

ATP

In 1959, L-α-amino acid (willardine) was isolated from seeds of Acacia willardiana and later from other species of Acacia.

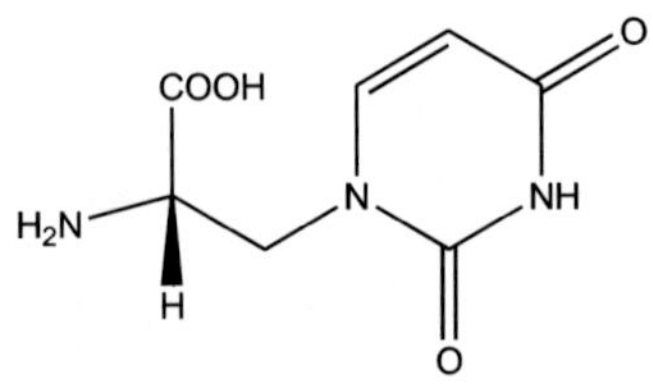

Willardine

Bacimethrin is chemically 4-amino-5-(hydroxymethyl)-2-methoxypyrimidine, an antibiotic which was isolated from Bacillus megatherium in 1961. It is active against bacterial infection both in *vivo* and *in vitro.*

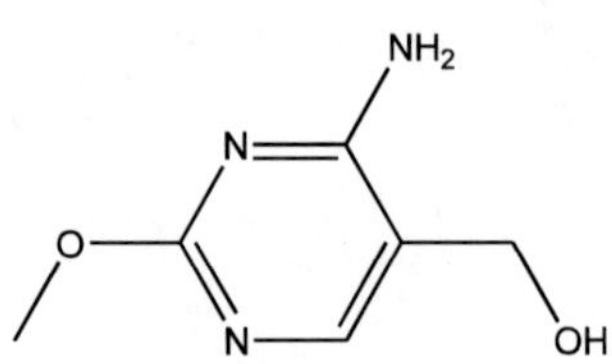

Bacimethrin

Febrifugine is a benzopyrimidine anlogue and is having antimalarial properties. It was isolated from Dichroa febrifuga. It's synthetic halogenated derivative halofuginone is used in veterinary medicine as a coccidiostat.

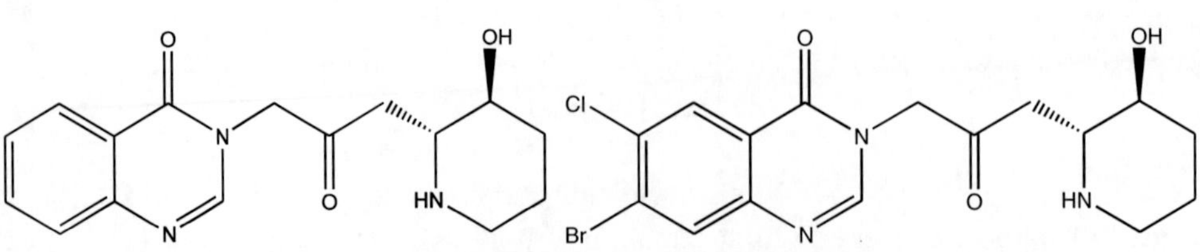

Febrifugine Halofuginone

Heteromines (F, G, H) are isolated from aerial parts of the plant *Heterostemma brownii* in 1997 and are found to possess cytotoxic effect on cancer cell line.

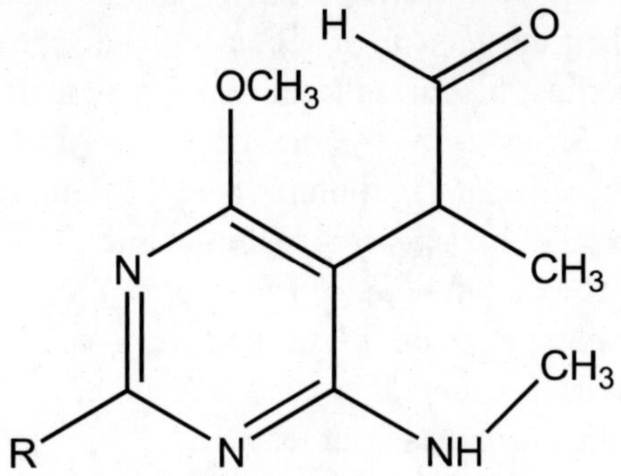

Heteromines-F R = Me_2N
G R = MeNH
H R = NH_2

Batzelladine is a pyrimidine derivative and is obtained from marine source which inhibits the binding of HIV gp-120 to CD4cells.

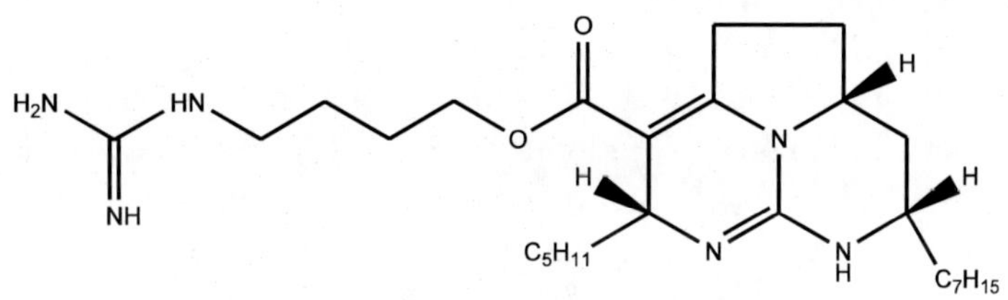

Batzelladine

Capreomycin contain pyrimidine ring and is produced from *Streptomyces capreolus.* Specifically it is a second line drug used for drug resistant tuberculosis.

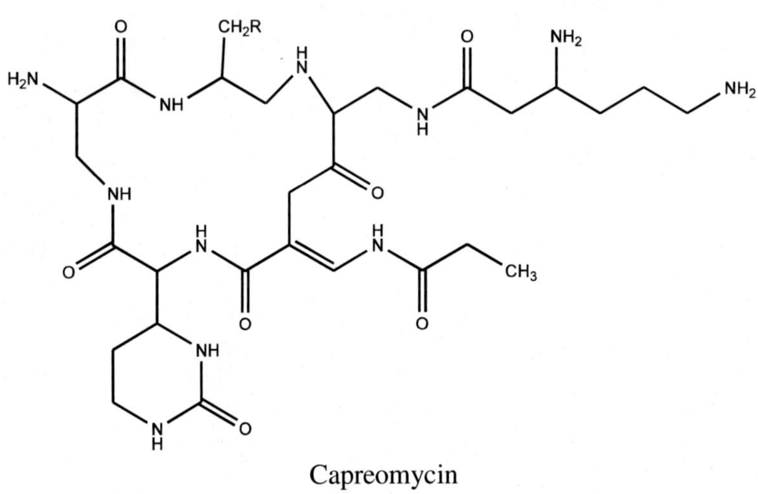

Capreomycin

CONCLUSION

The information collected from the research and development on the pyrimidine derivatives with respect to their mechanism of action, structure activity relationships, physical, chemical and spectral properties, have resulted in their potential use in the management of various diseases. In medicinal chemistry, pyrimidine derivatives have been familiar for their therapeutic potentials. The presence of a pyrimidine base in thymine, cytosine and uracil, which are the essential binding blocks of nucleic acids, DNA and RNA is one of the possible reasons for their biological activity. The literature indicates that compounds having pyrimidine nucleus possess wide spectrum of therapeutic activities, like 5-fluorouracil as anticancer, idoxuridine and trifluoridine as antiviral zidovudine and stavudine as antiHIV, trimethoprim, sulphamethiazine and sulphadiazine as antibacterial, sulphadoxin as antimalarial and antibacterial minoxidil and prazosin as antihypertensive, barbiturates eg. phenobarbitone as sedative, hypnotics and anticonvulsant propylthiouracil as antithyroid thionzylamine as H_1-antihistamine, toxoflavin and fervennuline as antibiotics. As a result of remarkable therapeutic efficacy of pyrimidine derivatives as demonstrated in this chapter, studies of pharmacological activity of drugs containing pyrimidine nucleus is necessary. There is significant scope for further research work to be done in this field to find out the potential drug candidates in relation to their safety and efficacy. This present book chapter will be helpful for the researchers who are continuing their research in this area. The chemistry of pyrimidines continues to be blossoming field and it would also be interesting to see development of pyrimidines as potentially active therapeutic compound. Despite significant research on the synthesis and studies of medicinal activities of pyrimidine derivatives, it appears that a focus of using pyrimidine as the starting material for further functionalization is ignored. As pointed out before that the electrophilic and nucleophilic substitution of pyrimidine is complicated. Therefore, methods to substitute pyrimidine ring with different other groups under mild conditions are necessary. Due to unique structure of these derivatives as described above, search for additional clinically active pyrimidines would continue in the future.

REFERENCES

[1] Katritzky, A. R., Ramsden, C. A., Scriven, E. F. V. and Taylor, R. J. K. 2008. "*Comprehensive Heterocyclic Chemistry*," Pergamon: Oxford, U.K., 1-13.

[2] Bansal, R. K. 1996. "*Heterocyclic Chemistry*," New Age International (P) Limited; 2nd ed.; 189-205.

[3] Bahl, B. S.; Bahl, A. 1996. "*Advanced Organic Chemistry*," S. Chand Group (P), 1st ed.; 15.

[4] Lagoja, I. M. 2005. "Pyrimidine as Constituent of Natural Biologically Active Compounds" *Chemistry & biodiversity* 2:1-50.

[5] Brown, D. J. 1984. "*Comprehensive Heterocyclic Chemistry*," Pergamon Press, Oxford, 3:84.

[6] Finar, I. L. 2008. "*Organic Chemistry: Stereochemistry and the Chemistry of Natural Products*"; 5th ed.; 2: 642-644.

[7] Adrien, A. 1982. "*Advances in Heterocyclic Chemistry*," Academic Press, London, 30.

[8] Agarwal, O. P. 2008. "*Organic chemistry: Reaction and reagents*" Krishna Prakashan Media (p) Ltd., 43rd ed.; 735-738.

[9] Michael, B. S., March, J. 2007. "*March's Advanced Organic Chemistry: Reactions, Mechanisms, and Structure*," Wiley-Interscience, 6th ed.102-110.

[10] Brown, D. J., Mason, S. F. 2008. "*Chemistry of heterocyclic compounds: the pyrimidines*" 16:31-81.

[11] Shishoo, C. J., Pathak, U. S., Jain, K. S., Devani, I. T. and Chabria, M. T. 1994. "Synthesis of some 2-substituted-6-phenyl- and 7-phenyl-thieno(3,2-d)-pyrimidin-4(3H)-ones" *ChemInform* 25(38):436-440.

[12] Pratyusha, C., Poornima, G., Sandhyarani, K., Krishnaveni, A., Brahmaiah, B. and Sreekanth N. 2013. "An overview on synthesis and biological activity of pyrimidines" *International Journal of Pharmaceutical Sciences Review and Research* 3(2):86-90.

[13] Corey, E. J., Cheng, X. 1989. "*The Logic of Chemical Synthesis*," Wiley-Interscience, 1st ed. *45-67.*

[14] Nadendla, R. R. 2005. "*Principles of Organic Medicinal Chemistry.*" New Age International (P) Limited; New Delhi, 2nd ed.; 61-76.

[15] Block, J. H., Beale, J. M. 2004. "*Wilson Gisvold's Textbook of Organic and Pharmaceutical Chemistry*" Lippincott Williams & Wilkins, 11th ed. 46-76.

[16] Lemke, T. L., Williams, D. A. 2012. "*Foye's Principles of Medicinal Chemistry*" Lippincott williams & wilkins, 7th ed. 32-45.

[17] Lednicer, D., Mitscher, L. A. and Georg, G.I. 1990. "*The organic chemistry of drug synthesis*" A Wiley-Interscience Publication, New York, 4:112-118.

[18] Tawari, N. R., Seema, B. and Mariam, D. S. 2008. "Pharmacophore mapping of a series of pyrrolopyrimidines, indolopyrimidines and their congeners as multidrug-resistance-associated protein (MRP1) modulators" *Journal of Molecular Modeling* 14(10):911-921.

[19] Bahekar, S. S. and Shinde, D. B. 2003. "Synthesis of some [2-amino-6-(4-substituted aryl)-4-(4-substituted phenyl)-1,6-dihydropyrimidin-5-yl]acetic acid derivatives and evaluated for antiinflammatory activity" *Acta Pharmaceutica* 53:223.

[20] Alagarsamy, V., Meena, S., Ramseshu, K. V., Solomon, V. R., Thirumuruganb, K., Dhanabal, K. and Murugan, M. 2006. "Synthesis, analgesic, anti-inflammatory, ulcerogenic index and antibacterial activities of novel 2-methylthio-3-substituted-5,6,7,8-tetrahydrobenzo (b)thieno[2,3-d]pyrimidin-4(3H)-ones" *European Journal of Medicinal Chemistry* 41:1293–1300.

[21] Dabiri, M., Nezhad, H. A., Khavasi, H. R and Bazgir, A. 2007. "A facile three components, one-pot synthesis of pyrimido[4,5-d]pyrimidine-2,5-dione derivatives under microwave assisted conditions" *J. Heterocyclic Chem.,* 44: 1009-1011.

[22] Agrawal, S. K., Ravi Kumar, T., Agrawal, P. and Savithri, S. 2004. "Preparation of pyrimidine derivatives for the treatment of inflammation and immunological disease" PCT. Int. Appl., WO, 2004 9,560; through *Chem. Abstr,* 140, 10: 146156m.

[23] Chhabria, M. T., Bhatt, H. G., Raval, H. G. and Oza, P. M. 2007. "Synthesis and biological evaluation of some 5-ethoxycarbonyl-6-isopropylamino-4(substitutedphenyl) amino pyrimidines as potent analgesic and anti-inflammatory agents" *Bioorganic & Medicinal Chemistry Letters,* 17:1022–1024.

[24] Venkatesan, J., Pandeya, S. N. and Selvakumar, D. 2007. "Synthesis and Biological Evaluation of 4,6-Diarylsubstituted-4,5-dihydro-2-aminopyrimidines" *Indian Journal of Pharmaceutical Sciences* 69 (4): 586-589.

[25] Sridhar, S., Prasad Y. R. and Dinda, S. C. 2011. "Synthesis and anticancer activity of some novel pyrimidine derivatives" *International Journal of Pharmaceutical Sciences and Research* 2(10): 2562-2565.

[26] Xie, F., Zhao, H., Zhao, L., Lou, L. and Hu, Y. 2009. "Synthesis of novel 2,4,5-substitutedpyrimidine derivatives and evaluation for inhibition against human hepatocellular carcinoma" *Bioorg Med Chem Lett.* 19:275–278.

[27] Rahaman, S. A., Pasad, Y. R., Kumar, P. and Kumar, B. 2009. "Synthesis and anti-histaminic activity of some novel pyrimidines" *Saudi Pharmaceutical Journal* 17:3.

[28] Ramesh, B., Swamy, S. Kilakarni, S. V. and Shivalinge G. K. P. 2010. "Design, synthesis and anti-inflammatory activity of some new pyrimidine derivatives" *Journal of Global Pharma Technology* 2(1):86-88.

[29] Rao, V. P., Prasad, Y. R., Kotra, V. and Rao, B. 2010. "Design, synthesis and anticancer activity of some new pyrimidine derivatives" *International Journal of Pharmacy & Technology* 2(4):1263-1269.

[30] Chhabria M. T. and Jani, M. H. 2009. "Design, synthesis and antimycobacterial activity of some novel imidazo[1,2-c]pyrimidines" *European Journal of Medicinal Chemistry* 44:3837-3844.

[31] Jain, S. K., Chaudhari, S. K., More, N. S. 2011. "A facile synthesis of 2-amino-5-cyano-4,6-disubstitutedpyrimidines under MWI" *International Journal of Organic Chemistry* 1:47-52.

[32] Trivedi, A. R., Bhuva, V. R., Dholariya, B. H., Dodiya, D. K., Kataria, V. B., Shah, V. H. "Novel dihydropyrimidines as a potential new class of antitubercular agents" *Bioorganic & Medicinal Chemistry Letters* 2010, 20, 6100–6102.

[33] Shanmugasundaram, P., Mohanarangan, J., Raj, R. K., and Aanandhi, M. V. 2009. "Synthesis and biological evaluation of pyrido(2,3-*d*)pyrimidine-carboxylate derivatives" *Rasayan journal of chemistry* 2(2):345-349.

[34] Virsodia, V., Raghuvir, R. S., Manvar, P. D., Dholakia, C., Adlakha, P., Shah, A. and Coutinho, E. C. 2008. "Synthesis, screening for antitubercular activity and 3D-QSAR studies of substituted N-phenyl-6-methyl-2-oxo-4-phenyl-1,2,3,4-tetrahydro-pyrimidine-5-carboxamides" *European Journal of Medicinal Chemistry* 43:2103-2115.

[35] Thore, S. N., Gupta, S. V., Baheti, K., Bora, R., Dekhane, D., Chhabria, M., Shingare, M., Pawar, S. and Shishoo, C.J. 2009. "Synthesis of novel bioactive derivatives of 3-(4-chlorophenyl)-2-hydrazino-5,6,7,8-tetrahydrobenzo(b)thieno[2,3-d]pyrimidine-4(3H)-ones" *European Journal of Medicinal Chemistry* 44:4721-4725.

[36] Chitre, T. S. and Bothara, K. G. 2011. "Pyrimidinedione: Pharmacophore optimization of selective thymidine monophosphate kinase inhibitors using group QSAR studies as potential antitubercular agents" *Journal of Chemical and Pharmaceutical Research* 3(2):479-488.

[37] Read, M. L., Brændvang, M., Miranda, P. O. and Gundersen, L. L. 2010. "Synthesis and biological evaluation of pyrimidine analogs of antimycobacterial purines" *Bioorganic & Medicinal Chemistry* 18:3885-3897.

[38] Dongarwar, A. S., Wanjari, B. E., Nema, M. V., Katolkar, P. P., Turaskar, A. O. and Nimbekar, T. P. 2010. "Synthesis, anti-inflammatory and anti-microbial activity of pyrimidine-5-carboxylate derivatives" *International Journal of Biomedical and Advance Research* 02(06):188-197.

[39] Shah, T. B., Gupte, A., Patel, M. R., Chaudhari, V. S., Patel, H. and Patel, V. C. 2010. "Synthesis and *in vitro* study of biological activity of heterocyclic *N*-Mannich bases of 3,4-dihydropyrimidine-2(1*H*)-thiones" *Indian Journal of Chemistry* 49B:578-586.

[40] Johar, M., Manning, T., Kunimoto, D. Y. and Kumar, R. 2005. "Synthesis and in vitro anti-mycobacterial activity of 5-substitutedpyrimidine nucleosides" *Bioorganic & Medicinal Chemistry* 13:6663-6671.

[41] Sawant, R., Sarode, V. 2011. "Synthesis, spectral characterization and analgesic activity of 2- methylthio-1,4-dihydropyrimidines" *Iranian Journal of Pharmaceutical Research* 10(4):733-739.

[42] Kathiravan, M. K., Shishoo, C. J., Kumar, K. G., Roy, S. K., Mahadik, K. R., Kadam, S. S. and Jain, K. S. 2007. "Synthesis and antihyperlipidemic activity of some novel condensed 2-chloroalkyl-4-chloro/hydroxy-5,6-disubstituted pyrimidines" *Arzneimittel-Forschung (Drug Research)* 57(9):599-606.

[43] Jain, K. S., Phoujdar, M. S., Kathiravan, M. K., Bariwal, J. B. and Shah, A. K. 2008. "Microwave-based synthesis of novel thienopyrimidine bioisosteres of gefitinib" *Tetrahedron Letters* 49:1269-1273.

[44] Shishoo, C. J., Shirsath, V. S., Rathod, I. S. and Yande, V. D. 2000. "Design, synthesis and antihistaminic (H1) activity of some condensed 3-aminopyrimidin-4(3*H*)-ones" *European Journal of Medicinal Chemistry* 35:351-358.

[45] Rathod, I. S., Chhabria, M. T., Chaudhari, A. S. and Jani, M. H. 2006. "Design, Synthesis and Diuretic Activity of Some Novel 2,4-Diamino-6-aryl-7-arylamino pyrimido[4,5-d]pyrimidin-5(6H)-ones" *Arzneim.-Forsch. (Drug Res.)* 56(6): 377-381.

[46] Choudhari, P. B., Jadhav, S. D. and Dhavale, R. P. 2012. "Development of pyrimidine derivatives as selective Cox-2 inhibitors" *American-Eurasian Journal of Scientific Research* 7(2):69-76.

[47] Mohamed, M.S., Awad, S. M. and Ahmed, N. M. 2011. Synthesis and antimicrobial activities of new indolyl-pyrimidine derivatives" *Journal of Applied Pharmaceutical Science* 01(05):76-80.

[48] Chikhalia, K. H. and Naik, T. A. 2007. "Studies on synthesis of pyrimidine derivatives and their pharmacological evaluation" *E-Journal of Chemistry* 4(1):60-66.

[49] Fathalla, O. A., Zeid, I. F., Haiba, M. E., Soliman, A. M. and El-Serwy, W. S. 2009. "Synthesis, Antibacterial and Anticancer Evaluation of Some pyrimidine derivatives" *World Journal of Chemistry* 4 (2): 127-132.

[50] Al-Abdullah, E. S., Elsayed, E. H. and El-Emam, A. A. 2011 "Synthesis of novel 6-phenyl-2,4-disubstituted pyrimidine-5-carbonitriles as potential antimicrobial agents" *European Journal of Medicinal Chemistry* 46:4642-4647.

[51] Anupama, B., Dinda, S. C., Prasad, Y. R. and Rao, A. V. 2012. "Synthesis and antimicrobial activity of some new 2,4,6-trisubstituted pyrimidines" *International Journal of Research in Pharmacy and Chemistry* 2(2):231-236.

[52] Sawant, R. and Sarode, V. 2011. "Synthesis, spectral characterization and analgesic activity of 2-methylthio-1,4-dihydropyrimidines" *Iranian Journal of Pharmaceutical Research* 10(4):733-739.

[53] Agarwal, N., Srivastava, P., Raghuwanshi, S. K., Upadhyay, D. N., Sinha, S., Shukla, P. K. and Ram, V. J. 2002. "Chloropyrimidines as a new class of antimicrobial agents" *Bioorganic & Medicinal Chemistry* 10:869-874.

[54] Singh, N., Kshirsagar, S. S., Nimje, H. M. and Chaudhari, P. S. 2011. "Microwave assisted synthesis of 4-substituted 1,2,3,4-tetrahydropyrimidine derivatives" *International Journal of Pharmacy and Pharmaceutical Sciences* 3(1):109-111.
[55] Borisagar, M., Joshi K., Ram H., Vyas K. and Nimavat K. 2012. "A one-pot microwave irradiation synthesis of 1,2,4-triazolo[1,5-a]pyrimidines" *Acta Chimica and Pharmaceutica Indica* 2(2):101-105.
[56] Chitre, T. S., Kathiravan, M. K., Chothe, A. S., Rakholiya, V. K., Asgaonkar, K. D., Patil, S. M. and Bothara, K. G. 2011. "Synthesis and antitubercular activity of some substituted pyrimidine derivatives" *Journal of Pharmacy Research* 4(6):1882-1883.
[57] Rajesh, S. M., Kumar, R. S., Libertsen, L. A., Perumal, S., Yogeeswari, P. and Sriram, D. 2011. "A green expedient synthesis of pyridopyrimidine-2-thiones and their antitubercular activity" *Bioorganic & Medicinal Chemistry Letters* 21:3012-3016.
[58] Chitre T. S., Bothara K. G., Patil S. M., Asgaonkar K. D., Nagappa S. and Kathiravan M. K. 2011. "Design, synthesis, docking and anti-mycobacterial activity of some novel thiouracil derivatives as thymidine monophoshate kinase (TMPKmt) inhibitors" *International Journal of Research in Pharmaceutical and Biomedical Sciences* 2(2):616-623.
[59] Pore, Y., Kuchekar, B., Bhatia, M. and Ingale, K. 2009. "Quantitative structure activity relationship (QSAR) studies on 5-cyano-N1,6-disubstituted,2-thiouracil derivatives as central nervous system depressants" *Digest Journal of Nanomaterials and Biostructures* 4(2):373-382.
[60] Basavaraja, H. S., Jayadevaiah, K. V., Hussain, M. M., Kumar, M. M. J. and Padmashali, M. 2010. "Synthesis of novel piperazine and morpholine linked substituted pyrimidine derivatives as antimicrobial agents" *Journal of Pharmaceutical Sciences and Research* 2(1):5-12.
[61] Mishra, R., Mishra, B. and Moorthy, N.S. 2008. "Synthesis and antimicrobial evaluation of some 3,4-dihydropyrimdin-2-ones Derivatives" *Trends in Applied Science Research* 3(2):203-208.
[62] Abdolmohammadi, S. and Balalaie, S. 2012. "An efficient synthesis of pyrido[2,3-d]pyrimidine derivatives via one-pot three-component reaction in aqueous media" *International Journal of Organic Chemistry*, 2:7-14.
[63] Rajasekaran, S., Rao, G. K., Pai, P. N. and Ajay, A. K. 2011. Design, synthesis and biological activity of substituted dihydropyrimidine-2-(1H)-thiones. *International Journal of PharmTech Research* 3(2):626-631.
[64] Morgan, J., Rachada, H. and Keller, P. A. 2003. "Anilinopyrimidines as novel antituberculosis Agents" *Bioorganic & Medicinal Chemistry Letters* 13:1755-1757.
[65] Bhatewara, A., Jetti, S. R., Kadre, T., Paliwal, P. and Jain, S. 2012. "An efficient one-pot multi component synthesis of pyrimidine derivatives in aqueous media" *Archives of Applied Science Research* 4 (3):1274-1278.
[66] Deshmukh, M. B., Anbhule, P. V., Jadhav, S. D., Jagtap, S. S., Patil, D. R., Salunkhe, S. M. and Sankpal, S. A. 2008. "A novel and environmental friendly, one-step synthesis of 2,6-diamino-4-phenylpyrimidine-5-carbonitrile using potassium carbonate in water" *Indian Journal of Chemistry* 47B:792-795.
[67] Hamouly, W., El-Khamry, A. M. and Abbas, E. M. H. 2006. "Synthesis of new 4-aryl-isoxazolo[5,4-d]pyrimidin-6-one(thione) and 4-aryl-pyrazolo[3,4-d]-pyrimidin-6-one

derivatives of potential antihypertensive activity" *Indian Journal of Chemistry* 45B:2091-2098.

[68] Mohamed, M. S., Awad, S. M. and Sayed, A. I. 2010. "Synthesis of certain pyrimidine derivatives as antimicrobial agents and anti-Inflammatory agents" *Molecules* 15:1882-1890.

[69] Ramesh, B. and Kulakarni, S. V. 2010. "Design, synthesis and anticancer activity of some new pyrimidine derivatives" *Journal of Global Pharma Technology* 2(4):110-112.

[70] Patil, P. A., Bhole, R. P., Chikhale, R. V. and Bhusari, K. P. 2009. "Synthesis of 3,4-dihydropyrimidine-2(1H)-one derivatives using microwave for their biological screening" *International Journal of Chem Tech Research* 1(2):373-384.

[71] Raju, B. C., Gangadasu, B., Narender, P. and Rao, V. J. 2006. "Calcium chloride catalyzed three component, one-pot condensation reaction: An efficient synthesis of 3,4-dihydropyrimidin-2(1*H*)-ones" *Indian Journal of Chemistry* 45B: 1259-1263.

[72] Sandhu, J. S. and Dhruv, K. 2010. "Microwave enhanced, solvent free green protocol for the production of 3,4-dihydropyrimidine-2-(1*H*)-ones using $AlCl_3.6H_2O$ as a catalyst" *Indian Journal of Chemistry* 49B:360-363.

[73] Nongkhlaw, R. L, Phucho, I. T., Nongpiur, A. and Nongrum, R. 2010. "Synthesis of 3,4-dihydropyrimidin-2(*1H*)-ones and 3,4,5,6,7,8-hexahydroquinazolin-2(*1H*)-ones *via* three component Cyclocondensation" *Indian Journal of Chemistry* 49B:346-350.

[74] Kumar, D., Mishra B. G. and Rao, V. S. 2006. "An environmentally benign protocol for the synthesis of 3,4-dihydropyrimidin-2(1*H*)-ones using solid acid catalysts under solvent-free conditions" *Indian Journal of Chemistry* 45B:2325-2329.

[75] Mohamad, Y. 2008. "Synthesis of pyrido[2,3-*f*]quinoxaline-3-carboxylic acid as antitumor drug" *Chem. Abstr.,* 148, 1:1382.

[76] Hood, P., Cao, J. J., Noronha, G. and Pathak, V. 2007. "Preparation of biaryl 2,4-pyrimidinediamines as inhibitor of kinase for treatment of cancer" PCT Int. Appl. WO 2007 84,017. through *Chem. Abstr.* 146, 25:1135.

[77] Stephen, T. 2006. "Preparation of phenyl substituted pyrimidines as kinase inhibitors for treating cancer" PCT Int. Appl. WO 2006 53,452. through *Chem. Abstr.*, 145, 11, 1245.

[78] Pass, M. 2007. "*Pyrimidine derivatives which are useful in producing an antiproliferative effect*" PCT WO 066099.

[79] Kozina, E. S. 2007. "Preparation of 4-indol-3-yl-N-phenyl pyrimidin-2 amines as tyrosine kinase inhibitors for treatment of cancer" PCT. Int. Appl. WO 2007 149,427 through *Chem. Abstr*., 2008, 148, 5:1134.

[80] Djung, J. 2007. "Preparation of 2-anilino-4-aminoalkyl amino pyrimidines as protein kinase C-alpha (PKC-α) inhibitors" US. Pat. 2007 293,525. through *Chem. Abstr.* 148, 4:7904.

[81] Fraley, Mark E., Hoffman, W. F. and Hartman, G. D. 2003. "*Preparation of 4-(thiazolyl)-2-pyrimidinamines as tyrosine kinase inhibitor*" PCT. Int. WO 2003 011,838.

[82] Mitchell, et al. "Preparation of dihydrothieno[3,4-*d*]pyrimidine derivatives as AKt kinase inhibitors" 2008. PCT. Int. Appl., WO 2008 6,039, through *Chem. Abstr.* 148, 7:1329.

[83] Nasser, S. M., Ismail, E. M. H. and Ali, D. A. I. 2016. "Pyrazolo[3,4-d]pyrimidine based scaffold derivatives targeting kinases as anticancer agents" *Future Journal of Pharmaceutical Sciences* 2: 20-30.

[84] Valarmathy, J., Joshua, L. S. and Senthilkumar, K. L. 2011. "Synthesis, characterization, antitubercular, anticonvulsant and anti-inflammatory activities of some novel 2-aminopyrimidine derivative" *International Journal of Chemistry* 2(2): 61-68.

[85] Pandeya, S. N., Srivastava, S. K., Rai, P. and Singh, A. 2012. "Synthesis and anticonvulsant activity of pyrimidine hydrazones" *Journal of Scientific Research in Pharmacy* 1(3):92-97.

[86] Sindhu, T., Mamatha, A., Reddy, K. S. and Venkateshwarlu, K. 2013. "Antidepressant activity of pyrimidine derivatives in mice" *Hygeia-Journal for Drugs and Medicines* 5(1):129-134.

[87] Gupta, J. K., Chaudhary, A., Dudhe, R. and Varuna, K. 2010. "A review on the synthesis and therapeutic potential of Pyrimidine derivatives" *International Journal of Pharmaceutical Sciences and Research* 1(5):34-49.

[88] Selvam, T. P., James, C. R., Dniandev, P. V. and Valzita, S. K. 2012. "A mini review of pyrimidine and fused pyrimidine marketed drugs" *Research in Pharmacy,* 2(4):01-09.

[89] Loukas, A. and Hotez, P. J. 2006. "*Chemotherapy of helminth infections: Goodman & Gilman's the pharmacological basis of therapeutics*" New York: McGraw-Hill, 11th ed.; 1073-1093.

[90] Kadatz, R. A. 1983. "Mechanism of action of Dipyridamole" *Thrombosis Research. Supplement* 4:39-46.

[91] Suzuki, Y., Yamaguchi, K., Shimada, S., Kitamura, Y. and Ohnishi, H. 1982. "Antithrombotic activity and the mechanism of action of trapidil (Rocornal)" *Prostaglandins, Leukotrienes and Medicine* 9(6):685-95.

[92] Mishra, R. and Tomar, I. 2011. "Pyrimidine: The molecule of diverse biological and medicinal importance" *International Journal of Pharmaceutical Sciences and Research* 2(4):758-771.

[93] Seth, A., Misra, A. and Umrigar, D. 2004. "Topical liposomal gel of idoxuridine for the treatment of herpes simplex: pharmaceutical and clinical implications" *Pharmaceutical Development and Technology* 9(3):277-289.

[94] Dua, R., Shrivastava, S., Sonwane, S. K. and Srivastava, S. K. 2011. "Pharmacological significance of synthetic heterocycles scaffold: a review" *Advanced Biomedical Research* 5(3):120-144.

[95] Brown, D. J. and Harper, J. S. 1965 "Neurotropic agent that promotes neurite outgrowth" *Journal of the Chemical Society* 5542-5551.

[96] Sriram, D. and Yogeeswari P. 2012. "Ultra short acting barbiturate-thiopental Sodium" *Medicinal Chemistry*. Pearson, 2nd ed. 99-100.

[97] Jain, K. S., Chitre, T. S., Miniyar, P. B., Kathiravan, M. K., Bendre, V. S., Veer, V. S., Shahane S. R. and Shishoo, C. J. 2006. "Biological and medicinal significance of Pyrimidines" *Current Science* 90(6):793-801.

[98] Mahadik, K. and *Bothra,* K. H. 1997. *Principles of Medicinal Chemistry.* Nirali Prakashan, 5th *ed.112-17.*

[99] Dudhe R., Sharma, P. K., Verma, P. and Chaudhary, A. 2011. "Pyrimidine as anticancer agent: A Review" *Journal of Advanced Scientific Research* 2(3):10-17.

[100] Vachala, S. D. 2012. "Fused pyrimidines: The heterocycle of diverse biological and pharmacological significance" *Der Pharma Chemica* 4(1):255-265.

[101] Kochkin, D. A. 1977. "Orotic acid and its derivatives: Preparation, properties, and uses" *Pharmaceutical Chemistry Journal* 11(8):1155-1160.

[102] Meisheri, K. D., Johnson, G. A. and Puddington, L. 1993. "Enzymatic and non-enzymatic sulfation mechanisms in the biological actions of minoxidil" *Biochemical Pharmacology* **45**:271-273.

[103] Neale, R., Fallon, S., Gerhardt, S. and Liebman, J. M. 1981. "Acute dyskinesias in monkeys elicited by halopemide, mezilamine and the antidyskinetic drugs, oxiperomide and tiapride" *Psychopharmacology,* 75(3):254-257.

[104] Goldberg, H. L. and Finnerty, R. J. 1979. "The comparative efficacy of buspirone and diazepam in the treatment of anxiety" *The American Journal of Psychiatry*, 136(9):1184-1187.

[105] Ruiu, S., Marchese, G., Saba, P. L., Gessa, G. L. and Pani, L. 2000. "The 5-HT_2 antagonist ritanserin blocks dopamine reuptake in the rat frontal cortex" *Molecular Psyciiatry* 5(6):673-677.

[106] Hiromi, K. and Minami, A. 1975. "Pipemidic Acid: a new antibacterial agent active against pseudomonas aeruginosa: in-vitro properties" *Antimicrobial Agents and Chemotherapy* 8:132-38.

[107] Vida, J. and Yevich, J. 2003. "*Sedative hypnotics: Burger's Medicinal Chemistry and Drug Discover*" John Wiley, New Jersy, 6th ed., 6:203-205.

[108] Schaffer, L. C., Schaffer, C. B. and Caretto, J. 1999. "The use of primidone in the treatment of refractory bipolar disorder" *Annals of Clinical Psychiatry* 11(2):61-66.

[109] Makarchikov, A. F., Lakaye, B., Gulyai, I. E., Czerniecki, J., Coumans, B., Wins, P., Grisar, T. and Bettendorff, L. 2003. "Thiamine triphosphate and thiamine triphosphatase activities: from bacteria to mammals" *Cellular and Molecular Life Sciences* 60(7):1477-1488.

[110] Lenzen, S. and Panten, U. 1988. "Alloxan: history and mechanism of action" *Diabetologia* 31:337-342.

[111] Gierski, F. and Peretti, C. 2007. "A. Effects of the dopamine agonist piribedil on prefrontal temporal cortical network function in normal aging as assessed by verbal fluency" *Progress in neuro-psychopharmacology and biological psychiatry* 31(1):262-268.

[112] Gupta, J. K., Chaudhary, A., Dudhe, R. and Varuna, K. 2010. "A review on the synthesis and therapeutic potential of Pyrimidine derivatives" *International Journal of Pharmaceutical Sciences and Research* 1(5):34-49.

[113] Pandeya, S. N. 2003. "*Pyrimethamine as antimalarial drug: Text Book of Medicinal Chemistry*" S. G. Publisher, 3rd ed. 622-24.

[114] App, E. M., Tomkiewicz, R. P., Hahn, H. L., Engler, H., Vergin, H. and King, M. 1997. "The effect of tasuldine, a bronchosecretolytic agent, on mucus rheology and clearability and the interaction with acetylcholine in ferrets" *Pulmonary Pharmacology and Therapeutics* 10(5):271-276.

[115] Kemp, J. P., Bernstein, I. L. and Bierman, C. W. 1992. "Pemirolast, a new oral nonbronchodilator drug for chronic asthma" *Annals of Allergy* 68(6):488-491.

[116] Nagasaka, A. and Hidaka, H. 1976. "Effect of antithyroid agent's 6-propyl-2-thiouracil and l-ethyl-2-mercaptoimidazole on human thyroid iodide peroxidase" *Journal of Clinical Endocrinology and Metabolism* 43:152.

[117] Verma, A., Sahu, L., Chaudhary, N., Dutta, T., Dewangan, D. and Tripathi, D. K. 2012. "A Review: Pyrimidine their chemistry and pharmacological potentials" *Asian Journal of Biochemical and Pharmaceutical Research* 1(2): 1-15.

[118] Lagoja, I. M. 2005. "Pyrimdine as constitutent of natural biologically acticve compounds" *Chemistry & Biodiversity* 2:1-50.

[119] Sharma, V., Chitranshi, N. and Agarwal, A. K. 2014. "Significance and biological importance of pyrimidine in the microbial world" *International Journal of Medicinal Chemistry*1-34. dx.doi.org/10.1155/2014/202784.

[120] Kumar, S., Deep, A. and Narasimhan, B. 2015. "Pyrimidine derivatives as potential agents acting on central nervous system" *Central Nervous System Agents in Medicinal Chemistry* 15:5-10.

In: Organic and Medicinal Chemistry, Volume 1
Editor: Bimal Krishna Banik

ISBN: 978-1-53614-453-6

Chapter 5

MEDICINAL AND PHARMACOLOGICAL IMPORTANCE OF THIAZOLIDINONE DERIVATIVES

Jnyanaranjan Panda[1,*]***, Biswa Mohan Sahoo***[2,†]
and Bimal K. Banik[3,‡]

[1]Roland Institute of Pharmaceutical Sciences, Berhampur, Odisha, India
[2]Department of Pharmacy, Vikas Group Of Institutions, Nunna, Vijayawada, Andhra Pradesh, India
[3]Community Health System of South Texas, Edinburg, TX, US

ABSTRACT

An important part of medicinal chemistry has been to set up a relationship between chemical and biological activity. It has been established that half of the therapeutic agents consist of heterocyclic compounds. The heterocyclic ring comprises the core of the active moiety or pharmacophore. An especially big attention is given to sulphur and nitrogen containing heterocyclic compounds, as they possess a broad spectrum of biological activity, and are used in various fields of pharmacy and medicine. The present chapter is designed with the main aim to give an importance of 4-thiazolidinone in the field of medicine. The introductory paragraph outlines the importance of organic chemistry in medicinal field followed by an overview of synthetic potential of 4-thiazolidinone scaffold. The other part of this chapter deals with the review of the work reported in various literature reports highlighting the pharmacological potential of 4-thiazolidinone nucleus.

Keywords: 4-thiazolidinone, broad spectrum, pharmacological potential, scaffold

* Corresponding Author Email: jrpanda77@gmail.com.
†Email: drbiswamohansahoo@gmail.com.
‡Email: bimalbanik10@gmail.com.

ABBREVIATIONS

ATCC	American Type Culture Collection
&	And
CNS	Central Nervous System
EC_{50}	Half Maximal Effective Concentration
FTIR	Fourier Transformation Infrared Spectroscopy
Gm	Gram
Hr	Hour
IR	Infra Red
IC_{50}	Half Maximal Inhibitory Concentration
i.p	Intra Peritoneal
MIC	Minimum Inhibition Concentration
MES test	Maximal Electroshock Seizure Test
μg	Micro Gram
MTCC	Microbial Type Culture Collection
Me	Methyl
M.P.	Melting Point
NMR	Nuclear Magnetic Resonance
%	Percentage
TLC	Thin Layer Chromatography
QSAR	Quantitative Structure–activity Relationship

1. INTRODUCTION

Heterocyclic chemistry has become a separate field of chemistry with long history and due to present society and future prospects. The earliest compounds known to mankind were of heterocyclic origin. Our life is totally dependent on the heterocyclic compounds. For example, birth depends on purine/pyrimidine bases and life depends on carbohydrates. Heterocycles act as medicines and cure our diseases.

Heterocyclic chemistry finds applications in biochemistry, polymers and material sciences as well. Significant number of compounds synthesized in the industrial sectors is heterocyclic in nature. A large number of synthetic heterocyclic compounds are predominant among all types of pharmaceuticals, agrochemicals and veterinary products. Some of the synthetic hetrocycles are used in photography and as rocket propellants. Heterocyclic chemistry is vastly expanding, because of the enormous amount of research work being done in this area. Among the heterocyclic compounds, thiazolidinones and their derivatives plays important role in the development of new drug molecules [1, 2].

2. 4-THIAZOLIDINONE

Thiazolidinones belong to an important group of heterocyclic compounds containing sulfur and nitrogen in a five member ring. Thiazolidinones are saturated form of thiazole, that have an atom of sulfur at position 1, an atom of nitrogen at position 3 and a carbonyl group at position 2, 4, or 5. The thiazole ring has been extensively studied and it forms a part of vitamin-B, Penicillins and the antibacterial thiazoles. Reduced thiazole serves in the study of polypeptide and proteins and occurs as structural units in compound of biological importance. Tetrahydro derivative of thiazole is known as thiazolidine and the oxo derivative of thiazolidine is known as thiazolidinone. For example, if the oxo group is at 4th position, it is called 4-thiazolidinone. Substituents in the 2-, 3-, and 5-position may be varied, but the greatest difference in the structure and properties is exerted by the group attached to the carbon atom in the 2- position. 4-thiazolidinones gained attraction by researchers due to their broad spectrum of biological and pharmacological activities [3].

The most reactive positions of 4-thiazolidone molecules are 2, 3, and 5, which determine the major directions of heterocyclic modifications of these types of molecules. The methylene group at 5-position of 4-thiazolidones is active enough; therefore, most of modification reactions of the 4-thiazolidone ring are carried out at this position. The carbonyl group of 4-thiazolidinone is highly unreactive. But in few cases, 4-thiazolidinone on reaction with Lawesson's reagent gives corresponding 4-thione derivatives. Substituents in the 2-, 3-, and 5-positions may be varied, but the greatest difference in structure and properties is exerted by the group attached to the carbon atom in the 2-position [4].

The 3-unsubstituted 4-thiazolidinones are usually solids, often melt with decomposition, but the attachment of an alkyl group to the nitrogen lowers the melting point. The 4-thiazolidinones that do not contain aryl or higher alkyl substituents are somewhat soluble in water [5].

The derivatives of 4-thiazolidinone nucleus have occupied a unique place in the field of medicinal chemistry due to wide range of biological activities like antibacterial, antifungal, antitubercular, anticancer, anticonvulsant, analgesic, anti-inflammatory, anthelmintic, antidiabetic, hypnotic, hypolipidemic, cardiovascular and anticonvulsant activities. They have interesting activity profiles mainly as cox-1 inhibitors, inhibitors of bacterial enzymes, non nucleoside inhibitors of HIVRT and antihistaminic agents. 4-Thiazolidinones are formed by the attack of sulfur nucleophile on imine carbon followed by intramolecular cyclization with elimination of water. According to Andres et al., 4-thiazolidinones may be considered as phosphate bioisosteres and therefore, they inhibit the bacterial enzyme MurB which is involved in the biosynthesis of peptidoglycan layer of the cell wall. In addition, some thiazolidinones were recently reported as novel inhibitors of mycobacterial rhamnose synthetic enzymes. This new approach is believed to be selective as rhamnose which is not found in humans. But, it is essential for mycobacterial cell wall synthesis [6].

Thiazolidinone is (I) the derivative of thiazolidine (II) with a carbonyl group at position 4. There may be substitution at 2, 3 and 5 position but group attached to the carbon atom in the 2-position of III brings about differences in structures and properties [7].

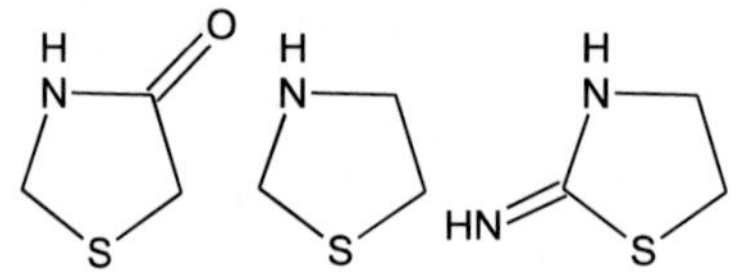

I II III

2.1. Synthetic Aspects of 4-Thiazolidinones

Synthesis of 4-thiazolidinones was reported either by cyclization of acyclic compounds or by interconversion among appropriately substituted-thiazolidinone derivatives. Several methods for synthesis was available in literature which involves conventional one-pot, two-pot and microwave method as well as by combinatorial synthetic methods. The main synthetic route to 1,3-thiazolidin-4-ones involved three components reactions with an amine, a carbonyl compound, and a mercapto-acid. The reaction took place by formation of an imine (the nitrogen of amine attacks the carbonyl of aldehyde or ketone), which undergoes attack by sulfur nucleophile in the media, followed by intramolecular cyclization through elimination of water (Scheme 1) [8-10].

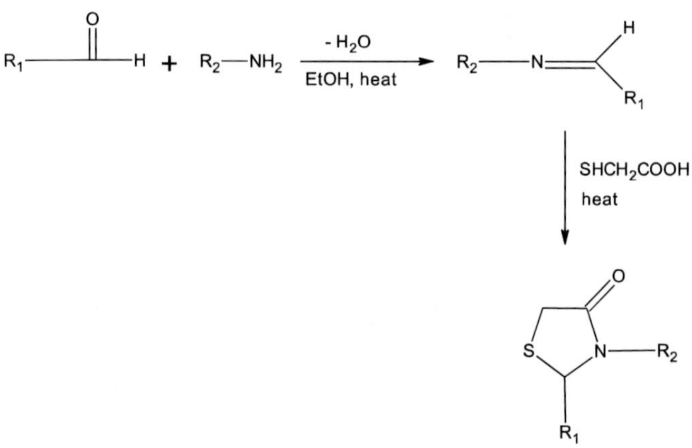

Scheme 1.

Quinazolinylazomethines (1) on treatment with mercaptoacetic acid in the presence of silica chloride produced 4-thiazolidinones (2) through intra-molecular cyclocondensation under solvent-free condition (Scheme 2) [11].

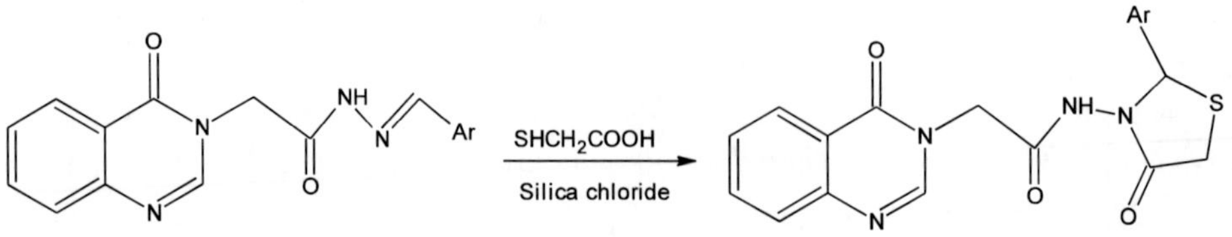

(1) (2)

Scheme 2.

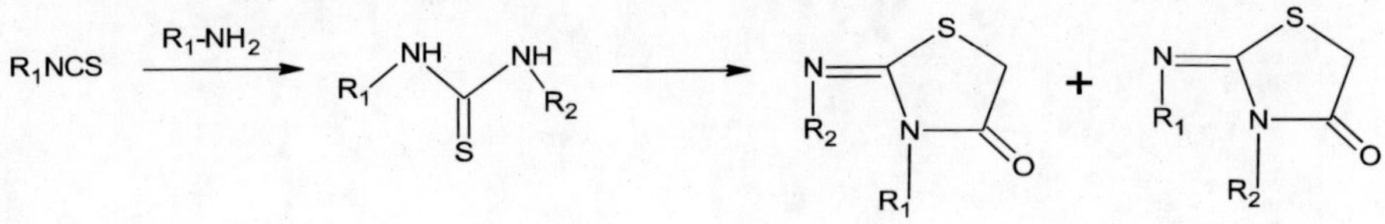

Scheme 3.

The reaction of aryl or alkyl isothiocyanate (3) with a primary amine provided the corresponding thiourea derivative (4), which was directly cyclized by treatment with halo acetic acid to the corresponding two isomeric 2-imino-thiazolidin-4-ones of the general structures 5 and 6 (Scheme 3) [12].

The dithiocarbamates formed by the reaction of primary amine with carbon disulfide in the presence of base on reaction with haloalkanoic acid in the presence of $NaHCO_3$ produced substituted 2-thiono-4-thiazolidinones as presented in the Scheme 4 [13].

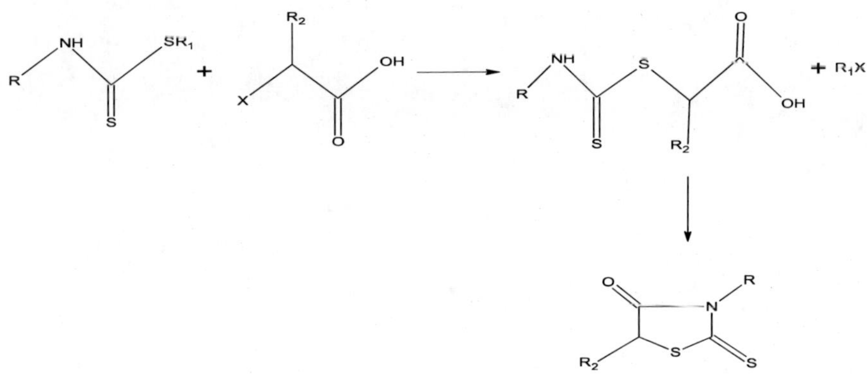

Scheme 4.

One-pot three component synthesis of 2-amino-4-thiazolidinone derivatives by reacting aldehyde, thiourea and chloroform (Scheme 5) was reported. Various iminothiazolidinones were developed by using different reagents and different reaction conditions [14].

Desai et al. carried out the microwave-assisted synthesis of thiazolidinone from the Schiff's bases (Scheme 6) by using thiolactic acid. The products were synthesized by conventional and microwave method and the yield were compared with each other. They concluded that the yield with the microwave-mediated synthesis was better than the conventional method [15].

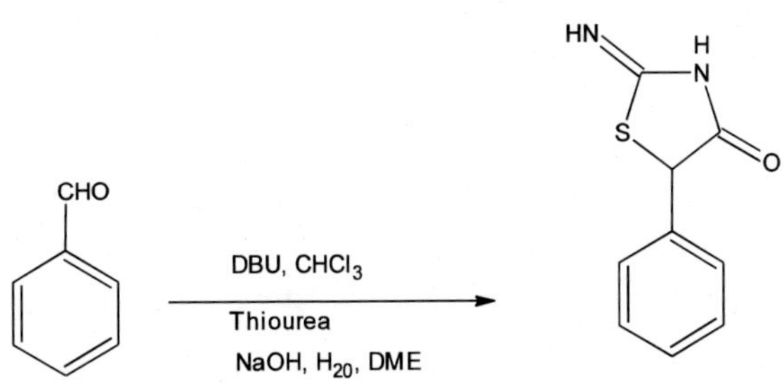

Scheme 5.

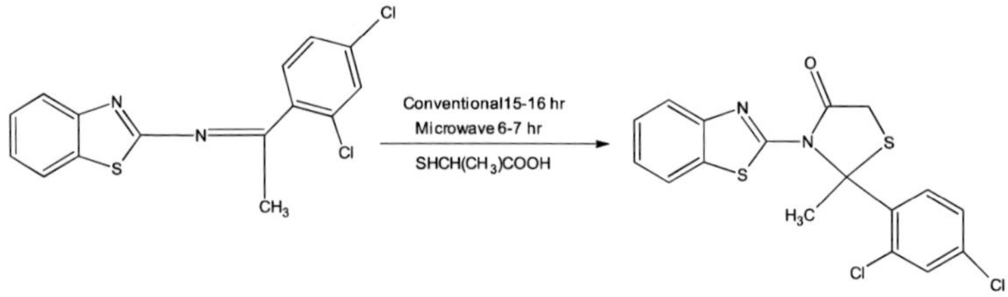

Scheme 6.

Synthesis of 4-thiazolidinone and its derivatives was performed by condensation of Schiff's bases with mercapto acid (Scheme 7). The key functional group required for the reaction was an amine, a carbonyl compound and mercapto acid. The reaction required about 12 hr by classical method, while the same reaction using microwave irradiation needed only 6-8 minutes [16].

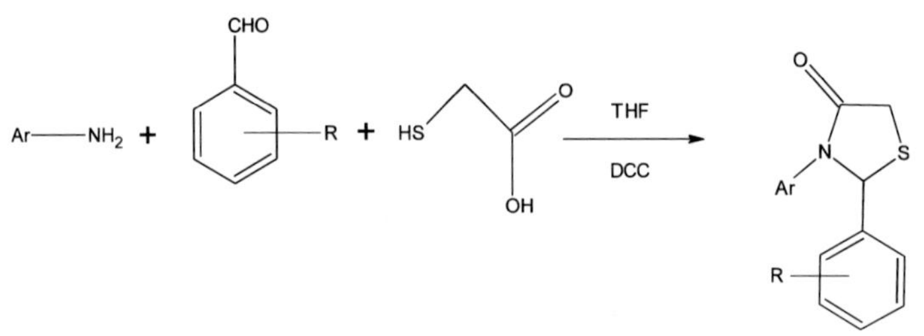

Scheme 7.

2.2. Synthetic Potential of 4-Thiazolidinones

Ralitoline, a thiazolidinone derivative chemically distinct from known antiepileptic drugs, possesses remarkable anticonvulsant properties as demonstrated in various animal models of epilepsy. The efficacy of this compound seemed to be comparable or even better than that of conventional antiepileptics (Scheme 8).

Another drug containing 4-thiazolidinone nucleus is etozoline. The active metabolite of etozoline is ozolinone. The effects of ozolinone (3-methyl-4-oxo-5-piperidino-thiazolidine-2-ylidene) on renal tubular and glomerular functions were studied in rats using clearance and micropuncture techniques. Ozolinone markedly increased urine volume, urinary sodium excretion and urinary potassium excretion at a minor extent. Renal blood flow, as measured by an electromagnetic flowmeter, increased considerably during the administration of the drug. Concerning the tubular site of renal action there is a striking similarity between ozolinone and the strong acting diuretic furosemide, although the chemical structures of these drugs are quite different. Differences between the two drugs exist concerning the end proximal tubular chloride concentration, which was decreased by furosemide but was not affected by ozolinone. These observations were due to different effects of the drugs on proximal tubular bicarbonate reabsorption. Ozolinone had similar renal actions to those of furosemide. Like furosemide, ozolinone increased renal blood flow, slightly decreased glomerular filtration rate, depressed tubular chloride reabsorption more than sodium reabsorption, increased potassium excretion, lowered the pH of urine, decreased urinary osmolarity towards isotonicity and depressed tubular PAH secretion. As regards to the effective doses and the maximal changes of tubular sodium excretion, ozolinone was somewhat less potent than furosemide (Scheme 8) [17].

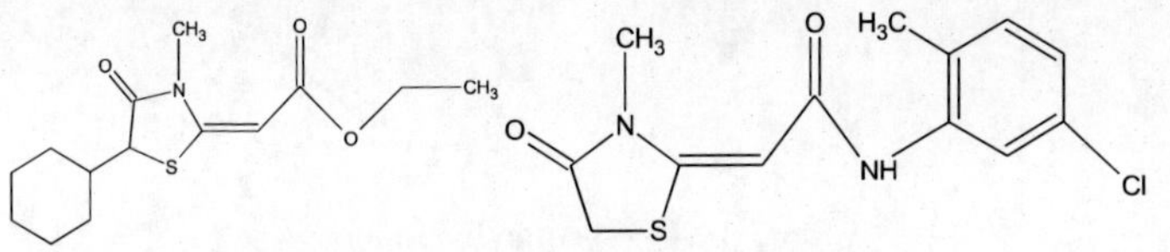

Scheme 8.

3. Biological Potential of 4-Thiazolidinone Derivatives

4-Thiazolidinone a mimic of bioactive β-lactam with additional sulfur atom is most privileged scaffolds among vast array of thiazolidine heterocycles. It has been widely recognized as wonder class in the field of medicinal chemistry and pharmaceutical use due its ability to accommodate a wide variety of bioactive motifs in its unique structural framework. Several 4-thiazolidinone analogs have been reported to show considerable biological activities as antibacterial, antifungal, anti-inflammatory, anticonvulsant, antitubercular, anthelmintic, cardiovascular, anticancer, antiviral and antihistaminic. The details of some important activities are given below.

3.1. Antimicrobial Activities

The dramatically rising prevalence of multi-drug resistant microbial infections in the past few decades has become a serious health problem. The search for new antimicrobial agents will consequently remain as an important and challenging task for the chemists. Thiazolidinones substituted at C-2 and N-3 positions, possess varied degrees of inhibition against bacteria and fungi. Some 4-thiazolidinone derivatives were prepared from ethyl(5-methyl-1-H-imidazole-4-carboxylate). All the synthesized compounds were evaluated against variety of pathogens for their antibacterial and antifungal activities. The results showed that the tested compounds possessed weak antibacterial and antifungal activities compared to standard drugs chloramphenicol and rifampicin for antibacterial activity and ketoconazole for antifungal activity (Scheme 9) [18].

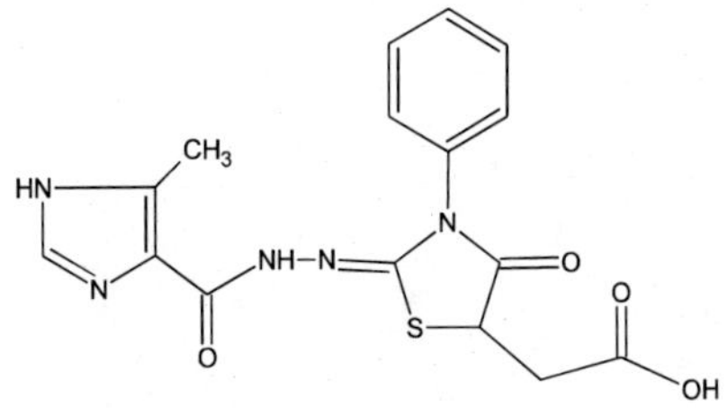

Scheme 9.

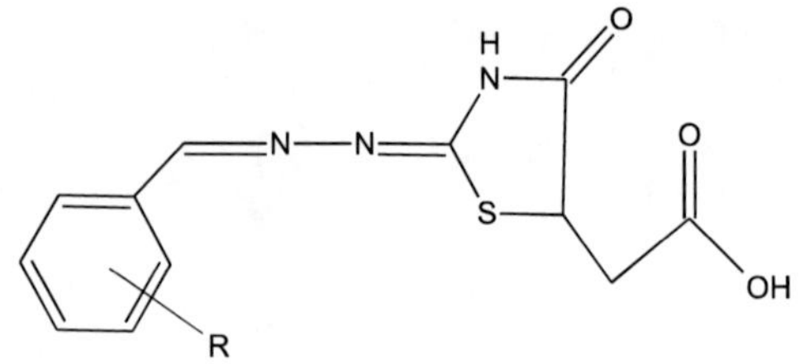

Scheme 10.

Some novel bioactive venlafaxine analogs such as 2,3-disubstituted-1,3-thiazolidin-4-ones were carried out by Kavitha et al. All the synthesized compounds were screened for their antimicrobials efficacy as by the disk diffusion and micro dilution method against pathogenic strains such as *B. subtilis, E. coli, P. fluorescens, X. campestris, X. oryzae, A. niger, A. flavus, F. oxysporum, T. species* and *F. monaliforme* species. Their results revealed that the presence of two fluorine atoms at 2nd and 6th positions is essential for the significant inhibitory activity (Scheme 10) [19].

Synthetic analogs of 2-phenyl-3-(4,6-diarylpyrimidin-2-yl)thiazolidin-4-ones were prepared by Gopalakrishnan et al. [20] and these compounds were investigated for their antibacterial activity against *Staphylococcus aureus*, *b-hemolytic Streptococcus*, *Vibrio cholera*, *Salmonella typhi*, *Shigellafelxneri*, *Escherichia coli*, *Klebsiella pneumonia*, and *Pseudomonas aeruginosa*. Results revealed that p-(OCH_3) and p-(CH_3) groups at phenyl ring attached to the pyrimidine ring exerted strong antibacterial activity against all the tested bacterial strains. On the other hand compounds with electron withdrawing groups such as para-chloro and para-fluoro functional group at phenyl ring attached to pyrimidine ring did not improve antibacterial activity (Scheme 11).

Substituted thiazolyl-thiazolidinylbenzothiazoles were studied by Singh et al. [21]. Their results showed that none of the compounds having 2-substituted-4-thiazolidinone ring showed any antibacterial activity but compounds were potent for insecticidal activity. Electron withdrawing group at phenyl ring such as p-OCH_3 enhanced its insecticidal activity. Compound containing the azetidinone moiety instead of thiazolidinone displayed antibacterial activity against *S. aureus* and *E. coli* (Scheme 12).

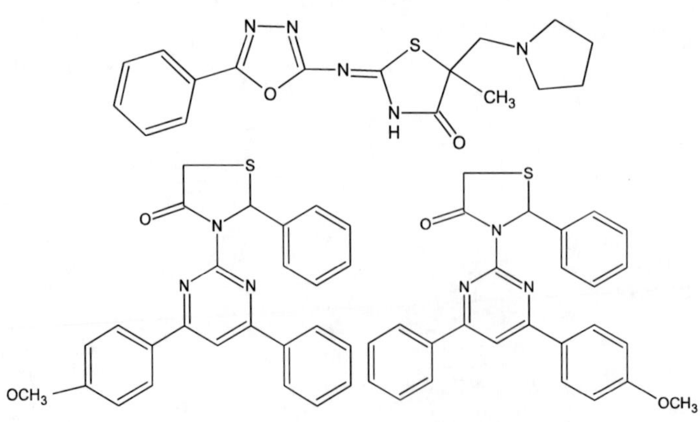

Scheme 11.

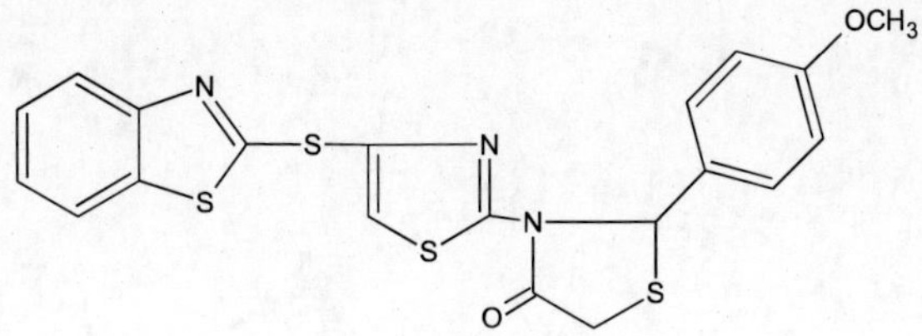

Scheme 12.

A series of 2-Imino-3-[carboxamido-p-hydroxyphenyl]-5-arylidene-4-thiazolidinone, unsubstituted or carrying hydroxy, nitro and chloro groups on the benzene ring, were synthesized and assayed *in vitro* for their antimicrobial activity against Gram positive and Gram negative bacteria and fungi by the cup-plate method by Sriram et al. [22]. The 5-arylidene derivatives showed an antibacterial efficacy considerably greater than that of the parent 2-imino-3-(carboxamido p-hydroxyphenyl)-thiazolidine-4-one, suggesting that the substituted and unsubstituted 5-arylidene moiety plays an important role in enhancing the antimicrobial properties of this class of compounds (Scheme 13).

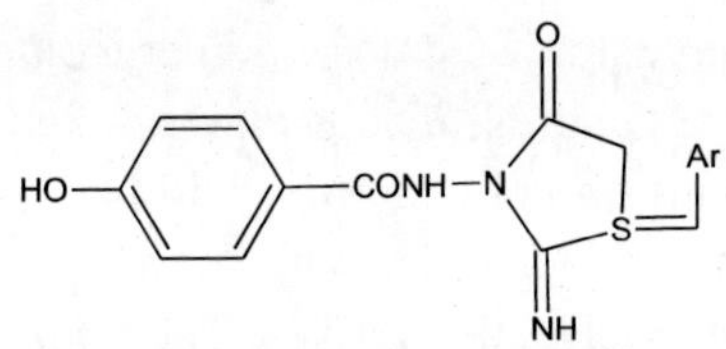

Ar = C_6H_5, 2-ClC_6H_4, 4-ClC_6H_4, 3-$NO_2C_6H_4$, 4-$NO_2C_6H_4$, 4-OHC_6H_4

Scheme 13.

Augusta et al. [23] synthesized 3-allyl-2-sulfanylamido-4-thiazolidinones after substitution of amino group of sulfanylamide with allylisothiocyanate and cyclization into 4-thiazolidinones. The synthesized compounds were tested for their antibacterial and antifungal activity (MIC) *in vitro* against microorganisms: *S. aureus, E. faecalis, E. coli, P. aeruginosa, K. pneumoniae, P. mirabilis, B. subtilis, B. cereus* and *C. albicans* taking sulfadimidine, sulfathiazole, sulfanilamide and sulfamethizole as standard drugs. Synthesized compounds demonstrated selective activity against *B. cereus*. New 4-thiazolidinones containing sulfanylamide were inactive against tested bacteria and fungi (Scheme 14).

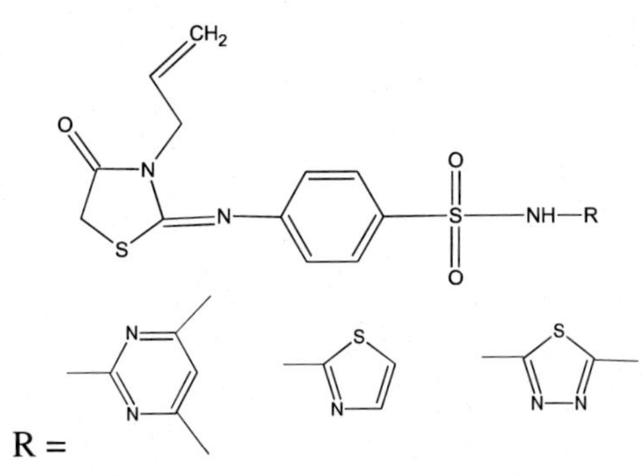

Scheme 14.

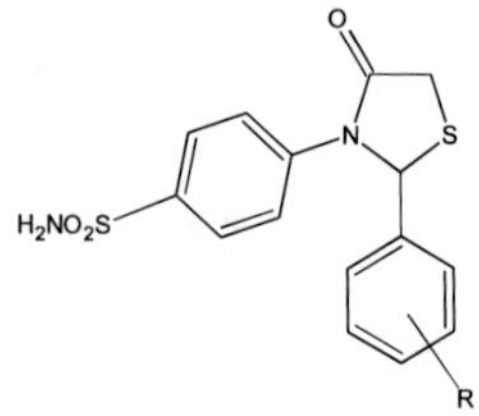

R = H, OH, Cl, NO_2, OCH_3

Scheme 15.

Some thiazolidine-4-one derivatives were prepared by Tribhuvan et al. [24] from Schiff's base. The compounds were screened against *Staphylococcus aureus*, *Bacillus pumilis*, *Bacillus subtilis*, *Escherichisa coli* and *Pseudomonas aeruginosa* by disk-diffusion method on nutrient agar media and Ampicillin was used as standard drug for the comparison at the concentration 50μg/mL and 100μg/mL against Gram-positive and Gram-negative bacteria used for the study. Some of the synthesized compounds displayed good activity (Scheme 15).

Condensation of 4-amino-2,3-dimethyl-1-phenyl-3-pyrazoline-5-one with different aromatic and heterocyclic aldehydes in dry toluene produced Schiff bases, which on reaction with mercaptoacetic acid and mercaptopropionic acid in dry toluene afforded the corresponding 2,3-disubstituted-4-thiazolidinones and 2,3-disubstituted-5-methyl-4-thiaz-olidinones. Further 2,3-disubstituted-4-thiazolidinones on condensation with 4-methoxybenzaldehyde in alcohol in the presence of sodiummethoxide gave its corresponding 2,3-disubstituted-5-[(4'-methoxy)benzylidene]-4-thiazolidinones derivatives. The prepared compounds were screened for their antimicrobial activity by Solankee et al. [25] (Scheme 16).

Novel 2-arylthiazolidin-4-one derivatives were synthesized in good-to-excellent yields (70-96%) by reaction of aromatic or aliphatic primary amines, aromatic aldehydes and thioglycolic acid in polypropylene glycol at 110°C by Davinder et al. [26]. The *in vitro* antimicrobial activity of the synthesized 2-arylthiazolidin-4-ones was investigated against a panel of six pathogenic fungal strains, a Gram-positive and three Gram-negative bacteria. Their results revealed that the compounds bearing 3-(4-(1H-imidazolylmethyl)phenyl)-substituent displayed significant antibacterial efficacy specifically against *Klebsiella pneumonia* (Scheme 17).

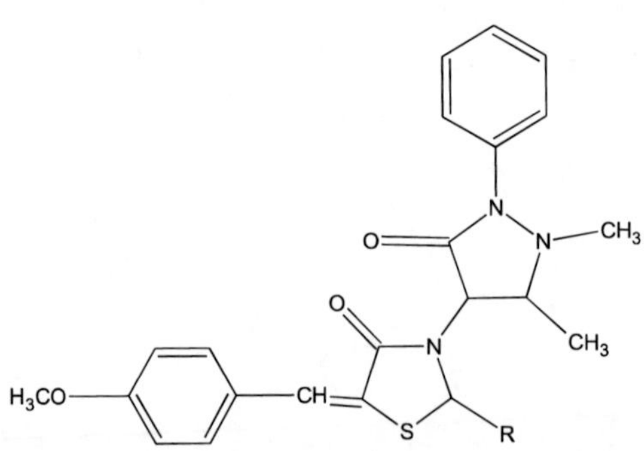

R = 2-Ethoxyphenyl, 2,3-Dichlorophenyl, 4-Hydroxyphenyl, 3-Ethoxy-4-hydroxyphenyl, Styryl, Pyridin-3-yl, 6-Methohynaphthalen-2-yl, 2-Ethoxyphenyl

Scheme 16.

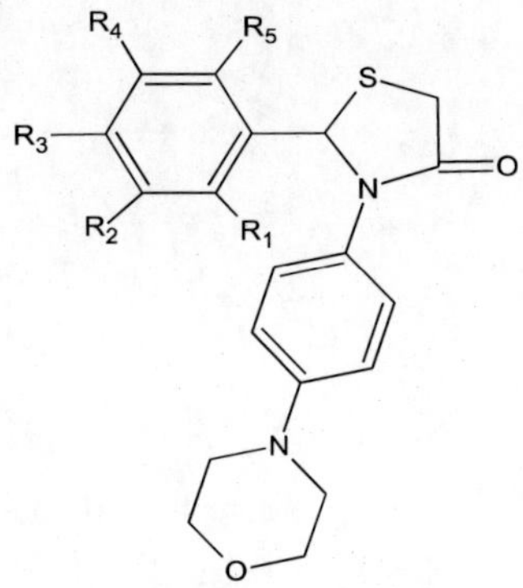

R_1 = H, CF_3; R_2 = H, F, Cl, Br; R_3 = H, F, Cl, Br, CF_3; R_4 = H, F; R_5 = H, F

Scheme 17.

2-(5-benzoyl-1H-benzo[d][1,2,3]imidazol-1-yl)acetohydrazide on facile condensation with aromatic aldehydes produced the corresponding 2-(5-benzoyl-1H-benzo[d]-[1,2,3]imidazol-1-yl)-N'-alkylideneacetohydrazide in good yield. Subsequent cyclo-condensation with thioglycolic acid yielded 3-(2-(5-benzoyl-1H-benzo[d][1,2,3]imidazol-1-yl)acetyl)-2-alkylthiazolidin-4-one [27]. The antibacterial activities of all the compounds were studied against gram-positive bacteria at a concentration of 50μg/ml by agar cup plate method. The fungicidal activity of all the compounds was studied at 1000 ppm concentration *in vitro*. Some of the tested compounds demonstrated significant antibacterial and antifungal activities (Scheme 18).

A series of novel thiazolidinones were synthesized by reaction of various Schiff bases of coumarin with thioglycolic acid by Divyesh et al. [28]. The reaction of 4-hydroxy coumarin with $POCl_3$ yielded 4-chloro coumarin 2 and 4-chloro-3, 4', 3', 4"-tercoumarin. Compound 2 was reacted with phenylenediamine to yield 4-[(4-aminophenyl)amino]-2H-chromen-2-one. Various Schiff bases of coumarin were synthesized by condensation of 4-[(4-aminophenyl)amino]-2H-chromen-2-one with different aldehydes. The thiazolidinone derivatives were evaluated for their antibacterial and antifungal activity by broth dilution method. Most of the compounds were more active against *E. coli*, *S. aureus* and *B. subtilis*. Some of the compounds were found equipotent to ampicillin and found less active than other standard drugs. Most of the compounds demonstrated antifungal activity for *C. albicans* similar to that of Griseofulvin, but, found less active than other fungal species (*A. niger*) (Scheme 19).

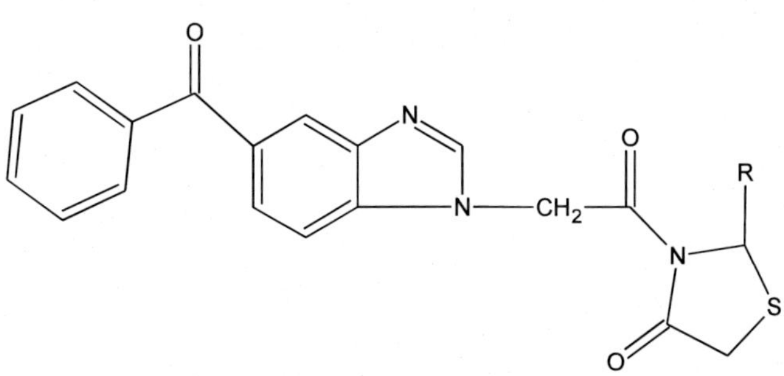

R = C_6H_5, 4-CH_3-C_6H_4, 4-OH-C_6H_4, 4-OCH_3-C_6H_4

Scheme 18.

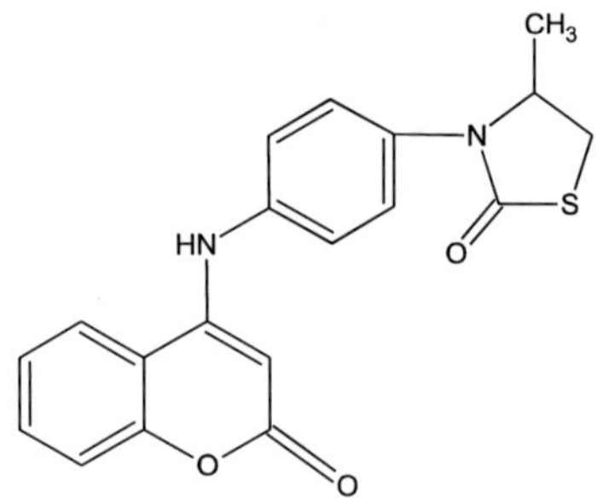

Ar = phenyl, 4-nitro phenyl, 3-nitro phenyl, 3,4-dimethoxy phenyl, 4-chloro phenyl, 4-methyl phenyl, Phenyl-2-carboxaldehyde, naphthyl, 2-chloro quinonyl

Scheme 19.

Some biologically active [1,2,4] triazolo [3,4-b][1,3,4] thiadiazole-2-aryl-thiazolidinone-4-ones were synthesized by Parmar et al. [29]. The structures of the synthesized compounds were established on the basis of IR, PMR, CMR and elemental analysis. The compounds were evaluated for antibacterial activity by Cup-plate agar diffusion method. Some of the tested compounds were shown significant activities against *B. subtilis*, *S. aureus*, *P. aeruginosa* and *E. coli* (Scheme 20).

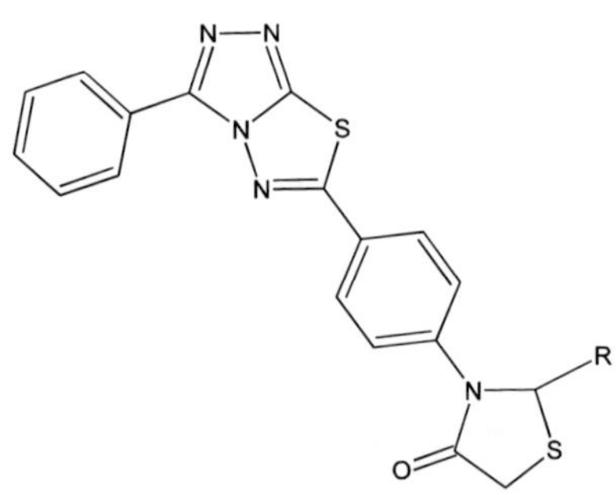

R = C_6H_6, 2-OH-C_6H_5, 4-OH-C_6H_5, 4-CH_3-C_6H_5, 4-OCH_3-C_6H_5, 4-OH-3-OCH_3-C_6H_5

Scheme 20.

A series of 2-[5-(arylidene)-2-imino-4-oxo-thiazolidin-3-yl]benzothiazole-6-carboxylic acid derivatives were prepared by Ameya et al. [30]. The synthesized compounds were screened for their antibacterial as well as antifungal activity. All the tested compounds showed slight to moderate activity against the selected microorganisms (Scheme 21).

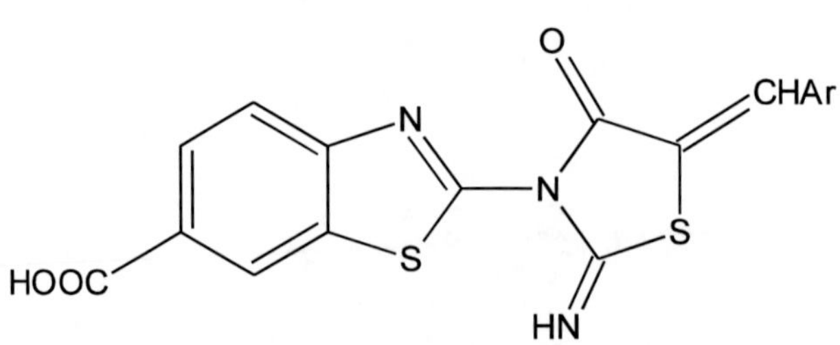

Ar = 2-Cl-C_6H_4, 4-Cl-C_6H_4, 4-OH-C_6H_4, 3-Br-C_6H_4, 3-NO_2-C_6H_4, 4-NO_2-C_6H_4

Scheme 21.

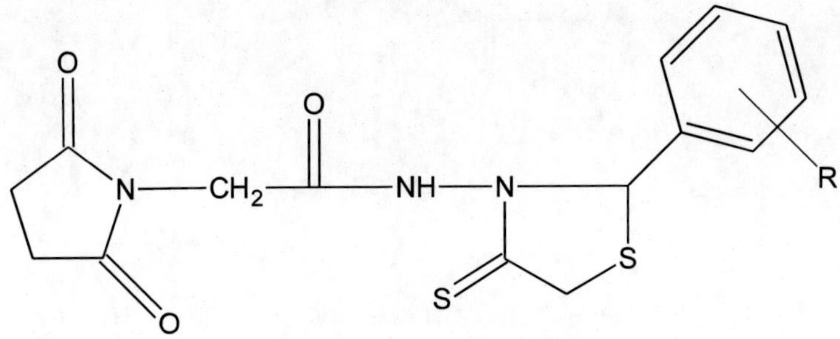

R = H, 2-OH, 2-NO_2, 4-Cl, 3-OCH_3, 3-OCH_3-4OH

Scheme 22.

Ethylchloroacetate and hydrazine hydrate were reacted with substituted aldehydes to get the corresponding substituted N-benzylidine-2-chloro acetohydrazides, which were further condensed with succinimide to obtain corresponding substituted N-benzylidine-2-(2,5-dioxopyrrolidin-1-yl) acetohydrazides. These compounds were cyclized with thioglycolic acid to get the corresponding thiazolidinone derivatives, which were evaluated for their antimicrobial activity by Vilas et al. [31]. All compounds demonstrated antibacterial activity against gram-negative bacteria (Scheme 22).

Synthesis and biological activity of several new 2-[2-(Substituedbenzyliden)-hydrazinoacetyl]-mercapto-5-methyl-1,3,4-thiadiazole, 2-[2-{2-(substitutedphenyl)-4-oxo-1,3-thiazolidin}-acetylimino]-mercapto-5-methyl-1,3,4-thiadizole and 2-[2-{2-(substituted phenyl)-4-oxo-5-(substituted aryliden)-1,3-thiazolidin}-acetylimino]-mercapto-5-methyl-1,3,4-thiadiazole were synthesized by Srivastava et al. [32]. All the synthesized compounds were evaluated *in vitro* for antibacterial activity by using filter paper disc method. Their results showed that some of the synthesized compounds possessed antibacterial and antifungal activities (Scheme 23).

N-(2-aryl-4-oxothiazolidin-3-yl)-2-(5-(phenoxymethyl)-2-thioxo-1,3,4-oxadiazol-3(2H)-yl) acetamide derivatives were synthesized and evaluated for their antibacterial and antifungal activities by Mishra et al. [33]. The fungicidal activity of all the compounds was studied at 1000 ppm concentration *in vitro.* The antibacterial activities of all the compounds were studied against gram-positive bacteria and gram-negative bacteria at a concentration of 50μg/ML by agar cup plate method. Some of the tested compounds exhibited significant activities (Scheme 24).

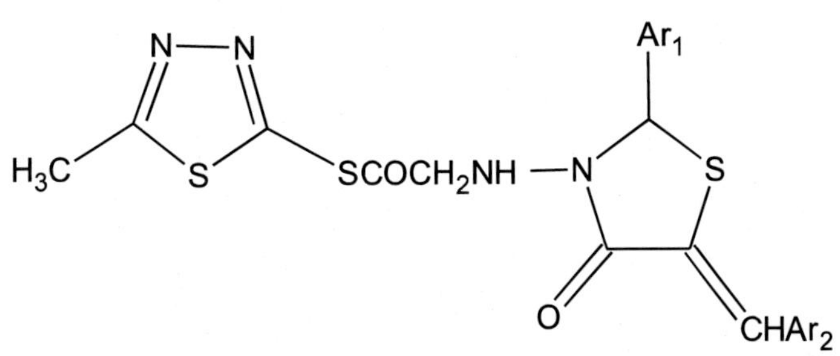

$Ar_1 = Ar_2$ = 2-Cl-C_6H_4, 3-Cl-C_6H_4, 4-Cl-C_6H_4, 2-Br-C_6H_4, 3-Br-C_6H_4, 4-Br-C_6H_4, 2-NO_2-C_6H_4, 3-NO_2-C_6H_4, 4-NO_2-C_6H_4, 2-OCH_3-C_6H_4, 3-OCH_3-C_6H_4, 4-OCH_3-C_6H_4

Scheme 23.

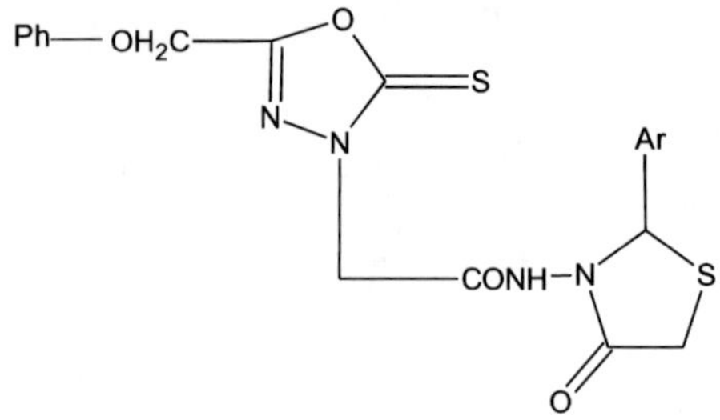

Ar = C_6H_5, 2-OH-C_6H_4, 4-OH-C_6H_4, 4-OCH_3-C_6H_4, 4-CH_3-C_6H_4

Scheme 24.

A new series of N-[3-(10H-phenothiazin-10-yl)propyl]-2-(substituted phenyl)-4-oxo-5-(substituted benzylidene)-3-thiazolidinecarboxamide were synthesized by Ritu et al. [34]. The reaction of thioglycolic acid with N-[3-(10H-phenothiazin-10-yl)propyl]-N'-[(substituted phenyl)methylidene]urea, in the presence of anhydrous $ZnCl_2$ afforded the new heterocyclic compounds N-[3-(10H-phenothiazin-10-yl)propyl]-2-(substitutedphenyl)-4-oxo-3-thiazolidine carboxamide. The latter product on treatment with several substituted aromatic aldehydes in the presence of C_2H_5ONa underwent the Knoevenagel reaction to yield the desired product. The structures of compounds were confirmed by IR, 1H-NMR, 13C-NMR and FAB mass spectroscopy and by chemical analysis. All the above compounds were screened for their antimicrobial activity against some selected bacteria and fungi and for their antituberculosis activity (*Mycobacterium tuberculosis*) (Scheme 25).

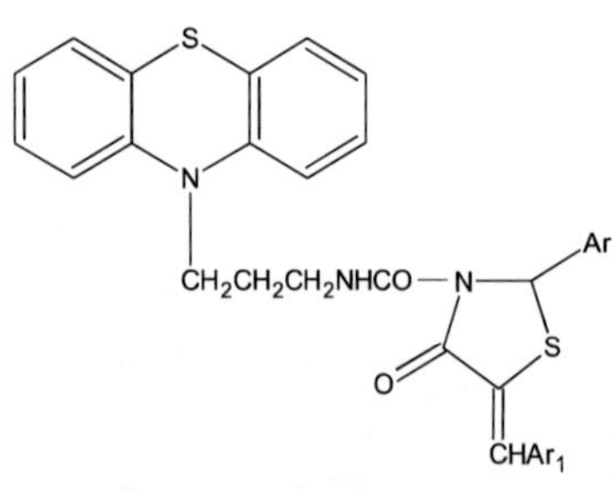

Ar = C_6H_5, 4-ClC_6H_4, 3-ClC_6H_4, 2-ClC_6H_4, 3-BrC_6H_4, 2-BrC_6H_4, 2-$NO_2C_6H_4$, 3-$NO_2C_6H_4$, 4-$NO_2C_6H_4$, 4-$CH_3C_6H_4$, 3-$CH_3OC_6H_4$

Scheme 25.

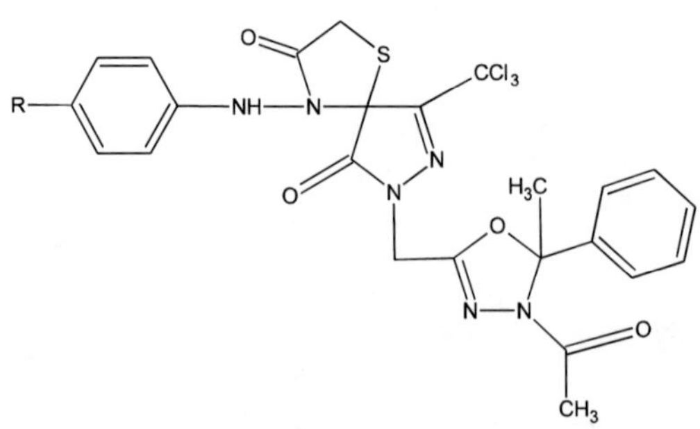

R = H, CH_3, OCH_3, OC_2H_5, Cl, Br

Scheme 26.

A series of six spiro compounds containing thiazolidinone and 1,3,4-oxadiazole moieties were synthesized and characterized by elemental analysis, IR spectral, 1H NMR spectral and 13C NMR spectral data by Ranga et al. [35]. The antimicrobial activity of the newly synthesized compounds was evaluated by cup plate method. The activity of the compounds was compared to that of standard agents. It was found that spiro compounds with -OCH_3, -Cl and -Br substituents, demonstrated better antimicrobial activity (Scheme 26).

Derivatives of methyl-4-thiazolidinone were synthesized by acid-catalyzed condensation of Schiff base and Thiolactic acid by Rathod et al. [36]. The prepared thiazolidinone derivatives were characterized by FTIR, NMR and element analysis. All the synthesized compounds were screened for their antimicrobial activities by using Broth dilution method. Antibacterial and antifungal studies of these compound indicated that some compound shows potential antimicrobial activity but overall the synthesized compound are less active than the standard drug (Scheme 27).

2-(5-(phenoxymethyl)-2-thioxo-1,3,4-oxadiazol-3(2H)-yl)acetohydrazide underwent facile condensation with aromatic aldehydes to afford the corresponding N'-aryl-2-(5-(phenoxymethyl)-2-thioxo-1,3,4-oxadiazol-3(2H)-yl)acetohydrazide in good yields. Cyclocondensation of these compounds with thioglycolic acid yielded N-(2-aryl-4-oxothiazolidin-3-yl)-2-(5-(phenoxymethyl)-2-thioxo-1,3,4-oxadiazol-3(2H)-yl) acetamide by Mishra et al. [33]. All the newly synthesized compounds were evaluated for their antibacterial and antifungal activities. The antibacterial activities of all the compounds were studied against gram-positive bacteria (*Staphylococcus aureus* and *Bacillus subtilis*) and gram-negative bacteria (*E.coli*, and *Klebsiella promioe*) at a concentration of 50μg/ML by agar cup plate method. The fungicidal activity of all the compounds was studied at 1000 ppm concentration *in vitro*. Plant pathogenic organisms used were *Nigrospora Sp*, *Aspergillus niger*, *Botrydepladia thiobromine*, and *Rhizopus nigricum*, *Fusarium oxyporium*. Some of the tested compounds exhibited significant activities at these test concentrations (Scheme 28).

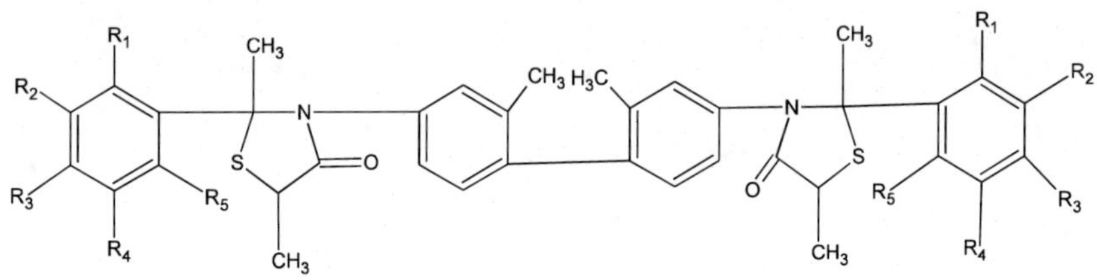

R_1 = H, OH R_2 =Cl, I, CH_3 R_3 = H, OH R_4 = Cl, I, CH_3 R_5 = H, OH

Scheme 27.

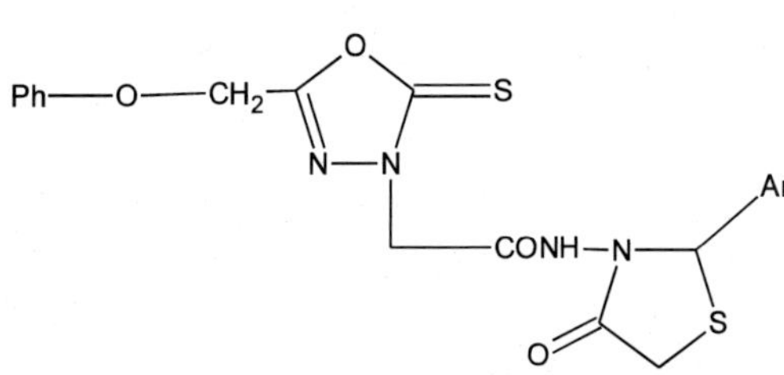

Ar = C_6H_5, 2-OHC_6H_4, 4-OHC_6H_4, 4-$CH_3C_6H_4$, 4-$OCH_3C_6H_4$

Scheme 28.

A new series of 1,3-benzothiazol-2-yl-1,3-thiazolan-4-one were synthesized by the reaction of 2-(4-methylphenyl)-3-(5,5,7-trimethyl-4,5,6,7-tetrahydro-1,3-benzothiazol-2-yl)-1,3-thiazolan-4-one with aryl aldehydes by Nagaraj et al. [37]. Chemical structures of all the new compounds were established by IR, 1H, 13C NMR, MS and elemental data. The synthesized compounds were evaluated for their antibacterial activity against Gram-positive bacteria *viz. Bacillus subtilis*(MTCC 441), *Bacillus sphaericus*(MTCC 11), *Staphylococcus aureus*(MTCC 96) and Gram-negative bacteria *viz. Pseudomonas aeruginosa*(MTCC 741), *Klebsiella aerogenes*(MTCC 39), *Chromobacterium violaceum*(MTCC 2656). Amongst them, compounds containing [(4-chlorophenyl)methylidene] moiety, [(3-nitrophenyl)methylidene] moiety and [(2-thienyl)methylidene] moietyshowed significant antibacterial activity, almost equal/more than the activity of the standard drug, Streptomycin. Further, the compoundswere also screened for their antifungal activity against *Candida albicans*(ATCC 10231), *Aspergillus fumigatus*(HIC 6094), *Trichophyton rubrum*(IFO 9185), and *Trichophyton mentagrophytes*(IFO 40996). Most of these tested compounds displayed appreciable activity against bacteria and fungi and emerged as potential molecules for further development (Scheme 29).

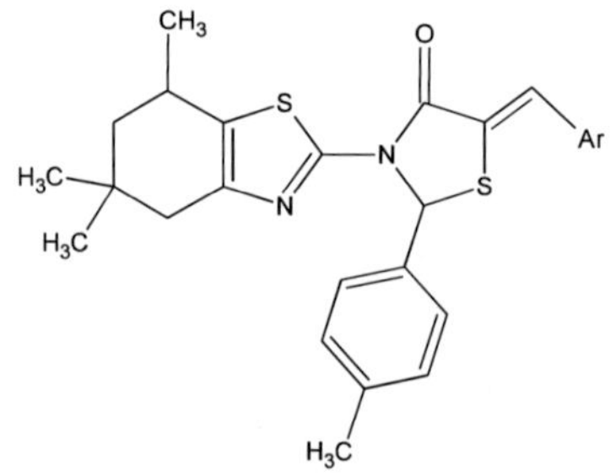

Ar = 4-$CH_3C_6H_4$, 4-ClC_6H_4, 4-$NO_2C_6H_4$, 3-$NO_2C_6H_4$, 2-OHC_6H_4, 2-furyl, 2-thienyl

Scheme 29.

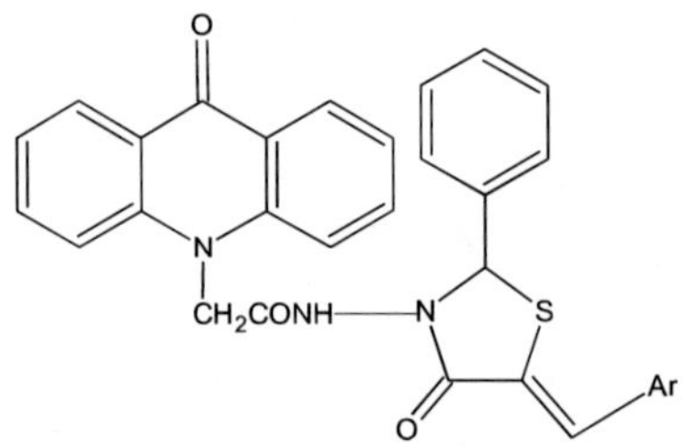

Ar = Benzaldehyde, 2,4-dichlorobenzaldehyde, 3,4-dimethoxybenzaldehyde, 3-methoxy-4-hydroxybenzaldehyde, 4-N,N-dimethylaminobenzaldehyde

Scheme 30.

A series of thiazolidinone derivatives with various aromatic aldehyde substitution were synthesized and characterized by physical (TLC and M.P.) and spectral data (IR, NMR and MASS) by Pankaj et al. [38]. They were evaluated for antimicrobial potential against Gram positive *Staphylococcus aureus, Bacillus subtilis,* gram-negative *Escherichia coli, Pseudomonas aeuroginosa, Shigella* and fungi *Candida albicans* and *Aspergillus niger.* The initial screening using zones of inhibition at 50 μl concentration revealed that they are moderately active against the tested strains. All the synthesized compounds showed mild to

moderate antimicrobial activity against Gram positive, Gram negative bacterium and fungal strains. So it was concluded that the substituted thiazolidinone derivatives with electron withdrawing groups showed potent activity compared to others (Scheme 30).

A series of new 4-thiazolidinone derivatives was synthesized, characterized by spectral techniques, and screened for antimicrobial activity by Amit et al. [39]. All the compounds were evaluated against five Gram-positive bacteria, two Gram-negative bacteria, and two fungi, at concentrations of 50, 100, 200, 400, 800, and 1600 μg/mL, respectively. Minimum inhibitory concentrations of all the compounds were also determined and were found to be in the range of 100-400 μg/mL. All the compounds showed moderate-to-good antimicrobial activity. Compounds [2-(4-fluoro-phenyl)-3-(4-methyl-5,6,7,8-tetrahydro-quinazolin-2-yl)-thiazolidin-4-one] and [3-(4,6-dimethyl-pyrimidin-2-yl)-2-(2-methoxy-phenyl)-thiazolidin-4-one] were the most potent compounds of the series, exhibiting promising antimicrobial activity against *Pseudomonas fluorescens*, *Staphylococcus aureus* and the fungal strains (Scheme 31).

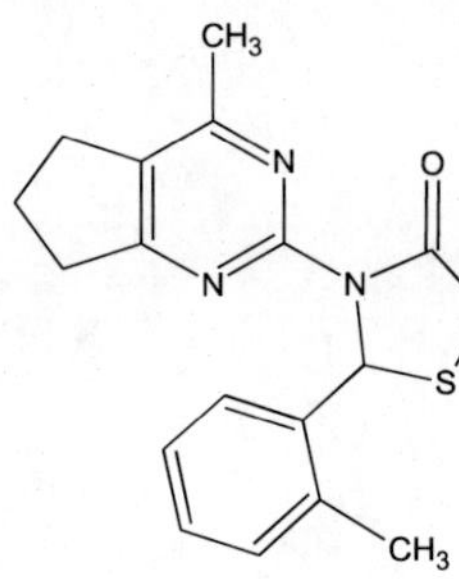

Scheme 31.

To develop novel antimycobacterial agents, a new series of thiazolidinone-azole hybrids were designed and synthesized by Barbaros et al. [40]. Thiazolidin-4-ones were obtained by the reaction of *Schiff* bases and hydrazones with mercaptoacetic acid. 5-Benzylidene derivatives were obtained by treatment of these with appropriate benzaldehydes according to Knoevenagel condensation. The structures of the synthesized compounds were confirmed by 1H NMR, IR, mass spectrometry and elemental analysis data. The target compounds were screened for their antimycobacterial activity against *M. tuberculosis* H37Rv strain using the microplatealamar blue assay method. Among them, few were found to be active compounds in the series. It was seen that both phenylamino and benzylidene substitutions on thiazolidin-4-one ring causes an improvement in the antimycobacterial activity (Scheme 32).

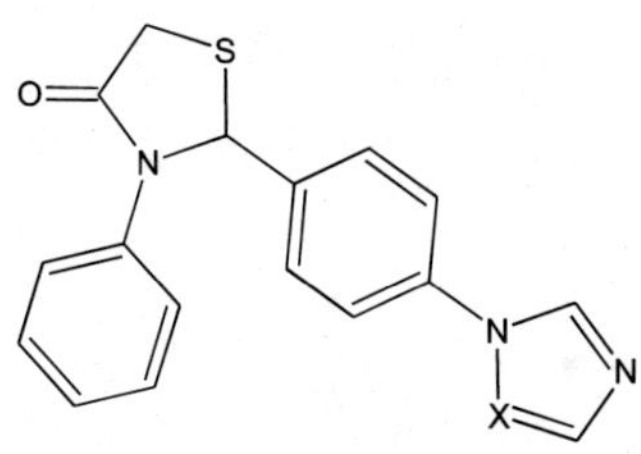

X = H, CH

Scheme 32.

3.2. Antitubercular Activity

The emergence of multi-drug resistant tuberculosis, coupled with the increasing overlap of the AIDS and tuberculosis pandemics has brought tuberculosis to the forefront as a major worldwide health concern.

Kucukguzel et al. [41] reported antitubercular activity of substituted 4-thiazolidinones and found that some of the tested compounds exhibited antitubercular activity (Scheme 33).

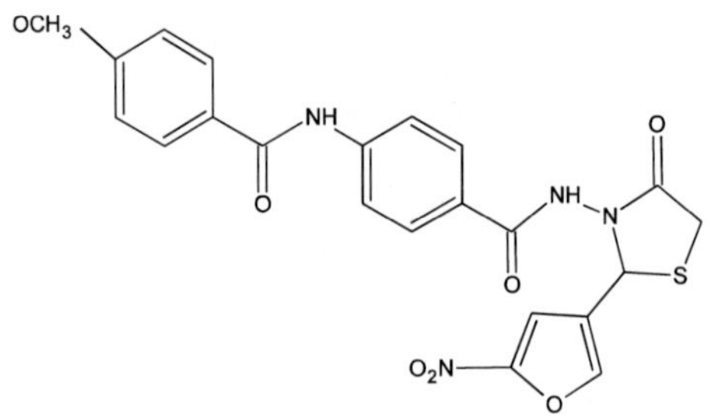

Scheme 33.

Isonicotinylhydrazide derivatives were prepared by Jaju and co-workers [42] and screened for antimycobacterial activity against *M. tuberculosis*. They found that the antitubercular activity was considerably affected by various substituents on the aromatic ring of 4-thiazolidinone and it was proved by the fact that compounds with no substitution at the aromatic ring did not show any considerable activity. Karali et al. [43] tested 4-(3-coumarinyl)-3-cyclohexyl-4-thiazolin-2-one benzylidenehydrazone derivatives for antitubercular activity (Scheme 34).

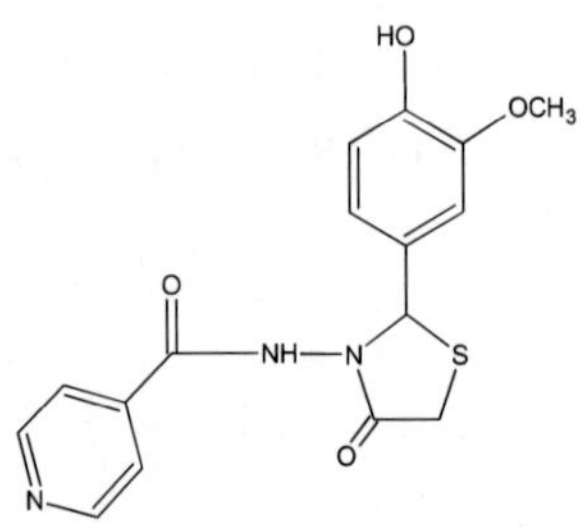

Scheme 34.

Dharmrajan et al. [44] synthesized some thiazolidinone derivatives and screened for their antitubercular activity against *Mycobacterium tuberculosis* H37Rv using resazurinmicroplate assay 25 (REMA) at the concentration of 1, 10, 50ìg/ml. All the compounds showed mild activity at 50mg/ml (Scheme 35).

A convenient, economic and a user-friendly process for the synthesis of substitute 4 thiazolidinones and phenolic derivatives [2 (substituted) 3-P-acetaamido phenoxy-4-thiazolidinones] of biological and medicinal importance was carried out by Salunke et al. [45]. The new drug like 2-Substituted-3-P-acetamindo phenoxy-4-thiazolidinones had antitubercular property. Derivatives were prepared by condensation reaction of Picric acid with Ethyl Chloro Acetate in the presence of Methanol in water. This method offered some advantages in terms of simplicity of performance, the choice of solvent, low reaction time, low cost and green chemistry conditions. The products were conveniently handled and isolated from the reaction mixture (Scheme 36).

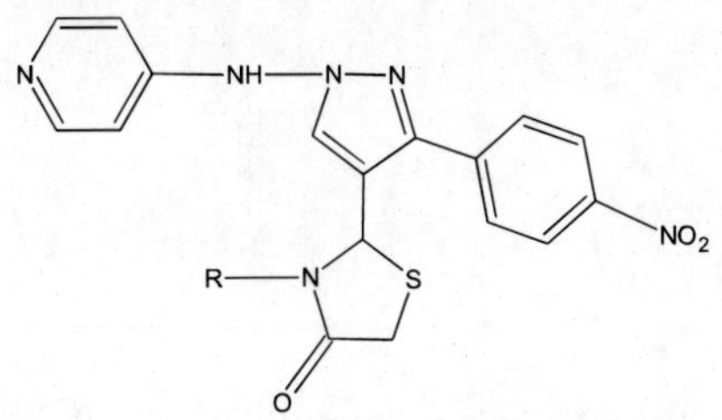

R = C_6H_5, 4-$C_6H_4NO_2$, 4-C_6H_4Cl, 4-C_7H_7O, 4-C_7H_7, 4-$C_7H_5O_2$, 4-C_6H_4F, 4-C_8H_7O

Scheme 35.

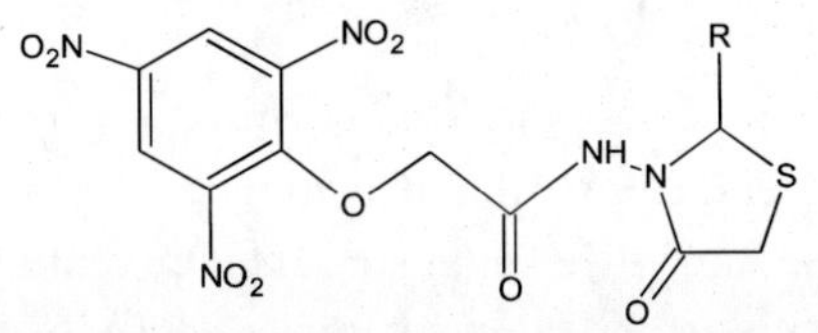

R = benzaldehyde, o-nitrobenzaldehyde, vanniline, cinnamaldehyde, p-dimethyl benzaldehyde

Scheme 36.

3.3. Anticancer Activity

Chandrappa et al. [46] prepared 2-(5-((5-(4-chlorophenyl)furan-2-yl) methylene)-4-oxo-2-thioxothiazolidin-3-yl) acetic acid derivatives and evaluated their anticancer activity. In searching for better anticancer agent Zhou and coworkers designed and synthesized, as well as studied cytoselective toxicity, structure-activity relationships and pharmacophore of thiazolidinone derivatives. Substitution on nitrogen atom in thiazolidinone ring and the -NMe_2 group at the 4-position of phenyl ring did not show any cytoselective toxicity (Scheme 37).

2-arylthiazolidine-4-carboxylic acid amide derivatives were synthesized by Gududuru et al. [47] for possible cytotoxic activity. The same research group designed new series of 2-aryl-4-oxo-thiazolidin-3-yl amides and all the synthesized compounds were evaluated against five human prostate cancer cell lines. They reported that increase in the alkyl chain enhanced the antiproliferative activity while replacement of the alkyl chain with aryl group reduced the biological activity (Scheme 38).

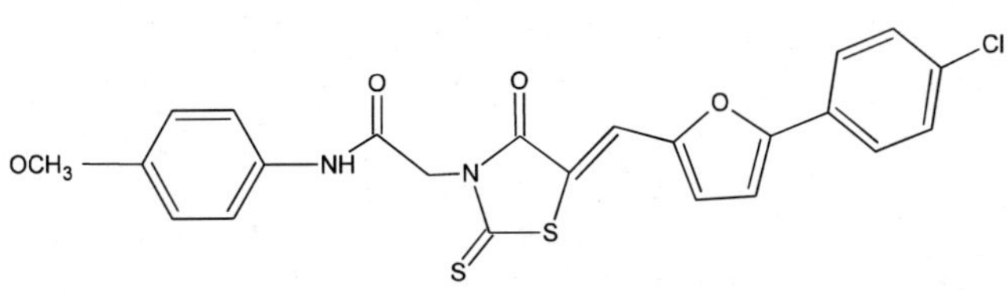

Scheme 37.

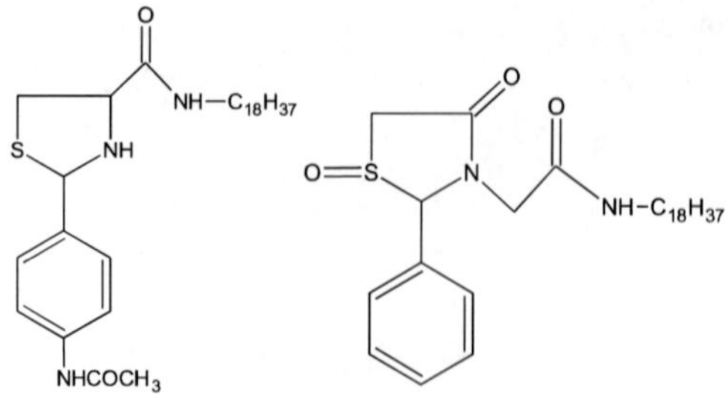

Scheme 38.

Some derivatives of 5-bromo-3-[(3-substituted-5-methyl-4-thiazolidinone-2-ylidene) hydrazono]-1H-2-indolinones were investigated by Karali et al. [48] for their primary cytotoxic activity against 3-cell line panel consisting of NCI-H460 (Lung), MCF7 (Breast), and SF-268 (CNS). It was observed that thiosemicarbazone derivatives of indolinones show promising cytotoxic activity.

Havrylyuk et al. [49] prepared a number of isatin-based thiazolidines conjugates. These were investigated for anticancer activity. Their affinity to tyrosine kinase, cyclin-dependent kinase and carbonic anhydrase isozymes suggested their potential as novel anticancer agents. None of the thiazolidinone conjugates showed greater activity than 1,3-dihydroindol-2-one conjugates with 3,5-diaryl-4,5-dihydropyrazolyne derivatives (Scheme 40).

Some derivatives of 5-benzylidene-2-[(pyridine-4-ylmethylene) hydrazono]-thiazolidin-4-ones were synthesized by Danniel et al. [50]. They investigated the antimicrobial and cytotoxic properties of the compounds. The compounds were evaluated *in vitro* against bacterial and fungal species. Some of them had weak antimicrobial activities. A few of them also displayed anticancer activities (Scheme 41).

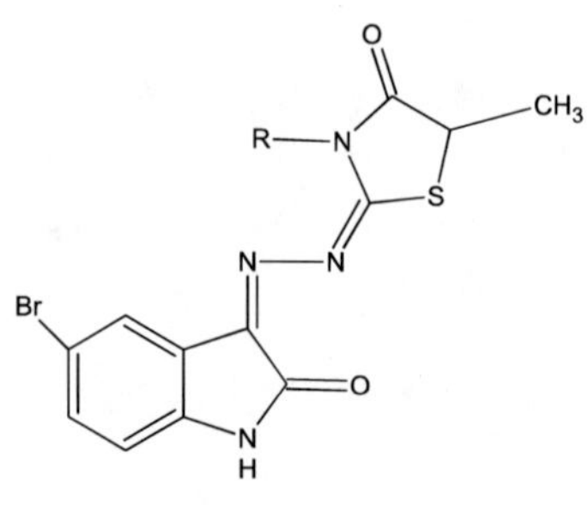

Scheme 39.

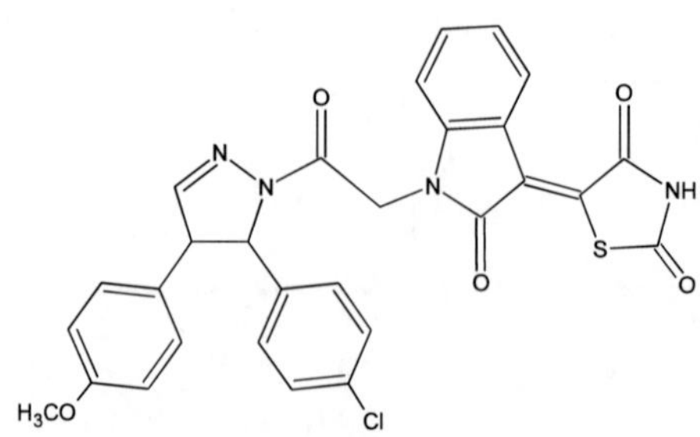

Scheme 40.

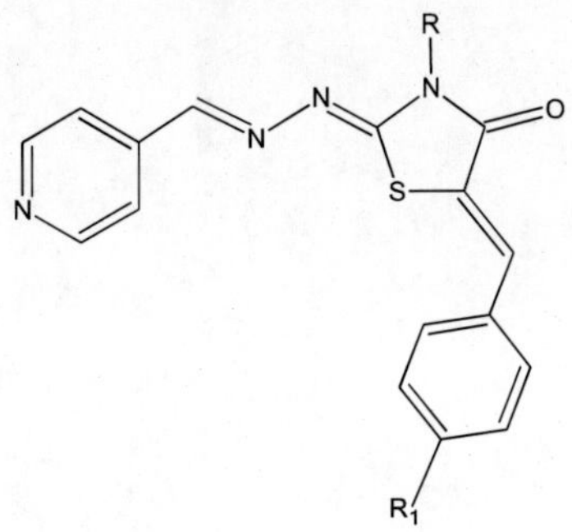

R = H, CH_3, C_2H_5, C_6H_5 R_1 = H, NO_2, $N(CH_3)_2$

Scheme 41.

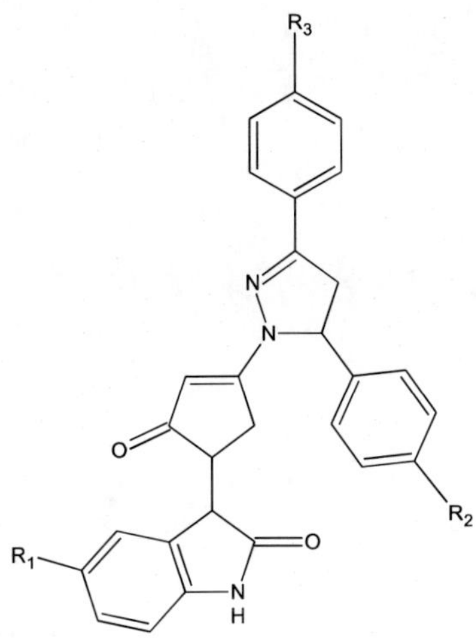

R_1 = H, Cl R_2 = Cl, NMe_2 R_3 = F, Cl, Br

Scheme 42.

In silico screening of virtual 4-thiazolidinones library was carried out with several self-developed QSAR and pharmacophore models of anticancer activity by Devinyak et al. [51]. Such approach allowed to choose 14 most probable anticancer hits-conjugates of 4-thiazolidinone, pyrazoline and isatin. These molecules were synthesized.

Half of these compounds were subjected to anticancer activity evaluation and biological experiments confirmed the predictions, although in a qualitative manner. The most sensitive to tested compounds cancer cell line is Mino (the lowest IC_{50} reached 0.17 μM), which represents mantle cell lymphoma-a rare but very aggressive tumor. This method identified further direction to study biological effects of the novel anticancer compounds (Scheme 42).

Monforte et al. [52] synthesized several series of 2,3-diaryl-1,3-thiazolidin-4-one derivatives and screened them for their possible cytotoxic activity. Among them the following compound was found to be most potent and reported as a new family of anti-viral agents acting as NNRTIs with minimal cytotoxicity. It was found that substitution by aryl group at nitrogen atom increases the cytotoxic activity of compound (Scheme 43).

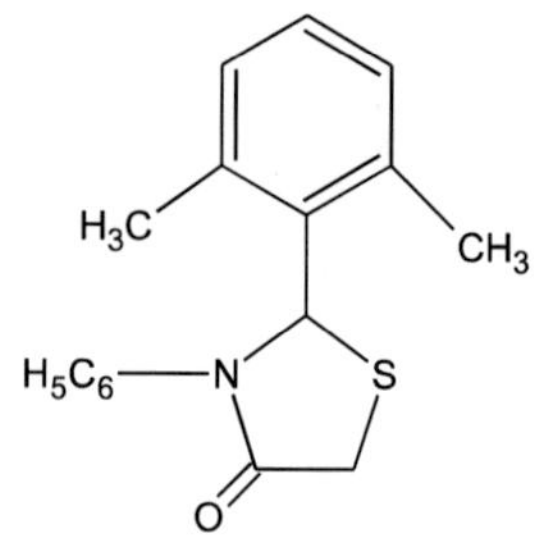

Scheme 43.

Rawal et al. [53] synthesized 2-(2,6-dibromophenyl)-3-heteroaryl-1,3-thiazolidin-4-one derivatives and studied their cytotoxic activity. Among the various derivatives the following compound showed the potent activity. A substitution of aryl group at nitro atom showed an increase in cytotoxic activity (Scheme 44).

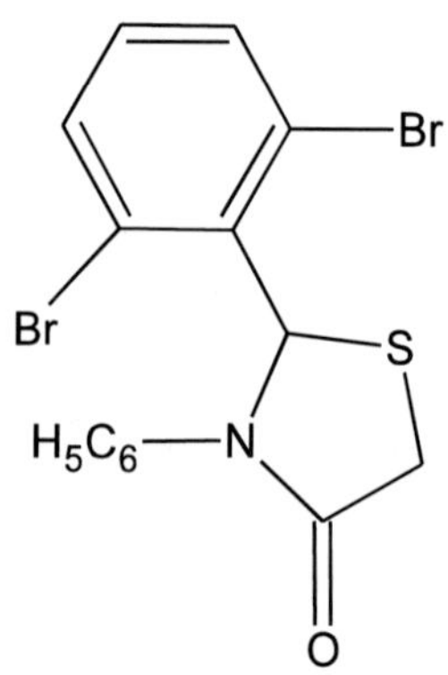

Scheme 44.

3.4. Anti-Inflammtory and Analgesic Activities

Study of anti-inflammatory and analgesic activity of new series of quinazolinone derivatives with thiazolidinone at 2nd position was reported by Kumar et al. [54]. In another study, Deep et al. [55] also reported biphenyl-4-carboxylic acid-5-(arylidene)-2-(aryl)-4-oxothiazolidin-3-yl amide derivatives that exhibit significant anti-inflammatory activity (Scheme 45).

Investigation of inflammatory/analgesic activities of 3,3'-(1,2-ethanediyl)-bis[2-aryl-4-thiazolidinone] derivatives was carried out by Ottana et al. [56]. Bulkiness at NH group of 2,4-thiazolidinedione ring either decreased or abolished the anti-inflammatory activity (Scheme 46).

A series of 2-(arylimino)-5-(pyridine-2-yl-methylidene)-1,3-thiazolidine-4-one were synthesized by Reetu et al. [57]. All the synthesized compounds were examined for their anti-inflammatory and analgesic activity using carrageenan induced paw edema method and acetic acid induced writhing method. The tested compounds showed significant anti-inflammatory and analgesic activities (Scheme 47).

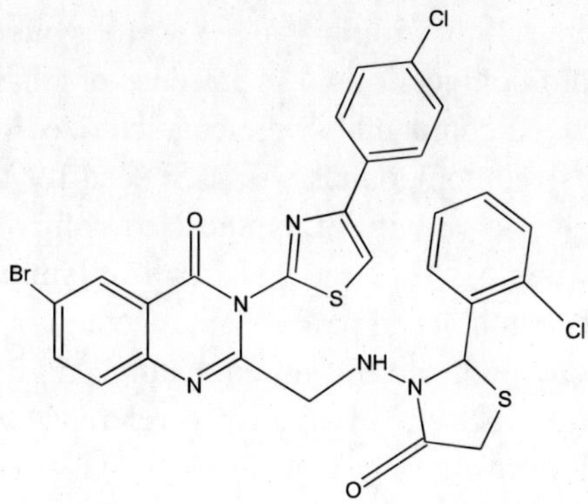

Scheme 45.

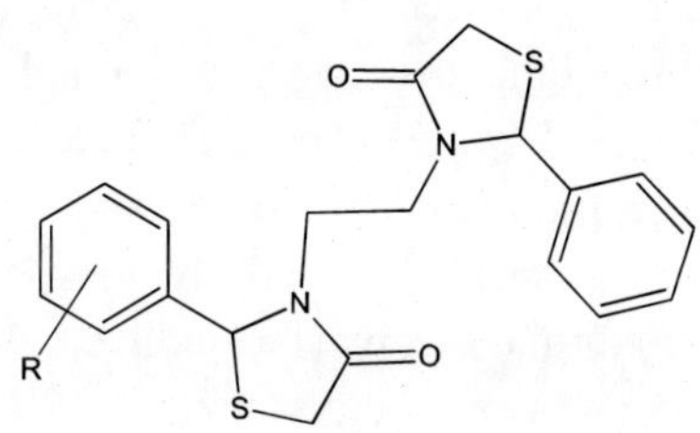

Scheme 46.

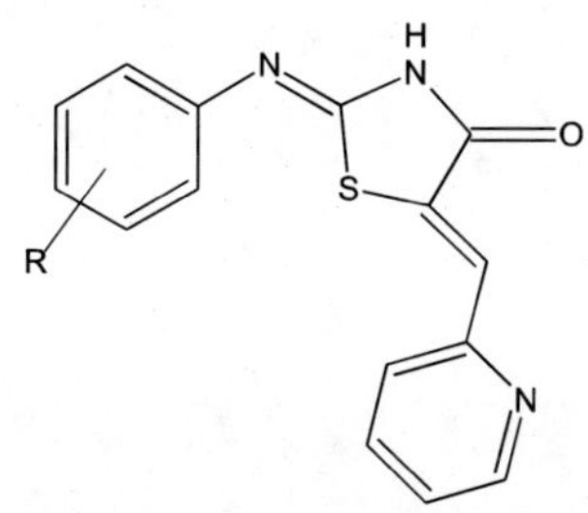

R = aniline, p-chloro aniline, m-toluidine, p-toluidine, 4-fluoro aniline, o-nitro aniline, m-nitro aniline, o-chloro aniline, o-bromo aniline

Scheme 47.

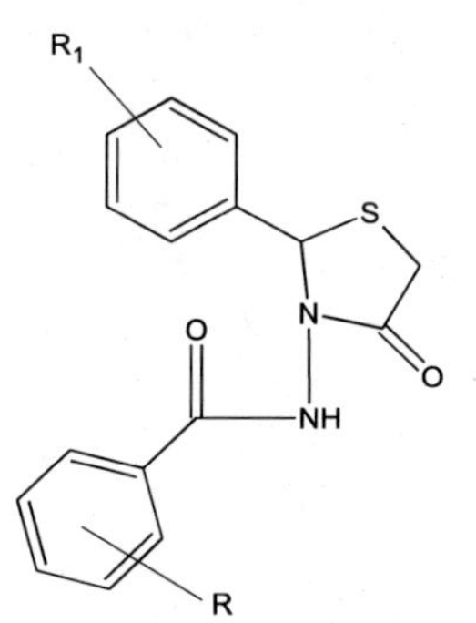

R = 2-Cl, 3-CH_3 R_1= 2-OH, 4-F, 4-NO_2, 4-OCH_3, 4-Cl, 2-NO_2

Scheme 48.

A series of substituted aryl thiazolidin-4-ones were synthesized by the reaction of substituted schiff's bases with thioglycolic acid in presence of ethanol by Kumudha et al. [58]. The structures of the synthesized compounds were characterized by FTIR, 1H-NMR spectral analysis. Some of the synthesized compounds were screened for their anti-inflammatory and analgesic activities. The analgesic activity of synthesized compounds was evaluated by tail immersion technique using albino wistar rats. The anti-inflammatory activity results were determined using the carrageenan induced paw oedema method in albino wistar rats. Some of the compounds showed weak to moderate anti-inflammatory activity and rest of the compounds showed good anti-inflammatory activity compared to reference drug diclofenac sodium. In general, all the compounds showed significant analgesic activity in comparison to the reference drug acetyl salicylic acid (Scheme 48).

Pravesh et al. [59] were prepared two new thiazolidinone derivatives by brominating 1-acetylnaphthalene in chloroform followed by condensation with substituted benzaldehydethiosemicarbazones using ethanol to get 4-naphthalen-1-yl-2-{2-[(substituted phenyl)methylidene] hydrazino}-1,3-thiazoles. These thiazole derivatives were then cyclized to title compounds by reacting with thiomalic acid in dioxane using $ZnCl_2$. All the synthesized compounds were characterized on the basis of their IR, melting point. The analgesic activity of thiazolidinones was performed on normal rats by acid-induced writhing method. The analgesic activity of new synthesized compound revealed that both of the compounds exhibit significant analgesic activity as compared to standard drug diclofinac sodium (Scheme 49).

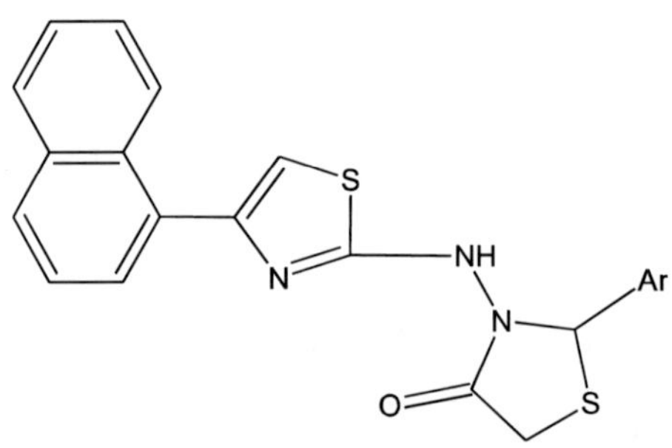

Ar = 2-nitrophenyl, 4-nitrophenyl

Scheme 49.

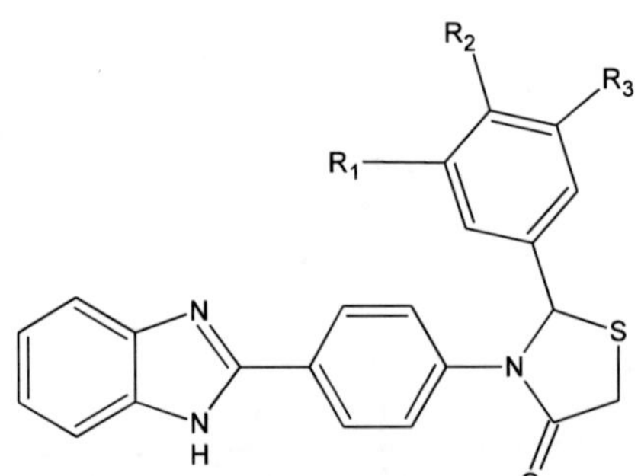

$R_1 = = R_2 = R_3 =$ H, Cl, OH, OCH_3, $N(CH_3)_2$

Scheme 50.

A new series of 2-[4-(azetidin-2-one)-3-chloro-4-phenyl]-1H-phenylbenzimidazoles and 2-(thiazolidin-4-one)-phenyl]-1H-phenylbenzimidazoles were synthesized by the reaction of schiff base [2-(4-aminophenyl)-benzimidazole and substituted benzaldehyde with chloroacetyl chloride and mercaptoacetic acid respectively by Anbalagand et al. [60]. The chemical structures of the synthesized compounds were confirmed by IR, 1H-NMR, mass spectral and C, H, N analysis. The synthesized compounds were screened for antibacterial (*Bacillus cereus, Escherichia coli, Micrococcus luteus, Klebssielap neumoniae, Staphylococcus aureus* and *Salmonella epidermidis*), antifungal (*Aspergillus niger* and *Candida albicans*), analgesic activity by writhing reflex method and anti-inflammatory activity by carrageenan induced paw edema method. The tested compounds exhibited significant activity of antibacterial, antifungal, analgesic and anti-inflammatory activity comparable to that of standard (Scheme 50).

A new series of 2-iminothiazolidin-4-ones and thiazolidin-2,4-diones were designed and synthesized by Waleed et al. [61]. The target compounds were evaluated for antimicrobial and anti-inflammatory activities, aiming to identify a candidate carrying dual activities. The antibacterial activity was determined by cup-diffusion technique and NCCLS broth dilution method. The MICs were determined using 96-well microtitre plates, while the anti-inflammatory activity was determined using carrageenan-induced rat paw edema model. The results revealed that target compounds show no antifungal activity towards tested Fungi, whereas the antibacterial activity was mainly on Gram-positive bacteria. The anti-inflammatory activity was observed in compounds belonging to the thiazolidin-2,4-dione series. The structures of the newly synthesized compounds were confirmed by physical and spectral data. The target compounds were screened for both antimicrobial and anti-inflammatory activity. Screening results revealed that the anti-inflammatory activity is superior to that of antimicrobial activity (Scheme 51).

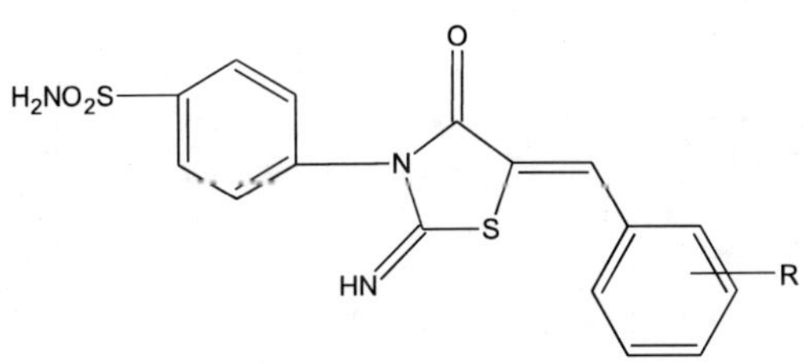

R = H, 4-OCH_3, 3,4-diOCH_3

Scheme 51.

Rheumatoid arthritis is an autoimmune inflammatory disease of yet unknown etiology. Tumor necrosis factor α (TNFα) is recognized as a regulatory substance that plays a central role in RA development and progression. On the other side, the bone morphogenetic protein (BMP) and Wnt signaling pathways are key mechanisms that induce and support cartilage and bone formation and maintenance. Previous studies showed that the pro-inflammatory cytokines TNFα and interleukin 1β (IL-1β) are central players in the inhibition of activity of skeletogenesis. Malyshev et al. [62] evaluated the anti-inflammatory activity of novel 4-thiazolidinone-based derivatives towards TNFα-induced pro-inflammatory effects during bone formation in different doses (0.02, 0.1, 0.3 and 1.0 μM) on the TNFα-mediated inhibition of the BMP-induced osteoblast differentiation in mouse mesenchymal precursor (stem) cells of C2C12 line. Treatment of these cells with TNFα completely inhibited their myogenic differentiation, as well as strongly inhibited the BMP-induced osteogenesis. Strikingly, the treatment of C2C12 cells with the synthesized compounds rescued the osteoblast differentiation from negative control of TNFα, and, moreover, converted this cytokine from the inhibitor of osteogenesis into its stimulator. Western-blot analysis of Inhibitory κBα (I-κBα) degradation was used to elucidate a mechanism of the anti-inflammatory effects. The following two tested compounds exhibited significant activity (Scheme 52).

There has been considerable interest for medicinal chemist in development of a promising target for new drug development. Rhodanine, thiazolidinedione (TZD) family pharmacophore has diverse biological activity. Chalcones are natural products belonging to flavanoid family possessed anti-cancer and anti-inflammatory activities. Phosphonates and phosphonic acids represent an important class of organophosphorus compounds. Substitutions with P-C-P linkage of the bisphosphonates can lead to extensive alterations in their physicochemical, biological, therapeutic and toxicological characteristics. Considering this biological importance, Khushbu et al. [63] aimed to synthesize a novel target which comprises the bisphosphonic acid and chalcone nucleuses with fused Rhodanine. Total 11 derivatives were synthesized and screened for *in vitro* cytotoxic assay on MCF-7 cell line (Breast Cancer) as well as for anti-inflammatory *in vivo* activity by carageenan induced Rat Paw Edema Method. The synthesized compounds were confirmed by physiochemical and spectral data. Among all 11 synthesized derivatives, three of them showed potent inhibition of inflammation when compared with Indomethacin as standard and promising results when screened over MCF-7 cell line against Letrozole (Scheme 53).

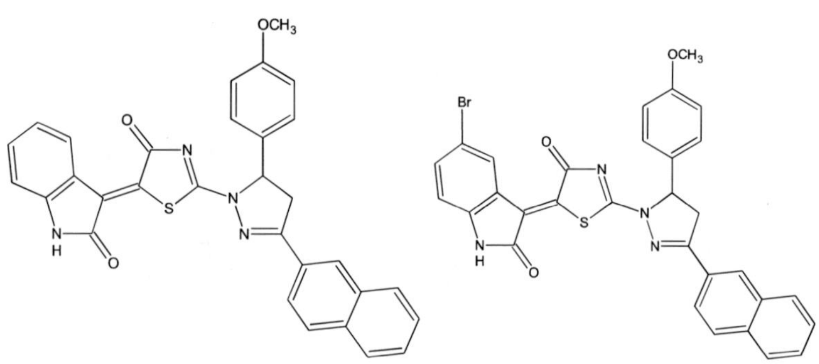

Scheme 52.

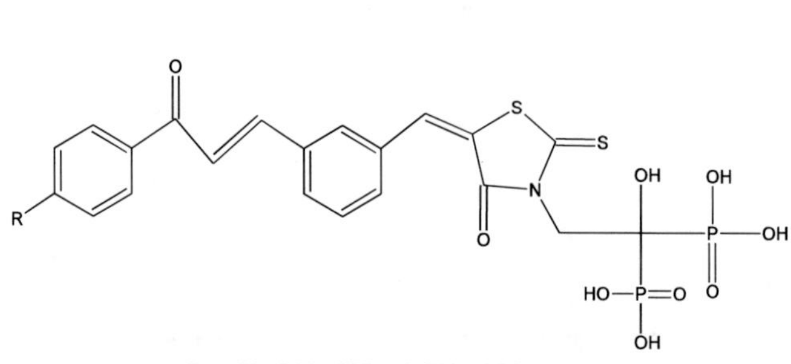

R = H, OH, CH_3, OCH_3, NO_2, F, Br

Scheme 53.

3.5. Antidepressant and Anticonvulsant Activities

A new series of thiazolidinone-barbituric acid and thiazolidinone-triazole derivatives were synthesized and studied for their antidepressant activity by Shiradkar et al. [64]. Agarwal, et al. [65] prepared a series of thiazolidinonyl-2-oxo/thiobarbituric acids derivatives. Compound with p-methoxyphenyl substitution at C-2 of thiazolidinone ring was found to be favorable for activity and exhibited more potent anticonvulsant activity as compared to standard drug sodium phenytoin (Scheme 54).

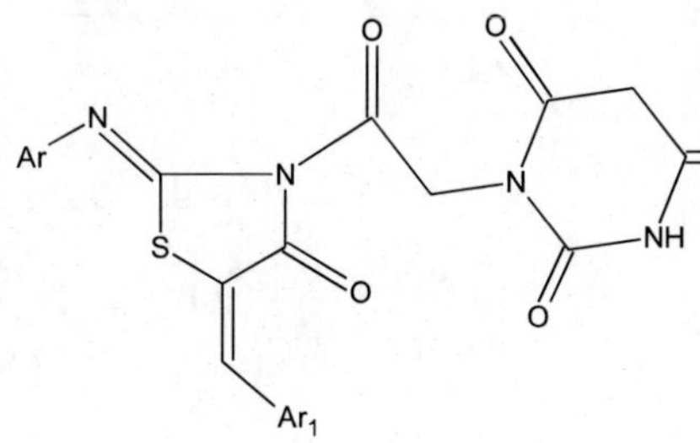

Scheme 54.

Amin et al. [66] reported some new substituted coumarinylthiazolines, coumarinyl thiazolidin-4-ones and substituted chromenothiazoles and evaluated their anticonvulsant activity. The following compounds were exhibited significant anticonvulsant activity (Scheme 55).

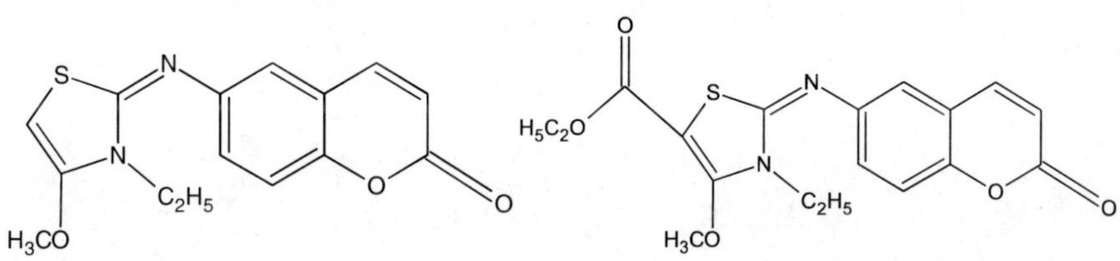

Scheme 55.

2-[3'-Aminoacetyl amino-2'-methyl-6-monosubstituted quinazolin-4'(3'H)-onyl-imino]-4-thiazolidinones were synthesised by cyclization of 1-[3'-Aminoacetyl-2'-methyl-6'-monosubstituted quinazolin-4'(3'H)-onyl]-thiosemicarbazide. These were further condensed with various substituted aldehydes to give the desired compounds. All the newly synthesized compounds were studied for their anticonvulsant activity by Archana [67]. The compounds were screened for their anticonvulsant activity against maximal electroshock induced seizures at a dose of 100mg/kg i.p., and were found to exhibit substantive anticonvulsant activity ranging from 50% to 90%. Compound 2-[3'-aminoacetyl amino-2'-methyl-6'-brom-osubstituted quinazolin-4'(3')-onylimino]-5-methyl-(*p*-N,N-dimethyl benzyl)-4-thiazo-lidinone showed maximum activity being more potent than the reference drug phenytoin sodium. The characteristic feature of this series was the incorporation of quinazolinone and thiazolidinone moieties into a single molecular framework. While evaluating the anticonvulsant activity, it was observed that compounds having bromo group at 6th position of quinazolinonyl moiety were found to be more potent than those without bromogroup. However compound having p-

N, N-dimethylbenzyl group at the 5th position of thiazolidinonyl moiety were shown most potent (activity of 90% against MES test which is more potent than standard drug phenytoin sodium, 8-0%). It was then studied at three graded doses of 25mg/kg i.p., 50mg/kg i.p., 100mg/kg i.p. and showed 30%, 50%, 90% inhibition of seizures respectively (Scheme 56).

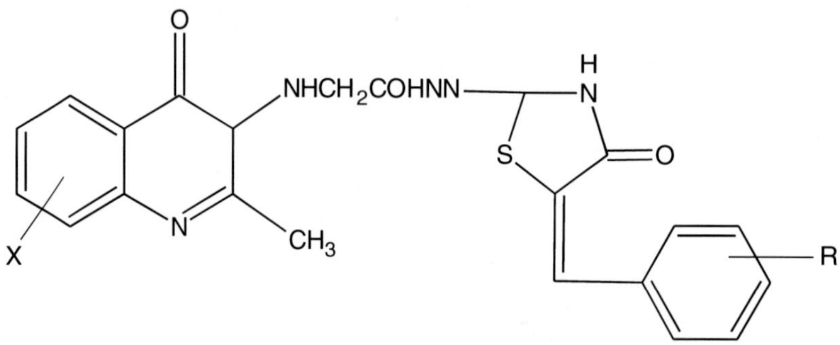

X = H, Br R = H, p-OH, m-OCH_3, p-OCH_3, p-$N(CH_3)$

Scheme 56.

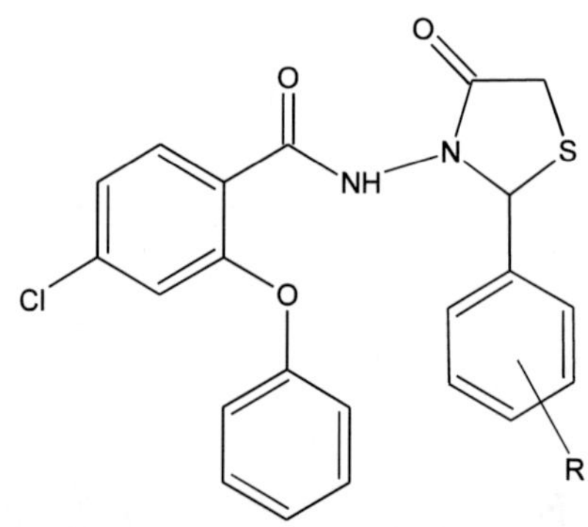

R = H, 4-OH, 2-Br, 4-Cl, 3-NO_2, 4-NO_2, 4-OCH_3, 4-CH_3

Scheme 57.

A new series of 4-chloro-N-(2-(substitutedphenyl)-4-oxothiazolidin-3-yl)-2-phenoxybenzamide derivatives were designed, synthesized and biologically evaluated as anticonvulsant agents by Mehrdad et al. [68]. The designed compounds have the main essential functional groups for binding to the benzodiazepine receptors and 4-thiazolidinone ring as an anticonvulsant pharmacophore. Some of the new synthesized compounds showed considerable anticonvulsant activity in electroshock and pentylenetetrazole-induced lethal convulsion tests. The compound 4-chloro-N-(2-(4-methoxyphenyl)-4-oxothiazolidin-3-yl)-2-phenoxy-benzamide, with the best activity was selected for evaluation of other benzodiazepine pharmacological effects. This compound induced significant sedative-hypnotic activity. However, it did not impair the learning and memory in the experimental condition. Flumazenil was able to antagonize the sedative-hypnotic and anticonvulsant effects of this compound indicating that benzodiazepine receptors are highly involved in the pharmacological properties of the novel compounds (Scheme 57).

Bis(4-thiazolidinone) derivatives were synthesized, characterized and evaluated for their anticonvulsant activity by Ulusoy et al. [69]. Some of them showed 90% protection in pentylenetetrazole-induced seizure. Substitution by phenyl group and alkyl group was found to shown potent activity (Scheme 58).

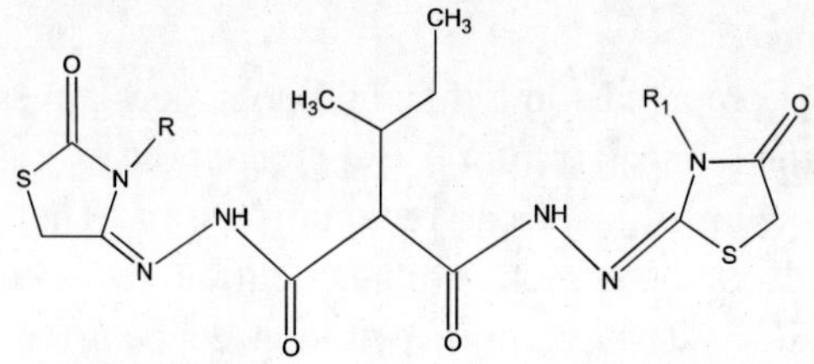

R = C_2H_5, $C_6H_5CH_3$ R1 = CH_3, OCH_3

Scheme 58.

Kaur et al. [70] synthesized novel substituted-thiadiazolylazetidinonyl compounds and screened their anticonvulsant activities. It was concluded that among the various derivatives the following molecule shows promising anti-convulsant activity (Scheme 59).

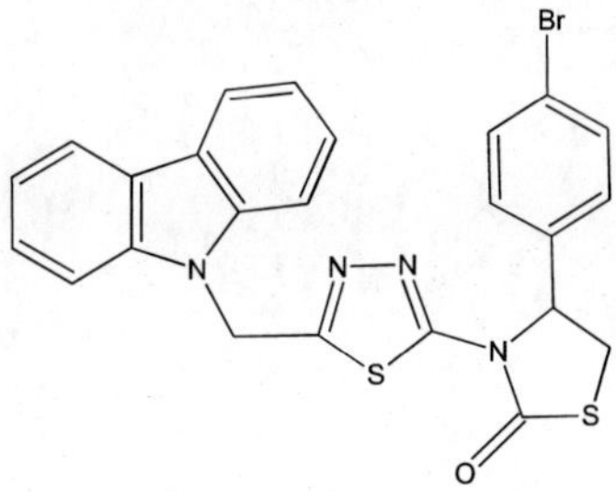

Scheme 59.

Shingalapur et al. [71] synthesized a group of 4-thiazolidinones containing 2-mercapto benzimidazole moiety and screened them for *in vivo* anticonvulsant activity by Maximal Electroshock model. Some of the tested compounds exhibited potent anticonvulsant properties (Scheme 60).

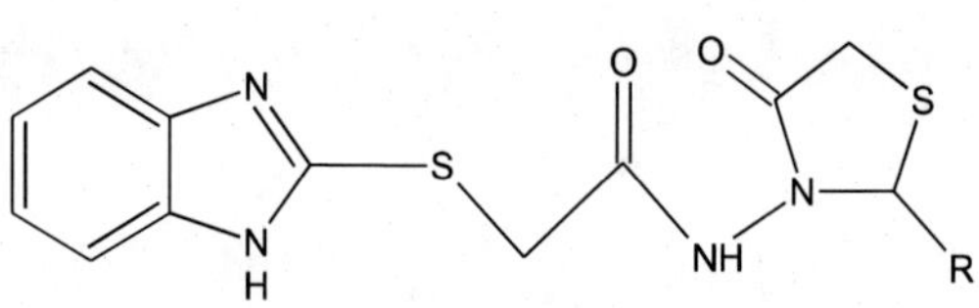

R = C_6H_5OH, 4-OHC_6H_4, 2-OH-3-$OCH_3C_6H_3$

Scheme 60.

3.6. Antiviral Activity

The activity of new 3-hydrazono-5-nitro-2-indolinone derivatives with various substituents at 3rd and 5th position of thiazolidinone ring was screened against the yellow fever virus and bovine viral diarrhea virus by Terzioglu et al. [72] (Scheme 61).

Chen et al. [73] prepared various 2-(2,6-dihalophenyl)-3-(4,6-dimethyl-5-(un)substituted-pyrimidin-2-yl)-thiazolidin-4-ones and studied their antiviral activity. Their findings suggested that overall hydrophobicity of the analogs and steric and electronic features of meta/para substituents of 3-hetero-aryl moiety on thiazolidin-4-one led to a substantial increase in antiviral activity (Scheme 62).

Dharmarajan et al. [44] prepared some thiazolidinone derivatives and evaluated them for their inhibitory effects on the replication of YFV in green monkey kidney (Vero) cells (ATCC CCL81), by means of a cytopathic effect reduction assay. The antiviral activity of the compounds was expressed as the effective concentration required to inhibiting the viral cytopathic effect by 50% (EC_{50}). The minimum cytotoxic concentration was expressed as CC_{50}, the concentration required to reduce cell growth by 50%. Some of them were found to prevent the YFV infection of the cells at concentrations that had no effect on cell growth. Some tested compounds were found to be more effective than ribaviri (EC_{50} values of 28.0 µm) (Scheme 63).

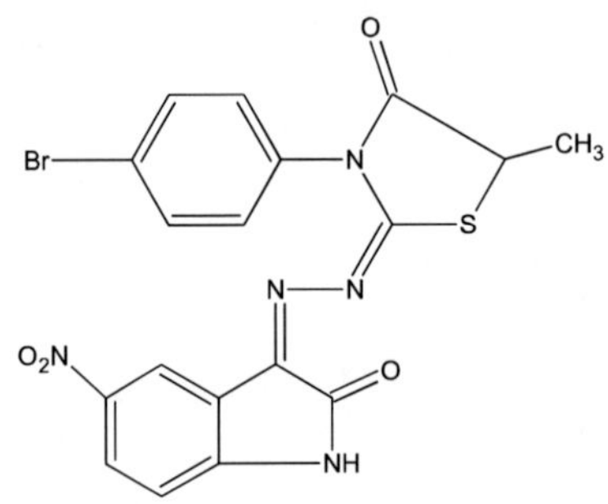

Scheme 61.

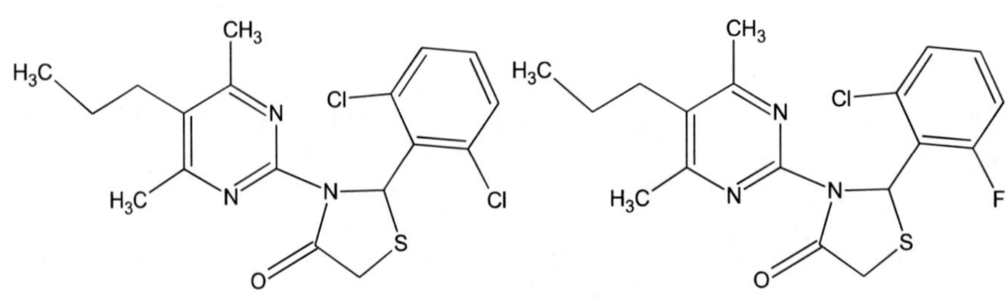

Scheme 62.

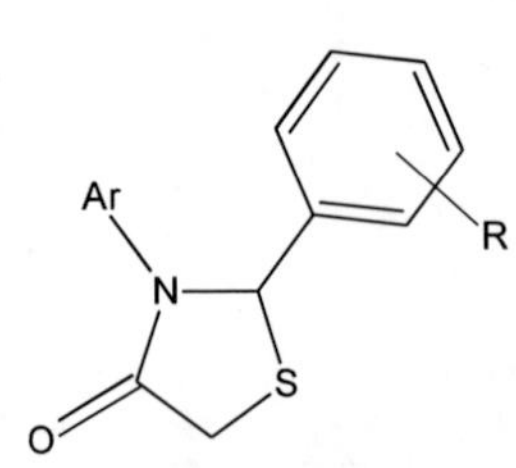

R = 2-Cl, 4-Cl, 3-NO_2, 4-NO_2, 4-OH

Ar = 4-FC_6H_4, 3-CH_3-2-C_5H_4N, 2,6-$(CH_3)_2C_6H_3$

Scheme 63.

3.7. Antidiabetic Activity

3-Subsituted-2-[p-(4-bromo-3-trifluoro-5-substituted-pyrazol-1)benzenesulfonylimino]-4-oxo thiazolidine derivatives were evaluated for their hypoglycemic activity using alloxan-treated female albino mice by Faidallah et al. [74]. Kini and Ghate [75] studied the oral hypoglycemic activity of 3-[5'-methyl-2'-aryl-3'-(thiazol-2''-yl amino) thiazolidin-4'-one]coumarin derivatives and reported that electronegativity of the groups seemed to be responsible for variation in hypoglycemic activity (Scheme 64).

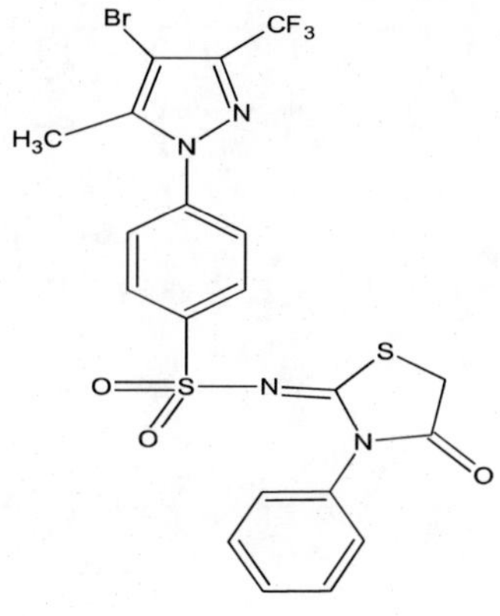

Scheme 64.

3.8. Antiulcer Activities

A series of thiazolidine-4-one derivatives were prepared by Taranalli et al. [76] from sulfanilamide and evaluated for anti-inflammatory, analgesic and antiulcer activities. Antiulcer activity was investigated by pylorus ligation ulcer model. The anti-inflammatory, analgesic and antiulcer activity was performed in 100 mg/kg b.w. rats. The nimesulide was used as standard drug for comparison. The following derivatives showed potential activity (Scheme 65).

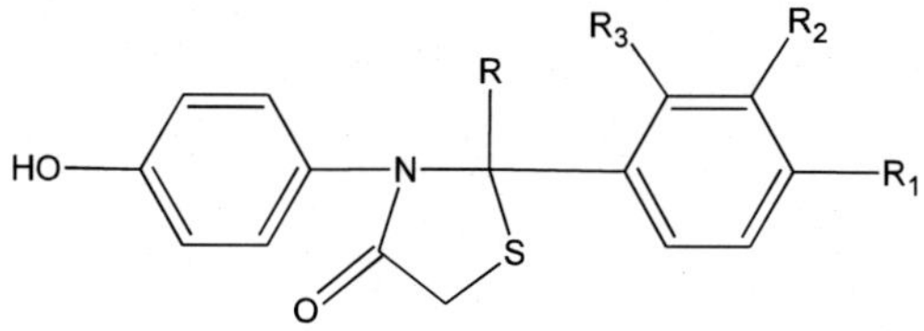

$R = R_1 = R_2 = R_3 = H, CH_3$

Scheme 65.

3.9. Antioxidant Activity

New coumarin derivatives, namely (2-(4-methyl-2-oxo-2*H*-chromen-7-yloxy)-*N*-(4-oxo-2-phenylthiazolidin-3-yl)acetamide, *N*-(2-(3-methoxyphenyl)-4-oxothiazolidin-3-yl)-2-(4-methyl-2-oxo-2*H*-chromen-7-yloxy)acetamide, 2-(4-methyl-2-oxo-2*H*-chromen-7-yloxy)-*N*-

(4-oxo-2-2,3,4trimethoxyphenyl)thiazolidin-3-yl)acetamide and *N*-(2-(4-bromophenyl)-4-oxothiazolidin-3-yl)-2-(4-methyl-2-oxo-2*H*-chromen-7-yloxy)acetamide) were synthesized starting from 4-methyl-7-hydroxycoumarin by Naceur et al. [77]. The structures of the obtained compounds were confirmed by analytical IR and NMR spectra to elucidate the different positions of protons and carbons and as well as theoretical studies. The new compounds were screened for antioxidant activity. Some of them were exhibited activity (Scheme 66).

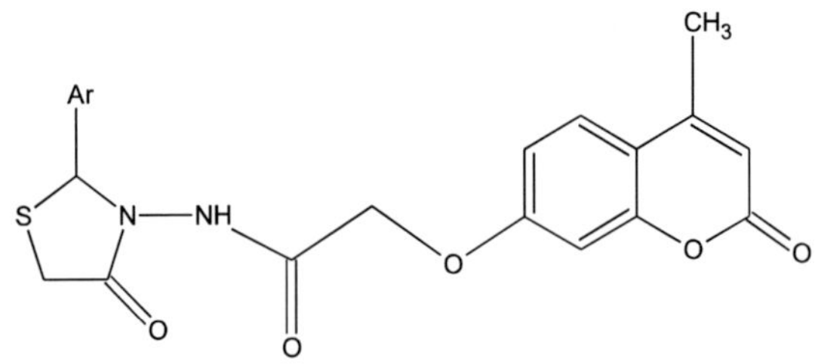

Ar = C_6H_5, p-$OCH_3C_6H_4$, 2,3,4-OCH_3C_6H2, p-BrC_6H_4

Scheme 66.

3.10. Anti-HIV Activity

Sriram et al. [78] synthesized several 1,3-thiazolidinone-4-one bearing various substituted diaryl ring at C-2 and N-3 position, utilizing microwave irradiation and evaluated for their anti HIV and anti YFV activities. The result of the *in vitro* anti HIV evaluation showed that 2-(2-chlorophenyl)-3-pyridin-2-yl-1,3-thiazolidin-4-one (12) proved to be an effective inhibitor of HIV-1 replication with EC_{50} of 10 im, CC_{50} of 120 im and percentage protection of 104%. These results indicated that the presence of halogen at C-2 and C-6 position on phenyl ring and the presence of 2-pyridyl substituent at N-3 position of thiazolidinone ring were important requirement for anti HIV activity (Scheme 67).

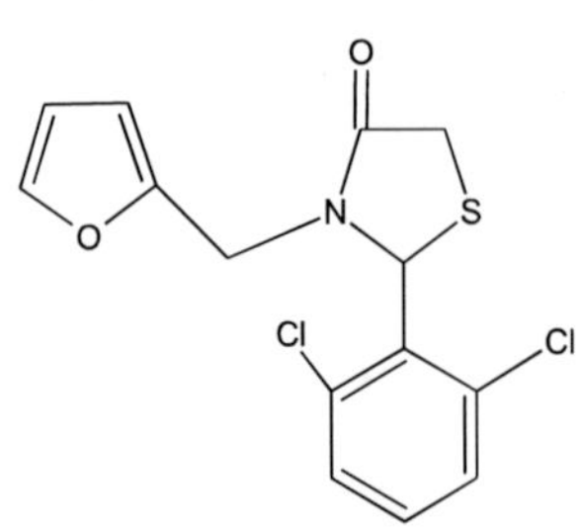

Scheme 67.

3.11. Antihistaminic Activity

Diurno et al. [79] prepared a series of 2-(Substituted-phenyl)-3-[3-(N,N-dimethylamino)propyl]-1,3-thiazolidin-4-ones. The H1-antihistaminic activity of the synthesized compounds was evaluated in vitro by measuring their ability to inhibit the histamine-induced contractions of isolated guinea-pig ileum and the activity reported, as pA2 values. The SAR study showed dependence of the potency of the H1-histamine antagonism on the m- and p-substituents suggesting that the aromatic moiety binds the receptor by a strong pi-interaction. Electron withdrawing substituents decreased the potency. While the electron-donating alkyl substituents, enhanced the aryl HOMO energy and increased the antihistamine activity. The m-substituents with the capability to form hydrogen bonds seemed to share an extra interaction with hydrogen accepting or donating groups of the histamine receptor and exhibited very high potency (Scheme 68).

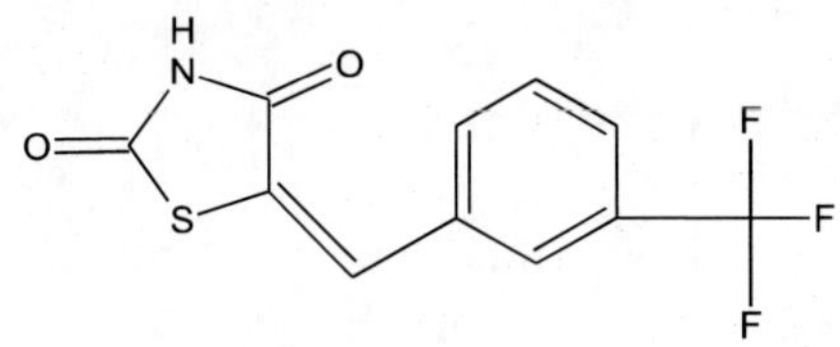

Scheme 68.

3.12. Cardiovascular Effect

Yoshiyuki et al. [80] carried out research on CP-060S a novel 4-thiazolidinone cardio protective drug and evaluated for its effect on cardiac function and myocardial oxygen consumption (MVO2) in anesthetized dogs. CP-060S (10-300 mg/kg IV) decreased heart rate, increased aortic flow and decreased mean blood pressure in a dose-dependent manner. The PR interval was significantly prolonged by administration of CP-060S (300 mg/kg IV). CP-060S (10-300 mg/kg IV) increased coronary blood flow in a dose dependent manner. Left ventricular end-diastolic pressure and maximal first derivative of left ventricular pressure were not significantly affected. CP-060S (10-300 mg/kg IV) increased coronary sinus blood flow and decreased arteriovenous oxygen difference and MVO2 in a dose-dependent manner. The effects of CP-060S on cardiac function and MVO2 were qualitatively similar to those of diltiazem, a typical Ca^{2+} antagonist (Scheme 69).

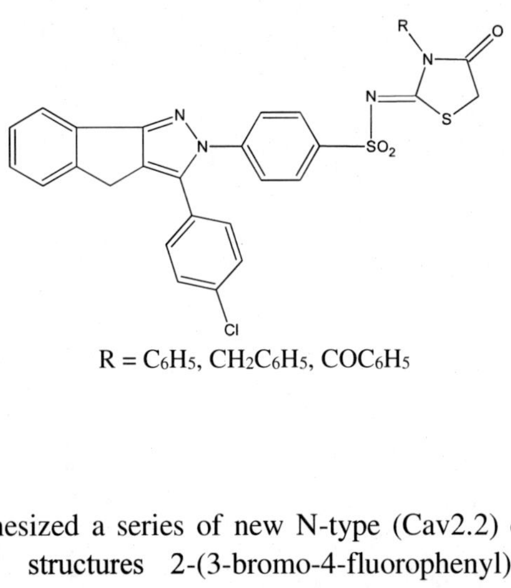

R = C_6H_5, $CH_2C_6H_5$, COC_6H_5

Scheme 69.

Lars et al. [81] synthesized a series of new N-type (Cav2.2) calcium channel blockers derived from the 'hit' structures 2-(3-bromo-4-fluorophenyl)-3- (2-pyridin-2-ylethyl) thiazolidin-4-one and its 2-[4-(4-bromophenyl)pyridin-3-yl]-3-isobutyl analogue. Extensive

SAR study of these series by using a range of synthetic approaches resulted in novel, patented compounds with IC_{50} values of up to 0.2 μM in an *in vitro* IMR32 assay, and selectivities for N/L of up to 30-fold. Compound 2-[4-(4-bromophenyl)pyridin-3-yl]-3-isobutylthiazolidine-4-one was found as the most potent, and single enantiomer 2-[4-(4-(tri fluro methyl) phenyl)pyridin-3-yl]-3-isobutyl-thiazolidine-4-one was found as the most selective compounds in this series. These compounds showed promise as lead structures in the quest for clinically effective N-type blockers in the treatment of pain.

3.13. Hypolipidaemic Activity

Gopalan et al. [82] prepared three 4-thiazolidinones, two with nicotinamide (NAT1 and NAT2) and one with 4-chlorophenoxyacetamide (PAT1) side chains. These compounds were evaluated for their hypolipidaemic and hypoglycaemic activity in Swiss albino mice fed a high-fat diet along with fructose administered in drinking water. NAT1 and PAT caused reduction of elevated triglycerides, cholesterol and glucose; NAT2 was effective only against triglycerides and has modest activity. None of them exerted any major adverse effects on the liver. Nicotinamide side chain contributed to the lipid lowering effect of both NAT1 and NAT2, but the bulky group of the latter affected proper binding to the receptor sites, making it ineffective against elevated cholesterol. On the other hand, the 4-chlorophenoxyacetamide side chain of PAT exerted powerful hypolipidaemic activity, despite the bulky substitution at C2. The thiazolidinone ring was responsible for the lipid lowering effect, which is however, modified by the type of substitutions at C2 and N of the ring. There was a structural resemblance between these molecules (4-thiazolidinones) and the glitazones (thiazolidindiones). The latter drugs acted through peroxisome proliferator activated receptor gamma (PPARγ). It was possible that the 4-thiazolidinone derivatives employed in this work may also have exerted some effect through PPARγ receptors (Scheme 70).

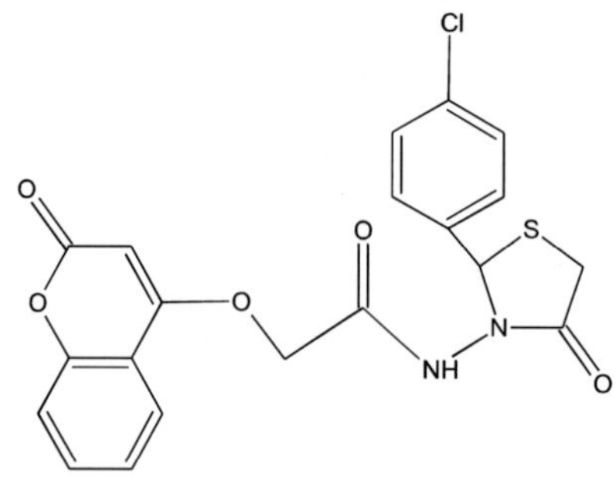

Scheme 70.

Jacob et al. [83] prepared a new molecule, incorporating a p-methoxy phenyl ring, a thiazolidinone-4-one and a nicotinyl moiety, and evaluated for its effect on serum total cholesterol, triglyceride and transaminases in rats. The effects were compared with those of nicotinic acid, a well known hypolipidemic drug. The compound 2-(4-methoxy phenyl)-3-nicotinamidothiazolidin-4-one, exhibited hypolipidemic effect with a significant reduction (19%) in total cholesterol, which was comparable with that of nicotinic acid (23.8%) and a significant reduction of triglycerides level (22.3%) while nicotinic acid caused a 26% reduction (Scheme 71).

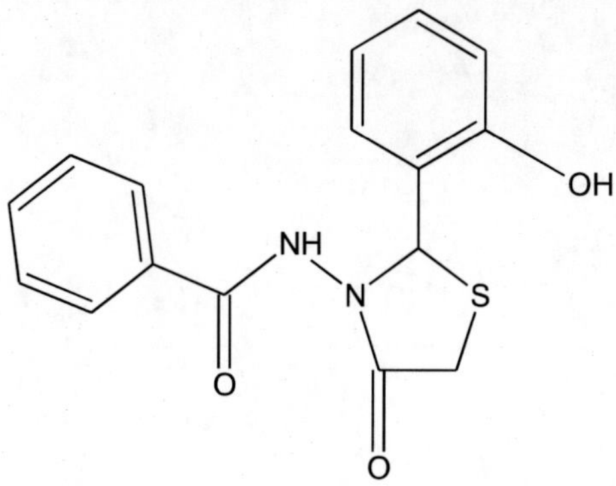

Scheme 71.

3.14. Anthelmintic Activity

Vipin et al. [84] synthesized a series of 2-aryl-3-substituted benzamido-1,3-thiazolidin-4-ones and all these compounds were evaluated for anthelmintic activity against *Pheritima posthuma* and Eudrilus sp. by Garg's method. Piperazine citrate was used as standard anthelminitic drug to compare the anthelminitic activity of synthesized compounds. The results showed that the compound N-[2-(2-hydroxyphenyl)-4-oxo-1,3-thiazolidin-3-yl]benzamide is a very potent and active against both the sepsis. All other compounds showed moderate to good anthelmintic activity (Scheme 72).

Brody et al. [85] prepared 3-Methyl-5-[(4-nitrophenyl)azo]rhodanin, nitrodan which was reported as a potent anthelmintic compound. This was effective when administered in feed against *Hymenolepsis nana* and *Syphaciaobvelata* infections in mice, *Asceridiagalli* infections in chickens, and *Toxoceracanis*, *Ancylostomacaninum*, and *Uncinariastenocephala* infections in dogs, pigs and horses (Scheme 73).

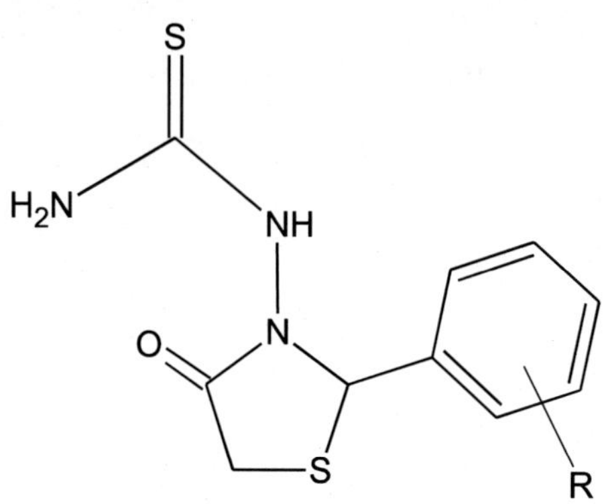

R = 4-N(CH_3)$_2$, 4-OCH_3

Scheme 72.

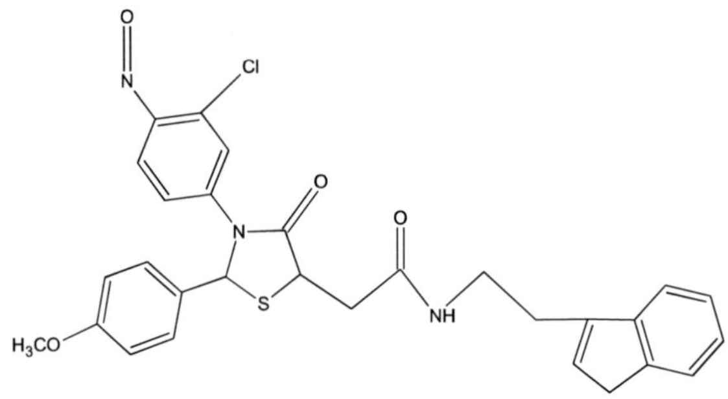

Scheme 73.

3.15. Antiarrhythmic Activity

The most common features of cardiac arrhythmias are atrial flutter and atrial fibrillation which are associated with an increase in heart failure, stroke, and mortality. Blockade of the Kv1.5 ion channel is a potentially artial-selective avenue for the treatment of atrial fibrillation and atrial flutter. Jackson et al. [86] reported synthesis and biological evaluation of thiazolidinone-based blockers of Kv1.5. The 3,4-dimethyl derivatives (IC_{50} = 0.069 lM) and (IC_{50} = 0.270 lM) were found to be the most potent compounds of this series. In 2009, Bhandari and co-workers discovered 2-(2-(4-(3-((5-substituted methylene)-4-oxo-2-(phenylimino)thiazolidin-3-yl)-2-hydroxypropylamino) benzoyl) hydrazinyl)-2-oxoethyl-nitrate derivatives and evaluated them for electrocardiographic, antiarrhythmic, vasorelaxing and antihypertensive activity as well as for *in vitro* nitric oxide (NO) releasing ability and found that some of these heterocycles are very potent. Compound 2-(2-(4-(3-(5-benzyliden-4-oxo-2-(phenylimino) thiazolidin-3-yl)-2-hydroxy propylamino) benzoyl)hydrazinyl)-2-oxoethyl nitrate, was found to be the most potent in this series (Scheme 74).

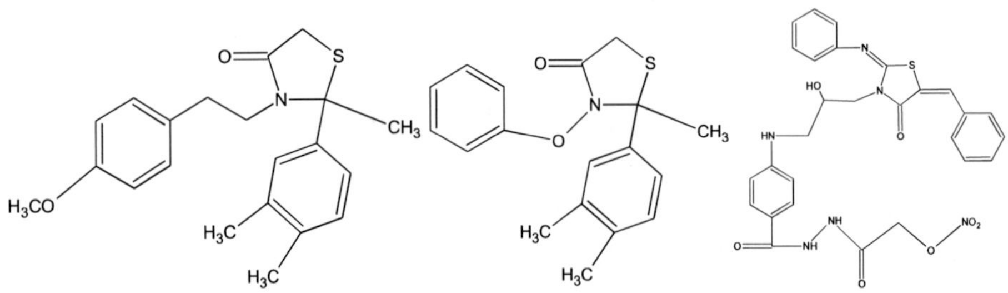

Scheme 74.

CONCLUSION

The 4-thiazolidinone is not only synthetically important scaffold but also possesses a wide range of promising biological activities. The present study gives an insight into the pharmacological profiles of 4-thiazolidinone-based heterocyclic frameworks. The literature reveals that 4-thiazolidinone has diverse medicinal potential, and therefore, syntheses of these types of compounds have attracted chemists, pharmacologists and researchers. This book chapter will motivate medicinal chemists and scientists to study structure activity relationship of 4-thiazolidinone-derived compounds in detail. Therefore, 4-thiazolidinone scaffold is competent in future drug design and development process. It is therefore, expected to undertake synthesis of novel 4-thiazolidinones that may improve specific medicinal properties with outstanding therapeutic potential.

REFERENCES

[1] Andrei, Y. 2014. "Introduction: Small Heterocycles in Synthesis" *Chemical Reviews* 114(16): 7783.

[2] Hassan, F. A. and Younus, K. W. 2012. "Biological Evaluation of Some Azole Derivatives in Cooling Fluids (Lubricant Oils)" *Research Journal of Biological Sciences* 7(1): 48–51.

[3] Horton, D. A., Bourne, G. T. and Smyth, M. L. 2003. "The combinatorial synthesis of bicycilc privileged structures or privileged substructures" *Chemical Reviews*103: 893.

[4] Laurent, D. R., Gao, Q. and Wu, D. D. 2004. "Regioselective synthesis of 3-(heteroaryl)-iminothiazolidin-4-ones" *Tetrahedron Letters* 45(9): 1907–1910.

[5] Frances, C. B. 1961. "4-Thiazolidinone" *Chemical Reviews* 61(5): 463–521.

[6] Davinder, P., Awanit, K., Praveen, K. S. and Mahendra, N. 2011. "Design, synthesis and antimicrobial evaluation of novel 2-aryl-thiazolidin-4-one derivatives" *Organic and Medicinal Chemistry Letters* 1(4): 1–7.

[7] Knott, E. B. 1954. "The electrophilic reactivity of alkoxyalkilydene derivatives of heterocyclic keto methylene compounds" *Journal of the Chemical Society* 1482-1490.

[8] Markovic, R., Stojanovic, M. 2005. "Stereo controlled synthesis of new tetrahydrofuro[2,3-d]thiazole derivatives via activated vinylogousiminium ions" *Heterocycles* 56: 2635.

[9] Pawar, R. B., Mulwad, V. V. 2004. "Synthesis of some biologically active pyrazole, thiazolidinone and azetidinone derivatives" *Chemistry of Heterocyclic Compounds* 40(2): 219–226.

[10] Ocal, N., Aydogan, F., Yolacan, C. and Turgut, Z. 2003. "Synthesis of some furo-thiazolidine derivatives starting from aldimines" *Journal of Heterocyclic Chemistry* 40 (4): 721–724.

[11] Mali, R. A., Pratap, J. R., Netankar, U. R. and Mane, P. D. 2009. "An efficient synthetic route for quinazolinyl 4-thiazolidinones" *Tetrahedron Letters* 50(35): 5025–5027.

[12] Bolli, M. H., Abele, S., Binkert, C., Bravo, R., Buchmann, S., Bur, D. and Weller, T. 2010. "2-imino-thiazolidin-4-one derivatives as potent, orally active S1P1 receptor agonists" *Journal of Medicinal Chemistry* 53: 4198.

[13] Cunico, W. 2007. "One-pot synthesis of 2-isopropyl-3-benzyl-1,3-thiazolidin-4-ones and 2-phenyl-3-isobutyl-1,3-thiazolidin-4-ones from valine, arenealdehydes and mercaptoacetic acid" *Tetrahedron Letters* 48: 6217–6220.

[14] Jieping, Z. and Blanchet, J. R. 2004. "Synthesis of 2-imino-4-thiazolidinone from alkyl (aryl) trichloromethylcarbinol revisited, a three-component process from aldehyde, chloroform and thiourea" *Tetrahedron Letters* 45: 4449–4452.

[15] Desai, K. R. and Mistry, K. 2006. "Microwave assisted synthesis of nitrogen and sulphur containing heterocyclic compounds and their pharmacological evaluation" *Indian Journal of Chemistry* 45B: 1762–1766.
[16] Bose, A. K., Manhas, M. S., Banic, B. K and Robb, E. W. 1994. "Microwave Induced organic reaction enhancement chemistry" *Research on Chemical Intermediates* 1: 20.
[17] Cameron, L., Maqual, G. and Julie, M. 2003. "A retrospective review of thiazolidinediones with development of a troglitazone conversion protocol" *The Scientific World Journal* 3: 477–483.
[18] Liesen, A. P., Aquino, T. M., Carvalho, C. S, Lima, V. T. and Araujo, J. M. 2010. "Synthesis and evaluation of anti-Toxoplasma gondii and antimicrobial activities of thiosemicarbazides, 4-thiazolidinones and 1,3,4-thiadiazoles" *European Journal of Medicinal Chemistry* 45 (9): 3685–3691.
[19] Kavitha, C. V., Basappa, S. and Nanjunda, S. 2006. "Synthesis of new bioactive venlafaxine analogs: Novel thiazolidin-4-ones as antimicrobials" *Bioorganic and Medicinal Chemistry* 14 (7): 2290–2299.
[20] Gopalakrishnan, M., Thanusu, J. and Kanagarajan, V. J. 2009. "Design, synthesis, spectral analysis and in vitro microbiological evaluation of 2-phenyl-3-(4,6-diarylpyrimidin-2-yl)thiazolidin-4-ones" *Journal of Enzyme Inhibition and Medicinal Chemistry* 24: 1088–1094.
[21] Singh, T., Srivastava, V. K., Saxena, K. K, Goel, S. L. and Kumar, A. 2006. "Synthesis of new thiazolylthiazolidinylbenzothiazoles and thiazolylazetidinylbenzothiazoles as potential insecticidal, antifungal, and antibacterial agents" *Arch Pharm Chem Life Sci* 339: 466–472.
[22] Sriram, D., Yogeeswari, P. and Ashok, K. T. G. 2005. "Microwave-Assisted Synthesis, And AntiHIV Activity Of 2,3-Dairyl-1,3-Thiazolidin-4-Ones" *Indian Journal of Pharmaceutical Sciences* 67 (4): 496–499.
[23] Augusta, Z., Andrejus, Z., Eduardas, T., Alvydas, P. and Vidmantas, D. 2012. *Acta Poloniae Pharmaceutica and Drug Research* 69 (5): 911–915.
[24] Tribhuvan, S. and Deepak, K. 2013. "Synthesis and evaluation of thiazolidine-4-one for their antibacterial activity" *Journal of Harmonized Research in Pharmacy* 2(1): 66–70.
[25] Solankee, A. N., Patel, K. P. and Patel, R. B. 2012. "Efficient synthesis and pharmacological evaluation of some new 4-thiazolidinones and 5-arylidenes" *Archives of Applied Science Research* 4(1): 72–77.
[26] Davinder, P., Awanit, K., Praveen, K. S. and Mahendra, N. 2011. "Design, synthesis and antimicrobial evaluation of novel 2-aryl-thiazolidin-4-one derivatives" *Organic and Medicinal Chemistry Letters* 1(4): 1–7.
[27] Deepak, K., Bux, F. B., Varsha, P. and Arun, S. 2012. "Thiazolidinone: Synthesis and biological studies" *Der Pharma Chemica* 4(1): 538–543.
[28] Divyesh, P., Premlata, K. and Navin, P. 2010. "Synthesis and characterization of some new thiazolidinones containing coumarin moiety and their antimicrobial study" *Archives of Applied Science Research* 2 (6): 68–75.
[29] Parmar, K., Prajapati, S., Patel, R. and Patel, R. 2011. "A simple and efficient procedure for synthesis of biologically active 1,2,4-triazolo-[3,4-b]-1,3,4-thiadiazole-2-aryl-thiazolidine-4-one derivatives" *Research Journal of Chemical Sciences* 1(1): 18–24.
[30] Ameya, A. C. and Nandini, R. P. 2007. "Synthesis and antimicrobial screening of 5-arylidene-2-imino-4-thiazolidinones" *ARKIVOC* 16: 148–155.

[31] Vilas, S., Ravi, T. and Venkataramana, C. H. S. 2010. "Synthesis and antimicrobial activity of some novel thiazolidinone derivatives" *Der Pharmacia Lettre* 2 (5): 146–154.

[32] Srivastava, S. K, Smita, V. and Srivastava, S. D. 2010. "Synthesis, characterization and biological activity of 1,3-thiazolidin-4-one derivatives of 2-mercapto-5-methyl-1,3,4-thiadiazole" *Journal of Chemical and Pharmaceutical Research* 2 (5): 270–276.

[33] Mishra, R. R., Nimavat, K. S and Vyas, K. B. 2012. "Synthesis and evaluation of thiazolidinone for antibacterial activity and antifungal activities" *Rasayan Journal of Chemistry* 5(2): 214–219.

[34] Ritu, S., Pushkal, S., Savitri, D. S. and Santosh, K. S. 2012. "Synthesis and biological activity of 4-thiazolidinone derivatives of phenothiazine" *Journal of the Serbian Chemical Society* 77(1): 17–26.

[35] Ranga, R. K., Raghavendra, G. P., Santosh, K. B. and Ravindranath, L. K. 2014. "Synthesis and antimicrobial evaluation of novel spiro compounds containing thiazolidinone and 1,3,4-oxadiazole moieties" *Journal of Chemical Technology and Metallurgy* 49 (3): 238–246.

[36] Rathod, M. S. and Jadhav, S. Z. 2015. "Synthesis, spectral and microbial studies of some methyl-4-thiazolodinone derivatives"*International Journal of Advances in Science Engineering and Technology* 1: 61–63.

[37] Nagaraj, A, Ravi, G., Sharath, K. S. and Nageswara, R. G. 2012. "Synthesis of new biologically active compounds containing linked thiazolyl-thiazolidinone heterocycles" *Organic Communication* 5 (4): 160–170.

[38] Pankaj, B., Girendra, G. and Gupta, S. K. 2017. "Synthesis of a series of thiazolidinone derivatives and evaluation of their antimicrobial properties" *International Research Journal of Pharmacy* 8 (4): 121–125.

[39] Amit, G., Rajendra, S., Pankaj, K. S. and Shailendra, K. S. 2016. "Novel 4-Thiazolidinone derivatives as anti-Infective agents: synthesis, characterization, and antimicrobial evaluation" *Biochemistry Research International* doi.org/10.1155/2016/8086762.

[40] Barbaros, E., Keriman, O., Oya, U., Sriram, D., Perumal, Y. and Ayla, B. 2016. "Novel thiazolidinone-azole hybrids: Design, synthesis and antimycobacterial activity studies" *Iranian Journal of Pharmaceutical Research* 15 (4): 783–790.

[41] Kucukguzel, S. G., Oruc, E. E., Rollas, S., Sahin, F. and Ozbek, A. 2002. "Synthesis, characterisation and biological activity of novel 4-thiazolidinones, 1,3,4-oxadiazoles and some related compounds" *European Journal of Medicinal Chemistry* 37 (3): 197–206.

[42] Jaju, S., Palkar, M., Maddi, V., Ronad, P. K., Mamledesai, S., Satyanarayana, D. and Ghatole, M. 2009. "Synthesis and antimycobacterial activity of a novel series of isonicotinylhydrazide derivatives" *Arch Pharm Chem Life Sci* 342: 723–731.

[43] Karali, N., Terzioglu, N. and Gursoy, A. 2002. "Synthesis and primary cytotoxicity evaluation of new 5-bromo-3-substituted-hydrazono-1H-2-indolinones" *Arch Pharm Pharm Med Chem* 335 (8): 374–380.

[44] Dharmrajan, S., Perumal, Y. and Ashok, K. T. G. 2005. "Microwave-assisted synthesis and anti YFV activity of 2,3-diaryl-1-3-thiazolidin-4-one" *Journal of Pharmaceutical Sciences* 8(3): 426–429.

[45] Salunke, M. S. 2015. "Synthesis and antitiuberculer activity of thiozolidenone derivatives" *Indian Streams Research Journal* 5(9): 1–7.

[46] Chandrappa, S., Kavitha, C. V., Shahabuddin, M. S., Vinaya, K., Anandakumar, C. S., Ranganatha, S. R., Raghavan, S. C. and Rangappa, K. S. 2009. "Synthesis of 2-(5-((5-(4-chlorophenyl)furan-2-yl)methylene)-4-oxo-2-thioxothiazolidin-3-yl)acetic acid derivatives and evaluation of their cytotoxicity and induction of apoptosis in human leukemia cells" *Bioorganic & Medicinal Chemistry Letters* 17 (6): 2576–2584.

[47] Gududuru, V., Hurh, H., Dalton, J. T. and Miller, D. D. 2005. "Discovery of 2-arylthiazolidine-4-carboxylic acid amides as a new class of cytotoxic agents for prostate cancer" *Journal of Medicinal Chemistry* 48: 2584–2588.

[48] Karali, N., Kocabalkanl, A., Gursoy, A. and Ates, O. 2002. "Synthesis and antitubercular activity of 4-(3-coumarinyl)-3-cyclohexyl-4-thiazolin-2-one benzylidenehydrazones" *II Farmaco* 57: 589.

[49] Havrylyuk, H., Kovach, N., Zimenkovsky, B., Vasylenko, O. and Lesyk, R. 2011. "Synthesis and anticancer activity of isatin-based pyrazolines and thiazolidines conjugates" *Arch Pharm Chem Life Sci* 344: 514.

[50] Danniel, D. F., Antonio, J. A. and Gildo, L. 2012. "Synthesis, antimicrobial and cytotoxic activities of 5-benzylidene-2-[(pyridine-4-ylmethylene) hydrazono]-thiazolidin-4-one and 2-[(pyridine-4-ylmethylene)hydrazono]-thiazolidin-4-one derivatives" *Quim Nova* 35(4): 694–698.

[51] Devinyak, O. T., Havrylyuk, D. Y., Avdieiev, S. S., Chumak, V. V., Panchuk, R. R., Stoika, R. S., Kavsan, V. M. and Lesyk, R. B. 2014. "Virtual screening and its experimental validation reveal novel compounds with promising anticancer activity among 4-thiazolidinone-pyrazoline- and isatin-based conjugates" *Austin Journal of Bioorganic and Organic Chemistry* 1 (1): 1–6.

[52] Monforte, P. and Verma, S. K. 2001. "Discovery of 2, 3-diaryl-1,3-thiazolidin-4-ones as potent anti-HIV-1 agent" *Bioorganic & Medicinal Chemistry Letters* 11: 1793–1796.

[53] Rawal, R. K., Katti, S. B., Kaushik-Basu, N., Arora, P. and Pan, Z. 2008. "Non-nucleoside inhibitors of the hepatitis C virus NS5B RNA-dependant RNA polymerase: 2-aryl-3-heteroaryl-1,3-thiazolidin-4-one derivatives" *Bioorganic & Medicinal Chemistry Letters* 18(23): 6110–6114.

[54] Kumar, A., Rajput, C. S. and Bhati, S. K. 2007. "Synthesis of 3-[4'-(p-chloro phenyl) thiazol-2'-yl]-2[(substituted azetidinone/thiazolidinone)-amino methyl]-6-bromo quinazolin-4-ones" *Bioorganic & Medicinal Chemistry* 15: 3089–3096.

[55] Deep, A., Jain, S. and Sharma, P. C. 2010. "Synthesis and anti-inflammatory activity of some novel biphenyl-4-carboxylic acid 5-(arylidene)-2-(aryl)-4-oxothiazolidin-3-yl amides" *Acta Poloniae Pharmaceutica Drug Research* 67 (1): 63–67.

[56] Ottana, R., Mazzon, E., Dugo, L., Monforte, F., Maccari, R., Sautebin, L., De Luca, G., Vigorita, M. G., Alcaro, S., Ortuso, F., Caputi, A. P. and Cuzzocrea, S. 2002. "Modeling and biological evaluation of 3,3'-(1,2-ethanediyl)bis[2-(4-methoxyphenyl)-thiazolidin-4-one], a new synthetic cyclooxygenase-2 inhibitor" *European Journal of Pharmacology* 448: 71–80.

[57] Reetu, V. K. 2011. "Synthesis, pharmacological evaluation & docking studies of some pyridine derivatives of thiazolidin-4-ones" *Asian Journal of Biochemical and Pharmaceutical Research* 4 (1): 219–233.

[58] Kumudha, D., Reddy, R. R., Kalavathi, T. and Leonard, J. T. 2012. "Synthesis and evaluation of substituted aryl thiazolidin-4-ones for analgesic and anti-inflammatory activities" *Der Pharmacia Lettre* 4 (4): 1149–1154.

[59] Pravesh, G., Pankaj, S., Birendra, S., Jaya, S. and Shikha, S. 2013. "Biological evaluation of some newly synthesized thiazolidinone derivatives for their analgesic activity" *International Journal of Current Trends in Pharmaceutical Research* 1 (3): 185–188.

[60] Anbalagand, N., Shanmugapandiyan, P., Denshing, K. S., Ilavarasan, R. and Nirmala, R. 2010. "Synthesis and biological activity of 2-(Thiazolidin-4-One)phenyl]-1h-phenylbenzimidazoles and 2-[4-(Azetidin-2-One)-3-chloro-4-phenyl]-1H-phenylbenzimidazoles" *International Journal of Pharmaceutical Sciences and Drug Research* 2 (2): 115–119.

[61] Waleed, A. B., Shaymaa, H. A. and Mohamed, E. S. 2014. "Synthesis and evaluation of new 2-iminothiazolidin-4-one and thiazolidin-2,4-dione derivatives as antimicrobial and anti-inflammatory agents" *Open Chemistry Journal* 1: 33–38.

[62] Malyshev, K. V. A., Finiuk, N. S., Pavlenko, O. K., Havr, D. Y., Le, R. B. S., Stoika, R. S. and Korc, H. O. G. 2017. "4-Thiazolidinone-based derivatives rescue TNAα-inhibited osteoblast differentiation in mouse mesenchymal precursor cells" *Ukrainian Biochemical Journal* 89: 111–122.

[63] Khushbu, S., Akhil, N., Monica, C. and Arjun, M. 2017. "Synthesis of chalcone and fused rhodanine derivatives of biphosphonates active as anti-inflammatory and over breast cancer" *Innovations in Pharmaceuticals and Pharmacotherapy* 4 (1): 33–61.

[64] Shiradkar, M. R., Ghodake, M., Bothara, K. G., Bhandari, S. B., Nikalje, A., Akula, K. C., Desai, N. C. and Burange, P. J. 2007. "Synthesis and anticonvulsant activity of clubbed thiazolidinone-barbituric acid and thiazolidinone-triazolederivatives" *ARKIVOC* XIV: 58–74.

[65] Agarwal, A., Lata, S., Saxena, K. K., Srivastava, V. K. and Kumar, A. 2006. "Chemistry and Biological Activities of 1, 3-Thiazolidin-4-ones" *European Journal of Medicinal Chemistry* 41: 1223.

[66] Amin, K. M., Rahman, A. D. E. and Al-Eryani, Y. A. 2008. "Synthesis and preliminary evaluation of some substituted coumarins as anticonvulsant agents" *Bioorganic & Medicinal Chemistry* 16: 5377–5388.

[67] Archana. 2015. "Synthesis of newer quinazolin-4(3H)-onylthiazolidinones as potent anticonvulsant agents" *Saudi Journal of Medical and Pharmaceutical Sciences* 1(2): 42–46.

[68] Mehrdad, F., Reza, J., Seyed, A. E., Sayyed, A. T., Elham, R., Mehrnaz, L., Mohsen, A. and Ali, A. 2017. "Novel 4-thiazolidinone derivatives as agonists of benzodiazepine receptors: design, synthesis and pharmacological evaluation" *Excli Journal* 16: 52–62.

[69] Ulusoy, N., Ergen, N., Ekinci, A. C. and Ozer, H. 1996. "Synthesis and anticonvulsant activity of some new arylidenehydrazides and 4-thiazolidinones" *Monatshefte fur Chemie* 127: 1197–1202

[70] Kaur, H., Kumar, S., Vishwakarma, S. P., Sharma, M., Saxena, K. K. and Kumar, A. 2010. *European Journal of Medicinal Chemistry* 45: 2777–2783.

[71] Shingalapur, R. V., Kallappa, M., Rangappa S. K. and Hugar, M. H. 2010. *European Journal of Medicinal Chemistry* 45: 1753–1759.

[72] Terzioglu, N., Karali, N., Gursoy, A., Pannecouque, C., Leysen, P., Paeshuyse, J., Neyts, J. and Clercq, E. 2006. "Synthesis and primary antiviral activity evaluation of 3-hydrazono-5-nitro-2-indolinone derivatives" *ARKIVOC* I: 109–118.

[73] Chen, H., Bai, J., Jiao, L., Guo, Z., Yin, Q. and Li, X. 2009. "Design, microwave-assisted synthesis and HIV-RT inhibitory activity of 2-(2,6-dihalophenyl)-3-(4,6-dimethyl-5-

(un)substituted-pyrimidin-2-yl)thiazolidin -4-ones" *Bioorganic & Medicinal Chemistry* 17 (11): 3980–3986.

[74] Faidallah, H. M., Khan, K. A. and Asiri, A. M. 2011. "Synthesis and biological evaluation of new 3-trifluoromethylpyrazolesulfonyl-urea and thiourea derivatives as antidiabetic and antimicrobial agents" *Journal of Fluorine Chemistry* 132: 131–137.

[75] Kini, D. and Ghate, M. 2011. "Synthesis and oral hypoglycemic activity of 3-[5'-Methyl-2'- aryl-3'-(thiazol-2''-yl amino)thiazolidin-4'-one]coumarin derivatives" *European Journal of Chemistry* 8: 386–390.

[76] Taranalli, A. D., Thimmaiah, N. V., Srinivas, S. and Saravanan, E. 2009. "Anti-inflammatory, analgesic and antiulcer activity of certain thiazolidinones" *Asian Journal of Pharmceutical Clinical Research* 2: 79–83.

[77] Naceur, H., Abdullah, S. A., Ridha, B. S. and Alary, F. 2012. "Synthesis and characterization of new thiazolidinones containing coumarin moieties and their antibacterial and antioxidant activities" *Molecules* 17: 9321–9334.

[78] Sriram, D., Pandeya, S. N., Nath, G. and DeClercq, E. 1999. "Synthesis, antibacterial, antifungal and anti-HIV activities of Schiff and Mannich bases derived from isatin derivatives and N-[4-(4′-chlorophenyl) thiazol-2-yl] thiosemicarbazide" *European Journal of Pharmaceutical Sciences* 9 (1): 25–31.

[79] Diurno, M. V., Mazzoni, O. and Correale, G. 1999. "2-(Substituted phenyl)-3-3-(N,N-dimethylamino)propyl.-1,3-thiazolidin-4-ones Acting as H1-Histamine Antagonists" *IL Farmaco* 54: 579–583.

[80] Yoshiyuki, S., Akima, M. and Tamura, K. J. 1999. "Novel 4-thiazolidinone and its Cardio protective action" *Pharmacology* 32: 57–63.

[81] Lars, J. S. K. and Christopher, J. H. 2007. "Synthesis and SAR of novel 2-arylthiazolidinones as selective analgesic N-type calcium channel blockers" *Bioorganic & Medicinal Chemistry* 17: 662–667.

[82] Gopalan, K. N., Suma, P. M., Vivek, K., Robby, T. Z., Suman, K. and Mallikarjuna, R. C.2008. "Assessment of hypolipidaemic activity of three thiazolidin-4-ones in mice given high-fat diet and fructose" *Chemico-Biological Interactions* 17: 363–368.

[83] Jacob, N. and Kutty, G. N. 2004. "Synthesis and hypolipidemic activity of a thiazolidinone derivative" *Indian Drugs* 41 (2): 76–79.

[84] Vipinm, K., Satyanarayana, D. and Subrahmanyam, E. V. S. 2002. "Synthesis and Biological activities of 2-Aryl-3-substituted Benzamido-1,3-thiazolidin-4-ones" *Indian Journal of Heterocyclic Chemistry* 11: 251–252.

[85] Brody, G. and Elward, T. E. 1971. "A brief review on recent developments and exploring activities of 4-thiazolidinone" *Journal of Parasitology* 57: 108.

[86] Jackson, C. M., Blass, B., Coburn, K., Janusz, J. M., Murawsky, M. and Ridgeway, J. M. 2007. "Crystal structure of (±)-3-[(benzo[d][1,3]dioxol-5-yl)meth-yl]-2-(3,4,5-tri-meth-oxy-phen-yl)-1,3-thiazolidin-4-one" *Bioorganic & Medicinal Chemistry Letters* 17: 282.

In: Organic and Medicinal Chemistry, Volume 1
Editor: Bimal Krishna Banik
ISBN: 978-1-53614-453-6

Chapter 6

SYNTHESIS, SYNTHONS AND MEDICINAL CHEMISTRY OF ISOXAZOLINES AND ANALOGUES

Sudipta Ray, Amiya K. Medda and Dilip K. Maiti*
Department of Chemistry, University of Calcutta, Kolkata, India

ABSTRACT

Isoxazole is an indispensible heterocyclic moiety that has been found in a wide variety of natural products and bioactive molecules. This makes it an important building block in the field of organic synthesis, synthons and medicinal chemistry. Some drugs such as Cloxacillin, Valdecoxib and Dicloxacillin having isoxazolines and analogue isoxazole motifs, and those are found in immense applications such as efficient drugs in the present perspective disease scenario. Therefore, synthesis of functionalized isoxazoles and analogues has wider therapeutic application and is a topic of great interest for medicinal chemists. This chapter sums up the reported synthetic strategies of isoxalines and analogues, their transformations to useful synthons and their major therapeutic fields exploited in the literature.

Keywords: isoxazolines, medicinal activities of isoxazolines, asymmetric synthesis, key synthons, chemistry of isoxazolines

INTRODUCTION

Isoxazolines (**1a-c**, Figure 1) are partially unsaturated indispensable heterocyclic motifs that are found in many marketed drugs. Isoxazolines (**1**) and its aromatic analogue isoxazole (**2**) are five membered heterocycles bearing oxygen and nitrogen atoms at adjacent positions. The completely saturated analogue is known as isoxazolidine (**3**). The isoxazolines (**1**), isoxazoles (**2**) and isoxazolidines (**3**) are widely used in pharmaceuticals and therapeutics because of their diverse biological activities. Isoxazolines and isoxazoles are considered as

* Corresponding Author Email: dkmchem@caluniv.ac.in

versatile synthons in organic synthesis, which have been extensively and efficiently transformed into various classes of medicinally useful molecules. For instance, the Δ^2-isoxazoline (**1a**) heterocyclic derivatives were known to exhibit insecticidal, antibacterial, antibiotic, antitumour, antifungal, antituberculosis, anticancer, ulcerogenic, COX-2 inhibitor and other pharmacological properties. In three ways, this book chapter describes the isoxazoline compounds: (i) Development of different methods to synthesize isoxazolines, isoxazoles, and isoxazolidines; (ii) transformation of the heterocycles to various synthons and (iii) their medicinal activities.

Structure and Nomenclature of Isoxazoline

The isoxazolines (**1a-c**, Figure 1) are similar to isoxazole but different in terms of the presence of an extra double bond to make the later one aromatic. In the isoxazole (**2**) the 'CH' group, adjacent to oxygen atom of furan, is replaced by 'N' atom making more polar in nature compare to furan. Isoxazoles are unsaturated aromatic heterocyclic compounds containing a ring with three carbon atoms, one oxygen atom and one nitrogen atom. The trivial name for the title five-membered fully unsaturated heterocycles as "isoxazole" was originally proposed by Hantszch as it was the isomer "oxazole" discovered first. The trivial name follows the Hantszch-Widman system of nomenclature: the prefix "iso" represents isomer, "oxa" represents the oxygen atom "aza" represents the nitrogen atom the suffix "ole" denotes the ring size as five-membered; altogether the derived name is "isoxazole" [1]. This name has been accepted in IUPAC and has been used in Chemical abstracts. In the Chemical Abstracts, the other systematic name 1,2-azole, is also used.

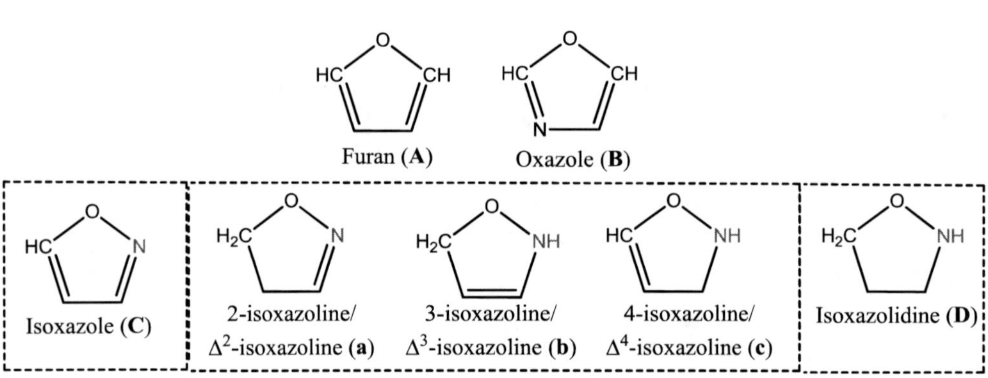

Figure 1. N-O bond-bearing five membered heterocycles.

CHEMISTRY OF ISOXAZOLE

The nitrogen hetero atom is more pronounced for electron withdrawing effect, while the oxygen atom for electron donating effect. As neutral molecules, isoxazoles undergo electrophilic substitution rather more readily at the position 4, than benzene. Effects of substituents can modify their behavior. Substituents at the position-5 apparently have more activating and deactivating effect than substituents at the position 3. In natural product synthesis, isoxazoles are used as latent synthons, such as masked new heterocyclic rings, masked fused rings, masked aromatic rings and masked aldol and related moieties [2]. The capability of isoxazole undergoing reaction is diverse: protonation, quaternization, complexation, oxidation, reduction, carbanionic condensations, thermolysis, photolysis, transformations into other heterocyclic ring systems and reaction with electrophiles, nucleophiles and Grignard reagents [3].

Like oxazole and isoxazole, isoxazoline is an immensely important heterocyclic molecule that has been found in many bioactive natural products [4]. Isoxazoline containing drug molecules are available in the market due to its antibacterial, antioxidant, antimicrobial, cytotoxicity and anti-inflammatory activity [5]. Isoxazoline is used as COX-2 inhibitor in the market [6]. Some of the bioactive natural products and drug molecules containing isoxazoline moiety are given below (Figure 2).

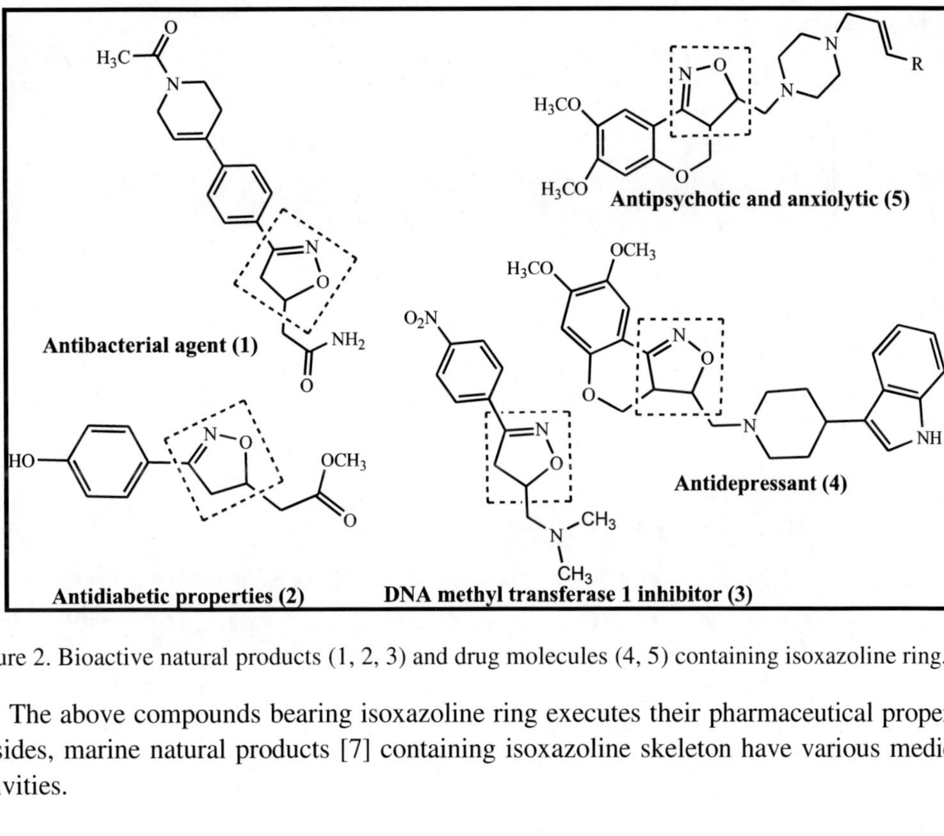

Figure 2. Bioactive natural products (1, 2, 3) and drug molecules (4, 5) containing isoxazoline ring.

The above compounds bearing isoxazoline ring executes their pharmaceutical properties. Besides, marine natural products [7] containing isoxazoline skeleton have various medicinal activities.

Synthetic Methods

Various synthetic approaches for isoxazoles and isoxazlines are reported in the literature: i) by cycloaddition reaction; ii) by metal catalyst; iii) by organocatalysis; iv) in aqueous organized media (AOM); v) nanoparticle mediated.

From the retrosynthetic point of view, as stated below, it is understood that as isoxazoline is a five member ring, therefore the ring can be constructed by a pricyclic reaction of the type [4+2] π-cycloaddition that is familiar as 1,3-dipolar cycloaddition reaction [8]. The 1,3-dipolar cycloaddition reaction either can be intermolecular or intramolecular. In both the situation dipole and dipolarophile are necessary. Two π-electrons of the dipolarophile (alkene or alkyne) and four electrons of the dipolar compound participate in a concerted manner. The addition is stereoconservative (suprafacial) and the reaction is therefore a $[\pi^4{}_s+\pi^2{}_s]$ cycloaddition. 1,3-Dipolar cycloaddition (DC) reaction has found versatile applications in chemistry and its allied branches of science and technology because this powerful atom-economical synthetic tool is capable of constructing 5-membered heterocyclic rings in a regio- and stereoselective way to furnish designed natural and synthetic functional molecules [9].

To synthesize isoxazoline ring, alkene is taken as a dipolarophile and nitrile oxide as 1,3-dipole (Figure 3).

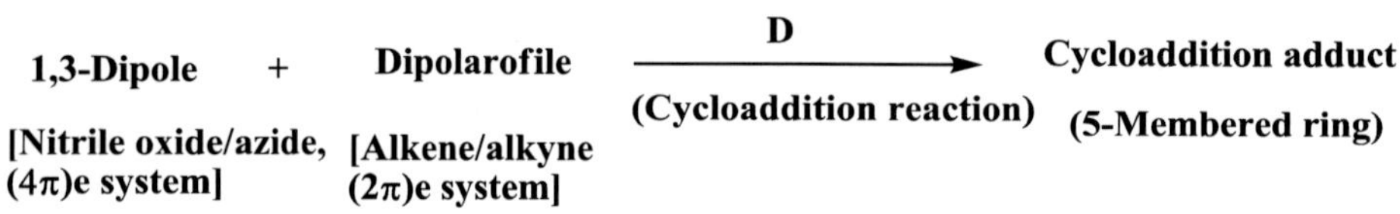

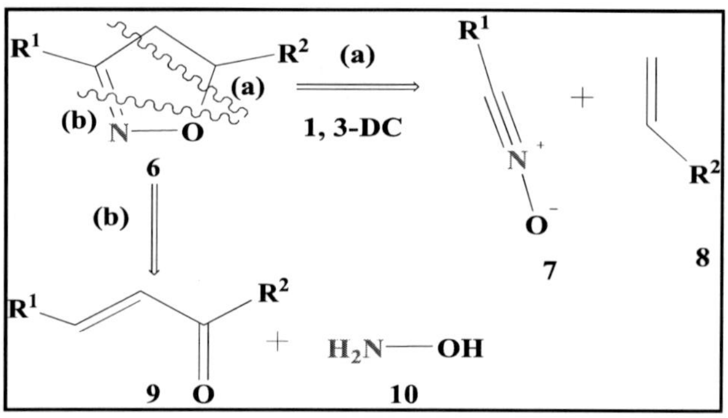

Figure 3. Probable retrosynthetic analysis of isoxazoline.

Retrosynthetic path '**a**' (Figure 3) describes the cycloaddition step and the other one is step '**b**' that is the reaction of α,β-unsaturated carbonyl compounds with hydroxyl amines [10]. Including the recent advances the overall methodology is given in short.

Vitale et al. synthesized isoxazole using enolate [11] ion of acetaldehyde **12** was quantitatively obtained at room temperature by the known cycloreversion of THF **11** in the presence of *n*-butyl lithium, and allowed to form 2-isoxazolines **14** by reaction with substituted benzonitrile oxides **13**. After quenching the first step of the reaction with NH_4Cl, the 2-isoxazolines were isolated in 83-97% yields; in the second step, they have been easily dehydrated to the corresponding 3-arylisoxazoles **15** after treatment with mild bases, by E1cB mechanism (Scheme 1).

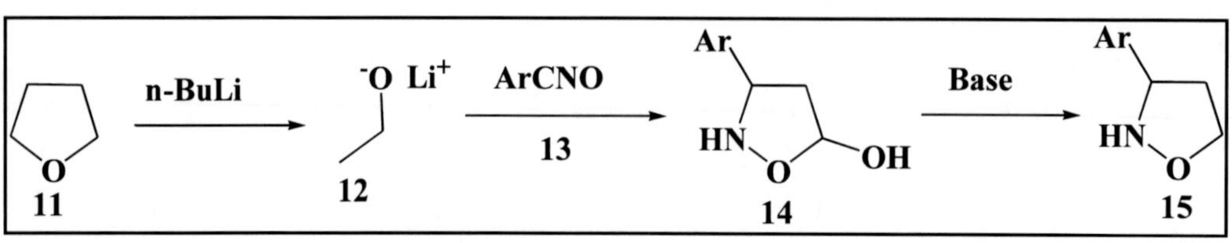

Scheme 1. Synthesis of isoxazole using enolate.

Sandeep et al. reported [12] 1,3-dipolar cycloaddition of alkenes **16** with nitrile oxides **17** generated *in situ* by treatment of aldoximes with Magtrieve (CrO_2) in either toluene or MeCN at 800°C affording isoxazolines in moderate yields (Scheme 2). It was observed that the formation of minor amount of de-oximination product along with isoxazolines **18** during the cycloaddition. The methodology found to be equally versatile for intramolecular nitrile oxide cycloaddition (INOC) reactions.

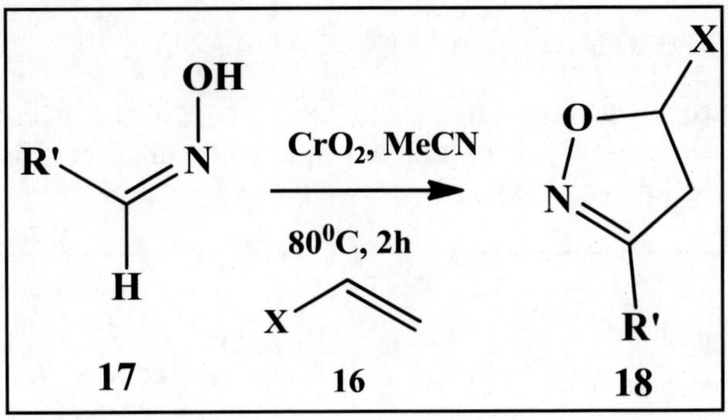

Scheme 2. Aldoximes forming isoxazolines.

However, most useful precursors are *N*-bromo-[13] and *N*-chlorosuccinimides [14] for both the two-steps and one-step (usually in the presence of Et_3N) procedures (Scheme 3). Other *N*-halogen-substituted compounds [15] such as chloramine-T (Scheme 4), trichloroisocyanuric acid (Scheme 5) [16], and *N*-(*t*-butyl)-*N*-chlorocyanamide [17] are also used for the oxidative dehydrogenation of aldoximes **19.**

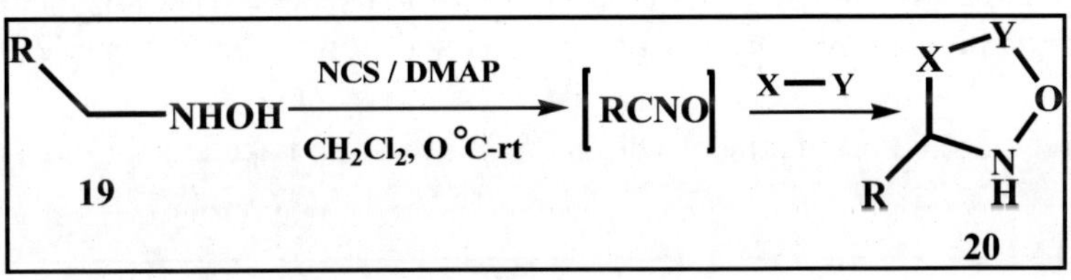

Scheme 3. Synthesis of isoxazolines using *N*-chlorosuccinimides.

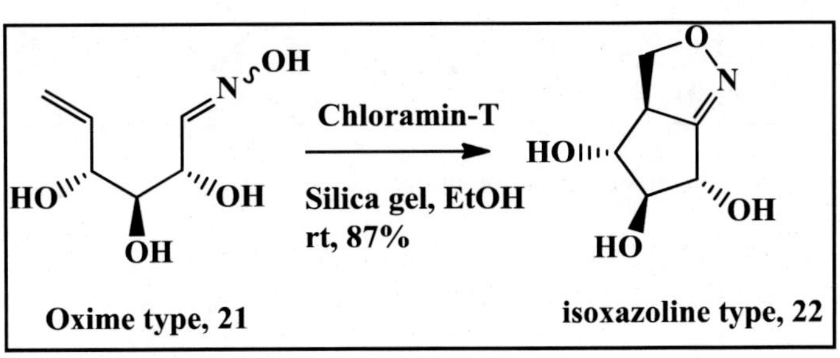

Scheme 4. Synthesis of isoxazolines using chloramine-T.

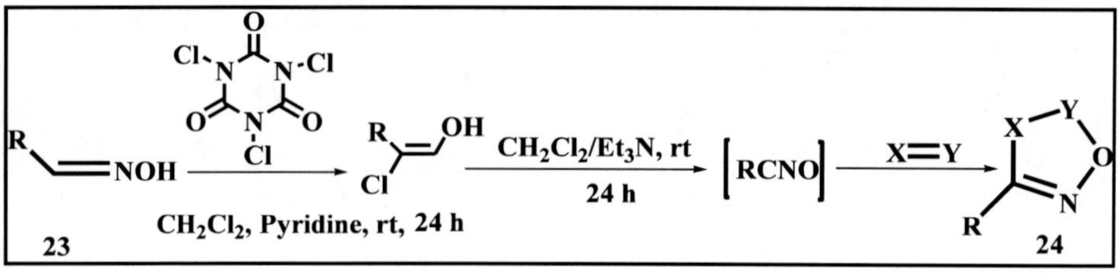

Scheme 5. Synthesis of isoxazolines using trichloroisocyanuric acid.

Generation of nitrile oxide **25** from aliphatic nitro compounds (**26**, Scheme 6) is generally done by Mukaiyama reaction by dehydration of primary nitroalkanes with a phenyl isocyanate in the presence of Et_3N as a base [18].

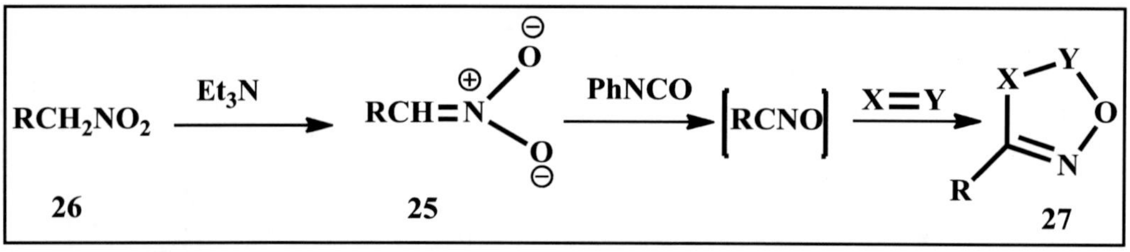

Scheme 6. Dehydration of primary nitroalkanes leads to the formation of isoxazolines.

Many reagents, other than arylisocyanates, have been tested for the dehydration of nitro alkanes, among them $POCl_3$, AcCl, Ac_2O, BzCl, and $MeSO_2Cl$ [19] Formation of nitrile oxides was also reported for the action of Mn(III) acetate on nitroacetate esters [20] and for the reaction of phosphorus trichloride with nitronate anion generated from β-nitrostyrene [21].

Formation of nitrile oxide can also be possible from alkyl bromide **28** (Scheme 7) and chloride on treatment with sodium nitrite and acetic acid in DMSO medium [22].

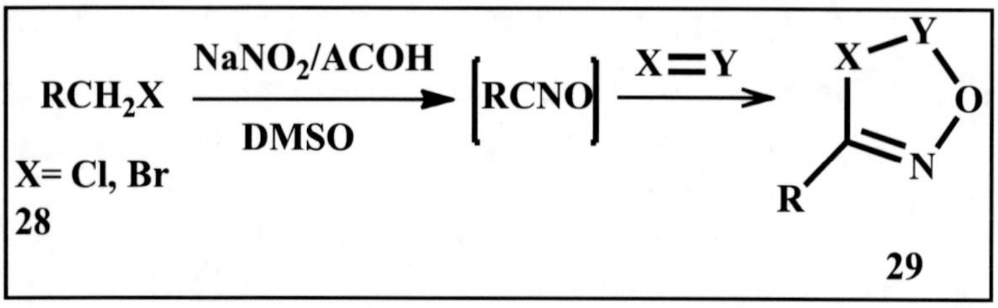

Scheme 7. Preparation of isoxazolines from alkyl bromide.

Dehydration of *O*-silylated hydroxamic acids **30** (Scheme 8) is used as a general method in the synthesis of nitrile oxides [23] in the presence of trifluoromethanesulfonic anhydride and triethylamine.

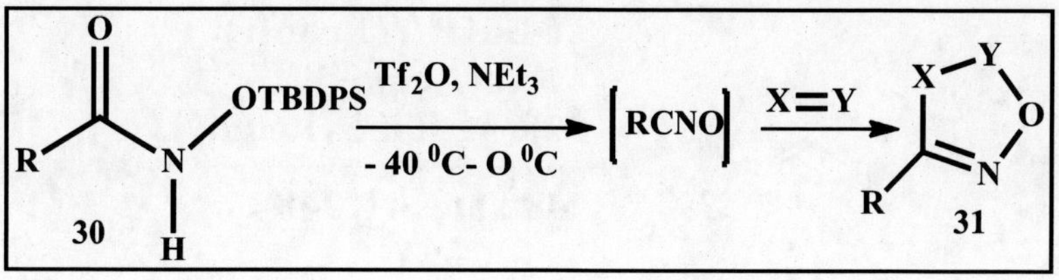

Scheme 8. Dehydration of *O*-silylated hydroxamic acids to produce nitrile oxides for synthesis of isoxazolines.

Some Recent Approaches to Δ²-Isoxazolines and Isoxazoles

A brief survey of the recent literature reveals that direct oxidation of aldoximes for the *in situ* generation of nitrile oxide and its 1,3-DC with alkenes/alkynes to synthesize isoxazolines/isoxazoles is becoming popular among synthetic chemists. Use of reagent $PhI(OAc)_2$ (DIB) [24], [hydroxy(tosyloxy)iodo]benzene (HTIB) [25], KI/I_2 [26], $PhICl_2$ [27], *t*-BuOI [28] and PhIO/CTAB [29] as oxidising agents are worth-mentioning. $PhI(OAc)_2$ is developed as an efficient organic Lewis acid-like oxidant to afford polyfunctional isoxazolines 32 (Scheme 9) from aldoximes in good yields.

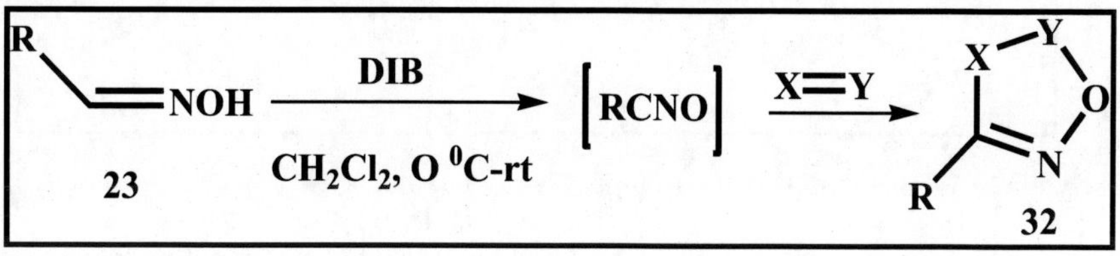

Scheme 9. Synthesis of polyfunctional isoxazolines using DIB.

Yao et al.[25] has synthesized tricyclic isoxazole **33** (Scheme 10) derivatives under 'on-water' conditions using [hydroxy(tosyloxy)iodo]benzene (HTIB or Koser's reagent) as the oxidizing agent.

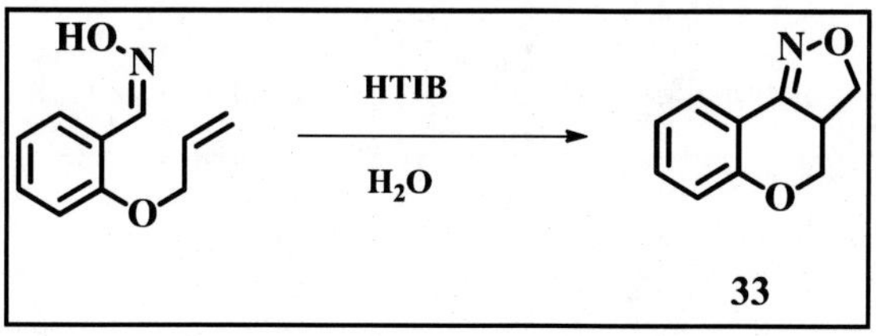

Scheme 10. Synthesis of tricyclic isoxazole using HTIB.

tert-Butyl hypoiodite (t-BuOI) was found to be a powerful reagent for generation of nitrile oxides from aldoximes and their cycloaddition with alkenes/alkynes leading to the formation of a variety of isoxazolines **34** (Scheme 11)under mild reaction conditions.

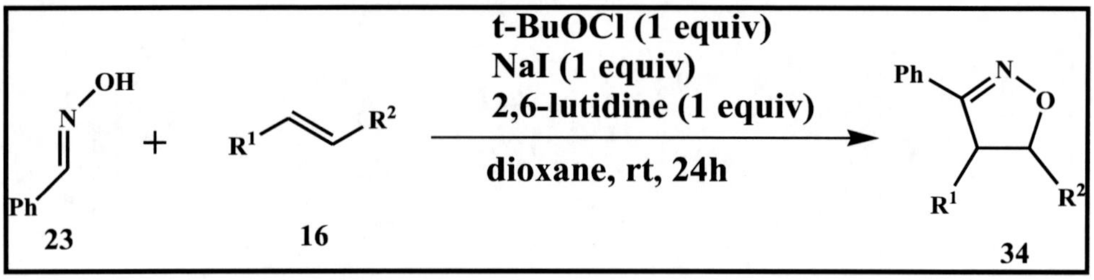

Scheme 11. Synthesis of isoxazoline using t-BuOI.

There are also reports on use of metal-based oxidizing agent such as MnO_2 [30], Magtrieve [31], CAN [32] and mercury (II) acetate [33] for the direct oxidation of aldoxime to generate nitrile oxide. These methods do not need any base for the 1,3-DC reaction. However, these methods need at least 7-10 equivalent of the metal oxidant to generate nitrile oxides at higher temperature.

Manganese (IV) oxide was found to oxidize aldoximes to nitrile oxides which successively condenses with the dipolarphile giving rise to the corresponding isoxazolines **35** (Scheme 12).

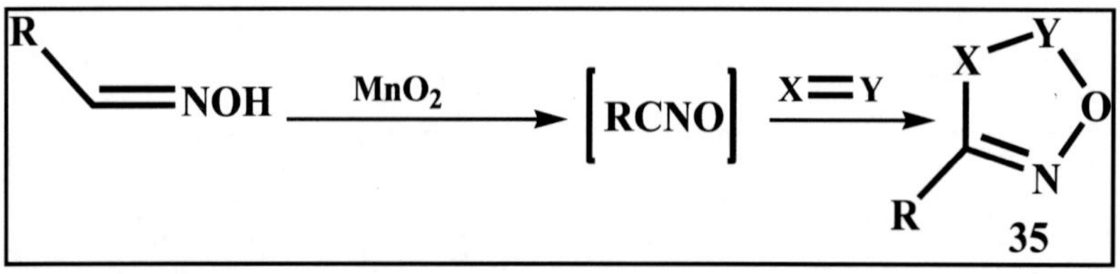

Scheme 12. Synthesis of isoxazoline using Manganese (IV) oxide.

Metal salt cerric ammonium nitrate (CAN) has been used as an efficient oxidant to convert aldoxime to nitrile oxide (Scheme 13) followed by cycloaddition with alkene to adduct **36** was established.

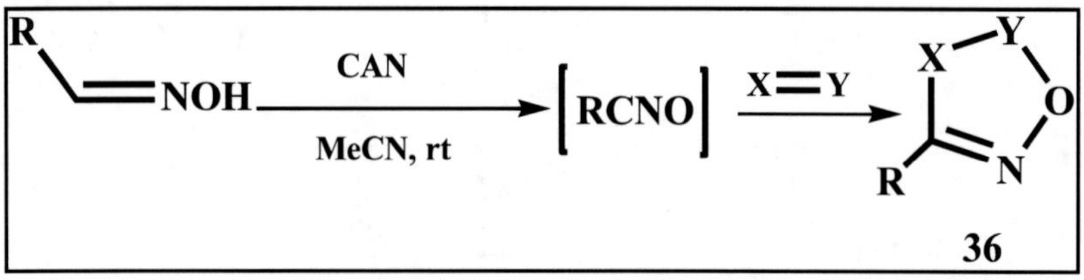

Scheme 13. Synthesis of isoxazoline using ceric ammonium nitrate (CAN).

Nitronates **37** can react with alkenes **38** in a 1,3-DC reaction to form *N*-alkoxy or*N*-silyloxyisoxazolidines **39** (Scheme 14). The 1,3-DC adduct **39**, obtained from acyclic nitronates, having a proton at the 3 position of the isoxazolidine ring, can easily eliminate the alkoxy or silyloxy group upon heating or treatment of acid to form 2-isoxazolines **40**. It should be noticed that isoxazolines are also obtained in the reaction of nitrile oxides with alkenes (vide infra), thus, nitronates can be considered as synthetic equivalents of nitrile oxides [34].

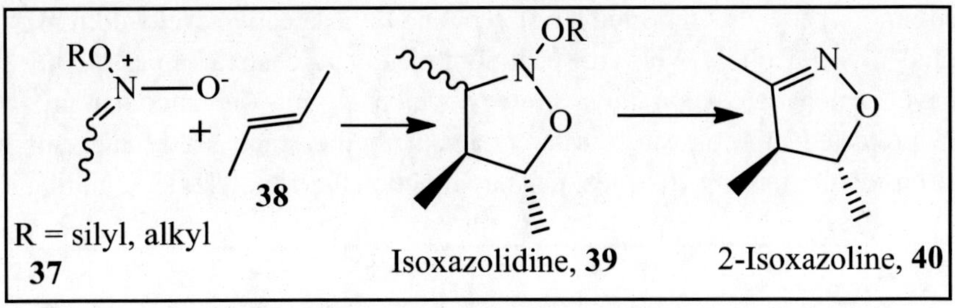

Scheme 14. Synthesis of isoxazoline upon reaction of nitrile oxides with alkenes.

Chiral Isoxazolines and Isoxazoles

Heterocyclic molecules have always attracted the attention of organic chemists due to their unique chemical, physical and biochemical properties [35]. Chirality plays an important role in several pharmacologically important molecules. Among them nitrogen-oxygen heterocycles constitute the largest class of biologically active compounds and several of these compounds have chiral isoxazoline and isoxazole building blocks [36]. Sandanayaka [37] et al. has prepared 6-(nitrileoxidomethyl) penam sulfone intermediate which underwent 1,3-dipolar cycloaddition (DC) reactions with various alkenes and alkynes to give several chiral isoxazolines **41** (Figure 4). A number of them have shown potent β-lactamase inhibition. Limonoids are degraded diterpenes, generally possessing a broad spectrum of biological activities as insect antifeedants, insect growth regulating properties, and antimalarial, anti-arthritic, anti-inflammatory and anticancer activities.

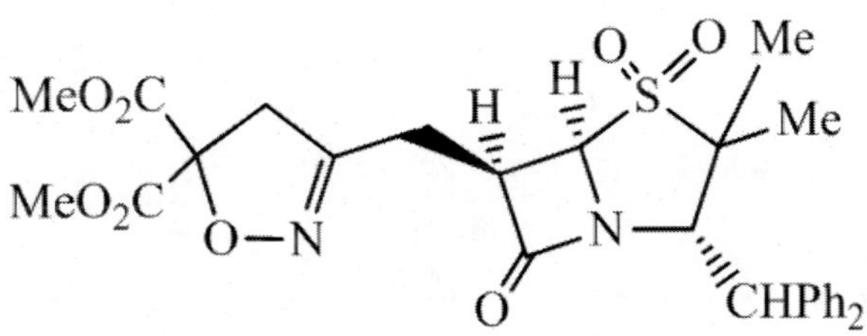

41

Figure 4. Isoxazolines 41 having potent *β*-lactamase inhibition. Synthetic Routes to Chiral Isoxazolines and Isoxazoles

Synthesis of chiral isoxazoline [38] can be achieved by exploiting

i. 1,3-DC of chiral nitrile oxide with achiral alkenes
ii. 1,3-DC of achiral nitrile oxide with chiral alkenes
iii. Intramolecular nitrile oxide cycloadditions (INOC) of chiral aldoxime
iv. Stereocontrolled cycloadditions using chirally modified metal catalysts.

Carreira [39] et al. has reported the stereo- and regioselective cycloaddition of chiral nitrile oxides and allylic alcohols to provide enantiomerically pure isoxazolines (**42-45**) (Scheme 15). They have demonstrated that their strategy which permits the modular preparation of all possible protected dipropionate diastereomers with the same set of reagents and a single reaction protocol affording complex, densely functionalized polyketide building blocks.

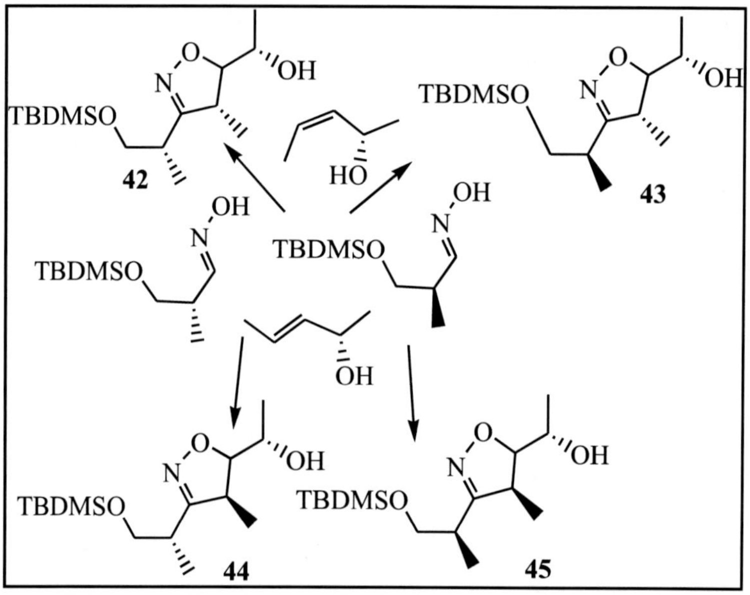

Scheme 15. Synthesis of enantiomerically pure isoxazolines.

In 2005, Jurczak et al. [40] reported synthesis of new chiral oximes and nitro alkanes, which were converted into nitrile oxides. They were subjected to 1,3-DC with (*E*)-3-hexene giving rise to corresponding 2-isoxazolines **46** (Scheme 16) in fair yields, but only with moderate stereoselectivities.

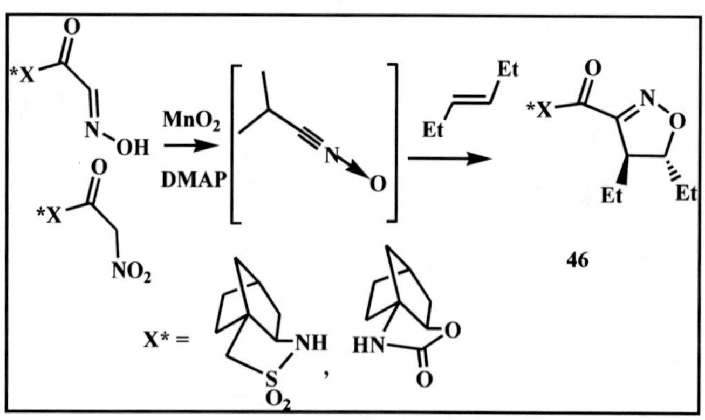

Scheme 16. Synthesis of chiral 2-isoxazolines from chiral oximes.

Contrary to the small number of publications dealing with 1,3-DC reactions of chiral nitrile oxides with alkenes several research groups have applied optically active alkenes in these reactions. Kamimura et al. [41] has synthesized optically active 4,5-dihydroisoxazoles **47**(Scheme 17) from chiral α-silyl allyl alcohols in almost diastereomerically pure form.

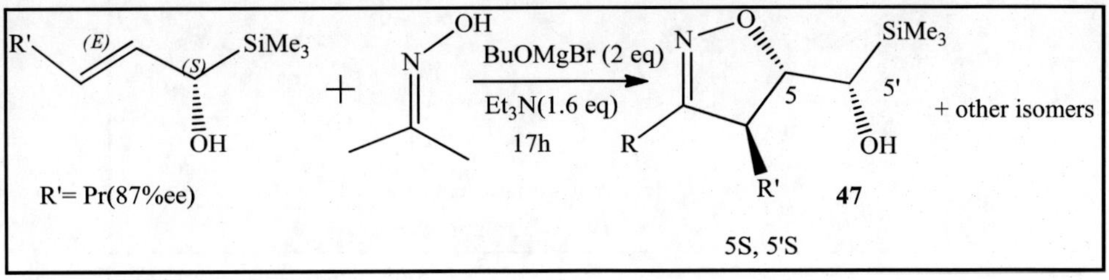

Scheme 17. Synthesis of optically active 4,5-dihydroisoxazoles.

Another interesting work was reported by Marsden et al. [42], where they synthesized chiral dioxazaborocines and applied them successfully for the asymmetric 1,3-DC reactions with achiral nitrile oxides to produce chiral isoxazoline **48** (ee 33%) (Scheme 18).

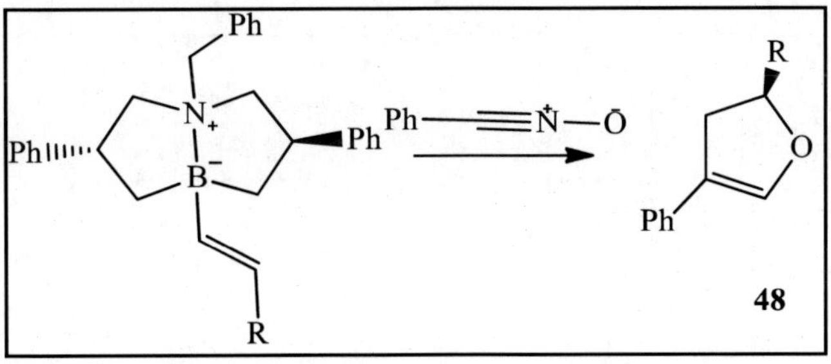

Scheme 18. Synthesis of chiral isoxazoline using asymmetric 1,3-DC reactions.

Aldoximes and L-proline esters were also utilized for stereoselective synthesis of spiro isoxazolinoprolinates **49-50** (Scheme 19), which was further converted into the corresponding unnatural spiro isoxazolinoproline-based amino acids. In particular, the *tert*-butyl spiro isoxazolinoprolinates were obtained with an 80:20 *trans*/*cis* ratio [43].

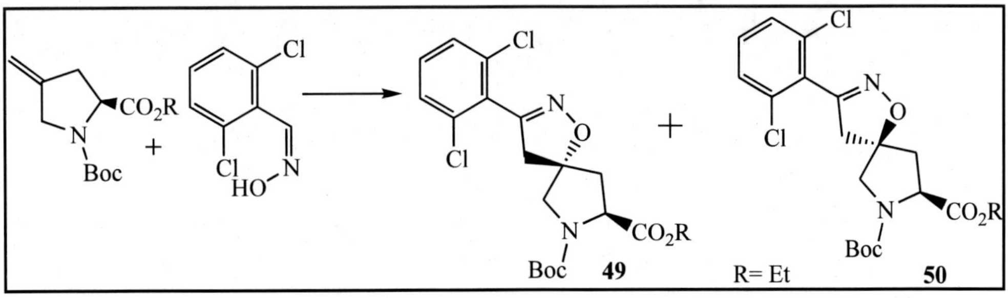

Scheme 19. Stereoselective synthesis of spiro isoxazolinoprolinates.

In 2005, Lassaletta et al. [44] reported the highly stereoselective 1,3-DC of a variety of aromatic and aliphatic nitrile oxides to 2,5-trans-2,5-diphenylpyrrolidine-derived acrylamide and cinnamamide, affording the corresponding 4,5-dihydroisoxazole-5-carboxamides as single diastereomers **51** (Scheme 20) in almost all cases.

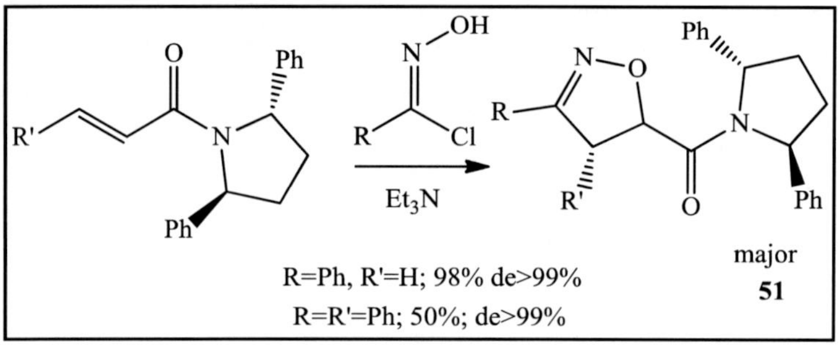

Scheme 20. Synthesis of 4,5-dihydroisoxazole-5-carboxamides as single diastereomers.

Chiacchio [46] et al. have stereoselectively synthesized enantiomerically pure annulated sultams by an INOC from the corresponding oximes. A total stereoselectivity was observed since only one of the eight possible stereoisomers **52** (Scheme 21) was produced.

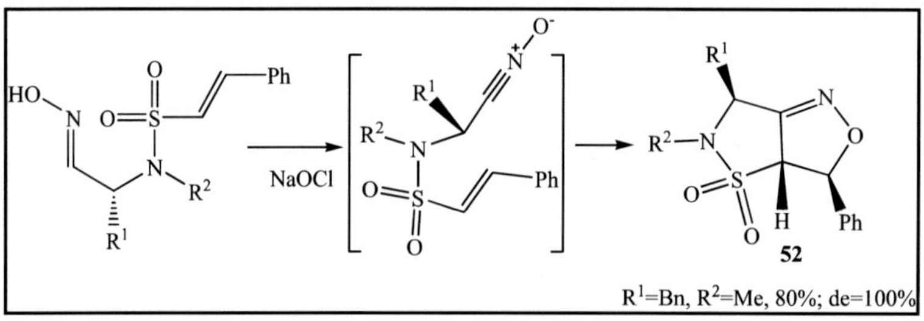

Scheme 21. Synthesis of enantiomerically pure annulated sultams.

Alcaide et al. (2004) reported [47] the regio- and stereoselective intramolecular 1,3-DC of novel chiral 2-azetidinone-tethered alkenyl nitrile oxides providing the corresponding chiral fused tricyclic β-lactams **53-54** (Scheme 22).

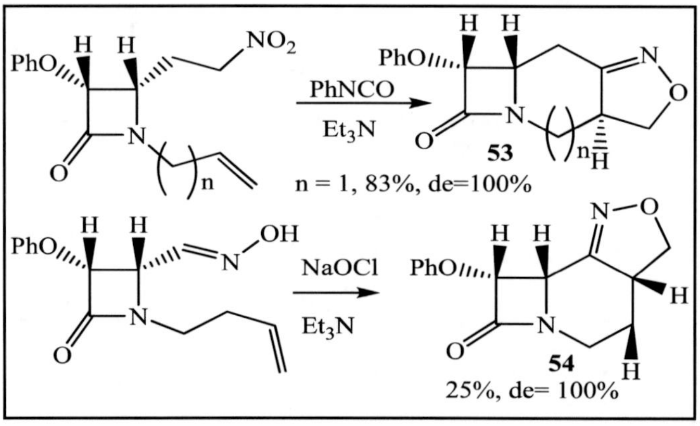

Scheme 22. Regio- and stereoselective intramolecular 1,3-DC of chiral nitrile oxides.

There are only a few publications on metal-assisted or metal-catalyzed 1,3-DC reactions ofnitrile oxides. Kanemasa et al. [48] used alkylmagnesium bromide or alkoxymagnesium bromide for the generation of the nitrile oxide instead of amines as bases in the reaction of chiral allylic alcohols with nitrile oxides generated by the conventional method. Low diastereoselectivity is mostly obtained. However, when the magnesium alkoxide is mixed with hydroximoylchloride, the resulting nitrile oxide reacts with the alkene to give the product **55** with a *syn*-selectivity of up to 98% de (Scheme 23).

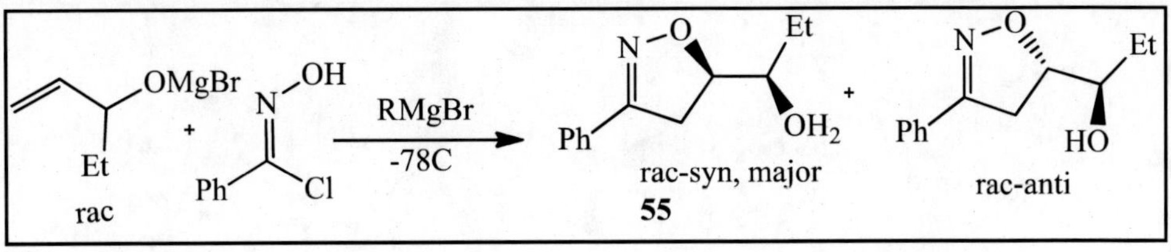

Scheme 23. Metal-assisted 1,3-DC reactions of nitrile oxides to produce isoxazolines.

Ukaji and co-workers [49] have applied an asymmetric metal-catalyzed 1,3-DC reaction of nitrile oxides with alkenes. Upon treatment of allyl alcohol with diethylzinc and (*R,R*)-diisopropyl tartrate, followed by the addition of diethylzinc and substituted benzoximoyl chlorides, the isoxazolines **56** (Scheme 24) are formed with impressive enantioselectivity (ee ~96%).

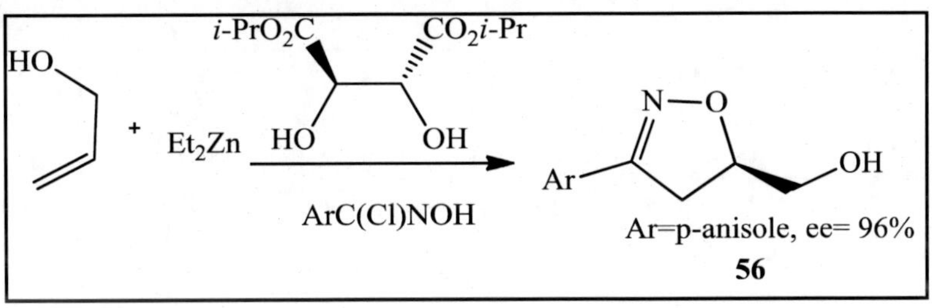

Scheme 24. Asymmetric metal-catalyzed 1,3-DC reaction of nitrile oxides.

Sibi et al. [50] has reported an enantioselective variant **57** (Scheme 25) of the 1,3-DC using a chiral Lewis acid prepared from magnesium iodide and a chiral bisoxazoline derived from amino indanol.

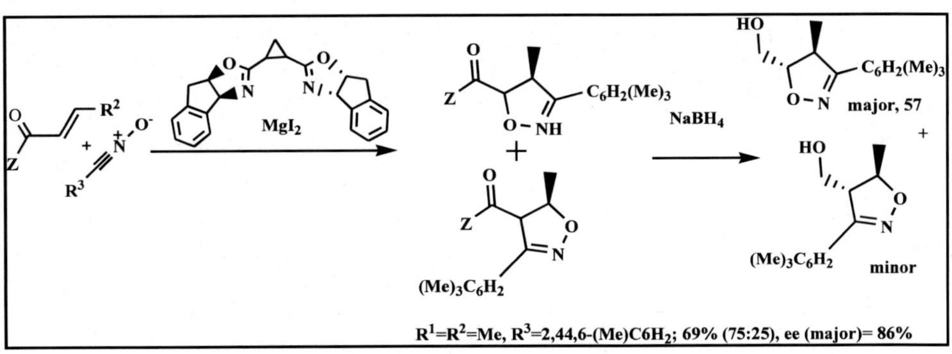

Scheme 25. Synthesis of chiral bisoxazoline derived from amino indanol.

In 2006, Porco et al. [51] has used Zr(IV) alkoxides to mediate a nitrile oxide cycloaddition which constituted the key step of a total synthesis of the spiroisoxazoline natural product (+)-calafianin **58** (Scheme 26).

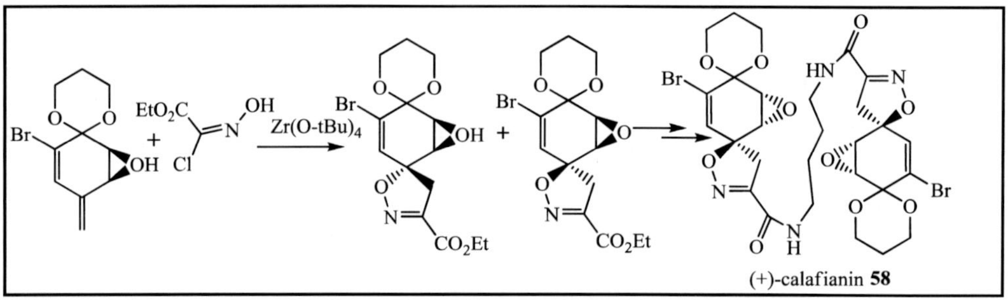

Scheme 26. Total synthesis of the spiroisoxazoline natural product (+)-calafianin.

A glycal-mediated synthesis of chiral 5-substituted isoxazoles **59**, (Scheme 27) bearing a differentially O-protected glycerol moiety in the side chain has been accomplished by Passacantilli et al. [52] dichloromethane at room temperature to give isoxazoles.

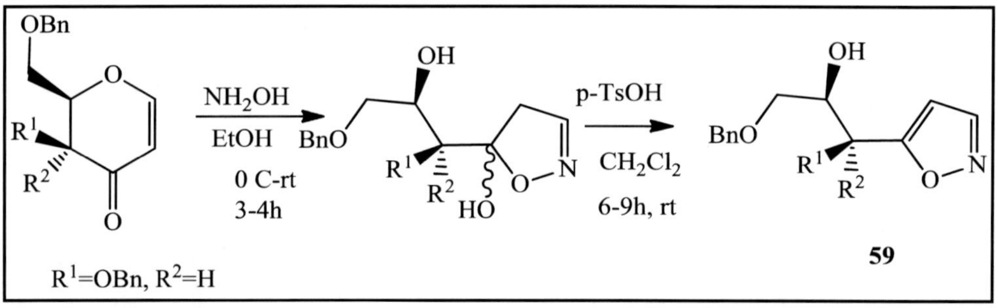

Scheme 27. A glycal-mediated synthesis of chiral 5-substituted isoxazoles.

Choice of Carbohydrate Derivatives as Precursors for Stereoselective Nitrile Oxide 1,3-DC Reaction

Carbohydrates are the most important members of the chiral pool. They are the main constituents of all major structural molecules in living systems. They occur alone in polymers such as starch, pectin, cellulose and chitin. They are also the key components of DNA and RNA in cell. Followings are the possible reasons for the extensive use of carbohydrates in the Nature:

(1) The cyclic structure guarantees an adequate conformational rigidity.
(2) The presence of multiple hydroxyl groups provides different positions for linkages and weak interactions with the receptor.

(3) The chirality provides different orientations of the hydroxyl groups and therefore different directions for the substituents linked to them.

Scientists are very keen to use carbohydrate as they are inexpensive, abundant, biocompatible and valuable precursor for installation of multiple chiral centers in the target molecules [53]. Sugar-derived chiral compounds are being employed in catalysis, asymmetric synthesis, construction of chiral natural products, new molecules for drug design and organic nanostructured materials [54].

Sugar-derived nitrile oxides are used for construction of pseudodisaccharides and higher carbon-chain monosaccharides [55], cyclization to chiral benzimidazole, benzoxazole and benzthiazole [56] multistep synthesis of bioactive natural products [57] and bioactive isoxazolines [58]. Sugar-derived isoxazolines and isoxazoles are used in the synthesis of naturally occurring [59] and artificial carbohydrate mimics. Kozikowski and Ghosh [60] reported the 1,3-DC reaction between the glyceraldehyde-derived alkene and nitrile oxide. The reaction proceeds with fair selectivity and the isoxazoline **60** (Scheme 28) is obtained as the major isomer with 60% ee.

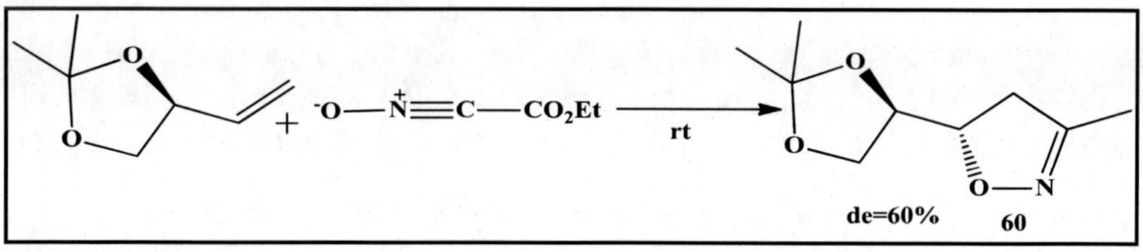

Scheme 28. 1,3-DC reaction of glyceraldehyde-derived alkenes and nitrile oxides to produce sugar-based chiral isoxazolines.

Rollin and colleagues [61] have reported the 1,3-DC of sugar-based ethenyl ethers with nitrile oxides providing the corresponding isoxazolines **61** (Scheme 29). Impressive properties in terms of stereoselective induction were obtained in the D-fructo series, whereas its epimeric β-D-psico counterpart was significantly less effective (up to 66% de), and the D-galactose-derived adduct gave no diastereoselectivity. Another example of a 1,3-DC performed with a sugar-derived chiral dipolarophile was reported by Tadano et al. [62] in 2003.

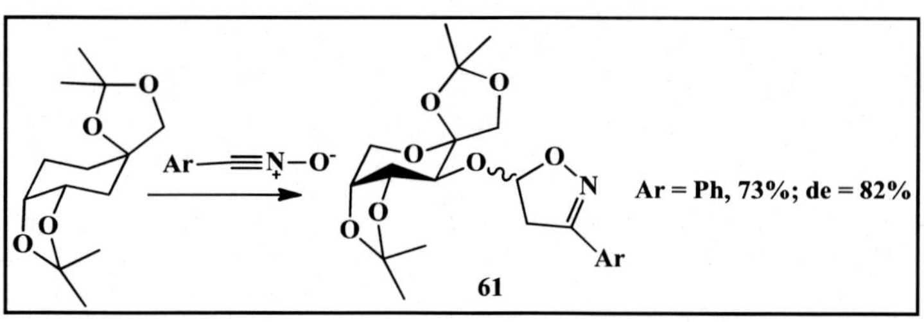

Scheme 29. 1,3-DC of sugar-based ethenyl ethers with nitrile oxides to isoxazolines.

Benzonitrile oxide and pivalonitrile oxide were reacted with methyl 4-*O*-acryloyl-6-deoxy-2,3-di-*O*-(*tert*-butyldimethylsilyl)-α-D-glucopyranoside, providing the respective cycloadducts **62** (Scheme 30) in excellent yields and each as a single diastereomer.

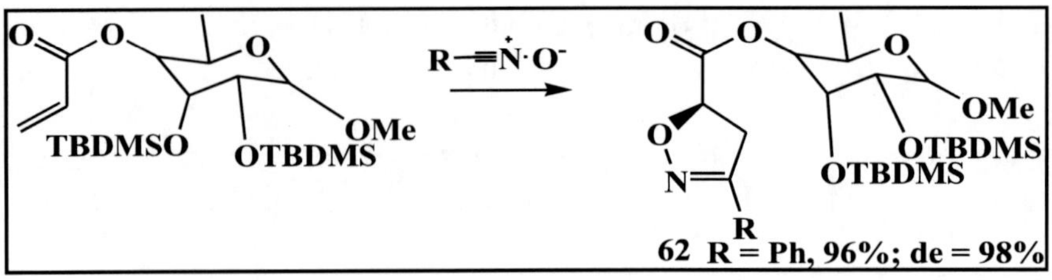

Scheme 30. 1,3-DC of benzonitrile oxide and methyl 4-*O*-acryloyl-6-deoxy-2,3-di-*O*-(*tert*-butyldimethylsilyl)-α-D-glucopyranoside to produce corresponding cycloadducts of isoxazolines.

Recently, Jäger et al. [63] has reported efficient routes to antibiotic L-(+)-Furanomycin and L-Carbafuranomycin from sugar-derived nitrile oxide utilizing commonly used *N*-chlorosuccinimide-HCl method. However, use of HCl is harmful for sugar moieties possessing ether linkages. So, easy access to sensitive sugar-derived nitrile oxides under neutral and benign reaction conditions is highly desirable. A highly efficient strategy (Scheme 33) for chemoselective oxidation of aldoximes to nitrile oxides by iodosobenzene in neutral aqueous media is reported by Maiti et al. [64]. The *in situ* intermolecular 1,3-DC of glycal-derived aldoximes **63** with olefins in nanometer aqueous micelles occurs with improved stereoselectivity and acceleration of reaction rate toward synthesis of new chiral synthons, 3-(2′-*C*-3′,4′,6′-tri-*O*-benzylglycal)-Δ^2-isoxazolines **64** and others. Construction of optically pure 2,8-dioxabicyclo[4.4.0]decene, (Scheme 31) skeleton is also performed by this green approach.

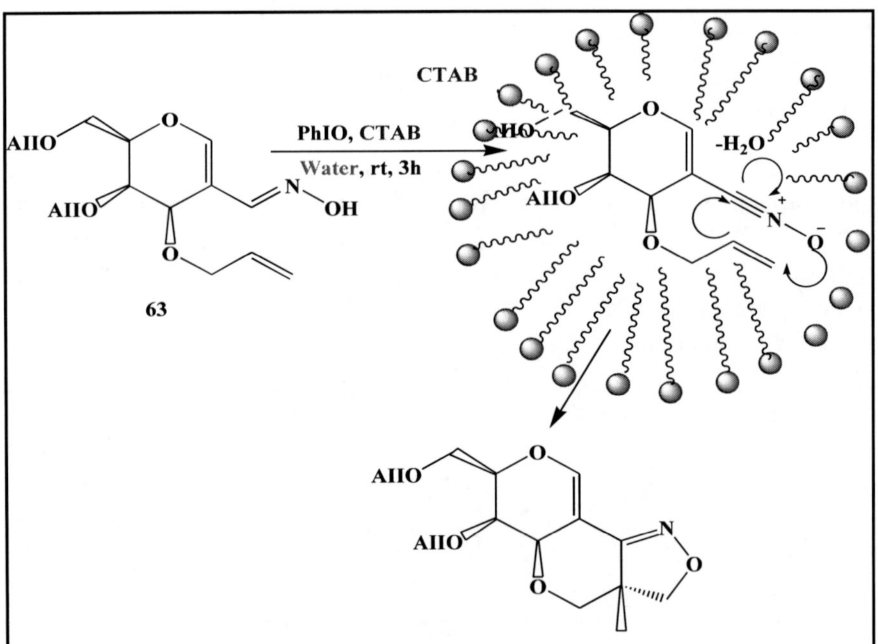

Scheme 31. A new *in situ* intermolecular 1,3-DC of glycal-derived aldoximes with olefins in nanometer aqueous micelles toward the synthesis of new chiral synthons, 3-(2′-*C*-3′,4′,6′-tri-*O*-benzylglycal)-Δ^2-isoxazolines.

The main features of this synthesis of isoxazoline in aqueous medium by 1,3-DC reaction in aqueous organized media(AOM) [64]. In the 1,3-DC reaction control of stereoselectivity and stereospecificity is very challenging. Two major obstacles are important to consider with respect to the stereo and regio selectivity. They are (a) generation of nitrile oxide in situ due to their short life time and b) the tertiary amine that is being used in synthesis can racemize the newly created chiral centre in isoxazoline. 1,3-DC reaction to synthesize varieties of 5-membered heterocyclic compounds including isoxazoline is studied in aqueous media by few groups. Maiti et al. had shown the development of new process to synthesize nitrile oxide in neutral aqueous media and trapped by alkene in 1,3-DC reaction and thereby avoids such challenges. They have successfully prepared aqueous organized media (AOM) which is a confined environment, called nanoreactor, built up by surfactant CTAB). It is sufficiently hydrophobic so that organic compound gets in contact quickly and execute the chemical transformation. Reaction of an α,β-unsaturated aldehyde or ketones and a hydroxyl amine (Scheme 32) is given below [65].

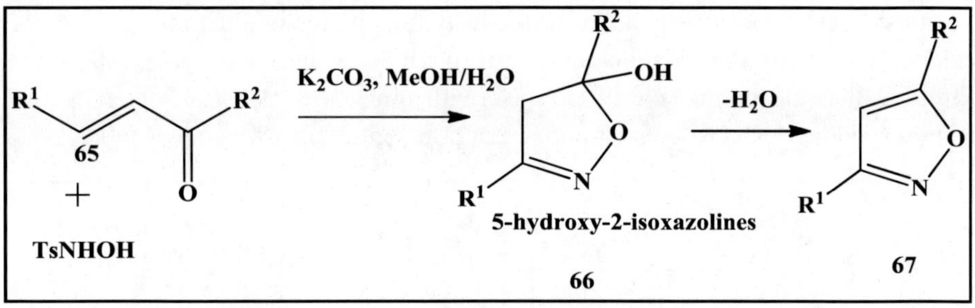

Scheme 32. Synthesis of 5-hydroxy-2-isoxazoline using a confined nanoreactor.

A series of novel substituted 1-amino-3-(5-phenyl-4,5-dihydroisoxazol-3-yl) quinolin-2(1*H*)-one (AJI1-AJI8) **68** have been synthesized by Kumar et al. [66] upon reaction with 1-amino-3-cinnamoyl-quinolin-2(1*H*)-one by using hydroxylamine hydrochloride as cyclising agent in alcohol medium (Scheme 33). The intermediate chalcones 1-amino-3-cinnamoyl-quinolin-2(1*H*)-one (AJC1-AJC8) were synthesized by condensing 3-acetyl-1-amino-quinolin-2-one with different substituted benzaldehyde in the presence of 40% ethanolic KOH.

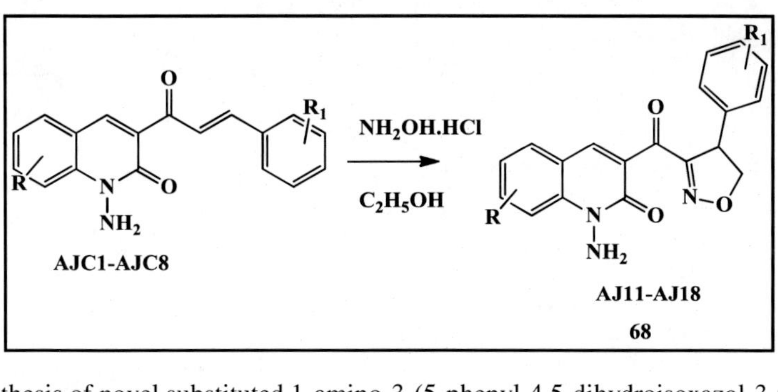

Scheme 33. Synthesis of novel substituted 1-amino-3-(5-phenyl-4,5-dihydroisoxazol-3-yl) quinolin-2(1H)-one (AJI1-AJI8).

A series of 3,5-disubstituted isoxazolines **69** (Figure 5) was synthesized by Ramón et. al. [67] to check their antiplatelet aggregation inhibitory potential. Most of the compounds in the same series lacked proaggregatory activity up to 1 mM, but inhibited, within the range from 1 to 1000 μM, both arachidonate-induced platelet aggregation and the second phase of ADP-induced platelet aggregation (primary ADP-induced aggregation was not affected).

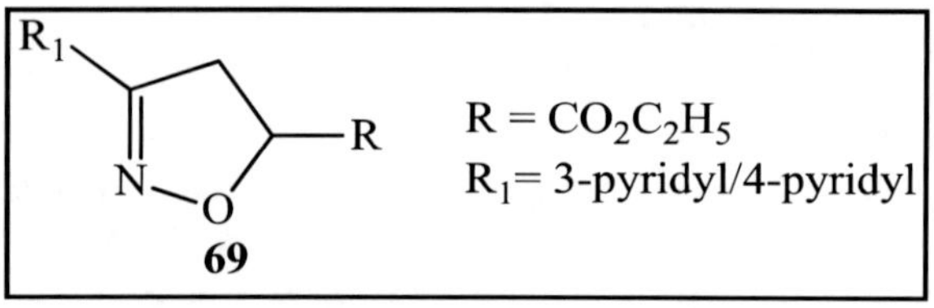

Figure 5. 3,5-Disubstituted isoxazolines (69).

Kumar et al. [67] emphasized on the synthesis of different substituted isoxazoline. The nitrile oxides generated *in situ* by the oxidative dehydrogenation of aldoximes with chloramine-T reacted with an α,β-unsaturated ketone to afford ethyl 3,5-diarylisoxazole-4-carboxylates. It was observed that the products were formed with unusual elimination of HCN from their expected cycloadduct isoxazolines **70** under reaction conditions (Scheme 34).

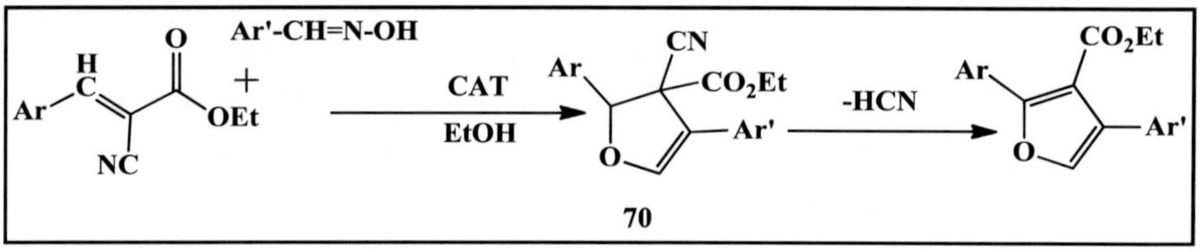

Scheme 34. Synthesis of differently substituted isoxazoline.

Rai and coworkers [68] reported the synthesis of series 3-Aryl-5*N*-aryl-4,6-dioxo-pyrrolo[3,4-*d*]-7,8-dihydroisooxazolines **71** via 1,3-dipolar cycloaddition of *in situ* generated nitrile oxides with *N*-aryl maleimides (Scheme 35). Later on they demonstrated the use of nitrile oxide as a dipolarophile in 1,3-dipolar cycloaddition with acetyl acetone and obtained the substituted isoxazolines **72** in good yield. Here the nitrile oxide gets added to enolic double bond of acetyl acetone [69].

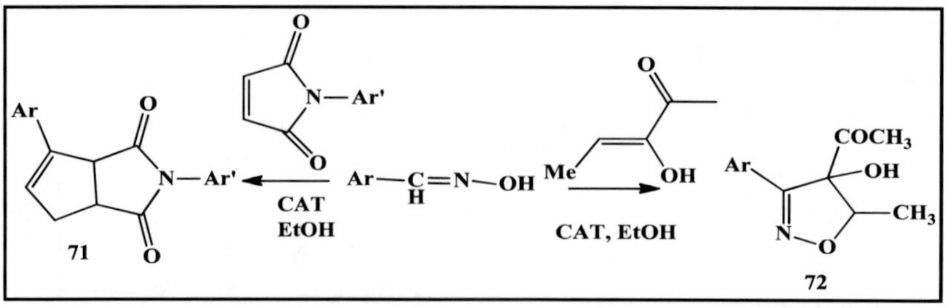

Scheme 35. Synthesis of 3-aryl-5*N*-aryl-4,6-dioxo-pyrrolo[3,4-*d*]-7,8-dihydroisooxazolines and its derivatives.

Fluorinated isoxazoline **73** were synthesized in one pot by the reaction of fluorinated chalcones and hydroxylamine in acetic acid medium under reflux conditions (Scheme 36) [69]. The products have been evaluated for their antibacterial activities. By introducing fluorine atom into specific position of organic molecule may cause significant changes in the stability, lipophilicity and biological activities of the resulting molecules. This is due to the high electro negativity of halogen, the strong C-F bond and the similar size of halogen and hydrogen atoms.

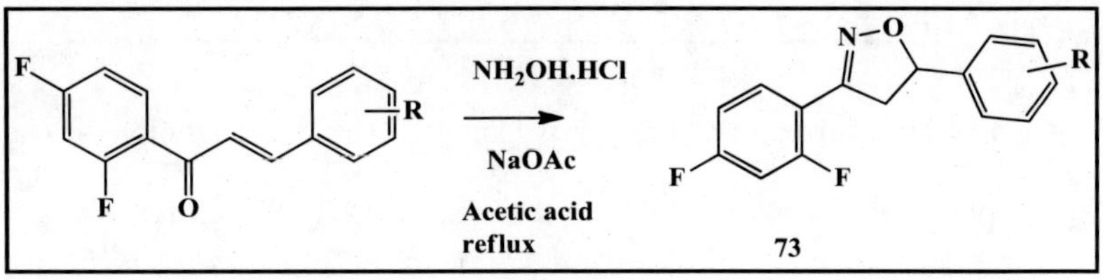

Scheme 36. Synthesis of fluorinated isoxazoline.

Merlo et al. [70] synthesized 3,5-disubstituted isoxazolines **74** by mild deselenenylation reaction of isoxazolinyl substituted phenyl selenide, which on treatment with the organic base 1,5-diazabicyclo[5,4,0]-undec-5-ene (DBU) or NaCN afford 5-methyl-3-substituted isoxazole (Scheme 37).

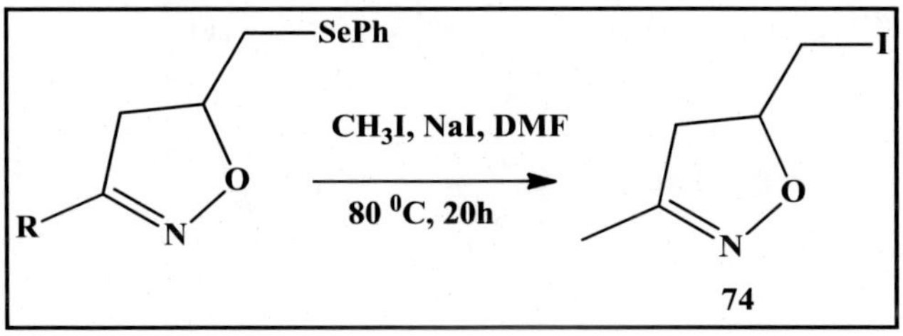

Scheme 37. Synthesis of 3,5-disubstituted isoxazolines by mild de-selenenylation reaction.

1,3-Dipolar cycloaddition reaction of nitrile oxides generated *in situ* from aldoximes with 4-methoxy cinnamonitrile afforded series of isoxazolines **75** in good yield. The dehydration of primary nitro compounds can be performed by bases in the presence of dipolarophiles. Among the organic bases examined, DABCO gave the best results. The reaction is applicable to activated nitro compounds and to phenylnitromethane and affords isoxazoline derivatives inhigher yields compared with those of other methods (Scheme 38). The reaction, however, is not compatible with nitroalkanes [71].

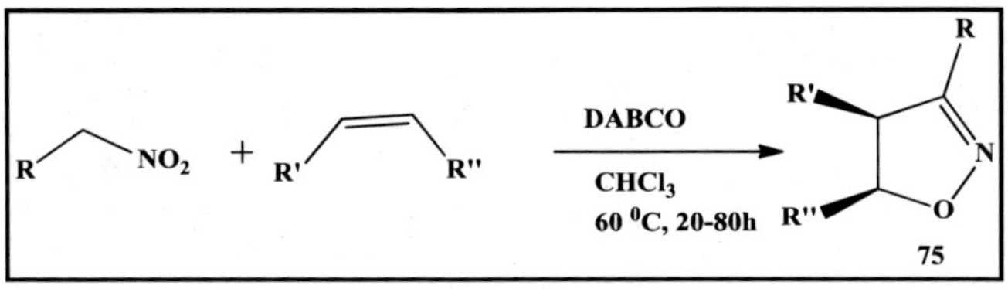

Scheme 38. Synthesis of isoxazoline derivatives using DABCO as a dehydrating agent of primary nitro compounds.

Unprotected *O*-propargylic hydroxylamines undergo cyclisations, when exposed briefly to silver nitrate adsorbed onto silica gel to give 4,5-dihydroisoxazoles [2-isoxazolines] (Scheme

39) **76** in very good yields, while *N*-acetyl derivatives give the corresponding 2,5-dihydroisoxazoles [3-isoxazolines] in similar yields [72].

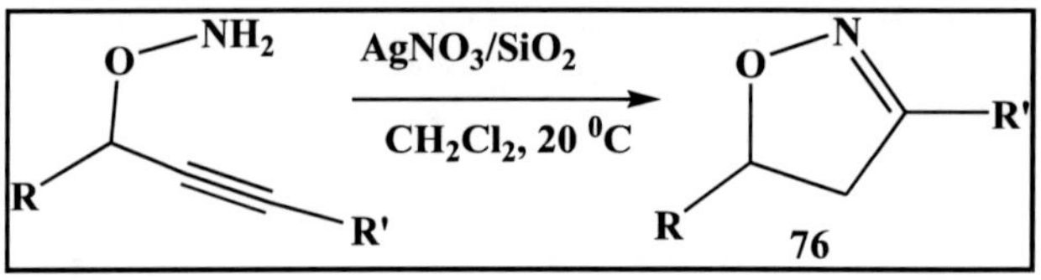

Scheme 39. Cyclisations of *O*-propargylic hydroxylamines to produce 2,5-dihydroisoxazoles.

3-Alkylpropargyl or 3-arylpropargyl hydroxylamines hydrochlorides were converted to 2-isoxazolines in good yields in refluxing methanol in the presence of K_2CO_3. Methods for 3-unsubstituted compounds and the direct transformation of *O*-propargyl phthalimide into 2-isoxazolines have been developed [73]. A mild and efficient copper-catalyzed trifluoromethylation reaction which involves the cyclization of oximes has been developed. The method provides a convenient access to a variety of useful CF_3-containing 4,5-dihydroisoxazoles **77** (Scheme 40) by constructing a C–CF_3 bond and a C–O bond in one step[73].

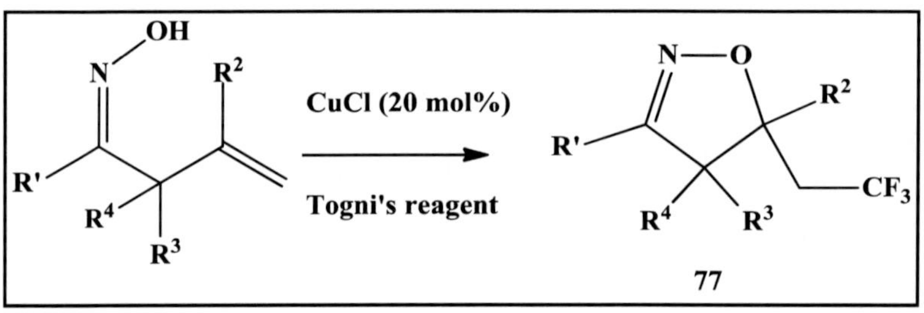

Scheme 40. Synthetic method provides a convenient access to a variety of useful CF3-containing 4,5-dihydroisoxazoles.

Metal Nanoparticles Mediated Isoxazoline Synthesis: $(TMSO)_2MnO_2$ Nanoparticles

$(TMSO)_2MnO_2$ mediated intermolecular 1,3-DC approach have been explored by the reaction of various functionalized aldoximes (**1'**) at ambient temperature to afford corresponding highly substituted Δ^2-isoxazolines **78** (Scheme 41). In the initial experiment, generation of nitrile oxides (**A**) is confirmed by trapping it with ethyl acrylate to obtain the known Δ^2-isoxazoline **78'**. Gratifyingly, other regioisomer **79** is not found in the benign synthetic process. This extraordinary reaction rate and regioselectivity can be explained in terms of *in situ* fabrication of highly active small nanoparticle and coordination of the metal during the cycloaddition step. Tolerance of the functional groups on the aldoximes (**1'**) and olefins (**2'**) are also examined. Both electron-rich and electron-deficient aromatic substituents are tolerated in this benign approach. The cycloaddition reaction is observed in large acceleration of reaction rate (3.5-5.0 h) with excellent isolated yields (71-85%). This is the benign synthetic protocol toward intramolecular nitrile oxide cycloaddition (INOC) reaction with olefin for synthesis of the fused isoxazoline **81** (Scheme 42) Intramolecular nitrile oxide cycloaddition strategy [74].

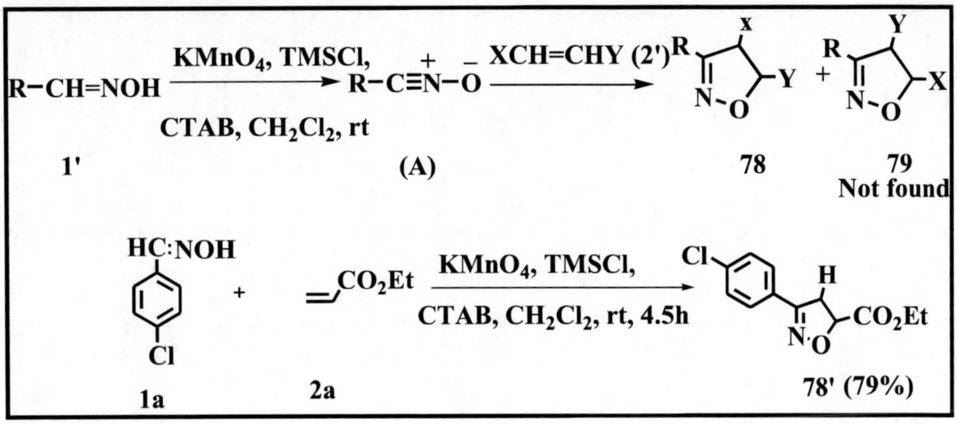

Scheme 41. $(TMSO)_2MnO_2$ mediated synthesis of Δ^2-isoxazoline.

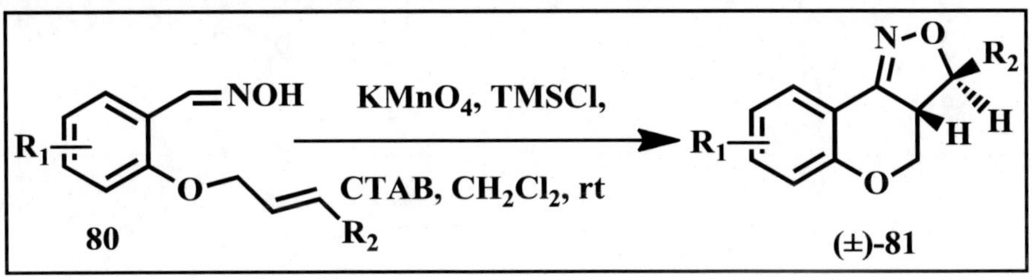

Scheme 42. Synthesis of the fused isoxazoline by Intramolecular nitrile oxide cycloaddition strategy using metal nanoparticle as active species.

The isoxazoline **82b-c** was isolated in 90% yield when the system made of benzaldoxime **87d**/sodium hypochlorite/THF-water (7:3) was stirred at room temperature for 18 h. An increasing in water content led to a yield drop. In pure water, the reaction yield was 25 and 90% for a reaction time of 18 and 72 h, respectively, and, performing the reaction in a 90mM SDS aqueous solution did not enhance the yield (24%). Moreover, no product was detected when the reaction was conducted in a 95:5 dichloromethane-water mixture because of the poor solubility of **82b** in dichloromethane [75].

However, this medium was favorable for benzaldoxime **82c** (Scheme 43) due to its good solubility, giving isoxazoline 228b in 90% yield after 18 h at room temperature. The latter isoxazoline was obtained in 40 and 92% yield in pure water after 18 and 48 h, respectively. The same observation was made as to the effect of SDS surfactant. The above experimental procedure when applied to the synthetic benzaldoxime **82a**, yielded diisoxazoline **82**, a cyclophane-type molecule, in 97% yield. About the same yield was obtained when the reaction was run in 95:5 DCM/water (Scheme 44).

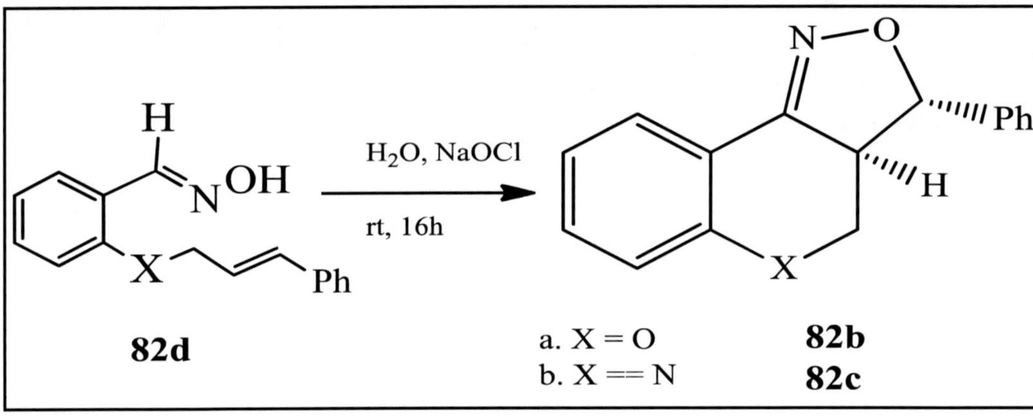

Scheme 43. Cyclization through isoxazoline formation.

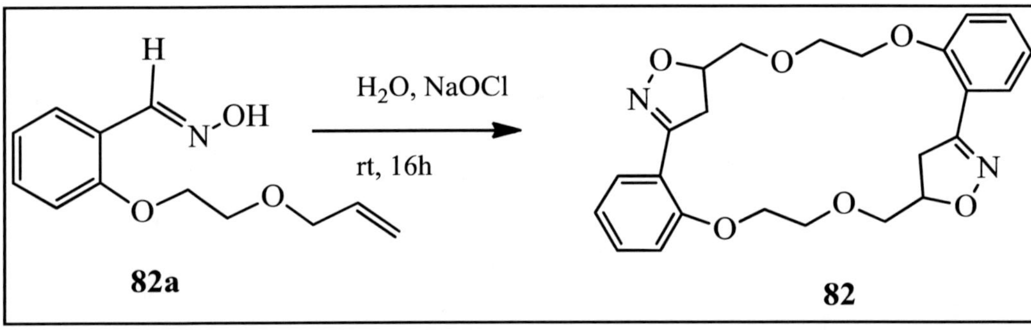

Scheme 44. Synthesis of diisoxazoline containing cyclophane-type molecule.

Singh et al. [76] found that the reaction of benzaldoxime **1a** (1 equiv.), maleimide **2a** (1.5 to 5 equiv.), and reagent **3** (1.2 equiv.) in dichloromethane at room temperature afforded pyrrolo-isoxazole **4a** in moderate to high yields of **83** (Scheme 45).

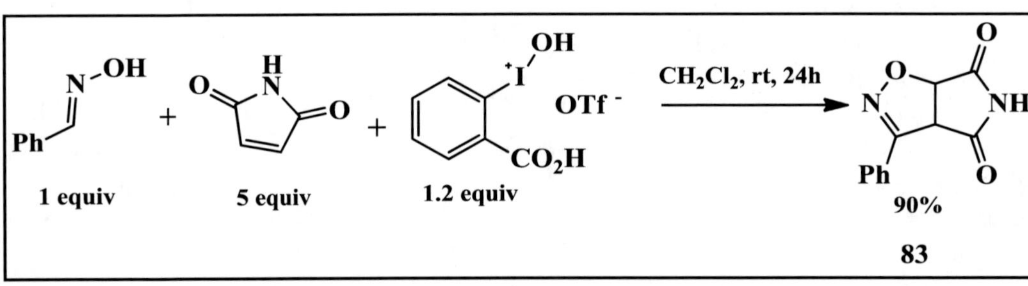

Scheme 45. Synthesis of pyrrolo-isoxazole.

Reaction mechanism is shown in Scheme 47. The active iodine (III) species, hydroxy(aryl)iodonium salt **84**, which is generated from 2-iodobenzoic acid **84a** and m-CPBA in the presence of TfOH, oxidizes the aldoxime **84b** *via* ligand exchange to furnish the corresponding nitrile oxide **84c** and the reduced 2-iodobenzoic acid **84a**. The generated nitrile oxide reacts with maleimide to afford pyrrolo-isoxazole **85**. Finally, the 2-iodobenzoic acid **84a** is reoxidized to continue the next catalytic cycle (Scheme 46).

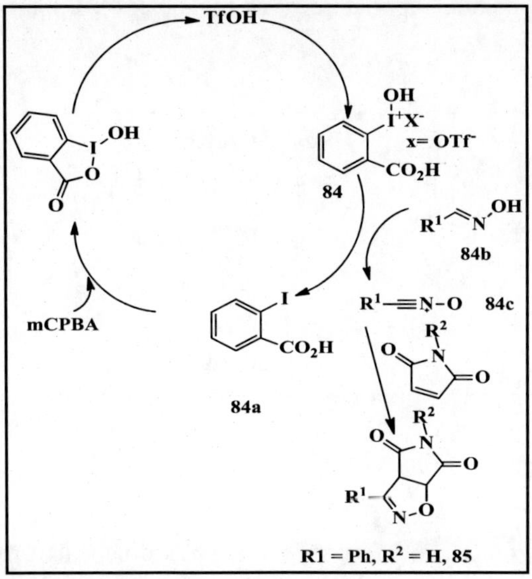

Scheme 46. Reaction mechanism of Synthesis of pyrrolo-isoxazole.

A green, sustainable, organocatalysis, and efficient synthesis of isoxazolines (Scheme 47) from allyloximes was developed by Kokotos et al. [77]. A 2,2,2-trifluoroacetophenone-catalyzed oxidation of allyloximes, utilizing H_2O_2 as the green oxidant, was taken advantage of in order to introduce a cheap and environmentally friendly protocol for the synthesis of substituted isoxazolines **86**. A variety of substitution patterns, both aromatic and aliphatic moieties, are well tolerated, leading to isoxazolines in moderate to excellent yields.

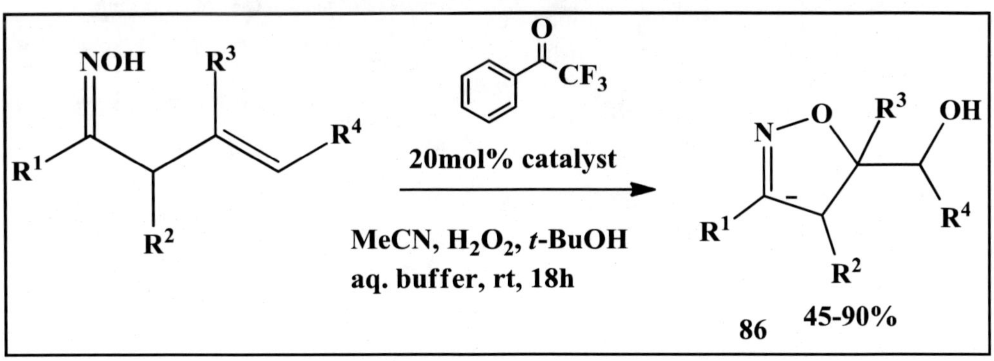

Scheme 47. Green, sustainable and efficient synthesis of isoxazolines.

Cu-catalyzed sp^{3}C–H bond nitration to form the nitrile oxides and subsequently to go through 1,3-dipolar cycloadditions with alkenes or alkynes in order to obtain isoxazoline derivatives. The sp^{3}C–H bonds of the methylazaarenes and substituted ethanones were successfully converted into the oxidative products combined with various alkenes. Generally, by using a cheap nitro source (KNO_3) and commercially available substrates, this method provided an efficient and concise approach to isoxazoline derivatives **87** (Scheme 48), which might possess great potential applications in the design of ligands as well as in tests of biological activity [78].

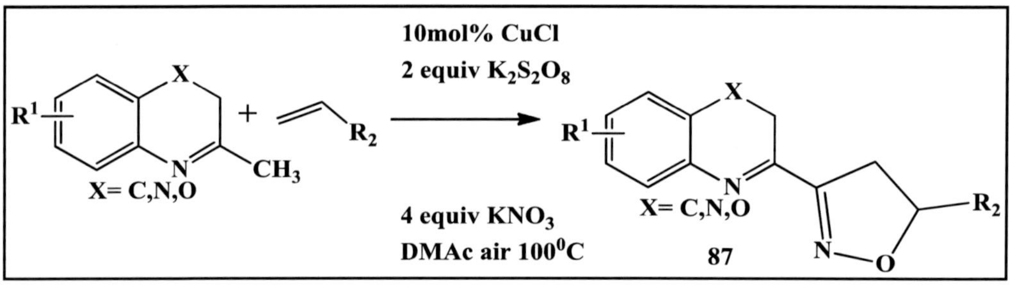

Scheme 48. Cu-catalyzed sp^3C–H bond nitration to form the nitrile oxides in order to obtain isoxazoline derivatives.

Zhdankin et al. reported the first hypervalent iodine catalyzed synthesis of isoxazole derivatives using iodoarene as a catalyst and oxone ($2KHSO_5$, 3 $KHSO_4$, $3K_2SO_4$) as an inexpensive and environmentally safe terminal oxidant. A new method was developed for the cyclization of aldoxime and alkene or alkyne using catalytic hypervalent iodine and oxone as a terminal oxidant in aqueous HFIP. This reaction proceeds via initial formation of nitrile oxides, which react with alkenes and alkynes to give corresponding isoxazolines **88** (Scheme 49). The reaction mechanism probably involves the electron-deficient hypervalent iodine species formed from the hydroxy(aryl)-iodonium ion and HFIP [79].

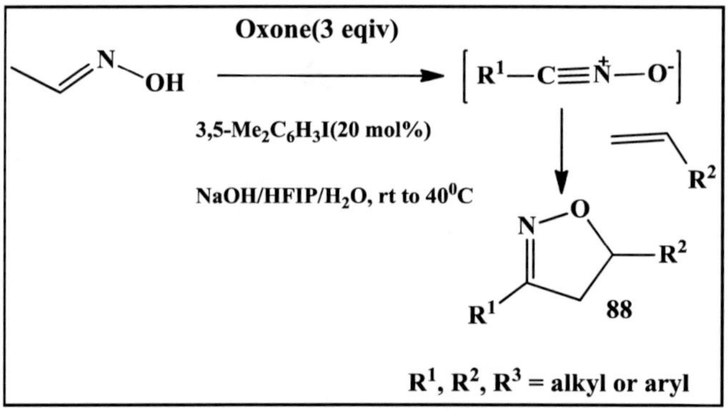

Scheme 49. Hypervalent iodine catalyzed synthesis of isoxazole derivatives.

An effective and environmentally benign procedure for the synthesis of isoxazolines and isoxazoles has been developed by a cycloaddition of nitrile oxides with alkenes or alkynes in water by Yan et al. [80]. In this approach, potassium chloride is first oxidized into chlorine in water by the environmentally friendly oxidant oxone, then aldoximes are oxidized into nitrile oxides by the in situ generated hypochlorous acid, finally a 1,3-dipolar cycloaddition between nitrile oxides and alkenes or alkynes occurs to provide the corresponding isoxazolines **89** in good yields (Scheme 50).

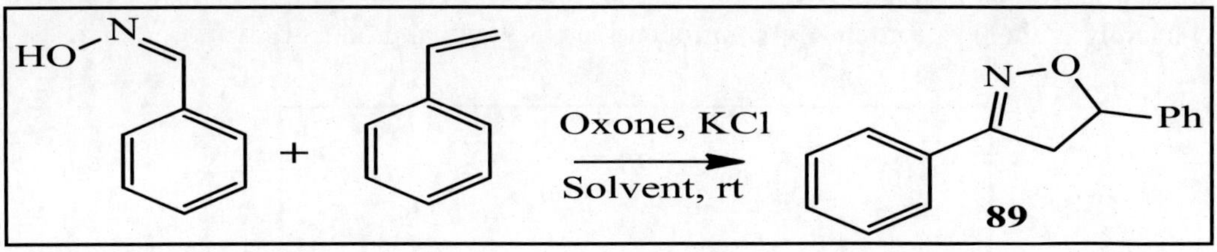

Scheme 50. An effective and environmentally benign procedure for the synthesis of isoxazolines and isoxazoles.

Zhu et al. [81] developed another method of oxime-mediated oxychlorination and oxybromination of unactivated olefins relying on palladium catalysis. A vast array of chlorinated and brominated isoxazolines has been divergently synthesized in moderate to good yields. The reaction has been achieved in good regio- and stereoselectivity and with high functional group tolerance. The halogenated isoxazolines **90** can be further transformed into diols, triols, and amino alcohols, which demonstrates the value of the method (Scheme 51).

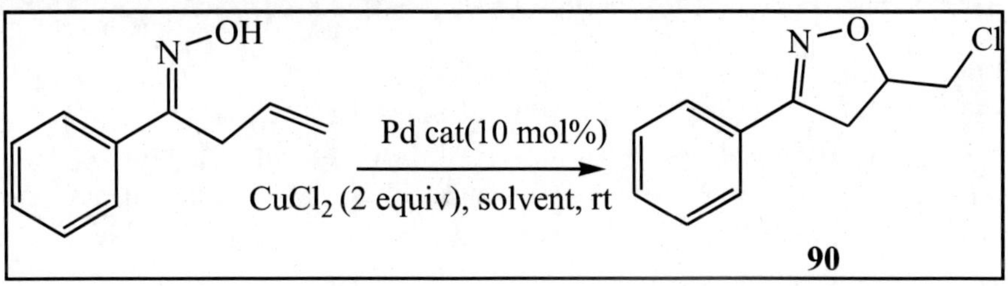

Scheme 51. Synthesis of chlorinated isoxazolines.

Isoxazolines (**91**) can be efficiently synthesized in good yields via a hypoiodite mediated catalytic oxidative cyclization of aldoximes and alkenes. This reaction involves active iodine species generated in situ from catalytic amounts of KI and oxone (Scheme 52).

Alkene + Aldoxime	KI(10mol%) Oxone (1..5 eq) → $MeOH-H_2O$(20:1) rt, 12h	Isoxazoline
1eq 2 eq		**91**

Scheme 52. Hypoiodite mediated catalytic oxidative cyclization of aldoximes with alkenes to produce isoxazoline.

The catalytic cyclization mechanism outlined in (Scheme 53). The active iodine species, hypoiodous acid, HOI, which is generated from KI and Oxone, reacts with aldoxime to give the iodonium ion I+, which is then opened by alkene to give N-iodohydroxy compound **91a**. Subsequently, the cyclization of **91a** affords the isoxazolidine **91b**, which undergoes β-elimination to give the isoxazoline **91** and iodide anion, and the regenerated iodide anion continues the catalytic cycle. The intermediate formation of iodonium ion is confirmed by the isolation of methyl ester of p-nitrobenzoic acid as a byproduct in the reaction of p-nitrobenzaldoxime. Indeed, nucleophilic ring opening of iodonium ion by methanol followed by hydrolysis should afford methyl p-nitrobenzoate as the final product [82].

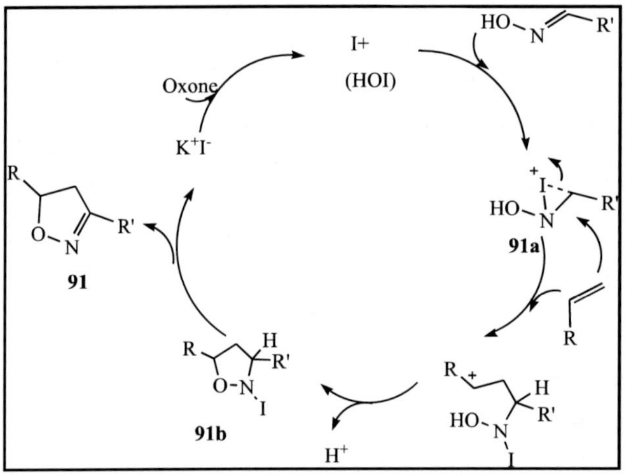

Scheme 53. Mechanism of hypoiodite mediated catalytic oxidative cyclization.

In 2016, Maiti et al. discovered a high-valent MoO_2(- HL)(H_2O)(DMF) complex with the rarely used H_3L ligand and found it to be a powerful catalyst for oxidative cyclocondensation using oxygen as a stoichiometric oxidant to furnish 2-substituted benzimidazoles selectively. In contrast to the common epoxidation catalysis by Mo(VI)-complexes, herein the MoO_2(HL)(H_2O)(DMF) complex has shown the diverse highly selective synthesis of ubiquitous 2-substituted benzimidazoles, 2-substituted isoxazolines **92** (Scheme 54) and their chiral analogues. Herein we demonstrated the MoVI(HL) as an attractive catalyst for developing benign, simple and general oxidative cyclization processes with excellent selectivity. It was anticipated that the discovery of a new high-valent catalyst, its novel properties and robust oxidative cyclization catalysis will have considerable application in chemical science towards devising new strategies for functionalizing molecules and discovering prospective new high-valent catalysts [83].

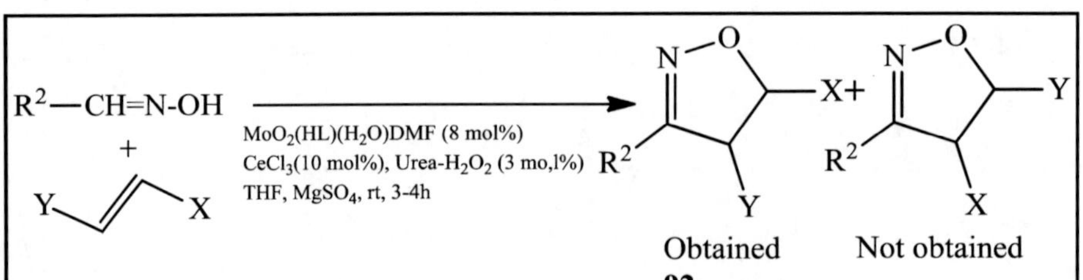

Scheme 54. Highly selective synthesis of ubiquitous 2-substituted benzimidazoles, 2-substituted isoxazolines.

It is established that **93a** treatment of styrene under inert atmosphere instead of open flask yielded corresponding isoxazoline derivative **93** (Scheme 55). The structure of **93** was derived from X-ray analysis [84].

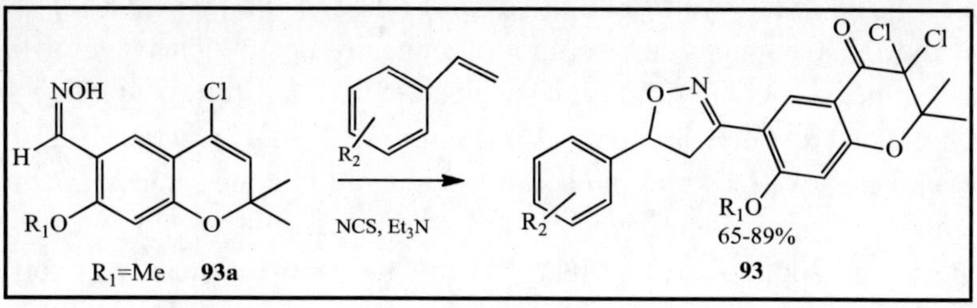

Scheme 55. Using styrene to form isoxazoline derivative.

IMPORTANCE OF Δ²-ISOXAZOLINES AS AN INTERMEDIATE AND APPLICATION AS A SYNTHONS

1,3-DC of nitrile oxides with alkenes followed by reductive cleavage of the heterocyclic ring has emerged as one of the most important strategies toward total synthesis of natural products [3,4].

Synthesis of biologically active natural products Bengazole A, (+)-paliciavine and (-)-vermiculline (**94-97,** Scheme 56) are some of the notable applications of this strategy [85]. Cycloaddition of nitrile oxide followed by cleavage of isoxazoline ring also form key steps in synthesis of epothilones, complex molecules mimicking the activity of anticancer drug Taxol.

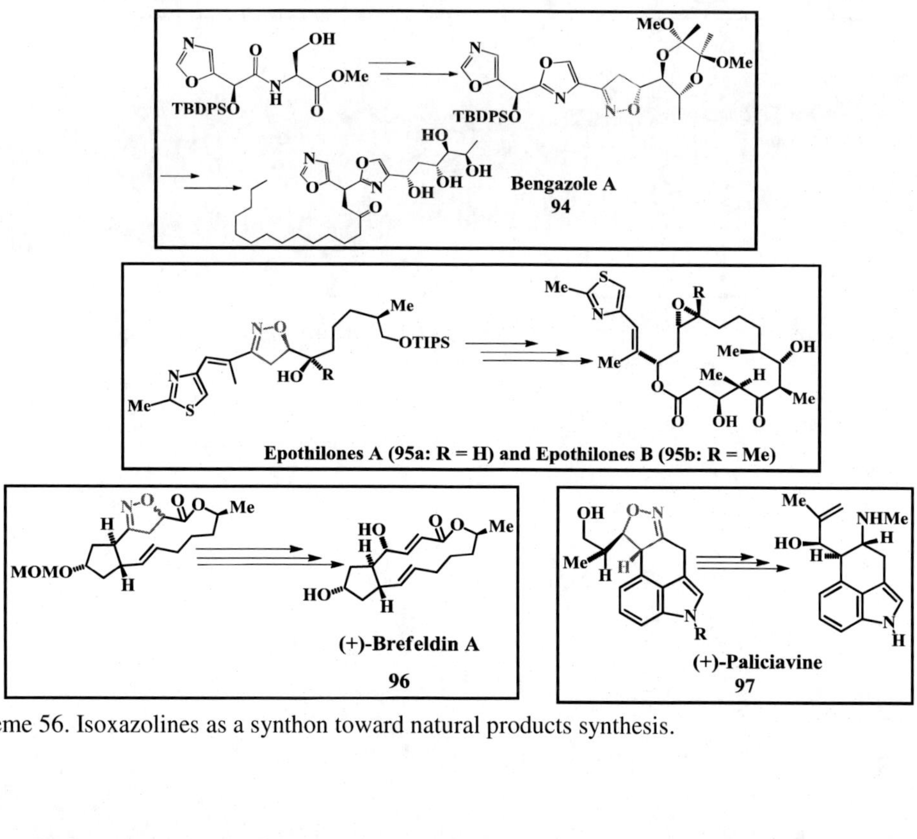

Scheme 56. Isoxazolines as a synthon toward natural products synthesis.

(+)-Brefeldin A is a 13-membered macrocyclic fugal metabolite containing diverse range of biological activities including cytostatic and antimitotic effects. The presence of a γ-oxygenated-α,β-unsaturated macrocyclic lactone unit in this moiety suggests the use of an intramolecular nitrile oxide cycloaddition (INOC) reaction (**98**, Scheme 57). The successful execution of the INOC strategy hinges upon the chemoselective cleavage of the N-O bond. There are also some isoxazolines molecules which are used as drugs possessing antidepressant (**I**), antipsychotic and anxiolytic activity (**II**) (Figure 4). 1,3-DC reaction is also an important strategy for synthesis of isoxazoles, a class of compound that are molecular components in a variety of natural products and drugs.

Application of appropriate reducing agents to the cycloadducts offers a facile stereoselective cleavage of the N-O and C=N bonds toward valuable chiral synthons [86]. Mateos [87] and coworkers have synthesized a limonoid analogue **99**, (Scheme 58) involving preparation of nitrile oxides, its intramolecular 1,3-DC and subsequent reductive transformation. It contains some of the key features of 12-acetoxy azadiradione.

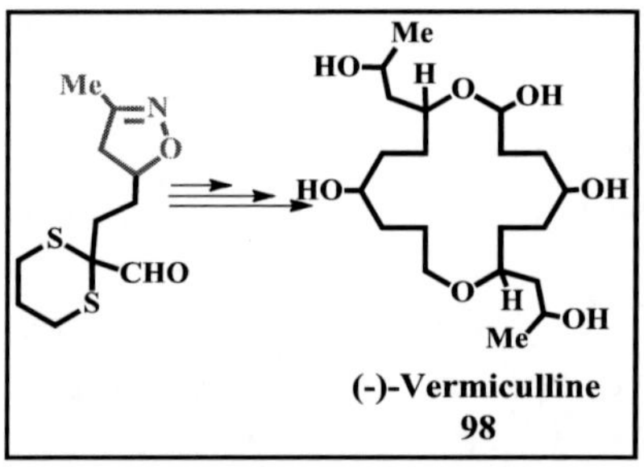

Scheme 57. Synthesis of 13-member heterocycles using isoxazoline.

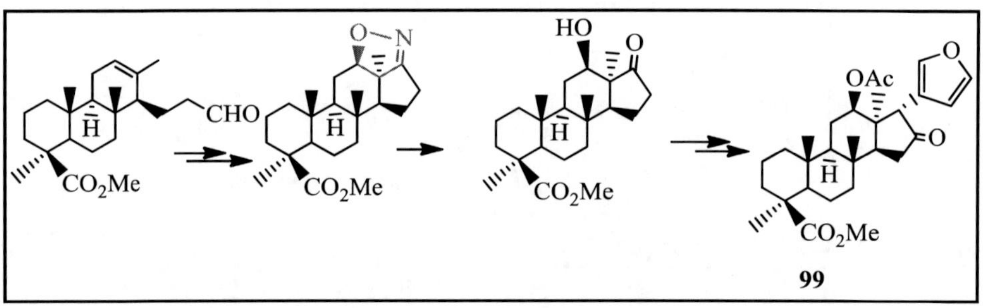

Scheme 58. The promising clinical synthesis of limonoid analogue from chiral isoxazolines.

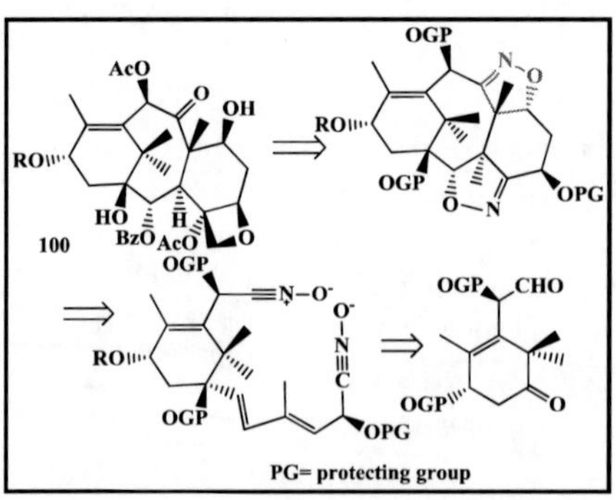

Scheme 59. Retrosynthetic approach to taxol 100.

Performance and unique molecular architecture of taxol **100** (Scheme 59) have evoked synthetic options for its total synthesis. The construction of the C-ring of taxol was accomplished by Mioskowski [88] and co-workers by the intramolecular [3+2] cycloaddition reaction of a nitrile oxide generated *in situ* and the involvement of the isoxazoline as an intermediate was observed (**101**, Figure 6).

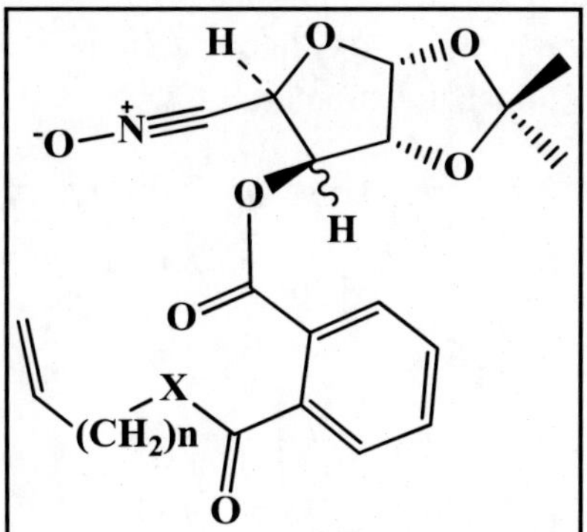

101

Figure 6. Phthalate the red substrate for INOC to produce macrocyclic lactone compound.

Depending on the value of n, and X (= O or N) and C3-H stereochemistry (α/β) different types of macrocycles will be obtained and they are on basic hydrolysis produces diol containing isoxazoline moiety **102** that on reductive cleavage results higher deoxy amino sugar **103** (Scheme 60) of biological and pharmaceutical properties. An instance is given below [89].

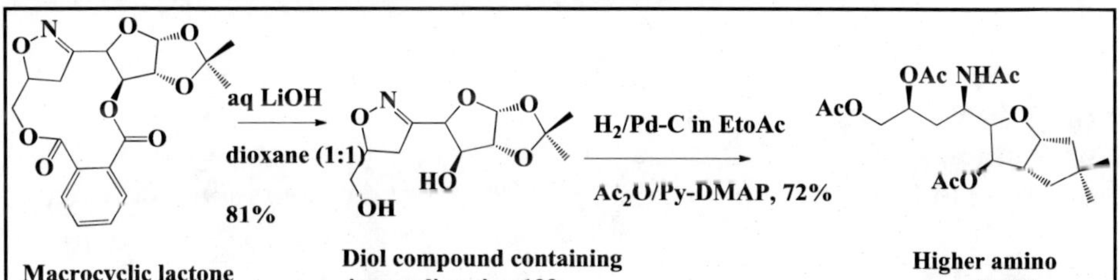

Scheme 60. Basic hydrolysis of macrocycles to produces diol containing isoxazoline an efficient precursor of higher deoxy amino sugar.

The highly selective synthesis of phthalate template bridged isoxazoline macrocyclic lactones is demonstrated using readily accessible carbohydrate precursors via INOC with alkene. The structure of macrocycle was analyzed by X-ray and 2D NMR. The orientation of bridged isoxazoline ring is dependent on the distance between the 1,3-dipole and dipolarophile. The macrocycles are amenable to further transformations to higher amino sugars by sequential removal of the phthalate template and reductive cleavage of the isoxaoline moiety **102**.

Erythromycin A, first isolated in 1952 by McGuire and co-workers at Lilly Research Laboratories from the actinomycete Sacharopolyspora erythraea. It consists of a 14-membered lactone ring with 10 stereogenic centers and two unusual sugars, L-cladinose at the C3 position and D-desosamine at the C5 position. Key steps of the approach include two magnesium-mediated nitrile oxide cycloadditions, a chelation-controlled Grignard reaction, and a Sharpless asymmetric dihydroxylation. In this synthesis the 1,3-dipolar cycloaddition reaction between nitrile oxides and allylic alcohol results isoxazolines which can serve as protected β-hydroxy ketone surrogates, the N-O bond reduction and hydrolysis can be effected at a late stage of the synthesis. The work we have delineated establishes the nitrile oxide cycloaddition reaction with allylic alcohols as a means of joining complex fragments and thus as a salient, powerful tactic for the synthesis of polyketide-derived natural products. The derived isoxazoline cycloadducts **104-108** (Scheme 61) can function as β-hydroxyketone surrogates that can be unmasked late in a synthesis within a densely functionalized structure [90].

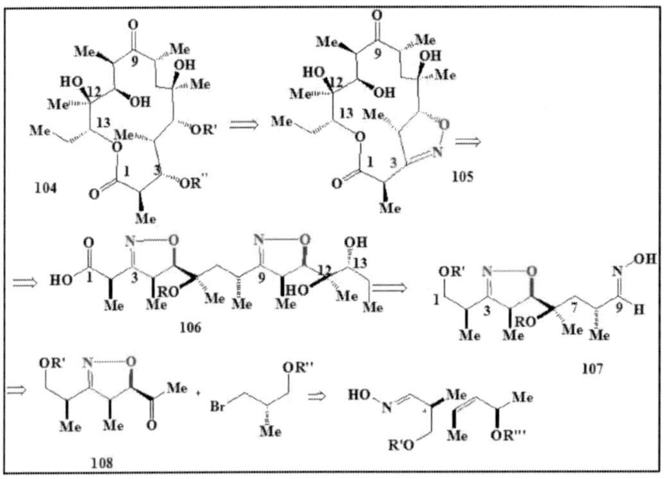

Scheme 61. Derived isoxazoline cycloadducts to produce polyketide-derived natural products.

With suitable reagent isoxazoline can produce: i) γ-amino alcohols**, 109**, ii) β-hydroxy ketones**, 110** (and thus α, β-unsaturated ketones, allylic alcohols, 1,3-diol, and 1,3-dienes), iii) β-hydroxy nitriles**, 111,** acids and esters and α,β and β,γ-unsaturated oximes, iv) as a ligand **112** (Scheme 62). Kozikowski et al. showed the possible paths of the isoxazoline ring opening [91].

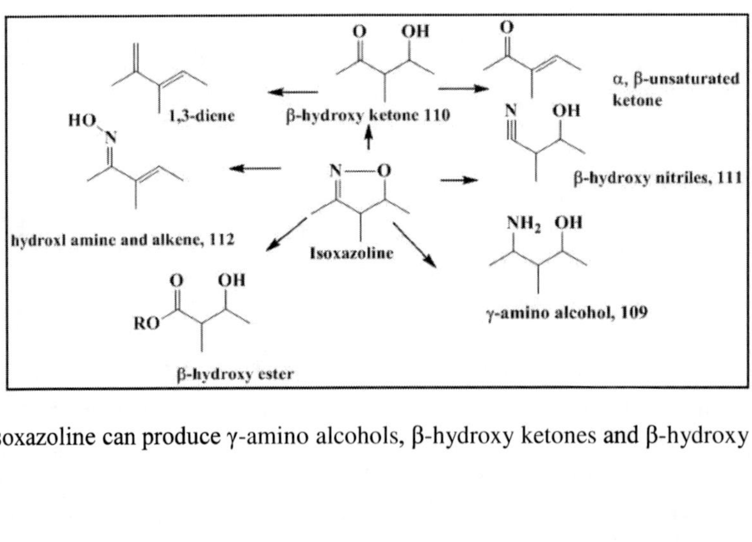

Scheme 62. Isoxazoline can produce γ-amino alcohols, β-hydroxy ketones and β-hydroxy nitriles.

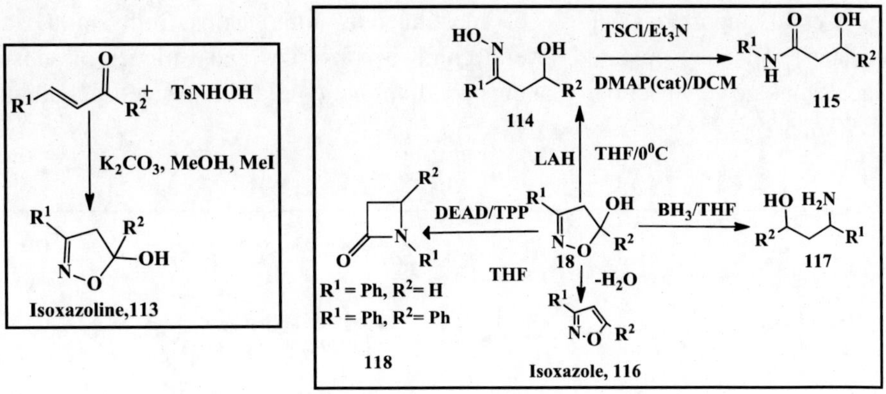

Scheme 63. Synthesis of 5-hydroxy-2-isoxazolines and its application in synthesis of isoxazoles, β-hydroxy oximes, γ-amino alcohols and β-Lactams.

The chiral 3-amino-3-phenylpropan-1-ol (**119**), a valuable intermediate for the synthesis of (S)-dapoxitene is reported (Scheme 64).

Chen et al. [93], developed a facile, economical, and efficient protocol for the chemoselective reduction of Δ^2-isoxazolines **121** to the corresponding *β*-hydroxy ketones **122** using Fe/NH_4Cl as the reducing agent (Scheme 65). The procedure revealed is general and is particularly useful for the transformation of both conjugated and nonconjugated Δ^2-isoxazolines to β-hydroxy ketones, a troublesome transformation with other known reagents. This finding should be of considerable interest for the construction of biologically active molecules.

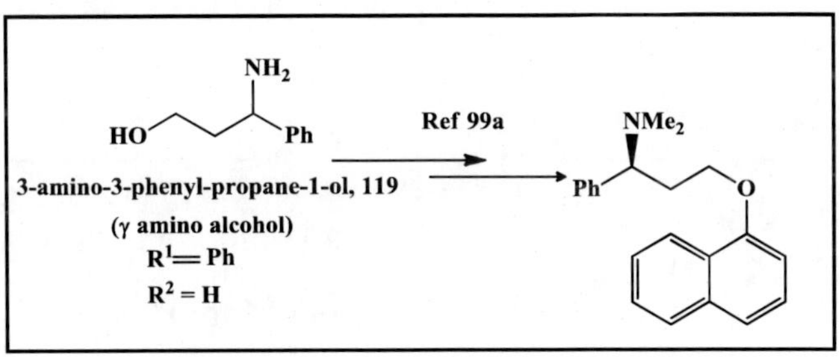

Scheme 64. Synthesis of (S)-dapoxetine.

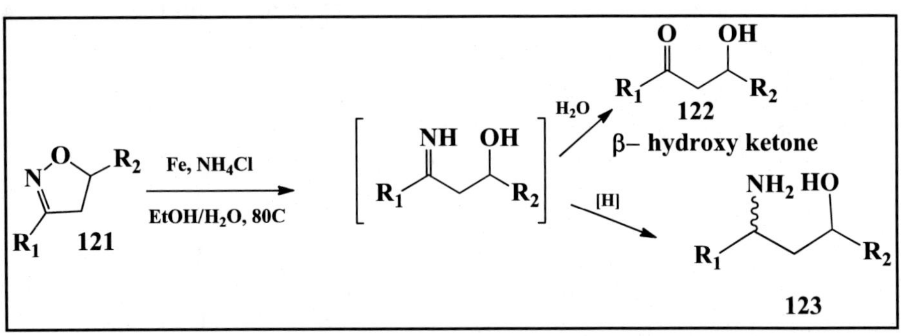

Scheme 65. Synthesis of conjugated and nonconjugated β-hydroxy ketones from Δ2-isoxazolines.

Mapp et al. reported synthesis of substituted β-amino acids from chiral isoxazoline intermediate [94]. They prepared several chiral isoxazolines and varieties of substituted β-amino acids that are very useful in making peptidomimetics and pharmaceutical building blocks (Scheme 66).

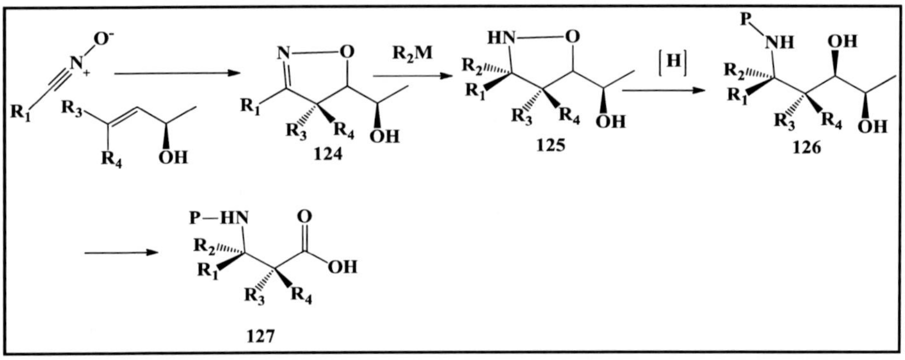

Scheme 66. Chiral isoxazolines are very useful in making peptidomimetics and pharmaceutical building blocks.

Wang et al. reported the first example of hypervalent iodine-promoted intromolecular oxy-fluorination of unsaturated oximes. By this method, the biologically interesting monofluoromethyl-substituted isoxazolines were obtained in good yield through fluorocylization. In the reaction, benign PIDP ($PhI(OPiv)_2$) was used as the oxidant and the cost effective HF–Py (HF–pyridine) was employed as the F-source. Furthermore, a synthetically useful building block, namely, monofluoromethyl-substituted β-hydroxyl ketone derivatives (**130**, Scheme 67), can now be synthesized via a simple ring-opening reaction from the isoxazoline product [95].

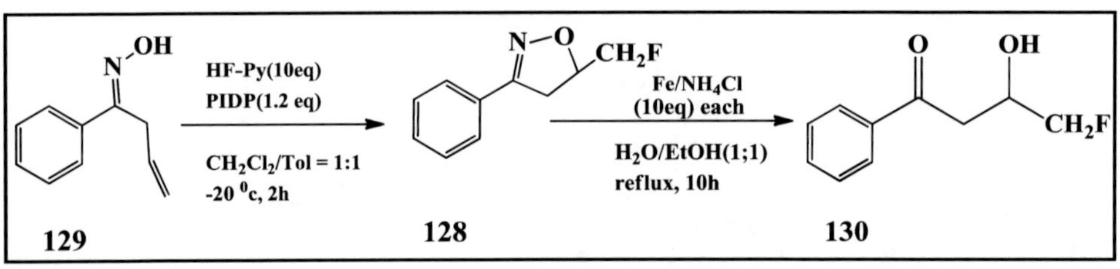

Scheme 67. Synthesis of monofluoromethyl-substituted β-hydroxyl ketone derivatives synthesized via a simple ring-opening reaction from the isoxazoline product.

The catalytic ring-expansive carbonylation of oxazolines, easily derived from α-amino acids, to yield β-amino acid derivatives is described in Scheme 68 [96].

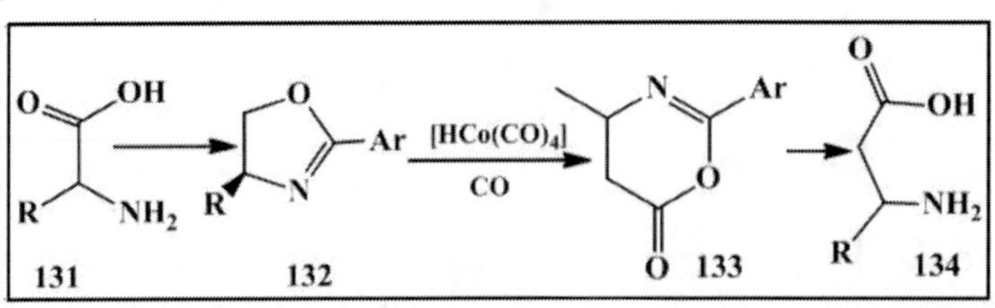

Scheme 68. Catalytic ring-expansive carbonylation of oxazolines.

Using different organometallics in the presence of etherate borane, Buhrlage et al. [97] performed a nucleophilic addition on isoxazolines **135** to obtain original isoxazolidines **136** in 73−80% yield with a dr between 7:1 and 20:1 (Scheme 69).

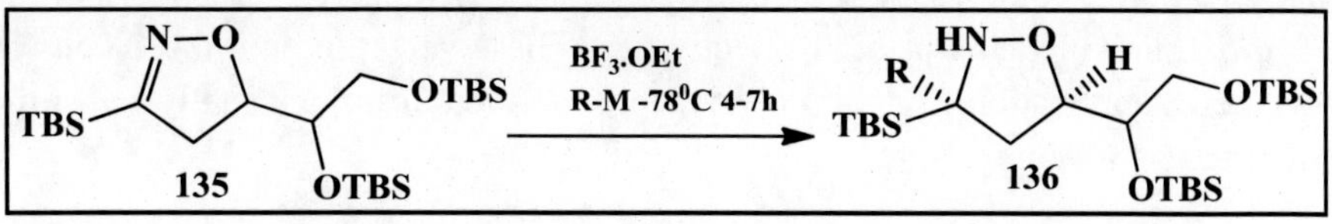

Scheme 69. Nucleophilic addition on isoxazolines to obtain original isoxazolidines.

Carreira et al. has described the mild, efficient, and provides the first general method for the selective reduction of this class of isoxazolines **137** and fills an important gap in the supporting methodology for the nitrile oxide approach to *β*-hydroxyl carbonyls [98]. Chemoselective reduction of unsaturated isoxazolines to the corresponding conjugated *β*-hydroxy ketones was performed using samarium iodide **138** (Scheme 70).

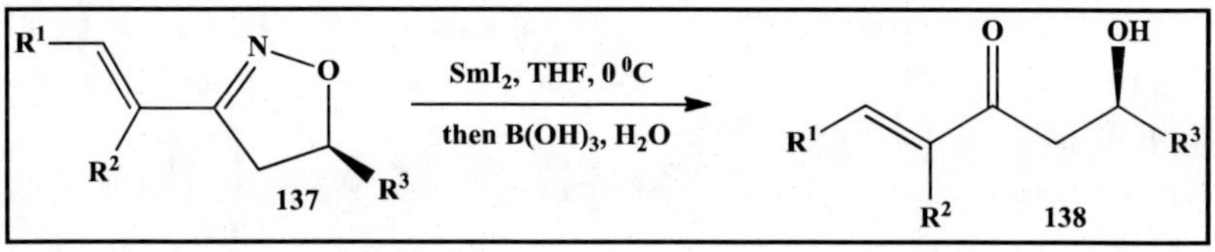

Scheme 70. Chemoselective reduction of unsaturated isoxazolines to conjugated *β*-hydroxy ketones.

L-(+)-Carbafuranomycin **139** is a novel analogue of L -(+)-furanomycin, an unusual antibiotic α-amino acid that attracted great interest due to its activity as an isoleucine antagonist. Jager et al. describes a concise and efficient asymmetric synthesis of this carba-analogue starting with the 1,3-dipolar cycloaddition of a chiral nitrile oxide **142** with cyclopentadiene **141**[99]. Notably, the methyl group was introduced by an SN2'cuprate substitution with high stereo- and regioselectivity (Scheme 71).

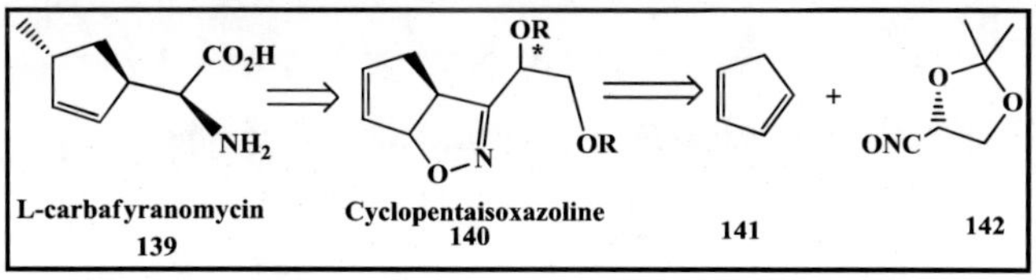

Scheme 71. Synthesis of L-(+)-Carbafuranomycin.

Mo-mediated cleavage [100] of cycloadduct *exo*-**143** underwent a tandem reductive N-O bond cleavage-retroaldol reaction followed by isomerization of the double bond (all three steps occurred in one pot) to provide cyclopentene **144** in 66% isolated yield (Scheme 72).

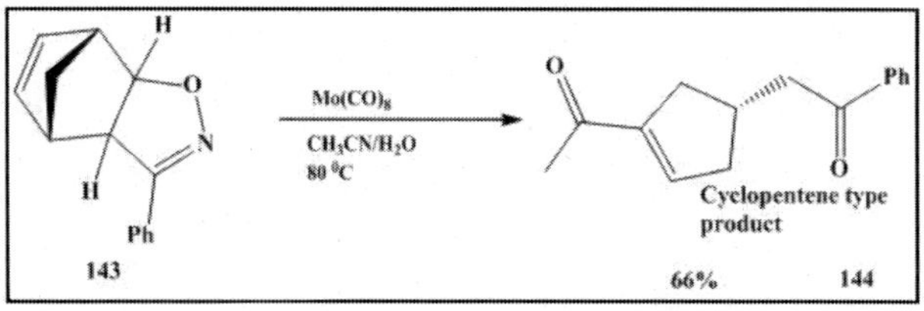

Scheme 72. Mo-mediated cleavage of cycloadduct to provide cyclopentene.

Carriera et al. [101] has shown that 2,3-dihydroisoxazoles **145** can serve as precursors for the preparation of either β-amino ketones or *β*-amino alcohols depending on their substitution pattern. The desired *β*-amino ketones can be obtained in preparatively useful yields. Additionally, a borohydride reduction followed by ring pening reaction provided a general entry to *syn-β*-amino alcohols in high selectivities. These versatile building blocks may find further synthetic applications, in particular when derived from unusually substituted 2,3-dihydroisoxazoles (Scheme 73).

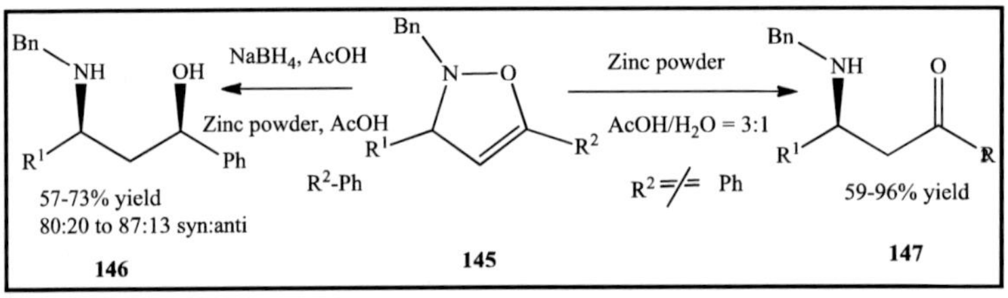

Scheme 73. Synthesis of syn-β-amino alcohols and β-amino ketones from 2,3-dihydroisoxazoles.

Maiti et al. [102] described use of small Cu(0)-nanoparticles (NPs) are fabricated utilizing $CuSO_4 \cdot 5H_2O$, surfactant (SDS) and ascorbic acid in aqueous medium. Its outstanding catalytic activity under low catalyst loading is developed toward reductive cleavage of isoxazoline, carbonyl azide and domino cyclization to furnish valuable 2-hydroxy-4-keto esters (**148**), primary amides and a new class of heterocycle, 4-hydroxy-2-pyrroline-5-one (**149**, Scheme 74).

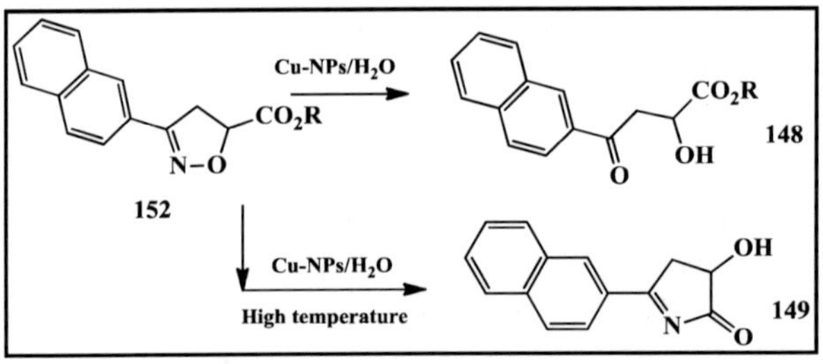

Scheme74. Reductive cleavage of isoxazoline, carbonyl azide and domino cyclization to furnish valuable 2-hydroxy-4-keto esters using small Cu(0)-nanoparticles.

In 2007, a diphenylamine-tethered bisoxazoline Zn(II) complex was used to add methoxyfuran to aromatic nitro olefins in an asymmetric Friedel-Crafts alkylation of 2-methoxyfuran with nitroalkenes with yield upto 96% (Scheme 75). The furan ring oxidatively gives an intermediate which is then treated with diazomethane to form the β-nitro ester from which the corresponding β²-amino acids are obtained.

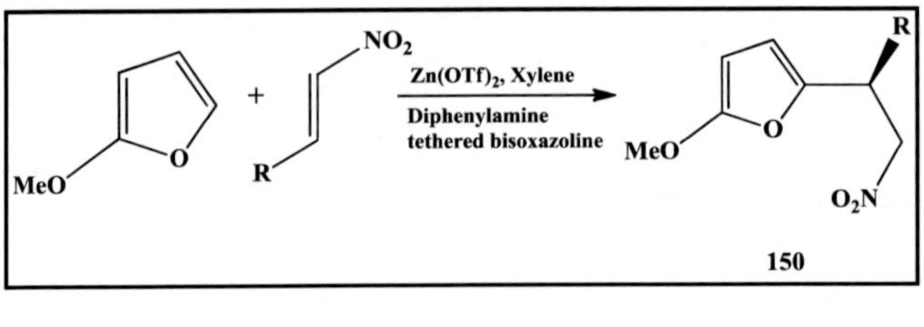

Scheme 75. Synthesis of methoxyfuran from diphenylamine-tethered bisoxazoline Zn(II) complex.

Emerging research interest on isoxazole moiety already has been proven by various search groups in the literature. Though many procedures are established for the synthesis of isoxazole core, but very few of them yield isoxazole with better percentage of product. But much more effort yet will be devoted in the future for developing new synthetic strategies.

MEDICINAL APPLICATION

Anticonvulsant Activity

The search for antiepileptic compounds with more selective activity and lower toxicity continues to be an area of investigation in medicinal chemistry [103]. In recent research it has been established GABA transport inhibitors and its particular astroglial uptake can act as anticonvulsant agents. Several isoxazole derivatives has been synthesised with proven anticonvulsant effect and to be so [104]. Compound **151** (Figure 7) is an example for this. Compound **152** is also an isoxazole derivative which affects the sodium channel to show its activity [105].

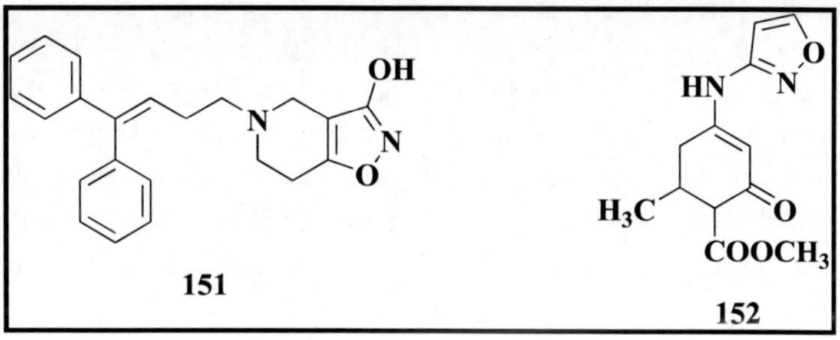

151

152

Figure 7. Structure of anti-convulsant isoxazoline compounds.

Cytotoxic Activity/Anti-Cancer Activity

Natural alkaloids containing isoxazolidinium compounds [106] show interesting cytotoxicity. The isoxazolidine scaffold is a potential candidate in the field of anticancer drug discovery. A promising approach can be observed with the isoxazolidinyl DNA intercalators or transcriptional activators.

Isoxaziline-DNA Intercalators

The development and design of novel cancer drugs and its therapies, mainly depends on the design of DNA intercalators and alkylating agent and this field is one of the most promising ones [107]. Lerman et al. [108] first proposed that DNA intercalation consists by inserting a planar polycyclic molecule between two base pairs without affecting the hydrogen DNA helix bonds. Rescifina et al. [109-112] has reported the generation of isoxazolidinyl polycyclic aromatic hydrocarbons (PAH), which exhibit moderate to good cytotoxicity toward various cell lines (compound **153-157**) as shown in Figure 8. They have also synthesized isoxazolidinyl PAH-calix[4]arene conjugates with higher cytotoxicity toward three human tumor cell lines

(IC50 95 nM). Molecular modeling studies along with circular dichroism analysis clearly give the evidence of intercalation of the compounds within the DNA double helix.

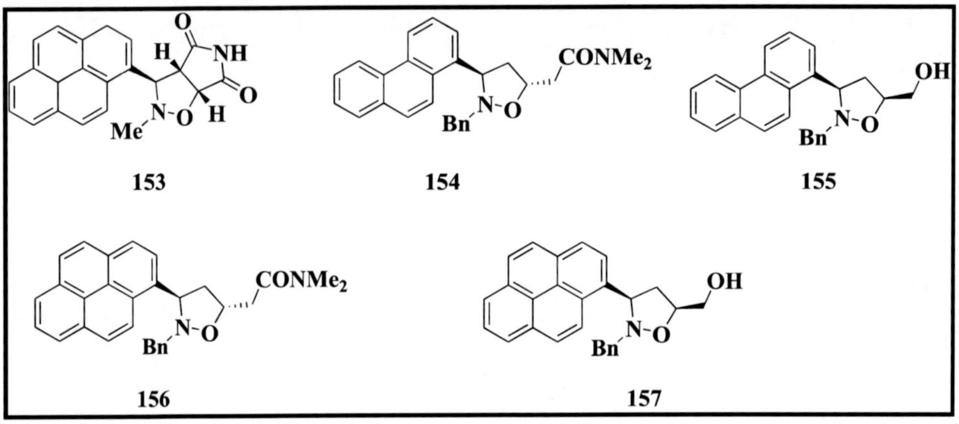

Figure 8. Isoxazolidine compounds as DNA intercalator.

Transcriptional Activators

Transcriptional activators are another promising way of DNA binding. Minter et al. reported [113] the synthesis and design of isoxazolidine **158** and **159** that exhibited transcription activation as potently as the peptide control ATF-4 (Figure 9). Key functional elements present in this molecule are isobutyl, carboxylic acid, hydroxyl, and phenyl groups were incorporated on the isoxazolidine ring to afford an adequate equilibrium between hydrophobicity and polarity that is crucial for overall DNA binding potentiality.

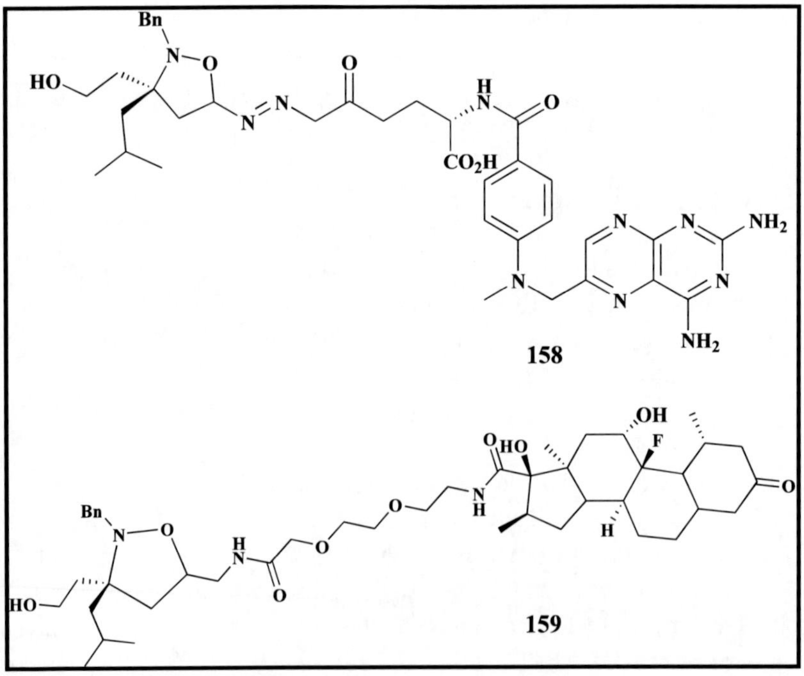

Figure 9. Isoxazolidinyl TAD.

Using Other Strategies

Different other strategies are followed by several groups of investigators towards the evaluation of cytotoxicity of isoxazolidine compounds toward cancer cell lines was observed (Figure 32). Khazir et al. [114] reported the spiro derivative of the sesquiterpenoid lactone (−)-α-santonin 400 that showed high potency against PC-3, MCF-7, and THP-1 cancer cell linesLast, Singh et al. [115] described chromano-piperidine fused isoxazolidines that showed a higher cytotoxic activity against COLO-05 than 5-fluorouracil. Again the effects of curcumin and of its isoxazole analogue in breastcancer cell line were already proved. The isoxazole analogue compound **160** has shown potent antitumor and molecular activities both in parental and in MDR tumor cells. Isoxazole derivatives also had significantly higher direct inhibition of the COX-2 catalytic activity than curcumin. The minimum metal chelation capability of isoxazole derivatives proved better than compared to curcumin [116]. The compound **161** also found highly effective against human tumor cell lines especially on renal cancer, CNS cancer cell and ovarian cancer cell lines [117]. Recently NO-NSAID was found as potent anti-cancer agents rather than their anti-inflammatory property [118]. Compound **162** is a NO donating compound used as anti-cancer agent (Figure 10).

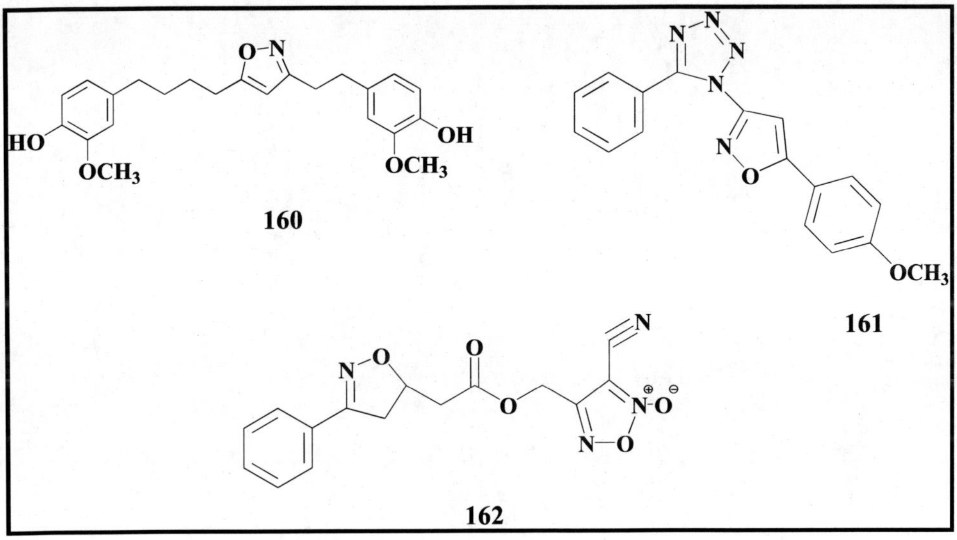

Figure 10. Structure of anti-cancer isoxazole compounds.

ANTIFUNGAL/ANTI-MICROBIAL ACTIVITY

A fungal and microbial infection also represents considerable health hazards for human life. The development of novel drugs is still a big challenge as well as to select the most potential drug for antifungal and antimicrobial from available drugs [119]. Drazoxolon (**163**) is a isoxaline derivative commonly used as anti-fungicidal agent. Acetyl Sulfisoxazole (**164**) is another important antimicrobial agent from the isoxazole family which is widely used in paediatric suspensions (Figure 11). Cycloserine (**165**) is a well-established molecule widely known as seromycin for its potency against mycobacterium tuberculosis [120-122].

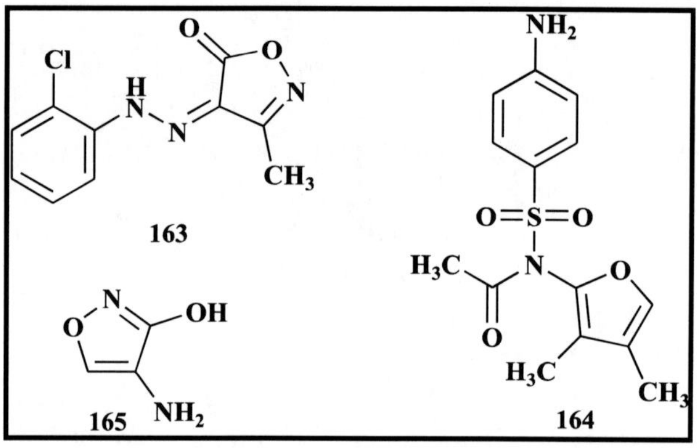

Figure 11. Structure of anti-microbial isoxazole compounds.

In 1997, China Shenyang Research Institute of Chemical Industry, discovered SYP-Z048, [123-125] a mixture of two diasteromers 3-[5-(4-chlorophenyl)-2,3-dimethyl-3-isoxazolidinyl]pyridine (**166**). It is an important fungicide and developed to control plant-pathogenic fungi (Figure 12) and showed high efficiency against a wide range of plant pathogens.

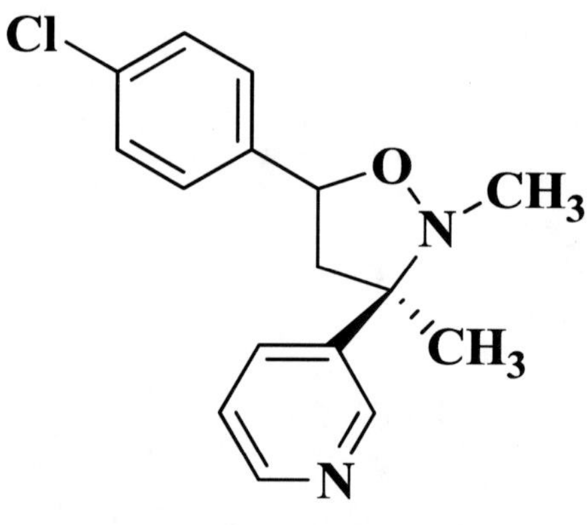

166

Figure 12. SYP-Z048 - an isoxazolidine based fungicide.

Rangappa group [126-128] also reported N-aryl-3−5-disubstituted isoxazoline (**167, 168**) rings, exhibiting antifungal or antimicrobial activity against three specific fungus and microorganisms. The activity of these compounds was similar to or even better than the reference Nystatin (for fungi) and Streptomycin (for bacteria) (Figure 13). Zelechowski et al. also described disubstituted isoxazolidines **169** that exhibited high fungicidal activity towards Rhizoctonia solani and Botrytis cinerea [129]. Nora et al. [130] reported 3,5-disubstituted isoxazolidines **170** as novel potential antibiotics (Figure 14). Spiroisoxazolidines **171** and **172** had been tested for antitubercular activity by Raunak et al. [131] the best inhibition was obtained with compound 164, especially when R_1 = H and R_2 = Cl (Figure 15).

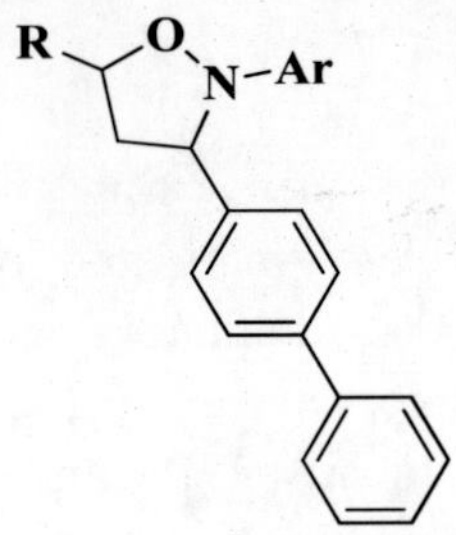

167

R = Ph, CO_2Et, CH_2OH
Ar = *p*-OMe, *p*-F-Ph

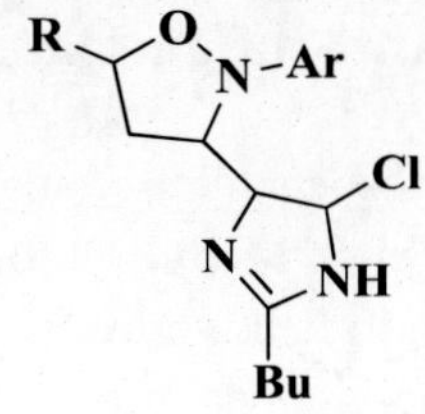

168

R = CO_2Ph, CO_2Me
Ar = Ph, *p*-Cl-Ph

Figure 13. N-Aryl-3–5-disubstituted isoxazolidines as antifungal and antimicrobial agents.

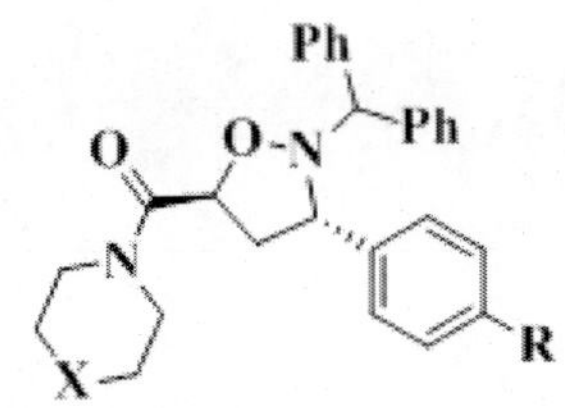

R = H, X = CH_2
R = CF_3, X = NCO_2Et

169

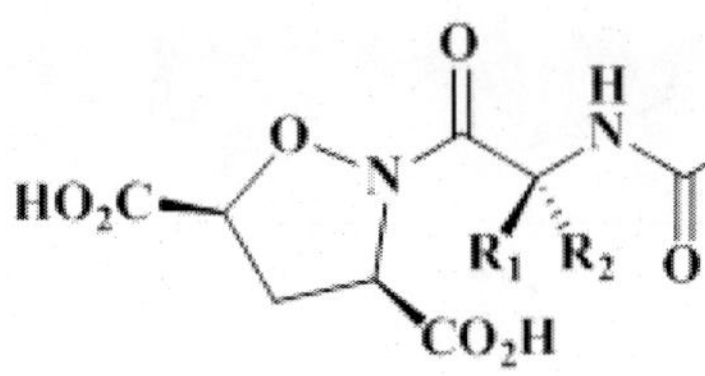

170

$R_1 = R_2 = H$
R_1 =Me, R_2 = H
R_1 = H, R_2 =Me

Figure 14. Example of disubstituted isoxazolidines with fungicidal or antibacterial activity.

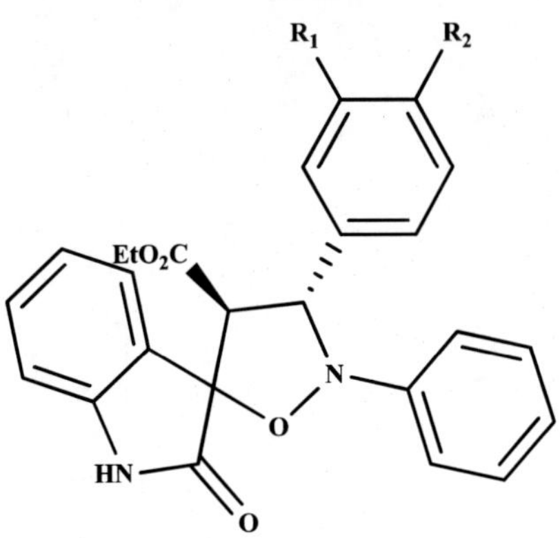

171

Anti-tubercular activity

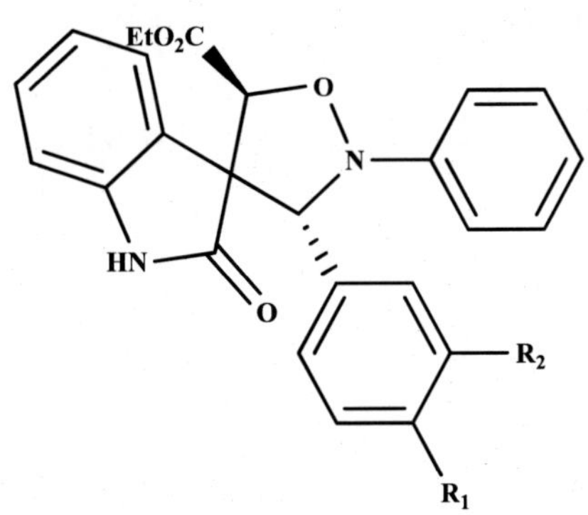

172

Anti-invasive activity

Figure 15. Two spiro-isoxazolidines isomers with different biological activities.

Analgesic and Anti-Inflammatory Activity

4,5-Diphenyl-4-isoxazolines containing one of the phenyl rings having a variety of substituents (H, F, MeS, $MeSO_2$) at the para position, were synthesized for its evaluation as a potent analgesic and selective cyclooxygenase-2 (COX-2) inhibitory and anti-inflammatory (AI) agents. The 4,5-phenyl-4-isoxazolines, which do not have a C-3 Me substituent, exhibit potent analgesic and AI activities in comparison with 2,3-dimethyl-5-(4-methylsulfonylphenyl)-4-phenyl-4-isoxazoline which exhibit excellent analgesic and AI activities, and it was a potent and selective COX-2 inhibitor [132]. Isoxazoline derivatives **173** were prepared from commercially available 2-hydroxy-aceto-naphthanone. Compound **173** exhibit potent antibacterial activity against four pathogenic bacteria, two Gram negative bacteria and two Gram positive bacteria (Figure 16). Majority of the isoxazoline analogues exhibited antibacterial activity comparable to ampicillin as reference drug [133]. Another example by Setoguchi et al. [134] is the isoxazolidine derivative **174** exhibited VLA-4 antagonist activity. The compound **174** showed an excellent IC50 and was orally efficacious (Figure 17).

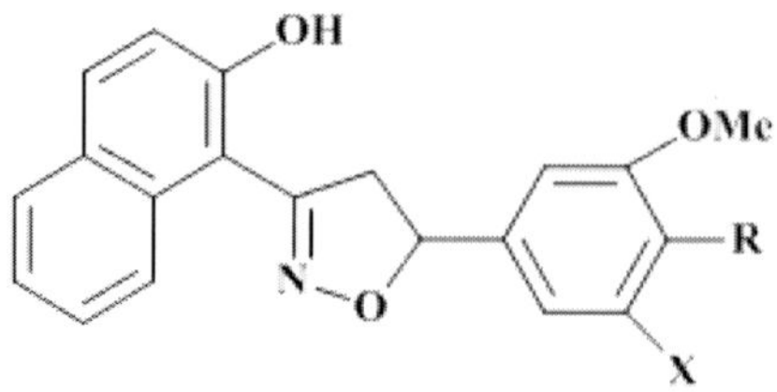

173

Figure 16. Structure of analgesic isoxazoline compounds.

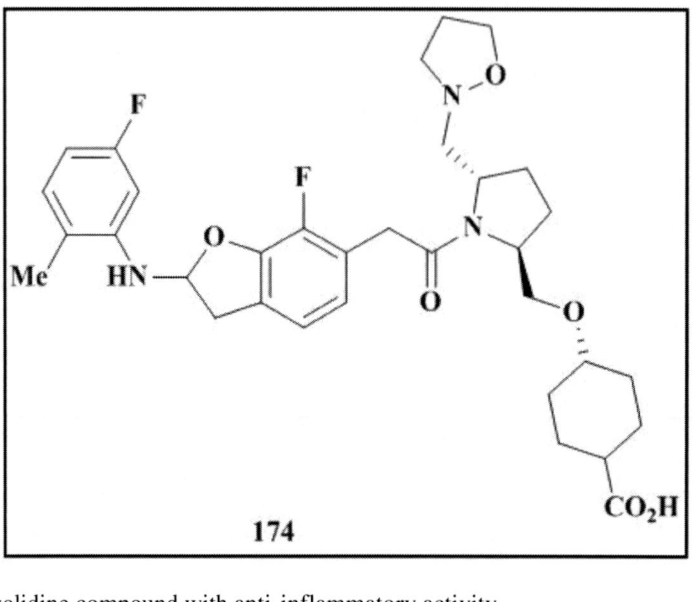

174

Figure 17. Isoxazolidine compound with anti-inflammatory activity.

Anti-Bacterial Activity

3-Aryl-5-(4-methoxyphenyl)-4,5-dihydroisoxazole-4-carbonitriles 175 [135] and 5-Acetyl-3-aryl-4-(2-furanoyl)-4,5-dihydroisoxazoles 176 [136], these two isoxazole based compounds were screened for their antibacterial and antifungal activity (Figure 18). A series of these isoxazole derivatives exhibited promising antibacterial and antifungal activity compared to standard drugs. The substitution of fluoro, chloro, bromo and cyano group at C3-substituted benzene ring of isoxazole ring resulted with potent antimicrobial activities. All these compounds also exhibited remarkable antioxidant activity and reducing power ability. A series of 2-(5-phenyl-4,5-dihydroisoxazol-3-yl)benzoic acids 177were synthesized and evaluated for their *in vitro* protein denaturation activity. The results showed that all these compounds possess significant anti-arthritic and anti-inflammatory action (Figure 18).

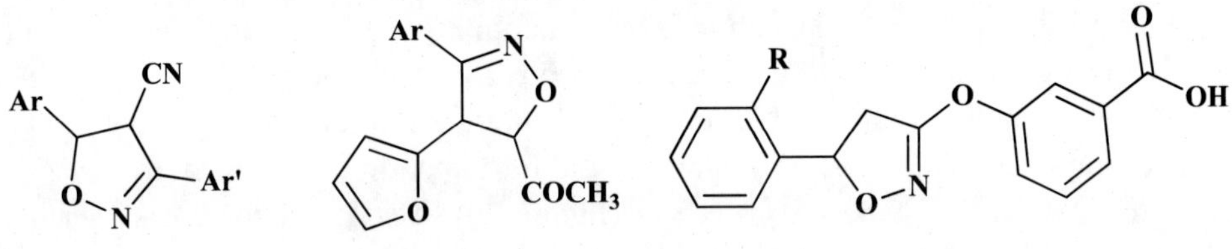

175 176 177

Figure 18. Structure of anti-bacterial isoxazaline compounds.

Curcumin-derived isoxazoles **178** [137] were minimize the metal chelation properties of curcumin. Curcumin analogue of isoxazoles became the potent ligands for fibrillar Aβ42 aggregates (Figure 19). Curcumin-derived isoxazoles inhibit Aβ secretion, bind to or inhibit the formation of fibrillar Aβ42 and tau aggregates. The enhancement in potency in comparison with curcumin is 10-100-fold. It is apparent from these data that curcumin-derived isoxazoles have multiple targets in Alzheimer's disease. The multifunctional curcumin–isoxazole **178** displayed interesting properties as an Aβ-modulating agent in primary neuronal cultures.

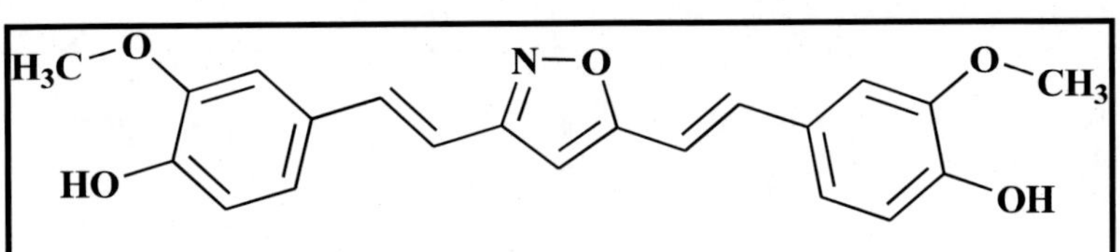

178

Figure 19. Structure of isoxazoline compounds for Alzheimer's disease.

A series of tetracyclic nitrofuran isoxazoline were designed to improve the pharmacokinetic properties of the initial compound, which had potent anti-tuberculosis activity but experienced hardship from poor solubility, high protein binding and rapid metabolism. The structural modifications were carried on the outer phenyl and piperidine rings to introduce solubilizing and metabolically blocking functional groups. Compounds with phenyl morpholine and pyridylmorpholine outer rings were found to be the most potent anti-tuberculosis agents. These compounds retained a narrow antibacterial spectrum of activity, with weak anti-Gram positive and no Gram negative activity, as well as good activity against non-replicating mycobacterium tuberculosis in a low oxygen model. Addition of solubilizing and metabolically blocked outer rings has improved solubility and decrease protein binding as designed [138]. A novel class of activators for chloride conductance in the cystic fibrosis transmembrane conductance regulator (CFTR) protein has been identified. These 3-(2-benzyloxyphenyl)isoxazolines were synthesized employing the 1,3-dipolar cycloaddition of nitrile oxides with various alkene dipolarophiles [139].

Antiviral Activities

A few medicinal organic compounds can be used against viruses. These are mainly nucleoside analogues. However, some very recent reference shows isoxazolidines based organic molecule had potential application towards inhibiting the HIV-1 replication. Loh et al. [140] showed isoxazolidine sulfonamides **179** and **180** had the potential to block the transcriptional activation of HIV-1. This is due to the presence of the halogen and the presence of aromatic rings seemed to have an important role for the antiretroviral activity towards the HIV-1 vector and the wild-type NL4.3 HIV-1 (Figure 20). Also the report from Lynch et al. [141] evaluated the ability of some isoxazolidines against HIV infection. They designed and tested the bicyclic isoxazolidine **181** as a CCR5 receptor antagonist. This compound blocked the MIP-α (chemokine CCL3) binding to the receptor, thus inhibiting the immune response.

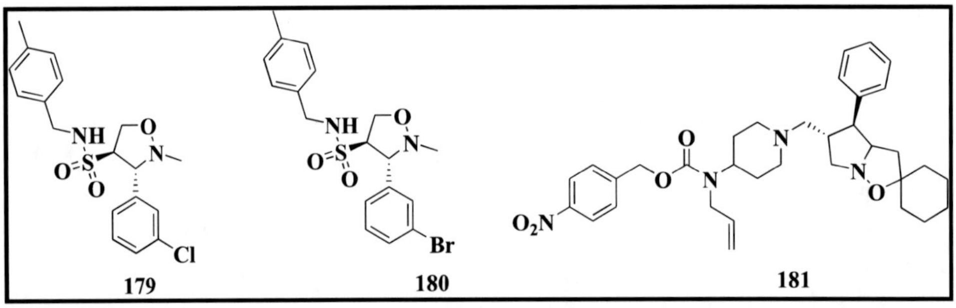

Figure 20. Isoxazolidine sulfonamides HIV-1 replication inhibitors and bicyclic isoxazolidine CCR5 receptor antagonist.

Conclusion

Isoxazoline is an important heterocyclic moiety. Emerging research interest on isoxazole moiety already has been proven by various research groups in the literature. Although many procedures are established for the synthesis of the isoxazole core, few of them yield isoxazole with a better percentage of product; hence, much more effort has yet to be devoted to develop new synthetic strategies. The isoxazoles derivative with biological activity with and new dimensions need to be explored for better medicinal understanding.

REFERENCES

[1] Lambein, Fernand, Kuo, Yu-Haey, Van Parijs, Roger. 1976. "Isoxazolin-5-ones, chemistry and biology of a new class of plant products" *Heterocycles* 4:567-593.

[2] Murai, Kenichi, Miyazaki, Shuji, Fujioka, Hiromichi. 2012. "Reactivityof the ester group attached isoxazoline, benzisoxazole, and isoxazole: a facial preparation of 3-acyl-substituted these heterocycles" *Tetrahedron Letters* 53:3746–3749.

[3] Triandafillidi, Ierasia, Kokotos, Christoforos G. 2017. "Green organocatalytic synthesis of isoxazolines via a one-pot oxidation of allyloximes" *Organic Letters 19:106-109.*

[4] Kaur, Kamalneet, Kumar, Vinod, Kumar Sharma, Anil, Kumar Gupta, Girish. 2014. "Isoxazoline containing natural products as anticancer agents: A review *European Journal of Medicinal Chemistry* 77:121-133.

[5] Kumar, K. Ajay, Govindaraju, M., Renuka, N., Kumar, G. Vasanth. 2015. "Isoxazolines: an insight to their synthesis and diverse applications" *Journal of Chemical and Pharmaceutical Research* 7:250-257.

[6] Bull, J. A., Balskus, E. P., Horan, R. A. J., Langner, M., Ley, S. V. 2006. "Stereocontrolled total synthesis of bengazole A: A marine bisoxazole natural product displaying potent antifungal properties" *Angewandte Chemie International Edition* 45:6714-6718.

[7] Huisgen, R. 1963. "Kinetics and mechanism of 1,3-dipolar cycloadditions"*Angewandte Chemie International Edition* 2: 633–645.

[8] Lee, Ja Young, Schiffer, Guido, Jäger, Volker. 2005. "Synthesis of L-Carbafuranomycin, an Unnatural Analogue of the Antibiotic Amino Acid Furanomycin" *Organic. Letters* 7:2317–2320.

[9] Han, Xiaoyu, Dong, Li, Geng, Caiwei, Jiao, Peng. 2015. "Catalytic asymmetric synthesis of isoxazolines from silyl nitronates"*Organic Letters* 17:3194–3197.

[10] Vitale, Paola, Scilimati, Antonio. 2013. "Five-membered ring heterocycles by reacting enolates with dipoles" *Current Organic Chemistry* 17:1986-2000.

[11] Bhosale, S., Kurhade, S., Prasad, U. V., Palle, V. P., and Bhuniya, D. 2009. "Efficient synthesis of isoxazoles and isoxazolines from aldoximes using Magtrieve CrO_2" *Tetrahedron Letters* 50:3948–3951.

[12] Dondoni, A., Giovannini, P. P. 2002. "Formyl*c*-glycosides as precursors to glycosyl nitrile oxides and nitrones" *Synthesis* 6:1701-1704.

[13] Katritzky, Alan R., Button, Martin A. C., Denisenko, Sergey N. 2000. "Efficient synthesis of 3,5-functionalized isoxazoles and isoxazolines *via* 1,3-dipolar cycloaddition reactions of 1-propargyl- and 1-allylbenzotriazoles" *Journal of Heterocyclic Chemistry* 37:1505–1510.

[14] (a) Rai, K. M. L., Hassner, A. 1997. "Intermolecular 1,3-dipolar cycloadditions of preformed nitrile oxides with phenyl vinyl sulphone" *Synthetic Commun*ications 27: 467-472. (b) Shing, T. K. M., Wong, W. F., Cheng, H. M., Kwok, W. S., So, K. H. 2007. "Intramolecular nitrile oxide-alkene cycloaddition of sugar derivatives with unmasked hydroxyl group(s)" *Organic Letters* 9:753-756.

[15] Rodrigues, R. D. C., Aguiar, A. P. D. 2001. *"A simple and efficient method for the synthesis of nitrile oxide from aldoxime using trichloroisocyanuric acid" Synthetic Communication* 31:3075-3080.

[16] Kumar, Vinod, Kaushik, M. P. 2006. "A novel one-pot synthesis of hydroximoyl chlorides and 2-isoxazolines using *N-tert*-butyl-*N*-chlorocyanamide" *Tetrahedron Lett*ers 47:1457-1460.

[17] Mukaiyama, Teruaki, Hoshino, Toshio. 1960. "The Reactions of Primary Nitroparaffins with Isocyanates" *Journal of American Chemical Society* 82:5339-5342.

[18] Feuer, Henry. 2007. *"Nitrile oxides, nitrones and nitronates in organic synthesis: novel strategies in synthesis"* John Wiley and Sons. Publication 129-130.

[19] Snider, Barry B., Che, Qinglin. 2002. "Oxidative and dehydrative cyclizations of nitroacetate esters with $Mn(OAc)_3$" *Tetrahedron* 58:7821-7827.

[20] Tu, Zhijay, Jang, Yaochung, Lin, Chunchi, Liu, Ju-Tsung, Hsu, Jianming, Sastry, M. N. V., Yao, Ching-Fa. 2005. "The study of reaction mechanism for the transformation of nitronate into nitrile by phosphorus trichloride" *Tetrahedron* 61:10541-10551.

[21] Zhang, Yaxing, Dong, Jianyu, Liu, Lixin, Liu, Long, Zhou, Yongbo, Yin, Shuang-Feng. 2017. "Manganese(III) acetate catalyzed oxidative amination of benzylic $C(sp^3)$–H bonds with nitriles" *Organic and Biomolecular Chemistry 15:*2897-2901.

[22] Muri, Dieter, Bode, Jeffrey W., Carreira, Erick M. 2000. "A novel, general method for the synthesis of nitrile oxides: dehydration of *o*-silylated hydroxamic acids" *Organic Letters* 2:539-541.

[23] (a) Das, Biswanath, Holla, Harish, Mahender, Gurram, Banerjee, Joydeep, Reddy, Majjigapu Ravinder. 2004. "Hypervalent iodine-mediated interaction of aldoximes with activated alkenes including Baylis–Hillman adducts: a new and efficient method for the preparation of nitrile oxides from aldoximes" *Tetrahedron Letters.* 45:7347-7350. (b) Jen, Tim, Mendelsohn, Brian A., Ciufolini, Marco A. 2011. "Oxidation of α-oxo-oximes to nitrile oxides with hypervalent iodine reagents" *Journal of Organic Chemistry* 76:728–731.

[24] Raihan, Mustafa J., Kavala, Veerababurao, Kuo, Chun-Wei, Raju, B. Rama, Yao, Ching-Fa. 2010. "On-water' synthesis of chromeno-isoxazoles mediated by [hydroxy(tosyloxy) iodo] benzene(HTIB)" *Green Chemistry* 12:1090-1096.

[25] Liaskopoulos, Theodoros, Skoulika, Stavroula, Tsoungas, Petros G., Varvounis, George. 2008. "Novel synthesis of naphthopyranoisoxazoles and versatile access to -naphthopyranoisoxazolines" *Synthesis* 5:711-718.

[26] Radhakrishna, A. S., Sivaprakash, K., Singh, B. B. 1991. "Iodobenzene dichloride - an efficient reagent for preparation of nitrile oxides from aldoximes" *Synthetic Communication* 21:1625-1629.

[27] Minakata, Satoshi, Okumura, Sota, Nagamachi, Toshiki, Takeda, Youhei. 2011. "Generation of nitrile oxides from oximes using*t*-buoi and their cycloaddition" *Organic Letters* 13: 2966-2969.

[28] Chatterjee, Nirbhik, Pandit, Palash, Halder, Samiran, Patra, Amarendra, Maiti, Dilip K. 2008. "Generation of nitrile oxides under nanometer micelles built in neutral aqueous media: synthesis of novel glycal-based chiral synthons and optically pure 2,8-dioxabicyclo[4.4.0] decene core" *Journal of Organic Chemistry 73*:7775 -7778.

[29] Kiegiel, Jarosław, Popławska, Magdalena, Jóźwik, Julita, Kosior, Małgorzata, Jurczak, Janusz. 1999. "New method ofin situgeneration of nitrile oxides by MnO2 oxidation of aldoximes" *Tetrahedron Letters* 40:5605-5608.

[30] Kim, Mingi, Hwang, Yoon Soo, Cho, Wansang, Park, Seung Bum. 2017. "Synthesis of 3,5-disubstituted isoxazoles containing privileged substructures with a diverse display of polar surface area" *ACS Combinatorial Science* 19:407–413.

[31] Das, Biswanath, Mahender, Gurram, Holla, Harish, Banerjee, Joydeep. 2005. "An easy access to 3,4-dihydro-3H-chromeno[4,3-c]isoxazoles and functionalized isoxazolines" Arkivoc 3:27-35.

[32] Lokanatha Rai, K. M., Linganna, N., Hassner, Alfred, Anjanamurthya, C. 1992. "Convenient method for the generation of nitrile oxide and its application to the synthesis of 2-isoxazolines" *Organic Preparations and Procedures International* 24: 91-93.

[33] Gothelf, Kurt V., Jørgensen, Karl Anker. 1998. "Asymmetric 1,3-dipolar cycloaddition reactions" *Chemical Review* 98:863-909.

[34] (a) Ezabadi, Iraj Rahavi, Camoutsis, Charalabos, Zoumpoulakis, Panagiotis, Geronikaki, Athina, Soković, Marina, Glamočilija, Jasmina, Ćirić, Ana. 2008. "Sulfonamide-1,2,4-triazole derivatives as antifungal and antibacterial agents: Synthesis, biological evaluation, lipophilicity, and conformational studies" *Journal of Bioorganic and Medicinal Chemistry* 16:1150-1161. (b) Ganesan, A. 1996. "The dimeric steroid-pyrazine marine alkaloids: Challenges for isolation, synthesis, and biological studies" *Angewandte Chemie International Edition* 35:611-613. (c) Ma, Chunrong, Liu, Xiaoxiang, Li, Xiaoyan, Flippen-Anderson, Judith,Yu, Shu, Cook, James M. 2001. "Efficient asymmetric synthesis of biologically important tryptophan analogues via a palladium-mediated heteroannulation reaction" *Journal of Organic Chemistry.* 66:4525-4542.

[35] (a) Casiraghi, Giovanni, Zanardi, Franca, Rassu, Gloria, Spanu, Pietro. 1995. "Stereoselective approaches to bioactive carbohydrates and alkaloids-with a focus on recent syntheses drawing from the chiral pool" *Chemical Review* 95:1677-1716. (b) Kwon, Taesoo, Heiman, Ann S., Oriaku, Ebenezer T., Yoon, Kyoungjin, Lee, Henry J. 1995. "New steroidal antiinflammatory antidrugs: steroidal [16.alpha.17.alpha.-d]-3'-carbethoxyisoxazolines" *Journal of Medicinal Chemistry* 38:1048-1051. (c) Gordaliza, Marina, Faircloth, Glynn T., Castro, Ma Angeles, del Corral, José M. Miguel, López-Vázquez, Ma Luisa, Feliciano, Arturo San. 1996. "Immunosuppressive Cyclolignans" *Journal of Medicinal Chemistry* 39:2865-2868.

[36] Sandanayaka, Vincent P., Yang, Youjun. 2000. "Dipolar cycloaddition of novel 6-(nitrileoxidomethyl) penam sulfone: an efficient route to a new class of β-lactamase inhibitors" *Organic Letters* 2:3087-3090.

[37] (a) Gothelf, Kurt V., Jørgensen, Karl Anker. 1998. "Asymmetric 1,3-dipolar cycloaddition reactions" *Chemical Review* 98:863-910. (b) Gallos, J. K., Koumbis, A. E. 2003. "1,3-Dipolar cycloadditions in the synthesis of carbohydrate mimics. part 1: nitrile oxides and nitronates" *Current Organic Chemistry*, 7:397- 426.

[38] Bode, Jeffrey W., Fraefel, Nina, Muri, Dieter, Carreira, Erick M. 2001. "A general solution to the modular synthesis of polyketide building blocks by kanemasa hydroxy-directed nitrile oxide cycloadditions"*Angewandte Chemie International Edition* 40:2082-2085.

[39] Kudyba, Iwona, Jóźwik, Julita, Romański, Jan, Raczko, Jerzy, Jurczak, Janusz. 2005. "The synthesis of oximes and nitroalkanes bearing a chiral auxiliary unit: convenient substrates for the preparation of enantiomerically pure nitrile oxides" *Tetrahedron: Asymmetry* 16:2257-2267.

[40] Litvinovskaya, Raisa P., Khripach, Vladimir A. 2001. "Regio- and stereochemistry of 1,3-dipolar cycloaddition of nitrile oxides to alkenes" *Russian Chemical Review 70*:405-424.

[41] Davies, Christopher D., Marsden, Stephen P., Stokes, Elaine S. E. 1998. "Chiral vinyl dioxazaborocines in synthesis: Asymmetric synthesis of 5-substituted Δ2-isoxazolines via nitrile oxide cycloaddition" *Tetrahedron Letters 39*:8513-8516.

[42] Park, Kyung-Ho, Kurth, Mark J. 2002. "Diastereoselective synthesis of hydantoin- and isoxazoline-substituted dispirocyclobutanoids" *Journal of Organic Chemistry* 65:3520-3524.

[43] Ros, Abel, Alvarez, Eleuterio, Dietrich, Hansjörg, Fernández, Rosario, Lassaletta, José M. 2005. "A practical synthesis of enantiopure 4,5-dihydroisoxazole-5-carboxylic acids" *Synlett* 19:2899-2904.

[44] (a) Trost, Barry M., Flemming, Ian, Semmelhack, Martin F. 1991. "*In Comprehensive Organic Synthesis*" Eds.; Pergamon Press: Oxford, 4:1111-1112. (b) Peet, Norton P., Huber, Edward W., Farr, Robert A. 1991. "Diastereoselectivity in the intramolecular nitrone, oxime, and nitrile oxide cycloaddition reactions. Synthesis of amino inositol derivatives as α-glucosidase inhibitors" *Tetrahedron* 47:7537-7550.

[45] Chiacchio, Ugo, Corsaro, Antonino, Rescifina, Antonio, Bkaithan, Majdi, Grassi, Giovanni, Piperno, Anna, Privitera, Tindara, Romeo, Giovanni. 2001. "Stereoselective synthesis of homochiral annulated sultams via intramolecular cycloaddition reactions" *Tetrahedron* 57:3425-3433.

[46] Alcaide, Benito, Almendros, Pedro, Sáeza, Elena. 2004. "Intramolecular 1,3-dipolar cycloaddition reaction of novel 2-azetidinone-tethered alkenyl nitrile oxides" *Arkivoc* IV:137-152.

[47] Kanemasa, Shuji, Nishiuchi, Masaki, Kamimura, Akio, Hori, Kenzi. 1994. "First successful metal coordination control in 1,3-dipolar cycloadditions. High-rate acceleration and regio- and stereocontrol of nitrile oxide cycloadditions to the magnesium alkoxides of allylic and homoallylic alcohols" *Journal of American Chemical Society* 116:2324–2339.

[48] (a) Ukaji, Yutaka, Sada, Kazunori, Inomata, Katsuhiko. 1993. "Enantioselective synthesis of 2-Isoxazolines*via* asymmetric 1,3-dipolar cycloaddition of nitrile oxide to achiral allyl alcohol" *Chemistry Letters* 22:111847-1850. (b) Shimizu, Makoto, Ukaji, Yutaka, Inomata, Katsuhiko. 1996. "Catalytic asymmetric 1,3-dipolar cycloaddition of nitrile oxides to an achiral allyl alcohol utilizing diisopropyl tartrate as a chiral auxiliary" *Chemistry Letters* 25:455-456.

[49] Sibi, Mukund P., Itoh, Kennosuke, Jasperse, Craig P. 2004. "Chiral Lewis acid catalysis in nitrile oxide cycloadditions" *Journal of American Chemical Society* 126:5366–5367.

[50] Bardhan, Sujata, Schmitt, Daniel C., Porco, John A. 2006. "Total synthesis and stereochemical assignment of the spiroisoxazoline natural product (+)-calafianin" *Organic Letters* 8:927–930.

[51] Weinig, Hans-Georg, Passacantilli, Pietro, Colapietro, Marcello, Piancatelli, Giovanni. 2002. "Glycal-mediated synthesis of enantiomerically pure 5-substituted isoxazoles containing a differentially *O*-benzylated glycerol moiety" *Tetrahedron Letters* 43:264613–4615.

[52] (a) Chapleur, Yves. 1998. "Carbohydrate mimics: concepts and methods" *Wiley-VCH: Weinheim*. (b) Hollingsworth, Rawle I., Wang, Guijun. 2000. "Toward a carbohydrate-

based chemistry: progress in the development of general-purpose chiral synthons from carbohydrates" *Chemical Review* 100:4267–4282. (c) Brasholz, Malte, Reissig, Hans-Ulrich, Zimmer, Reinhold. 2009. "Sugars, alkaloids, and heteroaromatics: exploring heterocyclic chemistry with alkoxyallenes" *Accounts of Chemical Research* 42:45–56. (d) Kurteva, Vanya B., Afonso, Carlos A. M. 2009. "Synthesis of cyclopentitols by ring-closing approaches"*Chemical Review* 109:6809–6857.

[53] (a) Marco-Contelles, José, Opazo, Elsa de. 2002. "Synthesis of enantiomerically pure, highly functionalized, medium-sized carbocycles from carbohydrates: formal total synthesis of (+)-calystegine B2" *Journal Organic Chemistry* 67:3705–3717. (b) Khiar, Noureddine, Mallouk, Siham, Valdivia, Victoria, Bougrin, Khalid, Soufiaoui, Mohammed, Fernández, Inmaculada. 2007. "Enantioselective organocatalytic oxidation of functionalized sterically hindered disulfides" *Organic Letters* 9:1255–1258.

[54] Smellie, Iain A. S., Moggach, Stephen A., Paton, R. Michael. 2011. "Synthesis of novel amidoxime-linked pseudodisaccharides" *Tetrahedron Letters* 52:95-97.

[55] Maiti, Dilip K., Halder, Samiran, Pandit, Palash, Chatterjee, Nirbhik, Joarder, Dripta De, Pramanik, Nabyendu, Saima, Yasmin, Patra, Amarendra, Maiti, Prabir K. 2009. "Synthesis of glycal-based chiral benzimidazoles by $VO(acac)_2$–$CeCl_3$ combo catalyst and their self-aggregated nanostructured materials" *Journal of Organic Chemistry* 74:8086-8097.

[56] Shing, Tony K. M., So, King H., Kwok, Wun S. 2009."Carbocyclization of carbohydrates: diastereoselective synthesis of (+)-gabosine f, (−)-gabosine o, and (+)-4-epi-gabosine o" *Organic Letters* 11:5070-5073.

[57] (a) Mishra, Ram Chandra, Tewari, Neetu, Verma, Shyam Sunder, Tripathi, Rama Pati, Kumar, Manish, Kumar Shukla, Praveen. 2004. "Diastereoselective synthesis and antifungal activity of glycosyl isoxazolines" *Journal of Carbohydrate Chemistry* 23:353-374. (b) Benltifa, Mahmoud, Hayes, Joseph M., Vidal, Sébastien, Gueyrard, David, Goekjian, Peter G., Praly, Jean-Pierre, Kizilis, Gregory, Tiraidis, Costas, Alexacou, Kyra-Melinda, Chrysina, Evangelia D., Zographos, Spyros E., Leonidas, Demetres D., Archontis, Georgios, Oikonomakos, Nikos G. 2009. "Glucose-based spiro-isoxazolines: a new family of potent glycogen phosphorylase inhibitors" *Bioorganic and Medicinal Chemistry* 17:7368-7380.

[58] ((a) Larsen, Karl Erik, Torssell, Kurt B. G. 1984. "An improved procedure for the preparation of 2-isoxazolines" *Tetrahedron* 40:2985-2988. (b) Schaller, Christophe, Vogel, Pierre. 2000. "Asymmetric syntheses of new polyhydroxylated quinolizidines: cross-aldol reactions of 7-oxabicyclo[2.2.1]heptan-2-one and 3a,4a,7a,7b-tetrahydro[1,3]dioxolo[4,5]furo[2,3-*d*]isoxazole-3-carbaldehyde derivatives" *Helva Chimia Acta* 83:193-232. (c) Berecibar, Amaya, Grandjean, Cyrille, Siriwardena, Aloysius. 1999. "Synthesis and biological activity of natural aminocyclopentitol glycosidase inhibitors: mannostatins, trehazolin, allosamidins, and their analogues" *Chemical Review* 99:779-844.

[59] Calabrese, Gary S., Buchanan, Robert M., Wrighton, Mark S. 1982. "Electrochemical behavior of a surface-confined naphthoquinone derivative. Electrochemical and photoelectrochemical reduction of oxygen to hydrogen peroxide at derivatized electrodes" *Journal of American Chemical Society* 104:5786-5788.

[60] Desroses, Matthieu, Chéry, Florence, Tatibouët, Arnaud, De Lucchi, Ottorino, Rollin, Patrick. 2002. "Sugar-based ethenyl ethers: stereoselective dipolar cycloadditions of nitrile oxides" *Tetrahedron: Asymmetry* 13:2535- 2539.

[61] Tamai, Tetsuo, Asano, Shingo, Totani, Kiichiro, Takao, Ken-ichi, Tadan, Kin-ichi. 2003. "Highly stereoselective [3+2] cycloadditions of nitrile oxides to methyl 4-*o*-acryloyl-6-deoxy-2,3-*o*-(*t*-butyldimethylsilyl)-α-*d*-glucopyranoside" *Synlett* 12:1865-1867.

[62] Lee, Ja Young, Schiffer, Guido, Jäger, Volker. 2005. "Synthesis of L-carba-furanomycin, an unnatural analogue of the antibiotic amino acid furanomycin" *Organic Letters* 7: 2317-2320.

[63] Pandit, Palash, Chatterjee, Nirbhik, Halder, Samiran, Hota, Sandip K., Patra, Amarendra, Maiti, Dilip K. 2009 "PHIO as a powerful cyclizing reagent: regiospecific [3+2]-tandem oxidative cyclization of imine toward cofacially self-aggregated low molecular mass organic materials" *Journal of Organic Chemistry*, 74:2581–2584.

[64] Tang, Shibing, He, Jinmei, Sun, Yongquan, He, Liuer, She, Xuegong. 2010. "Efficient and regioselective synthesis of 5-hydroxy-2-isoxazolines: versatile synthons for isoxazoles, β-lactams, and γ-amino alcohols" *Journal of Organic Chemistry* 75:1961–1966.

[65] Kumar, Abhishek, Fernandes, Jennifer, Kumar, Pankaj. 2014. "Synthesis and biological evaluation of some novel isoxazoline derivatives of carbostyril" *World Journal of Pharmacy and Pharmaceutical Sciences* 3:1267-1277.

[66] Pérez, Juana M., Ramón, Diego J. 2015 "Synthesis of 3,5-disubstituted isoxazoles and isoxazolines in deep eutectic solvents" *ACS Sustainable Chemistry and Engineering* 3:2343–2349.

[67] Umesha, K. B., Kumar, K. Ajay, Lokanatha Rai, K. M. 2002. "A novel synthesis of isoxazoles via 1,3-dipolar cycloaddition of nitrile oxides to acetyl acetone" *Synthetic Communication* 32:1841-1846.

[68] Kadnor, V. A., Pandhare, G. R., Gadhave, A. G., Uphade, B. K. 2011. "Synthesis and antibacterial activity of some fluorinated isoxazoline derivatives" *Rasayan Journal of Chem*istry 4:437-441.

[69] Tavares, Aline, Schneider, Paulo H., Merlo, Aloir A. 2009. "3,5-Disubstituted isoxazolines as potential molecular kits for liquid-crystalline materials" *European Journal of Organic Chmistry* 6:889–897.

[70] Cecchi, Luca,De Sarlo, Francesco, Machetti, Fabrizio. 2006. "1,4-Diazabicyclo[2.2.2]octane (DABCO) as an efficient reagent for the synthesis of isoxazole derivatives from primary nitro compounds and dipolarophiles: the role of the base" *European Journal of Organic Chmistry* 21: 4852–4860.

[71] Knight, David W., Proctor, Anthony J., Clough, John M. 2010. "New regiospecific catalytic approaches to 4,5-dihydroisoxazoles and 2,5-dihydroisoxazoles from*o*-propargylic hydroxylamines" *Synlett*. 4:628-632.

[72] (a) Pennicott, Lewis, Lindell, Stephen. 2006. "The preparation of 2-isoxazolines from*o*-propargylic hydroxylamines via a tandem rearrangement-cyclisation reaction" *Synlett*. 463-465. (b) He, Yu-Tao, Li, Lian-Hua, Yang, Yan-Fang, Wang, Yu-Qi, Luo, Jian-Yi, Liu, Xue-Yuan, Liang, Yong-Min. 2013. "Copper-catalyzed synthesis oftrifluoromethyl-substituted isoxazolines" *Chemical Communications* 49:5687-5689.

[73] Bala, Kason, Hailes, Helen C. 2005. "Nitrile oxide 1,3-dipolar cycloadditions in water: novel isoxazoline and cyclophane synthesis" *Synthesis* 19:3423–3427.

[74] Choe, Hyeonjeong, Pham, huy Trang, Leee, Joo Yun, Latif, Muhammad, Park, Haeil, Kang, Young Kee, Lee, Jongkook. 2016. "Remote stereoinductive intramolecular nitrile oxide cycloaddition: asymmetric total synthesis and structure revision of (−)-11β-hydroxycurvularin" *Journal of Organic Chemistry* 81:2612–2617.

[75] Anand, P., Singh, B. 2014 "Pyrrolo-isoxazole: a key molecule with diverse biological actions" *Medicinal Chemistry* 14:623-627.

[76] Triandafillidi, Ierasia, Kokotos, Christoforos G. 2017. "Green organocatalytic synthesis of isoxazolines via a one-pot oxidation of allyloximes" *Organic Letters* 19:106−109.

[77] Wang, Gang-Wei, Li, Shi-Xia, Wu, Quan-Xiang, Yang, Shang-Dong. 2015. "Cu-catalyzed sp^3C–H bond oxidative functionalization of alkylazaarenes and substituted ethanones: an efficient approach to isoxazoline derivatives" *Organic Chemistry Frontiers* 2:569–573.

[78] Yoshimura, Akira, Middleton, Kyle R., Todora, Anthony D., Kastern, Brent J., Koski, Steven R., Maskaev, Andrey V., Zhdankin, Viktor V. 2013. "Hypervalent iodine catalyzed generation of nitrile oxides from oximes and their cycloaddition with alkenes or alkynes" *Organic Letters* 15:4010–4013.

[79] Han, Liuquan, Zhang, Bijun, Zhu, Min, Yan, Jie. 2014. "An environmentally benign synthesis of isoxazolines and isoxazoles mediated by potassium chloride in water" *Tetrahedron Letters* 55:2308–2311.

[80] Dong, Kui-Yong, Qin, Hai-Tao, Liu, Feng, Zhu, Chen. 2015. "Oxime-mediated oxychlorination and oxybromination of unactivated olefins" *European Journal of Organic Chemistry* 7:1419–1422.

[81] Yoshimura, Akira, Zhu, Chenjie, Middleton, Kyle R., Todora, Anthony D., Kastern Brent J., Maskaevand Andrey V., Zhdankin, Viktor V. 2013. "Hypoiodite mediated synthesis of isoxazolinesfromaldoximesandalkenesusing catalytic KI and oxones the terminal oxidant" *Chemical Communications* 49:4800-4802.

[82] Pramanik, Nabyendu, Sarkar, Satinath, Roy, Dipanwita, Debnath, Sudipto, Ghosh, Sukla, Khamarui, Saikat, Maiti, Dilip K. 2015. "Synthesis and diverse general oxidative cyclization catalysis of high-valent $Mo^{VI}O_2$(HL) to ubiquitous heterocycles and their chiral analogues with high selectivity" *RSC Advances* 5:101959–101964.

[83] Dash, Ashutosh K., Jaladanki, Chaitanya K., Maiti, Dilip K., Singh, Deepika, Tripathi, Anil K., Gupta, Vivek K., Bharatam, Prasad V., Mukherjee, Debaraj. 2016. "Tandem*gem*–dichlorination and nitrile oxide generation from chlorochromene aldoximes: synthesis of a new class of room temperature fluxional 4-chromanone derivatives" *ChemistrySelect* 3:567 –571.

[84] (a) Bode, Jeffrey W., Carreira, Erick M. 2001. "Stereoselective syntheses of epothilones a and b via directed nitrile oxide cycloaddition" *Journal of American Chemical Society*, 123:3611-3612. (b) Andrés, J. Ignacio, Alcázar, Jesús, Alonso, José M., Alvarez, Rosa M., Bakker, Margot H., Biesmans, Ilse, Cid, José M., De Lucas, Ana I., Drinkenburg, Wilhelmus, Fernández, Javier, Font, Luis M., Iturrino, Laura, Langlois, Xavier, Lenaerts, Ilse, Martínez, Sonia, Megens, Anton A., Pastor, Joaquín, Pullan, Shirley, Steckler, Thomas. 2007, "Tricyclic isoxazolines: Identification of R226161 as a potential new antidepressant that combines potent serotonin reuptake inhibition and α_2-adrenoceptor antagonism" *Bioorganic and Medicinal Chemistry* 15:3649-3660.

[85] Curran, Dennis P. 1983. "Reduction of delta 2-isoxazolines: Raney nickel catalyzed formation of beta-hydroxy ketones" *Journal of American Chemical Society* 105:5826-5833.

[86] Behenna, Douglas C., Corey, E. J. 2008. "Simple enantioselective approach to synthetic limonoids" *Journal of American Chemical Society* 130:6720–6721.

[87] Alcaraz, L., Harnett, J. J., Mioskowski, C., Gall, T. L., Shin, D. S., Falck, J. R. 1995. "Taxol total synthesis: preparation of a chiral ring a moiety via biomimetic cyclization and evaluation of a tandem nitrile oxide strategy for rings b/c" *Journal of OrganicChemistry* 60:7209-7213.

[88] Majumdar, Swapan, Hossain, Jewel, Natarajan, Ramalingam, Banerjee, Ashish K., Maiti, Dilip K. 2015. "Phthalate tethered strategy: carbohydrate nitrile oxide cycloaddition to 12–15 member chiral macrocycles with alkenyl chain length controlled orientation of bridged isoxazolines" *RSC Advances* 5:106289–106293.

[89] Bode, Jeffrey W., Carreira, Erick M. 2009. "Stereoselective syntheses of epothilones a and b via nitrile oxide cycloadditions and related studies" *Journal of Organic Chemistry* 74:8695–8712.

[90] Kozikowski, Alan P. 1984. "The isoxazoline route to the molecules of nature" *Accounts of Chemical Research* 17:410-416.

[91] Nising, Carl F., Bräse, Stefan. 2012. "Recent developments in the field of oxa-Michael reactions" *Chemical Society Review* 41:988-999.

[92] Jiang, Dahong, Chen, Yuanwei. 2008. "Reduction of δ^2-isoxazolines to β-hydroxy ketones with iron and ammonium chloride as reducing agent" *Journal of Organic Chemistry* 73:9181–9183.

[93] Fuller, Amelia A., Chen, Bin, Minter, Aaron R., Mapp, Anna K. 2005. "Succinct synthesis of β-amino acids via chiral isoxazolines" *Journal of American Chemical Society* 127:5376-5383.

[94] Kong, Weidong, Guo, Quanping, Xu, Zhaoqing, Wang, Guoqiang, Jiang, Xianxing, Wang, Rui. 2015. "Iodine(iii)-mediated oxy-fluorination of alkenyl oximes: an easy path to monofluoromethyl-substituted isoxazolines" *Organic Letters* 17:3686−3689.

[95] Byrne, C. M., Church, T. L., Kramer, J. W., Coates, G. W. 2008. "Catalytic synthesis of β3-amino acid derivatives from α-amino acids" *Angewandte Chemie International Edition* 47: 3979- 3983.

[96] (a) Buhrlage, S. J., Chen, B., Mapp, A. K. 2009. "A flexible synthetic route to isoxazolidine β-proline analogs" *Tetrahedron* 65:3305−3313. (b) Berthet, Mathéo, Cheviet, Thomas, Dujardin, Gilles, Parrot, Isabelle, Martinez, Jean. 2016. "Isoxazolidine: a privileged scaffold for organic and medicinal chemistry" *Chemical Reviews 116*:15235-15283.

[97] Bode, Jeffrey W., Carreira, Erick M. 2001. "A mild and chemoselective method for the reduction of conjugated isoxazolines to β-hydroxy ketones" *Organic Letters* 3:1587–1590.

[98] Lee, Ja Young, Schiffer, Guido, Jäger, Volker. 2005. "Synthesis of L-carbafuranomycin, an unnatural analogue of the antibiotic amino acid furanomycin" *Organic Letters* 7:2317–2320.

[99] Tranmer, Geoffrey K., Tam, William. 2002. "Molybdenum-mediated cleavage reactions of isoxazoline rings fused in bicyclic frameworks" *Organic Letters* 4:4101–4104.

[100] Aschwanden, P. 2005."Reduction of 2,3-dihydroisoxazoles to β-amino ketones and β-amino alcohols" *Organic Letters* 7:5741-5742.

[101] Gayen, Krishnanka S., Sengupta, Tista, Saima, Yasmin, Das, Adita, Maiti, Dilip K., Mitra, Atanu. 2012. "Cu(0) nanoparticle catalyzed efficient reductive cleavage ofisoxazoline, carbonyl azide and dominocyclization in water medium" *Green Chemistry* 14:1589-1592.

[102] Lepage, F., Tombret, F., Cuvier, G., Marivain, A., Gillardin, J. M. 1992. "New N-aryl isoxazolecarboxamides and N-isoxazolylbenzamides as anticonvulsant agents" *European Journal of Medicinal Chemistry* 27:581–593.

[103] Bolvig, Tina, Larsson, Orla M., Pickering, Darryl S., Nelson, Nathan, Falch, Erik, Krogsgaard-Larsen, Povl, Schousboe, Arne. 1999. "Action of bicyclic isoxazole GABA analogues on GABA transporters and its relation to anticonvulsant activity" *European Journal of Pharmacology* 375:367–374.

[104] Eddington, Natalie D., Cox, Donna S., Roberts, Ralph R., Butcher, Raymond J., Edafiogho, Ivan O., Stables, James P., Cooke, Neville, Goodwin, Angela M., Smith, Carlynn A., Scott, K. R. 2002. "Synthesis and anticonvulsant activity of enaminones. 4. Investigations on isoxazole derivatives" *European Journal of Medicinal Chemistry* 37:635–648.

[105] Zhu, S. L., Wu, Y., Liu, C. J., Wei, C. Y., Tao, J. C., Liu, H. M. 2013. "Synthesis and in vitro cytotoxic activity evaluation of novel heterocycle bridged carbothioamide type isosteviol derivatives as antitumor agents" *Bioorganic and Medicinal Chemistry Letters* 23:1343−1346.

[106] Rescifina, A., Zagni, C., Varrica, M. G., Pistara, V., Corsaro, A. 2014. "Recent advances in small organic molecules as DNA intercalating agents: synthesis, activity, and modeling" *European Journal of Medicinal Chemistry* 74:95−115.

[107] Lerman, L. S. 1961. "Structural considerations in the interaction of dna and acridines" *Journal of Molecular Biology* 3:18−IN14.

[108] Rescifina, A., Chiacchio, U., Corsaro, A., Piperno, A., Romeo, R. 2011. "Isoxazolidinyl polycyclic aromatic hydrocarbons as dnaintercalating antitumor agents" *European Journal of Medicinal Chemistry* 46:129−136.

[109] Rescifina, A., Varrica, M. G., Carnovale, C., Romeo, G., Chiacchio, U. 2012. "Novel isoxazole polycyclic aromatic hydrocarbons as dna-intercalating agents" *European Journal of Medicinal Chemistry* 51:163−173.

[110] Rescifina, A., Chiacchio, M. A., Corsaro, A., De Clercq, E., Iannazzo, D., Mastino, A., Piperno, A., Romeo, G., Romeo, R., Valveri, V. 2006. "Synthesis and biological activity of isoxazolidinyl polycyclic aromatic hydrocarbons: potential DNA intercalators" *Journal of Medicinal Chemistry* 49:709−715.

[111] Rescifina, A., Zagni, C., Romeo, G., Sortino, S. 2012. "Synthesis and biological activity of novel bifunctional isoxazolidinyl polycyclic aromatic hydrocarbons" *Bioorganic and Medicinal Chemistry* 20:4978−4984.

[112] Minter, A. R., Brennan, B. B., Mapp, A. K. 2004. "A small molecule transcriptional activation domain." *Journal of American Chemical Society* 126:10504−10505.

[113] Khazir, J., Riley, D. L., Chashoo, G., Mir, B. A., Liles, D., Islam, M. A., Singh, S. K., Vishwakarma, R. A., Pilcher, L. A. 2015. "Design, synthesis and anticancer activity of michael-type thiol adducts of α- santonin analogue with exocyclic methylene"*European Journal of Medicinal Chemistry* 101:769−779.

[114] Singh, S., Chopra, A., Singh, G., Saxena, A. K., Ishar, M. P. S. 2013. "Synthesis and in-vitro cytotoxic evaluation of novel chromano-piperidine fused isoxazolidines: discovery of a potent lead" *Journal of Pharmaceutical Research* 7:337–341.

[115] Poma, Paola, Notarbartolo, Monica, Labbozzetta, Manuela, Maurici, Annamaria, Carina, Valeria, Alaimo, Alessandra, Rizzi, Michele, Simoni, Daniele, D'Alessandro, Natale. 2007. "The antitumor activities of curcumin and of its isoxazole analogue are not affected by multiple gene expression changes in an mdr model of the mcf-7 breast cancer cell line: analysis of the possible molecular basis" *International Journal of Molecular Medicine* 20:329-335.

[116] Bhaskar, V. H., Mohite, P. B. 2010. "Synthesis, characterization and evaluation of anticancer activity of some tetrazole derivatives" *Journal of Optoelectronics and Biomedical Materials* 2:249–259.

[117] Mijatovic, Sanja, Maksimovic-Ivanic, Danijela, Mojic, Marija, Malaponte, Graziella, Libra, Massimo, Cardile, Vera, Miljkovic, Djordje, Harhaji, Ljubica, Dabideen, Darrin, Cheng, Kai Fan, Bevelacqua, Ylenia, Donia, Marco, Garotta, Gianni, Al-Abed, Yousef. 2008. "Stanislavastosic-grujicic; ferdinandonicoletti novel nitric oxide-donating compound (s,r)-3-phenyl-4,5-dihydro-5-isoxazole acetic acid–nitric oxide (git-27no) induces p53 mediated apoptosis in human a375 melanoma cells" *Nitric Oxide* 19:177–183.

[118] Kathiravan, M. K., Salake, A. B., Chothe, A. S., Dudhe, P. B., Watode, R. P., Mukta, M. S., Gadhwe, S. 2012. "The biology and chemistry of antifungal agents: a review" *Bioorganic and Medicinal Chemistry* 20:5678–5698.

[119] Vyas, D. H., Tala, S. D., Dhaduk, M. F., Akbari, J. D., Joshi, H. S. 2007. "Synthesis, antitubercular and antimicrobial activities of some new pyrazoline and isoxazole derivatives" *Journal of Indian Chemical Society* 84:1140-1144.

[120] Anderson, J. R., Horsgood, R. K. 1971. "Studies on the degradation of an isoxazolone fungicide by pure cultures of soil bacteria" *Soil Biology Biochemistry* 3:271–280.

[121] Clark, D. G., McElligott, T. F. 1969. "Acute and short-term toxicity of drazoxolon(4-(2-chlorophenylhydrazone)-3-methyl-5-isoxazolone)" *Food and Cosmetics Toxicology* 7:481–491.

[122] Chen, F., Han, P., Liu, P., Si, N., Liu, J., Liu, X. 2014. "Activity of the novel fungicide syp-z048 against plant pathogens" *Scientific Report* 4:6473.

[123] Chen, F. P., Fan, J. R., Zhou, T., Liu, X. L., Liu, J. L., Schnabel, G. 2012. "Baseline sensitivity of monilinia fructicola from china to the dmi fungicide syp-z048 and analysis of dmi-resistant mutants" *Plant Disease* 96:416–422.

[124] Liu, P., Xu, Y., Li, J., Liu, J., Cao, Y., Liu, X. 2012. "Photodegradation of the isoxazolidine fungicide syp-z048 in aqueous solution: kinetics and photoproducts" *Journal of Agriculture and Food Chemistry* 60:11657–11663.

[125] Sadashiva, M. P., Mallesha, H., Hitesh, N. A., Rangappa, K. S. 2004. "Synthesis and microbial inhibition study of novel 5-imidazolyl substituted isoxazolidines" *Bioorganic and Medicinal Chemistry* 12:6389–6395.

[126] Ravi Kumar, K. R., Mallesha, H., Basappa, Rangappa, K. S. 2003. "Synthesis of novel isoxazolidine derivatives and studies for their antifungal properties" *European Journal of Medicinal Chemistry* 38:613–619.

[127] Sadashiva, M. P., Mallesha, H., Karunakara Murthy, K., Rangappa, K. S. 2005. "Enhancement in antimicrobial activity of 2-(phenyl)-3-(2-butyl-4-chloro-1*H*-

imidazolyl)-5-butylate isoxazolidine" *Bioorganic and Medicinal Chemistry Letters* 15:1811−1814.

[128] Zelechowski, K., Gołębiewski, W. M., Krawczyk, M. 2015. "Synthesis and fungicidal activity of 2-(diphenylmethyl)-3-arylisoxazolidine-5-carboxamides" *Monatsh. Chemistry*, 146:1895−1905.

[129] Nora, G. P., Miller, M. J., Möllmann, U. 2006. "The synthesis and in vitro testing of structurally novel antibiotics derived from acylnitroso diels−alder adducts" *Bioorganic and Medicinal Chemistry Letters* 16:3966−3970.

[130] Raunak, R., Kumar, V., Mukherjee, S., Poonam, P., Prasad, A. K., Olsen, C. E., Schäffer, S. J. C., Sharma, S. K., Watterson, A. C., Errington, W. 2005. "Microwave mediated synthesis of spiro-(indoline- isoxazolidines): mechanistic study and biological activity evaluation" *Tetrahedron* 61:5687−5697.

[131] Habeeb, Amgad G., Praveen Rao, P. N., Knaus, Edward E. 2001. "Design and synthesis of 4,5-diphenyl-4-isoxazolines: novel inhibitors of cyclooxygenase-2 with analgesic and antiinflammatory activity" *Journal of Medicinal Chemistry* 44: 2921–2927.

[132] Sailu, B., Mahanti, S., Srinivas, A., Balram, B., Ram, B., Taara, B., Vasudha, B. 2012. "Synthesis and antibacterial activity of novel isoxazoline derivatives" *Der Pharma Chemica* 4:2036-2041.

[133] Setoguchi, M., Iimura, S., Sugimoto, Y., Yoneda, Y., Chiba, J., Watanabe, T., Muro, F., Iigo, Y., Takayama, G., Yokoyama, M. 2013. "A novel, potent, and orally active vla-4 antagonist with good aqueous solubility: *trans*-4-[1-[[2-(5-fluoro-2-methylphenylamino)-7-fluoro-6-benzoxazolyl]acetyl]-(5s)-[methoxy(methyl)amino]-methyl-(2s)-pyrrolidinylmethoxy] cyclohexanecarboxylic acid" *Bioorganic and Medicinal Chemistry* 21:42−61.

[134] Prabhashankar, Jayaroopa, Kumar Govindappa, Vasanth, Renuka, N., Akshatha, K. N., Mahadevamurthy, S., Kumar Kariyappa, Ajay. 2012 *Der Pharma Lettre* 4:1685-1691.

[135] Ajay Kumar, K., Lokeshwari, D. M., Vasanth Kumar, G. 2012. "Isoxazolines: An insight to their synthesis and diverse applications" *International Journal of Pharmasutical Science and Drug Research* 4:236-239.

[136] Makarov, Vadim A., Riabova, Olga B., Granik, Vladimir G., Wutzler, Peter, Schmidtke, Michaela.2005. "Novel [(biphenyloxy)propyl]isoxazole derivatives for inhibition of human rhinovirus 2 and coxsackievirus B_3 replication" *J. Antimicrob. Chemother.* 55:483-488.

[137] Rakesh, David Bruhn, Madhura, Dora B., Maddox, Marcus, Lee, Robin B., Trivedi, Ashit, Yang, Lei, Scherman, Michael S., Gilliland, Janet C., Gruppo, Veronica, McNeil, Michael R., Lenaerts, Anne J., Meibohm, Bernd, Lee, Richard E. 2012. "Antitubercular nitrofuran isoxazolines with improved pharmacokinetic properties" *Bioorganic and Medicinal Chemistry* 20:6063-6072.

[138] Sammelson, Robert E., Ma, T., Galietta, Luis J. V., Verkman, A. S., Kurth, Mark J. 2003. "3-(2-Benzyloxyphenyl)isoxazoles and isoxazolines: synthesis and evaluation as CFTR activators" *Bioorganic and Medicinal Chemistry Letters* 13:2509-2512.

[139] Loh, B., Vozzolo, L., Mok, B. J., Lee, C. C., Fitzmaurice, R. J., Caddick, S., Fassati, A. 2010. "Inhibition of HIV-1 replication by isoxazolidine and isoxazole sulfonamides" *Chemical and Biological Drug Design* 75:461−474.

[140] Lynch, C. L., Gentry, A. L., Hale, J. J., Mills, S. G., MacCoss, M., Malkowitz, L., Springer, M. S., Gould, S. L., DeMartino, J. A., Siciliano, S. J. 2002. "CCR5

Antagonists: bicyclic isoxazolidines as conformationally constrained N1-substituted pyrrolidines" *Bioorganic and Medicinal Chemistry Letters* 12:677–679.

About the Authors

Sudipta Ray received her B. Sc. in chemistry from the Calcutta University in 1997 and completed her M. Sc in Organic Chemistry (major) from the Calcutta University in 1999. She was joined in the PhD program under the supervision of Professor Arindam Banerjee at Indian Association for the Cultivation of Science, India and received Ph. D. degree in 2007. Following postdoctoral studies as a JSPS fellow at Kumamoto University, Japan in 2011, she moved to the NIMS, Japan as a MANA postdoctoral fellow and currently associated with Professor Maiti's research group. Her current research interests are in synthetic organic chemistry of peptides as supramolecular building blocks and as templates for synthesis of valuable metal nanoparticles, their application in drug delivery, catalysis and smart electronics.

Amiya Kumar Medda received his BSc in Chemistry from the University of Calcutta in 2001 and completed his M. Sc. in Chemistry from IIT kharagpur, India in 2004. He was joint in the PhD programme under the supervision of Professor Hee-Seung Lee at Korea Advanced Institute of Science and Technology (KAIST), South Korea and received PhD degree in 2009. Following postdoctoral studies at the KAIST, he joined to the University of Texas at Dallas, Texas, USA as a postdoctoral researcher. At present, he is an Assistant Professor at the Department of Chemistry, Asutosh College, Kolkata, and associated with Professor Maiti's research group. His current research interest is in the area of development of chiral metal-ligand catalyst for enantioselective synthesis of bioactive natural products and drug like molecules, especially via sp^3C-H activated C-N and C-C coupling processess.

Dilip K. Maiti was born September 9, 1970, in West Bengal, India. He received his BSc in chemistry in 1991 and MSc (organic major) in 1993, from Calcutta University, India. He achieved his PhD on stereoselective synthesis, from Indian Institute of Chemical Biology in 1998. He had done his postdoctoral research in the School of Medicine, Wayne State University, USA. In 2005, he joined as a Reader faculty at Calcutta University and became full Professor in 2011. His major research activity is focused on organic synthesis, diverse catalysis and fabricating high-tech organic electronic devices.

In: Organic and Medicinal Chemistry, Volume 1
Editor: Bimal Krishna Banik

ISBN: 978-1-53614-453-6

Chapter 7

NOVEL SYNTHESIS OF BIOLOGICALLY ACTIVE PYRROLE

Ram Naresh Yadav[1,2], Indrani Banik[3] and Bimal Krishna Banik[1,3,4,*]

[1]Department of Chemistry, The University of Texas-Pan American, Edinburg, TX, US

[2]Department of Chemistry, Faculty of Engineering & Technology, Veer Bahadur Singh Purvanchal University, Jaunpur-220003, Uttar Pradesh, India

[3]The University of Texas, M. D. Anderson Cancer Centre, Department of Molecular Pathology Box-89, Houston, Texas, US

[4]Community Health System of South Texas, Edinburg, TX, US

ABSTRACT

Pyrrole is a five-membered heterocyclic aromatic organic compound that widely constitutes the naturally occurring molecules and chemotherapeutic agents. Pyrroles are also recognized as important biologically active compounds. The common naturally occurring molecules containing pyrroles nucleus includes Vitamin B_{12} bile pigments like bilirubin and biliverdin, and the porphyrins of heme, chlorophyll, chlorins, bacteriochlorins, and porphyrinogens. The present study focuses on the various synthesis of pyrroles of diverse interests which have been reported during last decades.

1. INTRODUCTION

Pyrrole and their derivatives are one of the most important hetrocycles, found in a broad range of natural products and chemotherapeutic molecules. They are also a valuable scaffold in material science. Pyrrole was first detected by F. F Runge in 1834 as a constituent of coal tar [1]. It was first isolated from the pyrolysate of bone in 1857.

[*] Corresponding Author Email: bimalbanik10@gmail.com and bimal.banik@chsst.org.

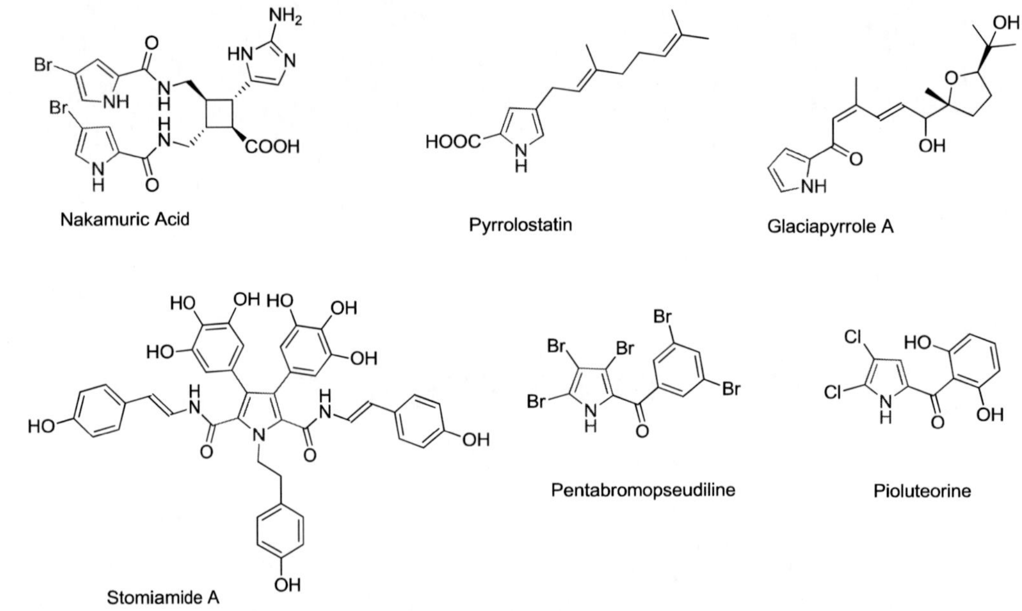

Figure 1. Some bio-active pyrrole-containing natural products.

It is recognized as the structural motif of the heame and chlorophyll pigment that are essential for life. Some important representative secondary metabolites containing pyrrole nucleus are summarized in (Figure 1).

This includes Nakamuric Acid isolated from marine sponge Agelas Nakamurai by Proksch in 1999. It is axially chiral marinopyrroles, which exihibit potential activity against metacillin-resistant staphylococcus aureus strains [2]. The Pyrrolostatin is a potential novel lipid peroxidation inhibitors in rat brain homogenate. It is isolated from the culture of Streptomyces Chrestomyceticus EC40 [3]. Glaciapyrrole A is another important pyrrolosesquiterpenes isolated from Streptomyces sp. strain from marine sediments of Alaska possess a potential antimicrobial activity.

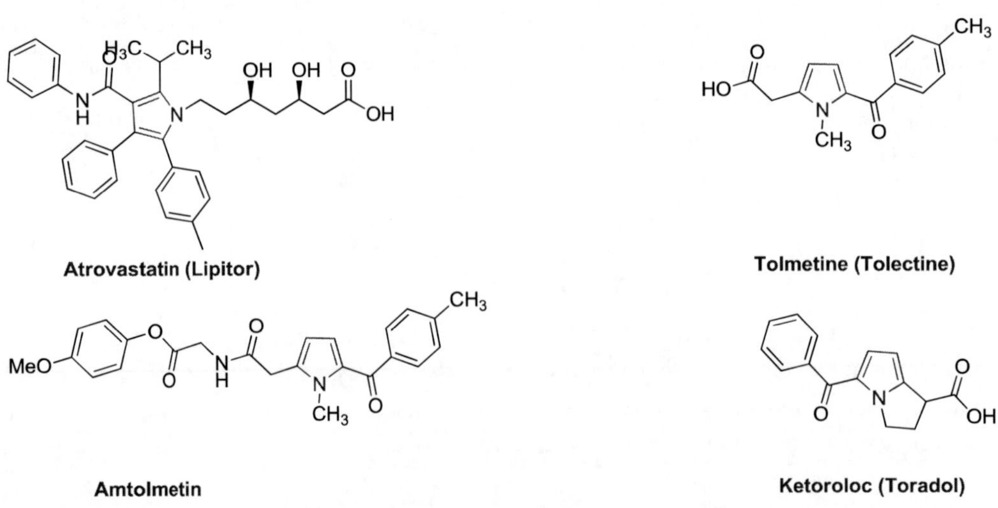

Figure 2. Some drugs contining pyroole nucleus.

The storniamide family is isolated from marine organisms (mollusks, ascidians, sponges) and containing 3, 4-diarylpyrrole substituted fragments. An O-methylated analogues of storniamide A have shown potent activity as inhibitors of the multidrug resistance (MDR) phenomenon, [4] which can be considered as the main obstacle to successful anticancer chemotherapy. For this reason, there is much current interest in the development of new MDR modulators [5–7]. The halo derivative pyrroles like penta bromopsudiline and pioleutiorine are isolated from bacterial source and show remarkable antibacterial properties [8]. A few pyrrole-containing chemotherapeutic drugs are shown in (Figure 2).

Atorvastatin is a famous drug having a pyrrole nucleus in its structure. It is a medicinally important drug used as a lipid-lowering agent and for prevention of events associated with cardiovascular disease and one of the top selling drugs worldwide under the trade name Lipitor [9]. Tolmetin is another non-steroidal anti-inflammatory drug (NSAID) used for the treatment to reduce the pain, swelling, and stiffness of arthritis [10-12]. Ketorolac (Toradol) is also an NSAID that is often used to treat eye pain and the itchiness of seasonal allergies [13-14]. Besides the above unique properties of pyrroles associated with biological activity and potential drugs candidates, they also play a very crucial role in the relevant of the material science research. In this context, polypyrroles are known to act as conducting polymers and used in solar cells [15-16] and batteries and gas sensor as well [17-18].

Another interesting property of natural and unnatural molecules containing pyrrole ring is found in their involvement in coordination and molecular recognition science. In this context the present study aims to provide the notable methods for the synthesis of pyrroles reported by the scientific communities and by us during the last decades.

2. Results and Discussions

Due its intriguing structural motif and remarkable biological activities the development of sustainable synthesis of pyrrole from renewable resources is highly desirable. Due to its biological activities and medicinal properties of this molecules is remain attracted attention of the scientist, researchers for the development new synthetic methods for the synthesis of this unique molecule. The classical methods for the pyrrole synthesis involve the Knorr [19], Pall-Knorr [20] and Hantzsch methods [21].

Apart from these methods, transition metal-catalyzed cyclization strategies and multicomponent coupling reactions (MCR) have extensively studied for the synthesis of functionalized pyrroles and its derivatives for their studies in chemical, biological and material sciences.

3. Pall-Knorr Synthesis

The Pall-Knorr is the classical method for the pyrrole synthesis. This method is based on the condensation of the 1, 4-dicarbonyl compounds with excess of ammonia or primary amine under protic or Lewis acid as catalyst. This reaction was initially reported independently by German chemists Carl Paal and Ludwig Knorr in 1884 as the method for the synthesis of furan and has been adapted for pyrroles and thiophenes.

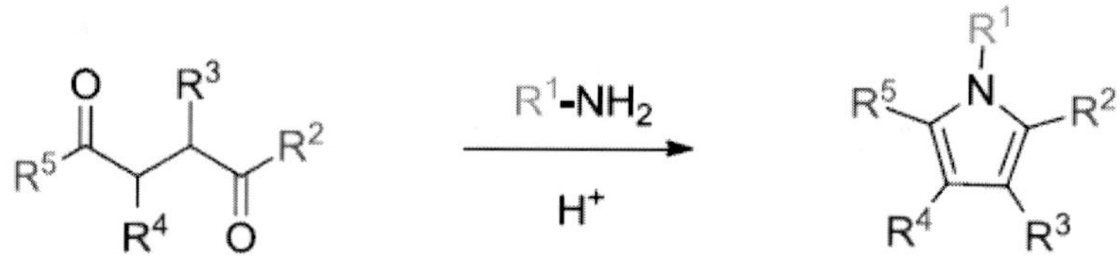

1 **2**

R^2 R^5 = H Aryl Alkyl
R^3 R^4 = H Aryl Alky / esters
R^1 = H Alkyl Aryl Amino or Hydroxy

Scheme 1. A classical Pall-Knorr cyclization of 1,4-dione with Amine derives to polysubstituted pyrrol(s).

Traditional Paal-Knorr conditions involved prolonged heating of strong acids to drive dehydration which occurred over a period of several hours. The common dehydrating agents include phosphorus pentaoxide, anhydrides, or zinc chloride. The Paal-Knorr method is considered limited by harsh reaction conditions, such as prolonged heating with acid, which may degrade sensitive functionalities as well as polymerisation in many potential pyrrole precursors [22]. (Scheme 1.)

It needs to modify the reaction conditions to make it more compatible with wide range of substrates, high yielding and reduce reaction time. A lot of study has been carried out in this area to improve the reaction conditions in order to get better yield with reduce reaction time. To substitute strong protic acid, there are several mild acids have been discovered for this reaction. The notable catalysts, Zeolite, $Sc(OTf)_3$, $In(OTf)_3$, $Ti(OiPr)_4$, Al_2O_3, p-TSA, montmorillnite-KSF has been employed under microwave conditions [23-29]. The used of these catalysts have improved the synthesis of pyrroles. Trost et al. reported the synthesis of the macrocyclic core of roseophilin, a potential antibiotic as drug candidates using Pall-Knorr methodology [30]. (Scheme 2).

The Pall-Knorr reaction has been also studied under microwave condition. This reaction was first employed for the synthesis of substituted pyrrol ring by Danks in a domestic microwave oven [31]. This method describes the synthesis of a wide range of N-substituted Pyrroles **7a-h** from 2, 5-hexadiones **5** and primary amine **6a-h** in excellent yield without using solvents (**Scheme 3, Table 3.1**).

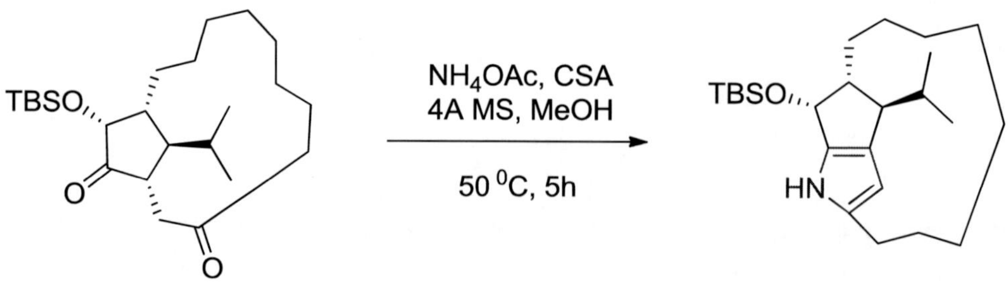

3 **4**

Scheme 2. Synthesis of macrocyclic core of Rosephilin via Pall-Knorr reaction.

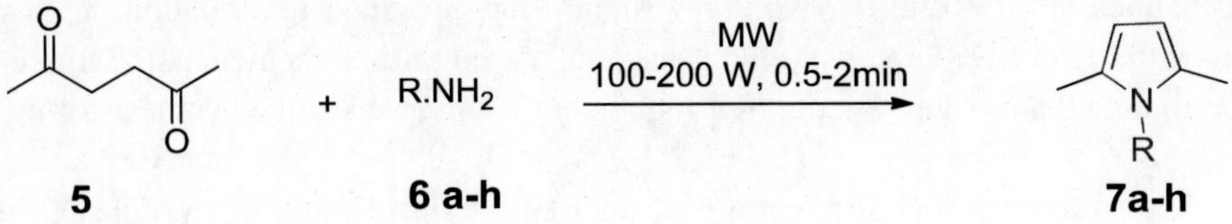

Scheme 3. Synthesis of N-substituted 2,5-dimethyl Pyrrole under microwave irradiation via Pall-Knorr reaction.

The electronic effect greatly influence the yield of the pyrrole. The bulky group in aromatic ring and electron withdrawing group drop the yield of the product significantly (entry **4** and **7** in Table 3.1).

Banik et al. reported the iodine-catalyzed solvent less synthesis of N-substituted pyrroles with monocyclic and polycyclic amine under microwave condition using Pall-Knorr reaction. This method involves a rapid reaction of 2, 5-dimethoxytetrahydrofuran (**8**) as masked 2, 5 dione with diverse amine (**9a-h**) in excellent yield of the N-substituted pyrrole (**10a-h**) [32] (Scheme 4, Table 3.2).

Table 3.1. Pall-Knorr results with different amine (a-h)

Entry	Amine (6) R[a-h]	Pyrrole (7) R [a-h]	Yield (%)
1	$C_6H_5.CH_2$	$C_6H_5.CH_2$	90
2	$C_6H_5.CH_2.CH_2$	$C_6H_5.CH_2.CH_2$	90
3	C_6H_5	C_6H_5	90
4	3-ClC_6H_4	3-ClC_6H_4	80
5	4- $CH_3O.C_6H_4$	4- $CH_3O.C_6H_4$	90
6	2- $CH_3C_6H_4$	2- $CH_3C_6H_4$	85
7	2,4- $(CH_3)_2C_6H_3$	2,4- $(CH_3)_2C_6H_3$	80
8	2,6- $(CH_3)_2C_6H_3$	2,6- $(CH_3)_2C_6H_3$	75

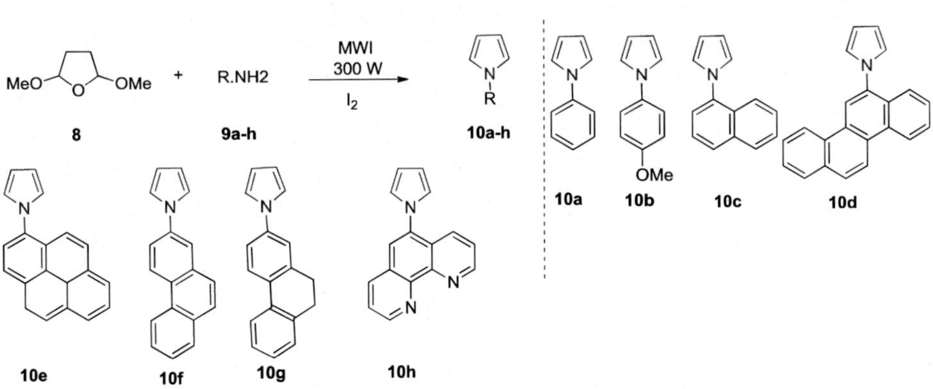

Scheme 4. Microwave assisted lodine catalyzed Pall-Knorr reaction for the synthesis of N-substituted Pyrrole.

As a part of our ongoing research on synthesis of anticancer agents, some of our trans-acetoxy β-lactams have demonstrated selective anticancer activity against a number of human cancer cell lines *in vitro* and *in vivo*. This finding has provided justification to develop an efficient synthesis of pyrrole-substituted β-lactams, based on the chemistry in Scheme 4. It is important to note that no deprotection/rearrangement is observed in the synthesis of pyrroles.

Table 3.2. Summary of Iodine catalysed microwave irradiation of pyrrole synthesis

Entry	Amine (9) 9 a-h (R)	Time (min)	Products 10 a-h	Yield (%)
a	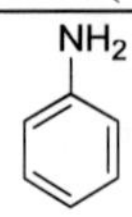	**5**	**10a**	**98**
b	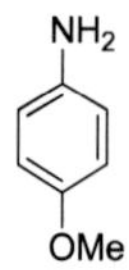	**5**	**10b**	**96**
c	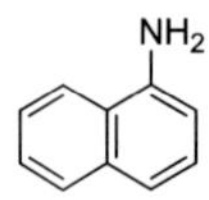	**5**	**10c**	**92**
d	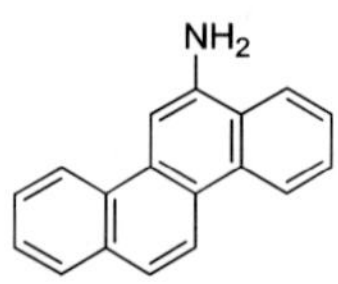	**20**	**10d**	**90**
e	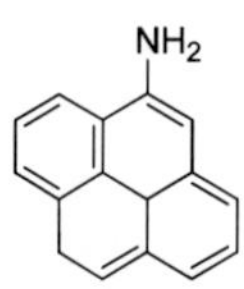	**20**	**10e**	**85**
f	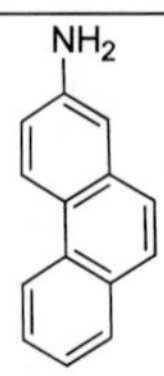	**5**	**10f**	**85**
g	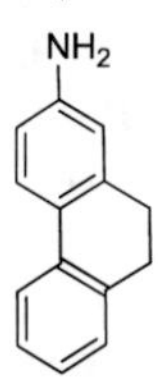	**40**	**10g**	**82**
h	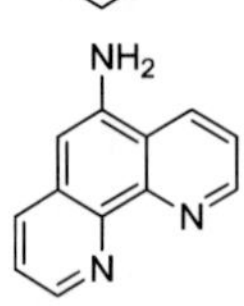	**55**	**10h**	**75**

Banik et al. had been studying the synthesis and biological evaluation of 3-amino-□-lactam for the past many years. Banik et al. reported a novel route for the synthesis of pyrrole using 3-amino-2-azetidinones **11a-c** and **11a-d** under microwave condition. This methodology was extended for the synthesis of enantiopure biologically important 3-substituted β-lactams **13a-b** (Scheme 5, Table 3.3).

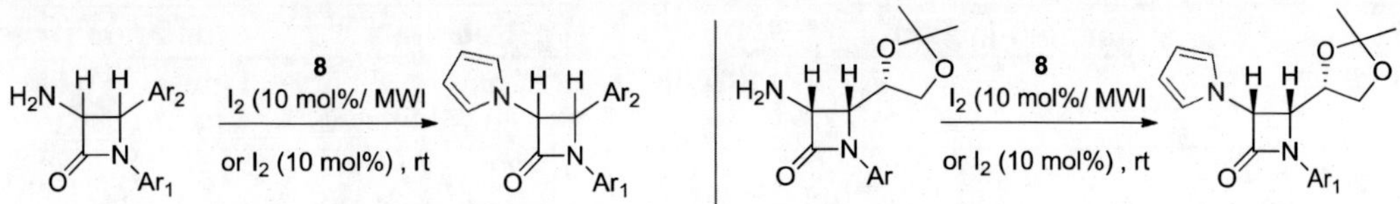

11 a-c Ar$_1$ = Phenyl, P-methoxyphenyl **14 a-i** (PMP), 6-Chrysenyl **11 d-i Ar$_2$** = Phenyl, PMP, 2-thiophenyl, 2-pyridyl, ferrocenyl

12 a-b Ar = Ph, PMP

13 a-b

Scheme 5. Synthesis of racimic and chiral Pyrrole substituted b-lactams under microwave irradiation using lodine as catalyst.

The whole series of reactions was repeated at room temperature (stirring) under solventless conditions but the reactions have produced lower yields after much longer times (Table 3.4). The reaction has proceeded equally well irrespective of the nature of the substituents at C-4 (aryl, heteroaryl, ferrocenyl or 2,2-dimethyl-1,3-dioxolanyl). No stereochemical change has been noticed in any examples. The results obtained from both the methods (Condition **A** and Condition **B**) are summarized in (Table 3.4).

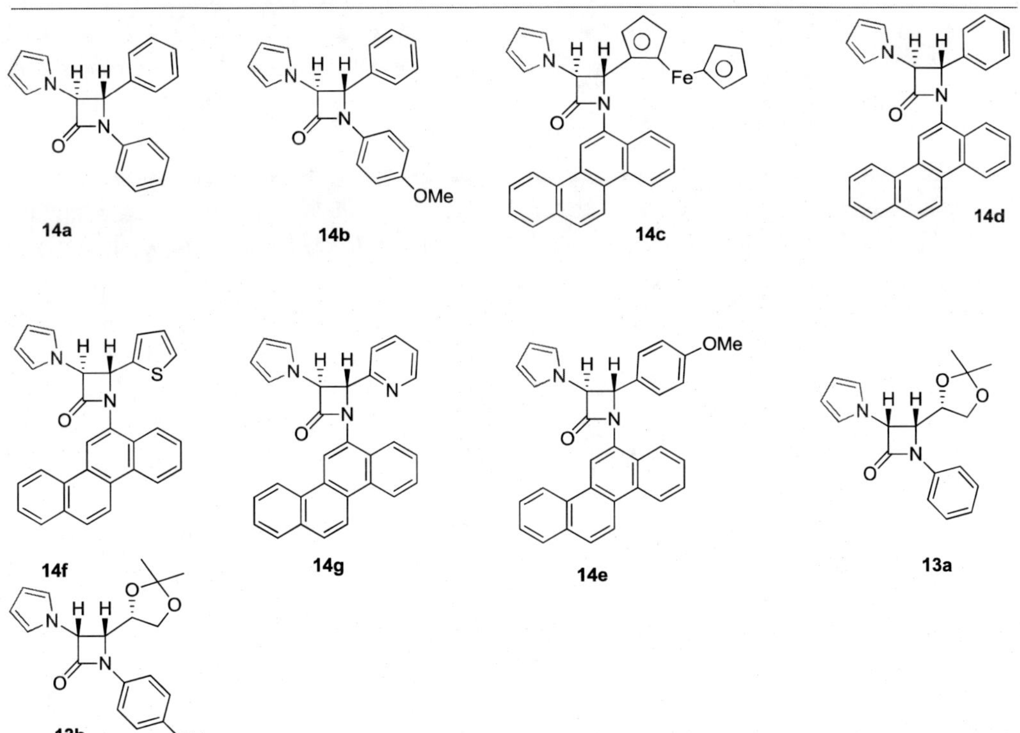

Table 3.3. Iodine-catalyzed microwave-assisted synthesis of 3-pyrrole substituted 2-β-lactam(s)

Entry	3-amino-β-lactam		Product(s)	Condition A		Condition B	
	Ar1	Ar2		Time (min)	Yield (%)[a]	Time (h)	Yield (%)[a]
1	Phenyl	Phenyl	**14a**	1	93	12	72
2	PMP	Phenyl	**14b**	1	94	12	73
3	Chrysenyl	Ferrocenyl	**14c**	3	90	24	63
4	Chrysenyl	Phenyl	**14d**	3	98	24	71
5	Chrysenyl	PMP	**14e**	3	95	24	67
6	Chrysenyl	Thiophenyl	**14f**	3	92	24	61
7	Chrysenyl	Pyridyl	**14g**	3	90	24	59
8	Phenyl	………	**13a**	1	93	12	70
9	PMP	……	**13b**	1	97	12	73

Condition A: (MWI) 300 Watts/90°C/24–50 psi; **Condition B:** Stirring at rt **a:** isolated yield

Mechanistically this reaction involve the deprotection of the methoxy groups in 2,5-dimethoxytetrahydrofuran takes place under mild acidic conditions and this process can be extremely accelerated by microwave irradiation. The intermediate dialdehyde as intermediate reacts with 3-amino-β-lactams to produce the corresponding 3-pyrrole-substituted β-lactams following a nucleophile addition and subsequent dehydration aromatization route. It is very convenient to conduct reactions with molecular iodine because of its stability in the presence of moisture and oxygen. In all the cases, the reactions are completed within 1–3 min and the products are obtained in excellent yield. This is convenient and environmentally benign route for the one-step synthesis of 3-pyrrole-substituted 2-azetidinones was developed.

The present simple methodology offers attractive features such as short reaction times and milder conditions, and produces the products in excellent yields. The procedure is equally effective with mono- and polyaromatic- 2-azetidinones pyrrole. This strategy should find application in the synthesis of various organic compounds of biological interest.

Karl A. Scheidt et al. made an important investigation for the synthesis of 2, 3, 4-substituted pyrroles in one-pot by Sila-Stetter/Paal-Knorr Strategy [33]. The reaction employed an acyl anion conjugate addition reaction of acyl silanes **15** and unsaturated ketones **16a-j** to generate 1, 4-dicarbonyl compounds *in situ.* Acyl silane employed for the generation of 1, 4 dicarbonyl compounds *in situ via* thiazolium salt (**17**) catalyzed Sila-Settrer reaction and then was directly converted to pyrroles **18a-j** in excellent yield simply upon the addition of an amine (Scheme 6, Table 3.4). The overall process is mildly dependent on acid composition, but good yield was obtained with *p*-toluenesulfonic acid (TsOH) and only with 20 mol % of thiazolium pre-catalyst,

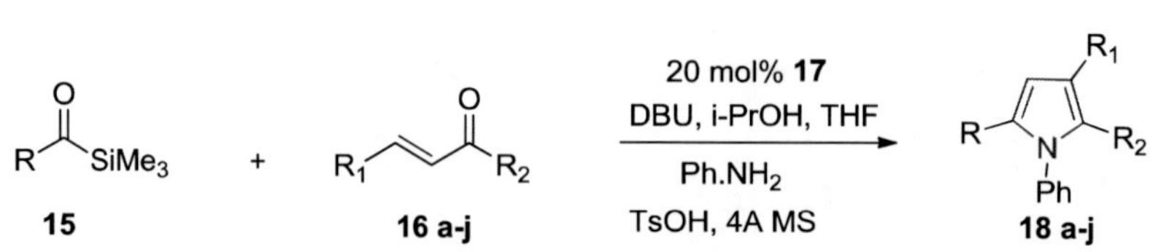

Scheme 6. Sila-Stetter/Paal-Knorr Reaction. A multicomponent approach to Pyrrole Synthesis[a].

Table 3.4. Summary of result of Sila-Stetter/Pall-Knorr Reaction

Entry	R	x	R_1	R_2	Yield[b] (%)	Pyrrole
1	Ph	SiMe3	Ph	Ph	66	**18a**
2	4-Me-Ph	SiMe3	Ph	Ph	71	**19b**
3	4-Cl-Ph	SiMe3	Ph	Ph	61	**18c**
4	Cyclohexyl	SiPhMe2	Ph	Ph	69	**18d**
5	CH_3	SiPhMe2	Ph	Ph	71	**18e**
6	Ph	SiMe3	4-Me-Ph	Ph	70	**18f**
7	Ph	SiMe3	3-Me-Ph	Ph	62	**18g**
8	Ph	SiMe3	4-OMe-Ph	Ph	69	**18h**
9	4-Cl-Ph	SiMe3	4-Cl-Ph	Ph	58	**18i**
10	Ph	SiMe3	Ph	4-Cl-Ph	80	**18j**

a: Reaction conditions: 20 mol % of 4 and 30 mol% of DBU, 4 equivalent of *i*-PrOH; 0.8 M at 70°C for 8h. Amine, TsOH, and 4 Å sieves were then added for an additional 8h. **b:** Isolated yield after purification.

Mechanistically this reaction involved in the addition of a neutral carbene/zwitterionic species generated *in situ* from the exposure of thiazolium salt to DBU) to an acylsilane. The complete mechanism of this transformation include the steps, desilylation, 1,4-michael type of addition and finally ejection of catalyst, delivered the desired products. The complete catalytic cycle is delineated in (Scheme 7.).

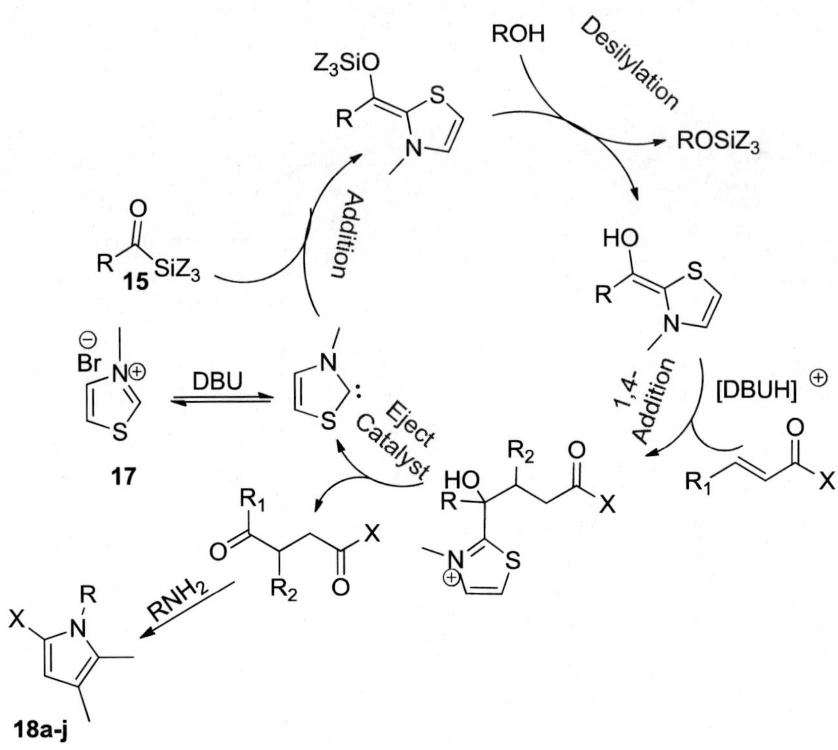

Scheme 7. Proposed mechanism for formation of 1,4-dicarbonyl compound *in situ*.

4. Hantzsch Pyrrole Synthesis

The Hantzsch pyrrole synthesis, is classical reaction named after Arthur Rudolf Hantzsch and this reaction is based on a condensation of β-ketoester **19** with ammonia or primary amine and α-halo ketone **20** to give substituted pyrroles (Scheme 8).

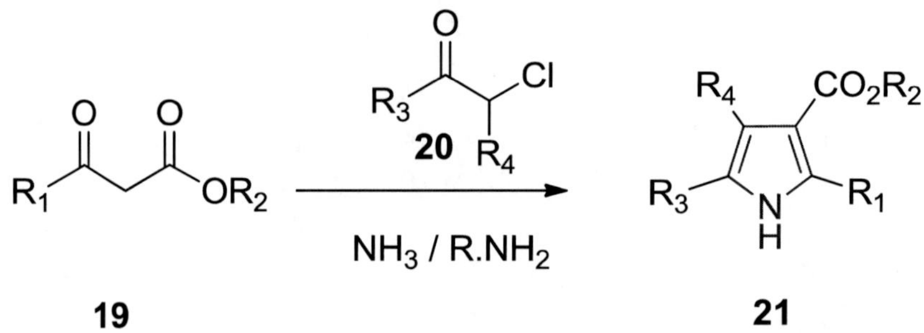

Scheme 8. Hantzsch pyrrole synthesis.

A generalization of the Hantzsch pyrrole synthesis was developed by Estevez et al. [34]. In this reaction highly substituted pyrroles was synthesized in a one-pot reaction, with relatively high yields. This reaction involved in the high-speed vibration milling (HSVM) of ketones with N-iodosuccinimide (NIS) and p-toluenesulfonic acid as a mild acid, to form a α-iodoketone *in situ* by the reaction of N-iodosuccinamide. This was followed by the addition of a aliphatic/aromatic and hetro-aromatic primary amine, a β-dicarbonyl compound, cerium (IV) ammonium nitrate (CAN) and silver nitrate, as the catalyst to afford the pyrrole **24** in 60-97% (Scheme 9).

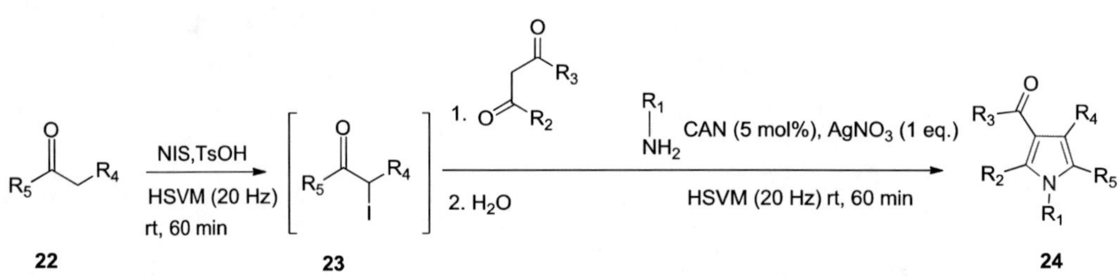

Scheme 9. Hantzsch pyrrole synthesis under high-speed vibration milling (HSVM).

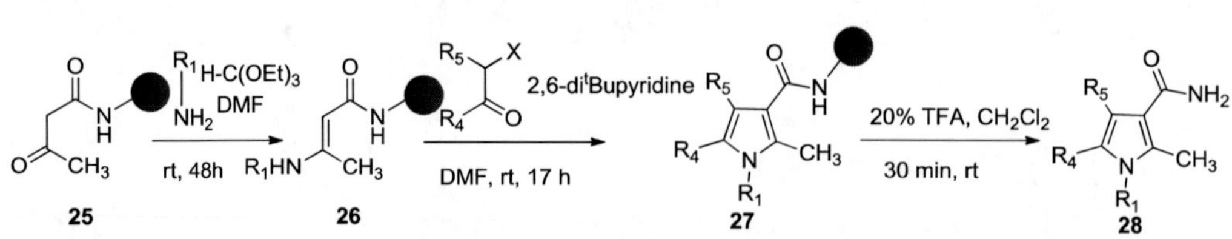

Scheme 10. Hantzsch pyrrole synthesis on solid support.

In 1998, Jung and co-workers demonstrated a three-component Hantzsch pyrrole synthesis on an acetoacetylated Rink resin as solid support. The reaction between this material and primary amines initially afforded the corresponding β-enaminones, which upon treatment with α-halocarbonyl compounds, followed by hydrolytic removal of the reaction products from the solid support afforded pyrroles (Scheme 10). This protocol produced pyrroles with excellent purities [35].

Another important investigation based on β-cyclodextrine as a supramolecular catalyst was used for the synthesis of 2,4,5-substituted pyrroles in good to moderate yield. This method had limitations and was not generalized for a wide variety of substrates. In this method, phenacyl bromide, acetylacetone and ammonium acetate or a variety of amines were reacted. Heating these starting materials and β-cyclodextrin in water at 60 - 70°C, the corresponding 1, 5-diaryl substituted pyrroles were obtained in good to excellent yields (Scheme 11) [36].

This methods offer the mild and efficient synthesis of series of N-aryl substituted pyrroles. This method especially good to install the keto moiety at the 4-postion of the pyrrole ring which could be further derivatized in to the diverse molecules of the biological importance.

5. Transition Metal-Catalyzed Annulation to Pyrroles

Metal-mediated synthetic routes for carbon-nitrogen bond are an important tool in synthesis of biologically relevant heterocyclic molecules [37]. It is estimated that over 90% of pharmaceutical have at least one nitrogen atom in their structures and about one reaction out of seven in the pharmaceutical industry involves the formation of a carbon-nitrogen bond. Nitrogen heterocycles occur in a wide-range of natural and biologically active compounds. Efficient methods for the synthesis of nitrogen-containing heterocycles are thus of fundamental importance and represent a major challenge in synthetic chemistry. In this context, we herein report the transition metal-catalyzed intermolecular annulation to aza heterocycles.

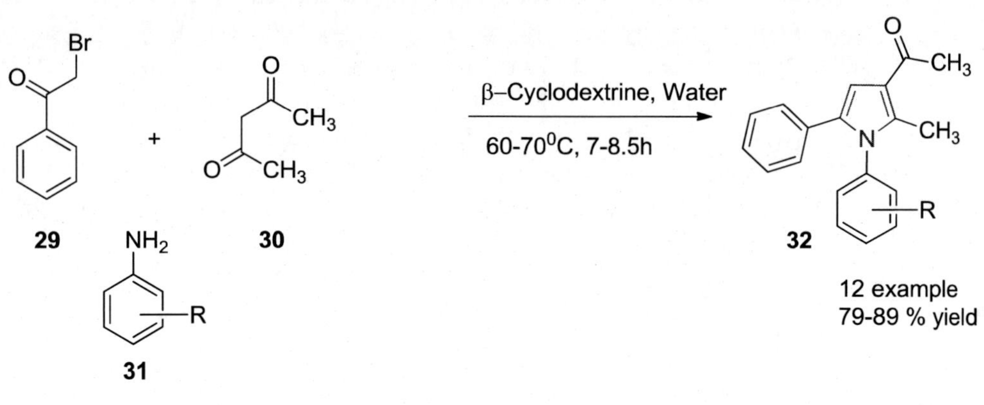

Scheme 11. Hantzsch type of pyrrole synthesis catalyzed by β-cyclodextrin.

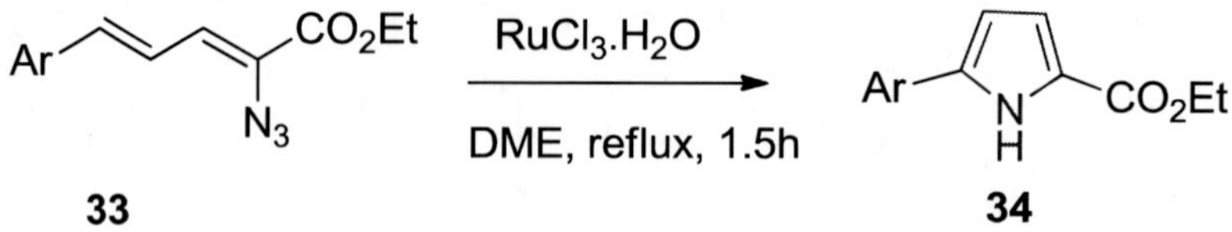

Scheme 12. $RuCl_3{\cdot}nH_2O$ catalyzed pyrrole synthesis.

Lin, Jia and co-workers reported the synthesis of pyrrole using commercially available $RuCl_3.nH_2O$-mediated intramolecular cyclization of aromatic substituted dienyl azide. This reaction required high temperature (85°C) and with 3 mol% catalyst loading. The authors of the papers investigated that the commercial available $RuCl_3.nH_2O$ exists in different oxidation states and influences the amination reaction greatly. The optimized solvent choice for the reaction was DME. A wide variety of pyrroles was obtained in excellent yield by refluxing the reactants in DME for1.5h. The authors also described the influence of electron withdrawing groups and donating groups in the dienyl azide in this reaction condition [38] (Scheme 12).

6. PtCl4-Catalyzed Cyclization of Homopropargyl Azide

Kou Hiroya and coworkers described the synthesis of pyrrole by platinum-catalyzed annulation of homopropargyl azide using ethanol as solvent with 2,6 ditert-butyl-4 methylpyridine as the base for a ring-closure reaction [39].

These reaction conditions were applied to the preparation of densely functionalized pyrrole derivatives with high tolerance of the functional groups (Scheme 13, Table 6.1).

The substrate bearing the phenyl ethynyl group was the most reactive and 5 mol% of PtCl4 and 20 mol % of 2, 6-di-*tert*-butyl-4-methylpyridine are enough to afford the product in good yield (Table 6.1, entry 1). Substrates with an alkyl group substituted alkyne (Table 6.1, entry 3), as well as compounds with silyl ether (Table 6.1, entry 4), hydroxyl (Table 6.1, entry 2), and ethoxycarbonyl groups (Table 6.1, entry 5), gave the corresponding pyrrole rings in reasonable yields under this reaction conditions (15 mol % of $PtCl_4$ and 60 mol % of 2,6-di-tert-butyl-4-methylpyridine, 0.1 M solution of the azide in ethanol). The amino group in the substrate needed to be protected in order to obstruct a complexation between the amino group and $PtCl_4$. N-Boc tosylamide was used as the protecting group. However, partial elimination of the Boc group in the product was observed under the reaction conditions. This side reaction was avoided by diluting the reaction mixture from 0.1 to 0.03 M solution. (Table 6.1, entry 6).

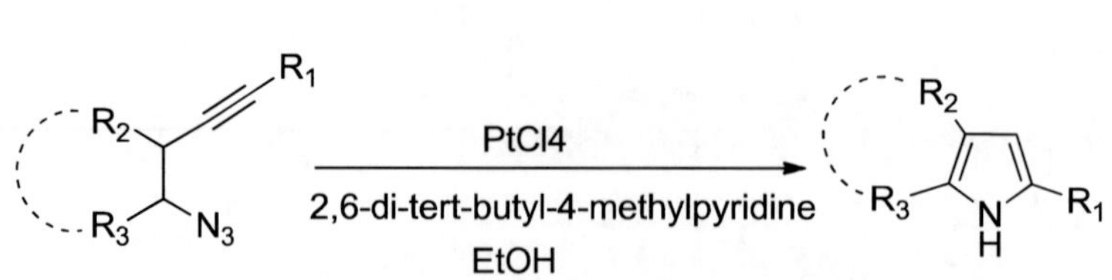

Scheme 13. Platinum catalysed cyclization reaction to pyrrole.

Table 6.1. Summary of Intramolecular annulation of homo propargyl azide

Entry	Homopropargyl azide	Products	T°C	Yield (%) (time)	
				Argon	Oxygen
1	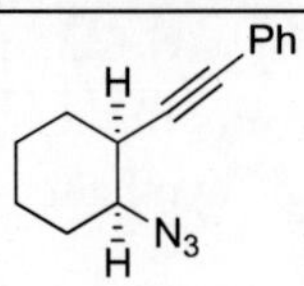 **35**	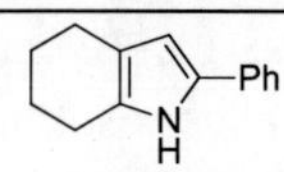 **36**	50	88 (**4h**)	78 (1h)
2	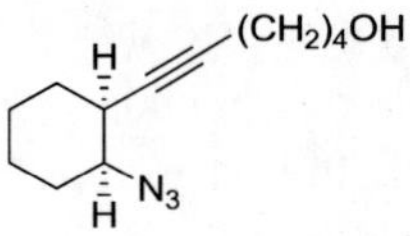 **37**	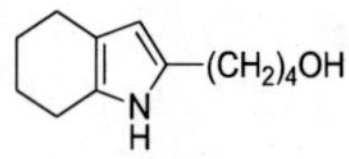 **38**	50	72 (**2h**)	—
3	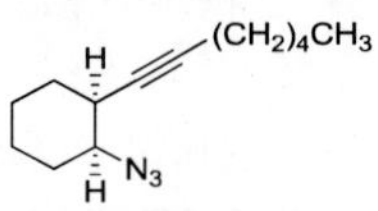 **39**	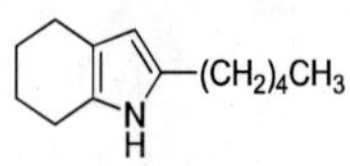 **40**	50	55 (**1.5h**)	24 (80 min)
4	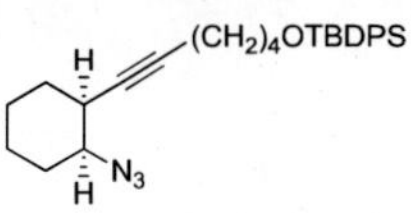 **41**	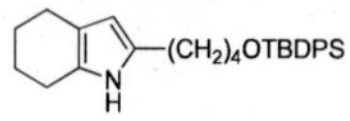 **42**	50	65 (**1h**)	—
5	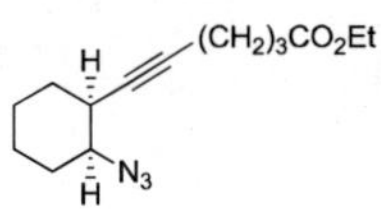 **43**	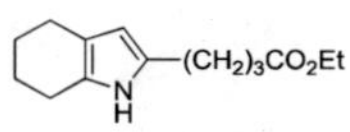 **44**	50	64 (**1h**)	—
6	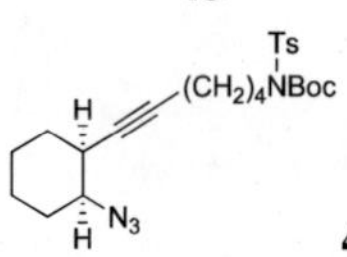 **45**	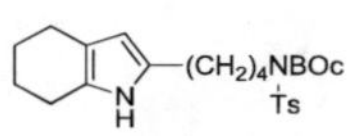 **46**	50	69 (**1/2h**)	23 (14)[a] (1h)
7	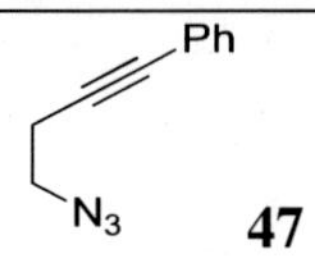 **47**	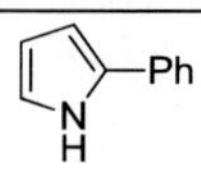 **48**	reflux	74 (**6h**)	70 (3)[a](2.5h)
8	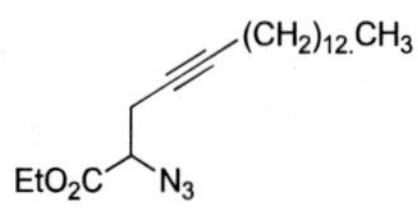 **49**	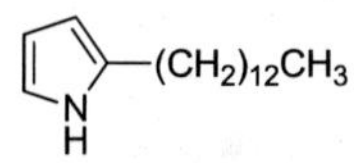 **50**	reflux	61 (**14h**)	—
9	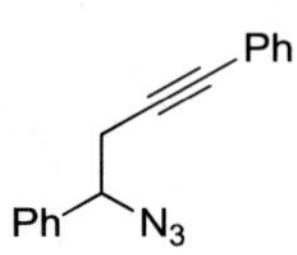 **51**	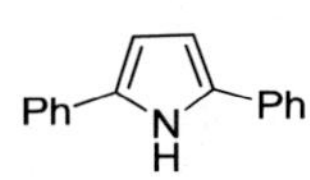 **52**	reflux	57 (**50min**)	65 (31)[a](1.5h)

[a] The number in parentheses was the yield of the recovered starting material.

7. Gold-Catalyzed Cyclization of Alkynylaziridines

Yuanhong Liu and his co-worker reported a facile gold-catalyzed cycloisomerization of N-phthlkynylazridines towards the noble synthesis of Nphth functionalized pyrroles [40]. Functionalized pyrroles including pyrrole-2-carboxylates or 2-pyrrolyl ketone were easily constructed in good yields by this method. The resulting pyrroles was further transformed to N-amino pyrrole or 2-acyl pyrrole, which serve as an important synthetic precursors for implementing the molecular complexity in their structures.

It is interestingly to note that only AgOTf failed to give any product (Table 7.1, entry 6). The reaction in presence of 5% $Ph_3PAuCl/AgOTf$ in THF solvent affords the products in excellent yield at room temperature (Table 7.1, entry 5). The title compounds with 2, 5 and 2, 3, 5 –substituted pyrroles has been delineated below. The phthalimido group can be easily deported to get the protecting group free pyrrole.

This method was successful with *cis*-aziridines. However, the authors observed different but interesting results. The *trans-* isomers failed to produce pyrroles. This method was particularly attractive for assembling multi-substituted pyrrole-2-carboxylates with high diversity through a regioselective manner. These compounds are potentially useful in pharmaceutical and material science.

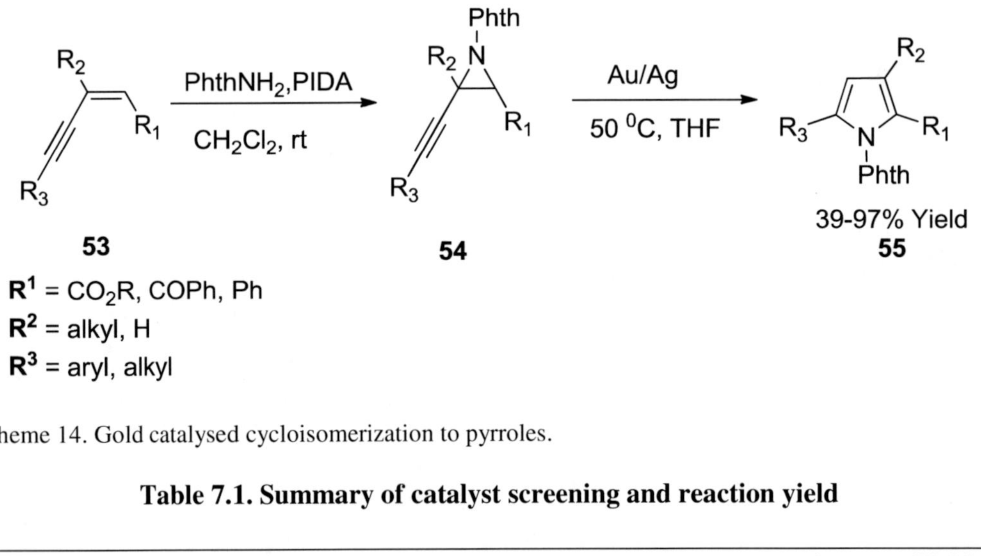

Scheme 14. Gold catalysed cycloisomerization to pyrroles.

Table 7.1. Summary of catalyst screening and reaction yield

Entry	catalyst	solvent	temperature (°C)	time (h)	Yield (%)55
1	5% $Ph_3PAuCl/AgOTf$	THF	50	1	97
2	5% $Ph_3PAuCl/AgOTf$	DCE	50	1.5	83
3	5% $Ph_3PAuCl/AgOTf$	MeOH	50	1	90
4	2% $Ph_3PAuCl/AgOTf$	THF	50	3	90
5	5% $Ph_3PAuCl/AgOTf$	THF	rt	6	96
6	5% AgOTf	THF	50	13	NR
7	10% $PtCl_2$	Toluene	80	21	61

$R^1 = CO_2Et$ $R^2 = H$ $R^3 = Ph$

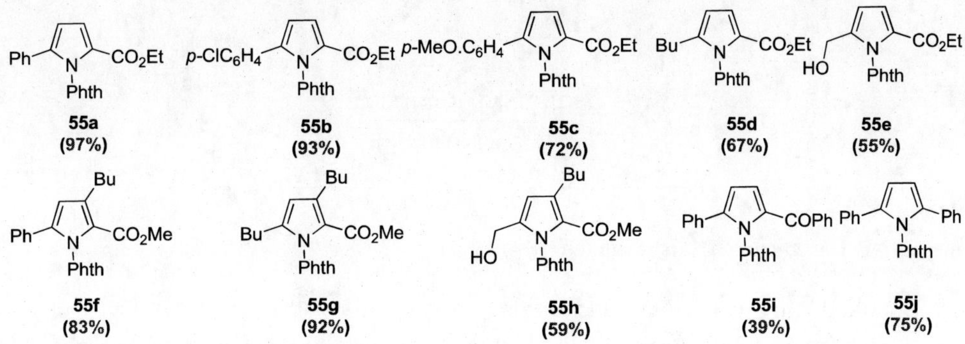

8. Copper/Nickel-Catalyzed Annulation to Pyrroles

Ning Jio et al. reported a novel and efficient copper or nickel-catalyzed selective denitrogenative annulation of vinyl azides with aryl acetaldehyde to synthesize 2,4- and 3,4-diaryl substituted pyrroles. This method was highly regioselective and required mild reaction condition without using any additive [41].

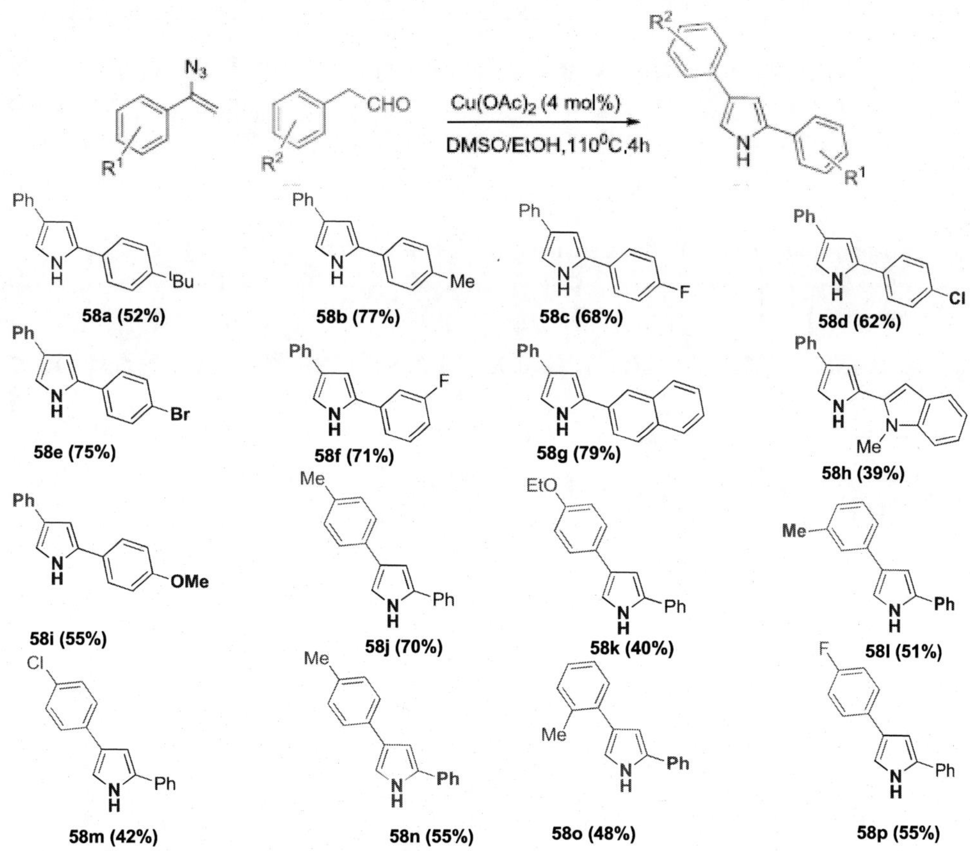

Scheme 15. Copper salt mediated 2,4-disubstituted pyrrole.

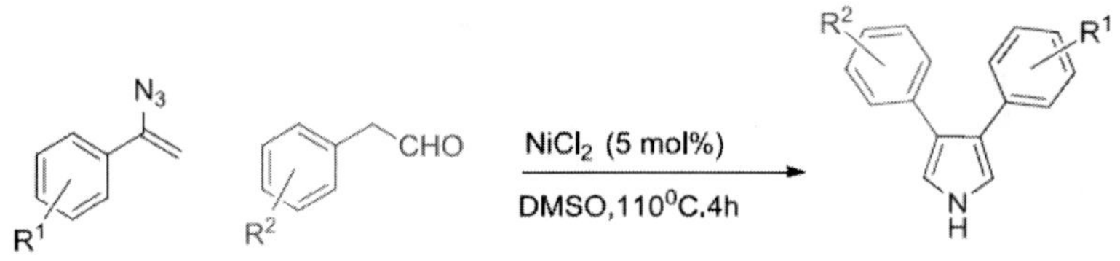

Scheme 16. **NiCl₂** mediated 3,4-disubstituted pyrrole.

This method provided a new and unique strategy to synthesize unsymmetrically 3,4-disubstituted pyrroles, which are difficult to achieve by other literature known protocols. This method involved the reaction of azidostyrene (**56**) and 2-phenylacetaldehyde (**57**) as the starting materials for the pyrrole synthesis. This reaction were carried out in 4 mol% copper (II) acetate at 110°C in DMSO under argon atmosphere in 56% yield. To optimize the reaction conditions, the authors screened various solvent systems for this reaction. The mixed solvent combination DMSO/EtOH was found to be a good combination and afforded the maximum yield of the products (Scheme 15).

A $NiCl_2$-catalyzed 3,4-disubsituted pyrrole formation was investigated. It was found that the substrate scope is broad with various substituted vinyl azides. (Scheme 16)

9. Palladium-Catalyzed Multicomponent Reactions

Dhawan and Arndtsen described the direct synthesis of pyrroles based on the multicomponent coupling of alkynes, imines and acid chlorides catalyzed by Palladium-based catalyst. This approach was based on 1,3-dipolar cycloaddition of alkyne to 1, 3-oxazolium-5-oxides (Mu¨nch-nones) to form the pyrroles [42] (Scheme 17).

10. Rhodium-Catalyzed Oxidative Pyrrole Synthesis

Glorius and co-workers investigated the novel pyrrole synthesis through an allylic Sp^3C-H activation of enamines followed by intramolecular coupling with unactivated alkynes under oxidative annulation strategies [43].

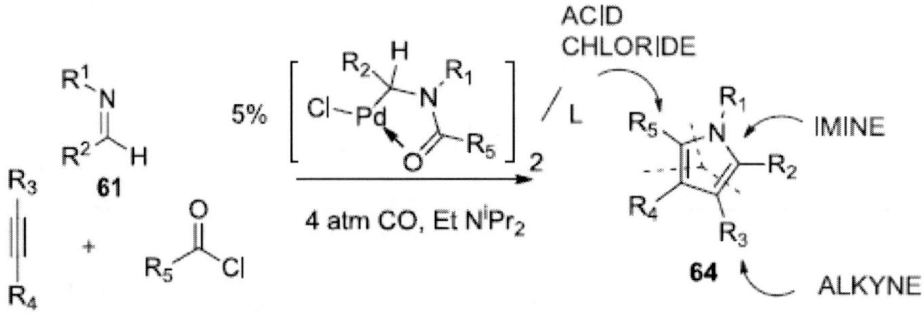

Scheme 17. Palladium catalyzed synthesis of pyrrole.

Table 10.1. Summary of Pd-catalysed pyrrole synthesis with different immine

Entry	imine	acid chloride	alkyne	products	Yield
1	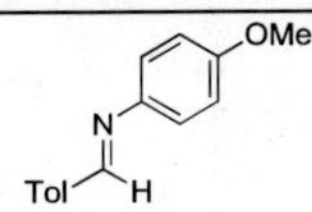	TolCOCl	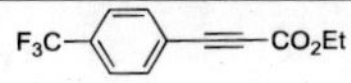	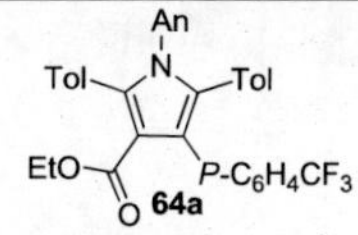 64a	81%
2	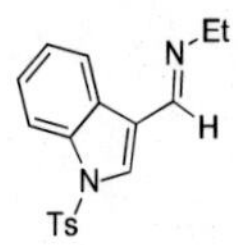	PhCOCl	MeO_2C—≡—CO_2Me	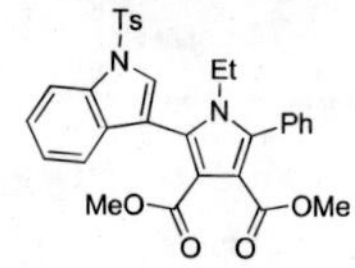 64b	71%
3	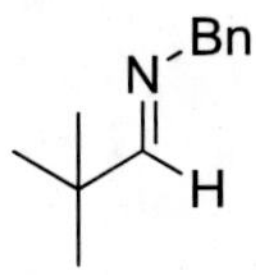	PhCOCl	Ph—≡—CO_2Me	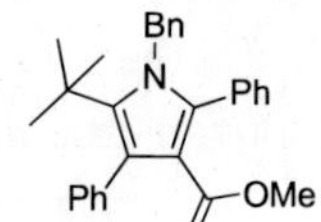 64c	80%
4	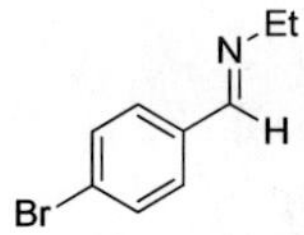	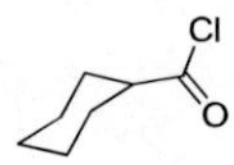	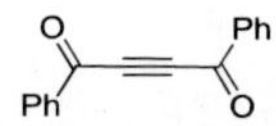	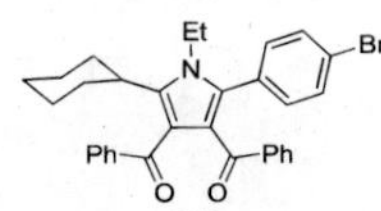 64d	63%
5	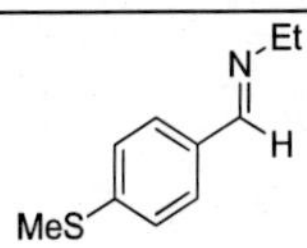	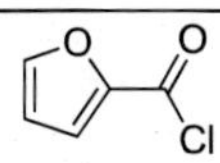	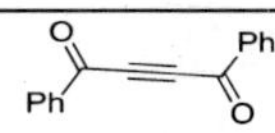	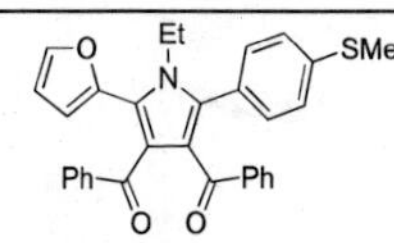 64e	73%
6	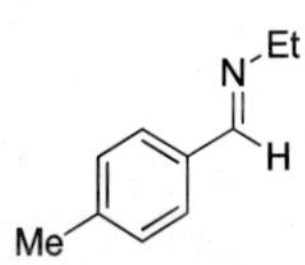	PhCOCl	H—≡—H	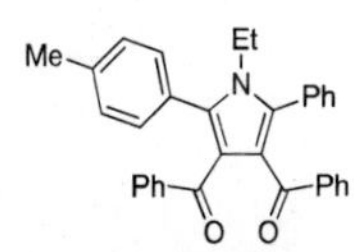 64f	77%
7	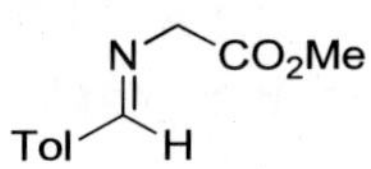	TolCOl	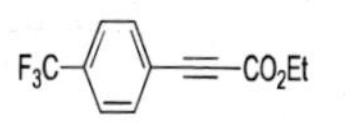	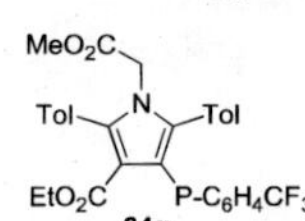 64g	65%
8	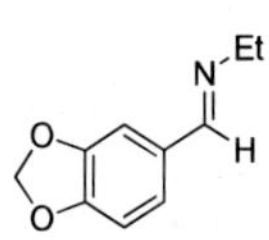	PhCOCl	MeO_2C—≡—CO_2Me	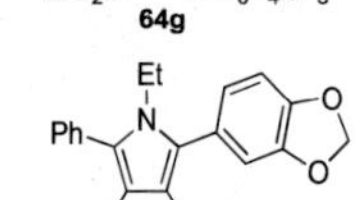 64h	81%
9	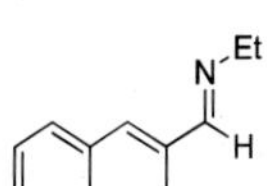	PhCOCl	MeO_2C—≡—CO_2Me	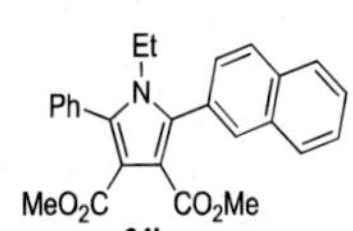 64i	88%

Enamides with methyl and ester group at and β–position respectively, underwent annulation with an akynes at N-H and methyl C-H bond, through Sp^3C-H bond activation. Interestingly, when β–substituent R_6 was H the rhodium catalyst appeared to reversibly activate the β-C-H position, and low yield of pyrrole was obtained. They investigated the effects of the N-substituent in this study and found the presence of acetyl group at N-position is very crucial for the success of reaction. The presence of an ester group at β-position also played a vital role for the reaction. A switch of ester to a keto group failed to deliver the products. The aryl disubstituted alkyne was found to be an excellent coupling partner with enamines and delivered excellent yield of the pyrroles (Scheme 18).

11. 1,3-Dipolar Cycloaddition to Pyrroles

11.1. Intermolecular 1, 3-Dipolar Cycloaddition

1, 3-Dipolar cycloaddition reaction is an important atom economy process and high yielding reaction in the synthesis of complex molecules in a single-pot operation. This strategy was exploited for constructing the diverse variety of bioactive organic molecules. This was also investigated for the synthesis of N-heterocyclic molecules. Azomethine ylides served as excellent component for 1,3-dipolar cycloaddition reaction for the construction the N-heterocyclic scaffolds [44].

Zhang and co-workers reported the synthesis of pyrrole ring using Ag (I)-catalyzed 1, 3-dipolar cycloaddition of azomethine ylides with *N*-alkyl maleimide, followed by dehydrogenation with DDQ. After the manipulation of the cycloadduct, it was transformed to a densely functionalized pyrrole scaffold. This method was based on the synthesis of azomethine ylide followed by cycloaddition and aromatization sequence.

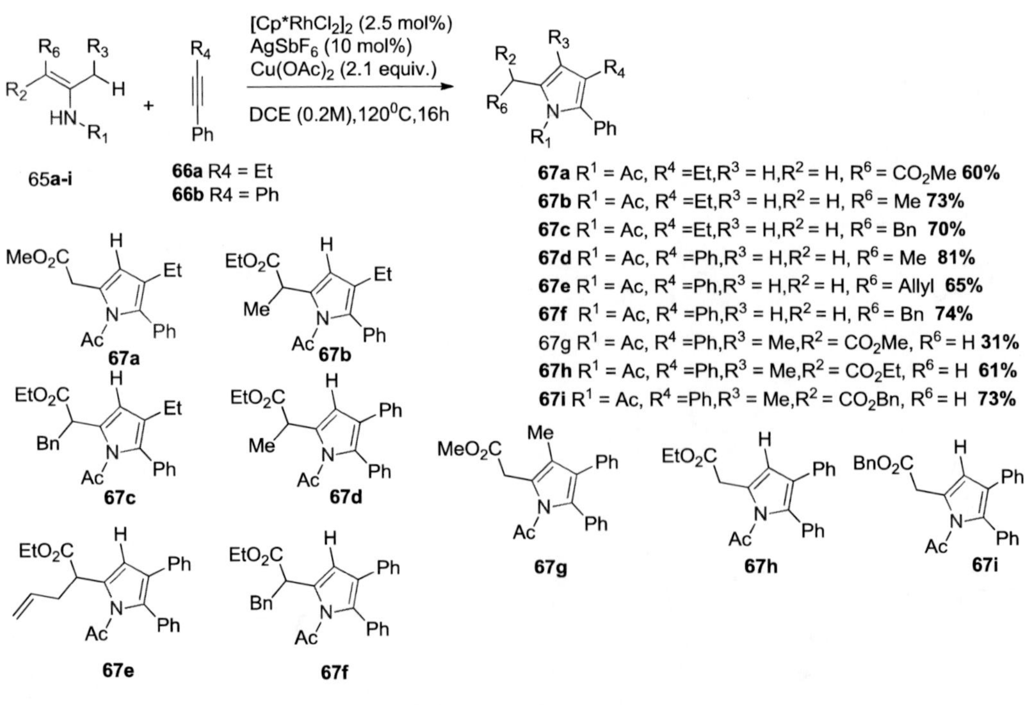

Scheme 18. Rhodium(III) -catalysed annulation of enamines and alkynes via allylic Sp^3 C-H bond activation.

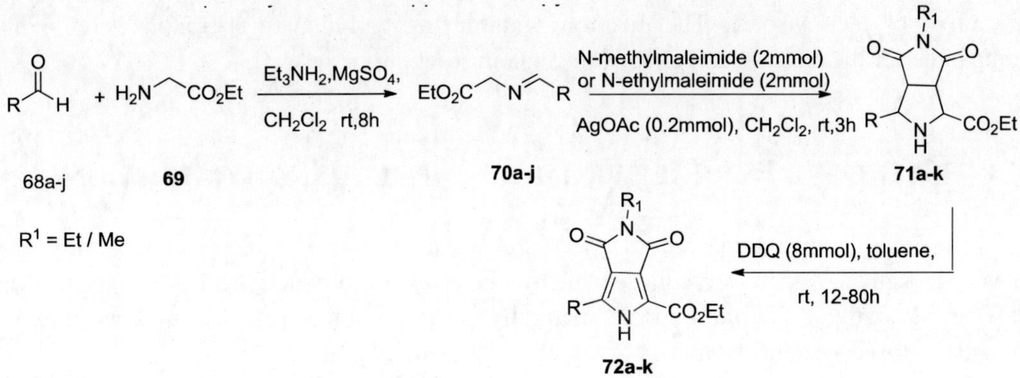

Scheme 19. 1,3-dipolarcycloaddition approache to pyrrole.

The diverse aldehyde was treated with ethyl glycinate hydrochloride in presence of triethyl amine in dichloromethane as a solvent to obtain the azomithine ylides. The resulting azomethine ylide was treated with maleimide as diploraphile in presence of 2.5 mol% of silver acetate and a triethyl amine as a base to achieve the cycloaddition reaction (Scheme 20).

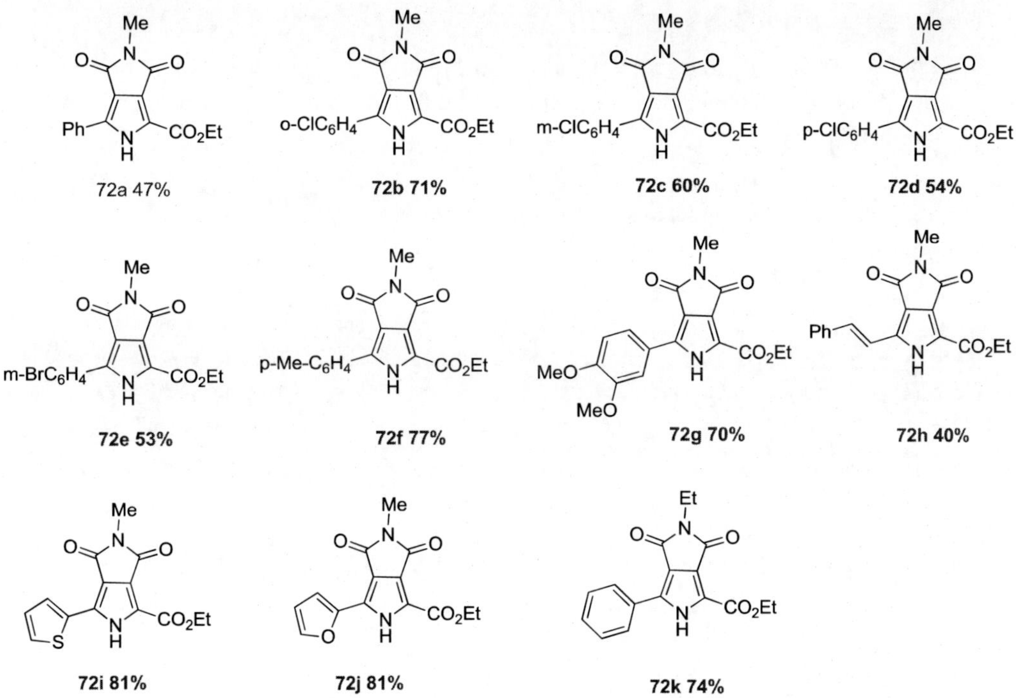

The cycloadducts were then employed dehydrogenation to get the pyrrole scaffold with DDQ (4 equivalent) in refluxing toluene. The effects of solvent were also investigated. Dichloromethane gave moderate yield of the desired product. The toluene and DDQ combination was found the optimized solvent and reagent for this reaction. A decrease of the amount of DDQ to 2 mol% lowered the yield of the reaction significantly. The maleimide ring at C_3-C_4-position open with reaction of series of amine which leads to novel 3,4 diamide derivative of the pyrroles. The electron withdrawing and donating groups were highly compatible for the reaction and underwent smooth reaction.

12. Indoyl Pyrrole from 1,3-Dipolar Cycloaddition

Biheteroaryl compounds are of great importance due to the broad applicability and privileged nature of these structures, which include natural products, and materials such as polymers and dyes. Pharmaceuticals, naturally occurring biheteroaryl structures exist as unfused or fused systems, examples of which are shown in Figure 2.

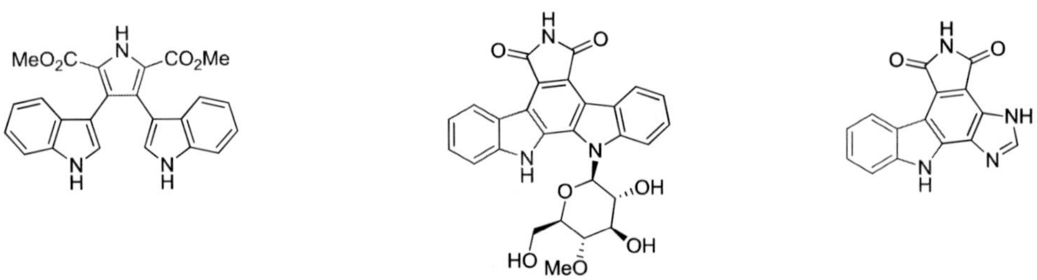

This protocol based on the reaction of 3-(2-nitrovinyl) indole with munchnones in THF under refluxing condition using N,N-diisopropylcarbodiimide as a base was performed without a transition metal catalyst.

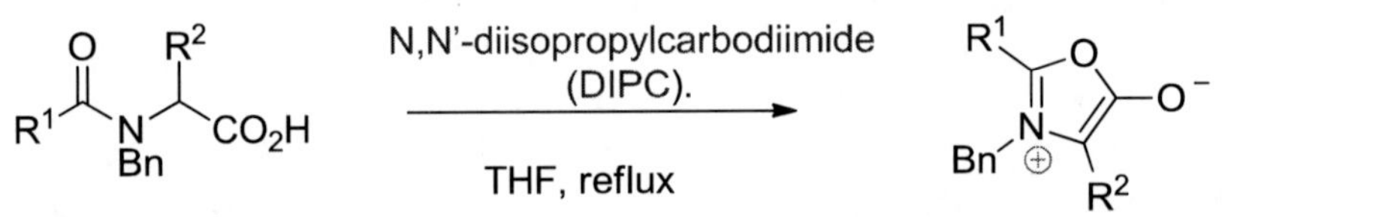

73 a $R^1 = R^2 = Ph$ | **74 a** $R^1 = R^2 = Ph$
73 b $R^1 = R^2 = Me$ | **74 b** $R^1 = R^2 = Me$
73 c $R^1 = Ph$ $R^2 = Me$ | **74 c** $R^1 = Ph$ $R^2 = Me$
73 d $R^1 = Me$ $R^2 = Ph$ | **74 d** $R^1 = Me$ $R^2 = Ph$

Scheme 20. Synthesis of Munchnone for 1,3-dipolar cycloaddition reaction.

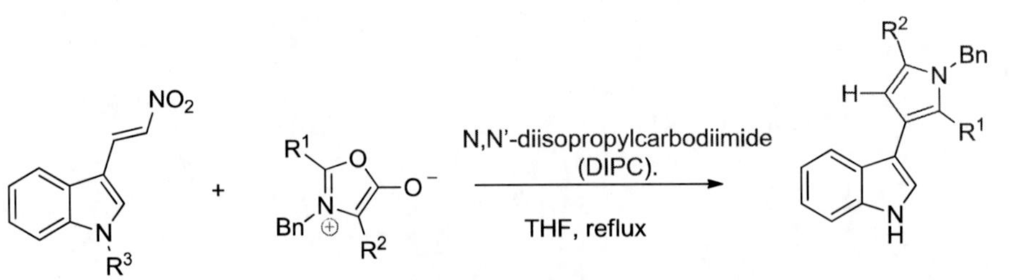

75 a-f

74 a $R^1 = R^2 = Ph$
74 b $R^1 = R^2 = Me$

76a-f

Scheme 21. Reaction of 3-Nitroalkenes with Symmetrical Münchnones.

Table 12.1. Summary of the synthesis of indoyl pyrrole via 1, 3-dipolar cycloaddition reaction

Entry	Indole	74	R^3	Product	Yield
1	**75a**	74a	H	76a	**78**
2	**75b**	74a	Me	76b	**62**
3	**75c**	74a	SO_2Ph	76c	**81**
4	**75d**	74b	H	76d	**84**
5	**75e**	74b	Me	76e	**75**
6	**75f**	74b	SO_2Ph	76f	**68**

The munchnones symmetrical and symmetrical **74a-d** were synthesize from amino acid derivatives **73a-d.** (Scheme 20). The reactions were tested with symmetrical munchnones (**74a-74b**) with N, N diisopropylcarbodiimide as base in refluxing THF in good to excellent yield (Scheme 21, Table 12.1). A series of 3-indoyl substituted pyrrole was synthesized and the result is summarized in Table 12.2. The authors were also studied the scope of this methodology with unsymmetrical munchnones (**74c-74d**) with high regioselectivity and good to moderate yield. This method proposed that 1,3-dipolarcycloaddition reaction proceeds through a bicyclic intermediate which decompose to release carbon dioxide and nitrous acid to the desired products.

The authors observed that the N-protecting group has no effects on the course of the product formation. The regiochemistry of the indoyl pyrrole was established by performing NOE interactions of the pyrrole ring protons with either the methyl or phenyl group of the pyrrole (**Scheme 22**). Phenyl sulfonyl protecting group provided the yield with low regioselectivity with unsymmetrical munchnones.

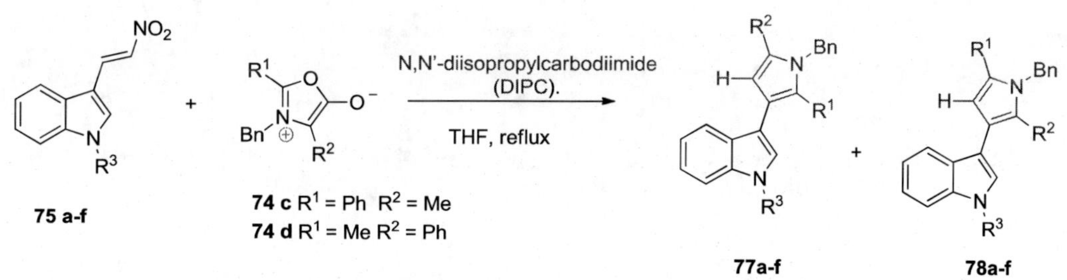

Scheme 22. Nitroalkenes with unsymmetrical münchnones regoselectivity of Indoyl pyrrole.

Table 12.2. Summary of the synthesis of indoyl pyrrole via 1, 3-dipolar cycloaddition reaction with unsymmetrical munchnone

Indole	Munchnone	R^3	Products	Ratio	Yield (%)[a]
75a	74c	H	77a:78a	77:23	**51**
75b	74c	Me	77b:78b	74:26	**86**
75c	74c	SO_2Ph	77c:78c	88:12	**71**
75d	74d	H	77d:78d	37:63	**20**
75e	74d	Me	77e:78e	44:56	**39**
75f	74d	SO_2Ph	77f:78f	49:51	**65**

The reaction protocol was allowed to access to a variety of substituted 2 and 3 indolyl pyrroles in moderate to high yields. In the case of unsymmetrical münchnones, the regioselectivity was varied.

The highly electron withdrawing group in indole ring greatly influence the products ratio with unsymmetrical munchnones. (Table 12.2, entry 75c and 75 f). The unprotected indole undergoes the cycloaddition reaction with poor yield and regeoselectivity comparatively (Table 12.2, entry 75a and 75 d).

13. 1,3-DIPOLAR CYCLOADDITION AND DECARBOXYLATION

Park and his co-workers demonstrated a novel regioselective synthesis of pyrrole through 1,3-dipolar cycloaddition-decarboxyllation strategy to tetra-substituted pyrroles. This protocol was based on the reaction of unsaturated benzofuron 3(3H)-one (**79**) and azalactone (**80a-b**). The authors confirmed the regioselectivity of the reaction by orthogonal synthesis and exhaustive NOE analysis of the products. The azalactone **80a-b** were derived from N-acetylated phenyl alanine and N-2-phenyl acetylated alanine respectively [46].

The reaction proceed best when 10 mol% of AgOAc in THF under microwave irradiation was performed. The protocol hypothesized that the pyrroles were formed by thermal extrusion of carbon dioxide from the cycloadducts. The authors thoroughly investigated the electronic effects on the regioselectivity of the product distribution. The substituent effect of both R1 and R2 group in azalactone was established extensively.

Indipolarophile (80) substituted R^2 with electron donating group led to increase the reaction time significantly. But, electron withdrawing group such bromo or nitro group in 1,3-dipolarophile enhanced the regioselectivity and yield of the product formation substantially. The complete investigation of the finding of the authors was described in (Scheme 23, Table 13.1).

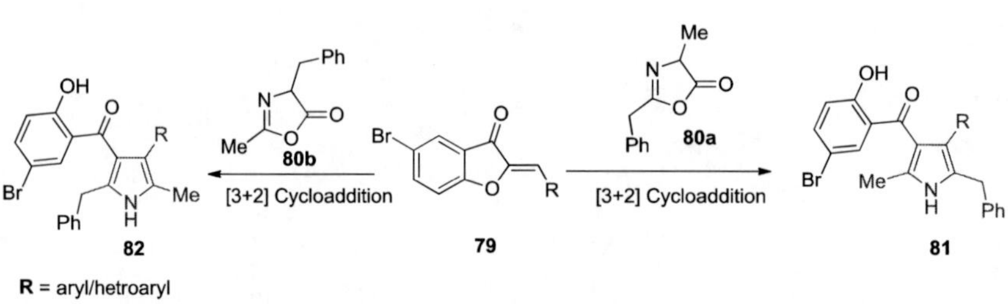

Scheme 23. Orthogonal synthesis of pyrrole via 2+3 cycloaddition reaction.

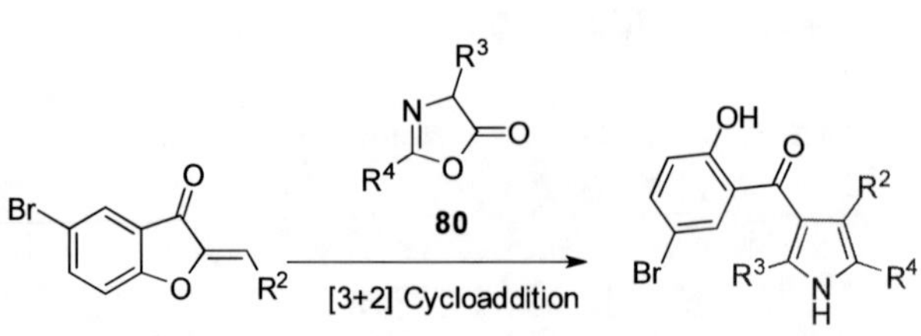

Table 13.1. Diversity of Pyrrole syntheses via [3+2] cycloaddition reaction

Compound	R^2(aldehyde)	R^3(AA)		R^4(acyl)	Yield
81a	4-pyridinyl	benzyl	(Phg)	methyl	**90**
81b	4-pyridinyl	methyl	(Ala)	benzyl	**75**
81c	4-pyridinyl	phenyl	(Phg)	methyl	**93**
81d	4-pyridinyl	isobutyl	(Leu)	methyl	**79**
81e	4-pyridinyl	2(methylthio)ethyl	(Met)	methyl	**70**
81f	4-pyridinyl	(1H-imidazolyl)methyl	(His)	methyl	**86**
81g	4-pyridinyl	phenyl	(Phg)	cyclohexyl	**97**
81h	4-pyridinyl	phenyl	(Phg)	phenyl	**94**
81i	4-pyridinyl	phenyl	(Phg)	benzyl	**91**
81j	4-pyridinyl	phenyl	(Phg)	n-butyl	**95**
81k	2-OMeC_6H_4	methyl	(Ala)	methyl	**91**
81l	4-OMeC_6H_4	phenyl	(Phg)	methyl	**97**
81m	3-$NO_2C_6H_4$	phenyl	(Phg)	methyl	**81**
81n	2-thiophenyl	phenyl	(Phg)	methyl	**84**

13.1. Silver-Catalyzed 1,3-Dipolar Cycloaddition

Kan and Wang described the tandem cycloaddition, oxidative dehydrogenative aromatization strategy for the synthesis of pyrrole. This method was relied on benzoyl peroxide-mediated oxidative tendem reaction for the construction the pyrrole. The reactions were catalyzed with well-defined silver-salt providing a practical method for the synthesis of multi-substituted pyrroles from easy available alkenes [47] (Scheme 24).

In this area, Arndtsen and his group developed a new one-pot strategy accompanied by phosphorus-based 1,3-dipolar cycloaddition reagents for the novel synthesis of the pyrrole scaffolds. This reaction was done by the coupling of the imines and acid chlorides followed by phosponite that resulted in 1,3-oxazolium-5-oxides (munchnone). The reaction was catalyzed by non nucleophic base DBU (equivalent) and lead to excellent yield of the products [48] (Scheme 25).

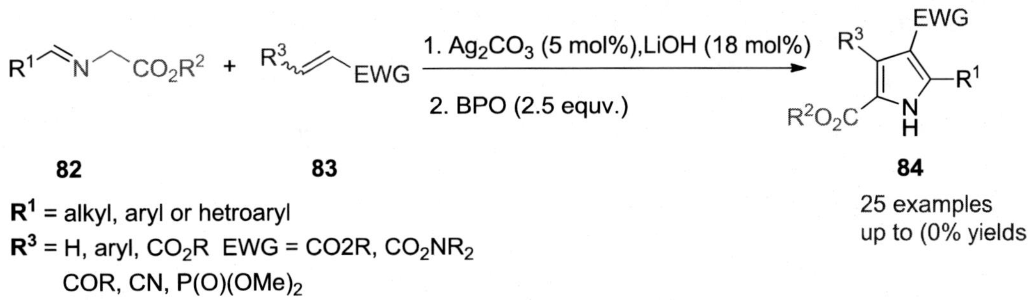

Scheme 24. 1,3-dipolar tandem reaction to multisubstituted pyrrole.

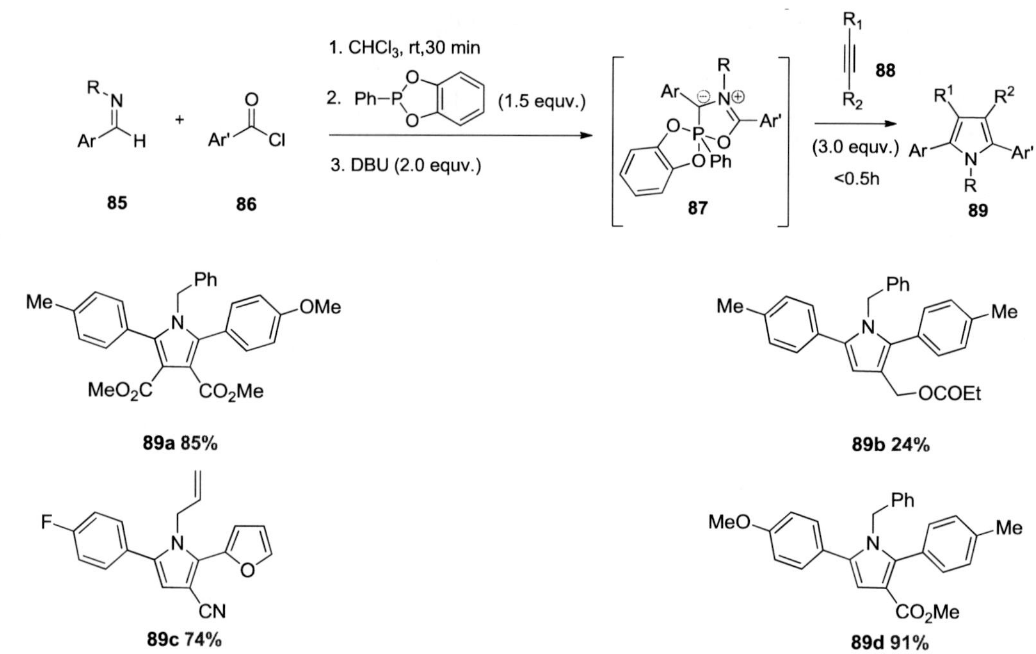

Scheme 25. Phosphonite basd 1,3-dipolar reaction to trisubstituted pyrroles.

14. Vinyl Oxime Rearrangement to Pyrrole

[3,3]-Sigmatropic rearrangement is a unique tool for forming new C-C bond in synthesis of complex molecules. A pioneering work to the synthesis of 2, 3, 5 and 2, 3,4 substituted pyrrole relevance of biological activity was carried out by Anderson et al. *via* rearrangement of O-vinyl oxime which can be obtained by iridium catalyzed isomerization of O-allyl oxime. The regiochemistry of pyrrole formation practically was controlled by either substituents or amine base additives. The reaction adopted either [3,3] or [1,3] sigmatoropic rearrangement followed by Pall-Knorr cyclization [49]. The authors further described the effects of the substituents. Electron-withdrawing substituents at the oxime and the addition of DBU as a amine base to the reaction mixture favored enolization, and the reaction followed the pathway of [3,3] rearrangement of O-vinyl oxime intermediate, and the formation of 2,3,4 trisubstituted pyrroles was realized.

On the other hand, electron-donating or electronically neutral substituents of the oxime and in absence of a amine base disfavored enolization, and adopted the pathways of [1,3] rearrangement of the O-vinyl oxime intermediate to occur followed by tandem tautomerization, cyclization, and elimination to give 2,3,5-trisubstituted pyrroles (Scheme 26). This method offered a simple, regioselective, and functional group tolerant synthesis of 2,5-disubstituted pyrroles and 2,3,4- or 2,3,5-trisubstituted pyrroles in two- or three steps sequential reactions from commercially available ketones and allyl hydroxylamine as starting materials. Moreover, this method worked well with both electron rich and poor substrates.

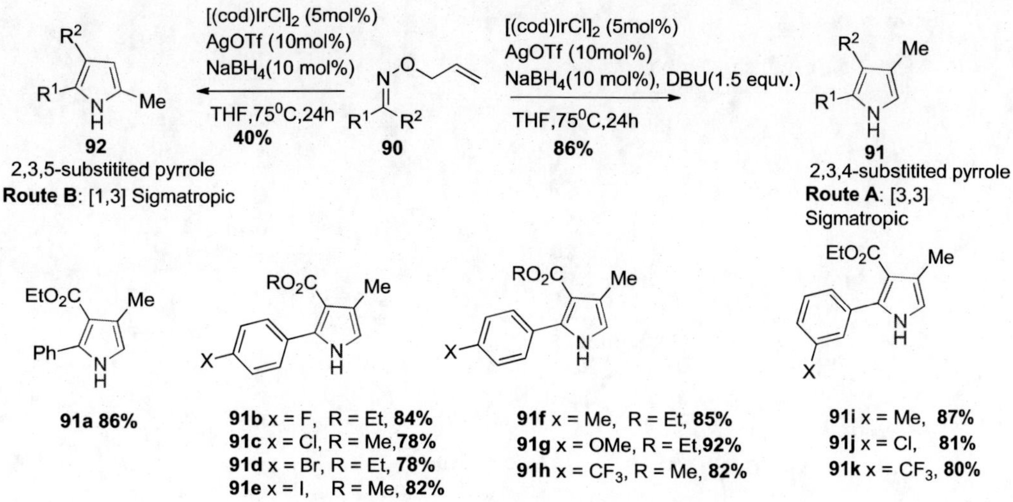

Scheme 26. Regioselective Synthesis of Trisubstituted Pyrroles from O-Allyl Oximes.

Camp and his co-workers developed the gold-catalyzed rearrangement of O-vinyl oxime to the highly substituted pyrrole in regiocontrolled manner under thermal conditions. The method was very efficient and mild to synthesize the densely functionalized pyrrole in a single pot operation. In this protocol authors used the diastereomeric mixture of O-vinyl oximes [50] (Scheme 27).

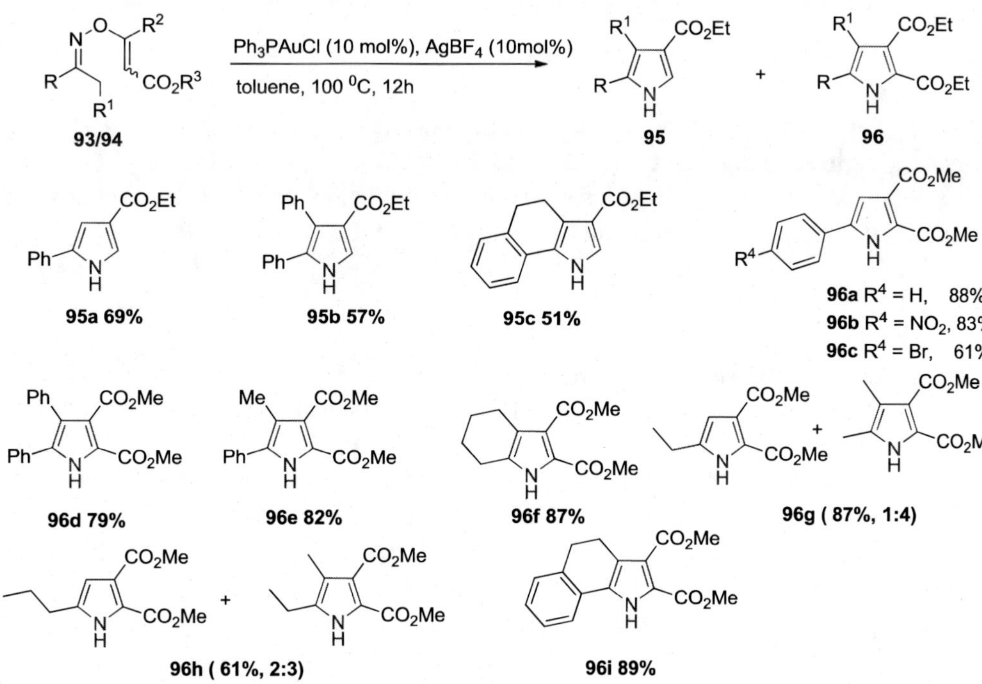

Scheme 27. Gold-catalysed rearrangement of O-vinyl oximes to pyrroles.

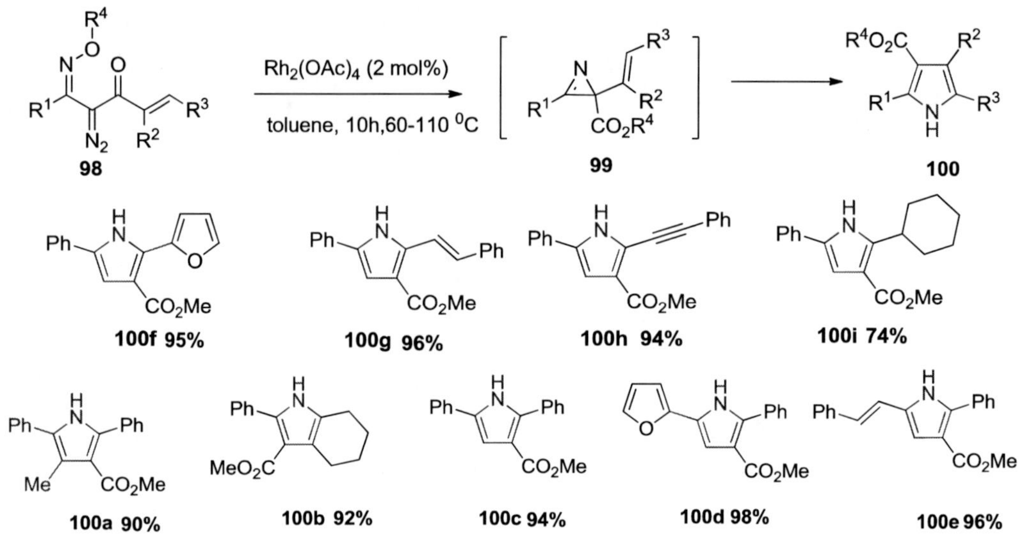

Scheme 28. Rhodium-catalysed rearrangement of a-diazo oxime ethes to pyrroles.

Park et al. developed efficient rhodium-catalysed synthesis of pyrrole via a novel rearrangement of α-oximino ketene derived from α-diazooxime followed by subsequent rearrangement to highly substituted pyrrole from oxime ethers in excellent yield [51] (Scheme 28).

15. Isocyanides in Pyrrole Synthesis

Isocyanides have been first described independently by Gautier [52]. A reaction by Hoffmann's carbyl ammine method was used for their preparation. The carbon atom in isocyanides have carbine like character which may function like both electrophilic and nucleiphilic which turned this compound one of the versatile building blocks in the synthesis of diverse organic molecules [x]. Isocyanides can participate in various transformations that include multicomponent reaction such as Ugi, Passerini reaction, Lewis acid-catalyzed isocynide-based cyclization and transition metal-catalyzed polymerization [53].

Isocynides are used towards the synthesis of various heterocyclic molecules of biological interests. One of the remarkable applications of isocynide was in the synthesis of N-heterocycle such pyrrole. Isocynide in combination with nitro alkene under basic condition underwent cyclization to offer 1,2-substituted pyrroles. This cyclization was first discovered by Barton and Zard and called Barton-Zard pyrrole synthesis [55]. This reaction was based on the treatment of isocynide (**101**) with nitro olefins (**102**) in non nucleophlic base such DBU/DABCO. Mechanistically the isocynide initially underwent base-induced enolization. The resulting enolate underwent Michael addition reaction with nitro alkene followed by 5-endo-dig cyclization and subsequent elimination and 1,5-sigmatropic H shift produced the pyrrole derivative (**106**) (Scheme 29).

The Barton-Zard methodology has been extensively exploited in synthesis of various natural products that include pyrrolosatatin and its analogue as well as the chromophores for biological system [56].

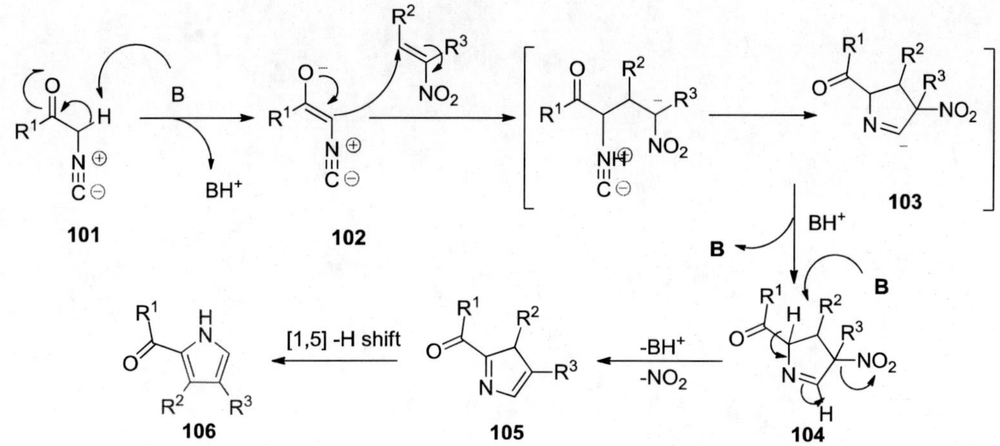

Scheme 29. Barton-Zard pyrrole synthesis.

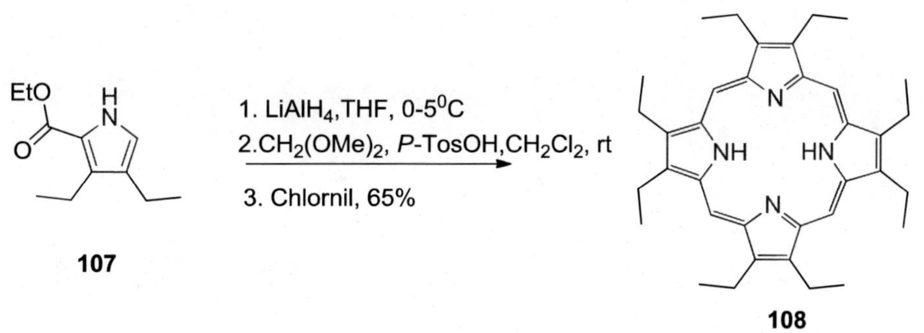

Scheme 30. Transformation of pyrrole in to porphyrin.

The pyrrole was synthesized using the Barton-Zard methodology and these molecules had substitution pattern to access the biologically active core structure of octaethyporphyrin ring [57] (Scheme 30).

Tosyl methyl isocynide (TosMIC) is a versatile reagent in synthesis of wide range of small heterocyclic molecules was first introduced by van Leusen and has become a classical reagent for the synthesis of 1,3-azole and pyrrole [58]. Upon treatment with electron deficient olefins under basic condition this gave pyrroles and this protocol is known as van Leusen pyrrole synthesis. The tosyl group was prone to eliminate under basic conditions makes this reagent high potential in pyrrole synthesis.

Zhang and his co-workers reported a facile synthesis of 3-(isoxazol-5-yl)-substituted pyrroles through [3+2] cycloaddition of tosylmethyl isocyanide (TOSMIC) and styrylisoxazole under strong alkaline condition at ambient temperature [59]. The authors screened the various bases such as DBU, K_2CO_3, KO^{t}Bu, TMG and NaOH in combination with different solvents such as acetonitrile, DMF, THF and ethanol. Acetonitrile and potassium hydroxide combination was found to be best to perform this reaction. The electron withdrawing or electron donating groups in aryl ring attached to the alkene of the Michael accepters underwent smooth addition of the nucleophilic reagent TOSMIC and produced high pyrroles with high yields (Scheme 31).

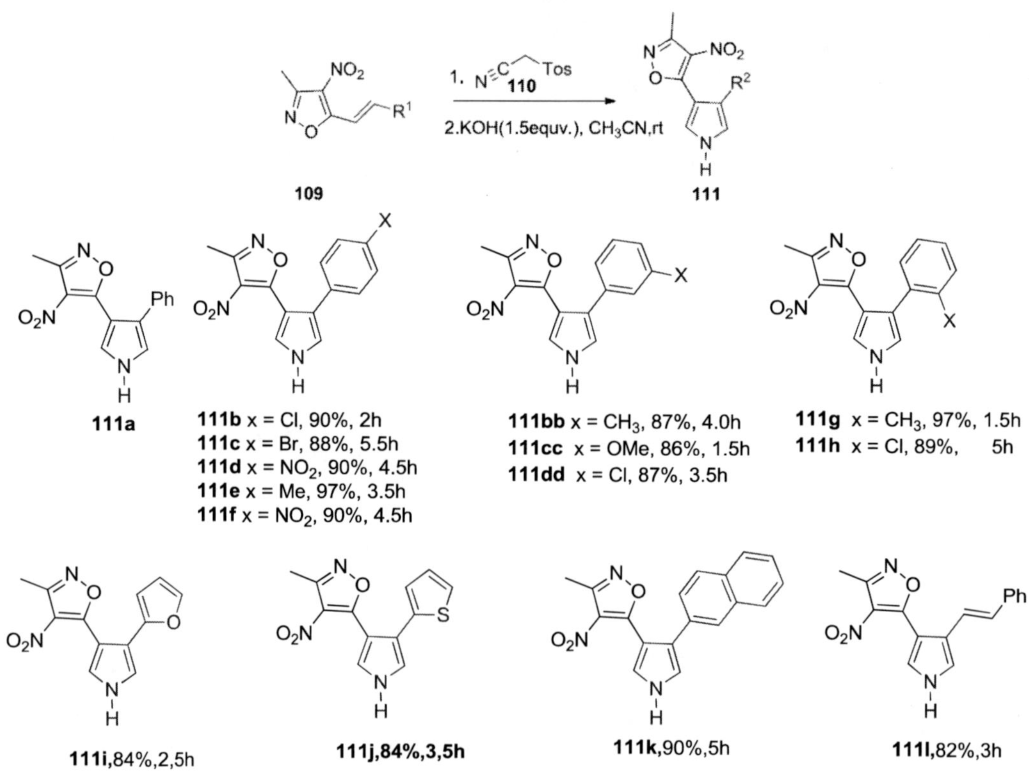

Scheme 31. Synthesis of 3-isoxazole bisubstituted pyrrole derivatives via [3+2] cycloaddition.

This reaction was also explored with substituted TOSMIC to synthesize the 2, 3, 4-substituted pyrroles in good yields (Scheme 32 Table 15.1). The confirmation of the products was established by the NMR analysis and the X-ray crystallographic analysis.

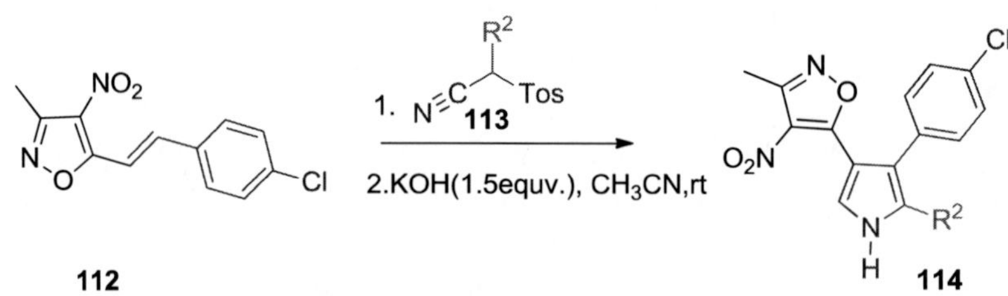

Scheme 32. Synthesis of 3-isoxazole trissubstituted pyrrole derivation via [3+2] cycloaddition.

Table 15.1. Summary of substituted N-TosMIC, 1, 3-dipolar cycloaddition reaction

Entry	R^1	Time (h)	Pyrrole (114)	Yield (%)
1	Et	8.0	**114a**	**67**
2	Allyl	9.0	**114b**	**56**
3	Phenyl	4.0	**114c**	**81**
4	Benzyl	7.0	**114d**	**78**
5	4-Me.$C_6H_4CH_2$	5.0	**114e**	**83**

Yamamoto et al. reported an efficient synthesis of pyrrole (**117**) by phosphene-mediated formal cycloaddition strategy of isocynide (**115**) across the electron deficient alkyne (**116**). This interesting organocatalytic transformation was found to give best yields in dioxane at 100°C with bidentate phosphine such as dppp as catalyst.

This protocol initially involved in the nucleophic addition of phosphine to electron deficient π bond of an alkyne (**116**). This led to a dipolar ionic intermediate which on deprotonation of the isocynide (**115**) produced alkene followed by the formal cycloaddition, and 1,5-H shift to the pyrrole. The presence of electron withdrawing group in methyl isocynide accelerated the reaction significantly. This methods did not work with electron rich methyl isocynid [60] (Scheme 33).

In this context the rhodium carbonyl complex catalyzed reaction of β-cyanoacetate was another excellent method to offer the pyrrole ring efficiently $Rh_4(CO)_{12}$ as catalyst. The formamide intermediate was decomposed under thermal conditions to give imine which underwent cyclodehydration to the products (**122**) in high yield (Scheme 34).

The cycoldehydration of unsymmetrical 1, 3-dicarbonyl compound gave the pyrroles regioselectively based on the electronic effects and favourable stabilizing factors [61]. The reaction of β-cyanoacetate (**119**) with excess of 1, 3-dicarbonyl compound (**118**) initially produced the unsaturated formamide in presence of $Rh_4(CO)_4$.

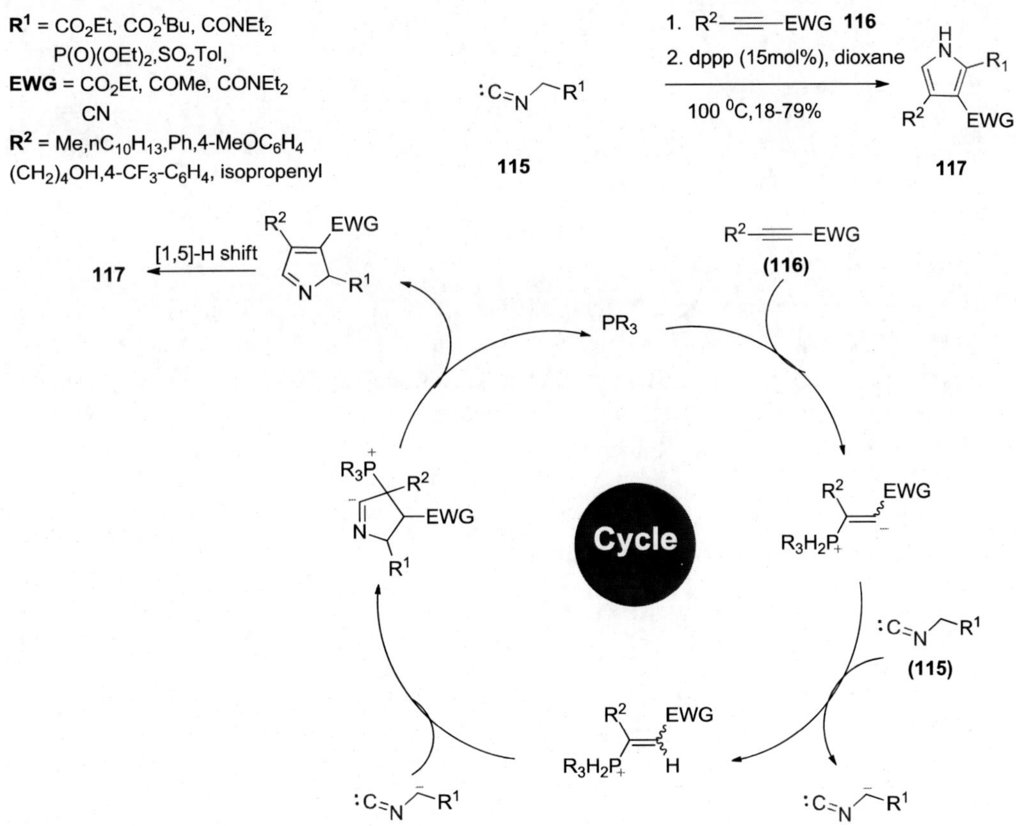

Scheme 33. Mechanistic patway for the phospine catalyzed synthesis of pyrrole.

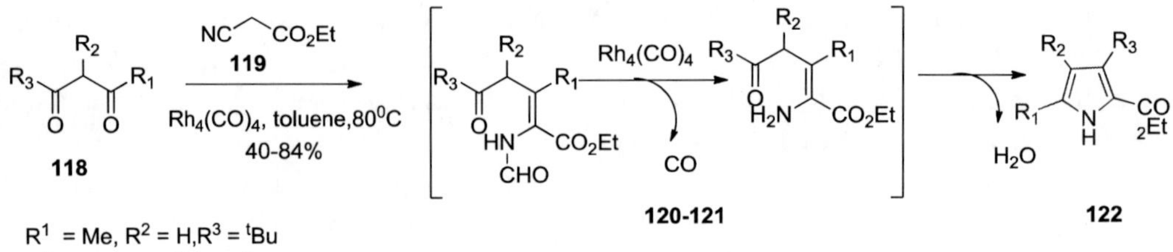

Scheme 34. Rhodium catalysed synthesis of pyrrole from 1,3-dicarbonyl compounds.

16. Ring Closing Metathesis to Pyrrole

Olefin metathesis is an important organic reaction that entails the redistribution of fragments of two alkenes (olefins) by the scission and regeneration of carbon-carbon double bonds. This reaction was first developed by Grubbs, Schrock and Chauvin who were awarded with the Nobel Prize in Chemistry in 2005. This discovery had opened a revolution in the synthesis of many organic molecules and natural products.

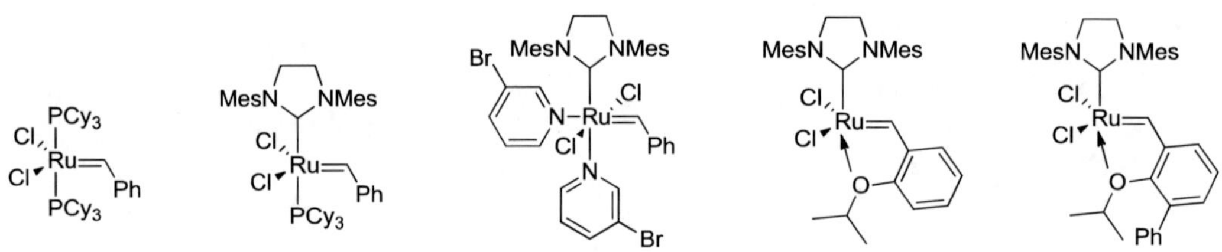

The reactions are catalyzed by molybdenum-carbene (Schrock catalyst) or ruthenium-carbene (Grubbs' type catalysts) complexes. In particular, the Ru-carbenes (Grubbs' 1st, 2nd and 3rd generation **Ru1-3,** Hoveyda **Ru4** and Blechert **Ru5** catalysts, (Figure 3) are very crucial.

Ruthenium chloride and Grubbs 2nd generation catalyst combination was used for the synthesis of pyrroles. Ruthenium was responsible for the oxidative aromatization leading to the pyrrole (Scheme 35). A variety of amines linked to olefin moiety in a pressure tube was subjected to a tandem catalytic system of the 'standard' second generation metathesis catalyst Ru2 in combination with $RuCl_3$ under ultrasound conditions and this method resulted in the construction of pyrroles. The ultrasound was used to obtain a fine dispersion of the $RuCl_3$ additive which was less soluble in reaction medium. The electron withdrawing groups such as Cl and ester at vinylic position were failed to yield the desired pyrrole and this was explained due to dimerization of the olefins. A reasonable conversion to the pyrrole was observed but the yield of the products was dropped during purification [62].

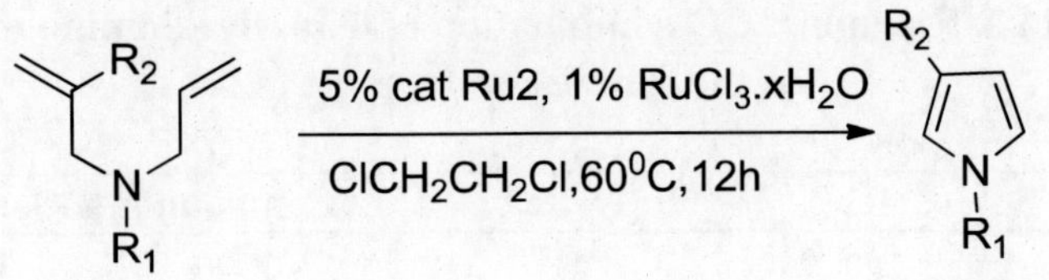

123 124

Scheme 35. Synthesis of pyrrole in using Grubb's 2nd generation catlyst and ruthenium chloride.

Table 16.1. Summary of Ru2 and $RuCl_3.xH_2O$ for pyrrole synthesis.

Entry	R1	R2	Conversion	Compound	Yield
1	Bn	H	74%	124a	**55**
2	$OMeC_6H_4$	H	95%	124b	**74**
3	CH_2CN	H	30%	124c	**0**
4	CH_2CH_2CN	H	44%	124d	**30**
5	Bn	Me	69%	124e	**50**
6	Bn	CH_2moarpholine	82%	124f	**0**
7	Bn	Cl	0	124g	**0**
8	CO_2Me	Br	0	124h	**0**
9	Bn	CH_2Cl	56%	124i	**0**
10	Bn	CO_2Et	0	124j	**0**
11	CH_2CO_2Me	H	91%	124k	**63**
12	$CH(CH_3)CO_2Et$	H	78%	124l	**60**
13	$CH_2P(O)(OEt)_2$	H	71%	124m	**60**

The authors explored the transformation of pyrroline to pyrrole by the addition of hydrogen acceptor to the reaction mixture which enhances the conversion of dihydo ruthenium (RuH_2) to ruthenium species (Ru). The oxidation of pyrroline to pyrrole was achieved by DDQ and TCQ. The latter compound was found to be best oxidizing agent for this transformation.

The mechanism suggested by the authors was summarized in **Scheme 36.** The metathesis products pyroline underwent aromatization *in situ* by hydroquinone assisted oxidative-dehydrogenative pathways.

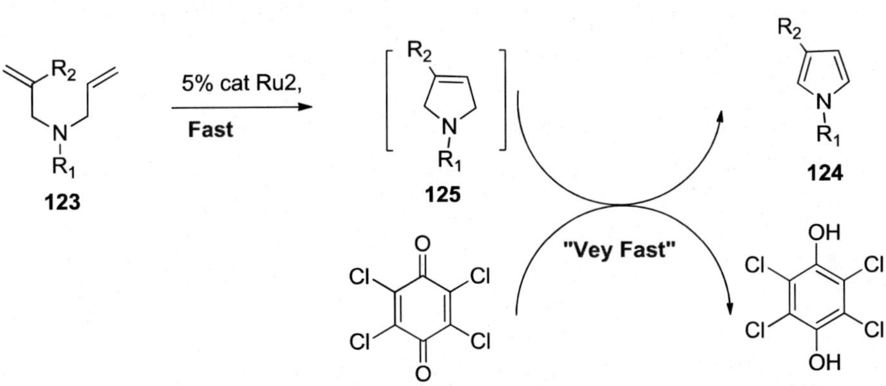

123 125 124 126 127

Table 16.2. Ru2 and TCQ combination of oxidative domino metathesis for the synthesis of pyrrole (124)

Entry	R1	R2	Compounds	Conversion
1	Bn	H	124a	**100%**
2	CH_2CO_2Me	H	124k	**96%**
3	$CH_2P(O)(OEt)_2$	H	124m	**95%**
4	$OMeC_6H_4$	H	124b	**100%**
5	Bn	CH_3	124e	**93%**
6	Bn	CH_2moarpholine	124f	**90%**

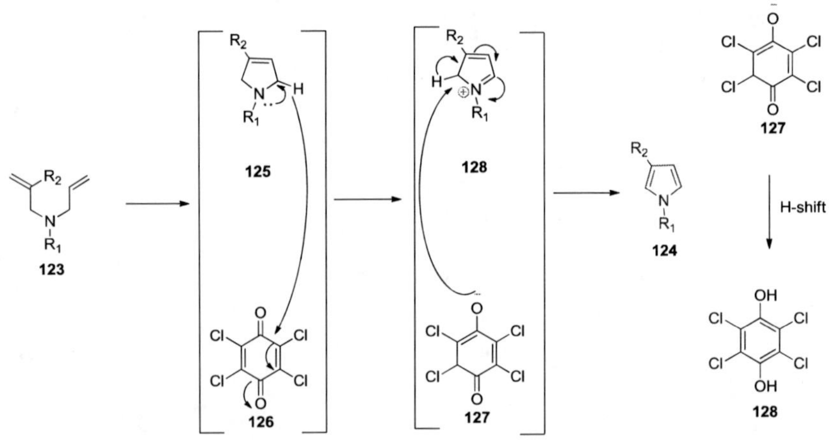

Scheme 36. Plausible reaction mechanism.

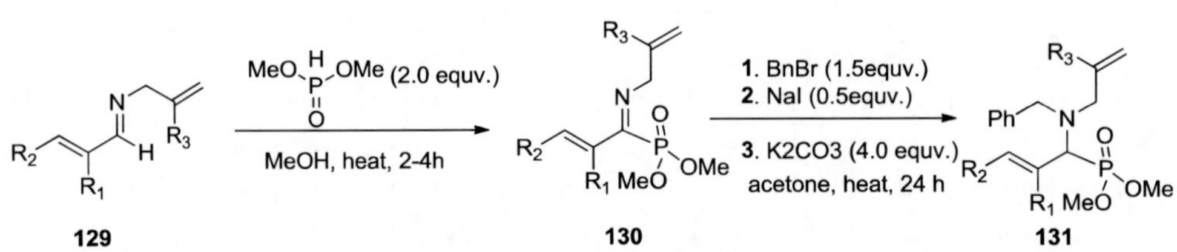

Scheme 37. Synthesis of 2-phosphonylated N-diallyl immine.

Table 16.3. Summary of the synthesis of 2-Phosphonylated N-diallyl immine

Entry	R1	R2	R3	130(%)	131(%)
a	Ph	H	H	95	61
b	Ph	Me	H	90	50
c	Me	Bn	H	27	50
d	Ph	Isoamyl	H	88	54
e	Me	Ph	H	80	35
f	Me	CH_2CH_2Ph	H	44	92
g	Ph	H	Me	74	86

The authors next attempted the synthesis of 2-phosphonopyrroles (**130a-f**) by standardized RCM/oxidation strategy. The metathesis substrate was obtained by regioselective 1,2-addition of phosphorus nucleophile to imine (**129**) followed by protection of the N-atom of the imine with benzyl group (**131**) (**Scheme 37**).

2-Phosphonoacetate N-diallyl was then subjected to Grubbs 2nd generation catalyst through domino reaction sequence. This methods offered the good yield of phosphono pyrrole at room temperature (**Scheme 38**).

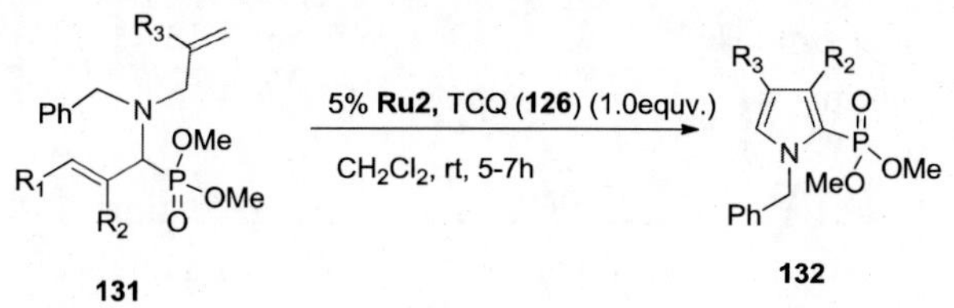

Scheme 38. Synthesis of 2-phosphonylated pyrrole by oxiadtiv-RCM domino reaction by TCQ.

Table 16.4. Summary of the results of 2-Phosphonopyrrole

Entry	R1	R2	R3	132(%)
a	Ph	H	H	**75**
b	Ph	Me	H	**84**
c	Me	Bn	H	**72**
d	Ph	Isoamyl	H	**70**
e	Me	Ph	H	**75**
f	Me	CH_2CH_2Ph	H	**71**
g	Ph	H	Me	**--**

17. RCM AND RCEYM APPROACH TO PYRROLE

Metathesis reactions were well exploited in the synthesis of pyrrole from alkene. A very few report on pyrrole synthesis was known by enyne metathesis. Stevan and coworkers describe the synthesis of pyrrole (**135**) by RCEYM-aromatization approach. The authors reported the synthesis of a series of phosphono-3-pyrrole from enyne.

The RCEYM initially gave 3-substituted pyrrolines witch was then oxidised *in situ* by the addition of TCQ (tetrachloroquinone) to pyrrole. Interestingly the addition of DDQ to the reaction mixture decomposed the ruthenium catalyst and failed to give the products. (Scheme 39).

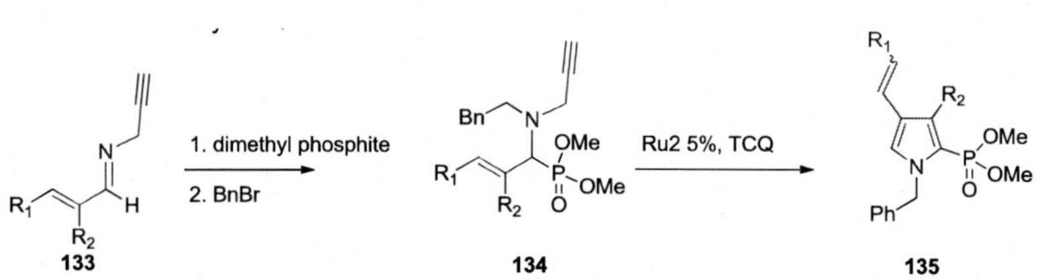

Scheme 39. Synthesis of 2-phosphonylated pyrrole by oxiadtiv-RCEYM domino reaction by TCQ.

Castagnolo and coworkers explored the EYCM reaction for the synthesis of pyrroles in single pot operation from N-protected propargyl amine (**141**) and ethyl vinyl ether (**137**) with Grubbs 2nd generation catalyst. Terminal alkyne on reaction with ethyl vinyl ether in the presence of Grubbs 2nd generation catalyst under microwave-induced conditions in aqueous medium led to the formation of crotanaldehyde. The ehoxy diene hydrolyzed to enol derivative followed by tautomerization to crotanldehyde [63] (Scheme 40).

The authors envisaged if the reaction through the formation diene ether can be cleaved by the nucleophile present in the terminal alkyne. The N-protected propargyl amine and EVE (ethyl vinyl ether) under microwave-mediated condition with **Ru2** in aqueous medium produced diene ether intermediate which knocked out the OEt and afforded pyrrole in one-pot operation (Scheme 41).

The reaction took place only when $CuSO_4$ is added to the reaction medium. The assumption behind the success of this reaction was that Cu^{+2}coordinate with ethoxy group and leads to the aromatization to furnish the pyrrole.

The method proved to be versatile leading to a variety of pyrroles substituted at C-2 with both aromatic and aliphatic groups as well as bearing a variety of aromatic and EWG-substituents on the nitrogen [64] (Scheme 42).

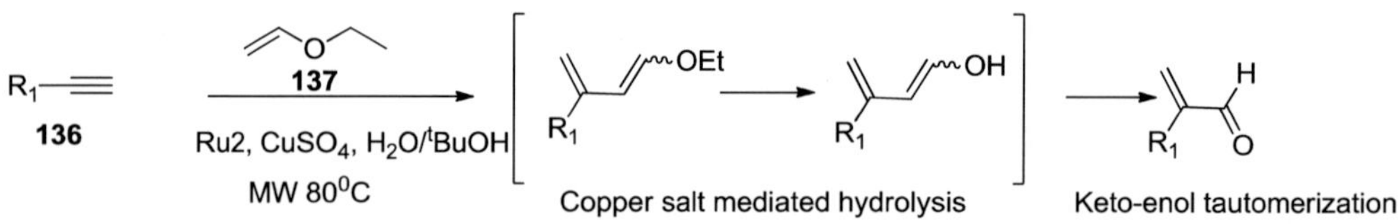

Scheme 40. EYCM approch for crotanaldehyde.

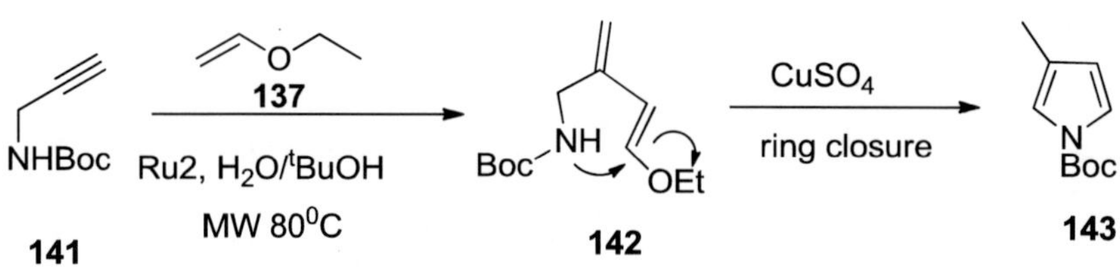

Scheme 41. EYCM approch for pyrrole.

Table 17.1. Optimization of EYCM reaction conditions

Solvent	Ru2 (mol%)	$CuSO_4$	T°C	Time	Yield
$H_2O/^tBuOH$	10	2	80°C	20 min	25%
Toluene	5	2	80°C	30 min	36%
Toluene	5	2	120°C	30 min	56%
Toluene	5	...	120°C	30 min	0
Toluene	5	1	120°C	30 min	18%
Toluene	10	2	120°C	30 min	55%

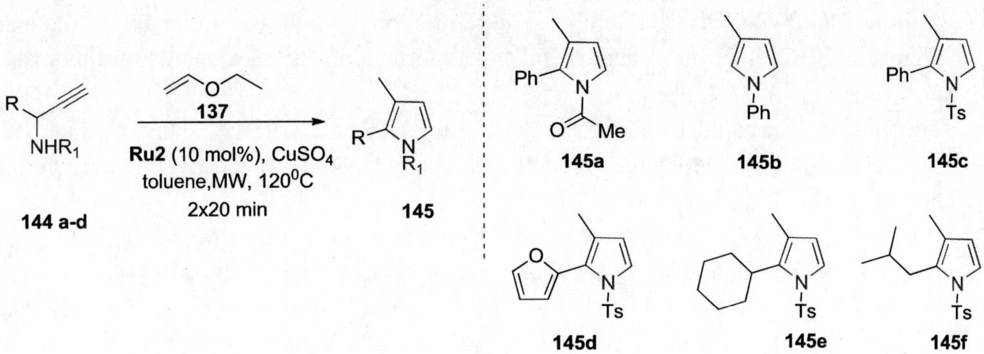

Scheme 42. EYCM approch for 2,3-substituted pyrrole.

CONCLUSION

Diverse novel pyrroles are prepared using different methods. These method have expanded the scope of pyrrole synthesis. It is our understanding that synthesis of many other medicinally active pyrroles is possible using the novel chemistry in this book chapter. Some of the methods as described herein are extremely new and conceptually bright. We believe that these methods of pyrrole synthesis will bring considerable values to the researchers.

REFERENCES

[1] Runge, F. F. (1834). "Ueber einige Produkte der Steinkohlendestillation" *Annalen der Physik und Chemie*, *31*, 65-78.

[2] O'Malley, D. P., Li, K., Maue, M., Zografos, A. L. & Baran, P. S. (2007). "Total Synthesis of Dimeric Pyrrole-Imidazole Alkaloids: Sceptrin, Ageliferin, Nagelamide E, Oxysceptrin, Nakamuric Acid, and the Axinellamine Carbon Skelton" *J. Am. Chem. Soc.*, *129*, 4762-4775.

[3] Kato, S., Shindo, K., Kawai, H., Odagawa, A., Matsuoka, M. & Mochizuki, J. (1993). "Pyrrolostatin, a novel lipid peroxidation inhibitor from Streptomyces chrestomyceticus. Taxonomy, fermentation, isolation, structure elucidation and biological properties" *J. Antibiot.*, *46*, 892-9.

[4] Boger, D. L., Boyce, C. W., Labroli, M. A., Sehon, C. A. & Jin, Q. (1999). "Total Syntheses of Ningalin A, Lamellarin O, Lukianol A, and Permethyl Storniamide A Utilizing Heterocyclic Azadiene Diels–Alder Reactions" *J. Am. Chem. Soc*, *121*, 54-62.

[5] Avendan˜o, C. & Mene´ndez, J. C. (2002). "Inhibitors of multidrug resistance to antitumor agents (MDR)" *Curr Med Chem*, *9*, 159-93.

[6] Avendan˜o, C. & Mene´ndez, J. C. (2004) "Recent Advances in Multidrug Resistance Modulators" *Med. Chem. Rev.–Online.*, *1*, 419.

[7] Raub, T. J. (2006). "P-Glycoprotein Recognition of Substrates and Circumvention through Rational Drug Design" *Mol. Pharm.*, *3*, 3.

[8] Varun, Bhardwaj., Divya, Gumber., Vikrant, Abbot., Saurabh, Dhiman. & Poonam, Sharma. (2015). "Pyrrole: a resourceful small molecule in key medicinal hetero-aromatics" *RSC Adva.*, *5*, 15233-15266.

[9] Teixeira, C., Barbault, F., Rebehmed, J., Liu, K., Xie, L., Lu, H., Jiang, S., Fan, B. & Maurel, F. (2008). "Molecular modeling studies of N-substituted pyrrole derivatives--potential HIV-1 gp41 inhibitors" *Bioorg Med Chem.*, *16*, 3039.

[10] *Cordrey, L. J. (1976).* "Tolmetin sodium, a new anti-arthritis drug: double-blind and long-term studies". *Journal of the American Geriatrics Society*, 24, 440–446.

[11] Cardoe, N. & Steele, C. E. (1976–1977). "A double-blind crossover comparison of tolmetin sodium and phenylbutazone in the treatment of rheumatoid arthritis". *Current medical research and opinion.*, *4*, 688–694.

[12] Liyanage, S. P. & Steele, C. E. (1977–1978). "Tolmetin in osteoarthrosis of the hip and knee: double-blind crossover trials". *Current medical research and opinion*, *5*, 299–305.

[13] Mallinson, Tom. (2017). "A review of ketorolac as a prehospital analgesic". *Journal of Paramedic Practice. London: MA Healthcare*, *9*, 522–526.

[14] Vallerand, April H. (2017). *Davis's Drug Guide for Nurses.* Philadelphia: F. A. Davis Company, p. 730. ISBN 9780803657052.

[15] Seth, C., Rasmussen, Sean. & Evenson, J. (2013). "Dithieno[3,2-b:2′,3′-d]pyrrole-based materials: Synthesis and application to organic electronics" *Progress in Polymer Science.*, *38*, 1773.

[16] Sean, J. Evenson., Matthew, J. Mumm., Konstantin, I. Pokhodnya. & Seth, C. Rasmussen. (2011). "Highly Fluorescent Dithieno[3,2-b:2′,3′-d]pyrrole-Based Materials: Synthesis, Characterization, and OLED Device Applications" *Macromolecules*, *44*, 835-841.

[17] Sudam, S. Pandule., Suresh, U. Shisodia., Rajendra, P. Pawar. & Vasant, V. Chabukswar. (2016). "Synthesis, Properties, and Ammonia Gas Sensing Applications of Poly-[1-(4-nitronaphthalen-1-yl)-2,5-di(thiophen-2-yl)-1H-pyrrole]" *Polymer-Plastics Technology and Engineering*, *56*, 268-275.

[18] Jonathan, M. Slater., Esther, J. Watt., Neville, J. Freeman., Iain, P. May. & Donald, J. Weir. (1992). "Gas and vapour detection with poly (pyrrole) gas sensors" *Analyst.*, ***117***, 1265-1270.

[19] Knorr, L. (1884). "Synthese von Furfuranderivaten aus dem Diacet bernsteinsäureester" [Synthesis of Furfuranderivaten from diacetic acid ester] *Berichte der deutschen chemischen Gesellschaft*, *17*, 2863.

[20] Paal, C. (1884). "Ueber die Derivate des Acetophenonacetessigesters und des Acetonylacetessigesters" [About the derivatives of the acetophenone acetoacetic ester and the Acetonyl acetoacetate], *Berichte der deutschen chemischen Gesellschaft*, *17*, 2756.

[21] Hantzsch, A. (1890). *Chem. Ber. 23*, 1474.

[22] Holman, R. W. (2005). "Strategic Applications of Named Reactions in Organic Synthesis: Background and Detailed Mechanisms (Kürti, László; Czakó, Barbara)", *Journal of Chemical Education*, *82*, (1780).

[23] Sreekumar, R. & Padmakumar, R. (1998). "Simple, efficient and convenient synthesis of pyrroles and pyrazoles using zeolites" *Synth. Commun*, *28*, 1661-1665.

[24] Chen, J., Wu, H., Zheng, Z., Jin, C., Zhang, X. & Su, W. (2006). "An approach to the Paal–Knorr pyrroles synthesis catalyzed by $Sc(OTf)_3$ under solvent-free conditions" *Tetrahedron Lett.*, *47*, 5383-5387.

[25] Chen, J. X., Liu, M. C., Yang, X. L., Ding, J. C. & Wu, H. Y. (2008). "Indium (III)-catalyzed synthesis of N-substituted pyrroles under solvent-free conditions". *J. Braz. Chem. Soc.*, *19*, 29-34.

[26] Yu, S. X., Le Quesne, P. W. & Quararibea. (1995). "Total synthesis of (±)-funebral, a rotationally restricted pyrrole alkaloid, using a novel Paal-Knorr reaction" *Tetrahedron Lett.*, *36*, 6205-6208.

[27] Ballini, R., Barboni, L., Bosica, G. & Petrini, M. (2000). "2,5-Dialkylfurans and nitroalkanes as source of 2,3,5- trialkylpyrroles". *Synlett.*, 2000, 391-393.

[28] Raghavan, S. & Anuradha, K. (2003). "Solid-phase synthesis of heterocycles from 1,4-diketone synthons" *Synlett.*, *5*, 711-713.

[29] Samajdar, S., Becker, F. F. & Banik, B. K. (2001). "Montmorillonite KSF-mediated facile synthesis of pyrroles" *Heterocycles.*, *55*, 1019-1022.

[30] Trost, B. M. & Doherty, G. A. (2000). "An Asymmetric Synthesis of the Tricyclic Core and a Formal Total Synthesis of Roseophilin via an Enyne Metathesis" *Journal of the American Chemical Society.*, *122*, 3801.

[31] Danks, T. N. (1999). "Microwave assisted synthesis of pyrroles" *Tetrahedron Lett.*, *40*, 3957.

[32] Debasish, Bandyopadhyay., Jessica, Cruz., Ram, N. Yadav. & Bimal, K. Banik. (2012). "An Expeditious Iodine-Catalyzed Synthesis of 3-Pyrrole-substituted 2-Azetidinones" *Molecules.*, *17*, 11570-11584.

[33] Ashwin, R. Bharadwaj. & Karl, A. Scheidt. (2004). "Catalytic Multicomponent Synthesis of Highly Substituted Pyrroles Utilizing a One-Pot Sila-Stetter/Paal–Knorr Strategy" *Org. Lett.*, *6*, 2465-2468.

[34] Estevez, V., Villacampa, M. & Menendez. (2012). "Three-component access to pyrroles promoted by the CAN–silver nitrate system under high-speed vibration milling conditions: a generalization of the Hantzsch pyrrole synthesis" *Chem. Commun.*, *49*, 591-593.

[35] Trautwein, A. W., Su¨ssmuth, R. D. & Jung, G. (1998). "Hantzsch pyrrole synthesis on solid support" *Bioorg. Med. Chem. Lett.*, *8*, 2381.

[36] Murthy, S. N., Madhav, B., Kumar, A. V., Rao, K. R. & Nageswar, Y. V. D. (2009). "Multicomponent Approach Towards the Synthesis of Substituted Pyrroles under Supramolecular Catalysis Using β-Cyclodextrin as a Catalyst in Water Under Neutral Conditions" *Helvetica.*, *92*, 2118.

[37] Halfen, Jason A. (2005). "Recent Advances in Metal-Mediated Carbon-Nitrogen Bond Formation Reactions: Aziridination and Amidation" *Current Organic Chemistry.*, *9*, 657.

[38] Shou, W. G., Li, J., Guo, T., Lin, Z. & Jia, G. (2009). "Ruthenium-Catalyzed Intramolecular Amination Reactions of Aryl- and Vinylazides". *Organometallics*, *28*, 6847-6854.

[39] Kou, Hiroya., Shigemitsu, Matsumoto., Masayasu, Ashikawa., Kentaro, Ogiwara. & Takao, Sakamoto. (2006). "Cyclization Reactions of Homopropargyl Azide Derivatives Catalyzed by PtCl4 in Ethanol Solution: Synthesis of Functionalized Pyrrole Derivatives" *Org. Lett.*, *8*, 5349.

[40] Xiangwei, Du., Xin, Xie. & Yuanhong, Liu. (2010). "Gold-Catalyzed Cyclization of Alkynylaziridines as an Efficient Approach toward Functionalized N-Phth Pyrroles" *J. Org. Chem.*, *75*, 510.

[41] Feng, Chen., Tao, Shen., Yuxin, Cui. & Ning, Jiao. (2012). "2, 4- vs 3, 4-Disubsituted Pyrrole Synthesis Switched by Copper and Nickel Catalysts" *Org. Lett.*, *14*, 4926-4929.

[42] Rajiv, Dhawan. & Bruce, A. Arndtsen. (204). "Palladium-Catalyzed Multicomponent Coupling of Alkynes, Imines, and Acid Chlorides: A Direct and Modular Approach to Pyrrole Synthesis" *J. Am. Chem. Soc.*, *126*, 468-469.

[43] Souvik, Rakshit., Frederic, W. Patureau. & Frank, Glorius. (2010). "Pyrrole Synthesis via Allylic sp3 C−H Activation of Enamines Followed by Intermolecular Coupling with Unactivated Alkynes" *J. Am. Chem. Soc.*, *132*, 9585-9587.

[44] Meng, Gao., Wenting, Zhao., Hongyi, Zhao., Ziyun, Lin., Dongfeng, Zhang. & Haihong, Huang. (2018). "An efficient and facile access to highly functionalized pyrrole derivatives" *Beilstein J. Org. Chem.*, *14*, 884.

[45] Gribble, G. W., Pelkey, E. T., Simon, W. M. & H. A. (2000). "Regioselective 1,3-Dipolar Cycloaddition Reactions of Unsymmetrical Münchnones (1,3-Oxazolium-5-olates) with 2- and 3-Nitroindoles. A New Synthesis of Pyrrolo[3,4-b]indoles" *Tetrahedron.*, *52*, 10133.

[46] Yongju, Kim., Jonghoon, Kim. & Seung, Bum Park. (2009). "Regioselective Synthesis of Tetrasubstituted Pyrroles by 1,3-Dipolar Cycloaddition and Spontaneous Decarboxylation" *Org. Lett.*, *11*, 17-20.

[47] Yan, Liu., Huayou, Hu., Xiang, Wang., Sanjun, Zhi., Yuhe, Kan. & Chao, Wang. (2017). "Synthesis of Pyrrole via a Silver-Catalyzed 1,3-Dipolar Cycloaddition/ Oxidative Dehydrogenative Aromatization Tandem Reaction" *J. Org. Chem.*, *82*, 4194-4202.

[48] Daniel, J. St. Cyr. & Bruce, A. Arndtsen. (2007). "A New Use of Wittig-Type Reagents as 1,3-Dipolar Cycloaddition Precursors and in Pyrrole Synthesis" *J. Am. Chem. Soc.*, *129*, 12366-12367.

[49] Heng-Yen, Wang., Daniel, S. Mueller., Rachna, M. Sachwani., Hannah, N. Londino. & Laura, L. Anderson. (2010). "Carbon−Carbon Bond Formation and Pyrrole Synthesis via the [3,3] Sigmatropic Rearrangement of O-Vinyl Oxime Ethers" *Org. Lett.*, *12*, 2290-2293.

[50] Simbarashe, Ngwerume. & Jason, E. Camp. (2011). "Gold-catalysed rearrangement of O-vinyl oximes for the synthesis of highly substituted pyrroles" *Chem. Commun.*, *47*, 1857.

[51] Yaojia, Jiang., Wei, Chuen Chan. & Cheol-Min, Park. (2012). "Expedient Synthesis of Highly Substituted Pyrroles via Tandem Rearrangement of α-Diazo Oxime Ethers" *J. Am. Chem. Soc.*, *134*, 4104-4107.

[52] Gautier, A. (1868 and 1867). Ann. Chem. Pharm. "Isocyanides as new isomers of cyanides are mentioned and announced previously (but not described) in: A. Gautier, *Ann. Chem. Pharm.*, *146*, 119−124 and 142, 289−294.

[53] Suginome, M. & Ito, Y. (2004). "*In Science of Synthesis*', Vol. *19* (Ed.: S.-I. Murahashi), Thieme, Stuttgart, pp. 445–530, and references cited therein.

[54] Derek, H. R. Barton. & Samir, Z. Zard. (1985). "A new synthesis of pyrroles from nitroalkenes" *J. Chem. Soc., Chem. Commun.*, *0*, 1098-1100.

[55] Fumoto, Y., Eguchi, T., Uno, H. & Ono, N. (1999). "Synthesis of pyrrolostatin and its analogues" *J. Org. Chem.*, *64*, 6518-6521.

[56] Jiansheng, Tang. & John, G. Verkade. (1994). "Nonionic Superbase-Promoted Synthesis of Oxazoles and Pyrroles: Facile Synthesis of Porphyrins and .alpha.-C-Acyl Amino Acid Esters" *J. Org. Chem.*, *59*, 7793–7802.

[57] van Leusen, A. M., Boerma, G. J. M., Helmholdt, R. B., Siderius, H. & Strating, J. (1972). "Chemistry of sulfonylmethylisocyanides. Simple synthetic approaches to a new versatile chemical building block" *Tetrahedron Lett.*, *23*, 2367-2368.

[58] For reviews, see: a) van Leusen, A. M. & van Leusen, D. (1995). "In Encyclopedia for Organic Synthesis; L. A. Paquette Ed.; Wiley: New York, *7*, 4973–4979; b) van Leusen, A. M. (1980). "*Lect. Heterocycl. Chem*," *5*, 111–122.

[59] Xueming, Zhang., Xianxiu, Xu. & Dawei, Zhang. (2017). "[3+2] Cycloaddition of Tosylmethyl Isocyanide with Styrylisoxazoles: Facile Access to Polysubstituted 3-(Isoxazol-5-yl)pyrroles" *Molecules.*, *22*, 1131.

[60] Shin, Kamijo., Chikashi, Kanazawa. & Yoshinori, Yamamoto. (2005). "Copper- or Phosphine-Catalyzed Reaction of Alkynes with Isocyanides. Regioselective Synthesis of Substituted Pyrroles Controlled by the Catalyst" *J. Am. Chem. Soc.*, *127*, 9260-9266.

[61] Hikaru, Takaya., Sachiko, Kojima. & Shun-Ichi, Murahashi. (2001). "Rhodium Complex-Catalyzed Reaction of Isonitriles with Carbonyl Compounds: Catalytic Synthesis of Pyrroles" *Org. Lett.*, *3*, 421-424.

[62] Nicolai, Dieltiens., Christian, V. Stevens., David, De Vos Bart Allaert., Renata, Drozdzak. & Francis, Verpoort. (2004). "Pyrrole synthesis using a tandem Grubbs carbene–$RuCl_3$ catalytic system" *Tet. Lett.*, *45*, 8995-8998.

[63] Daniele, Castagnolo., Lorenzo, Botta. & Maurizio, Botta. (2009). "Alkyne-Enol Ether Cross-Metathesis in the Presence of CuSO4: Direct Formation of 3-Substituted Crotonaldehydes in Aqueous Medium" *J. Org. Chem.*, *74*, 3172-3174.

[64] Chachignon, H; Nicolò, Scalacci., Elena, Petricci. & Daniele, Castagnolo. (2015). "Synthesis of 1,2,3-Substituted Pyrroles from Propargylamines via a One-Pot Tandem Enyne Cross Metathesis–Cyclization Reaction" *J. Org. Chem.*, *80*, 5287-5295.

In: Organic and Medicinal Chemistry, Volume 1
Editor: Bimal Krishna Banik
ISBN: 978-1-53614-453-6

Chapter 8

POLYAROMATIC COMPOUNDS AS NOVEL ANTICANCER AGENTS

***Indrani Banik*[1], *Ram Naresh Yadav*[2,3], *Frederick F. Becker*[1] *and Bimal Krishna Banik*[1,2,*]**

[1]The University of Texas, M. D. Anderson Cancer Centre, Department of Molecular Pathology Box-89, Houston, Texas, US

[2]Department of Chemistry, The University of Texas-Pan American, Edinburg, TX, US

[3]Department of Chemistry, Faculty of Engineering & Technology, Veer Bahadur Singh Purvanchal University, Jaunpur-222003, Uttar Pradesh, India

ABSTRACT

Synthesis of polyaromatic hydrocarbons by various methods is studied. The metabolic activation of these types of hydrocarbons is investigated. However, methods to prepare new anticancer compounds from polyaromatic hydrocarbons are not investigated. We report here synthesis and biological evaluation of a few derivatives derived from polyaromatic molecules. Several interesting results are obtained and therefore, the use of these agents for further anticancer research is identified.

Keywords: polyaromatic amine, anticancer, chrysene, dibenzofluorene, glycosylation

1. INTRODUCTION

Synthetic and biochemical research on polyaromatic hydrocarbons has received excellent recognitions from scientific community [1]. But there has been poor attention on the use of the derivatives of these molecules as anticancer agents. Bair et al. had identified a correlation

* Corresponding Author Email: bimalbanik10@gmail.com and bimal.banik@chsst.org; Community Health System of South Texas, 3135 S. Sugar Road, Edinburg, TX.

between the effectiveness of a few of these derivatives against cancer cell lines [2]. As a result of these efforts, benzylic aminopropanediols were investigated. Some of these compounds were believed to be active against numerous cancer cell lines. These compounds were also found to interact with DNA by intercalation and showed topoisomerase inhibitory activities. The intercalative properties was confirmed for the potent napthalimides, amonafide and mitonafide [3-6].

Studies by Denny et al. raised doubts regarding the validity of the target by poly aromatic-based antitumor compounds. These crucial studies prompted our group to offer excellent possibilities to expand research on poly aromatic compounds to demonstrate their effective synthesis and selective anticancer activities *in vitro* and *in vivo* [7-12].

2. Results and Discussions

2.1. Substituted Chrysene

A limited number of studies have been conducted with polyaromatic compounds as anticancer agents. Our group has continued to synthesize polyaromatic compounds and studied their biological activity as novel anticancer agents. Notably, we do not find other similar studies in the literature. This is due to the hypothesis and belief that polyaromatic compounds may not create anticancer agents due to their mutagenic activities. However, mutagenic activations depend on a number of factors. The mutagenic character of benzene is investigated. However, many benzene derived organic compounds are extremely useful drugs. Therefore, in science the best conclusions can be obtained from performing the actual work. On this basis, we report here our investigations on the synthesis and biological studies that occurred exclusively in our research laboratory. This paper may encourage other scientists to think more carefully about polyaromatic derivatives from different perspectives and evaluate these derivatives for different purposes.

In previous investigations, a series of mono substituted chrysene derivatives were synthesized and some of them demonstrated specific antitumor activity. These studies suggested possible interactions at the tumour cell membrane as an alternate target for anticancer activity. Preliminary structure-activity study indicated that nature and length of the appended chain is crucial as well as the terminal group [13].

The results obtained from this investigation indicated that the anticancer properties of these molecules are also dependent on the location of a side chain in the polyaromatic ring compounds. This was confirmed by the fact that the 2-substituted chrysene molecules have superior anticancer activity than the 6-substituted compounds within a similar group of compounds. In this paper, we mainly focus on compounds as shown in Figure 1.

These are amines with a polyaromatic ring and a side chain. The side chain can be inserted into the specific position of polyaromatic rings using varieties of known and unknown reactions. Electrophillic reactions were conducted on these hydrocarbons. To obtain functionalized derivatives, a new nitration reaction was attempted. This study culminated in a facile aromatic nitration reaction of aromatic compounds by bismuth nitrate impregnated with solid support (**2** to **3**) (Scheme 1) [14].

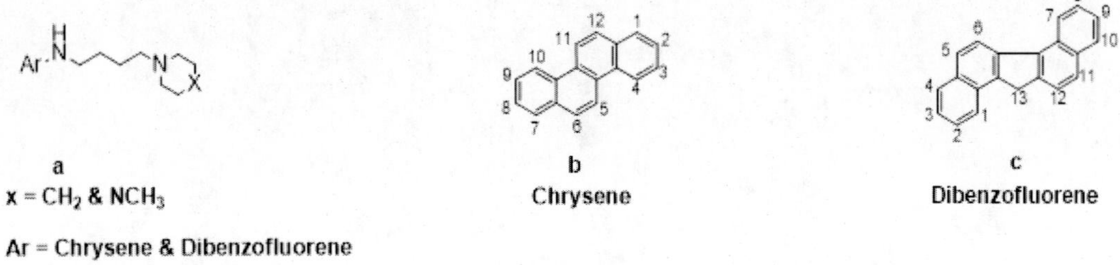

Figure 1 (a-c)

Figure 1.

New methods for the reduction of nitro compounds present in the aromatic groups (**3** to **4**) by samarium and indium metal were also developed by us (Scheme 1) [15, 16]. The amine **4** was then condensed with acid **5** to produce the diamide **6**. Amide **6** on reduction afforded corresponding amine **8a** and **8b** (Scheme 1). The acid **5** was synthesized through a reaction of succinic anhydride with corresponding heterocyclic bases (piperidine and N-methyl piperazine) in organic solvent. The amide **7b** that has N-methyl piperazine side chain was considerably more active than that of piperidine side chain **7a** against every cancer cell lines tested. A 4-carbon chain was found to be superior to a **3** or **5**-carbon chain as the spacer that connects the polyaromatic ring and the heterocyclic ring.

The amines **8a** and **8b** are much more active than the amides **7a** and **7b** regardless of the terminal units present in the system. Considering the activity of these series of compounds, our next attempts were to prepare other analogues for biological evaluation.

6-Aminochrysne (**9**) was converted to amide **7a** and **7b**. On nitration with nitric acid-sulphuric acid mixture or bismuth nitrate-impregnated with clay, the amides **7a** and **7b** produced 12-nitro-6-amido derivative **10**. The nitro group in **10** was reduced to amine **11** which was diazotized and methylated to **12**.

In another route, **9** was acetylated to acetamide **13** and it was nitrated to 12-nitroacetamido derivative **14**. The nitro amide was the hydrolyzed followed by the coupling reaction with acid 5 to afford 12-nitro derivative **10a-b**. The 12-nitro-6-acylamide compound **10a-b** was reduced to the corresponding amino derivative by hydrazine hydrate and 10% Pd/C to 12-amino-6-acylchrysne **11a-b**. The amino group was then diazotized and converted to methoxy derivative **12a-b** (Scheme 2).

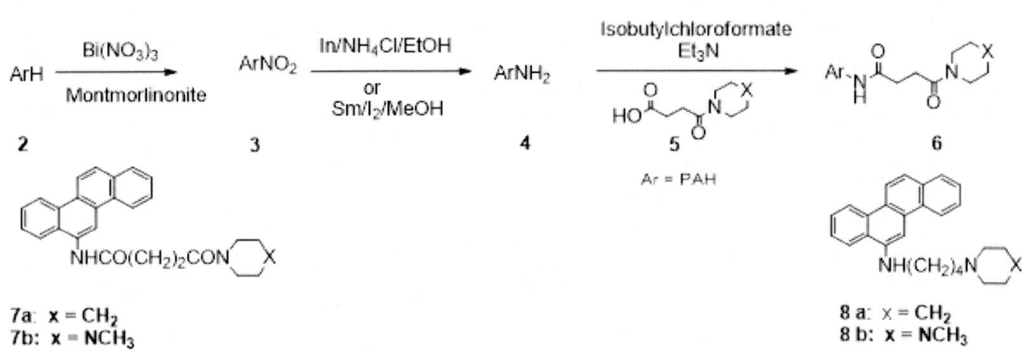

Scheme 1.

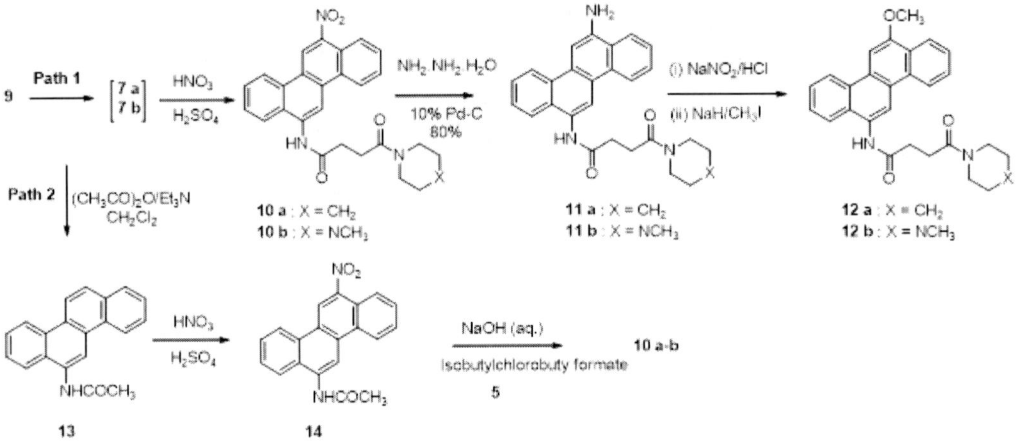

Scheme 2.

The N-acetamide derivative **13** was converted to metyl ketone **15** by Fridel-Crafts acylation reaction. The methyl ketone derive **15** was hydrolyzed to amino ketone **16** in a good yield. A coupling reaction then produced the corresponding compounds **17a-b** (Scheme 3). The amide **7a-b** on bromination produced the bromo derivative **18a-b**.

The 12-bromo-6-acylamidochrysene **18a-b** was obtained in excellent yield. The bromo amide **18** obtained from 7 on hydrolysis with strong alkali produced the **12**-bromo-**6**-aminochrysene **18** which was coupled with the side chains to afford identical bromodiamide **17** as described earlier to produce the amides (Scheme 4).

In vitro cytoxicity of the bromo, acyl, nitro and methoxy substituted chrysenes with the side chains was performed against numerous cancer cell lines. The results were interesting since substituents at the aromatic group at the 12-position of the chrysene have influence on the anticancer activities. Electron withdrawing groups (bromo, nitro and acetyl) decreased the activity against all cancer cell lines tested compared to the 12-unsubstituted isomers. However, electron donating methoxy group at 12-position increased the activity compared to the 12-unsubstituted compound indicates that the basicity of these products may have some influence on their anticancer activities.

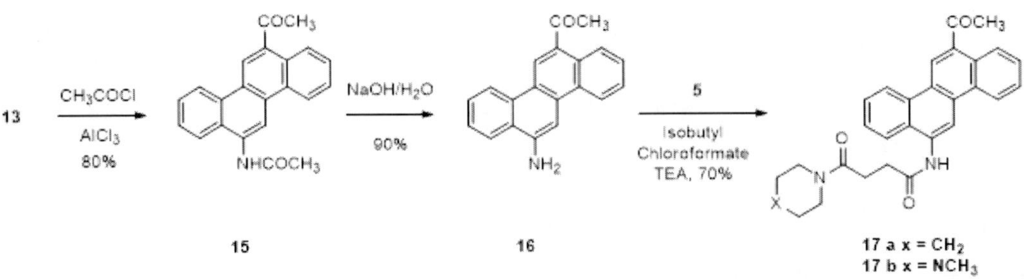

Scheme 3.

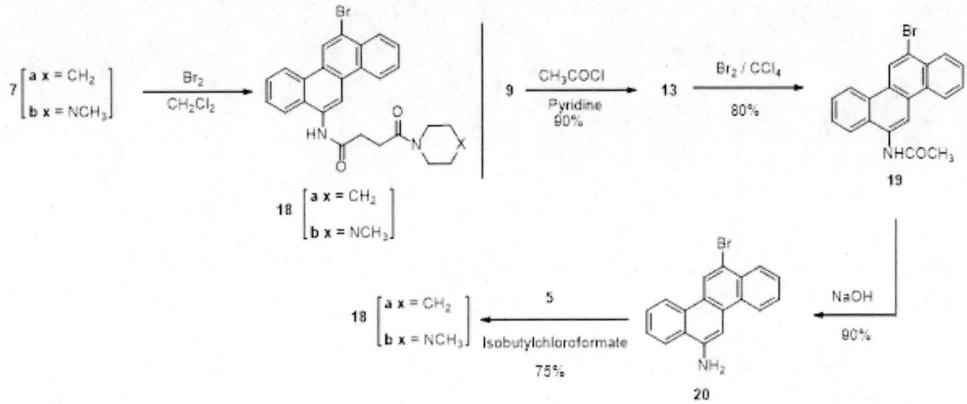

Scheme 4.

Examination of the structures of the compounds indicated a general trend of *in vitro* cytotoxicity. The more basic amido compounds had superior anticancer activity while the di amino compounds in a specific series had similar activities. These results indicated that a polar and basic group at the terminal side chain may maintain or improve the activities of the derivatives. Therefore, we decided to prepare some of the compounds that have a polar or basic side chain at the terminal site of these types of molecules.

Piperazine-2-ethanol **21** was protected as the silyl derivative **22** and reacted with succinic anhydride to produce the acid **23**. The acid **23** on coupling reaction with 6-aminochrysene **9** afforded the desired product **24**. Upon deprotection of the silyl group, the alcohol **25** was prepared (Scheme 5).

The benzyl protected piperazine **26** was condensed with succinic anhydride and the resulting acid **27** was condensed with 6-aminochrysene **9** to produce the diamide **28**. The benzyloxy protecting group was then removed by hydrogenolysis to obtain a secondary amino compound **29**. This amino compound **29** was alkylated with ethyl bromoacetate to produce the alkylated product **30** (Scheme 6).

The *in vitro* cytotoxicity of these ethers **23**, alcohol **24** and amino compounds was evaluated against numerous cancer cell lines. It was notable to observe better activity of the compounds that have alcohol and amino groups at the terminal site of these derivatives. But, the ether derivative was totally inactive. These experimental results suggested that the nature of the terminal site of these molecules is important. The basic terminal sites had interaction with negative component of the phospho lipid bilayer in the cell membrane. The terminal polar group helped to bind the molecule to the cell surface and the amino group present in the molecule was good enough for this binding and therefore, to enhance the *in vitro* cytotoxicity.

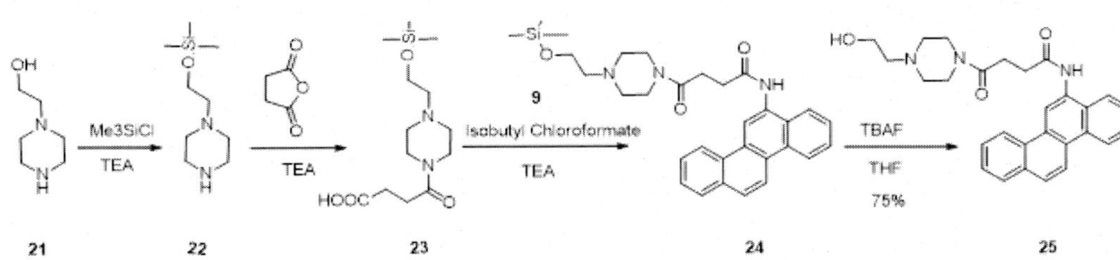

Scheme 5.

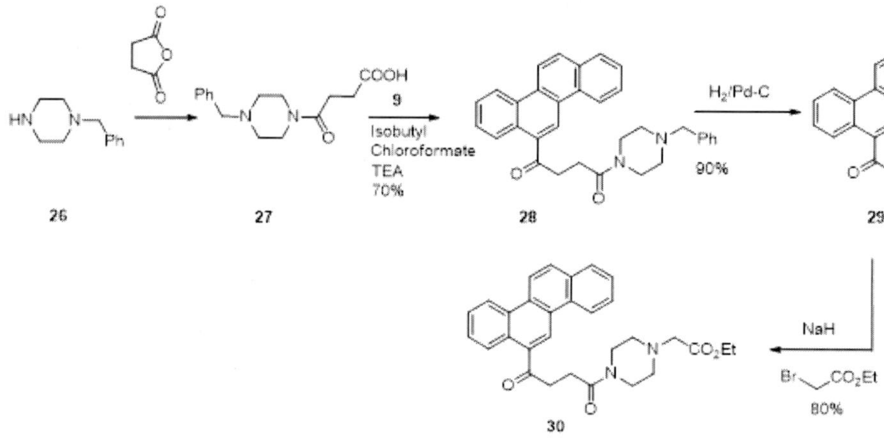

Scheme 6.

2.2. Disubstituted Chrysenes

To extend the previous studies of these biologically active aromatic compounds, synthesis of numerous 6, 12-disubstitued chrysenes were performed. In addition, these molecules were also tested against numerous human and animal tumor cell lines. Specifically, three amide di substituted derivatives were prepared through a sequence of reactions (Scheme 7). Synthesis of symmetrical amide (same terminal heterocyclic rings on the aliphatic side chain) was prepared from 6-aminochrysene.

6-Aminochrysnene (**9**) on standard acetylation reaction produced 6-acetamidochrysene **13**. 6-Acetamidochrysene **13** on nitration with concentrated nitric acid produced the 12-nitro derivative **14**. Bismuth nitrate-impregnated with clay also produced the same nitro compound **14**. The nitro amide **14** on reduction and hydrolysis reaction afforded the 6, 12-diaminochrysene **31**. Since we prepared a number of derivatives with 6-aminochrysene, availability of the new 6, 12-diamino compound of chrysene is helpful for huge possibilities. A coupling reaction with the side chain afforded the amide **32a-b** which was then reduced to produce the amine **33a-b**. The amine was converted to the hydrochloride salt **34a-b**. It was important that coupling reaction with the acid hindered no problems with 6, 12-diamnochrysene and proceeded smoothly (Scheme 7).

This method although worked well, however, it has a number of steps. Therefore, it was an important goal to simplify the synthesis. After a few experimentations, a simple synthetic route was also developed. For example, it was found that chrysene on nitration with nitric acid produced 6, 12-dinitrochrysene **35**. The dinitro compound **35** was reduced by hydrazine hydrate/Pd-C (catalytic transfer hydrogenation method) to obtain 6, 12-diaminochrysene (**31**).

The 6, 12-diaminochrysene **31** was coupled with the acid side chain in the presence of isobutyl chloroformate to produce the amide **32** in excellent yield (**Scheme 7**). After the preparation of 6, 12-disubstituted symmetrical chrysene amide **32**, it was necessary to synthesize 6, 12-disubstituted asymmetrical chrysene amides [**14**].

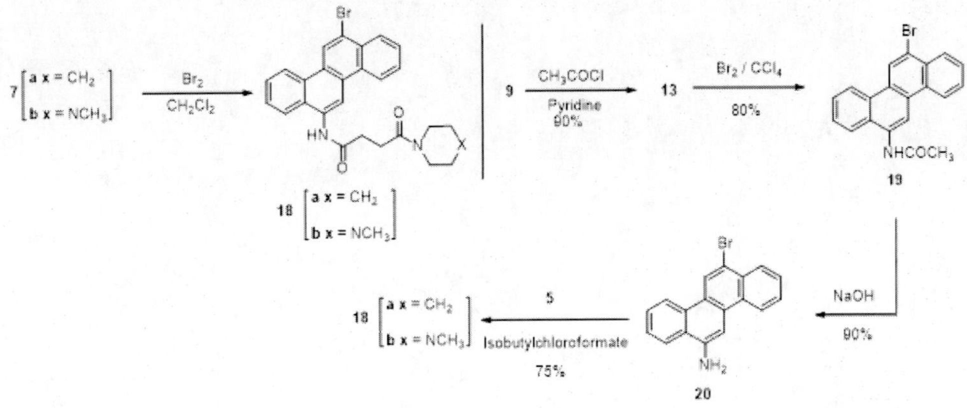

Scheme 4.

Examination of the structures of the compounds indicated a general trend of *in vitro* cytotoxicity. The more basic amido compounds had superior anticancer activity while the di amino compounds in a specific series had similar activities. These results indicated that a polar and basic group at the terminal side chain may maintain or improve the activities of the derivatives. Therefore, we decided to prepare some of the compounds that have a polar or basic side chain at the terminal site of these types of molecules.

Piperazine-2-ethanol **21** was protected as the silyl derivative **22** and reacted with succinic anhydride to produce the acid **23**. The acid **23** on coupling reaction with 6-aminochrysene **9** afforded the desired product **24**. Upon deprotection of the silyl group, the alcohol **25** was prepared (Scheme 5).

The benzyl protected piperazine **26** was condensed with succinic anhydride and the resulting acid **27** was condensed with 6-aminochrysene **9** to produce the diamide **28**. The benzyloxy protecting group was then removed by hydrogenolysis to obtain a secondary amino compound **29**. This amino compound **29** was alkylated with ethyl bromoacetate to produce the alkylated product **30** (Scheme 6).

The *in vitro* cytotoxicity of these ethers **23**, alcohol **24** and amino compounds was evaluated against numerous cancer cell lines. It was notable to observe better activity of the compounds that have alcohol and amino groups at the terminal site of these derivatives. But, the ether derivative was totally inactive. These experimental results suggested that the nature of the terminal site of these molecules is important. The basic terminal sites had interaction with negative component of the phospho lipid bilayer in the cell membrane. The terminal polar group helped to bind the molecule to the cell surface and the amino group present in the molecule was good enough for this binding and therefore, to enhance the *in vitro* cytotoxicity.

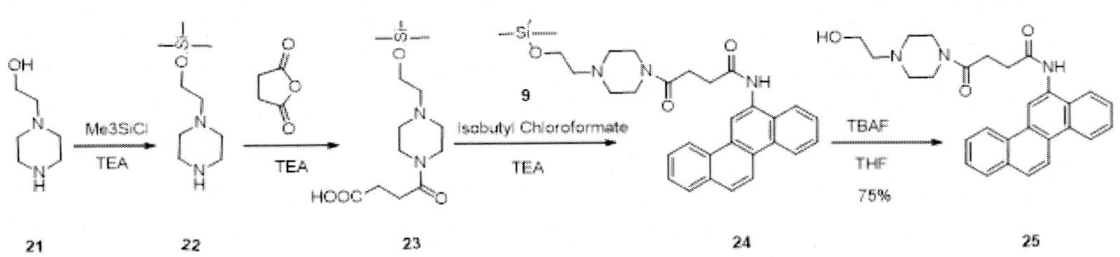

Scheme 5.

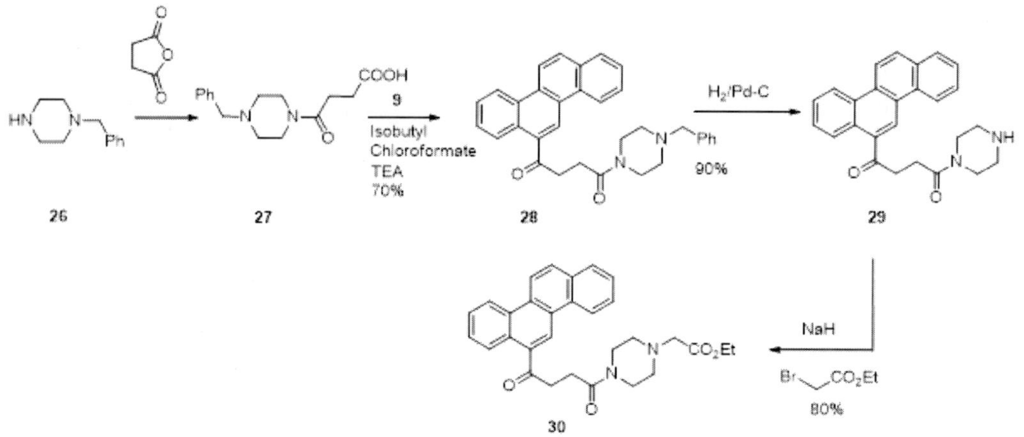

Scheme 6.

2.2. Disubstituted Chrysenes

To extend the previous studies of these biologically active aromatic compounds, synthesis of numerous 6, 12-disubstitued chrysenes were performed. In addition, these molecules were also tested against numerous human and animal tumor cell lines. Specifically, three amide di substituted derivatives were prepared through a sequence of reactions (Scheme 7). Synthesis of symmetrical amide (same terminal heterocyclic rings on the aliphatic side chain) was prepared from 6-aminochrysene.

6-Aminochrysnene (**9**) on standard acetylation reaction produced 6-acetamidochrysene **13**. 6-Acetamidochrysene **13** on nitration with concentrated nitric acid produced the 12-nitro derivative **14**. Bismuth nitrate-impregnated with clay also produced the same nitro compound **14**. The nitro amide **14** on reduction and hydrolysis reaction afforded the 6, 12-diaminochrysene **31**. Since we prepared a number of derivatives with 6-aminochrysene, availability of the new 6, 12-diamino compound of chrysene is helpful for huge possibilities. A coupling reaction with the side chain afforded the amide **32a-b** which was then reduced to produce the amine **33a-b**. The amine was converted to the hydrochloride salt **34a-b**. It was important that coupling reaction with the acid hindered no problems with 6, 12-diamnochrysene and proceeded smoothly (Scheme 7).

This method although worked well, however, it has a number of steps. Therefore, it was an important goal to simplify the synthesis. After a few experimentations, a simple synthetic route was also developed. For example, it was found that chrysene on nitration with nitric acid produced 6, 12-dinitrochrysene **35**. The dinitro compound **35** was reduced by hydrazine hydrate/Pd-C (catalytic transfer hydrogenation method) to obtain 6, 12-diaminochrysene (**31**).

The 6, 12-diaminochrysene **31** was coupled with the acid side chain in the presence of isobutyl chloroformate to produce the amide **32** in excellent yield (**Scheme 7**). After the preparation of 6, 12-disubstituted symmetrical chrysene amide **32**, it was necessary to synthesize 6, 12-disubstituted asymmetrical chrysene amides [**14**].

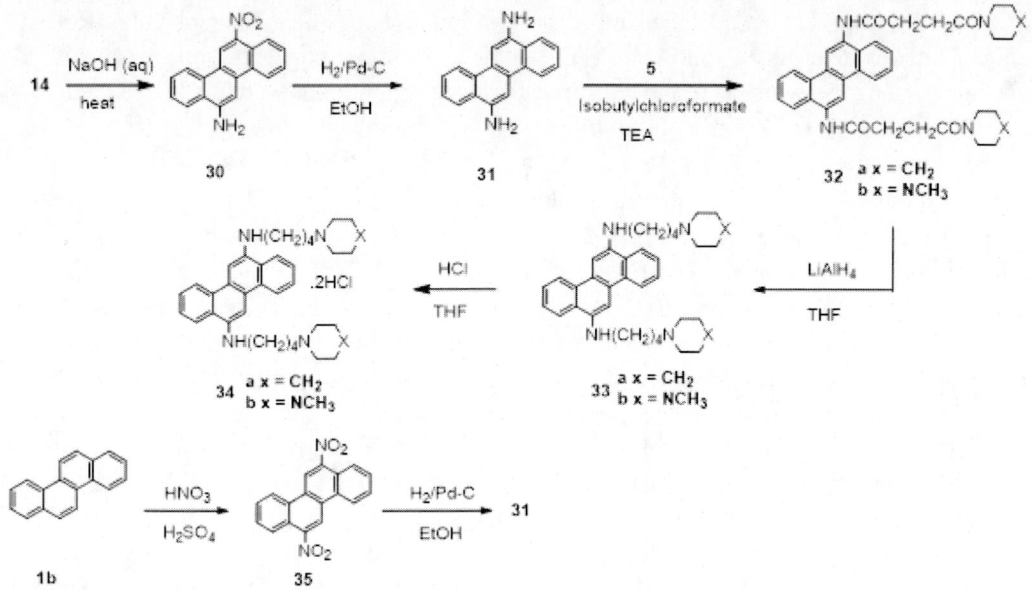

Scheme 7.

Asymmetric chrysene were not chiral products. Asymmetric structure meant by the presence of different terminal heterocyclic rings. It was obvious that different sequences are necessary to prepare 6, 12-disubstituted asymmetrical chrysene amides. 6-Aminochrysene **9** on acetylation reaction with acetic anhydride and subsequent nitration under standard conditions produced 12-nitro-6-acetamidochrysene **14**. Nitroamide **14** on hydrolysis of the ester group afforded 12-nitro-6-aminochrysene **36** in excellent yield. The side chain acid **5a** was then condensed with 12-nitro-6-aminochrysene **36** to produce the desired product **37**. The nitro group in **36** was then reduced by catalytic transfer hydrogenation method to the amino derivative **37**. This amino compound was well suited for another coupling reaction with a different side chain. Condensation of the amino compound **37** with acid **5b** produced the amide **38** following our earlier method. The amide bond in **38** was reduced to obtain non-symmetrical amine **39** in excellent yield (Scheme 8).

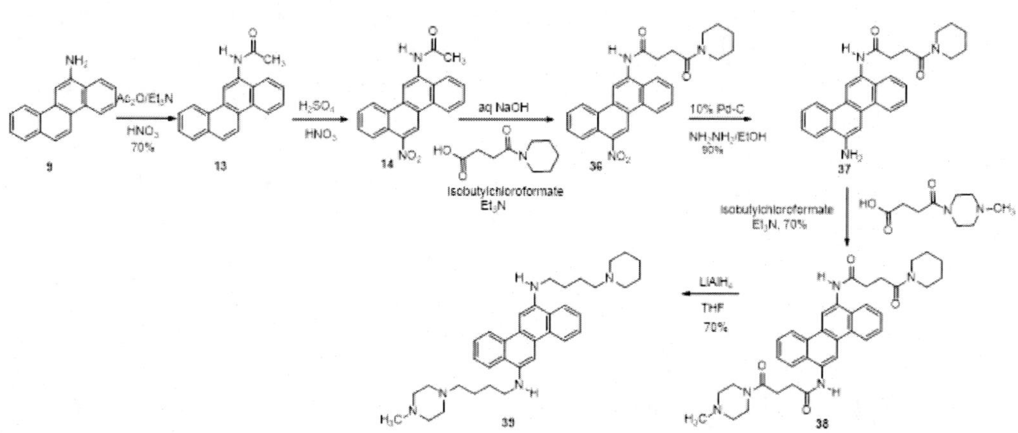

Scheme 8.

In vitro assays were conducted using 72h of continuous exposure to the molecules. Relative cell growth was determined following MTT method. The final concentration of solvent was kept below <0.625%, which is found non-toxic to the cells. Dilutions were made in RPMI 1640 with 10% FBS. Cancer Cell growth inhibition data was expressed in IC_{50} [15].

To demonstrate activity or non-activity and for a comparison, the following terms were used. Compounds that demonstrated IC_{50} values above 10.0 μM was described as inactive; 5.0 μM < 10 μM as limited activity; 2.5 μM < 5.0 μM as modest activity, and below 2.5 μM as active. However, some authors described a compound is active even the cell growth inhibition data is determined to be around 10.0 μM. The controls for the MTT assay were maintained including in every run two known anticancer compounds including a well-known drug cisplatin or adriamycin. To obtain consistent and reliable results, in the majority of the experiments, an inactive polyaromatic molecule was used as a negative control, but the experiment was not counted valid unless at least one new synthesized molecule expressed positive and another negative inhibition data. This consistency experiment of the *in vitro* test was performed repeatedly comparing all molecules (particularly the promising groups) over 20 months. There was minor changes of the inhibition data in the result of a given compound and cisplatin against a single tumor line. But it was remarkable to note that in the vast majority of the examples, the active or inactive compound's status remained within its original activity classification as described above. The activities of the mono substituted chrysenes were identical with previously reported observations.

Interestingly **21a**, the amide with a terminal piperidine moiety of chrysene was inactive against all cancer cell lines, a similar molecule **21b** with a terminal piperazine group demonstrated limited activity against 4/7 cancer cell lines. The molecule **22a**, the amino derivative of chrysene with a terminal piperidine group demonstrated modest activity against three/seven and was active against four/seven cancer cell lines. The considerable changes of activity were found for all related amino analogues in this series regardless of the terminal moiety.

The *in vitro* cytotoxicity data of 6, 12-disubstituted amide derivatives demonstrated interesting variations between symmetrical and asymmetrical molecules. Surprisingly, the symmetrical *bis* molecules with terminal piperidine group **12a** and with terminal piperazine ring **12b** were all inactive against the tested cancer cell lines. But, the molecule that had non-symmetrical piperidine/N-methylpiperazine terminal ring **19** demonstrated limited activity against three/seven cancer cell lines. Two of these cancer cell lines were identical to those affected by **21b**, MCF-7 (human breast) and HL-60 (human leukemia). Compound **13a** was active against all seven tumor cells and similarly **13b** was active against six/seven cancer cell lines. The hydrochloride salt of **13b** and **14b** remained active against all cancer cell lines. Importantly, the hydrochloride salts were soluble in water despite the presence of a large hydrophobic aromatic ring present in these molecules.

Some results of amide compounds were believed to mention. It was not surprising that the amide variant of terminal piperidine in *bis* compounds was as inactive. However, the lack of any reasonable activity by the *bis* compounds with N-methyl piperizine was unexpected because it was observed that those bearing a piperazine ring as the terminal moiety demonstrated activity against a number of tumor cell lines. These surprising results of the *bis* compounds with symmetrical structures were unexpected and no explanation was advanced to support these experimental results. This conclusion again supported by the fact that the molecule which possessed non-symmetrical terminal piperidine and piperazine terminal groups demonstrated activity similar to the compound that had only one side chain with N-methyl piperazine. It was extremely difficult to explain the results. However, it was proposed that in this example the piperidine group is in a null mode and that the *bis* molecule behaved like a molecule that has only one side chain with the active component. This conclusion was supported by the hypothesis that the terminal piperidine group containing molecules bound to chrysene have no activity in association with amide chain compositions regardless of their locations.

The *bis* amino compound, the amino variant of the amino compound with one side chain demonstrated similar results. It was found that conversion of the amide to amino compounds was helpful to increase the anticancer activity of the resulting molecules irrespective of the linker and the nature of the terminal groups. An excellent activity of the amino compounds was observed against almost every cancer cell lines tested, regardless of its structural alterations. To support this claim, the data showed that each of the amino compounds **20**, **13a**, **13b** and **14b**, the hydrochloride salt of **13b** behaved similarly and they were active in majority of the cancer cell lines. Probably, the basic nature of the amino compounds was responsible for their effects. However, it was not certain to explain the properties of these molecules that cause these significant differences. The exact mechanism of action of these new molecules was not certain.

Because of the excellent *in vitro* activity of **14b**, it was tested *in vivo* against human ovarian cancer line, SKOV-3. The compound was dissolved in PBS sand the solution was adjusted to pH 7.4 prior to injection. The cancer cells were harvested at 80% confluence and viability was measured by trypan blue exclusion. Cancer cells were used for this experiment if viability was found to be approximately 90%. Six to ten weeks old nu/nu female mice were used for this test. The cancer cells were washed in PBS, re-suspended in PBS, injected intra-peritoneally at 0.65 x 10^6 per mouse and all mice were sacrificed after a standard period of time (at seven weeks after injection). Administration of **14b** was started 72 hours after the injection of the tumor at 40 mg/k i.p. The regimen was followed for two weeks. Interestingly, a weight loss of approximately between 5% and 7% was calculated in the treated mice relative to the untreated mice. Moreover, complete restoration of the weights of the treated mice was observed to those of control mice at the end of the approximately third week. It was also crucial to note that in untreated mouse multiple nodules were found throughout the abdominal cavity and these varied in diameter generally from 1 to 10 mm. It was gratifying to observe that administration of **14b** reduced the tumor number per tumor bearing mouse by 35% and the tumor volume was reduced by 47%. This number against SKOV-3 tumor line was considered to be fascinating because of the notorious nature of this tumor. In addition, the salt of these amino compounds also demonstrated activity *in vivo* against the human ovarian cancer cell line SKOV-3.

The *in vivo* data of this compound was certainly promising but not curative in the limited regimen used. However, this tumour was resistant to numerous known antitumor clinical drugs. Therefore, based upon the *in vivo* activity, studies of this compound with various regimen and conditions were warranted.

It was remarkable to state this compound also demonstrated reasonable activity *in vivo* against colon cancer cell lines. Tumor growth was evident in the majority of mice at 11 days with a rapid increase in average tumour volume through the 25th day. After injection of 1.25 x10^6 cells, growth of this colon tumour in control animal study consistently was between 85 and 100%. The mice was sacrificed when tumours achieve a volume of 1 cm^3. An inhibition of the growth of colon tumour with this molecule with respect to tumour growth was proved consistently. It was found that the inhibition of growth of the colon tumour is best achieved at the 20th day of the treatment when each regimen inhibited the average colon tumour volume by 50%. Interestingly, the hydrochloride salt of this compound was soluble in water and this excellent property opens up further significant research in this area.

The results as described above were also fascinating since NCI Program investigated this molecule further on additional human colon cancers and better sensitivity against numerous colon cancer cell lines were observed. However, the NIH-Compare Program reported that this compound has an "unidentified mode of action".

The monoamine with piperidine **22a** caused extensive cytotoxicity while the amide version **13a** was far less active against human leukemic Jurkat T cells *in vitro*. This much lesser activity of the diamide **13a** was also proved in a significantly less degree of apoptotic damage, one of the possible mechanisms of cell death. It was difficult to predict if this compound may have an identical mechanism against another cell line because selectivity of action against different tumors is more specific in the antitumor compounds. The apoptotic pathways were involved in cell death induced in Jurkat cells by the diamide and diamine polyaromatic compounds described here. Importantly, no effects were demonstrated in a normal, non-transformed line of human YT cells. The activation of caspase-3 during apoptosis was measured. The results indicated that the amino compounds demonstrated excellent caspase-3 activity with respect to amides that are bound to chrysenes. Jurkat cells incubated with various monoamines with chrysenes were much more TUNEL-positive than the amides with identical group. The results of *bis* diamino compounds in chrysenes did not follow a similar activation. This was due to a different mechanism which is responsible for their high activity. An increase of significant caspase-3 activation, PARP-cleavage and DNA damage showed by the monoamino compound proved the apoptotic pathway was involved in the overall cell killing process.

Our studies in this area prompted us to synthesize new 2, 8-disubstituted chrysene derivatives. However, methods were not available in the literature to prepare 2, 8-disubstituted chrysenes. Therefore, a new method was necessary for this purpose. Samarium-mediated reductive dimerization reaction of methyl cinnamate was developed for the preparation of 2, 8-diamino chrysene as one of the crucial steps.

Electrophilic substitution is the main reaction used to add a functional group to an aromatic system but, chrysene on electrophilic produced 6-substituted isomer because this is the most electron rich center. A photocyclization route was available to obtain 2-substituted chrysene. In reality, realizing the conditions of this method, it appeared there would be considerable difficulty to obtain 2, 8-disubstituted chrysenes by this procedure. We realized that metal-mediated reaction can produce a suitable intermediate that can be used for the synthesis of valuable new 2, 8-disubstituted chrysene compounds.

The use of organic lanthanides in synthesis of molecules is known [16].The use of samarium and indium metal in synthesizing organic compounds was demonstrated by our group [**17**]. Many other scientists were also actively pursued different types of new reactions with samarium. For example, samarium metal was used for the reductive dimerization of imines and carbonyl molecules. However, methyl cinnamate and ethyl cinnamate did not dimerize following similar conditions to obtain the dimeric compound **41**. This reaction produced 3-phenylpropionate because of the reduction of the alkene bond.Samarium iodide-mediated methods failed in the dimerization of ethyl cinnamate. But, dimerization of conjugated acid compounds was obtained with samarium diiodide/THF/amine. In contrast, this reaction was not reported with cinnamate ester or acid. Ytterbium iodide produced reduced compounds as a result of the saturation of the alkene group. Indium metal failed to induce dimerization completely. Based upon these series of observations and following our experience on samarium metal-mediated reactions, we tested the dimerization of methyl cinnamate because we realized that the dimeric compound can be used for the synthesis of unusual chrysene derivatives.

From our study, it appeared that the success of samarium-induced reactions and product distribution were controlled with the selection of appropriate solvents [7]. Therefore, our aim was to promote dimerization of methyl cinnamate **40**. Single-electron transfer across the carbon-carbon unsaturated bond by samarium produced the ion radical **A** in the first step of the process. In turn, the generated ion radical **A** followed two different pathways leading to two different products. The first pathway was the transfer of additional electron from the metal to create dianion **B** and subsequent protonation of the dianion **B** toward the reduction product **42**. It was obvious that this pathway was not our choice. The second pathway was a coupling process of the ion radical **A** to **C** and subsequent protonation of **C**. This mechanism clearly indicated that if a reaction follows the first pathway, the reduction product **40 A** to **B** would be favourable **Scheme 9**. Alternatively, if the reaction proceeds through the second pathway **A** to **C** the product would be dimeric **41**. With this mechanistic rationale in mind, it was conceivable that the stability of the ion radical dictates the product distribution. An increase in the stability was required to prepare or to increase the yield of the dimer **41** and achieve our goal.

To test this hypothesis, we investigated a few reactions using samarium metal in the presence of iodine, allyl bromide and methyl iodide in methanol and THF. However, the dimeric product was not formed. A low yield of the monomer was obtained. However, a reaction of methyl cinnamate **40** with samarium in the presence of aluminum foil in methanol produced the dimer 42 in good yield. The use of aluminum chloride/samarium gave identical results. In contrast, samarium iodide failed to produce the dimer, since the product was a monomer. Samarium diiodide in the presence of aluminum foil or aluminum chloride produced monomer as the major product. Aluminum alone did not yield the dimer 4**1**. These results confirmed that the stability of the ion radical was high enough in the presence of the samarium-aluminum combination. As a result, coupling of the ion radical occurred, which was the key factor for dimer formation **A** to **C**. It was hypothesized that aluminum or aluminum chloride coordinates to the ion radical **A**, and this process increased its stability, and favored the coupling process. In another route, there was a possibility of dimer formation *via* a Michael addition of the ketyl with the cinnamate, not through the dimerization of the two ketyls [18].

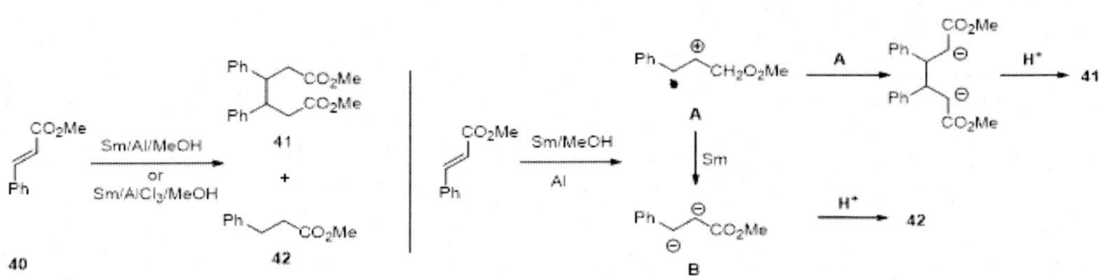

Scheme 9.

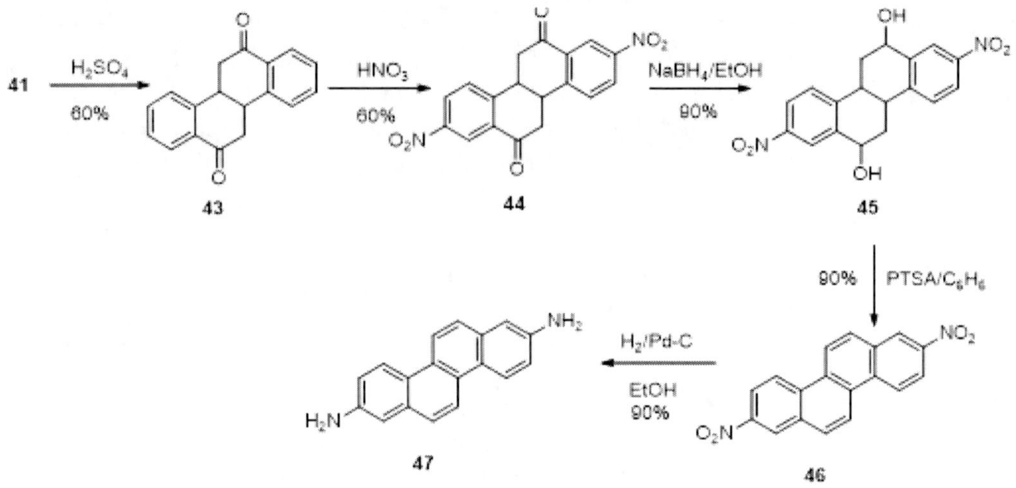

Scheme 10.

After identification of a suitable route for the dimerization reaction, next step was a cyclization process of the diester **41**. The cyclization was performed using sulfuric acid to afford the tetracyclic compound. Aromatic nitration and reduction-aromatization were then performed to obtain the 2, 8-diamino chrysene in excellent yield (Scheme 10).

The important feature of this method was the functionalization of chrysene at the uncommon **2** and 8 positions. No methods were available for the preparation 2, 8-didusbtituted chrysene [21]. The *in vitro* cytotoxicity data of several derivatives as described for the 6, 12-isomeric chrysene demonstrated excellent antitumor properties against a number of cancer cell lines. Notably, the anticancer activity of the 2, 8-disubstituted isomers were superior to those obtained from 6, 12-disubstitued compounds.

2.3. Dibenzofluorenes

Functionalization of benzene and naphthalene derivatives by electrophilic reaction is routine organic chemistry. The orientation of the electrophile in such monocyclic or bicyclic derivatives is predictable. A similar substitution reaction in a polycyclic aromatic system is extremely difficult, and for polycyclic non-alternate hydrocarbon is poorly predictable. In fact, little is known about the electrophilic substitution reaction in polycyclic aromatic system [19]. We planned to link a 4-carbon side chain with a heterocyclic base at the end to the aromatic ring through nitrogen of amino dibenzofluorene. Therefore, our task was to prepare amino dibenzofluorene **53** for the subsequent derivatization. Towards this goal, we reacted dibenzofluorenone with nitric acid in acetic acid under different conditions but failed to produce the desired nitro derivative **52**. However, the hydrocarbon **51** produced a single nitro compound **52** with nitric acid-acetic acid at 0-5°C in 80% yield. The location of the nitro group in the aromatic system was determined by nmr spectra. Based on the homonuclear decoupling and COSY nmr studies, all of the protons in **52** were assigned**.** These assignments were supported by the data reported by Jones et al.

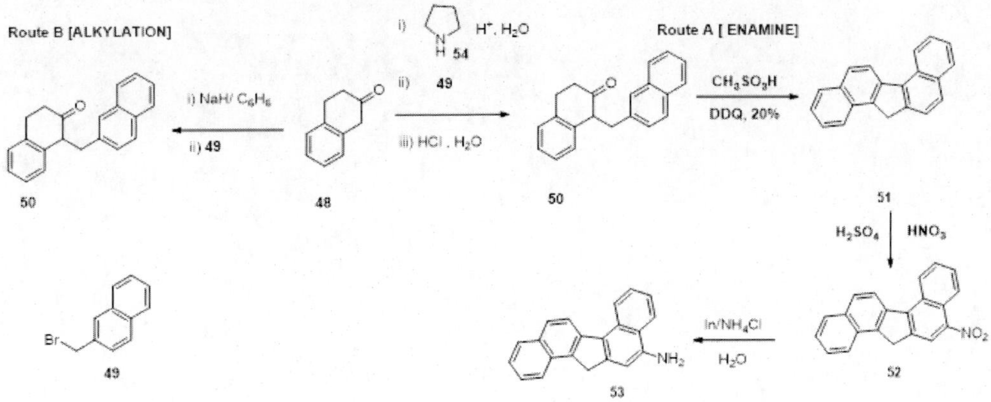

Scheme 11.

The NMR spectrum revealed that the AB quartet present in the hydrocarbon was absent from the nitro compound. The spectrum of the nitro compound showed a new singlet at δ 8.39 and a new doublet at δ 8.44 (J = 8.75 Hz). We eliminated positions C_1, C_4, C_7 and C_{10} for the nitro group because of the singlet at δ 8.44. The positions C_5, C_6, C_8 and C_9 were eliminated based on the homonuclear decoupling and COSY experiment. That region (δ 7.5-7.6) of the hydrocarbon remained unaffected clearly ruled out positions C_2 and C_3 for the nitro group. We eliminated position C_{12} because of the downfield doublet at δ 8.44. The ^{13}C NMR spectrum showed the presence of nine quaternary carbons. The signal at δ 123.58 due to the C_{11} carbon (verified by HETCOR study) was not present in the spectrum of **52**. A new peak at δ 145.24 appeared because of thc nitro group. Thus, we assigned C_{11} as the site of the nitro group in **52** based on extensive NMR study. The site of electrophilic attack was further verified by computer-assisted charge density calculation of the Whelands intermediates The relative energies of the Wheland intermediates of each of the carbons were calculated as C_1 (6.947), C_2 (21.278), C_3 (13.68), C_4 (11.008), C_5 (14.403), C_6 (16.919), C_7 (0.0), C_8 (10.558), C_9 (8.591), C_{10} (0.894), C_{11} (2.325), C_{12} (19.638), and C_{13} (7.358). These values indicated that preferential attack can take place at positions 7, 10, and 11 positions. On the basis of NMR data we eliminated positions 7 and 10.

Reduction of the nitro compound **52** to the amino compound **53** was carried out by following our own methodology. We demonstrated that samarium-iodine was highly effective for the transformation of nitro to the amino compound. The amine **53** was then condensed with the side chains **5a** and **5b** by mixed anhydride method. Many other condensing agents, such as DCC and HOBt, failed to give the amides **54a-b** (Scheme 12).

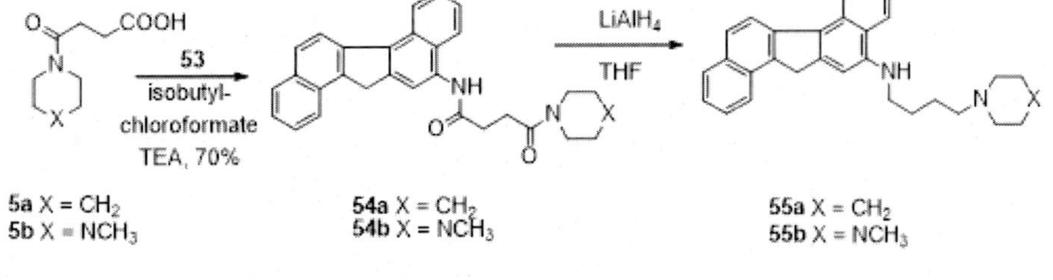

Scheme 12.

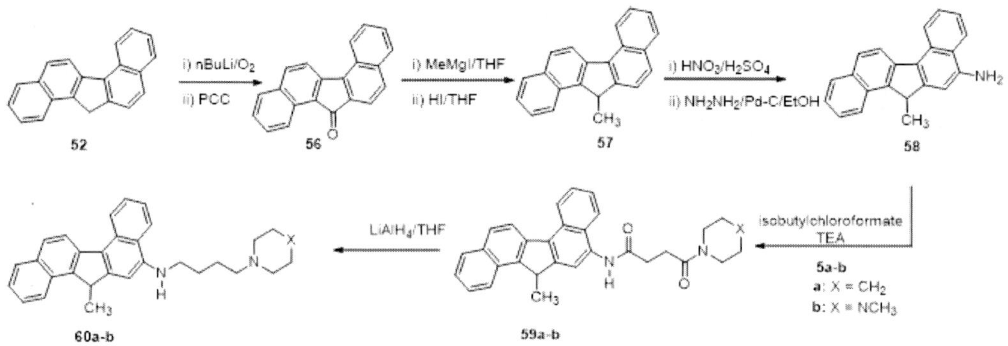

Scheme 13.

The desired diamides **54a-b** were isolated by column chromatography. Reduction of the amide functionality to the amino group by lithium aluminum hydride is a standard chemical transformation. Therefore, in order to produce the diamino compounds, reduction of **55a-b** was carried out by this reagent under a variety of conditions. However, the desired diamino compounds **55a** or **55b** were isolated from the complex reaction mixtures in very poor yield. After many experiments, diborane was found to be the reagent of choice for this reaction and it resulted in a good yield of the amino compound **55**.

These dibenzofluorene derivatives were tested against 3 tumor lines of animal origin and 6 of human origin. These cancer lines were used in the NCI panel for the testing of chemotherapeutic agents. The antitumor cytotoxicity of the new compounds was compared to cisplatin and adriamycin. The didenzofluorene with N-methylpiperazine side chain was much more active than the piperdine analogues. It was important to observe that compound derived from dibenzofluorene system was considerably more active than chrysene series in each and every cancer cell lines tested. In the amino series, all compounds were equally active. Moreover, in the amino series, compounds with N-methylpiperazine or piperidine remained equally active.

The antitumor activity of these newly synthesized dibenzofluorene derivatives, **52**, **53, 54** and **55** was tested and a comparison with respect to cisplatin was performed. The diamides with N-methyl piperazine moieties were more potent than the diamides with piperidines. However, all the diamines were potent irrespective of the terminal groups.

The keto compound **56** prepared by an oxidation experiment on the treatment with methylmagnesium iodide and dehydration in the presence of acid afforded the methylated dibenzofluorene **57**. Nitration with concentrated nitric acid, reduction of the nitro group by catalytic transfer hydrogenation to the amino group and coupling reaction as described above afforded the amide **59a-b** in excellent yield. The amide was successfully reduced by lithium aluminum hydride to the amine **60a-b** (Scheme 13).

Interestingly, the *in vitro* activity of these methyl derivatives had followed identical results like the 13-unsubstituted dibenzofluorene derivatives. The amino series were more potent against all cell lines tested compared to amide series. The activity of the 13-methyl derivatives was found to be better than the unsubstituted derivative in the amide and amine series against seven cancer cell lines. The methyl group is an electron donating group and it has a power to increase the basicity of the resulting compounds compared to the non-methylated analogues. Therefore, it was concluded that even minor changes of the structure of the dibenzofluorenes may alter the anticancer activity.

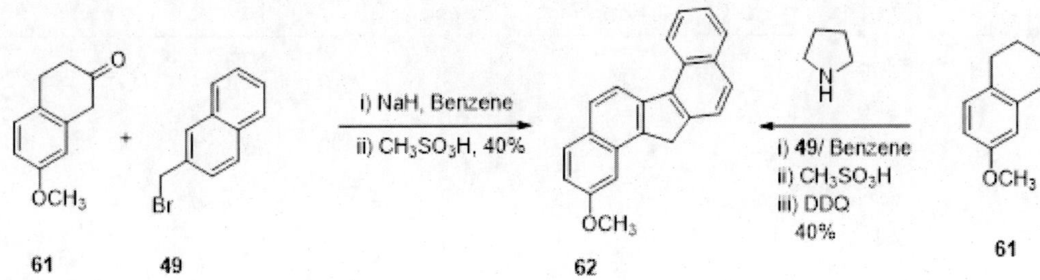

Scheme 14.

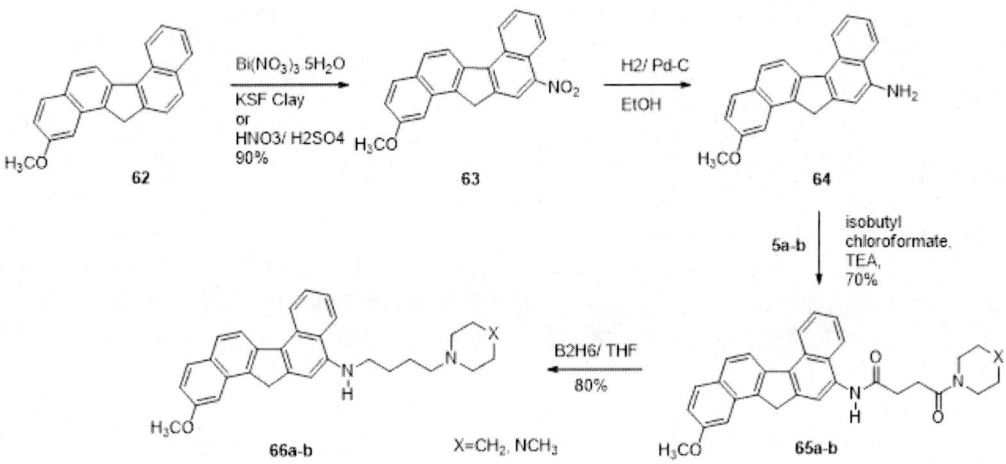

Scheme 15.

The success of the above methods had prompted us to consider the synthesis and biological evaluation methoxy dibenzofluorene derivatives. Following an identical route, the methoxy derivative was prepared. For example, the enamine of 7-methoxy beta-tetralone **61**was alkylated with 2-naphthyl bromide in the presence of a strong base and the resulting ketone was cyclized and aromatized to the methoxy hydrocarbon **62**. This methoxy dibenzofluorene on nitration with nitric acid/sulfuric acid mixture afforded the nitro derivative which was reduced to the amino derivative by indium-induced method. Alternatively, the methoxy dibenzofluorene was prepared through a one-pot method. In this method, preparation of enamine was not necessary.

Table 2.1. IC_{50} (micro molar) of compounds to 10 MTT assay (72h continuous exposure)[a]

Cell lines	Cisplatin	54a	54b	55a	55b
B-16	7.33	128.50	4.30	4.05	1.67
BRO	5.66	103.54	3.89	4.26	3.54
HL-60	-	33.33	3.48	-	-
L-1210	-	-	53.70	7.05	-
MCF-7	15.99	50.63	4.32	4.82	4.56
OVCAR-3	-	94.32	4.11	3.12	1.80
PC-3	1.66	32.52	3.78	4.49	4.39

HT-29	15.99	>100	3.86	4.11	3.27

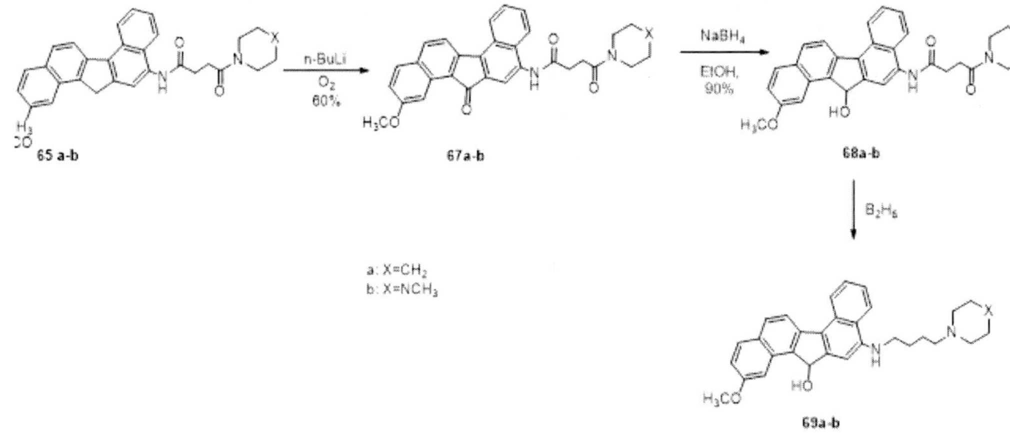

Scheme 16.

Alkylation of 7-methoxy beta tetralone with the corresponding naphthyl bromide in the presence of a strong base and acid-induced cyclization-dehydration produced the hydrocarbon. Oxidation of the methoxy hydrocarbon was possible to the keto derivative. But, the ketone failed to produce the nitro compound, probably due to deactivation of the aromatic ring system. On the basis of the deactivation hypothesis of the keto compound, nitration of the hydrocarbon **63** with nitric acid/sulfuric acid and bismuth nitrate-induced clay impregnated reactions was conducted. The reaction produced a single nitro compound **63**. The location of the nitro group in the aromatic ring was assigned by NMR spectra and comparison with our previous compounds that have no methoxy groups. It was important to note that nitration took place at the unsubstituted aromatic ring. Several methods were attempted to reduce the nitro group in **63**. Thus reduction of **63** with hydrogen gas/Pd-C, hydrazine/Pd-C, samarium/iodine, and indium/ammonium chloride were effective in producing the corresponding amino compound **64**. The amine **64** was then reacted with the acids described earlier in the presence of isobutylchloroformate and triethylamine [4]. The diamides **65a** ($X = CH_2$) and **65b** ($X = NCH_3$) were obtained in 70% yield (Scheme 15). On oxidation, the methylene group in **65a-b** produced the ketone **67a-b**. The ketone **67a-b** was reduced to the alcohol **68a-b**. Diborane was used to reduce the amide groups in **69a-b** (Scheme 16).

These compounds were tested in the University of Texas M. D. Anderson Cancer Center's Core Analytical Laboratory against eight tumor lines all of which have been used in the NCI panel for the testing of chemotherapeutic agents. The antitumor activity of these new methoxy dibenzofluorene derivatives **54** to **55** were performed and compared with cisplatin. The results of these tests are interesting, and indicate the significance of altering structures in terms of antitumor activities. Some compounds were more active than cisplatin in many of these cancer cell lines (Table).

In fact, **55a** and **55b** were some of the most active molecules that we have derived from the study of polyaromatic compounds. The keto group at position 13 created the fluorenone with reduced activity (data not shown). It was clear that the piperidine group at the terminus of the alkyl chain decreases activity compared with those that terminate with the N-methyl piperazine. Clearly, this was observed in compounds in **13**-keto and **13**-alchol substituted derivatives. The alcohol group at position 13 in produced a reduced activity when compared with **54b**. Reduction of **54** resulted in their amine derivative **55**. An impressive increase was seen in the amine derivatives **55a** and **55b** regardless of the terminus units. The cause of the spectacular differences of the activity of the two series of compounds in all cell lines is not known. The introduction of at one basic group (tertiary amine) at the terminal site of these molecules was probably responsible for the increased activity. The basic units will help to increase the pKa value so that these molecules can be protonated at physiological pH. These type of polyaromatic highly lipophilic and cationic molecules were expected to bind to several anionic structures inside cells such as to phosphate groups of nucleic acids (DNA and RNA). These lipophilic molecules were also expected to interact strongly with the lipids components in cell membrane. The *in vitro* cytotoxicity of these new molecules was similar or better than that of cisplatin which acts through a different mechanism of action primarily through cell damage and death. However, it was also known that activity *in vitro* is not proportional to clinical applications in many instances. For example, adriamycin is one of the most potent anticancer drugs *in vitro* and *in vivo*, but it is also established that it is not a successful clinically active agent [20-21].

2.4. Pyrrole-Substitued Polyaromaic Compounds

Pyrroles are important heterocycles that have diverse medicinal activities. For this reasons, a number of procedures for the preparation of pyrrole derivatives were performed by scientists. Conjugate reactions, transition metal-mediated reaction, [4] reductive coupling reaction [5] aza-Wittig method, and several multistep procedures [7] were developed for the synthesis of substituted pyrrole derivatives. Despite many crucial methods, Paal-Knorr [8] reaction was still considered as the most efficientfor the preparation of numerous pyrroles. An acidic clay-induced [9] reaction and microwave-mediated irradiation method were used for the preparation of pyrrole systems following Paal-Knorr conditions. These procedures helped synthesis of pyrrole in a number of ways, but efficient reaction with low nucleophilic aromatic amines was not studied. Specifically, high power microwave irradiation and high temperature or large excess of acids was useded for a successful Paal-Knorr reaction. Thus, mild and simple conditions were necessary to overcome the shortcomings of the current methods.

Simple methods for the synthesis of substituted pyrroles by iodine-catalyzed and montmorillonite-KSF-induced Paal-Knorr reactions were developed. Based upon the anticancer activities of polyaromatic amino compounds, synthesis of pyrrole-containing polyaromatic compounds were necessary. Considering the reaction conditions of Paal-Knorr method, iodine-catalyzed and clay-induced reactions were investigated. It was gratifying that pyrroles were synthesized by clay-mediated reactions following Paal-Knorr's method.

Although clay worked well, however, silica gel and alumina failed to react with the substrates to produce the products. Montmorillonite clay under identical reaction conditions reacted and produced the products in high yields. The great success of montmorillonite-mediated procedure for the synthesis of pyrroles indicated the significance and crucial role of the nature and acidity of the solid support. It was important to realize that o strong acids and alkali, high temperature, microwave, extraction, stirrer, and dry solvents were necessary [22-25].

Table 2.2. Paal-Knorr's method, condition and Yields of 2,5-disubstituted pyrrole

Entry	Amine	Product(s)	Time(h)	Yield (%)
1	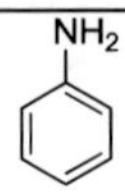 70a	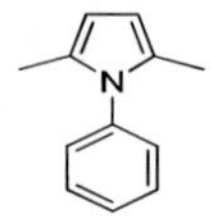 72a	10	96
2	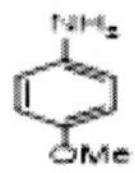 70b	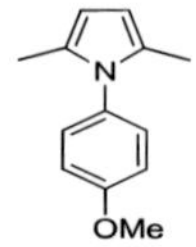 72b	15	81
3	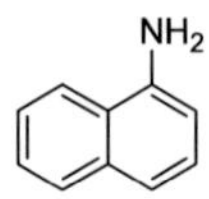 70c	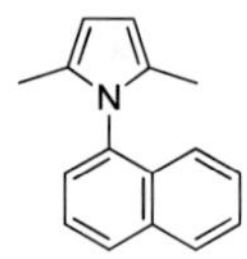 72c	11	83
4	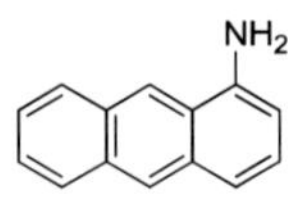 70d	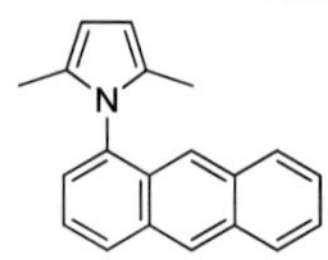 72d	19	96
5	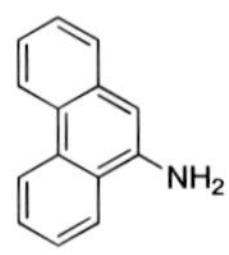 70e	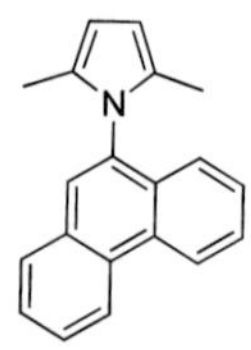 72e	18	94
6	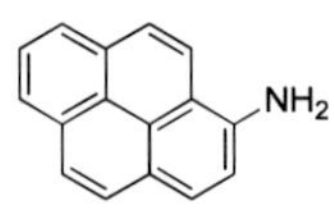 70f	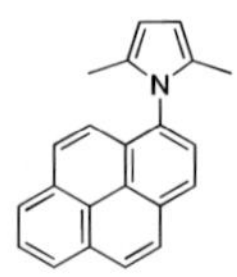 72f	20	88
7	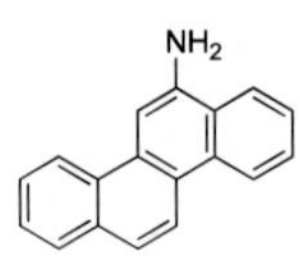 70g	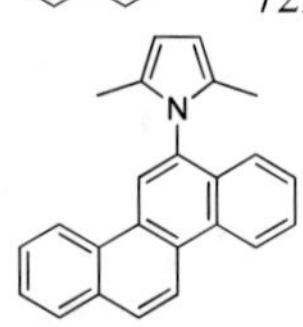 72g	22	85
8	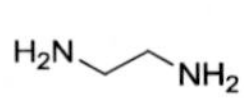 70h	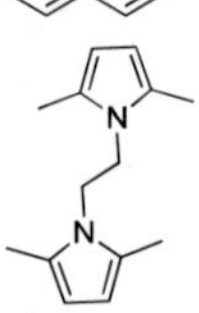 72h	10	85
9	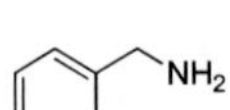 70i	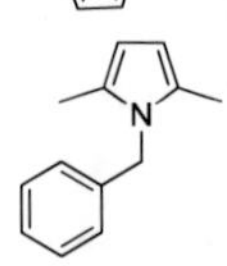 72i	10	95
10	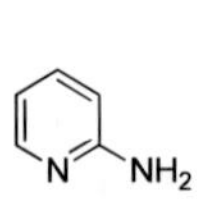 70j	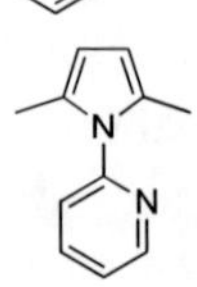 72j	25	70

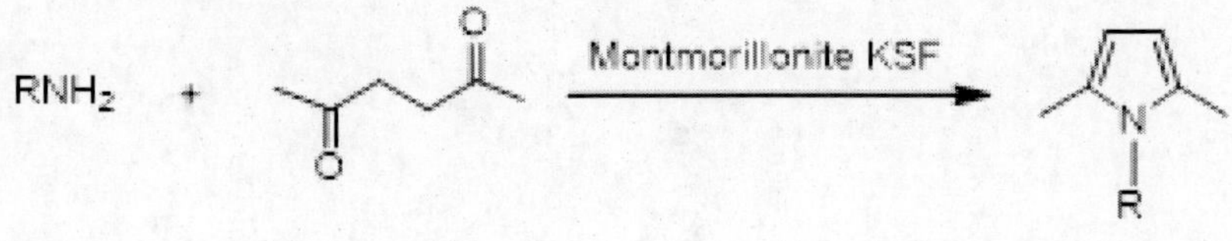

Scheme 17.

All types of primary amines including polyaromatic amines were successfully employed for this study. The diketone, 2, 5-hexanone **71** reacted very smoothly (Scheme 17).

The experimental procedure of this reaction was straight forward. The clay, amine and the ketone in dichloromethane were mixed together in solvent. Solvent was then evaporated under reduced pressure and the reactant mixture was stored at room temperature for some time depending upon the basicity and nucleophilicity of the amines. The yields of pyrroles and time for completion of the reactions were given in Table 2.2. The more basic and superior nucleophilic aromatic amines needed less time to react and produced pyrroles quickly. Although polyaromatic amines are much less basic, the yields of the products from them were excellent (Entry 4-7 in Table 2.2).

Preparation of substituted pyrroles was also accomplished by iodine-catalyzed reaction [15] (Scheme 18). It was demonstrated that the reactions conducted by iodine was more convenient than the clay-mediated process. The amines and ketones were taken with solid molecular iodine in dichloromethane and the reactants were kept at room temperature. For liquid starting materials, no solvent was necessary. But, the reaction required a small amount of solvent with sterically hindered polyaromatic amines. In addition, small amount of solvent was necessary to make the reactants soluble.

Importantly, different types of amines were tested for this investigation. The results were shown in Table 2.3. The polyaromatic amines needed much more time to form the products, but finally they demonstrated similar yields like monocyclic amines.

Table 2.3. Iodine-Catalyzed Synthesis of Substituted Pyrroles

Entry	R_1	R_2	R_3	R_4	Yield (%)	Time (h)
1	phenyl	Me	Me	H	90	0.5
2	benzyl	Me	Me	H	92	0.5
3	1-naphthyl	Me	Me	H	85	1
4	1-anthryl	Me	Me	H	88	1
5	9-phenanthryl	Me	Me	H	91	1
6	1-pyrenyl	Me	Me	H	87	1.5
7	6-chrysenyl	Me	Me	H	80	1.5
8	4-anisyl	Me	Me	H	84	1.5
9	2-pyridyl	Me	Me	H	78	1
10	phenyl	Me	Ph	H	91	7
11	benzyl	Me	Ph	H	89	9
12	1-butyl	Me	Ph	H	93	8
13	cyclopentyl	Me	Ph	H	76	10
14	phenyl	Ph	Ph	Ph	15	20

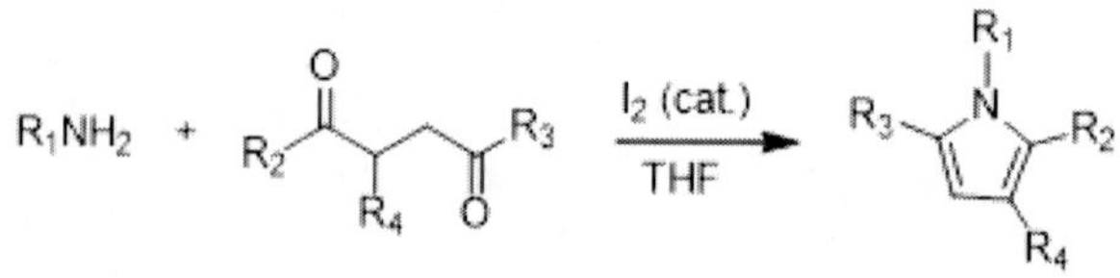

73 74 75

a: $R2 = R_3 = Me,\ R_4 = H$
b: $R_2 = Me, R_3 = Ph,\ R_4 = H$
c: $R_2 = R_3 = R_4 = Ph$

Scheme 18.

The yields of pyrroles with primary and polyaromatic amines established that iodine was an excellent catalyst. These reactions were performed at room temperature. In contrast, high temperature was necessary to obtain pyrroles using other methods.

To extend the variety, this procedure was tested using substituted diketones **74b** and **74c** and pyrrole formation was observed with **74b**. Numerous 1, 2, 5-trisubstituted pyrroles were synthesized by this procedure (entries 10-13,). This method did not work with **74c**. Using a higher proportion of iodine, the yield of the pyrrole was increased. The steric hindrance by the three aromatic rings present in **74c** was not supportive for this reaction. Clearly, the reaction had no problem if there is a single aromatic ring in the ketone component.

The amine and diketone did not react in the absence of iodine. Catalytic amounts of iodine (5-10 mol%) was essential for the reaction to take place successfully. It was assumed that hydroiodic acid is the real catalyst in this process. The formation of hydroiodic acid in the reaction media was highly possible with iodine in the presence of non-anhydrous organic solvents and reactants. The reaction of an aromatic amine and the diketone with in the hydroiodic acid (1 equivalent) was conducted and lower yields of the pyrroles were obtained. Therefore, it was concluded that iodine-catalyzed and clay-induced procedures were superior to acid-induced synthesis of pyrroles.

2.5. Sugar-Containing Polyaromatic Compounds

The synthesis of sugar containing polyaromatic substrates is a difficult task. In this area, we developed two important methods for the preparation sugar-containing polyaromatic compounds *via* glycosylation [26].

Glycosylation of alcohols is a significant area of research due to solubility of the glycosides in water and their medicinal activities of *O*-glycosides. As a result, many methods of glycosylation reactions were devoped. The glycosyl fluoride [3a] and thioglycoside [1], and Ferrier rearrangement [3b] were prominent methods for effective glycosylation. Lewis acids [4] and solid acidic materials [5] were also used for this procedure. The principal shortcomings of these procedures were the nonstereoselectivity nature of the reactions. A mixture of α- and β-glycosides were synthesized in different ratios by these myhods. A few stereospecific methods were identified to improve the significance of these procedures. Lewis acid-mediated synthesisof β-glycosides using glycosyl phosphates was studied. These phosphates were synthesized as a mixture of compounds through many reactions. These methods required a complicated oxidation of the alkene bonds in glycals by dimethyl-dioxirane [8].

Trichloroacetamidate procedure worked well, but it does not have selectivity [9]. The limitations of these procedures required to develop a new activator. Our research resulted in an indium metal-induced method of stereoselective synthesis of β-glycosides by a reaction of alcohol and β-D-bromoglucose. A similar method of β-D-bromoglucoses with zinc, samarium diiodide, and titanium reagents produced the glycals.

We developed a few samarium-and indium-induced reactions Indium metal-mediated synthesis was investigated by many researchers. Indium has a high reducing power because of its low ionization potential value (5.8 eV). Indium is inert in aqueous solution and mild acid conditions. These unique characteristics of indium metal are exploited in chemical research for the synthesis of organic compounds. In our investigations with polyaromatic compounds, indium had become a powerful promoter. For example, we demonstrated indium-medaited synthesis of various 3-unsubstituted β-lactams. The success of the indium-medaited reaction to obtain β-lactam by the reaction of ethyl bromoacetate and Schiff base was attractive. It was believed that reaction between indium and a bromo compound forms an oganometallic intermediate that reacts with the imines. This process failed to follow radical coupling route to form the dimeric products of bromo compound or imine. This idea was supported because a facile addition reaction can really take place with bromo-indium species and carbonyl molecules. This, bromo-indium species took part in a nucleophilic reaction.

On reaction of methanol, ethanol, isopropanol, 3-methylbutanol, 9-fluorenol, and pentenol with 2,3,4,6-tetra-*O*-acetyl-D-glucopyranosyl bromide **76a** was performed with indium metal powder. This method generated a single glycoside stereospecifically regardless of the nature of the alcohols **77a-g**. The anomeric carbon stereochemistry of these glycosides was found to be β from NMR analysis of the crude reaction mixtures (**Scheme 19**).This method was also used for the preparation of a disaccharide 78g. It was gratifying to observe that this method produced a single product with controlled stereochemistry with multicyclic aromatic alcohol in which nucleophilicity of the molecule was too weak.

The mechanism of indium-mediated sugar addition was not known. The formation of β-glycosides was noteworthy. The indium-mediated reaction was mechanistically different than the Lewis acid-induced reactions, glycosyl phosphate-induced methods and triflate-catalyzed methods since generation of an ionic intermadiate is not possible by following this method. Moreover, indium-induced method produced β-glycosides in the absence of any acidic reagents with sugar 76. A neighbouring group participation of the C-2 functional moiety present in the sugar in promoting the formation of β-glycosides was not a favourable pathway by the indium-mediated reaction. The reaction did not go to yield any product in the absence of indium metal. This new chemistry was adavnced by a substitution nucleophilic bimolecular reaction of alcohols with sugar derivatives that have bromine at the anomeric center of the carbohydrate.

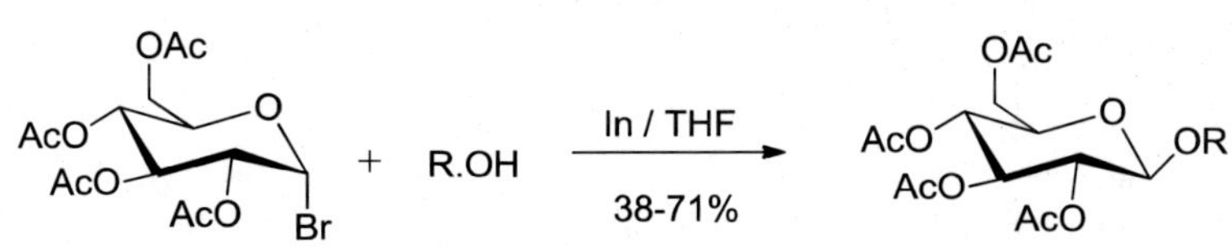

76 **77a-g** **78 a-g**

Scheme 19.

Table 2.4. Indium-mediated Glycosylation of alcohols by 2, 3, 4, 6-Tetra-O-acetyl-α-D-glucopyranosy bromide

Entry	Alcohol	Sugar	Glycosides	Time (h)	Yield (%)
1	MeOH 77a	76	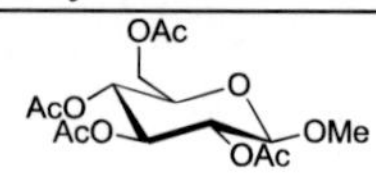 78a	10	47
2	EtOH 77b	76	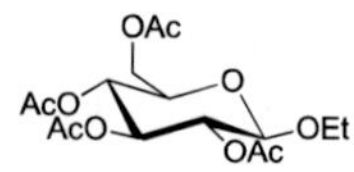 78b	9	45
3	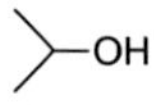 77c	76	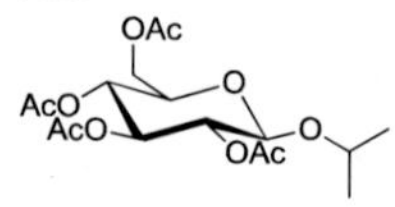 78c	12	44
4	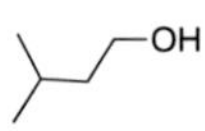 77d	76	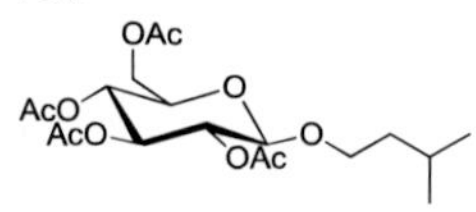 78d	11	40
5	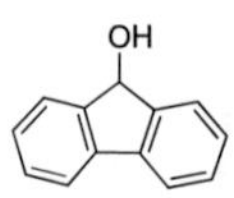 77e	76	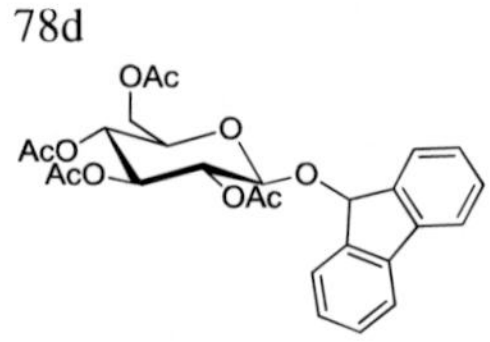 78e	10	43
6	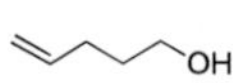 77f	76	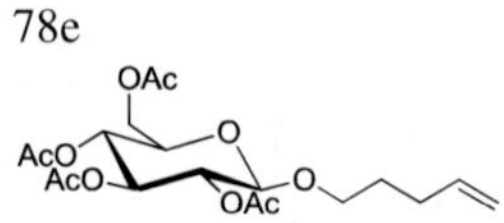 78f	8	38
7	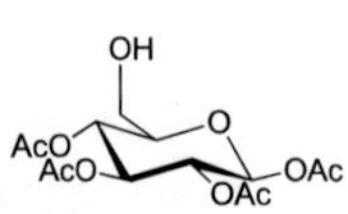 77g	76	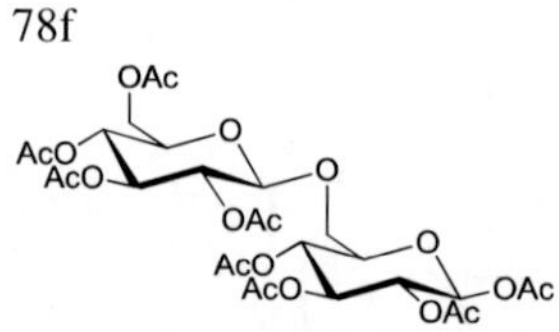 78g	14	40

A reaction of the alcohol 77e with the glycal 79 in the presence of borontrifluoride in anhydrous THF produced two glycosides 80a and 80b in 60% yield. The purification of one of them was possible. But, purification of the another isomer was difficult. The second compound, however, was isolated in 90% pure form through numerous recrystallization. The unsaturated glycosides 80a and 80b were reacted with acid and the products (+)-77e and (-)-77e were obtained in good yield (Scheme 20).

The optical rotation of the product alcohol was found to be almost equal and opposite to each other. But, it was difficult to determine the absolute stereochemistry of the alcohols. A plus and minus sign was used to assign the chiral nature of the alcohols.

Most of the glycosides prepared following Ferrier rearrangement conditions showed to possess α-stereochemistry. Therefore, an α-stereochemistry was indicated. However, single crystal X-ray crystallography study is required to determine the absolute stereochemistry of the product glycosides **80a** and **80b**. The data demonstrated a successful resolution of complex benzylic polyaromatic alcohols for the first time.

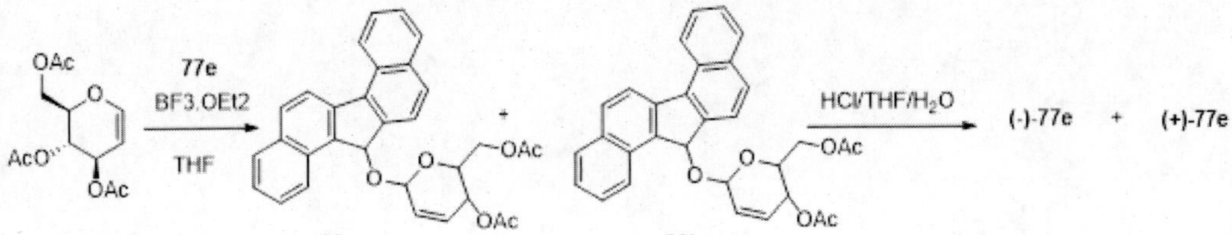

Scheme 20.

2.6. Unusual Dehydrogenation and Novel Polyaromatic Amines

Bismuth nitrate-induced nitration of several aromatic compounds was reported in our previous publication. Reaction of 9,10-dihydrophenanthrene 81 with bismuth nitrate pentahydrate impregnated with clay produced a single 2,7-dinitro compound 82 under the influence of microwave irradiation. The dinitro product 83 was then attempted to transform to the corresponding diamino derivative 83 under the catalytic hydrogenation reaction over Pd/C in anhydrous ethanol at room temperature. Surprisingly, an unusual aromatic compound corresponding to 2, 7-dinitro-phenanthrene derivative 84 was obtained as the minor product (Scheme 21, Table 2.5).

The formation of **84** was explained and it was due to a catalytic oxidative dehydrogenation or aromatization reaction. Our goal was to improve the yield of **84** by changing catalyst loading and the reaction conditions. An increase in the catalyst amount to 10-20 mol% resulted in a substantial improvement in the yield of compound **84**. Moreover, it was the exclusive product with excess of catalyst and longer reaction time. The usual diamine **84** was obtained under controlled conditions.

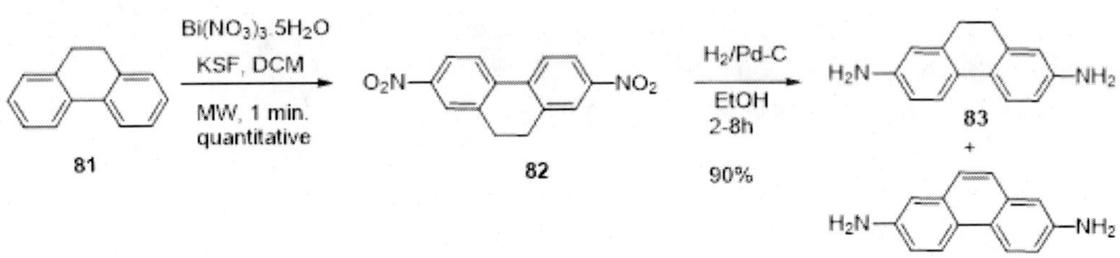

Scheme 21.

Table 2.5. Synthesis of Phenantherene and 9, 10-dihydrophenantherene-2, 7-diamine 83 and 84

Entry	Time((h))	Catalyst loading	Compound (83)[a]	Compound (84)[a]
1	2	5-10 mol %	80	80
2	3	5-10 mol %	70	70
3	5	10-15 mo 1%	30	30
4	8	10-15 mol %	10	10

a: Isolated after column chromatography.

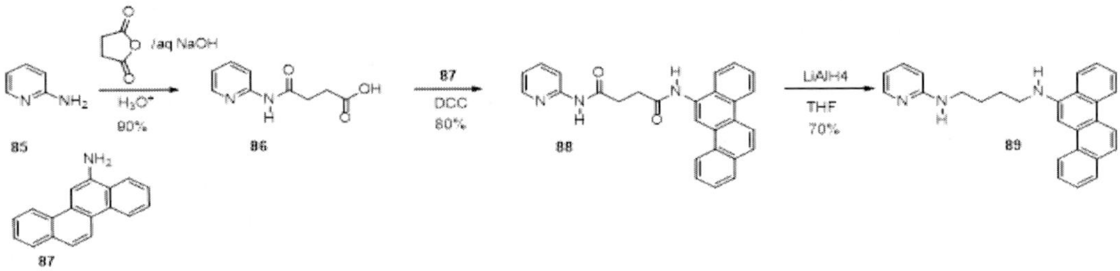

Scheme 22.

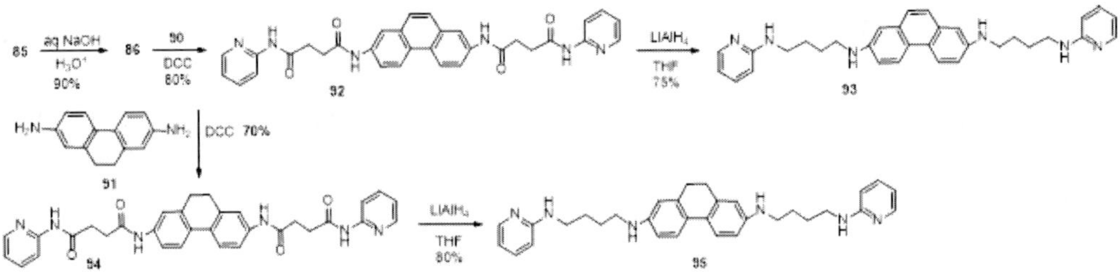

Scheme 23.

Piperidine and N-methyl piperazine were used as the terminal heteroycyclic rings extensively as shown above. After careful analysis of the results and availability of unknown diamino phenanthrene, prompted us to investigate the synthesis further. For example, it was decided to prepare compounds that have a pyridine ring at the terminal site of a few similar molecules. To introduce a pyridine ring at the side chain, the best starting material was 2-aminopyridine **85** which can produce a similar acid by reacting with succinic anhydride. The acid was then condensed with 6-amino chrysene **87** diaminodihydrophenanthrene **90** and diaminophenanthrene **91** (Schemes 22-23). The resulting diamides **88, 92** and **94** were formed in excellent yields. Reduction of the diamides to amino derivatives **89**, **93**, and **95** produced amino compounds with a pyridine ring at the terminal site of these molecules. The *in vitro* cytotoxicity and other biological studies with the new compounds described in this series are in progress.

2.7. Mechanism of Actions

Since polyaromatic compounds were used less frequently than other types of organic molecules as anticancer agents, little is known regarding the mechanism of action of these types of molecules. Because of the presence of multiple aromatic rings, this area of research was studied on the interaction with the DNA systems and associated enzymes. From this study, it was also believed that other sites of activity have a substantial role. For example, interaction of these molecules against cell membrane was heavily recognized. This was also exemplified by an experiment with well-known cancer drug, adriamycin. Agarose linked with adriamycin showed significant activity against mouse leukemia L1210 confirmed serious interaction at the plasma membrane. It was necessary for the linked-adriamycin to enter into the cell nucleus. Probably, adriamycin exerted its activity through a covalent type of bonding with the lipid or other types of interaction and this physiological conditions of the cells were altered. Cytotoxic action of adriamycin was therefore, possible from its interaction with the membrane of the cell. The *in vitro* cytotoxicity data of the amino derivatives as reported herein seemed to support the interaction of these compounds with the cell membrane. The membrane interaction also depended on varieties of factors. For example, the charge density, the nature of the functional groups and lipophilicity and hydrophobicity of a molecule may selectively altered the function of membrane. The interaction of organic compounds with cell components in the plasma membrane was very complex. A variety of reactions, for example, compound's interference with surface receptors, intercalation into the membranes, alteration of ion distribution, and lipid disruption were the most common. The active amino compound that interacts with caspase-3 had also interaction with the membrane of the red blood cells like a membrane stabilizing agent. Possibly this molecule intercalated into the cell membrane by hydrophobic interaction.

CONCLUSION

We demonstrated various new methods including but not limited to bismuth nitrate-mediated nitration of polyaromatic hydrocarbons, indium-induced reduction of polyaromatic nitro compounds and imines to amino derivatives, samarium-mediated reductive dimerization, one-pot synthesis of dibenzofluorenes, oxidation of activated methylene group to ketone by sodium bismuthate, stereospecific glycoside formation and optical resolution of polyaromatic alcohol with carbohydrates. These methods were used in combination with other available pathways for the preparation of numerous polyaromatic compounds. Structure-activity relationships studies of many of these molecules were conducted against numerous cancer cell lines *in vitro* and useful selectivity was observed. Some of the most active compounds were tested against specific cancer cell lines and a reduction of tumor growth was observed.

Many scientists may argue to use these compounds against cancer because of their close relationships with carcinogens and mutagens. However, results from these investigations were encouraging because mutagenicity and carcinogenicity of the molecules are not observed. It was believed that structurally altered polyaromatic derivatives are capable of mitigate the carcinogenic and mutagenic properties. In contrast, they exerted cytotoxicity by interacting with specific cell organelles. Some anticancer agents had specific components that many scientists believe to be carcinogenic/mutagenic. Despite the apparent contradiction, many phenothiazines, mustards, anthracenes and carbazoles were undergoing clinical trials or in use clinically. Moreover, anticancer agents were not administered for a long period of time that is required for carcinogenic effects. This study had identified apoptotic route through caspase activation as a major pathway in killing the cancer cells.

Many compounds with unique structures were prepared and useful studies were conducted with the lead molecules based upon the current findings. These studies were required to identify clinically active agents because of their initial effectiveness. The adverse clinical situations of many anticancer drugs justified to study new and effective anticancer agents with high activity and low toxicity.

ACKNOWLEDGMENTS

BKB is grateful to NCI and the University of Texas M. D. Anderson Cancer Center to support this work.

REFERENCES

[1] (a) Harvey, R. G. (1997). Polycyclic Aromatic Hydrocarbons, Wiley-VCH, (b) Rice, J. E., Cai, Z-W. (1993). An Intramolecular Arene-Triflate Coupling Reaction for the Regiospecific Synthesis of Substituted Benzofluoranthenes, *J. Med. Chem.*, *58*: 1415-1424. (c) Rice, J. E., Czech, A., Hussain, N., La Voie, E. J. (1988). Synthesis of Fluorinated Derivatives of Benzo[k]fluoranthene and Indeno[1,2,3-cd] pyrene and 8,9-Dihydro-8,9-epoxybenzo[k]fluoranthene, *J. Org. Chem.*, *53*, 1775-1779. (d) Chung, Y.-S., Kruk, H., Barizo, O. M., Katz, M., Lee-Ruff, E. (1987). Syntheses of Cyclopentene-Fused Polynuclear Aromatic Hydrocarbons, *J. Org. Chem.*, *52*, 1284-1288. (e) Lee, H., Harvey, R. G. (1988). New Synthetic Approaches to Cyclopenta[a]phenanthrenes and their Carcinogenic Derivatives., *J. Org. Chem.* *53*, 4253-4256. (f) Harvey, R. G., (1985). Polycyclic Hydrocarbons and Carcinogenesis; American Chemical Society: Washington, DC., (g) Clar, E. (1964). Polycyclic Hydrocarbons; Academic Press; New York,

[2] (a) Bair, K. W., Andrews, C. W., Tuttle, R. L., Knick, V. C., Cory, M., McKee, D. D. (1991). 2-(Arylmethyl) amino-2methyl-1,3-propanediol DNA Intercalators. An Examination of the Effects of Aromatic Ring Variation on Antitumor Activity and DNA Binding, *J. Med. Chem.*, 34, 1983-1990. (b) Bair, K. W., Tuttle, R. I., Knick, V. C., Cory, M., McKee, D. D. (1990). (1-Pyrenylmethyl) amino Alcohols, a New Class of Antitumor DNA Intercalators. Discovery and Initial Amine Side Chain Structure-Activity Studies, *J. Med. Chem.*, *33*, 2385-2393. (a) Malviya, V. K., Liu, P. Y., Alberts, D. S., Surwit, E. A., Craig, J. B., Hanningan, E. V. (1992). Evaluation of Amonafide in Cervical Cancer Phase II., *Am. J. Clin. Oncol.*, *15*, 41-44. (b) Rosell, R., Carles, J., Abad, A., Ribelles, N., Barnadas, A., Benavides, A., Martin, M. (1992). Phase I Study of Mitonafide in 120h Continuous Infusion in Non-Small Cell Lung Cancer., *Invest. New Drugs,* 10, 171-175.

[3] (a) Sami, S. M., Dorr, R. T., Alberts, D. S. and Remers, W. A (1993). 2-Substituted 1,2-Dihydro-3H-dibenz[de,h]isoquinoline-1,3-diones. A New Class of Anti-tumor Agent., *J. Med. Chem.*, 36, 765-770. (b) Sami, S. M., Dorr, R. T., Solyion, A. M., Alberts, D. S., Remers, W. A. (1995). Amino-Substituted2-[2′-(Dimethylamino) ethyl]-1,2-Dihydro-3H-dibenz[de,h]isoquinoline-1,3-diones. Synthesis, Anti-tumor Activity and Quantitative Structure-Activity Relationship, *J. Med. Chem.*, *38*, 983-993.

[4] Fukushima, T., Kawai, Y., Nakayama, T., Yamaguchi, T., Yoshida, A., Urasaki, Y., Imamura, S., Kamiya, K., Tsutani, H., Ueda, T., Nakamura, T. (1996). Superior Cytotoxic Potency of Mitoxantrone in Interaction with DNA: Comparison with that of Daunorubicin, *Oncol. Res.*, *8*, 95-100.

[5] Denny, W. A., Rewcastle, G. W., Ganguley, B. C. (1990). Potential Antitumor Agents. 59. Structure-Activity Relationships for 2-Phenylbenzimidazole-4-carboxamides, a New

Class of Minimal DNA-Intercalating Agents Which May Not Act via Topoisomerase II., *J. Med. Chem.*, 33, 814-819.

[6] (a) Leon, P., Garbay-Jaureguiberry, C., Lambert, B., Le Pecq, J. B., Roques, B. P. (1988). Asymmetrical Bisintercalators as Potential Antitumor Agents, *J. Med. Chem., 31*, 1021-1026. (b) Cherubim, P., Deady L. W., Dorkos, M., Qazi, N. H., Baguley, B. C., Denny, W. A. (1993). Synthesis and Biological Evaluation of Phenanthrene-Derived Carboxamides as Cytotoxic Agents, *Anti-cancer Drug Design, 8*, 429-438. (c) Lee, H. H., Palmer, B. D., Boyd, M., Baguley, B. C. Denny, W. A. (1992). Potential Antitumor Agents. 64. Synthesis and Antitumor Evaluation of Dibenzo[1,4]dioxin-1-carboxamides: a New Class of Weakly Binding DNA-Intercalating Agents., *J. Med. Chem., 35*, 258-266. (d) Agbandje, M., Jenkins, T. C., McKenna, R., Reszka, A. P., Neidle, S. (1992). Anthrecene-9,10-diones as Potential Anticancer Agents. Synthesis, DNA-Binding, and Biological Studies on a Series of 2,6-Disubstituted Derivatives. *J. Med. Chem., 35*, 1418. (d) Venitt, S., Crofton-Sleigh, C., Agbandje, M., Jenkins, T. C., Neidle, S. (1988). Anthracene-9,10-diones as Potential Anticancer Agents: Bacterial Mutation Studies of Amido-Substituted Derivatives Reveal an Unexpected Lack of Mutagenicity, *J. Med. Chem.*, 41, 3748-3752. (e) Perni, R. B., Wentland, M. P.' Huang, J. I., Powles, R. G., Aldous, S., Klinbeil, K. M. Peverley, A. D., Robinson, R. G., Corbett, T. H., Jones, J. L., Mattes, K. C., Rake, J. B., Coughlin, S. A. (1998). Synthesis and Antitumor Activity of 4-Aminomethylthioxanthenone and 5-Aminomethylbenzothiopyranoindazole Derivatives, *J. Med. Chem.*, 41, 3645-3654. (f) Perry, P. J., Gowan, S. M., Reszka, A. P. Polucci, P., Jenkins, T. C., Kelland, L. R., Neidle, S. (1998). 1,4 and 2,6-Disubstituted Amidoanthracene-9,10-dione Derivatives as Inhibitors of Human Telomerase. *J. Med. Chem.*, 41, 3253-3260. (g) Palmer, B. D., Rewcastle, G. W., Atwell, G. J., Baguley, B. C., Denny, W. A. (1988). Potential Antitumor Agents. 54. Chromophore Requirements for *in vivo* Antitumor Activity Among the General Class of Linear Tricyclic Carboxamides., *J. Med. Chem.*, 31, 707-712.

[7] (a) Tritton, T. R., Yee, G. (1982). The Anticancer Agent Adriamycin can be Actively Cytotoxic Without Entering Cells., *Science, 217*, 248-250, (b) Tritton, T. R. (1991). Cell Surface Action of Adriamycin, *Pharmacol. Ther., 49*, 293-309.

[8] (a) Born, G. V., Housely, G. M. (1983). Effects of Modification of the Membranes of Intact Erythrocytes on the Anti-Haemolytic Action of Chlorpromazine. Br. *J. Pharm.*, 79: 481-487. (b) Becker, F. F., Banik, B. K. (1998). Polycyclic aromatic Compounds as Anticancer Agents: Synthesis and Biological Evaluation of Some Chrysene Derivatives, *Bioorganic & Med. Chem. Lett., 8*, 2877-2880. (c) Banik, B. K., Becker, F. F. (2001). Synthesis, Electrophilic Substitution and Structure-Activity Relationship Studies of Polycyclic Aromatic Compounds for the Development of Anticancer Agents, *Curr. Med. Chem., 8*, 1513-1533. (d) Banik, B. K., Becker, F. F. (2001). Polycylic Aromatic Compounds as Anticancer Agents: Structure-Activity Relationships of New Chrysene and Pyrene Derivatives, *Bioorg. & Med. Chem., 9*, 593, (e) Becker, F. F., Mukhopadhyay, C., Hackfeld, L., Banik, I., Banik, B. K. (2000). Polycyclic Aromatic Compounds as Anticancer Agents: Synthesis and Biological Evaluation of Dibenzofluorene Derivatives, *Bioorg. & Med. Chem.*, 8, 2693-2699. (f) Banik, I., Becker, F. F., Banik, B. K. (2003). Stereoselective Synthesis of □□Lactams with Polyaromatic Imines: Entry to New and Novel Anticancer Agents. *J. Med. Chem.,* 46, 12-15. Also see: (g) Becker, F. F. and Banik, B. K. *US Patent*, 6015811, 2000. (h)

Becker, F. F., Banik, B. K., *US Patent*, 6184224, 2001. (i) Becker, F. F. Banik, B. K. *US Patent*, 6362200, 2002. Samajdar, S., Becker, F. F., Banik, B. K., Surface-Mediated Highly Efficient Regioselective Nitration of Aromatic Compounds by Bismuth Nitrate, *Tetrahedron Lett.*, 2000, *41*, 8017-8020.

[9] (a) Banik, B. K., Mukhopadhyay, C., Venkatraman, M. S., Becker, F. F. (1998). A Facile Reduction of Aromatic Nitro Compounds to Aromatic Amines by Samarium and Iodine, *Tetrahedron Lett.*, *39*, 7343-7346. (b) Basu, M. K., Becker, F. F., Banik, B. K.(2000). Ultrasound-Promoted Highly Efficient Reduction of Aromatic Nitro Compounds to the Aromatic Amines by Samarium/Ammonium Chloride, *Tetrahedron Lett.*, *41*, 6551-6554, (c) Banik, B. K., Suhendra, M., Banik, I., Becker, F. F. Indium/Ammonium Chloride Mediated Selective Reduction of Aromatic Nitro Compounds: Practical Synthesis of 6-Aminochrysene, *Synth. Commun.*, 2000, *30*, 3745-3754, (d) Banik, B. K., Banik, I., Becker, F. F. Indium/Ammonium Chloride-Induced Selective Reduction of Aromatic Nitro Compounds, *Organic Synthesis*, 2004, *81*, 188.

[10] (a) Banik, B. K., Zegrocka, O., Banik, I., Hackfeld, L., Becker, F. F. (1999). Samarium-Induced Iodine-Catalyzed Reduction of Imines: Synthesis of Amine Derivatives, *Tetrahedron Lett.*, *40*, 6731-6734. (b) Ghatak, A., Becker, F. F., Banik, B. K. (2000). Samarium Induced Alkyl Halide Mediated Reductive Coupling of Ketones. *Tetrahedron Lett.*, *41*, 3793-3796.

[11] Piwowar-Landis, K. R., Chen, D., Cui, Q. C., Minic, V. Becker, F. F. Banik, B. K., Dou, Q. P. (2006). Apoptotic-Inducing Activity of Novel Polycyclic Aromatic Compounds in Human Leukemic Cells, *Int. J. Mol. Med*, 17, 931-935.

[12] (a) Lainton, J. A. H., Hoffman, J. W., Martin, B. R., Compton, D. R. (1995). 1-Alkyl-3-(1-naphthoyl) pyrroles: A new class of cannabinoid. *Tetrahedron Lett.* 36, 1401. (b) De Leon, C. Y., Ganem, B. A. (1997). New approach to porphobilinogen and its analogs. *Tetrahedron*, 53:7731.

[13] Banik B. K., Becker F. F. (2011). "Polycyclic aromatic compounds as anticancer agents: structure-activity relationships of chrysene and pyrene derivatives." *Bioorg Med Chem.* 9:593-605.

[14] Banik, Bimal K., Basu, Manas K., Becker, Fredrick F. (2010). "Novel disubstituted chrysene as a potent agent against colon cancer" *Oncol Lett.* 1, 1033–1035.

[15] Becker, Frederick F. and Banik, Bimal K.(2014). "Polycyclic aromatic compounds as anticancer agents: synthesis and biological evaluation of methoxy dibenzofluorene derivatives." *Front. Chem.,* 2, 55.

[16] Molander, Gary A. (1992). "Application of lanthanide reagents in organic synthesis." *Chem. Rev.* 92, 29–68.

[17] Bimal K Banik. (2018). Indium Metal-Induced Reactions: Synthesis of Biologically Active Molecules. *Asian Journal of Chemistry.*30, 1-4.

[18] For samarium-induced iodine-catalyzed reactions see: (a) Banik, B. K., Samajdar, S., Ghatak, A. (2001). A Convenient Samarium-Mediated Reduction of Ferrocenyl Imines. *Heterocycles,* 55, 1957-1961; (b) Banik, B. K. (2003). Samarium-Induced Reductive Dimerization of Aryl Ketones in Aqueous Alcohol. *Chemistry-An Indian Journal,* 1, 149, (c) Samajdar, S., Banik, B. K. (2003). Samarium-Induced Reductive Dimerization of Ketimines. *Chemistry-An Indian Journal*, 1, 230 (d) Banik,B. K., M. S., Banik, I., Basu, M. K. (2004). Samarium-Induced Reductive Dimerization of Methyl Cinnamate: Synthesis of 2, 8-Diamino Chrysene *Tetrahedron Lett.* 45, 4737-4739; (e) Banik, B. K.,

Banik, I., Aounallah, N., Castillo, M. (2005). Samarium-Induced Convenient Reductive Dimerization of Aromatic Ketones: A Mechanistic Approach." *Tetrahedron Lett.* 46, 7065-7068.

[19] Banik, B. K. and Becker, F. F. (2001). Synthesis, Electrophilic Substitution and Structure-Activity Relationship Studies of Polycyclic Aromatic Compounds for the Development of Anticancer Agents. *Current Medicinal Chemistry*, 8:1513.

[20] Banik, B. K., Mukhopadhyay C., Becker, F. F. (2010). Synthesis and biological evaluation of novel dibenzofluorene derivatives as anticancer agents. *Oncol. Lett.* 309–311.

[21] Becker, F. F., Mukhopadhyay, C., Hackfeld, L., Banik, I., Banik, B. K. (2000). Polycyclic aromatic compounds as anticancer agents: synthesis and biological evaluation of dibenzofluorene derivatives. *Bioorg. Med. Chem.* 8, 2693–269.

[22] Banik, B. K., Garcia, I., Moarles, F., Aguilar, C. (2007). Novel synthesis of substituted pyrrole bound to indolinone via molecular iodine-catalyzed reaction. *Heterocycl. Commun.* 13, 109–112.

[23] Dieter, R. K., Yu, H. (2000). A facile synthesis of polysubstituted pyrroles. *Org. Lett.* 2, 2283–2286.

[24] Arcadi, A., Rossi, E. (1998). Synthesis of functionalized furans and pyrroles through annulation reactions of 4-pentynones. *Tetrahedron,* 54, 15253–15272.

[25] Periasamy, M., Srinivas, G., Bharathi, P. (1999). Conversion of Aryl methyl ketimines to 2,5- diarylpyrroles using $TiCl_4/Et_3N$." *J. Org. Chem.* 64: 4204–4205 and references cited therein.

[26] (a) Banik, B. K., Manhas, M. S. and Bose, A. K. (1997). EnantiopureHydroxy β-Lactams via Glycosylation. *Tetrahedron Lett.* 38, 5077-5080; (b) Banik, B. K., Zegrocka, O., Manhas, M. S., Bose, A. K. (1997). Enantiomerically Pure β-Lactams with the Thienamycin Side Chain via Glycosylation. *Heterocycles* 46, 173-176; (c) Banik, B. K., Manhas, M. S., Bose, A. K. (1994). Stereospecific Glycosylation via Ferrier Rearrangement for Optical Resolution. *J. Org. Chem.*, 59, 4714-4716 (d) Banik, B. K., Banik, I., Becker, F. F., (2010). Asymmetric Synthesis of Anticancerβ-Lactams via Staudinger Reaction: Utilization of Chiral Ketene from Carbohydrate, *Eur. J. Med. Chem.,* 45, 846-848.

In: Organic and Medicinal Chemistry, Volume 1
Editor: Bimal Krishna Banik
ISBN: 978-1-53614-453-6

Chapter 9

The Ferrier Rearrangement: A Versatile Methodology for the Preparation of O, S, N and C-Glycosides and Application for the Synthesis of Biologically Active Compounds

***Ram N. Yadav*[1], *Indrani Banik*[2], *Amrendra K. Singh*[1] and *Bimal Krishna Banik*[2,3,4,5,*]**

[1]Department of Chemistry, Faculty of Engineering and Technology, Veer Bahadur Singh Purvanchal University, Jaunpur, Uttar Pradesh, India
[2]Department of Molecular Pathology, The University of Texas-M. D. Anderson Cancer Center, Houston, TX, US
[3]Department of Chemistry, The University of Texas-Pan American, Edinburg, TX, US
[4]Department of Chemistry and Chemical Engineering, Stevens Institute of Technology, Castle Point on the Hudson, Hoboken, NJ, US
[5]Current Address: Vice President of Research and Education Development, Community Health Systems of South Texas, Edinburg, TX, US

Abstract

Ferrier rearrangement is one of the most fascinating reactions for installing diverse nucleophiles with high degree of stereoselectivity at the anomeric centre of the carbohydrate system toward the synthesis biologically active natural products. In carbohydrate chemistry, a reaction of glycosyl donor and acceptor catalyzed by various

[*] Corresponding Authors Emails: bimalbanik10@gmail.com and bimal.banik@chsst.org.

promoters is called glycosylation. There are numerous methodology developed to improve the reaction and streoselectivity at the amomeric center of the carbohydrates moiety. To form the glycosidic bond, a number of methods like Koenigs and Knorr(glycosyl iodide), Fischer (O-glycosides methods), Schmidt Mukaiyama (glycosyl fluride), Fraser-Reid (n-pentyl glycosides methods), Lönn, Garegg, van Boom (S-glycosides methods) and Lemieux, Thiem, and Danishefsky (glycosylation of glycals) are performed. The most notable and well-studied reaction in the literature towards this subject is the Ferrier rearrangement of the glycal. This reaction is used for the synthesis of biologically important diverse glycosides, oligosaccharides and glycocongugates which are integral structural motif in many medicinally active natural products such as antibiotics or cardiac glycosides as well as numerous synthetic intermediates. The ready availability and diverse stereogenic centre makes the glycal as one of most important starting materials for this reaction. In this chapter we cover a comprehensive report on the Ferrier glycosylation that is reported in past one decade.

Keywords: glycosides, glycosidic bond, anomer, natural products, ferrier rearrangement, stereoselectivity

1. Introduction

Ferrier has reported two important reactions concerning the transformation of carbohydrates to important biologically active molecules and intermediate precursors for the various chemical transformations. Specifically, the acid-catalyzed rearrangement of glycal in which various nucleophiles are introduced at C-1 position of the carbohydrate to give 2,3-unsubstituted glycosylated products is called Ferrier rearrangement [1, 2]. Although the allylic rearrangement of glycal with water has been investigated by Emil Fiescher, Ferrier has generalized the synthetic utility of this reaction in 1960. This chapter has focused on the Ferrier rearrangement of the glycal for the synthesis of diverse C, O, and S-glycosides.

The most notable glycosides are Fluoride, Nitrogen, Sulphur, and Carbon linked products and these are synthesized with relatively high stereoselectivity. One of the most important synthetic applications of this rearrangement reaction is the synthesis of C-glycosides of diverse interests.

1.1. The General and Mechanistic Aspects of Ferrier Rearrangement

The Ferrier glycosylation reaction involves the allylic coupling of a nucleophile to a glycal having a leaving group at C-3 position, which leads to the corresponding 2,3-unsaturated glycoside [3]. Initially the delocalized allyloxocarbenium ion (**2**) is formed by acid activators. This intermediate trapped by *in situ* with nucleophiles which leads to the anomeric mixtures of α and β **3** and **4** glycosides followed by the double bond migration to the position 3, 4 (Scheme 1).

All major classes of glycosidic bonds have been successively synthesized using this protocol. The following important general aspects and factors which greatly influence the course of the reaction are delineated below.

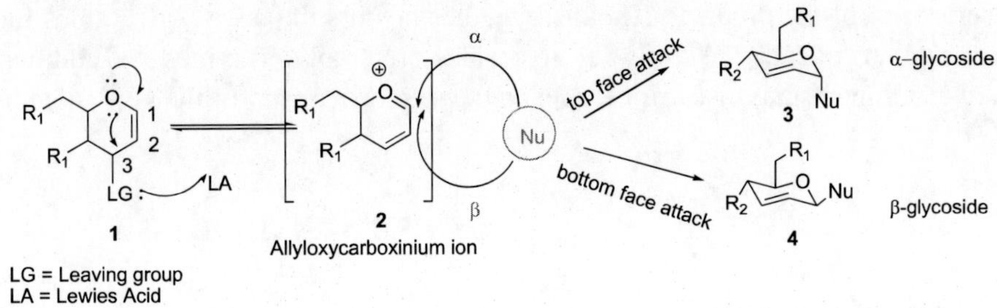

Scheme 1.

1.2. Formation of Glycosidic Bond

The glycosidic bond is usually formed by the nucleophilic attack of the nucleophile or other partially protected sugars moiety to the electrophilic center of a glycal. The compound that gives the glycosyl moiety is called the *glycosyl donor*, and the nucleophile that receives it is known as *glycosyl acceptor*. The reaction generally is performed in the presence of an activator called "promoter." The role of the promoter is to assist the departure of the leaving group from the anomeric postion of the sugar moiety. Promoters are often used in catalytic amounts, although in some instances they are used stoichiometrically. In some cases, other additives such as molecular sieves or base as acid scavenger are used.

1.3. Challenge and Problem with Glycosidic Bond Formation

The success of a coupling reaction between glycals and nucleophiles depends on the reactivity of the donor and acceptor, and the nature of the protecting groups present in the sugars structures. It greatly influences the preferred selectivity of the reaction towards the α and β-anomeric form. The mode of reaction condition and experience of the person conducting the reaction also plays a role. The choice of solvent and temperature also play a crucial role in products distribution.

1.4. Promotor, Solvent and Experimental Conditions

The promoters are generally a Lewis acid, transition metal, ionic liquid, organocatalyst and solid supported resin. These have a great influence in the reaction as it favours the departure of the leaving group. The solvent has also an influence on the overall rate of the processes and on the stereochemistry of the glycosylated products. Solvents of the low polarity such as dichloromethane or diethyl ether are frequently used for the glycosylation reaction. The polar aprotic solvents, such as acetonitrile or nitromethane have also been used for this purpose. It known that the coordinating properties of the solvents to form a complex with sugar oxonium cation intermediate influence the approaches of the nucleophiles and influence the anomeric ratio of the products. The diethyl ether favours the *cis* 1,2-glycosides formation while the acetonitrile leads the *trans*-1,2 glycosides preferably. The mode of addition of the reagents in experiments has also critical role on the stereochemistry of the glycosidic bond.

2. Glycal in Glycosides Synthesis

2.1. Synthesis of Glycal Derivatives

Several protected glycals are available commercially, including the simple and important glycals 3,4,6-tri-O-acetyl-D-glucal (**1**) and 3,4,6-tri-O-acetyl-D-galactal (**2**). These as well as others serve as precursors to a wide range of useful reactions in organic, medicinal and carbohydrate chemistry. The synthetic value of acetylated glucals has been enhanced by the availability of simple one-pot procedure for the conversion of acyl into alkyl protected glucThe common O-acetylated pyranoid glycals can be obtained using traditional Fischer–Zach method [4] of zinc-promoted elimination of glycosyl halides in acetic acid. The scope of this method has been extended with the development of efficient aprotic and non-acidic conditions for this reaction using zinc dust and 1-methyl imidazole in ethyl acetate [5].

Acetylated glycosyl halide on treatment with catalytic amounts of Vitamin B (as source of Co (III)) together with zinc and ammonium chloride or with **6** in THF solutions of the Ti (III) dimer (Cp_2TiCl) are the other methods [7].

This elimination methods are, however, limited to the readily available glucosyl and galactosyl halide precursors, 2 and therefore, are primary sources of D-glucal and D-galactal derivatives. The most attractive route to D-allal and D-gulal derivatives **11** and **12** (Scheme 2) is *via* oxidation and rearrangement of S-phenyl-2,3-dideoxy derivatives **7** and **8** derived from acetylated D-glucal and D-galactal, respectively, using the Ferrier reaction on glycals.

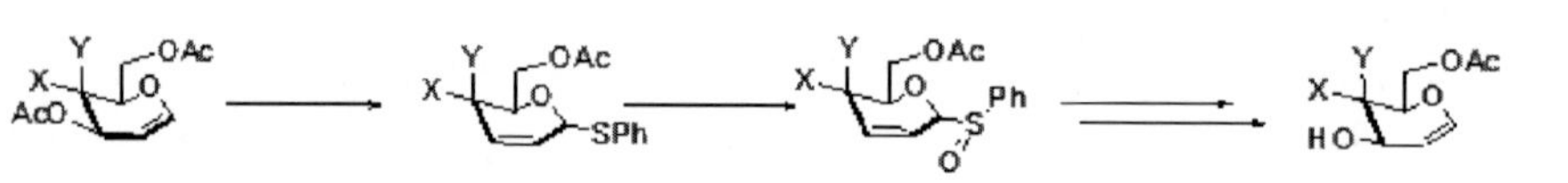

5 (X = H Y = OAc) **6** (X = H Y = OAc) — **7** (X = H Y = OAc) **8** (X = H Y = OAc) — **9** (X = H Y = OAc) **10** (X = H Y = OAc) — **11** (X = H Y = OAc) **12** (X = H Y = OAc)

Scheme 2. Synthetic route to D-alla and D-glucal derivatives.

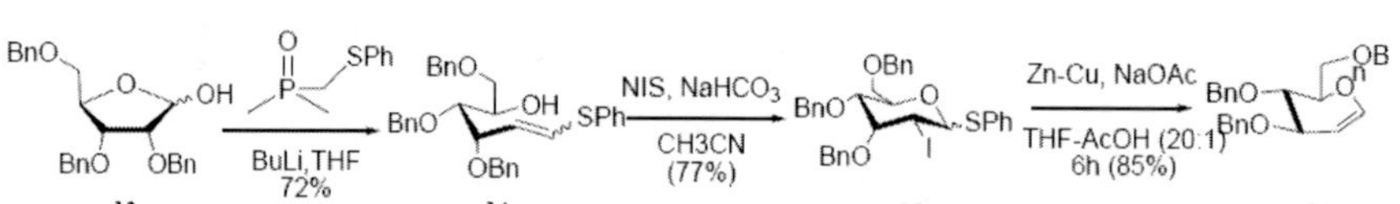

13 14 15 16

Scheme 3. Synthesis of a D-allal derivative.

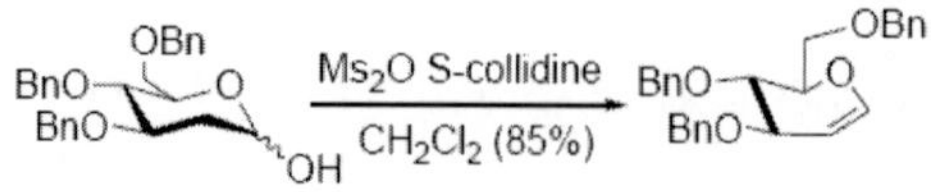

17 18

Scheme 4. Preparation of glycals via base induced elimination.

The most promising methodology to access the wide range of glycals and its derivatives is illustrated in Scheme 3. This transformation describes conversion of benzyl protected D-ribose into a D-allal *via* olefination-cyclization-elimination sequences.

Glycals are also available from 2-deoxy sugars by acid or base-induced eliminations of anomeric substituents. These methods are limited by the availability of the 2-deoxy sugars, for which the glycals themselves are the most obvious synthetic precursors. However, examples of these methods (Scheme 4) include the direct preparation of tri-O-benzyl D-glucal (**18**) from 2-deoxy-tri-O-benzyl-D-glucopyranose (**17**) via its 1-O-mesylate in excellent yield [8].

Tri-O-acetyl and pivaloyl protected glycals are the most widely used as a starting material for the study and synthesis of the various glycosidic bonds.

2.2. O-Glycosides from Glycal

Perhaps, oxygen-glycosides are common and these can be synthesized from glycals by nucleophilic addition of alcohols in the presence of a catalyst. A wide range of O-glycosides are prepared following this method. The most common catalysts include $BF_3{\cdot}Et_2O$ [9], BCl_3 [10], Ph3P-HBr [11a,b], p-TsOH, CSA and $Pd(OAc)_2$ [12]. Ferrier rearrangement is also a powerful approach for the synthesis of pseudoglycals and their derivatives, starting from glycals. This process is usually performed by reacting the glycal with a Lewis acid catalyst or with redox reagent, such as $InCl_3$ [13], ceric ammonium nitrate [14], $HClO_4$-SiO_2, $SnCl_4$, $Sc(OTf)_3$, [16] and $Fe_2(SO_4)_3.xH_2O$ [17].

Kinfe et al. reported a temperature-controlled selective reaction. Direct addition of alcohols to the alkene of tri-Obenzyl-D-glucal (**18**) to generate 2-deoxy-O-glycoside sugars and the Ferrier rearrangement reaction to produce the corresponding pseudoglycals are mediated by Al $(OTf)_3$ [18].

Interestingly, at 0°C, the mechanistic course of the reaction switched to produce, the Ferrier rearranged pseudoglycal products instead of the 2-deoxy-1-O-glycoside (Scheme 5, Table 2.1, entries 6–11).

This reaction proceeded with numerous types of alcohols and the products were obtained predominantly as the α-anomers, in good yields. The alteration of the mechanistic route with respect to temperature changes was very attractive part of this method. Importantly, THF or acetonitrile as solvent failed to demonstrate this switch in mechanism which was observed only with DCM and 1, 2-dichloroethane. In both instances (i.e., for THF or acetonitrile as solvent), only the Ferrier rearranged products were obtained.

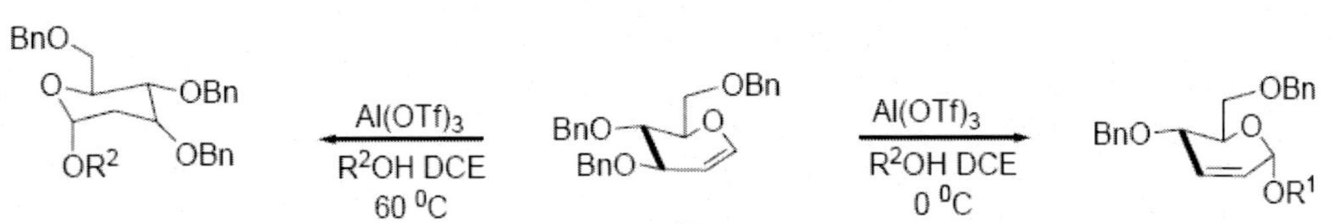

20 **18** **19**

DCE : 1,2- Dichloroethane

Scheme 5. Aluminium-triflate catalysed, temprature controlled Ferrierrier rearrangement vs selective addition of nucleophile.

Table 2.1. Al $(OTf)_3$-catalysed addition of alcohol to glucal 1[a]

Entry	Alcohol (R^2)	T°C	Products	Ratio[b] (α:β)	Yield[c] (%)
1	CH_3	60	20a	01:01	56
2	Benzyl	60	20b	03:02	69
3	Propargyl	60	20c	07:03	74
4	Allyl	60	20d	02:03	72
5	CH3	0	19a	07:01	74
6	Benzyl	0	19b	07:01	69
7	Propargyl	0	19c	08:01	74
8	Allyl	0	19d	03:01	83
9	4-C6H4Br	0	19e	04:01	63
10	4-C6H4OCH3	0	19f	09:01	76
11	Cyclohexyl	0	19g	08:01	86

[a] 5mol% of $Al(OTf)_3$, 0.48 mmol substrate, 0.72 mmol alcohol. (1.5 eq), 2 mL DCE; [b]calculated using ^1H NMR spectroscopy.

Rafiee et al. explored the alkali metals, potassium dodeca-tungstocobaltate, $K_5CoW_{12}O_{40}{\cdot}3H_2O$ HO (POM), to recognize the role of potassium as the promoter. Various glycosylations with different acetylated glycals **21** and POM as the catalyst in various solvents produced the α-glycosides **23** as the major product in good-to-excellent yields (Scheme 6). Extensive optimization revealed acetonitrile as the best choice of solvent for this method. The catalysts showed high yields, even after being recovered four times, thus proving the high efficiency of the alkali metal as the catalyst [18].

The catalytic efficiency of iron-based Lewis acid catalyzed Ferrier rearrangement of peracetylated glucal with diverse alcohol was successful employed by Wang and Chen et al. A comparative study for the glycosylation reaction with various iron-based Lewis acid catalysts was performed. They found that ferric triflate is one of the most versatile Lewis acid promoter for performing the glycosylation of per acetyl glycals in terms of both selectivity and yields. The use of 10 mol% of the catalyst in dichloromethane as the solvent gave a maximum of 87%yield with the α-anomer as the major product (Scheme 7, Table 2.2) [19].

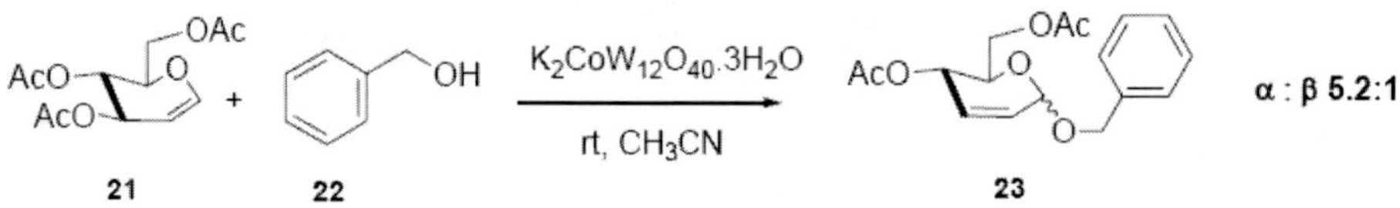

Scheme 6. Potassium dodecatungstocobaltate catalysed Ferrier rearrangement of per acetylated D-Glucal.

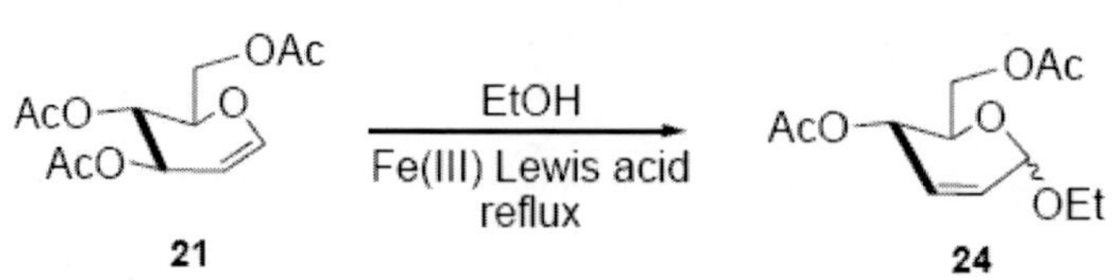

Scheme 7. Comparative study of Fe(III) catalyst in Ferrier rearrangement.

Table 2.2. Streoselectivity in Fe (III) catalysed Ferrier glycosylation

Entry	Promoters	Time(h)	Yield (%)	Anomeric ration	
				α	β
1	$FeCl_3$	2	31	5	1
2	$Fe_2(SO4)_3$	2	Negligible	Negligible	
3	$Fe(NO_3)_3$	2	41	9	1
4	$Fe(OTf)_3$	1/2	48	10	1

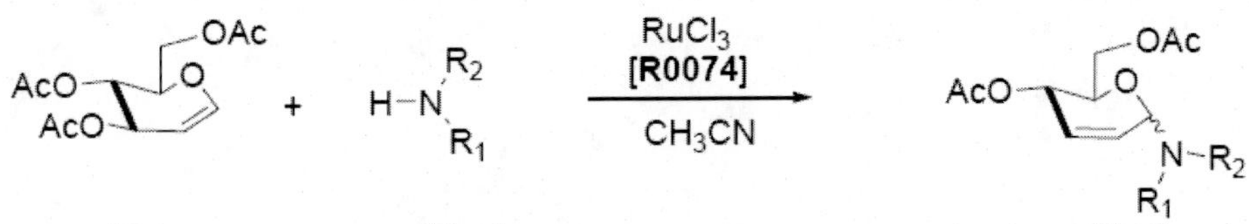

21 25a-d 26a-d

Scheme 8. Ruthenium-catalyzed synthesis of 2,3-unsaturated-N-glycosides via Ferrier rearrangement.

Zinc chloride was also proven an effective promoter for the glycosylation of the glucal with allylic, benzylic and tertiary alcohols through a SN^1 nucleophilic pathway. Liu and co-workers demonstrated $ZnCl_2$ impregnated with activated alumina was most effective catalyst for the synthesis of 2, 3-unsaturated glycosides. They also performed the glycosylation reaction in the presence of solvent and in the absence of any solvent.

Kashyap et al. reported the ruthenium-catalyzed synthesis of 2,3-unsaturated-N-glycosides *via* Ferrier azaglycosylation. According to their results, in the presence of ruthenium (III) chloride catalyst, glycals smoothly reacted with various N-nucleophiles such as sulphonamides, benzamides, and carbamates to afford the corresponding 2,3-unsaturated-N-glycosides in high yields. High α-anomeric selectivity was observed with N-benzyl or N-phenyl group-substituted sulfonamides. It was expected that this protocol may become successful for constructing N-glycosidic linkages in neoglycoconjugates (Scheme 8) [20].

More recently, $Zn(OTf)_2$ was developed as a cheaper, milder and efficient catalyst for the stereo selective synthesis of α-glycosides with a variety of O- and S-nucleophiles. Many functional groups were tolerated well under reaction conditions [21].

Srinivas et al. reported $ZnBr_2$-catalyzed glycosylation of the peracetylated glucal with variety of primary, secondary, tertiary, allylic and benzylic alcohols under microwave conditions [22].

Diethyl zinc is an efficient reagent for the cyclopropanation reaction. The catalytic properties of this reagent were studied for activating the alcohol during glycosylation reaction. The synthesis of the 2, 3 unsaturated glycosides was achieved by treating the zinc salt-activated alcohol with glucal in presence of Pd catalyst. The special emphasis of this catalytic protocols was the anomeric selectivity of the Ferrier products which mainly governed by the reagent and the solvent system rather than the nature of the protecting groups of the sugar moiety. Using the ligand such as di(tert-butyl)-2-biphenyl phosphine exclusively delivered the α-anomer while the use of $P(OMe)_3$ ligand furnished α-anomer predominantly as glycosylated products (Scheme 9). Nguyen et al. reported the direct synthesis of the α-glycosides from the glycal *via* glucal trichloroacetimidates using Pd-II/L [23].

Table 2.3. N-glycosylation through Ferrier rearrangement

Entry	Nucleophile	N-glycosides	Time	Yield (%)	α : β ratio
1	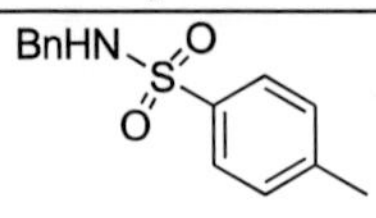 25a	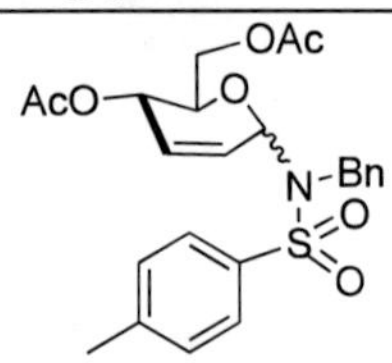 26a	30 min	81	88:12
2	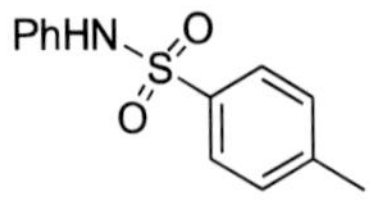 25b	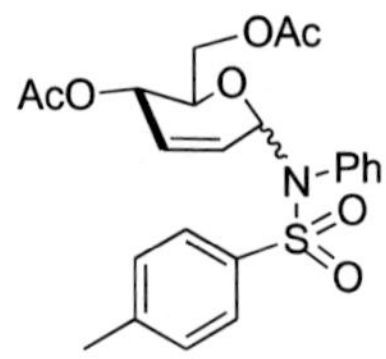 26b	15 min	65	α only
3	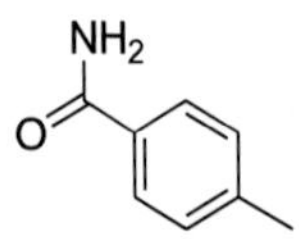 25c	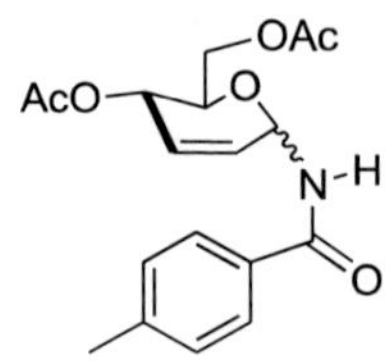 26c	3h	82	75:25
4	H_2N-CBz 25d	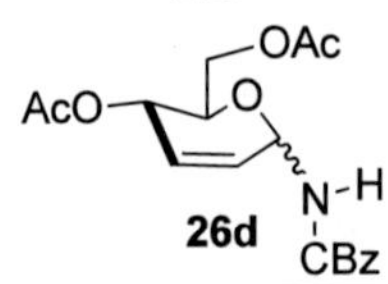 26d	1h	76	71:29

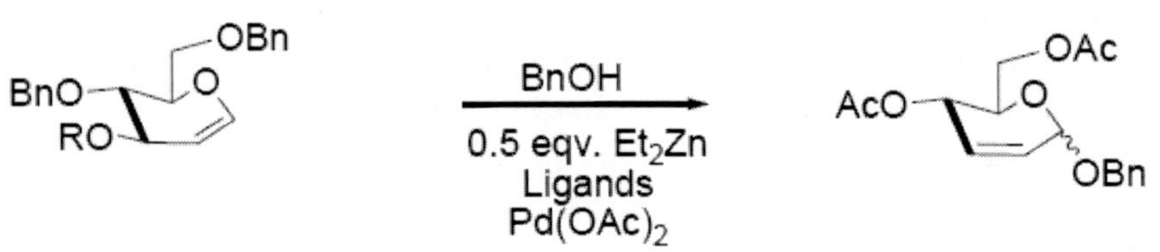

27 a R = OAc
27 b R = tBoc

28 a R = OAc
28 b R = ^{t}Boc

Scheme 9. Pd-catalysed glycosylation of benzyl protected glycals.

Table 2.4. Pd-catalysed glycosylation of protected benzyl alcohol with glycals

Entry	Compound	Ligand	Yield (%)	α/β
1	28a	15%DTBBP	92	<1:25
2	28a	30%$P(OMe)_3$	90	7:1
3	28b	15%DTBBP	92	<1:25
4	28b	30%$P(OMe)_3$	90	5:1

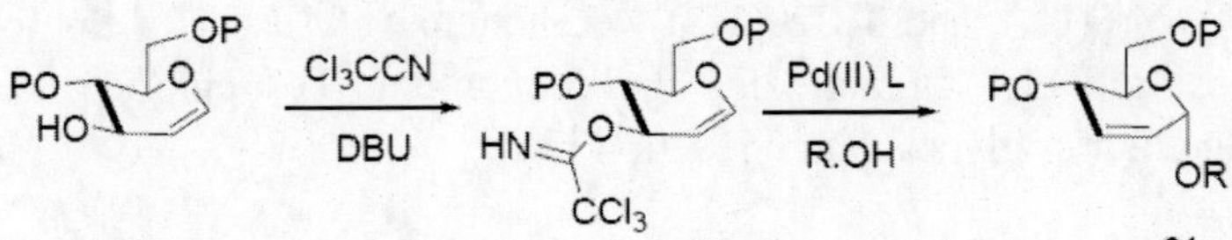

29 30 31

Scheme 10. Pd II catalysed transformation of glycal to glycosides.

The Pd (II)-catalyzed-ligand strategy was extensively exploited in the activation of the pi system of the glycals which mainly controlled the selectivity of the products formation during glycosylation reaction. Recently, Liu and co-workers developed the decarboxylative strategy for the synthesis of unsaturated O-glycosides using $Pd_2(dba)_3$ catalyst and DtBPF [1,1'-bis(di-tert-butylphosphino)ferrocene] as ligand in presence of base $CsCO_3$. The anomeric selectivity was depended upon the concentration of the base employed. A higher amount of the base favored the formation of β-glycosides selectively. The higher α-anomer selectivity was achieved using Re IV-Oxo complex.

Galan et al. [23] explored the additive-free palladium-catalyzed Ferrier glycosylation of glycal for the synthesis of 2, 3-unsaturated O-glycoside from acetylated glycal in high α-anomeric selectivity. The mechanistic study envisaged that the reaction proceed through the formation of alkoxy-palladium intermediate that enhances the proton acidity and nucleophilicity of the oxygen in the glycosyl donor. This method was explored for the synthesis of wide range of glycoside and glycoconjugate of synthetic efficacy (Scheme 11, Table 2.5).

The gold metal-based Lewis acid catalyzed Ferrier rearrangement was employed by Balamurgan et al. In this reaction, $AuCl_3$ was used as the catalyst under mild reaction condition for the synthesis of 2, 3 unsaturated glycosides with high degree of α-stereoselectivity. The reaction of glycal with propargyl alcohol in presence of $AuCl_3$ in acetonitrile led to the formation of the intermediate (**32**) *in situ*. This intermediate upon treatment with a variety of the glycosyl acceptors produced α-glycosides predominantly in good to moderate yields. An extensive study was performed to establish the versatility of this method.

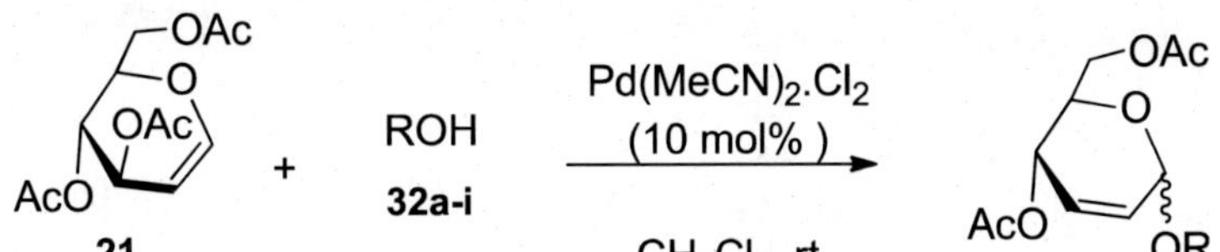

21 32a-i 33a-i

Scheme 11. Stereoselective Ferrier rearrangement of peracetylated glucal **21** with O- nucleophiles.

Other transition metal-based Lewis acid-catalyzed Ferrier glycosylation toward the synthesis of unsaturated glycosides was also studied. In this context, the notable experiments was performed using $Cu(OTf)_3$-mediated synthesis of O-glycosides under mild reaction conditions. Kashyap et al. used $RuCl_3.3H_2O$ as an efficient catalyst for the synthesis of O-glycosides in good to moderate yield from glycols [24]. Niobium (V) chloride was also implemented as efficient catalyst for the synthesis of 2, 3 unsaturated glycosides from per acetylated D-glucal and variety of glycosyl acceptors. In addition to other transition metal triflates, $Sc(OTf)_3$, $Y(OTf_3)$, and $TiCl_3(OTf)$ were extensively used for the formation of O-glycosides stereoselectively. Organo zinc and zirconium (IV) chloride were effectively investigated for the Ferrier glycosylation reaction [25].

Table 2.5. Ferrier rearrangement of peracetylated glucal 21 with O- nucleophiles.

Entry	ROH		Time (h)	Yield (%)	α/β
1		32a	3	97	9:1
2	BnOH	32b	3	97	9:1
3	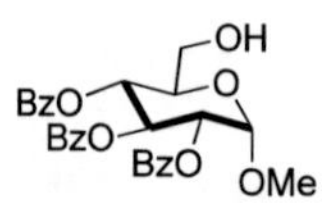	32c	7	90	9:1
4	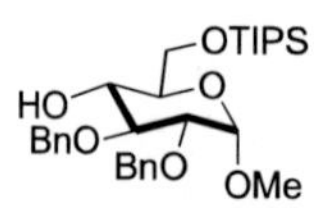	32d	16	68	9:1
5	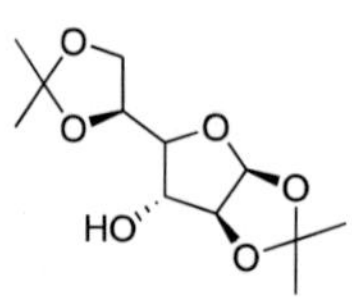	32e	17	70	4:1
6	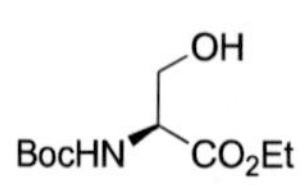	32f	17	68	3:1
7	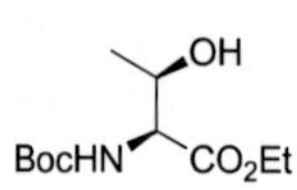	32g	17	64	>99:1
8	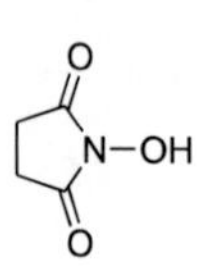	32h	15	90	>99:1
9	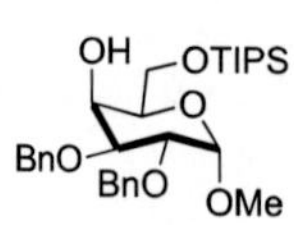	32i	17	65	20:1

[a] Yield of isolated product; [b] Determined by crude ^{1}H NMR; [c] Reaction carried out at 50°C.

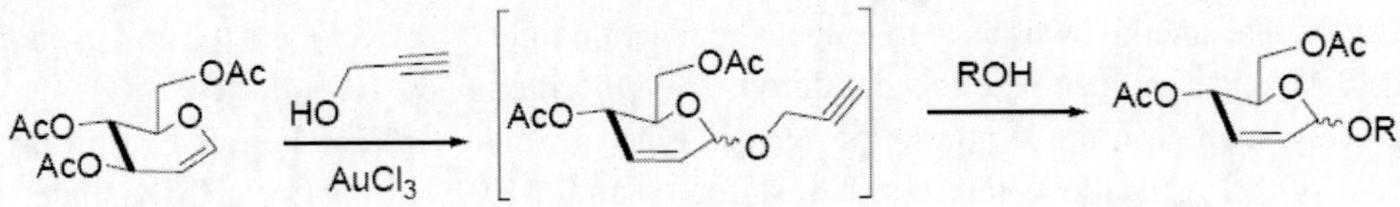

21 32 33

Scheme 11. $AuCl_3$ - catalysed through proporgylation-glycosylation strategy for Ferrier rearrangement.

In order to explore the catalytic efficacy of the rare earth elements as glycosylating promoters for Ferrier glycosylation, Yadav et al. reported the synthesis of 2, 3 unsaturated glycosides from the glycal using $Dy(OTf)_3$ as the catalyst with various alcohols, phenols and amino acids in ionic liquid medium [26]. Lanthanum (III) nitrate was also used for stereoselective synthesis of α-glycosides from glycals under solvent free conditions. However, this catalytic protocol failed to give the products in dichloromethane, THF and diethyl ether. In contrast, nitromethane was found as the best solvent for this reaction and the yield of the reaction improved significantly. The high stability towards moisture and heat and milder reaction condition made $Er(OTf)_3$ as one of the most efficient catalysts for promoting Ferrier glycosylation reaction. In almost all the reactions investigated, notable stereoselectivity was observed. The rate of reaction was also fast. Zirconium catalyst generated by electrochemical method acted as a novel catalyst for the effective glycosylation of per acetylated glycals [27].

The catalytic property of aluminium trichloride was extensively used in organic synthesis for the synthesis of various useful compounds. It was used to trigger the Ferrier glycosylation reaction. In this reaction semi-protected glactal was treated with diethyl phosporochlorodite and triethyl amine that furnished 3-O-diethoxyphosphanyl-D-glactal derivative which underwent Ferrier rearrangement with glycosyl acceptor in presence of aluminium trichloride yielding the desired 2, 3-unsaturated glycosylated products. Indium was frequently employed as the glycosylation promoter for Ferrier rearrangement reaction. In this context Boga et al. reported an interesting result on Ferrier glycosylation reaction. They treated the peracetylated glucal using 20 mol% of anhydrous indium trichloride with a variety of nucleophiles to accesses the 2, 3 unsaturated O-glycosides with high α-stereochemistry at the anomeric center. They allowed the reaction for longer time under this condition. Interestingly, the O-glycosides were converted to the corresponding C-glycosides. In similar strategy boron metal-based catalyst was played a pivotal role in Ferrier glycosylation reaction. $BiCl_3$ and $Bi(OTf)_3$ with or without silica gel bed-supported glycosylation of glycals with diverse nucleophiles was explored extensively. Various type of aliphatic and aromatic alcohols produced corresponding 2, 3 unsaturated glycosides with high yield of α-glycosides.

The heavier metals have an adverse effect on environment and biological systems. Therefore, it is demanding to establish protocols without using metals as the catalyst. De et al. developed a protocol for the synthesis of 2, 3 unsaturated O-glycosides without using any metal catalyst. They showed that reaction of excess nucleophiles (20 mol%) and glycal proceeded smoothly and produced 2, 3 unsaturated glycosides in good yield without using any metals or Lewis acids as a promoter. Montmorillo K-10 was established a highly efficient catalyst for rearrangement reaction under solvent-free reaction condition. Application of microwave heating also reported advantageous over traditional heating in the formation of glycosides. This method was used to reduce the reaction time effectively and to form α-anomer exclusively under milder reaction condition. Montmorillonite K-10 suported $Bi(OTf)_3$ was extensively studied in the synthesis of various type of Sialic acid *via* Ferrier rearrangement protocol. Other versatile catalyst systems, for example $HClO_4$-Silica, $NaHSO_4$-silica, and H_2SO_4-Silica, CF_3SO_4-Silica were used for the synthesis of a wide range of pseudo glycosides. In the context of organic acid catalysts, acetic acid failed to yield any desired unsaturated glycosides. However, Liu and co-workers decided to test the effectiveness of camphorsulfonic acid (CSA) as probable catalyst for triggering the formation of pseudo glycosides. Both (R) and (S)-CSA, were reported to give the desired product, the (S)-isomer was the more efficient one.

Various other catalysts were developed over the decades for the formation of 2, 3-unsaturated O-glycosides from glycals. The most notable catalysts were cyanuric chloride-catalyzed Ferrier transformation, BDMS-mediated rearrangement and DNBA-mediated synthesis of O- glycosides. $BF_3.OEt_2O$-catalyzed glycosylation Ferrier glycosylation was also studied.

Freitas et al. investigated the tellurium (IV) tetrabromide catalyzed O-glycosylation of peracetylated glycal with diverse alcohols in good to excellent α-selectivity.

Tellurium tetrahalides ($TeCl_4$, $TeBr_4$ and TeI_4) were extensively used for research activity because of their structural versatility. The chemical behaviors of these halides were due to partially ionic Te-X bonding [28]. These were well suited to react with both Lewis bases and Lewis acids.

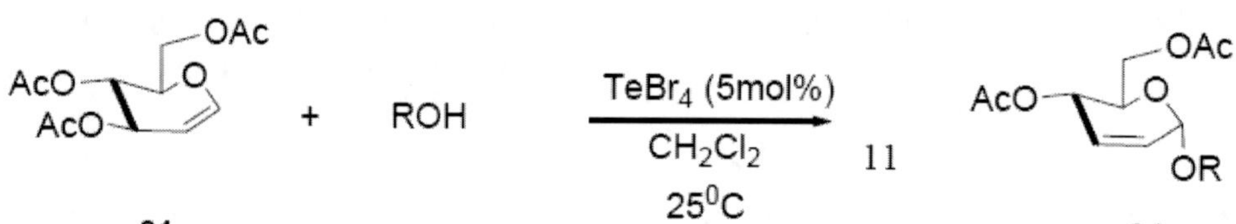

21 **34**

Scheme 12. $TeBr_4$ promoted synthesis of 2,3 unsaturated O-glycosides through Ferrier glycosylation.

The reaction of glycal **21** with t-BuOH was described to be difficult to proceed. But, under these catalytic conditions the reaction gave a clean product in high yield and excellent anomeric selectivity (Table 2.6, entry 10). More reactive alcohols, such as allylic, benzylic and propargylic alcohols gave better yields and reactions were completed in shorter reaction times when compared with aliphatic and alicyclic alcohols (Table 2.6, entries 1, 2, 3 and 5). The reaction time increased when a homoallylic alcohol was used (Table 2.2, entry 4). A lower yield was observed when phenol was used as the nucleophile (Table 2.6, entry 13). This effect was also observed before for other catalytic systems [29].

Vankar et al. investigated a new catalyst system comprising of $AuCl_3$ and phenylacetylene that promotes the Ferrier rearrangement of glycals and 2-acetoxymethylglycals with different nucleophiles, and also the O-glycosylation of 1-O-acetyl sugars to obtain a variety of useful glycosides at room temperature through relay catalysis. Good anomeric selectivity was observed for the Ferrier rearrangement, whereas the O-glycosylation of 1-O-acetyl sugars gave mixtures of dia-stereomers with moderate to excellent selectivity (Scheme 13).

Table 2.6. Influence of temperature and anomeric ratio of TeBr4 catalysed Ferrier O-Glycosylation of D-Glucals

Entry	Alcohol	Glycosides	Time	Ratio [a] α : β	Yield (%) [b]
1	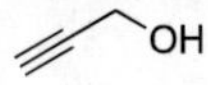	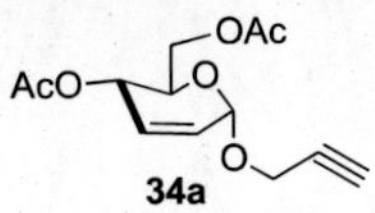 34a	0.1	89:11	92
2	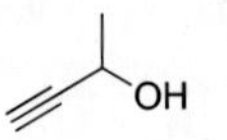	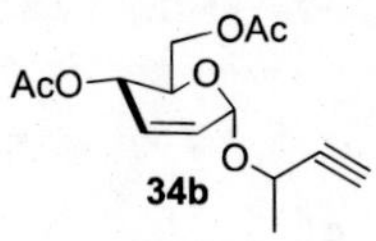 34b	0.1	88:12	94
3	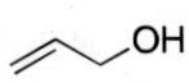	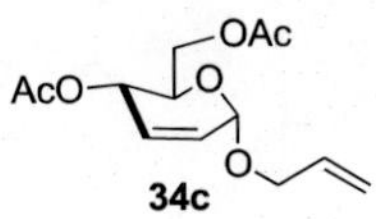 34c	0.25	88:12	86
4	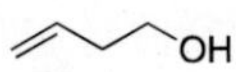	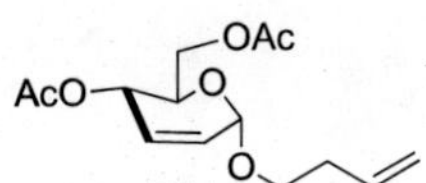 34d	1.0	89:11	87
5	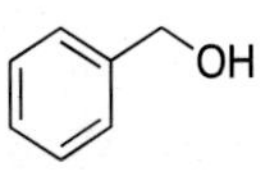	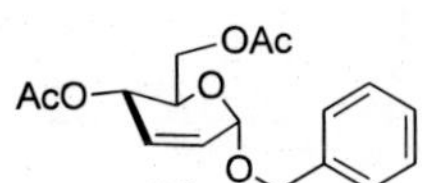 34e	0.25	86:14	86
6	MeOH	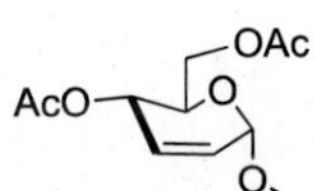 34f	1.5	78:22	86
7	EtOH	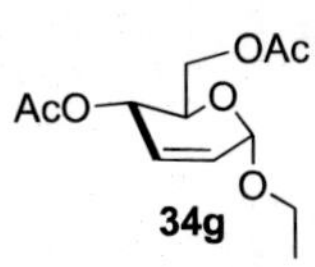 34g	5.0	83:17	84
8	^{i}PrOH	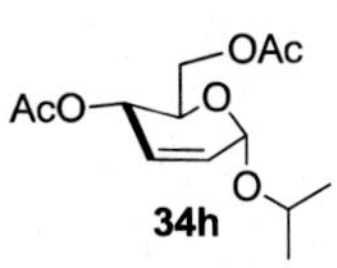 34h	1.5	90:10	92
9	n-PrOH	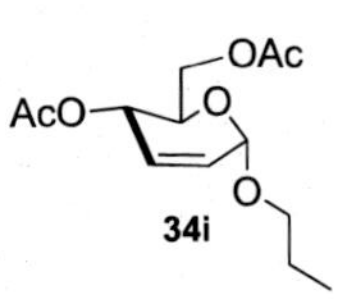 34i	2.5	88:12	85

Table 2.6. (Continued)

Entry	Alcohol	Glycosides	Time	Ratio [a]α : β	Yield (%)[b]
10	^{t}BuOH	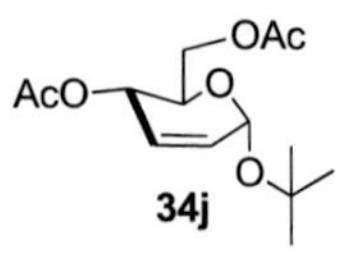 34j	2.0	93:7	86
11	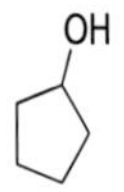	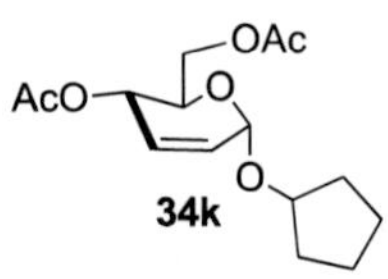 34k	1.0	89:11	85
12	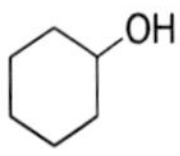	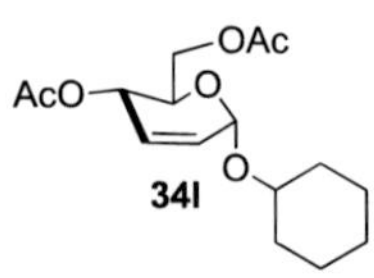 34l	1.0	91:9	90
13	PhOH	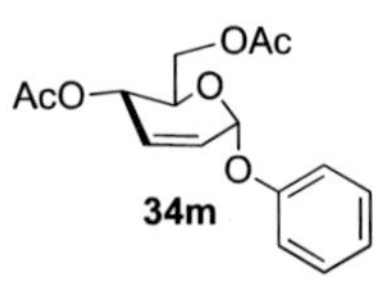 34m	1.0	88:12	70

[a]The anomeric ratios were obtained by ^{1}H NMR and confirmed by gas chromatography for all reactions.
[b] Isolated yield.

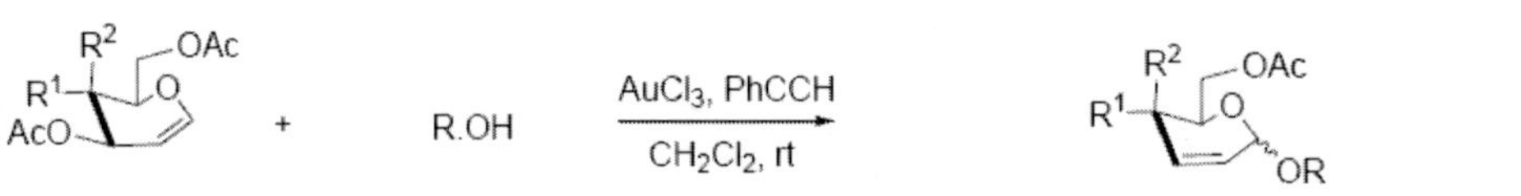

35 R^1 = OAc R^2 = H
36 R^1 = H R^2 = OAc

37 a-k R^1 = OAc R^2 = H
38 a-d R^1 = H R^2 = OAc

Scheme 13. $AuCl_3$/phenylacetylene-catalysed Ferrier rearrangement of glycal derivative of **35** and **36.**

Lignans are widely distributed in angiosperms and gymnosperms. The range of their structures and biological activities is broad. Various lignans are known to have anti-tumor, antimitotic and antiviral activity and they specifically inhibit certain enzymes. Toxicity to fungi, insects and vertebrates is observed for some lignans and a variety of physiological activities are documented.

Due to its complex structure and densely functionalized core, it can be modified to therapeutic agents. The few lignans such as epodophyllotoxin, epipodophyllotoxin & picropodophyllin has been investigated as anticancer agent, investigated the utility of $Yb(OTf)_3$ as a Lewis acid catalyst for Ferrier rearrangement, to access stereoselective O-glycosylation of these lignans. The O-glycosides demonstrated better bioavailabity, reduced toxicity and improved cytotoxicity to generate semi synthetic analogues, i.e., O-pseudoglycals. The synthesis of 2, 3-unsaturated glycopyranosides was attempted *via* Ferrier rearrangement of tri-O-acetyl glucal (**21**) using an excess of O-nucleophiles (**39a-c**) in the presence of catalytic amounts of $Yb(OTf)_3$ using anhydrous CH_2Cl_2 as solvent (Scheme 14) [30].

Table 2.7. $AuCl_3$/phenyl acetylene- catalysed Ferrier rearrangement of D-glucal (37 a-k) in dichloromethane at room temperature

Entry	Nucleophile	(O S C N)-glycosides	Time	Yield[b] (%)	α:β [c]ratio
1		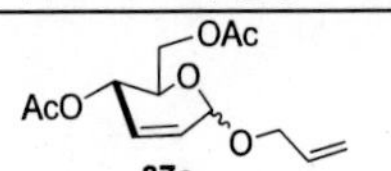 **37a**	15 min	92	10:1
2		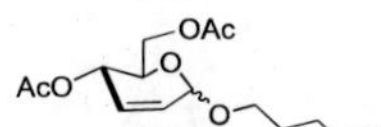 **37b**	15 min	91	8:1
3	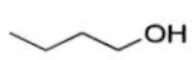	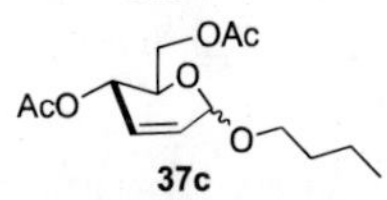 **37c**	1h	90	7.5:1
4	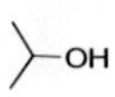	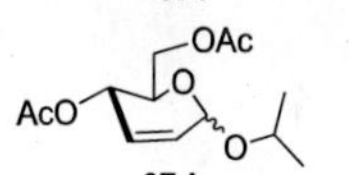 **37d**	75 min	82	5:1
5		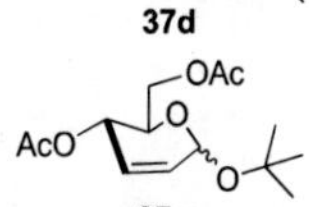 **37e**	8 h	72	>19:1
6	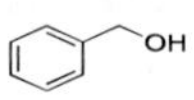	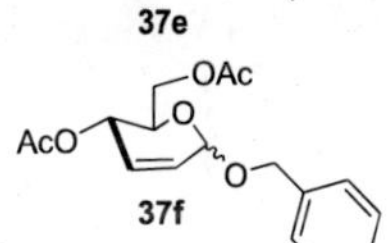 **37f**	5 min	96	7:1
7	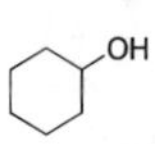	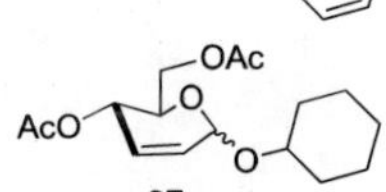 **37g**	1h	96	>19:1
8	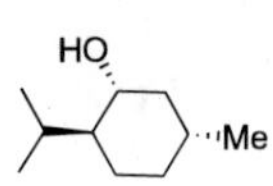	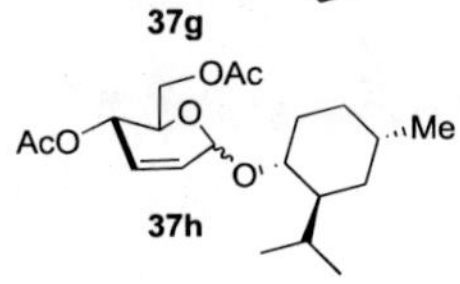 **37h**	15 min	85	11:1
9	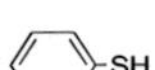	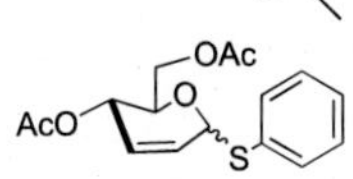 **37i**	5 min	93	5:1
10	$TsNH_2$	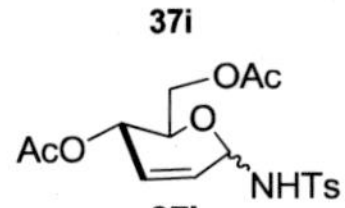 **37j**	3 min	92	3:1
11	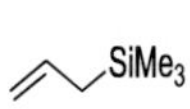	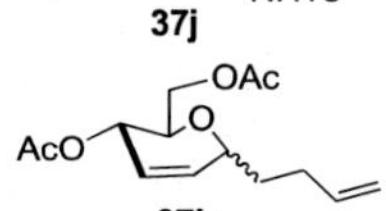 **37k**	1h	94	10:1

[a]AuCl3 (1 mol%) and phenylacetylene (1 mol%) were used at room temperature (35-37 °C). [b]Isolated yields [c] α/β ratios were determined from ^{1}HNMR spectral analysis.

Table 2.8. AuCl₃/phenylacetylene-catalysed Ferrier rearrangement of galactal (38 a-d) at room temperature in dichloromethane

Entry	Nucleophile	(O S)-glycosides	Time	Yield (%)[b]	α:β ratio[c]
1	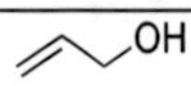	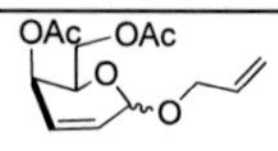 **38a**	5 min	75	α only
2	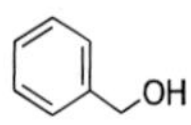	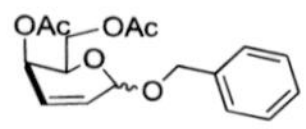 **38b**	3 min	75	α only
3	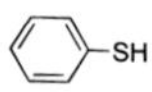	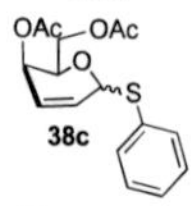 **38c**	4.5 min	75	α only
4	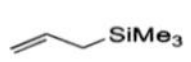	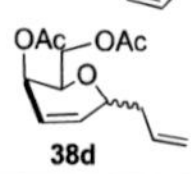 **38d**	3 min	96	8:1

[a]$AuCl_3$ (1 mol%) and phenylacetylene (1 mol%) were used at room temperature (35-37 °C). [b]Isolated yields [c] a/b ratios were determined from ^{1}HNMR spectral analysis.

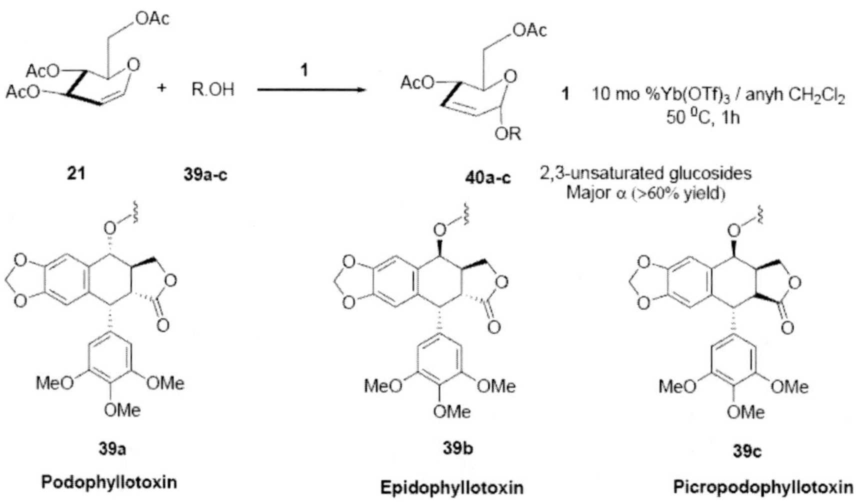

Scheme 14. Ferrier rearrangement for the synthesis O-pseudoglycosides.

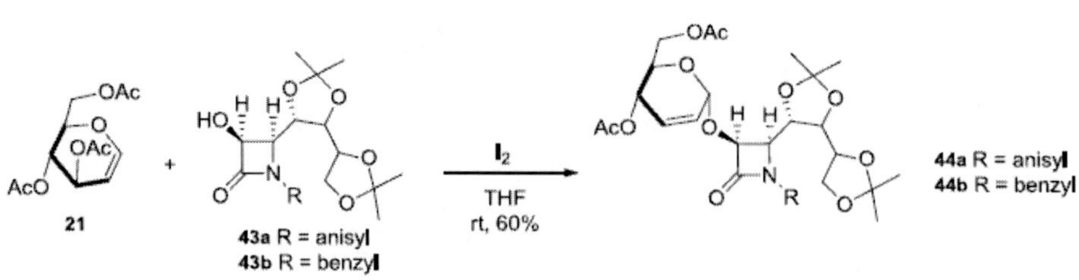

Scheme 15. Ferrier rearrangement of 3-hydroxy β –lactams with diacetonide.

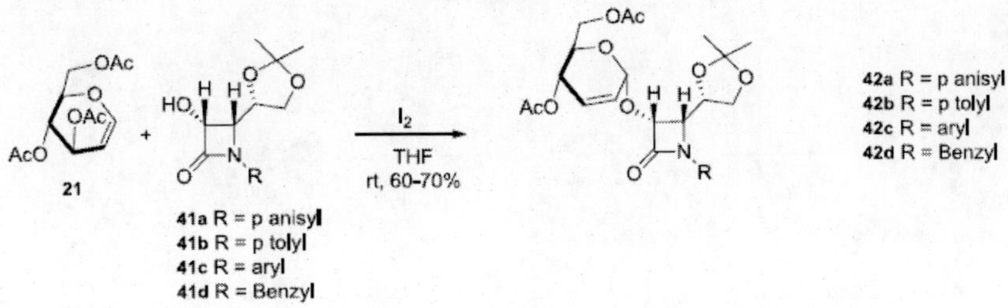

Scheme 16. Ferrier rearrangement of 3-hudroxy β-lactams with monoacetonide.

Banik et al. reported a highly streospecific iodine-catalyzed Ferrier glycosylation of enantiopure 3-hydroxy β-lactam (41a-d) with glycal (21) in excellent yield (Scheme 16) [31].

Table 2.9. Ferrier rearrangement for the synthesis of Anticancer Lignans

Entry	Nucleophiles	Glycosides	Yield
1	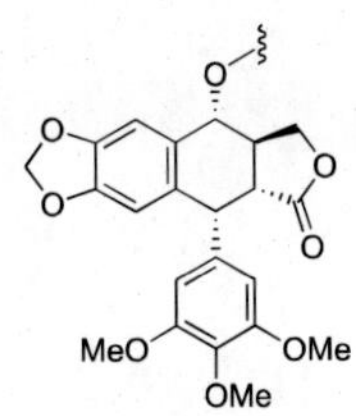 **39a** **Podophyllotoxin**	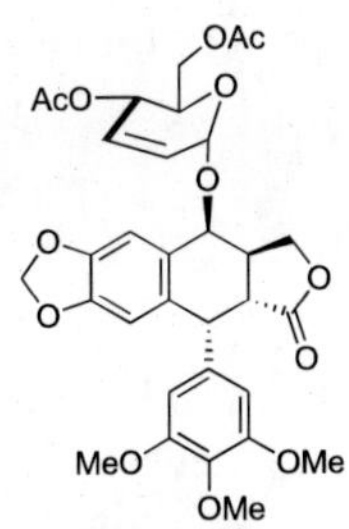 **40a**	63
2	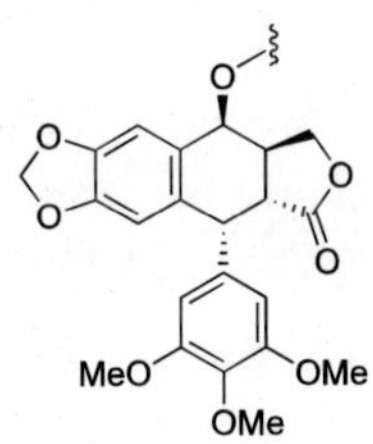 **39b** **Epidophyllotoxin**	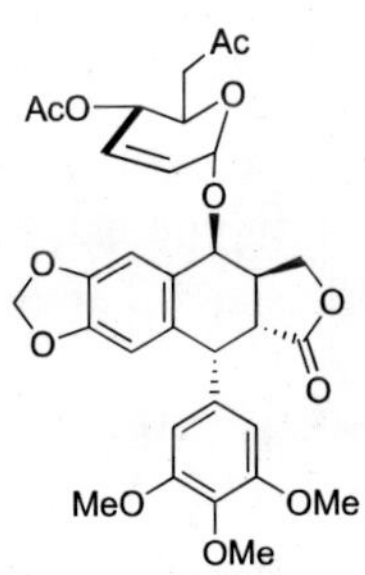 **40b**	64
3	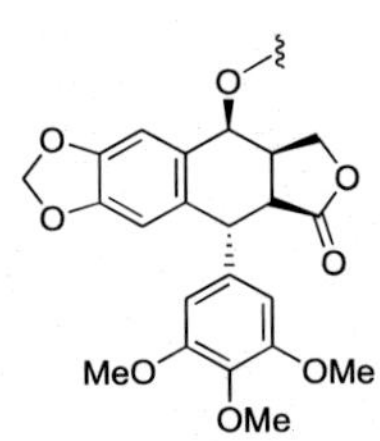 **39c** **Picropodophyllotoxin**	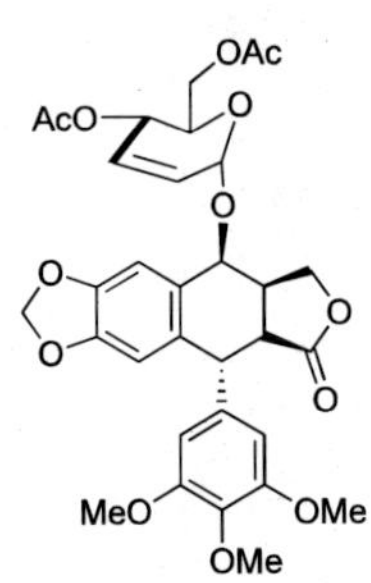 **40c**	65

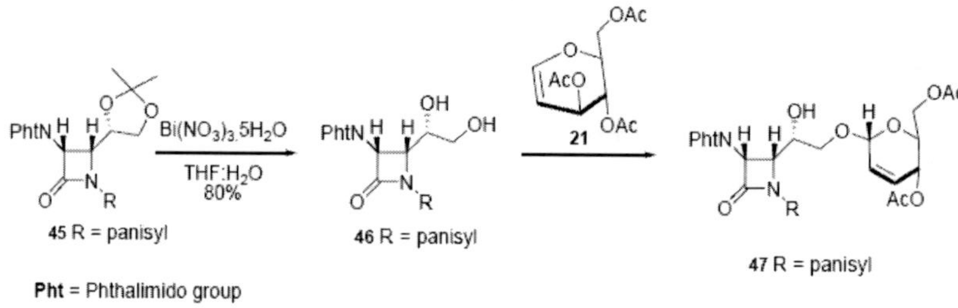

Scheme 17. Ferrier rearrangement of vicinal-diol of 3-amino β-lactams.

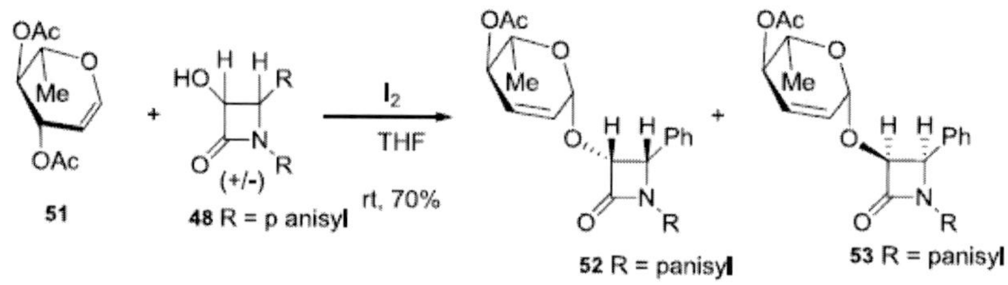

Scheme 18. Ferrier rearrangement of L-rhamnal acetate with racemic 3-hydroxy β-lactam.

3. S-GLYCOSIDES FROM GLYCAL

In recent years, acid zeolites have been extensively used for various organic transformations with great advantage of being catalytically eco-friendly materials. Acidic zeolites are well known catalyst for industrial process like etherification of benzylic alcohol for the synthesis of insecticides, verbutin and acetylation of dimethoxyarenes such as veratrole.

In carbohydrate chemistry zeolite was efficiently applied to catalyze the isomerization of diasacchrides and its hydrolysis. Rauter et al. [31] reported a one step synthesis of thioglycosides of the biologically interest. In this investigation the benzyl protected D-glucal was treated with O- nucleophiles and S nucleophiles in presence of acidic zeolite as promoter. (Scheme 19).

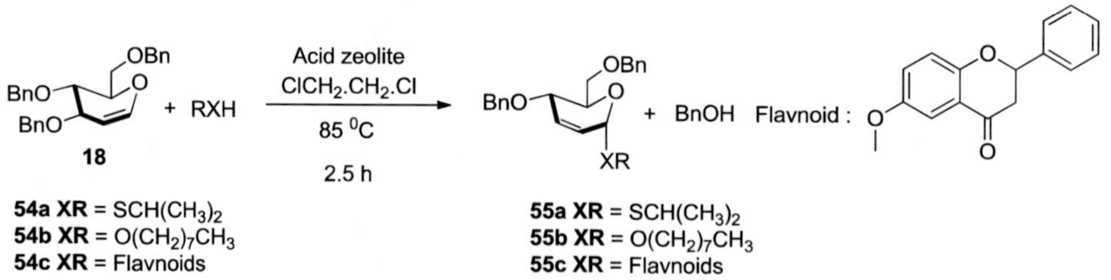

Scheme 19. Synthesis of 2,3-unsaturated O- and S-glycosides by Ferrier rearrangement of a benzyl protected glycal catalysed by acid zeolites as glycosyl promoter.

The flavonoid glycosides, known to possess a diverse bioactivities and exhibit a remarkable array of biochemical and pharmacological actions, on inflammation, heart diseases and cancer.

The stereoselective synthesis of 2, 3-unsaturated S-glycosides was accomplished by using mild and convenient protocol employing a catalytic amount of $RuCl_3$. Various acceptors comprising of aromatic, aliphatic, alicyclic, and naphthyl mercaptans were reacted smoothly with different glycals to provide regioselectively C-1 S-linked glycosides in good yields (72–92%) with high α-anomeric selectivity (72:28–99:1). This protocol was mild enough for constructing S-linkages at anomeric position of lactose derived glycal (Scheme 20).

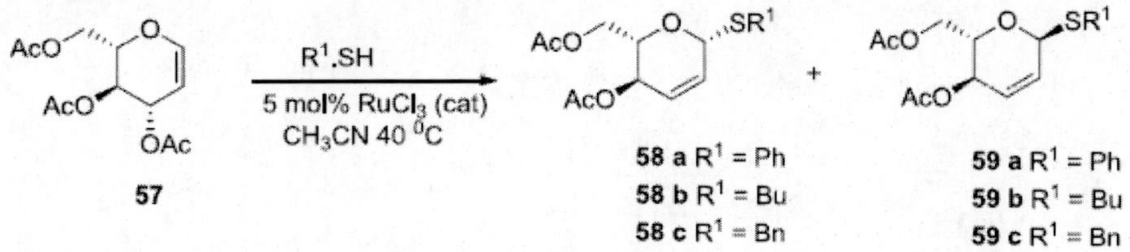

Scheme 20. Synthesis of 2,3-unsaturated S-Glycosides via Ruthenium catalysed Ferrier rearrangement of Glycals.

4. C-GLYCOSIDES

C-glycosides are carbohydrate analogues in which the glycosidic oxygen is replaced with a methylene substituent at anomeric position. Because of their high stability under acidic, basic or enzymatic reaction conditions, they are widely used in medicinal chemistry. These types of glycosides have attracted much attention, considering that many of them have demonstrated their effectiveness as therapeutic agents. Structurally C-glycosides can be constituted by aliphatic or aromatic systems, and the sugars can be pyranose or furanose-containing ring systems.

4.1. Allyl Trimethyl Silane as C-Nucleophile

Toshima and co-workers demonstrated a highly stereoselective synthesis of allyl α-*C*-glycosides with high α-anomeric selectivity in excellent yield by treating diverse acetate protected 1,2-glycals **61a-d** with allyltrimethylsilane (ATMS) as a source of C-nucleophile in the presence of 30-50 mol% of 2,3-dichloro-5,6-dicyano-*p*-benzoquinone (DDQ) as a glycosyl activator in acetonitrile at low temperature (Scheme 21 Table 4.1) [32-33].

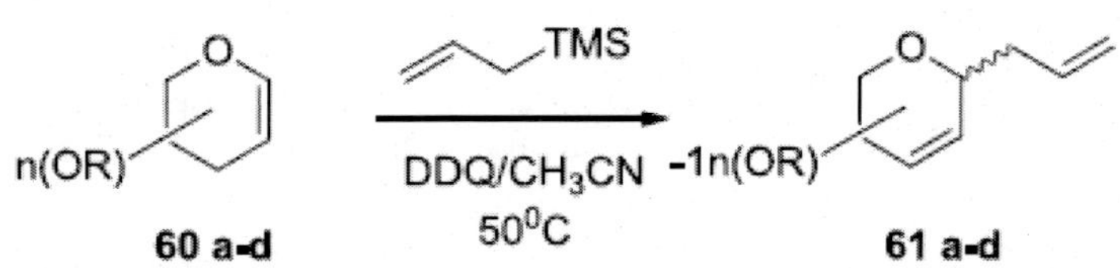

Scheme 21. C-glycosylation of 1,2-glycal with allytrimethylsilane and DDQ as a promoter.

Table 4.1. C-glycosylation of 1,2-glycal with allyltrimethylsilane and DDQ.

Entry	Glycosyl Donar (Glycal)	Protecting group (R^1)	T (^{0}C)	Yield (%)	Anomeric Distribution	
					α	β
1	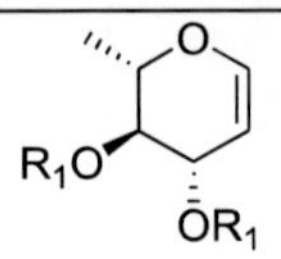 60a	Ac	50	90	15	1
2	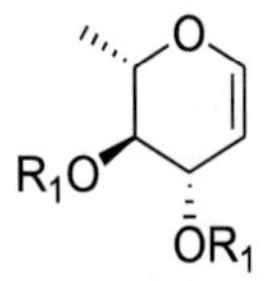 60 b	Bz	70	77	10	1
3	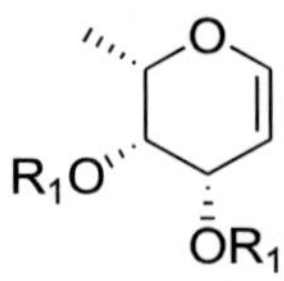 60c	Ac	70	74	>99	1
4	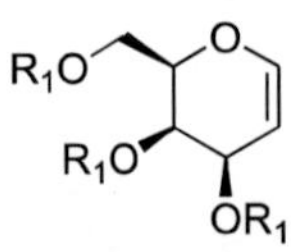 60d	Ac	70	76	>99	1
5	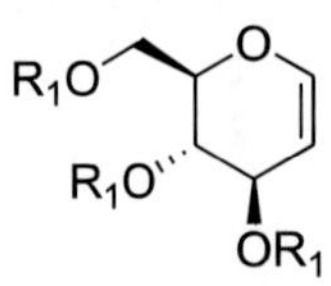 60e	Ac	50	85	16	1

4.2. Silyl Cyanide as a Source of C-Nucleophile

Glycosyl cyanides are important precursor in the synthesis of C-glycosides and its derivatives. The cyano group can be transformed into various functional groups and can be further exploited in organic synthesis. The first Lewis-catalysed addition of trimethylsilyl cyanide (TMSCN) on glycals was reported by De Las Heras et al. in 1983 [34]. The installation of cynide group to the anomeric position of the glucal was achieved by various Lewis acid catalysts (Scheme 22).

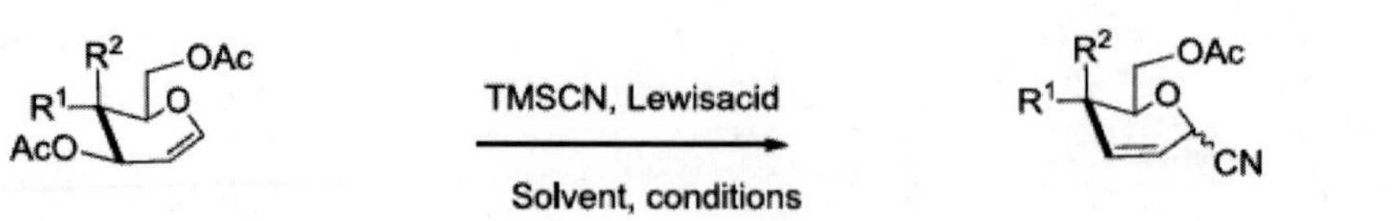

62 R^1 = OAc R^2 = H
63 R^1 = H R^2 = OAc

64 R^1 = OAc R^2 = H
65 R^1 = H R^2 = OAc

Scheme 22. Lewis acid mediated C-glycoside synthesis via Ferrier rearrangement.

Table 4.2. Lewis acid mediated C-glycoside synthesis via Ferrier rearrangement

Entry	Glycal	Promoter	Catalyst loading	Solvent	Yield (%)	α:β	Glycosides
1	62	$Sc(OTf)_3$	3 mol%	CH_2Cl_2	90	6:4	64
2	62	I_2	8 mol%	CH_2Cl_2	80	6:4	64
3	63	I_2	8 mol%	CH_2Cl_2	72	7:3	65
4	62	$InBr_3$	5 mol%	CH_2Cl_2	90	6:4	64
5	63	$InBr_3$	5 mol%	CH_2Cl_2	87	7:3	65
6	62	$Bi(OTf)_3$	2 mol%	CH_2Cl_2	92	7:3	64

4.3. Acetylene iodide as source of C-nucleophile

Lubin-Germain et al. [35] investigated the indium metal-catalyzed glycosylation of alkynated pseudoglycosides and alkyne iodide with high α-anomeric selectivity. The iodide functionality is essential for the Ferrier rearrangement. The iodine functionality in alkyne C-glycosides was established by the reaction of silver nitrate and N-iodosuccinimide. Indium metal was found to be superior catalyst over other metals like manganese, zinc and other salt in term of both yield and selectivity (Scheme 23 Table 4.3).

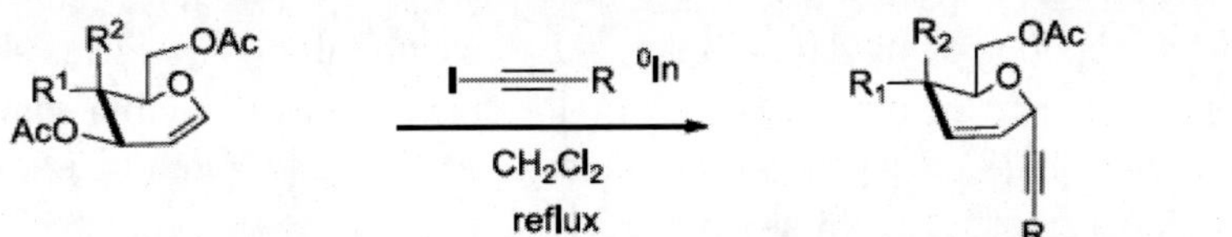

62 R^1 = OAc R^2 = H
63 R^1 = H R^2 = OAc

66 R^1 = OAc R^2 = H
67 R^1 = H R^2 = OAc

Scheme 23. Indium metal catalyzed C-Ferrier rearrangement.

Table 4.3. Indium metal mediated glycosylation of acetylene derivatives

Entry	Glycal	R	Major Glycosides	Time	Yield (%)	α/β
1	62	Ph	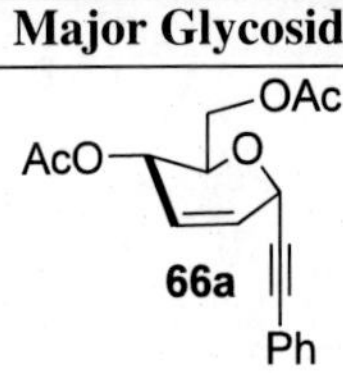 **66a**	4	79	90:10
2	62	$SiMe_3$	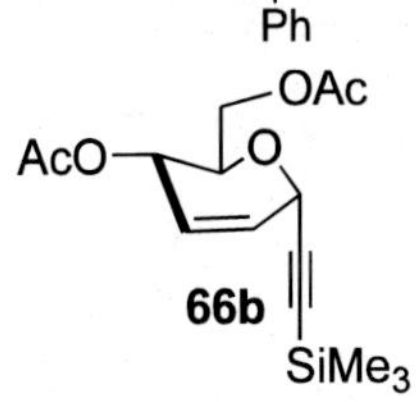 **66b**	24	82	85:15
3	62	α-glycal	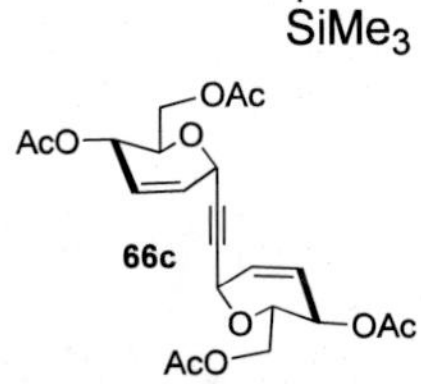 **66c**	24	75	84:16

Table 4.3. (Continued)

Entry	Glycal	R	Major Glycosides	Time	Yield (%)	α/β
4	62	β-gluco	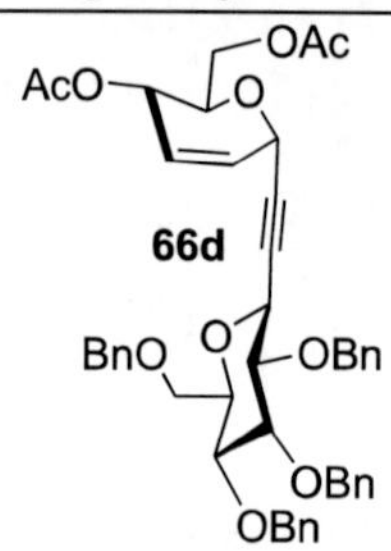 66d	3	94	95:5
5	63	Ph	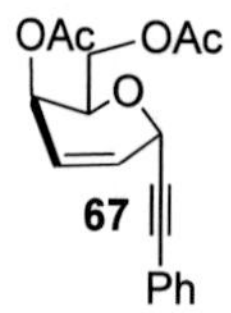 67	16	95	83:17

4.4. Organometallics as C-Nucleophiles

Nicolaou et al. reported trialkylaluminium and alkylaluminium chloride as sources of the carbon nucleophile in Ferrier rearrangement of the triacetylated 1-methyl glucal **68** using titanium tetrachloride ($TiCl_4$) as a glycosyl promoters successfully to synthesize 1,1-dialkylated pseudoglycals **69ab** stereospecifically (Scheme 24).

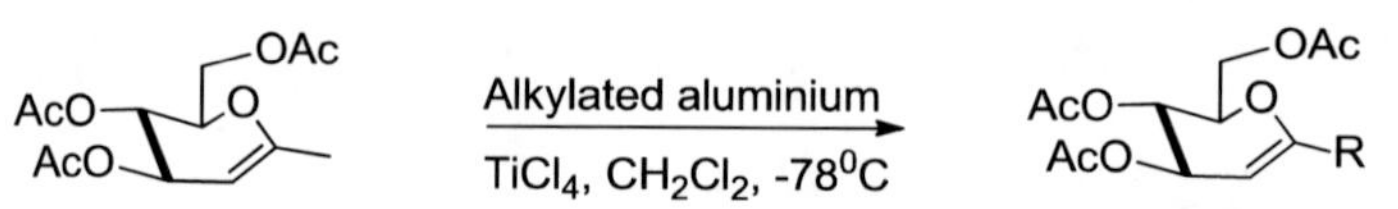

68

69 a R = Me
69 b R = Et

Scheme 24. Aluminium alkyls as carbon nucleophiles.

Table 4.4. Aluminium catalyst in glycosylation

Regents	R	Yield (%)	Regents	R	Yield (%)
$AlMe_3$	Me	92	$AlEt_3$	Et	82
$AlEt_2Cl$	Et	85	$AlEtCl_2$	Et	81

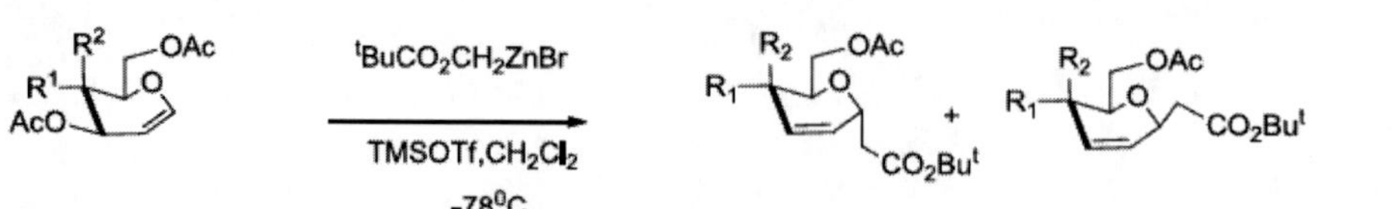

62 R^1 = OAc R^2 = H
63 R^1 = H R^2 = OAc

68 R^1 = OAc R^2 = H
69 R^1 = H R^2 = OAc

70 R^1 = OAc R^2 = H
71 R^1 = H R^2 = OAc

Scheme 25. TMSOTf-catalysed nucleophilic addition of Reformatsky regents on glycals.

Orsini and Pelizzoni carried out the study of carbon Ferrier rearrangement using tert-butoxycarbonylmethylzinc bromide (**50**) as Reformatsky reagent with peracetylated glycal derivatives to synthesize the pseudoglycosides **68, 69** and **70, 71** in presence of trimethysilyl triflate (TMSOTf) as catalyst at 0^0C in good to moderate yield. The D-glucal gave the excellent α-anomeric selectivity while with D-galactal and D-rhamnose, reversal of anomeric selectivity was observed (Scheme 25).

Table 4.5. Reformatsky reagent for glycosylation reaction

Entry	Glycal	Glycosides(α)	R^1	R^2	Yield (%)	α/β
1	62	68	OAc	H	48.6	1:2
2	63	69	H	OAc	16.1	2:1

Synthesis of α-C-aryl glycosides *via* palladium metal-catalyzed Heck-type of coupling was also investigated. Yang and his group disclosed the palladium-catalyzed Heck-type coupling of 3,4,6-tri-O-benzyl-Dglucal **18** with various bromo aryls in the presence of 5 mol %, TBAB (tetra-n-butylammonium bromide) (1.0 equiv.), and $Pd(OAc)_2$ and K_2CO_3 (3.0 equiv.) in DMF under microwave irradiation at 170°C afforded C-aryl glycosides **72–74** in good yields and excellent anomeric selectivity (Scheme 26) [36].

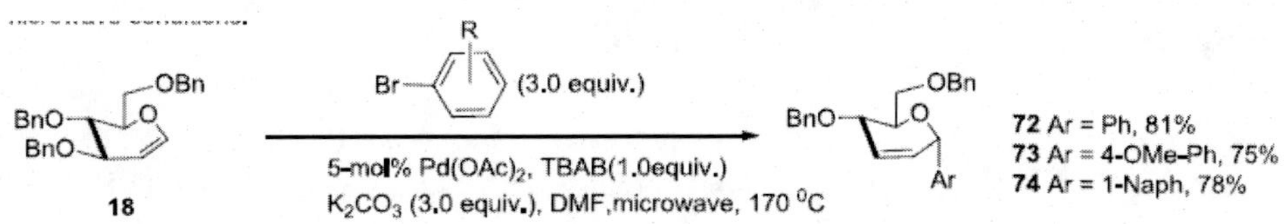

Scheme 26. Palladium-catalysed Heck-type C-glycosylation of glycal with substituted bromo aryl halides under microwave conditions.

CONCLUSION

Ferrier rearrangement is extremely versatile to prepare numerous glycosides mostly by a stereoselective pathway. Various catalysts are able to promote the glycosylation reaction between acceptor-donor. Because of the biological activity of the glycosides, this method as described herein will find a wide range of application in chemistry, biology, medicine and pharmacy.

REFERENCES

[1] Ferrier, R. J. (1979). Unsaturated Carbohydrates. Part 21. A Carboxylic Ring Closure of a Hex-5-enopyranoside Derivative. *J. Chem. Soc., Perkin Trans.* 1, 1455–1458.

[2] Ferrier, R. J.; Zubkov, O. A. (2003). Transformation of Glycals into 2,3-Unsaturated Glycosyl Derivatives. *Org. React.* 62.

[3] Fraser-Reid, B. (1996). Some Progeny of 2,3-Unsaturated Sugars They Little Resemble Grandfather Glucose: Twenty Years Later." *Acc. Chem. Res.*, 29 (II), 57–66.

[4] Zhao, J., Wei, S., Ma, X., Shao, H. (2010). *A simple and convenient method for the synthesis of pyranoid glycals. Carbohydrate Research,* 345, 168–171.

[5] Franz, A., Wei, Yi., Samoshin, V., Gross, P.H. (2002). Mild Synthesis of Disaccharidic 2,3-Enopyranosyl Cyanides and 2-C-2-Deoxy Pyranosyl Cyanides with $Hg(CN)_2/HgBr_2$/TMSCN. *The Journal of Organic Chemistry*, 67, 7662-9.

[6] Csuk, R., Alois, F., Brigitte, B. I. and Weidmann, H. (1986). A new, zinc-promoted synthesis of 1, 4-(1, 5)-anhydro-2-deoxy-pent-(hex)-1-enitols (furanoid and pyranoid glycals). *Chemical Communications,* 0, 1149-1150.

[7] Spencer, R. P., Cavallaro, C. L., and Schwartz, J. (1999). Rapid Preparation of Variously Protected Glycals Using Titanium(III). *Journal of Organic Chemistry, 64*, (11) 3987-3995.

[8] Fraser-Reid, B. J., Lopez, J. C, Gammon, D. W., and Sels, B. F. (2008). Other Methods for Glycoside Synthesis, In Handbook of Chemical Glycosylation. *Wiley-VCH Verlag GmbH and Co. KGaA,* 103, 104, 105, 129, 130, 381–448.

[9] Agarwal, A., Rani, S., and Vankar, Y.D. (2004). Protic Acid (HClO4 Supported on Silica Gel)-Mediated Synthesis of 2,3-Unsaturated-O-glucosides and a Chiral Furan Diol from 2,3-Glycals, *Journal of Organic Chemistry*, *69* (18), 6137–6140.

[10] Gómez, A. M., Miranda, S., López, J. C. (2017) Ferrier rearrangement: an update on recent developments. *Carbohydrate Chemistry*, 42, 2010-247.

[11] Morris, W. J., Shair, M. D. (2008). Stereoselective synthesis of 2-deoxy-β-glycosides using anomeric O-alkylation/arylation. *Organic letters*, 11(I) 9-12. (b) Rauter, A. P, Almeida, T, Vicente, A.I, Ribeiro, V, Bordado, J. C, Marques, J. P., Ramôa-Ribeiro, F., Ferreira, M. J., Oliveira, C., Guisnet, M. (2006) Reaction of N-, S- and O-Nucleophiles with 3,4,6-tri-O-Benzyl-D-Glucal Mediated by Triphenylphosphine Hydrobromide versus HY Zeolite. *Eur. J. Org. Chem.* (X), 2429-2439.

[12] Kim, H., Men, H., Lee, C. (2004). Stereoselective Palladium-Catalyzed O-Glycosylation Using Glycals. *Journal of the American Chemical Society*, 126 (V) 1336-1337.

[13] Fugedi, P. (2006). Glycosylation Methods. The Organic Chemistry of Sugars. *CRC Press; Boca Raton*, 89–179.

[14] Reddy, Y.S., John Pal, A. P., Gupta, P., Ansari, A. A., Vankar, Y. D. (2011). Ceric Ammonium Nitrate-Catalyzed Azidation of 1,2-Anhydro Sugars: Application in the Synthesis of Structurally Diverse Sugar-Derived Morpholine 1,2,3-Triazoles and 1,4-Oxazin-2-ones. *Journal of Organic Chemistry*, 76 (15), 5972-5984.

[15] Das, R. and Mukhopadhyay, B. (2016). Chemical O-Glycosylations: An Overview. *Chemistry Open,* 5 (5), 401-433.

[16] Li, X., Zhu, J. (2016). Glycosylation via Transition-Metal Catalysis: Challenges and Opportunities, *European Journal of Organic Chemistry*, 28, 4724-4767.

[17] Williams, D. B. G., Simelane, S. B. and Kinfe, H. H. (2012). Aluminium triflate catalysed O-glycosidation: temperature-switched selective Ferrier rearrangement or direct addition with alcohols. *Organic & Biomolecular Chemistry,* 10, 5636-5642.

[18] Letton, J. C., Ohio, F. P (1987). Process for Preparing Alkyl Glycosdes. *United States Patent*, 19, 4,713,447

[19] Zhang, G., Liu, Q. (2007). Ferric Sulphate Hydrate–Catalyzed, Microwave-Assisted Synthesis of 2,3-Unsaturated O-Glycosides via the Ferrier Reaction. *Synthetic Communications,* 3485-3492.

[20] Routier, F. H., Nikolaev, A. V., Ferguson, M. A. J. (1999). The preparation of neoglycoconjugates containing inter-saccharide phosphodiester linkages as potential anti-Leishmania vaccines. *Glycoconjugate journal,* 16, 12, 773-780. (b) T.R Reddy, S Chittela, S Kashyap, (2014). Ruthenium trichloride catalyzed synthesis of 2, 3-unsaturated-N-glycosides via Ferrier azaglycosylation. *Tetrahedron,* 70, 48, 9224-9229.

[21] McKay, M. J., Nguyen, H. M. (2012). Recent advances in transition metal-catalyzed glycosylation. *ACS catalysis* 2 (8), 1563-1595

[22] Murakami, T., Hirono, R., Sato, Y., Furusawa, K. (2007). Efficient synthesis of ω-mercaptoalkyl 1, 2-trans-glycosides from sugar peracetates. *Carbohydrate research*, 342 (8), 1009-1020.

[23] Mercer, G. J., Yang, J., McKay M. J., and Nguyen, H. M. (2008) "Palladium(II)-Catalyzed Rearrangement of Glycal Trichloroacetimidates: Application to the Stereoselective Synthesis of Glycosyl Ureas. *J. Am. Chem. Soc.* 130 11210–11218.

[24] Sau, A., Galan, M. C. (2017). Palladium-Catalyzed α-Stereoselective O-Glycosylation of O-(3)-Acylated Glycals. *Organic letters*, 19, (11), 2857-2860.

[25] Swamy, N. R., Srinivasulu, M., Reddy, T. S., Goud, T. V. (2004). Zirconium(IV) Chloride Catalyzed Synthesis of 2,3-Unsaturated C, N, O, S, and Heteroaromatic Glycosylation in the Ferrier Rearrangement. *Journal of Carbohydrate Chemistry,* 23:6-7, 435-441,

[26] Augé, J., Sizun, G. (2009). Ionic liquid promoted atom economic glycosylation under Lewis acid catalysis. *Green Chemistry*, 11, 1179-1183.

[27] Smitha, G., Reddy, C. S. (2004). ZrCl4-catalyzed efficient Ferrier glycosylation: A facile synthesis of pseudoglycals. *Synthesis*, 6, 834-836.

[28] Freitas, J. C. R., Couto, T. R., Paulino, A. A. S. (2012). Stereoselective synthesis of pseudoglycosides catalyzed by TeCl4 under mild conditions. *Tetrahedron,* 68,41, 8645-8654.

[29] Freitas, Juliano C. R., Couto, Túlio R., Filho, João R. Freitas, Menezes, Paulo Henrique, (2011). Synthesis of O-Glycosides Dimers Catalysed by Tellurium (IV) Tetrachloride. *Brazilian Meeting on Organic Synthesis,* (1), 1.

[30] Daley, L., Guminski, Y., Demerseman, P., Kruczynski, A., Etiévant, C., Imbert, T., Hill, Bridget T. and Monneret, C. (1988). Synthesis and Antitumor Activity of New Glycosides of Epipodophyllotoxin, Analogues of Etoposide, and NK 611. *Journal of Medicinal Chemistry,* 41, 4475-4485.

[31] Banik, B.K., Manhas, M. S., Bose, A. K. (1997). Enantiopure α-hydroxy-β-lactams via stereoselective glycosylation, *Tetrahedron letters*, 38, 29, 5077-5080. (b) B.K Banik, M.S Manhas. (2012). Stereospecific novel glycosylation of hydroxy β-lactams via iodine-catalyzed reaction: a new method for optical resolution, *Tetrahedron*, 68, 52, 10769-10779. (c) Chandra, S., Yadav, R. N., Paniagua, A., Banik, B. K. (2016). Indium salts-catalyzed O and S-glycosylation of bromo sugar with benzyl glycolate: an unprecedented hydrogenolysis. *Tetrahedron Letters,* 67 (13), 1425-1429.

[32] Xavier, N. M., Lucas, S. D., Rauter, A. P. (2009). Zeolites as efficient catalysts for key transformations in carbohydrate chemistry. *Journal of Molecular Catalysis* A 305, 84-89.

[33] Toshima, K. (2006). Novel glycosylation methods and their application to natural products synthesis. *Carbohydrate research*, 341, 1282-1297.

[34] Federico G. De L. H., Ana S., Felix, P., Fernández, R. (1983). Cyanosugars—II: Synthesis of hexopyranosyl cyanides by reaction of glycals with trimethylsilyl cyanide. *Tetrahedron*, 39, (9), 1617-1620.

[35] Tsukiyama Takahiro, Isobe, Minoru. (1992). C-Glycosidation with Silylacetylenes to D-glucals. Tetrahedron Letters, 33, 7911-7914. (b) Ayed, C., Palmier, S., Lubin-Germain, N., Uziel, J., Augé, J. (2010). Indium-mediated alkynylation of sugars: synthesis of C-glycosyl compounds bearing a protected amino alcohol moiety. *Carbohydrate Research,* 345, 2566-2570.

[36] Doyle, D. G. Jr. (**1990**). "C-glycoside synthesis by palladium-mediated glycal-aglycon coupling reactions." *Acc. Chem. Res.*, 23, 201–206.

About the Editor

Bimal Krishna Banik

Professor Banik has served as a Tenured Full Professor in Chemistry and First President's Endowed Professor at the University of Texas-Pan American. At present, he is the Vice President of Research & Education Development of Community Health Systems of South Texas.

The principal areas of Dr. Banik's research are divided among several disciplines: (a) new β-lactams as anticancer agents; (b) synthesis and studies of mechanism of action of novel polyaromatic compounds as anticancer agents; (c) catalysis.

Dr. Banik's teaching skills is exceptionally strong and this is proved by students' confidential evaluations and thousands of comments. He has mentored approximately 453 students, 20 postdoctoral fellows, and 7 PhD level research scientists. He has acted as the advisor of two major scientific societies.

Dr. Banik has been awarded $7.25 million in grants from the NIH, NCI, and Kleberg Foundation. He has published more than 341 papers and 460 presentation abstracts. The number of citations of his research papers is close to 6000. He has presided over 20 research sessions at the ACS National Meetings. He is an editorial board member of 26, editor-in-chief of 11, founder of 4, associate editor of 4, and guest editor of 4 journals.

Professor Banik has received numerous awards. Some of these include: Mahatma Gandhi Pravasi Samman; Dr. Bose endowment medal; Dr. M. N. Ghosh medal; University of Texas Board of Regents' Outstanding Teaching award; five top-cited papers awards by Elsevier; best advisor award in USA; ACS Member Service award.

INDEX

#

A

B

C

D

E

F

G

H

I

K

L

M

N

O

P

Q

R

S

T

U

V

W

X

Y

Z

α

β

Modern Advances in Organic and Inorganic Chemistry

Editors: Balkose, Devrim (Ýzmir Institute of Technology, Turkey); Zaikov, Gennady E. (N. M. Emanuel Institute of Biochemical Physics, Moscow, Russia)

Hardcover ISBN: 978-1-59454-402-6
Retail Price: $180

Organic Chemistry, Biochemistry, Biotechnology and Renewable Resources. Research and Development. Volume 1 - Today and Tomorrow

Editors: Gennady E. Zaikov (N.M. Emanuel Institute of Biochemical Physics, Russian Academy of Sciences, Moscow, Russia) Frank Pudel (Pilot Pflanzenoltechnologie, Magdeburg e.V., Magdeburg, Germany), Grzegorz Spychalski (Institute of Material Fibres and Material

Series: Biochemistry Research Trends

Book Description: This book includes the latest information in the field of organic chemistry, biochemistry, biotechnology, material science (synthesis, properties, applications) and renewable resources.

Hardcover ISBN: 978-1-62081-155-9
Retail Price: $209

Organic Chemistry, Biochemistry, Biotechnology and Renewable Resources. Research and Development. Volume 2- Tomorrow and Perspectives

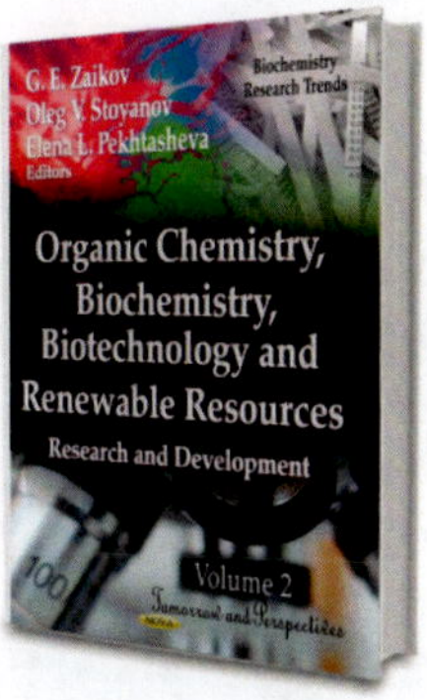

Editors: G.E. Zaikov, Oleg V. Stoyanov, Elena L. Pekhtasheva (Institute of Biochemical Physics, Russian Academy of Sciences, Moscow, Russia)

Series: Biochemistry Research Trends

Book Description: This book includes the latest information in the field of organic chemistry, biochemistry, biotechnology, material science (synthesis, properties, applications) and renewable resources.

Hardcover ISBN: 978-1-62081-159-7
Retail Price: $209

Emerging Topics in Organic Chemistry

Editors: Rattaporn Thonggom (Mahidol Univ., Bangkok, Thailand) , Thevarak Rochanapruk (PTT Phenol Co. Ltd, Chatuchak, Bangkok, Thailand) and Gennady E. Zaikov (N.M. Emanuel Institute of Biochemical Physics, Russian Academy of Sciences, Moscow, Russia)

Book Description: This book is a first attempt of scientific cooperation between Mahidol University and PTT Phenol Company from Thailand and Institute of Biochemical Physics, Russian Academy of Sciences. The authors selected for their cooperation the topics from organic chemistry (quantitative level).

Hardcover ISBN: 978-1-60456-633-8
Retail Price: $150

From Medicinal Chemistry to Food Science: A Transfer of In Silico Methods Applications

Editor: Pietro Cozzini (Department of Food Science, University of Parma, Parco Area delle Scienze, Parma, Italy)

Series: Food Science and Technology

Book Description: The aim of this book is to illustrate the big opportunity to transfer computational techniques, well accepted and used in medicinal chemistry to food science; through this, we hope to discover new pollutants, possible dangerous food additives, contaminants from industries, xenoestrogens, etc.

Hardcover ISBN: 978-1-63483-587-9
Retail Price: $230

New Developments in Medicinal Chemistry

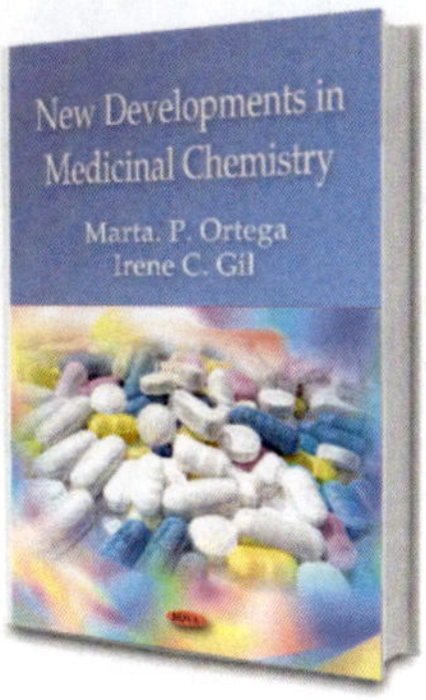

Authors: Marta P. Ortega and Irene C. Gil

Book Description: Medicinal or pharmaceutical chemistry is a scientific discipline at the intersection of chemistry and pharmacology involved with designing, synthesizing and developing pharmaceutical drugs. Medicinal chemistry involves the identification, synthesis and development of new chemical entities suitable for therapeutic use.

Hardcover ISBN: 978-1-60456-810-3
Retail Price: $130